CHEMISTRY AND WORLD FOOD SUPPLIES: THE NEW FRONTIERS

CHEMRAWN II

INTERNATIONAL UNION OF PURE AND APPLIED CHEMISTRY

in conjunction with

INTERNATIONAL RICE RESEARCH INSTITUTE

CHEMISTRY AND WORLD FOOD SUPPLIES: THE NEW FRONTIERS

CHEMRAWN II

Invited Papers Presented at the International Conference on
Chemistry and World Food Supplies, Manila, Philippines, 6 – 10 December 1982

Edited by

L.W. SHEMILT

McMaster University

Hamilton, Ontario, Canada

PERGAMON PRESS

OXFORD • NEW YORK • TORONTO • SYDNEY • PARIS • FRANKFURT

U.K.	Pergamon Press Ltd., Headington Hill Hall, Oxford OX3 0BW, England
U.S.A.	Pergamon Press Inc., Maxwell House, Fairview Park, Elmsford, New York 10523, U.S.A.
CANADA	Pergamon Press Canada Ltd., Suite 104 150 Consumers Rd., Willowdale, Ontario M2J 1P9, Canada
AUSTRALIA	Pergamon Press (Aust.) Pty. Ltd., P.O. Box 544, Potts Point, N.S.W. 2011, Australia
FRANCE	Pergamon Press SARL, 24 rue des Ecoles, 75240 Paris, Cedex 05, France
FEDERAL REPUBLIC OF GERMANY	Pergamon Press GmbH, 6242 Kronberg-Taunus, Hammerweg 6, Federal Republic of Germany

First edition 1983

Library of Congress Cataloging in Publication Data

International Conference on Chemistry and World Food Supplies (1982: Manila, Philippines) Chemistry and world food supplies.
(IUPAC symposium series)
1. Agricultural chemistry – Congresses. 2. Food industry and trade – Congresses. I. Shemilt, L.W. II. International Union of Pure and Applied Chemistry. III. International Rice Research Institute. IV. Title. V. Series.
S583.2.I57 1982 630'.2'4 83-8218
ISBN 0-08-029243-7
ISBN 0-08-029242-9 (soft)

British Library Cataloguing in Publication Data

International Conference on Chemistry and World Food Supplies *(1982: Manila)*
Chemistry and world food supplies – (IUPAC symposium series)
1. Agricultural chemistry – Congresses. I. Title. II Shemilt, L.W. III. International Union of Pure and Applied Chemistry. IV. International Rice Research Institute. V. Series.
540'.2463 S 584
ISBN 0-08-029243-7
ISBN 0-08-029242-9 Pbk

In order to make this volume available as economically and as rapidly as possible the authors' typescripts have been reproduced in their original forms. This method unfortunately has its typographical limitations but it is hoped that they in no way distract the reader.

CONTENTS

The Role of Chemistry and Biochemistry in Improving Animal Production Systems

Contributions of Chemistry and Biochemistry to Developing New and Improved Food Sources

Chemistry and Biochemistry in the Processing and Storage of Food

Chemistry in the Assessment and Control of the Food Supply

The Forward Edge

CHEMRAWN II COMMITTEE

General Chairman	B.W. Rossiter
Vice-Chairman – Local Arrangements	M.R. Vega
Vice-Chairman – Finance	R.L. Hall
Secretary	Joyce C. Torio
Developing Country Participation Chairman	E.S. Ayensu
Future Actions Committee Chairman	C. Ponnamperuma
IUPAC	S. Nagakura
Poster Session Chairman	E. del Rosario
Program Committee Chairmen	N.E. Borlaug
	M. Calvin
	P.A. Oram
Scientific Editors	L.W. Shemilt
	G. Bixler
Publications Committee Chairman	W.G. Schneider
Publicity Committee Chair	Carol L. Rogers
Treasurer	F.J. Marin-Price

INTERNATIONAL UNION OF PURE AND APPLIED CHEMISTRY

IUPAC Secretariat: Bank Court Chambers, 2-3 Pound Way
Cowley Centre, Oxford OX4 3YF, UK

PREFACE

The International Union of Pure and Applied Chemistry (IUPAC) is an international, non-governmental organization dedicated to the advancement and application of chemical science and technology. Through its broad and world-wide scientific and engineering connections, IUPAC was in a strong position to carry through the concept of CHEMRAWN, first enunciated in 1975. As the acronym for "Chemical Research Applied to World Needs", CHEMRAWN was conceived as a focus for identifying and addressing world problems amenable to solutions through chemistry. It recognized the interdependent nature of world communities, expressing a concern for problems facing both developed and developing nations, and at the same time, noting the limitations of a purely technological approach, endeavoured to address simultaneously scientific, social, economic and environmental factors.

The general purpose of CHEMRAWN was further seen as being achieved through providing an effective international forum, through a series of world conferences, each on a selected problem area, bringing together experts, leaders and decision-makers from governments, industries and universities in developed and developing countries. Through such meetings, the identification and publication of medium- and long-term goals for research and development were visualized as giving further effect to the objectives of CHEMRAWN. The first such world conference, CHEMRAWN I, held in Toronto, Canada, in July 1978, fully exemplified this international and interdisciplinary approach as it considered the problem of "Future Sources of Organic Raw Materials".

The second international meeting invoking this CHEMRAWN concept was the International Conference on Chemistry and World Food Supplies - the New Frontiers, held in Manila, Philippines, in December 1982 as CHEMRAWN II. Recognizing the problem of achieving adequate food and nutrition for the burgeoning populations in developing nations, it was fitting that IUPAC was joined in a sponsorship of such a topic by the International Rice Research Institute (IRRI). Renowned throughout the world for the significant role it has played and is playing in the advancement of agriculture in the developing nations, IRRI was both sponsor and, in many respects, a co-host, as its facilities are located in the Philippines at Los Banos.

CHEMRAWN II provided discussion of critical needs and solutions relating to chemistry, agriculture and world food supplies. Treating chemical-based needs as a system, and including the social, economic and environmental factors, it was an interdisciplinary meeting bringing together modern aspects of the fields of chemistry, biochemistry and microbiology with the critical needs of agriculture production and food processing. Special attention was given to strengthening the indigenous capacity of developing countries through the involvement of their scientists and leaders in all phases of the conference.

CHEMRAWN II had the following objectives:

- First, to identify and put into perspective those areas of research and development having the potential to increase significantly food production and improve food storage and processing.

- Second, to strengthen scientific research in developing nations, particularly in those fields which require professional competence and initiative without excessive capital and human resources.

- Third, to accelerate implementation of research priorities and objectives by fostering cooperation among governments, industries, and universities.

The first objective was met by a series of 56 invited lectures, which are now the contents of this volume of scientific proceedings, and a further group of plenary lectures. The latter, along with a well-considered compilation of recommendations, are contained in a companion volume on "Perspectives and Recommendations", being published by IRRI. The presentation at the conference of over one hundred poster session papers (mostly from the developing countries), their publication by edited abstract in a Conference Handbook, and the above two CHEMRAWN II volumes are designed together to meet the remaining objectives.

In editing the papers presented in this volume, I have been greatly assisted by the cooperation and understanding of the authors. An effort has been made to achieve both clarity and some uniformity of presentation, and the acceptance by authors of suggestions for abbreviation and re-wording has been much appreciated. If there is textual deviation from the ideas they presented, it is the editor's responsibility and I request their further understanding.

I gratefully acknowledge the invaluable assistance in the editorial task (not only of these papers but also of the poster session abstracts) of Dr. Clay M. Switzer, Dean of the Faculty of Agriculture of the University of Guelph, of Dr. Leon J. Rubin, Professor of Food Engineering, University of Toronto, and of Dr. Ross H. Hall of the Department of Biochemistry, McMaster University, Hamilton. Dr. Hall, both by propinquity as well as by dedication and experience, gave guidance and support throughout the assembly of the papers into this volume. I am also indebted to the Publications Committee Chairman, Dr. W.G. Schneider and to the General Chairman, Dr. Bryant Rossiter, for their continuing support; and am equally grateful to the CHEMRAWN II Local Arrangements Committee under Dr. Marcos Vega for providing editoral office facilities and support during the conference. Dr. Joyce Torio, as Secretary and Co-ordinator for CHEMRAWN II, was always cheerful and effective in the numerous communication efforts required. The excellent work of the staff of the McMaster University Engineering Word Processing Centre deserves both praise and extensive thanks - with special thanks to Mrs. Betty Petro who remained efficient and enthusiastic throughout countless revisions. I am also most grateful to my secretary, Mrs. Rita Donaldson, for her interest, many hours of proofreading and general support. Mr. Don Crawley of Pergamon Canada Limited has lent expert advice and full support throughout.

L.W. Shemilt

Hamilton, Canada
April 1983

ACKNOWLEDGEMENTS

We are grateful to the following organizations and companies for their sponsorship and/or generous financial assistance.

Co-sponsors:

International Union of Pure and Applied Chemistry
International Rice Research Institute

Affiliate Sponsors:

Chemical Society of the Philippines
International Fertilizer Development Center
International Food Policy Research Institute
International Society for Horticultural Science
International Union of Biochemistry
International Union of Biological Sciences
International Union of Food Science and Technology
International Soil Science Society
International Union of Nutritional Sciences

Financial Contributors:

Agency for International Development, USA
Ajinomoto Co., Japan
Asahi Chemical Industry Co. Ltd., Japan
Australian Development Assistance Bureau, Australia
Carnation Company, USA
Castle & Cooke, Incorporated, USA
Canadian International Development Agency, Canada
Chemical Society Located in Taipei, China
Chiyoda Chemical Engineering and Construction Co., Japan
Ciba Geigy, Ltd., Switzerland
The Coca Cola Company, USA
The Commemorative Association for the Japan World Exposition, Japan
Department of Agriculture, USA
E.I. Du Pont de Nemours & Company, USA
Eastman Kodak Company, USA
Exxon Corporation, USA
The Ford Foundation, USA
French Ministry of Research and Industry, France
Fujisawa Pharmaceutical Co. Ltd., Japan
General Foods Corporation, USA
H.J. Heinz Company Foundation, USA
Hershey Foods Corporation, USA
Hoffman-La Roche Incorporated, USA
International Council of Scientific Unions
International Development Research Centre, Canada
International Union of Pure & Applied Chemistry
Kajima Foundation, Japan
Kikkoman Shoyu Ltd., Japan
Kyowa Hakko Kogyo KK, Japan
Dr. R. Mag Ltd., Switzerland
McCormick & Company Incorporated, USA
Meiji Seika Kaisha Ltd., Japan
The Meito Sangyo Co. Ltd., Japan
National Science Foundation, USA
Nestle S.A., Switzerland
The Quaker Oats Company, USA
Ralston Purina Company, USA
The Rockefeller Foundation, USA
Rohm & Haas Company, USA
Sandoz AG, Switzerland
Shionogi and Co. Ltd., Japan
Sumitomo Chemical Co. Ltd., Japan
Suntory Ltd., Japan
Swift & Company, USA
Takara Shuzo Co. Ltd., Japan
Takeda Chemical Industries, Japan
Teijin Ltd., Japan
Thomas J. Lipton, Incorporated, USA
Toray Industries, Inc., Japan
Toyo Jozo Co. Ltd., Japan
Treuhandstelle IM Verb. der Chemischen Industrie, FRG
UNESCO
Wm. Underood Company, USA
United National Development Programme
Velsicol Chemical Corporation, USA
Xerox Corporation, Canada and USA

In addition to the sponsors and financial donors, the following contributed significantly to CHEMRAWN II and its success.

Philippine Government
American Association for the Advancement of Science
American Chemical Society
International Food Policy Research Institute
National Academy of Sciences/National Research Council, USA

CHEMRAWN II COMMITTEE MEMBERS:

General Chairman	Dr. Bryant W. Rossiter, Eastman Kodak Company, USA
Vice Chairman for Local Arrangements	Dr. Marcos R. Vega, International Rice Research Institute, Philippines
Vice Chairman for Finance	Dr. Richard L. Hall, McCormick and Company, Inc., USA
Secretary	Dr. Joyce C. Torio, International Food Policy Research Institute, USA
Developing Country Participation Chairman	Dr. Edward S. Ayensu, Smithsonian Institution, USA
Future Actions Committee Chairman	Dr. Cyril Ponnamperuma, University of Maryland, USA
International Union of Pure and Applied Chemistry (IUPAC)	Dr. S. Nagakura, IUPAC, President, Institute for Molecular Science, Japan
	Dr.Heinrich Zollinger, IUPAC, Past President, Technische Hochschule Zurich, Switzerland
Poster Session Chairman	Dr. Ernesto del Rosario, University of the Philippines, Philippines
Program Committee Chairmen	Dr. Norman E. Borlaug, Nobel Laureate, International Maize and Wheat Improvement Center, Mexico
	Dr. Melvin Calvin, Nobel Laureate, University of California-Berkeley, USA
	Mr. Peter A. Oram, International Food Policy Research Institute, USA
Scientific Editors	Dr. L.W. Shemilt, McMaster University, Canada
	Dr. Gordon Bixler, American Chemical Society, USA
Publications Committee Chairman	Dr. William G. Schneider, IUPAC Vice President, Canada
Publicity Committee Chair	Ms. Carol L. Rogers, American Association for the Advancement of Science, USA
Treasurer	Mr. Francisco J. Marin-Price, American Chemical Society, USA

Organizing Committee:

Dr. Bryant W. Rossiter Eastman Kodak Company, USA, Chairman

Dr. Adnan Al-Ageel, Kuwait Foundation for the Advancement of Sciences, Kuwait

Dr. Eduardo Alvarez-Luma, International Council for the Development of Underutilized Plants, Mexico

Dr. William O. Baker, Bell Laboratories, USA

Sir John G. Crawford, The Australian National University, Australia

Mr. Charles S. Dennison, Council on Science & Technology for Development, USA

Dr. Jacques Diouf, Secretary of State for Scientific & Technical Research, Senegal

Dr. Nicholi Emmanuel, Academy of Sciences, USSR

The Honorable He Kang, Ministry of Agriculture, China

Dr.Kenzo Hemmi, University of Tokyo, Japan

Dr. W. David Hopper, The World Bank, USA

Mr. Joseph H. Hulse, International Development Research Centre, Canada

Dr. Franklin A. Long, Cornell University, USA

Dr. Jamal T. Manassah, City College of New York, USA

Dr. William G. Padolina, University of the Philippines at Los Banos, Philippines

Dr. Bukar Shaib, Special Advisor to the President, Nigeria

Dr. M.S. Swaminathan, The International Rice Research Institute, Philippines

Dr. Werner Treitz, Federal Ministry for Economic Cooperation, FRG

Dr. Moriya Uchida, TEIJIN Ltd., Japan

Program Committee:

Dr. Norman E. Borlaug, Nobel Laureate, International Maize and Wheat Improvement Center, Mexico, Chairman

Dr. Melvin Calvin, Nobel Laureate, University of California-Berkeley, USA, Chairman

Mr. Peter Oram, International Food Policy Research Institute, USA, Chairman

Dr. B.A. Abeywickrama, University of Ceylon, Sri Lanka

Dr. W.K. Agblé, Crops Research Institute, Ghana

Dr. C.C. Balch, University of Reading, England

Dr. Ricardo Bressani, Instituto de Nutricion de Centro Americana y Panama, Guatemala

Dr. Ralph W.F. Hardy, E.I. du Pont de Nemours & Company, USA

Dr. Rogelio O. Juliano, University of the Philippines, Philippines

Dr. Mohammed El-Khash, The Center for the Studies of Arid Zones and Dry Lands, Syria

Dr. Mohammed Nour, International Center for Agricultural Research in Dry Areas, Lebanon

Dr. Thomas R. Odhiambo, The International Center of Insect Physiology & Ecology, Kenya

Sir Charles Pereira, Peartrees, England

Dr. David Pimentel, Cornell University, USA

Dr. R. Scott Russell, The Grange, East Hanney,Wantage, Oxon, England

Dr. George W. Cooke, Rothamsted Experimental Station, England

Dr. Masao Fujimaki, Ochanomizu University, Japan

Dr. Kurt F. Gander, Unilever Forschungsgesell-schaft MbH, FRG

Dr. Hans Geissbuhler, Ciba-Geigy Corporation, Switzerland

Dr. Bernard S. Schweigert, University of California-Davis, USA

Dr. Gerald G. Still, U.S. Department of Agriculture, USA

Mr. R.V.K.M. Suryarau, Coromandel Fertilizers Limited, India

Dr. Lee Kum Tatt, Singapore Institute of Standards and Industrial Research, Singapore

Future Actions Committee:

Dr. Cyril Ponnamperuma, University of Maryland, USA, Chairman

Dr. Bansi Amla, The World Bank, USA

Dr. Norman E.Borlaug, Nobel Laureate, International Maize & Wheat Improvement Center, Mexico

Dr. Nyle C. Brady, Agency for International Development, USA

Ms. Teresa Salazar de Buckle, Junta del Acuerdo de Cartegena, Peru

Dr. Melvin Calvin, Nobel Laureate, University of California-Berkeley, USA

Dr. F. Egami, Mitsubishi-Kasei Institute of Life Sciences, Japan

Dr. Khaled El-Shazly, Alexandria University, Egypt

Dr. W. David Hopper, The World Bank, USA

Dr. Herwig Hulpke, Bayer, A.G., FRG

Dr. Emil O. Javier, University of the Philippines at Los Banos, Philippines

Dr. Kwoh-Ting Li, Magsaysay Awardee, Academia Sinica located in Taipei, China

Dr. P. Mahadevan, Food and Agriculture Organization, Italy

Dr. John W. Mellor, The International Food Research Institute, USA

Dr.Louis G.Nickell, Velsicol Chemical Corporation, USA

Dr. Thomas R. Odhiambo, The International Centre of Insect Physiology & Ecology, Kenya

Dr. Raul N. Ondarza, Consejo Nacional de Ciencia y Technologia, Mexico

Dr. Thomas Parton, Ciba-Geigy Corporation, Switzerland

Sir Charles Pereira, Peartrees, England

Sir George Porter, Nobel Laureate, The Royal Institution, England

Dr. Abdus Salam, Nobel Laureate, International Centre for Theoretical Physics, Italy

Dr. Glenn T. Seaborg, Nobel Laureate, University of California-Berkeley, USA

Professor Li Su, Academia Sinica, Bejing, China

Dr. M.S. Swaminathan, The International Rice Research Institute, Philippines

CONTRIBUTIONS OF CHEMISTRY TO REMOVING SOIL CONSTRAINTS TO CROP PRODUCTION

R. Scott Russell
(formerly Director, Letcombe Laboratory)
East Hanney, Wantage, Oxfordshire, OX12 0HQ, England
and
G.W. Cooke
Rothamsted Experimental Station
Harpenden, Hertfordshire, AL5 2JQ, England

ABSTRACT

Chemistry provided the background for the scientific control of soil fertility by providing fertilizers and other agrochemicals. It is closely involved also in multidisciplinary studies of the soil-plant system which have created our understanding of how the growth and function of root systems is regulated. New opportunities have thus been provided for identifying constraints to crop production and for devising methods to overcome them. Fertilizers make it possible to remove all nutrient constraints, but information is needed on their interaction with other production factors, such as the supply of available water, before recommendations for applying them can have a sound scientific basis. Fertilizers are rarely used with the maximum efficiency and more research is needed to specify fertilizers to meet local needs in developing countries, to provide nitrogen fertilizers by methods which consume less energy, and to determine the amounts and methods of application which fit local conditions of soil, climate and cropping system. Considerable increases in crop yields should be possible with adequate research, development and advisory services.

KEYWORDS: Crop yields, Cultivation, Fertilizers, Nitrogen, Plant nutrients, Rhizosphere, Roots, Soil, Yield gap.

INTRODUCTION

The major part of the food which man eats has its origin in plants which grow in the soil. Thus, the topic of this paper is basic to any discussion of world food supplies, especially when emphasis is given to the contribution of chemistry. This began with the identification of the elements which are essential plant nutrients and was followed by the development of industrial chemical processes to provide additional supplies of them as fertilizers, as well as other agrochemicals which benefit crop production. These pioneer contributions of chemists paved the way for modern studies of the plant/soil system, which requires inputs also from plant physiologists and pathologists, microbiologists and physicists. This interdisciplinary approach has in the past 30 years, much increased our understanding of the basic principles which govern crop growth in the soil, and enlarged the opportunities to manage crops and use inputs, including fertilizers, more efficiently and economically. Some of the background information is reviewed in the first part of this paper, attention being given

Table 1: Comparison of Average Yields of Three Major Crops With the "Potential Yield" Based on Experiments or Farmers' Results.

	Yield, tonnes per hectare		
	Wheat	Rice	Maize (corn)
Potential Yield:	14	14	22
Average Yield (from FAO):			
Developed countries - 1961-65	1.7	4.9	3.5
1979	2.3	5.9	5.8
Developed countries - 1961-65	1.0	1.6	1.1
1979	1.4	2.1	1.3
Best country with large production - 1979	5.9	6.6	6.9

- derived from Cooke (1982).

the situation. Major constraints to crop growth, which often cannot be adequately overcome by the traditional methods, are commonly caused by stresses which root systems experience in the soil (Table 2). To relieve them efficiently we require an understanding of the principles which govern the growth and function of root systems, and their interaction with the soil. This subject is considered here in relation to physical stresses, although some of this information is relevant also to relieving stresses from other causes, which is discussed later in these proceedings.

Roots and the Regulation of Plant Growth

Until recently, the roots of plants were thought to fulfil only three essential functions - the absorption of water, the absorption of nutrients and anchoring the plants in the soil. A fourth equally essential function is now known: roots participate in the hormonal control processes which regulate plant growth. Detailed discussion of these complex and still partially understood mechanisms is given in these proceedings by Bruinsma (1983) and Nickell (1983), who also consider possibilities for benefitting crop production through the manufacture of synthetic substances. However, for two reasons, account must be taken of exogenous hormones in discussing the effect of stresses in the soil on plant growth. Roots, especially their apices, are major sources of four of the main classes of growth regulator which are now recognized, viz., cytokinins, gibberellins, ethylene, and abscisic acid. These interact with substances produced in shoots, for example, auxins, to regulate growth throughout the plant. Many stresses in the soil affect hormonal synthesis in roots and this is sometimes the primary cause of injury to shoots. It is of practical importance to realize this as some hormonal effects on shoots could otherwise be mistakenly attributed to an inadequate supply of water or nutrients.

It is hoped that further research on hormonal mechanisms will explain responses of roots, which are at present little understood, but are of far-

Table 2. Stresses in the Soil Which Can Restrict the Ability of Plants to Use Nutrients Which Are Present In It.

Physical stresses	Water stress Mechanical impedance Anaerobiosis (Waterlogging) Unfavourable temperature
Chemical stresses	Toxic substances and unfavourable pH Nutrient inbalance
Biological stresses	Pests Diseases Competition from weeds

reaching significance in their behaviour under stress. Compensatory growth (Russell, 1977) serves as an example. If one part of the rooting zone is more favourable for root growth than another - and this is the usual situation when stresses arise - the favourably placed roots can grow better than if the whole root system experienced that environment, thus compensating for reduced growth of roots elsewhere. The localization of nitrogen and phosphate fertilizers in the soil can evoke this response, which is here illustrated by an experiment in an artificial system. A young cereal plant, three weeks old, illustrated in Figure 1, had been grown in solution culture with a deficient supply of nitrate to the root system (0.01 M), except for a short segment which grew through a compartment 4 cm deep which received an ample supply (1.00 M) and was then returned to the dilute solution (Drew and Saker, 1975). Within a few days, the weight of the "well fed" zone increased and subsequently there was a considerable localized proliferation of lateral roots. Increased uptake of nitrate in that zone had caused the preferential translocation to it of metabolites from the shoots; the mechanism responsible is as yet unknown but, seems explicable only in hormonal terms.

When these plants (the L treatment) were compared with those for which the entire root systems had received the ample nitrate supply (the H treatment), it was found that, although the rate of growth (g/g.d) of the L plants was less in the early part of the experiment, there was no difference between L and H plants when the photograph was taken. Moreover, the nitrogen content of the two types of plant differed little. Five times as much had been taken up by the "well fed" zone in the L treatment than by the corresponding zone in the H treatment; this was due both to the numerous laterals and to the rate of uptake; absorption per unit weight of root, from the same external supply, can increase if less root is present to absorb it or if the requirement for metabolism elsewhere increases (see also; Pitman, 1972). Again the detailed mechanism is unknown but a hormonal response is suspected.

The scope for further research by biochemists and plant physiologists is evident but, for the present discussion, the salient facts are clear: the performance of the whole plant is closely coordinated; changes in the environment which root experience can affect the pattern of growth throughout the plant in many ways; moreover metabolism in shoots can influence absorption of nutrients by roots.

particularly to the effects of physical constraints in soil and opportunities for mitigating them. The second part of the paper deals with the history and present state of crop nutrition involving fertilizers. Although fertilizers have been with us for nearly a century and half, there are still many problems in securing their efficient use, and in integrating them into improved and more intensive agricultural systems in both developed and developing countries. These problems are outlined. The need is stressed for governments to recognize that these scientific problems exist, and that they must provide the support for research and development to investigate soil/root relationships more fully so that more food is available, and at lower cost through the more efficient use of fertilizers and other inputs.

Many results of past research have not yet been fully applied. Experiments, and trials on farmers' fields in many countries, show that average yields are now much below those which are possible. The realization of this fact has been much encouraged by estimates, based both on field experiments and theory, of what could be achieved if we eliminated all constraints which we are now capable to control. While the potential yield estimated in this way is hardly attainable on the world-wide average, the comparison made in Table 1 for three major crops is nonetheless a stimulus. A very considerable "yield gap" between average and potential yields is evident; so also are the achievements of the last two decades. In some developed countries where soil and climate are most favourable, a fairly close approach to the theoretical potential is now possible. Generally, in the developing world greater environmental constraints and fewer resources make the situation different. For example, the International Rice Research Institute concluded that the "practical" potential yield of rice in the Philippines is about half that shown in Table 1 (IRRI, 1978). But this still leaves a considerable "yield gap". More importantly, IRRI and its associates have demonstrated, dramatically, how interdisciplinary effort can reduce this gap. In 1972, the average rice yield of farmers in the Philippines was 1.6 t/ha and it was not clear why it was much below that obtained by research workers (Chandler, 1975). However, a few years later, an interdisciplinary "constraints" program of IRRI and its associates on about 400 sites in six Asian countries, identified constraints many of which arose in the soil, and led to average yields of 3.6 and 4.3 t/ha for wet and dry season rice respectively (Herdt, 1979; De Datta, 1981). It was also shown that greater inputs could increase yields by a further 25 to 30%; this would make them more than twice the average for developing countries shown in Table 1.

No other crop in the developing world has yet been studied as widely as rice and there is no reason to believe that it is the only one of which the yield could be doubled by the fuller use of existing knowledge. Closing the "yield gap" is one of the most important tasks for agricultural science in the immediate future. Progress will depend primarily on the effort devoted to experimental work under the varied field conditions which farmers experience.

SOME ASPECTS OF ROOT/SOIL RELATIONSHIPS

Faced with our present problems, it may seem surprising that the manner in which roots interact with the soil received so little attention at an earlier time. But there is a simple explanation. In those favourable soils of temperate regions, where modern agricultural research developed, it often seems that the soil imposed little constraint on plant growth, provided that adequate nutrients were available, the land being well drained and elaborately cultivated. Fertilizers and manure could satisfy the first requirement; manual labour and machinery could deal with the other two. A detailed understanding of how roots grow and interact with the soil thus seemed unnecessary. The extension and intensification of agriculture to less favourable soils and climates have changed

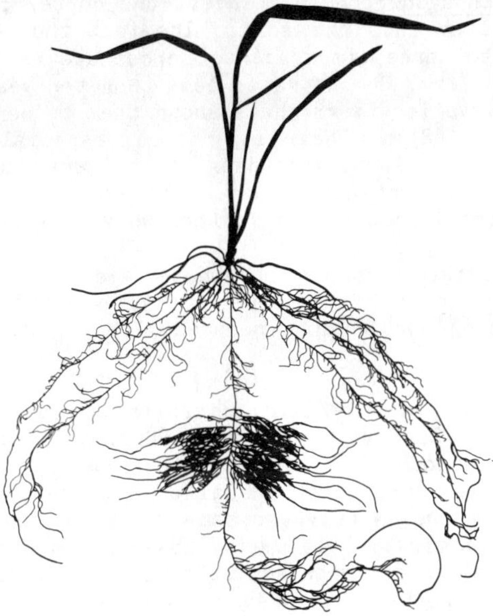

Figure 1: Modification in the form of roots of barley (Hordeum vulgare) caused by providing an enhanced supply of nitrate (0.1 mM) to the mid-part of one root axis; the remainder of the root system received one-hundredth of that concentration. Photographs taken 15 days after treatment (Drew and Saker, 1975).

Movement of Nutrients Through the Soil Into Roots

The quantity of roots produced by crop plants varies widely depending both on species and the environment, but they always occupy only a small fraction of the total volume of the soil - often less than 5% in the zones where they are densest (Wiersum, 1961). Thus, if nutrients diffuse so slowly in soil that little reaches the surface of roots, they will be largely inaccessible to the crop for which they are provided; those planted in later years may, however, benefit. The properties of ions which cause them to move slowly can also lead their fixation, which is discussed later in this paper. On the other hand, if nutrients move very freely through the soil, at a rate little different from that of water, the efficiency with which they are used can be much reduced by leaching.

Nitrogen fertilizers in general move sufficiently rapidly for local depletion to be prevented and the main problem is often to avoid loss by leaching or in other ways. Phosphate fertilizers are in sharp contrast. They diffuse through moist soil at much less than one-hundredth of the rate of nitrate, and they may be virtually exhausted in the soil adjacent to roots even though the supply a few mm away is unaffected (Nye and Tinker, 1977). There is good evidence that, when abundant root hairs penetrate the soil, phosphate is readily taken up in the zone they penetrate. But it remains uncertain whether this is due to absorption by the hairs themselves or because the abundant microorganisms and exudates in this zone increase the mobility of phosphate (Russell, 1977). The consequence of practical importance is that the restraint on absorption imposed by slow diffusion is reduced because the diameter of the root hair cylinder is much greater than that of the root itself.

Recently, it has been realized that a further mechanism can facilitate the movement to roots of some nutrients, especially phospate, when it is in deficient supply - a common situation in many tropical regions. Vesicular arbuscular mycorrhizas of the genus Endogone commonly infect the cortex of roots and develop hyphae which can extend 2 cm into the soil. The fact that they are capable of providing roots with two to three times as much phosphate as could reach them by diffusion through soil was first demonstrated less than ten years ago (Sanders and Tinker, 1973). Field surveys in Nigeria have shown them to be present on roots of many crop plants (Sanders, 1981). Their importance, especially for cassava, has also been indicated in other regions, including South America (Howeler, 1981). At present, adequate quantitative results from field experiments are lacking, but there can be no doubt of the importance of further work on this subject.

Potassium is intermediate between phosphate and nitrate in its movement through the soil. In some soils, it moves freely, but when 2:1 clay minerals occur, its rate of diffusion can be much reduced, leading to localized depletion close to roots.

Knowledge of the parts of the roots systems which are able to absorb nutrients is also important for considering the efficient passage of nutrients to them. Improved research methods have lately shown that all parts of the entire root systems of cereals, and some other annual crops, can absorb the three main nutrients, N, P and K, at approximately the same rates if the external supply of water and nutrients does not vary. Calcium is, however, an exception. Its uptake is confined to the apical parts of roots and it does not move downwards within the roots to the young growing parts. An adequate concentration of calcium in the zone where they extend is therefore essential.

Physical Soil Stresses Which Restrict Root Growth and Function

The size and distribution of pores in the soil are its most important structural characteristic which affect the occurrence of physical stresses to roots. The following simplified classification of pores is used here:

> Transit pores of channels, exceeding cir. 50 μm in diameter. Water moves freely through them under gravity; air enters and roots commonly grow down them.
>
> Storage pores, cir. 5 - 50 μm in diameter. They hold water against gravity but at a potential (free energy) sufficiently high for roots to extract it.

Abundant smaller pores are also present, but do not enter into this discussion as the water in them is at too low a potential to be readily available to plants. Both the pedological origin of the soil and its management can greatly influence its pore characteristics.

Water Stress: When there is a complete ground cover of vegetation, water can be lost into the atmosphere at about three quarters of the rate from a free water surface; solar radiation incident on leaves is the driving force. This considerable use of water is an inescapable consequence of the structure of leaves which provides for gas exchange between leaf cells and the atmosphere, on which photosynthesis depends. Stomatal closure, which is influenced by abscisic acid, one of the hormones produced in roots, can restrict water loss during the daylight hours when insolation is highest. Plant species differ much in the efficiency of this process and also in other characteristics which can reduce water use. Plant breeding may improve their water economy but a considerable loss is inescapable in rapidly growing plants.

Water moves along a gradient of decreasing water potential from soil through the plant and, if transpirition losses are not balanced by absorption, injury can be rapid - much more so than would be often caused by a similar interruption in the supply of nutrients. Some decrease in root extension can occur when the soil water potential is still above -1 bar; nutrient uptake and yield can be rapidly affected as the potential falls and death can occur if the entire root system experiences -10 to -15 bar. However the dessication of the older parts of roots, which are nearest to the soil surface may have little effect, provided that sufficient of their apical growing regions receive adequate water. This characteristic is often of prime survival value during drought because water is lost much more rapidly from the surface soil than from the deeper layers; the water potential at the depth of 20 - 25 cm may fall from close to zero to below -5 bar within a week, but at the depth of about 1 m the water potential decreases much more slowly. Continuing growth then depends on the downward extension of roots being sufficiently rapid for an adequate part of the root system to be in soil where water remains available. Compensatory growth can much assist this; decreasing water potential in the surface can cause roots to be more numerous, sometimes by a factor of more than four, in the deeper soil (Klepper et al., 1973).

The main characteristics of soil which facilitate survival of drought are: (1) the conservation of water in the rooting zone; this depends on there being sufficient stable and continuous transit pores for rain from periodic storms to enter, and adequate storage pores to retain it, and (2) conditions which permit deep root penetration. Major requirements for this are an adequate depth of soil and the absence of stresses which restrict root penetration; mechanical impedance can be among the most important constraints which soil management can mitigate.

Mechanical Impedance: Modern understanding of the effects of this stress in field conditions can be considered to date from the demonstration, by Wiersum (1957), that roots cannot decrease in diameter to enter pores smaller than themselves; indeed mechanical constraint commonly causes their diameter to increase. If they have to resist pressures as low as 0.2 bar to expand pores their rate of elongation can be much reduced (Figure 2) though it can continue at a slow rate when pressures are considerably greater. Species differ but present evidence suggests that root growth is always retarded if roots have to resist external pressures of 0.5 bar or more. Reports that a few species have a special ability to grow in heavily compacted soil may be due to them being able to survive with a very slow rate of root extension. The sensitivity of roots to very small external pressures is due to physiological mechanisms in the apical meristem which are presumably either hormonal, or the result of changes in fine structure (Russell and Goss, 1974; Goss and Russell, 1980).

If the extension of root axes is impeded, but there are pores large enough for the finer laterals to enter freely, these latter show compensatory growth; laterals proliferate immediately above the zone where compaction occurs. When adequate nutrients and water are present in this restricted zone the yield of plants may be little affected, but, as explained above, shallow rooted plants are very susceptible to injury from even slight water stress. Low water potential in the soil also reduces root elongation and increases soil strength, i.e. its resistance to deformation, so that the forces roots must resist in enlarging pores are increased. Thus the effects of water stress and mechanical impedance can interact positively in several ways.

Roots can penetrate the soil by two strategies and increasing interest in reduced cultivation or "no till" systems makes it important to distinguish between them. If the soil is relatively homogeneous, with few continuous pores larger in diameter than roots, the extension of roots depends primarily on their ability to expand pores. Because pore size and the bulk density of soil vary inversely, the measurement of this property may then indicate when it is desirable to increase porosity by soil disturbance. The use of penetrometers can also help by showing

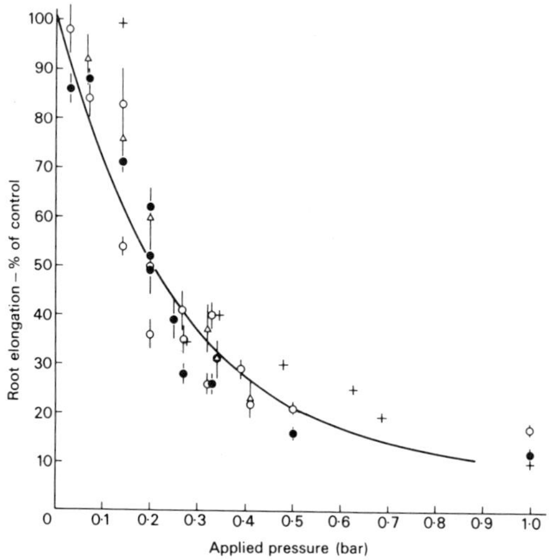

Figure 2: Effect of pressure which the roots of barley (Hordeum vulgare) have to overcome to penetrate artifical porous systems. The diameter of the barley roots was about 400 μm and the diameter of the pores ranged from 15 μm (triangles) to 150 μm (open circles). For sources, see Russell (1977).

changes in soil strength. By contrast, in heavy soils with massive structure, for example clays, the penetration of roots can depend primarily on the existence of continuous transit pores or fissures, caused often by soil shrinkage, burrowing organisms or dead roots. These can persist in undisturbed soil but are largely destroyed by ploughing. Because of the small fraction of the soil volume which roots occupy, quite limited numbers of these channels can be sufficient for their downward penetration so that measurements of the bulk density of the soil do not provide guidance on their occurrence. The same is true with "gravel horizons", which are common in some regions; even though the bulk density of the intervening fine material may suggest that roots can penetrate freely, this may be prevented if the gravel aggregates are so close together that the fine material can "lock" between them when roots are expanding pores (Wiersum, 1957; Harrison-Murray, 1978). These are examples of responses which at present can be adequately detected only by examining living roots in situ and with knowledge of the morphological effects of mechanical stress on roots.

Anaerobiosis: When roots and the microorganisms of the soil are respiring rapidly they can, within a day, consume about as much oxygen as is normally present in a well aerated soil. Because oxygen diffuses very slowly in water - at about 1/10,000th of the rate in air - anaerobic conditions can develop rapidly unless sufficient transit pores in the soil are filled with air. Anaerobiosis is thus caused by waterlogging and it can occur transiently in normally well drained soils if their surface is degraded by intense rain so that all pores are occluded with water for a limited time (Wien et al. 1979).

Except in plants such as rice, in which aerenchyma allows oxygen to pass downwards from shoots, a continuing supply of oxygen at the surface of roots is essential for normal growth. Plants differ much in their sensitivity to waterlogging, which can also vary with their age. Some legumes die after 3-4 days waterlogging. The absorption of water and nutrients is affected, nodulation can be inhibited; attack by microorganisms after the original injury may hasten death (Wien et al., 1979). However, in at least some cases effects on hormonal processes are the primary cause of injury; epinasty in tomato plants, which superficially can suggest wilting of leaves, occurs rapidly and is known to be due to ethylene formed from a precurser, which is translocated from roots (Jackson and Campbell, 1976; Bradford and Yang, 1980). In contrast ethylene can sometimes assist survival of waterlogging. Drew et al. (1979) have shown that it is responsible for the development of cortical air spaces, comparable to the aerenchyma of rice, in the crown roots of maize (corn) when oxygen is in limited supply.

In addition to depriving roots of oxygen, anaerobic conditions can give rise to many toxic products in the soil. For example in root apices can be rapidly killed by fatty acids, especially acetic acid, produced by the decomposition of water-saturated plant material. This is well known in wet rice soils (Wang et al., 1967) and can also occur in temperate conditions (Lynch, 1978).

It is evident that some of the soil physical characteristics which discourage anaerobiosis may increase the susceptibility of plants to water stress. Very rapid drainage through the soil, which shorten periods of waterlogging, could reduce the amount of water retained in the rooting zone for subsequent use by plants, and while numerous storage pores are beneficial during water stress, they can decrease the drying of waterlogged soils. Thus, since water stress and mechanical impedance also interact, it may be necessary to consider these two stresses and waterlogging conjointly in assessing soil conditions for plant growth.

Unfavourable Temperature: Temperatures in excess of $40^{\circ}C$ are not uncommon a short distance below the soil surface in the tropics (Lal, 1974). This is appreciably above the optimum mean temperature for root growth in any species, which appears to be about $33^{\circ}C$ (Cooper, 1973). However there is inadequate field data to show whether high temperature, as opposed to low water potential in surface soil, is ever a major cause of injury to young plants. The likelihood that high temperatures affect well established root systems, which have reached below the depth of cir. 20 cm, seems remote on the basis of Lal's measurements, especially as shading by foliage is likely to reduce soil temperature in the immediate vicinity of stems. If, nonetheless, the mitigation of the effects of high temperature on roots did become important the obvious solution would be to shade the soil by mulches or in other ways. Since there are already strong reasons for mulching to increase the infiltration of water and reduce soil erosion (Lal, 1983) no separate discussion of mitigating the effects of high temperatures on root systems is necessary.

LIVING ROOTS AND SOIL CONDITIONS

The prime importance of a water-stable pore regime for minimizing stresses to root growth from water stress, mechanical impedance and anaerobiosis, is obvious; it is equally important in the control of soil erosion. Organic carbon compounds and especially polysaccharides are involved in stabilizing the soil but the mechanisms responsible are not fully understood. Roots and microorganisms, especially fungi, are believed to be mainly responsible for preserving aggregates greater than 250 μm in diameter, on which desirable soil structure largely depends (Tisdall and Oades, 1982). Materials released on the death of roots has often been thought to be their main contribution, but the direct involvement of living ones now seems also to deserve attention; they can exude considerably more organic materials than was formerly suspected. Studies using ^{14}C have shown that about 15% of the carbon fixed in photosynthesis can be released into the soil during a few weeks by young plants, in which time root death is negligible (Barber and Martin, 1976). Soil microorganisms can stimulate the release of exudates and also rapidly respire a considerable fraction of them. There seems no doubt that these ephemeral substances make a contribution to stabilizing the soil; particles often adhere firmly to living roots and an increase in soil stability is sometimes, though not always, observed when young plants have been grown in pot culture for a few weeks (Reid and Goss, 1981). Although the extent to which products from living roots contribute to soil stability in field conditions is not known, it is beyond doubt that the destruction of root systems by soil disturbance, and the enhanced soil respiration to which it often leads, can make soil conditions less suitable for subsequent crops, especially when the soil is left bare for an appreciable period.

Despite the limitation of present knowledge of plant-soil interactions, the developments here referred to have increased our understanding of ways for mitigating stresses in the soil which can restrict the extent to which the nutrients are used by preserved plants. More is known also of how desirable characteristics of the soil can be preserved or improved by its management (see Lal, 1983).

CROP NUTRITION

The greatest contribution of chemistry to removing constraints imposed by soil on crop growth has been in plant nutrition. The identification of the simple chemical ions that are plant nutrients (which occurred early in the last century for major nutrients and, in this century, for micronutrients), was the greatest breakthrough in the whole history of agricultural science. Chemical industry responded by making these ions available to farmers as fertilizers. The outcome was that crops need no longer be limited in their growth by lack of nutrients; the amounts of elements required to supplement deficient supplies in soil could be supplied in forms immediately available to plants. This simple concept is, however, at variance with the considerable complexities involved in plant nutrition, and the difficulties that are now recognized in using fertilizers efficiently.

History and Present World Use of NPK Fertilizers

Fertilizers were being used in Europe by the 1830's - bones were used to supply P and nitrates were imported from South America. Superphosphate was first made in the 1840's and the fertilizer industry steadily developed from that time. Ammonium sulfate, made from the waste liquor from gas-making, supplied N until the Haber-Bosch process for fixing atmospheric N was established in 1912. The mining of potassium deposits began late in the last century. Developments in this century have included a wider range of chemical forms of fertilizers, the provision of liquid fertilizers and a great improvement in the physical condition of solids, as well as a vast expansion of the industry.

Information on the amounts of N, P and K used as fertilizer is now published annually by FAO (1981). Estimates of the amounts applied at intervals since 1913 are in Table 3. In the last 40 years use has increased by 22 times for N, by 9 times for P and by 8 times for K. In total the developed countries use 50% more N, nearly 3 times as much P, and nearly 6 times as much K as is used in developing countries. Relative to N, developed countries use about one-third as much P and one-half as much K by weight; in the developing countries these fractions are one-sixth as much P, and one eighth as much K. These small proportions of P and K relative to the N used in developing countries cannot continue if production is to be increased. Soils of the developing countries have small reserves of all nutrients which just suffice to maintain yields in traditional farming systems. When fertilizers are used to supply more N and to raise yields the supplies of P, and particularly the supplies of K, in the soil are rapidly depleted and the amounts of these nutrients that are available then limits yield. Consequently as developing countries increase their production by using more N-fertilizer and improved varieties of crops the amounts of P and K used will have to be increased if a full return is to be obtained from other inputs. The need for using more K in intensified systems in India has already been shown (Roy et al. 1978).

Table 3 also shows the great variations in rates of application of fertilizers in the continents of the world, using data given by FAO (1981). The small amounts of N, P and K used in Africa, and the large amounts used in Europe, are striking features; so also is the fact that in the period 1967-78, the percentage increases in the use of fertilizers were greater in developing than developed countries (von Peter, 1980). FAO has estimated that, if the needs for food of the World's expanding population are to be met, the fertilizers used in developing countries would have to increase by seven times between 1975 and the year 2000. More modest estimates are that the use of NPK must increase by four or five times in the next 20 years.

Factors Affecting Future Use of NPK Fertilizers

The constraints that impede the increased use of fertilizers have been discussed by von Peter (1980) and also by Mathieu and de la Vega (1978). They include 1) lack of fertilizer; 2) shortage of credit; 3) poor advisory services and badly informed farmers; 4) poor quality of seeds; 5) inadequate water for irrigation; 6) patterns of tenancies; 7) government action in setting ratios between prices of crops and fertilizers; and 8) lack of basic information on the effects of fertilizers on crop yields. The last two constraints require further comment.

The most effective way of increasing fertilizer use, and therefore of increasing food production, is to establish ratios for value of crop (v), to the cost of fertilizer (c), that make the use of fertilizer attractive to the farmer. Examples of such relationships are shown below (IFC, 1981 and FAO, 1982).

	$\frac{v}{c} = \frac{\text{price of paddy/kg}}{\text{price of N/kg}}$	N used on all arable crops 1979 average kg/ha	Yield of paddy 1979-1981 avge t/ha
Nepal	21	6	1.8
Philippines	30	23	2.2
Sri Lanka	55	40	2.5
Indonesia	65	32	3.3
Korean Republic	101	199	5.5

Governments must recognize that, if they have a food deficit, the remedies are in their hands. The surest way to have more food produced is to remove financial uncertainty from farming by establishing v/c ratios which make it profitable for farmers to use the inputs that research and development have shown to be necessary to increase production. Three-quarters of the increase in food production in developing countries in recent years has come from achieving larger yields on existing cultivated land, and only a quarter by bringing new land into cultivation (von Peter, 1980). This trend will continue; both administrative

Table 3: World and Regional use of NPK Fertilizers

	N	P	K
Total Amounts Applied in the World, Millions of tonnes			
1913 (approx.)	(1.3)	(0.8)	(0.7)
1939	2.6	1.6	2.3
1954	5.6	2.8	4.7
1960	9.7	4.2	7.1
1970	28.7	8.2	12.9
1975	38.9	9.9	16.5
1980	57.2	13.6	19.5
Total Amounts, 1979/80, Millions of tonnes			
All Developed Countries	34.7	10.0	16.6
All Developing Countries	22.6	3.6	2.9
Ratios Applied 1979/80			
World	1	0.24	0.34
All Developed Countries	1	0.29	0.48
All Developing Countries	1	0.16	0.13
Rates of Application to Agricultural Land in 1978, kg/ha			
World	11.3	2.9	4.5
Africa	1.4	0.4	0.3
North-Central America	18.8	4.2	8.3
South America	2.2	1.4	1.7
Asia	14.1	2.3	2.1
Europe	61.1	17.5	31.5
Oceania	0.5	1.1	0.4
All Developed Countries	17.3	5.1	9.0
All Developing Countries	6.8	1.3	1.2
Annual Rates of Growth, 1967-78, in Percent Per Annum			
World	7.5	5.1	5.8
Developed countries	6.5	3.7	5.3
Developing countries	10.3	12.0	10.4

action relating to the profitability of inputs, and research and development on the efficient use of these inputs, must concentrate on raising the productivity of land that is now cultivated.

Effects of Fertilizers on Crop Yields

When crops have an environment that provides adequate water, and gives complete freedom from pests, diseases and competing weeds, their growth is directly proportional to the supply of nutrients, up to the limit that genetic capability and solar radiation make possible. These ideal circumstances rarely exist in practical framing, but in modern research on the causes of yield variation such conditions must be established by the use of inputs (fertilizers, irrigation if needed, and agrochemicals to control pests, diseases and weeds) so that the true potential of crops are established in the field. Yields are built up by the interactions which remove constraints; these must be investigated and understood.

Before a fully scientific basis for advice on the use of fertilizers can be provided the constraints affecting yield in an area must be identified and their interactions with fertilizers established. These facts are commonly ignored and no nation has a complete understanding of the effects of inputs, and their interactions, on crop yields for all the regions of the country. These circumstances have arisen because fertilizers were first used, and methods for advising on their use were developed, in the well-watered regions of temperate countries. It was considered sufficient to make field trials of the effects of fertilizers on crop yields and to apply economic techniques to the results of these trials. Simple trials which test the effects of increasing amounts of fertilizers, and sometimes compare different forms and methods of application, do no more than sample the environment with its constraints on crop growth resulting from climate, soil properties, and the limited capabilities of the farmers who grow the crops. While such trials help provide immediate advice to the farmers,they do not satisfy the longer-term objective of having full scientific control of crop growth and thus the ability to obtain yields nearer to established potentials. These limitations to the value of field experimentation with fertilizers cannot be accepted indefinitely.

The remedy can only be to establish research and development programs which identify constraints to the growth of major crops in particular regions, and then investigate the interactions between the fertilizers and the other inputs, which remove these constraints. The work can begin with simple experiments such as described by Widdowson and Penny (1965) who identified the constraints on yield of wheat imposed by shortage of N, by root disease and pests, and by drought, and investigated the interactions of the inputs applied to overcome these constraints. Multidisciplinary factorial experiments are the next step. Lester (1980, 1981, 1982) has described elaborate studies of this type, and the work of IRRI illustrates the application of this method in developing countries. Establishing such research and development programs is the best investment that governments can make in implementing policies for agricultural development, for the knowledge acquired of local resources and problems has permanent value despite any changes in economic or social conditions.

The essential role of soil surveys in trasferring information from experimental sites to farmers' fields must be stressed. The soil classification and mapping which result from the surveys take account of parent materials, climate, and soil-forming processes which are the factors that determine the nutrient status of soils in their unimproved condition. Soils that are similarly classified, and with similar past histories, will have similar needs for fertilizers. Governments must recognize the need for soil surveys as the basis for agricultural improvement and subsequent management in a region.

Other Nutrients Required as Fertilizers

Most discussions on fertilizers relate to N, P and K, but when the other major nutrients (Ca, S and Mg) and the micronutrients (B, Cu, Fe, Mn, Mo and Zn) must be applied to supplement the reserves in the soil and so grow a larger crop, they too are fertilizers. There are no published statistics to show the amounts of these elements applied as fertilizers to the World's crops. The only assessments of current needs that can be made are from the results of field experiments testing these fertilizers, and soil analyses, made by local workers.

Calcium: Many tropical soils are very acid and contain much soluble aluminium which is toxic to many crops; some contain so little calcium that they cannot supply sufficient of this nutrient. Crops that have traditionally been grown in these areas tolerate the acid conditions and high concentrations of Al which are serious obstacles to the introduction of other species of plants. Much may be done by breeding crops to develop varieties which tolerate the chemical stresses caused by soil acidity, for example, wheat varieties that tolerate very acid soils have been bred in Brazil. However this remedy cannot be applied to all crops, e.g., no variety of soybean can thrive on very acid soils. Assessments must be made of the need for lime to precipitate soluble Al, and to supply Ca, wherever improvement of acid soils is planned; the value that plant breeding may have in adapting crops to tolerate soil acidity must be explored further. Work will also be required on methods of diminishing subsoil acidity to encourage the deep rooting of crops that helps in overcoming drought, and on the assessments of lime needed to precipitate soluble Al and on the timing of the dressings, so that losses of Ca by leaching are minimised.

Sulfur: Deficiencies of sulfur are common in many developing countries (IFDC 1979). In most temperate and industrial regions the atmosphere supplies more than sufficient S for crop growth by direct deposition and in precipitation. In the tropics the atmosphere supplies much less S, but too few data are available for any adequate assessment of the need to use S as a fertilizer. Studies on the S-cycle, on S-reserves in soils, and on crop responses to S-fertilizers are urgently needed in most developing countries. S is certain to be needed in larger quantities as natural sources will be insufficient to support the larger yields when systems are intensified by applying other fertilizers and supplying other inputs. Research will also be required on the materials and methods to secure high efficiency in the use of S-fertilizers.

Magnesium: Deficiency of Mg does not appear to be a common cause of poor yields in developing countries at present. Mg is supplied by rain, also by dolomitic liming materials. Little has been done on the Mg-cycle in tropical countries. Although the need for investigations does not appear to be urgent, the situation must be kept under review. Experimental work on crop responses to Mg-fertilizers, and tests of alternative sources of Mg are desirable.

Micronutrients: These elements will be needed in increasing quantities as efforts are made to increase production in developing countries. For example, half of the soils of India are reported to be deficient in zinc (Tandon et al., 1981). In most other countries little has been done to assess the reserves of micronutrients in soils and the likely need to apply them as fertilizers. More studies similar to that of Sillanpaa (1982), sponsored by the Government of Finland and FAO, are needed. Even this latter comprehensive work said the findings could only "be considered as general guidelines to draw attention to the nature of problems likely to arise and to the future work necessary to resolve them". All new work must be related to soil classification since parent material and weathering processes determine the amounts of trace elements present in soils. The need for research on the most efficient methods and materials to supply micronutrients is also urgent. There are large differences in the amounts and costs of these elements needed to correct deficiencies by 1) applications to soil, and 2) foliar sprays. We know little of the value to later crops of the large residues left by soil applications. There is controversy on the rival merits of applying soluble salts or chelates of the elements to both soil and leaves.

Efficiencies of NPK Fertilizers

The efficiency of fertilizers has received far too little attention in the past. The objective of all scientific work on the subject must be to secure 100 percent uptake of the applied nutrients by the crop, or series of crops, for which the fertilizers have been applied.

Nitrogen: Often no more than 30% of the fertilizer nitrogen applied is recovered by a crop. Much of the remainder is lost by leaching or dentrification in wet periods, volatilisation of ammonia also occurs; but most early investigations failed to account for a considerable proportion of the N applied. In developed countries little interest was taken in this serious inefficiency of an input, either by fertilizer manufacturers (whose sales would logically be diminished if the N sold was more efficient), or by farmers who were more concerned with buying fertilizer cheaply than with using it efficiently. Discussions on environmental pollution began to change these attitudes in the late 1960's. Leached nitrate damages the quality of natural waters by eutrophication, and it may pose a problem to health when much is present in drinking water. Nitrous oxide, evolved by dentrification, may interfere with the ozone layer of the stratosphere. These concerns, reinforced by the large increases in the price of oil, and hence in the cost of fixed nitrogen, have given high priority to research on the efficiency of N-fertilizers.

Much success has been achieved. Work with ^{15}N has made it possible to account for about 90% of the N-fertilizer applied (Jenkinson 1982a, 1982b). Modelling of the transformations and movement of forms of N in soil, in relation to soil type, weather and crop, has lead to more precise recommendations for applying fertilizer (Greenwood, 1982). Growing a larger crop by the use of their inputs to high-yielding varieties can double the amounts of N recovered by the crop. Research on the production of forms of N-fertilizers, and on methods of application, that lead to higher efficiency have been successful, particularly for flooded rice culture. IFDC (1982) has taken a leading role in this research to improve the efficiency of N. McCune (1981, 1982) has described research on the production and use of large granules placed deeply in soil, on coating of granules to delay their solution, and on nitrification and urease inhibitors; all of these techniques have a place in raising the efficiency of nitrogen.

Phosphate: The reactions by which soluble phosphates are precipitated in soil ('fixation' of phosphate) result commonly in not more than a quarter of the applied phosphate being recovered by the first crop. While there has been considerable advance in understanding of the processes involved, the only practical advance has come from placing the phosphate close to seeds or roots of the crop which has increased short-term efficiency. The residues of phosphate fertilizers in soil retain some solubility and can benefit later crops, and they enhance soil fertility. When account is taken of recoveries over very long periods, phosphate fertilizers can have very high efficiencies. More research is needed on phosphate economy in developing countries to determine whether near-complete recoveries can be expected in the long-term on most soils where residues have built up the level of soil phosphate. Research, such as that done by IFDC (1982), is required on forms of phosphate, and particularly on the use of indigenous rock phosphates processed in various ways. The problem in developing countries is whether to seek for short-term efficiency of P by using special forms of soluble phosphate and placement methods, or to regard phosphate dressings as a long-term improvement of the soil, and use the cheapest effective source to build up residues that enhance phosphate status. This problem has been discussed by McCune (1982) who states that "it may not be advisable for developing countries to adopt developed country fertilizers and concepts" and "more work is needed to tailor fertilizers to meet the special needs of developing country agriculture". These statements are particularly relevant to phosphate.

Potassium: This cation is retained in soil in the exchangeable form and is not readily leached except from soils with very little clay and very small exchange capacity. In soils with 2:1 clay mineral lattices reserves of K are accumulated in 'fixed' forms which can be slowly used by later crops; again on a long-term basis K can be a very efficient nutrient almost all of a fertilizer-K dressing being recovered by a sequence of crops. Tropical soils commonly have only 1:1 lattice clay minerals which lack this capacity to accumulate fixed K as a reserve protected from leaching. The relatively small amount of work done on the efficiency of K in tropical soils indicates that there should be no serious difficulty in securing the high efficiency of this nutrient applied as a fertilizer, providing that attention is given to applying the correct amounts at the right times which will be judged by the rainfall, the root development of crop and the rate of K uptake.

FUTURE SUPPLIES OF NITROGEN

In many regions the supply of nitrogen determines agricultural productivity and this element is certain to be a limiting factor in most schemes for development that involve cereals, root crops and pastures. therefore the source of future supplies of N require careful consideration and planning, as well as appropriate research work. A major factor is the large amount of energy needed to manufacture N-fertilizers through the synthesis of ammonia and alternative processes should be sought.

Alternative Chemical Fixation Systems

In 'natural' (i.e. biological) fixation systems the enzyme nitrogenase operates at normal temperatures and pressures, taking dinitrogen from the air and combining it with hydrogen from water. If this process could be understood so well that it could be operated in a factory, considerable savings in energy would result. This has been the objective of research done in the 1970s and reviewed by Chatt (1976). Ammonia can be produced from compounds having molecular-N attached to elements such as molybdenum or tungsten at ordinary temperatures and pressures in an aqueous medium supplying protons. Cyclic catalytic systems in which the metal compound is not destroyed are now being sought. Chatt concluded - "Obviously the goal is still far away, but at last one can see it as a definite possibility". Work on this topic is a very important challenge to chemists.

Biological Fixation Systems

Chatt (1976) estimated that 176 million tonnes of N were fixed annually by biological processes - 3 times as much as chemical industry provided for world agriculture in 1980. Research is being done on the natural systems of understand them better, to increase their efficiency, and to develop new systems. The potential of this work was discussed by Chatt, and, more recently, by Postgate (1982). The latter concluded - "Genetically-screened, and later genetically improved, inocula for established legumes are likely to become available in future, as are organisms genetically-altered so as to be able to form agronomically significant biocoenoses with non-legumes such as cereals". Beringer and Day (1981) reviewed the improvement of legumes, reporting annual fixation ranging up to 550 kg/ha of N by beans, 460 kg/ha by alfalfa, and 670 kg/ha by clover. Improved clovers have fixed up to 470 kg/ha of N in English experiments. Such work points to the path that must be taken in developing countries where the badly-needed improvement of pasture cannot afford to rely on the treatment of grass with fertilizer-N, the practice which is so common in Europe. The use of clover to fix N will also become more important in much pasture land of temperate countries where the high cost of N-fertilizer deters its use on grasslands. Other possibilities are the transfer of genes for nitrogen fixation to other organisms which may form new symbioses with plants. Work on nitrogen-fixing bacteria should be encouraged; some of these have already been shown to be associated with the roots of grasses and cereals; others fix N while decomposing organic matter added to soil.

CONCLUDING REMARKS

Increased food production will depend on our success in providing information on the working materials of cropping systems - soil, plant, water and nutrients. This demands research on soils and crops, and on the fertilizers essential to increase soil fertility. This information must be used to guide both the advice on using fertilizers in cropping systems that is given to farmers, and also the advice that is given to fertilizer manufacturers. The high efficiency which is so essential to repay the cost of inputs will come from the right fertilizers, applied in the right place and at the right time.

In developing countries many farmers still need to be introduced to fertilizers as vital aids to higher crop yields. The most serious problems will be in formulating and manufacturing materials to suit the constraints set by soil, crop and climate, and by the farmer's capabilities and equipment. The fertilizers that suit agriculture in temperate developed countries may not fit these requirements; modification of their physical form and chemical composition may be required to suit methods of application that secure high efficiency. Full use should be made of indigenous raw materials, notably mineral rock phosphates which may prove superior in the long-term to the water-soluble and citrate-soluble forms of phosphate used in temperate countries. These are challenges which should be solved by the soil chemists and the chemical engineers working together.

In planning to increase food production by agricultural development governments must establish the financial incentives that encourage farmers to modernize and intensify their systems, and give full support to research and development programs to support these changes. The most rewarding investment that governments can make will be to support scientific investigations of the constraints to yield under local conditions; the information acquired will have immediate practical application and it will also be of permanent value to the country's agriculture. Lines of communication must also be examined to ensure that all the information acquired from investigations on the efficient use of inputs is transferred smoothly from research workers to advisers, and then to farmers. Advice based on sound scientific knowledge must not be set aside to suit some temporary economic advantage.

The farming systems of the future, which will ensure freedom from hunger for the world's expanding population, will be based on the results of scientific investigations, and will be controlled by scientific methods. The magnitude of the work that must be done should not be under-estimated. The sequence will always be, firstly to identify the constraints in the production systems, and secondly to devise means of overcoming them by research. It is essential to identify the hierarchy of limiting factors, to know their relative importance, and then to estimate the effects of progressively removing constraints. From the ideal package of improved practices that will be proposed the governments and farmers concerned will then have to choose those components that are practicable and which give the best economic returns. Finally, it must be stressed that the ability to make sound policy decisions at all management levels, from an individual farm, to a country, depends on being able to make reliable predictions of the likely results of changes in production systems. The ability to make these accurate predictions can only come from the results of research and development investigations on the cropping systems of the individual regions of a country.

REFERENCES

Barber, D.A., and J.K. Margin (1976). The release of organic substances by cereal roots into soil. New Phytol. 76, 69-80.

Beringer, J.E. and J.M. Day (1981). The role of biological nitrogen fixation in U.K. agriculture. Proc. Fertil. Soc. No. 205.

Bradford, K.J. and S.F. Yang (1980). Xylem transport of 1-aminocyclopropane-1-carboxylic acid, an ethylene precursor, in waterlogged tomato plants. Plant Physiol. 65, 322-326.

Bruinsma, J. (1983). Modifying crop performance with growth-regulating chemicals. Chemrawn II Tech. Proc., Pergamon Press Canada, Ltd.

Chandler, R.F., Jr. (1975). Case history of IRRI's research management during the period 1969 to 1972. Asian Vegetable Research Center, Taiwan.

Chatt, J. (1976). Nitrogen fixation - future prospects. Proc. Fertil. Soc., No. 155.

Cooke, G.W. (1982). Fertilizing for Maximum Yield. Third edition, Granada, London, New York.

Cooper, A.J. (1973). Root Temperature and Plant Growth. Research Review No. 4, Commonwealth Bureau of Horticulture and Plantation Crops, Maidstone, England.

De Datta, S.K. (1981). Principles and Practices of Rice Production. Wileys, New York, 546-591.

Douglas, J.T., and M.J. Goss (1982). Stability and organic matter content of surface soil aggregates under different methods of cultivation and in grassland. Soil and Tillage Res. 2, 155-175.

Drew, M.C., M.B. Jackson and S. Gifford (1979). Ethylene-promoting adventitious rooting and development of cortical air spaces (aerenchyma) in roots may be an adaptive response to flooding in Zea mays L., Planta 147, 83-88.

Drew, M.C. and L.R. Saker (1975). Nutrient supply and the growth of the seminal root system in barley. II Localized, compensatory increases in lateral root growth and rates of nitrate uptake when nitrate is supplied to one part of the root system. J. Exp. Bot. 26, 79-80.

Goss, M.j. and R. Scott Russell (1980). Effects of mechanical impedance on root growth in barley (Hordeum volgare L), III Observation on the mechanism of response. J. Exp. Bot. 31, 577-588.

FAO (1981). FAO Fertilizer Yearbook. Vol. 30 (and earlier issues) FAO, Rome, Italy.

FAO (1982). 1981 Production Yearbook. Vol. 35 (and earlier issues), FAO, Rome, Italy.

Greenwood, D.J. (1982). Modelling of crop response to nitrogen fertilizer. Phil. Trans. R. Soc. Lond. B, 296, 351-362.

Harrison-Murray, R. (1978). Crop responses to mulching in a tropical environment with special reference to high soil temperature. Ph.D. Thesis, University of Reading, Reading.

Herdt, R.W. (1979). An overview of the constraints project results. Farm-level constraints to high rice yields in Asia: 1974-77. IRRI, Manila, 395-411.

Howeler, R.H. (1981). The effect of mycorrhizal inoculation on the phosphate nutrition of cassava (Manihot esculenta Crantz). The soil/root system in relation to Brazilian agriculture. IAPAR< Parana, 243-258.

IFC (1981). Urea/paddy price ratios in various Asian countries. Int. Fertil. Corresp. 22 (no. 6), 3.

IFDC (1979). Sulfur in the tropics. Technical Bulletin IFDC-T12 International Fertilizer Development Center, Muscle Shoals, Alabama.

IFDC (1982). Annual Report 1980. International Fertilizer Development Center, Muscle Shoals, Alabama.

IRRI (1978). Annual Report of the International Rice Research Institute for 1977. Manila.

Jackson, M.B. and D.J. Campbell (1976). Waterlogging and petiole epinasty: the role of ethylene and low oxygen. New Phytol. 76, 21-29.

Jenkinson, D.S. (1982a). The nitrogen cycle in long-term field experiments. Phil. Trans. R. Soc. Lond., B. 296, 563-571.

Jenkinson, D.S. (1982b). Work with ^{15}N. Rep. Rothamsted Exp. Stn for 1981, Part 1, 253-4.

Klepper, B., M.H. Taylor, M.G. Huck and E.L. Fiscus (1973). Water relations and growth of cotton in dry soil. Agron J. 65, 307-310.

Lal, R. (1974). No-tillage effects on soil properties and maize (Zea mays L.) production in Western Nigeria. Plant and Soil, 40, 321-331.

Lal, R. (1983). Chemical approaches towards increasing water availability to crops. Chemrawn II Tech. Proc., Pergamon Press Canada, Ltd.

Lester, E. (1980). Multidisciplinary activities. Rep. Rothamsted Exp. Stn. for 1979, Part 1, 17-25.

Lester, E. (1981). Multidisciplinary activities. Rep. Rothamsted Exp Stn. for 1980, Part 1, 17-33.

Lester, E. (1982). Multidisciplinary activities. Rep. Rothamsted exp. Stn. for 1981, Part 1, 19-26.

Lynch, J.M. (1978). Production and phytotoxicity of acetic acid in anaerobic soils containing plant residues. Soil Biol. Biochem. 10, 131-135.

Mathieu, M. and J. de la Vega (1978). Constraints to increased fertilizer use in developing countries and means to overcome them. Proc. Fertil. Soc. 173.

McCune, D.L. (1981). Fertilizers for tropical and sub-tropical agriculture. Proc. Fertil. Soc. 196.

McCune, D.L. (1982). Tailoring fertilizers to developing country agriculture. Fm. Chem. June 1982, pp. A10-11.

Nickell, L.G. (1983). The role of growth regulators and hormones in enhancing food production. Chemrawn II Tech. Proc., Pergamon Press Canada, Ltd.

Nye, P.H. and P.B. Tinker (1977). Solute movement in the soil-root system. Blackwell, Oxford.

Pitman, M.G. (1972). Uptake and transport of ions in barley seedlings. III Correlation between transport to the shoot and relative growth rate. Aust. J. Biol. Sci., 25, 905-919.

Postgate, J.R. (1982). Biological nitrogen fixation: fundamentals, Phil Trans. R. Soc. Lond., B296, 375-385.

Reid, J.B. and M.J. Goss (1981). Effect of living roots of different plant species on the aggregate stability of two arable soils. J. Soil Sci., 32, 521-541.

Roy, R.N., S. Seetharaman and R.N. Singh (1978). Soil and fertilizer potassium in crop nutrition. Fertil. News, 23 (6), 3-26.

Russell, R. Scott (1977). Plant root systems: their function and interaction with the soil, McGraw-Hill, London, New York.

Russell, R. Scott and M.J. Goss (1974). Physical aspects of soil fertility - the response of roots to mechanical impedance. Neth. J. Agric. Sci., 22, 305-318.

Sanders, F.E. (1981), Microbiological factors. B. Vesicular-arbuscular mycorrhiza. Characterization of soils in relation to their classification and management for crop production: examples from some areas of the humis tropics. University press, Oxford, 177-187.
Sanders, F.E. and P.B. Tinker (1973). Phosphate flow into mycorrhizal roots. Pestic. Sci., 4, 385-395.
Sillanpaa, M. (1982). Micronutrients and the nutrient status of soils. FAO Soils Bulletin No. 48. FAO, Rome.
Tandon, H.L.S., V. Kumar, P.O. Meelu, Rajendra Prasad, D.N. Puri, N.S. Randhawa, G.S. Sekhon and Ramendra Singh (1981). Soil fertility and fertilizer use in India. Quarter Century of Agronomic Research in India (1955-1980). Indian Society of Agronomy, New Delhi, 9-25.
Tisdale, M.J. and J.M. Oades (1982). Organic matter and water-stable aggregates in soils. J. Soil Sci. 33, 141-163.
von Peter, A. (1980). Fertilizer requirements in developing countries. Proc. Fertil. Soc. No. 188.
Wang, T.S.C., S-Y Cheng and H. Tung (1967). Dynamics of soil organic acids. Soil Sci., 104, 138-144.
Widdowson, F.V. and A. Penny (1965). Effects of formalin on the yields of spring wheat. Rep. Rothamsted Exp. Stn. for 1964, 65-66.
Wien, C., R. Lal and E.L. Pulver (1979). Effects of transient flooding on growth and yield of some tropical crops. Soil physical properties and crop production in the tropics. Wiley, Chichester and New York, 235-245.
Wiersum, L.K. (1957). The relationship of the size and structural rigidity of pores to their penetration by roots. Plant and Soil, 9, 74-85.
Wiersum, L.K. (1961). Utilization of soil by the plant root system. Plant and Soil, 15, 189-192.

DEVELOPING MORE EFFICIENT FERTILIZERS THROUGH FORMULATION, MANUFACTURING, AND DISTRIBUTION TECHNOLOGY

Donald L. McCune, Managing Director
International Fertilizer Development Center
Muscle Shoals, Alabama 35660, USA

ABSTRACT

Chemicals are being evaluated for their capacity to improve the agronomic efficiency of urea, the most widely used nitrogen fertilizer. Examples are inhibitors, coatings, metal ions to reduce losses. Solid and liquid forms of conditioners are reducing losses through improved product quality. New manufacturing routes, being evaluated for their usefulness in processing raw materials, especially phosphate ores, more efficiently include direct acidulation of ore as mined, use of the ore directly as fertilizer on acid soils, and conversion to partially soluble forms without total dissolution. New flotation reagent chemicals are used to upgrade phosphate ores that cannot be used without beneficiation. Improvements in the manufacturing, handling, storage, and distribution of fertilizers, raw materials, and intermediates are being made through better materials of construction, a wider variety of good quality bags, more distribution in bulk, and other means. Thus, chemistry is playing an important role in providing "cost effective" fertilizers for farmers with the ultimate goal of increasing the world food supply.

KEYWORDS: Improved fertilizer efficiency, fertilizer manufacturing, fertilizer distribution.

INTRODUCTION

The need for more intensive agriculture is illustrated by the decline in arable land per person from about 0.41 ha in 1962 to 0.31 ha in 1979 for the world as a whole. The developing countries had an average of 0.21 ha of arable land per person in 1979. Although more arable land may be brought into production, population growth is expected to greatly exceed arable land growth. So more intensive agriculture -- higher yields and more crops per year where practical -- will be necessary with intensive and efficient fertilizer use as an essential component. Modern fertilizers supply one or more of the primary nutrients -- nitrogen, phosphorus, and potassium. Scientists now recognize that secondary nutrients -- calcium, magnesium, and sulfur -- may be needed, and many commercial fertilizers are formulated to supply them. Further, small quantities of such micronutrients as boron, chlorine, copper, iron, manganese, molybdenum, zinc, and cobalt may also be needed for optimum yields. The chemical form of these elements is likely to influence their agronomic effectiveness, but fortunately most are available in several chemical and physical forms. Manipulation of chemistry in the formulation, manufacturing, and distribution of fertilizers must be considered if the challenge to make these nutrients more "cost effective" is to be achieved. This paper discusses some of the current research challenges and opportunities for the fertilizer sector, as well as how chemistry is affecting the world food supply through various fertilizers.

FERTILIZER FORMULATION

In fertilizer terminology, formulation indicates the manner, types of materials, and proportion in which nutrients are supplied in fertilizers. Nitrogen is derived from air and may be fixed in the ammonia, nitrate, or amide form, or a combination of these. Examples are anhydrous ammonia (82.3% N), ammonium sulfate (21% N), ammonium nitrate (33-34% N), urea (45-46% N), calcium cyanamide (20-40% N), and sodium nitrate (15% N). Phosphate rock, the basic raw material for all forms of phosphate fertilizer, is a complex mineral, with apatite ($Ca_{10}(PO_4)_6X_2$, where X may be F, OH, or Cl) being the major constituent. Numerous other elements are often substituted for Ca and P in the apatite structure. Decomposition of the rock with a mineral acid (H_2SO_4, H_3PO_4, or HNO_3) determines the form of the resulting product. Forms such as single superphosphate (SSP, 16-22% P_2O_5), triple superphosphate (TSP, 44-48% P_2O_5), ammonium phosphates (MAP, 11% N-55% P_2O_5; DAP, 18% N-46% P_2O_5), and nitric phosphates (14-20% P_2O_5) are widely used. The predominant form of potash used today is high-grade potassium chloride (60-62% K_2O), but such nonchloride types as potassium sulfate (50% K_2O) and potassium nitrate (13% N-44% K_2O) are also available. Compound fertilizers are products that contain two or more nutrients. Compounding may be achieved by chemical reactions and/or physical mixing. Formulation denotes the proportion of nutrients in compound fertilizers as well as the sources from which they are derived.

Chemicals for Increasing Fertilizer Agronomic Efficiency

Efficiency of a fertilizer indicates the degree of utilization of a nutrient by a growing plant and depends on how it is used. Specific types and/or management systems are being studied in an attempt to achieve higher efficiency. Efforts are underway to determine the mechanisms and magnitudes of losses of nutrients, especially nitrogen applied to rice and various upland crops. In some cases only about one-third of the applied nitrogen is actually utilized by the crop. Losses may occur thorugh ammonia volatilization, nitrification-denitrification, leaching, and runoff. Soil type, temperature, and moisture as well as the form, timing, and placement of nitrogen fertilizers have a significant effect on the extent and mechanism of the nitrogen losses.

Inhibitors -- Ammonia volatilization is a prime cause of nitrogen loss, particularly when urea is used to supply nitrogen for flooded rice. Urea is converted to ammonia and carbon dioxide by urease which is present in the soil. The ammonia can escape to the atmosphere through the floodwater. Ammonia volatilization losses also can be important for surface-applied urea on upland (unflooded) soils. Urease inhibitors are chemicals that are added to urea to prevent or delay hydrolysis of urea and thus decrease nitrogen losses by ammonia volatilization. Such chemicals, added to urea in the range of 0.5-5.0%, include phenol phosphorodiamidate (PPD), catechol, thiourea, dimethylphenol, and benzoquinone. In preliminary greenhouse tests, PPD was spectacularly successful in decreasing ammonia losses from flooded soils from 30% (without PPD) to about 4%. The addition of 1% PPD was as effective as that of 5%, and incomplete data indicate that 0.5% may be sufficient. In further greenhouse tests with rice, urea containing 5% PPD gave improved nitrogen recovery, but plant growth was not improved (Byrnes and coworkers, 1982). Plants cannot effectively utilize urea as such, so the degree of inhibition must be adjusted to supply ammonia nitrogen at a rate adequate for plant growth. Further study is needed to understand the mechanism, define the optimum dosage, and find lower cost urease inhibitors. At this time, only experimental quantities of these materials are available.

Other compounds are being evaluated as nitrification inhibitors. Nitrification is a two-step biological process whereby the ammonium ion (NH_4^+) is first converted to nitrite (NO_2^-) by Nitrosomonas or nitrite-forming bacteria and then the nitrite is conveted to (NO_3^-) by Nitrobacter or nitrate-forming bacteria. In flooded soils lacking air circulation, the nitrate ion can be converted to gaseous elemental nitrogen (N_2) and nitrous oxide (N_2O) by microbes that destroy the nitrate ion to obtain oxygen and thus allow the nitrogen to be lost to the atmosphere. Such compounds as a sulfathiazole-formaldehyde reaction product, 2-chloro-6-(trichloromethyl) pyridine (N-Serve), dicyandiamide (DCD), and nitrapyrin are being tested as nitroficiation inhibitors. In India, nitrification inhibition for urea is being claimed for neem cake, although the active ingredient has not been isolated (Ketkar, 1976). Some results with urease and nitrification inhibitors are shown in Tables 1 and 2. Some of the results are promising, but much more work is needed to evaluate these materials.

Table 1: Apparent Fertilizer Recovery From Various Fertilizer Sources on Rice (Source: IFDC Greenhouse data, 1981, unpublished.)

Material Type, Broadcast and Incorporated	Apparent Fertilizer Recovery, %
Continuously Flooded	
Urea	27
Urea + PPD (urease inhibitor)	42
Urea + DCD (nitrification inhibitor)	39
Urea + nitrapyrin (nitrification inhibitor)	34
39-Day Dry Period, Followed by Flooding	
Urea	21
Urea, split application	24
Urea + PPD	32
Urea + DCD	22
Urea + nitrapyrin	18
Urea + neem cake (nitrification inhibitor)	11
Coated urea, rubber latex (controlled release type)	19

Table 2: Effect of Phenyl Phosphorodiamidate (PPD) on ^{15}N Recovery by Sorghum from Urea as Affected by Rainfall. (Source: IFDC Greenhouse data, 1981, unpublished.)

Material	Simulated Rainfall, mm		
	Low (245)	High (465)	Intense (500)
	(Apparent ^{15}N Recovery, %)		
Urea, Broadcast	47.2	52.3	-
Urea + PPD,[a] broadcast	52.5	54.7	-
Urea, Incorporated	56.5	59.4	49.1

a. Urease inhibitor.

The use of inhibitors with urea offers promise for improved nitrogen efficiency, but two basic problems must be solved. The inhibitors must be made "cost effective", and a practical method must be developed to incorporate them into nitrogen fertilizers.

Coating of Soluble Fertilizers -- Another potential method for improving the efficiency of soluble nitrogen fertilizers is to coat them with a material that will control the rate of nutrient release (Powell, 1968). The concept is that the nutrient will be released at the proper time and in the amount that the plant can effectively use. Researchers have evaluated a wide range of materials, including acrylic latex, sulfur, rice bran wax, urea-formaldehyde, cement, natural rubber latex, and magnesium ammonium phosphate. Sulfur-coated urea (SCU) and a water-permeable, resinous film-coated urea (Osmocote) are available commercially, but because of the cost they have very limited use on food crops. Coating would appear to minimize losses by ammonia volatilization, nitrification-denitrification, and leaching. Surface-applied coated urea for rice in Asia has provided significant yield increases compared with surface-applied, split applications of uncoated urea. Table 3 gives some yield data for SCU versus uncoated urea. In some cases SCU would be a cost effective nitrogen fertilizer if there were a commercial supply in Asia. However, the results are highly variable, and on the average it is doubtful whether the improved efficiency justifies the additional cost. Further study is needed to define the conditions under which SCU is likely to be cost effective and to lower the manufacturing cost. Larger granules of SCU, such as forestry grade (about 4-6 mm average diameter), require less sulfur for an effective coating and thus reduce the cost. One advantage of SCU is that no change in farmers' practices is required, no machinery is needed for application, and less labor is needed since a single application can replace split applications.

Table 3: Effects of Particle Size and Coating of Urea at Two Nitrogen Rates on the Grain Yield of IR36, IRRI, 1981 Dry Season.

Source of Urea	Grain Yield[1], ton/ha 58 kg, N/ha	87 kg, N/ha	F-means
No fertilizer nitrogen	-	-	2.7 c
Urea, prilled	3.1 c	3.9 b	3.5 b
Urea, forestry grade	3.5 bc	4.3 ab	3.9 b
SCU,[2] forestry grade	4.9 a	4.8 a	4.8 a
Urea, supergranule (1 g)	4.2 b	4.7 a	4.4 a
Mean	3.9	4.4	

[1] In a column, means followed by a common letter are not significantly different at the 5% level.
[2] SCU = sulfur-coated urea.

Source: De Datta, S.K., E.T. Creaswell, I.R.P. Fillery, and C. Menguito, IRRI, 1981, unpublished data.

Metal Salt-Urea Complexes -- Another approach to reduce urea hydrolysis or ammonia volatilization loss is to combine the urea with a metallic ion which can form a complex with the ammonia. Iron and aluminum have been found to be effective in reducing ammonia loss when the molar ratio of urea to metal does not exceed 40:1 (Lewis, 1979). Other ions, such as copper, zinc, and manganese, which are also micronutrients are worthy of study. The concept, therefore, offers promise as a means of adding micronutrients and an ammonia volatilization inhibitor in a single application.

Deep Placement -- An alternative means for improving the efficiency of nitrogen fertilizers, especially for rice, is placement in the soil at a depth of about 10 cm. This method has been extensively investigated in Asia. The usual method is to place large granules called "supergranules" or briquettes, each containing 1-3 g of urea, about 10 cm deep between rice hills after transplanting. This method is effective, but as with SCU the effectiveness varies from one location to another. The main disadvantage is the labor required for hand placement or the expense of machine placement. Flinn and O'Brien (1982) have analyzed the economic advantages of deep placement and SCU under Philippine conditions concluding that in each case the higher resulting efficiency would cover the extra costs involved.

Phosphate and Potash Fertilizers

Phosphate and potash fertilizers are not considered lost in the same sense as nitrogen fertilizers. Phosphate fertilizer applied to certain high P-fixing soils, however, may become immobilized so that the plant cannot efficiently utilize it (Sanchez, 1976). Most phosphate fertilizers were developed for temperate agriculture and may not be both economically and agronomically optimum for the humid tropics and acid soils (pH 4.5-5.5). In sandy soils with high rainfall, potash fertilizers are subject to leaching losses, and in extreme cases soluble phosphate fertilizers may also be leached. Through chemical reactions it is possible to manipulate the solubility of phosphate fertilizer. (Solubility of phosphate (P_2O_5) can be expressed by several methods -- solubility in water and in solvents such as neutral ammonium citrate, citric acid, or formic acid.) A range of experimental products is emerging, and these range from (1) the least soluble form -- directly applied ground phosphate rock; to (2) controlled solubility products -- partially acidulated phosphate rock and nitrophosphates; to (3) highly soluble, ammonium phosphates -- (MAP/DAP), superphosphates (SSP/TSP), and certain nitrophosphates. Such manipulations offer an opportunity to better tailor or formulate the phosphate product to suit the soil, crop, and rainfall conditions, the market demand, the particular characteristics of the raw materials, any lack of foreign exchange, and the utilization of less skilled manpower for production. Furthermore, the opportunity exists for reducing the investment and production cost and, as discussed under "Phosphate Manufacturing", for making use of locally available raw materials that otherwise would not be exploited.

In general, agronomists agree that potash utilization efficiency is not a serious problem at this time. Because of the relative abundance and low cost of high-grade potassium chloride, it is less likely to attract much research effort compared with nitrogen and phosphate fertilizers. Coating of soluble potassium chloride is one obvious way of reducing leaching losses if and when this problem becomes significant. No low cost alternative forms of potassium, such as potassium phosphate, potassium nitrate, or potassium sulfate, are readily available. However, double sulfates of potassium and magnesium, such as langbeinite ($K_2SO_4 \cdot 2MgSO_4$), may be economic fertilizers when all three nutrients -- K, S, and Mg -- are needed.

IMPROVED EFFICIENCY THROUGH MANUFACTURING

Improved manufacturing efficiency can be achieved by (1) reduced consumption of raw materials and (2) reduced processing cost. Some examples are given below to show the ways that chemistry and a better understanding of it are contributing to higher efficiency. Improved efficiency is also being achieved through equipment modifications that are not discussed in this paper.

Nitrogen Manufacturing

Most nitrogen fertilizers today are derived from ammonia whose hydrogen source may be natural gas (methane), liquified petroleum gas (butane and propane), naphtha, coke-oven gas, heavy oil, coal, methanol, ethanol, or water, and whose nitrogen source is air. About 70% of the world's ammonia production is based on natural gas. With this feedstock and most of the others, the hydrogen is associated with carbon which can be utilized indirectly to produce hydrogen by partial oxidation or steam reforming followed by a shift reaction. For example, the reactions using methane as feedstock are:

Steam reforming -- $CH_4 + H_2O = 3H_2 + CO$ (1)

Shift reaction -- $CO + H_2O = H_2 + CO_2$ (2)

Overall reaction -- $CH_4 + 2H_2O = 4H_2 + CO_2$ (3)

The carbon dioxide is separated from the hydrogen in modern plants by such solvents as mono-ethanolamine, potassium carbonate, methanol propylene carbonate. Such processes can usually operate at a relatively low temperature and the gas is released without a high heat input. The pure carbon dioxide is often used for urea production in an adjacent plant. In the People's Republic of China, carbon dioxide is utilized by reaction with ammonia in small plants to produce ammonium bicarbonate. The reaction of hydrogen and nitrogen to form ammonia is carried out catalytically at high pressure and temperature. Highly reactive metal powders that would be effective catalysts at low temperature and pressure would greatly simplify the ammonia synthesis step (Rieke, 1977). While most such work is proprietary, some has been reported from Japan (Ozaki, 1981). Although other energy-saving steps have been proposed and some of them are in use, the increased investment may offset the improved efficiency (Mudahar and Hignett, 1982). However, substantial energy savings are possible by more efficient operation of present facilites.

Except in developed countries anhydrous ammonia is rarely used directly as a fertilizer. Instead it is converted to solid products: urea (ammonia plus carbon dioxide), ammonium nitrate, ammonium sulfate or ammonium chloride. In terms of world consumption for agricultural use, urea and ammonium nitrate are most important. The most common method of producing the solid product is to prill an anhydrous melt by letting droplets of it fall through an upward-moving stream of air in a prilling tower. Some newer plants are converting to granulation where the anhydrous melt is sprayed onto a rotating or fluidized bed of granules. Solid particles, prills or granules, may not be sufficiently strong to be handled, distributed, and mechanically applied. Weak granules that form dust, tend to cake, or absorb moisture may become a big problem for humid tropical countries. In some cases deterioration in physical condition may represent significant manufacturing and distribution losses. Small amounts (0.25%-1.0%) of additives, called conditioners, combined with improved operating techniques have been found to prevent one or more of the above problems.

Conditioners are basically of two types -- those such as formaldehyde and paraformaldehyde that are incorporated into the granules to impart strength, and those that are applied to the surface to retard caking. Solid conditioners, such as powdered diatomaceous earth, siliceous dust, and clays, are added to the surface primarily as anticaking agents but are not intended to inhibit moisture absorption. Some surface-applied liquid conditioners are Uresoft for urea and alkylamines for ammonium nitrate. Both types of conditioners are rendered ineffective if the moisture content of the urea or ammonium nitrate is not carefully controlled during production and distribution. Ammonium nitrate undergoes changes in crystal state at 84.2^{o} and $32.1^{o}C$ which cause crystal expansion, possibly resulting in particle degradation. This lower temperature can easily be reached in storage in tropical areas. About 1% by weight of chemical additives will reduce or eliminate this problem. Magnesium nitrate and "Permalene" are used for this purpose in Europe and the United States, respectively.

There is a continuing interest in nitrogen fixation processes such as the arc process that may be applicable to developing countries. Through an electrical arc, nitrogen and oxygen in the air are converted to nitrogen oxides which then form dilute nitric acid by reaction with water. The nitric acid is neutralized with lime or phosphate rock and applied as a solution fertilizer. Alternatively, the calcium nitrate solution can be dried and granulated. The process consumes a large quantity of electrical energy per ton of fixed nitrogen (~74,000 kWh/mt N). Unless electrical energy is relatively low in cost and the byproduct heat can be utilized, the process is economically questionable (Livingston, 1977). Electrolysis of water to produce hydrogen for ammonia is more efficient (~12,000 kWh/mt N fixed) and may be more easily justified for remote areas (Grundt, 1982).

Phosphate Manufacturing

Because of the wide variation in the chemical and mineralogical analysis of phosphate rock "as mined", the technologist is challenged to develop processes for handling the difficult ores. Two approaches can be taken: (1) use the ore "as is" in nonconventional processes, or (2) upgrade the ore to make it suitable for conventional processes. This research is gaining in importance since many known deposits are yielding lower grade (lower P_2O_5 content) ores having a wider range of impurities. Also, many developing countries want to make better use of their indigenous resources. Before any research program can be planned, it is necessary to study or completely characterize a representative ore sample to establish possible research alternatives. Problem ores generally have one or more of the following impurities in relatively high quantities: (1) iron and aluminum, (2) carbonates (magnesium and/or calcium), (3) silica, (4) chloride, (5) organics, and (6) sulfides (Table 4). An excessive amount of any of these impurities seriously affects the use of the ore in conventional processes (Lehr, 1982).

Some ores can be benficiated merely by washing to remove a clay-like fraction called slimes. In most cases a significant part of the original phosphate is lost in the slimes. After removal of slimes, some ores still contain excessive amounts of impurities and may be treated by flotation to separate the phosphate rock (apatite) from the impurity minerals. This is only possible when the impurity minerals are present as separate particles or can be separated by grinding. Separation of silica by flotation is widely practiced. Tall oil, a mixture of long chain fatty acids, is a commonly used flotation reagent; it is available as a byproduct from the paper industry. Studies are underway to identify more effective or less expensive or less expensive reagents.

Only recently has it been possible to remove calcium or magnesium carbonate by flotation. The U.S. Bureau of Mines has developed a carbonate-silicate flotation process that has been successfully used in the western United States to upgrade the quality and increase the recovery of the phosphate rock concentrate from a low-grade ore. In bench-scale tests a sample of ore from China was upgraded to 32% P_2O_5 with 94% recovery (Stowasser, 1981). Similar processes have been developed to remove both carbonates and silicates from south Florida ore (Dufour and coworkers, 1980). In Finland a special flotation chemical is being used to upgrade an ore containing only 4% P_2O_5 to produce a 35% P_2O_5 concentrate with 85% recovery (Kuikkola, 1980).

A better understanding of the chemistry and mineralogy of phosphate rock could lead to its utilization without benefication in some cases. It has been demonstrated in pilot-scale work that wet-process phosphoric acid can be made directly from siliceous ores if the equipment is specially designed to handle the larger quantity of byproduct solids and to elimiante corrosion/erosion problems. By direct acidulation of phosphate ore, P_2O_5 recovery is 95% or higher compared with the usual recovery of 50%-70% through beneficiation followed by acidulation. Thus, it may be more economical in some cases to use a high-silica ore directly to make phosphoric acid than to beneficiate the ore and use the concentrate.

Another use of ore directly involves partial acidulation with sulfuric or phosphoric acid to produce a solid fertilizer. Since the ore is not dissolved completely, it is likely that impurities will be less of a problem. Depending on the chemistry of the apatite structure, phosphate ores vary widely in reactivity. Highly reactive ores -- having a high degree of CO_3^{-2} substitution in the apatite lattice -- are relatively effective fertilizers on acid soils. Use of the ore directly results in essentially no processing losses. In addition it provides certain secondary elements; also, micronutrients are usually present in the ore. These and other processes now make it possible to utilize certain ores that previously were not considered exploitable and that could have a significant impact on the food supply in developing countries.

Table 4: Chemical Composition of Selected Phosphate Ores "As Mined" or Beneficiated (IFDC unpublished data).

	P_2O_5	CaO	F	CO_2	Al_2O_3	Fe_2O_3	SiO_2	NAC Soluble P_2O_5 [a]
	(wt %)							
Africa								
1. Upper Volta (Kodjari)								
Pit No. 2	26.5	37.1	2.4	1.1	2.3	1.6	26.0	2.3
Pit No. 3	17.5	23.2	1.7	0.6	2.7	2.6	48.3	2.4
2. Mali (Tilemsi Valley)	27.6	39.2	2.5	2.0	2.7	7.1	13.9	2.7
3. Abu Tartur, Egypt	26.5	43.3	2.8	5.6	0.99	4.0	4.1	3.5
4. Angola, Africa	37.0	51.0	4.0	2.2	2.0	0.44	1.1	5.5
Latin America								
1. Baja California, Mexico	23.6	37.8	2.6	3.4	1.6	1.3	24.4	5.3
2. Tapira, Braxil	35.9	50.1	1.5	0.52	0.57	2.5	2.7	1.6
3. Patos de Minas, Brazil	17.0	21.8	1.3	0.50	9.1	3.8	37.8	1.4
4. Sechura, Peru	30.2	46.5	2.9	4.4	0.84	0.81	3.2	6.0
5. Sardinata, Colombia	25.8	42.2	2.9	5.3	2.1	0.89	16.6	2.4
6. Pesca, Colombia	20.4	28.9	2.0	1.4	1.1	0.85	42.2	3.8
7. Napo Formation, Ecuador	27.8	49.6	3.4	7.4	0.87	1.50	7.1	4.4
Asia								
1. Mussoorie, India (Maldeota mine)	21.2	38.5	2.3	13.8	0.73	6.3	6.6	2.2
2. Guizhou, China	17.6	40.0	2.2	25.0	0.54	0.33	3.5	1.0
3. Sri Lanka	36.4	48.5	2.6	1.2	1.9	4.8	0.8	2.8
4. Hazara, Pakistan	31.0	42.7	3.2	1.0	1.4	1.0	17.0	2.5
5. Lamphun, Thailand	33.2	48.0	0.1	6.2	2.5	0.78	2.5	7.1
U.S.								
1. Tennessee[b]	31.4	44.2	3.5	1.6	2.4	4.2	7.8	2.9
2. North Carolina (uncalcined)	30.0	49.0	3.8	6.0	0.43	0.72	1.4	6.7
3. Central Florida[b]	33.4	48.9	3.9	3.0	1.3	0.9	4.5	4.0

a. P_2O_5 soluble in neutral ammonium citrate.

b. Beneficiated phosphate rock.

Table 4 shows the chemical analysis of some phosphate rocks "as mined", or in some cases as beneficiated. The relative reactivity of these rocks is indicated by the percentage of P_2O_5 that is soluble in neutral ammonium citrate[1]. When these rocks are finely ground and applied directly to acid soils in humid tropical areas, the short-term response is proportional to their reactivity as shown in Figure 1. Highly reactive rocks are about as effective as some water soluble phosphates such as TSP, whereas rocks of low or medium reactivity range from 50% to 75% as effective as TSP.

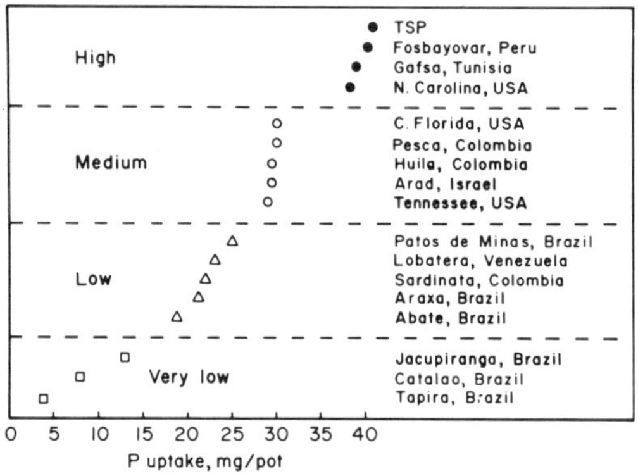

Figure 1. Relative Availability of Phosphorus Sources as Measured by P-Uptake by *Panicum maximum* Grown on an Acid Oxisol (Las Gaviotas, Colombia) Under Greenhouse Conditions. Sum of Three Cuttings.

Source: León, L. A., W. E. Fenster, and L. L. Hammond. "Relative Agronomic Potential of Finely Ground Phosphate Rocks From Brazil, Colombia, Peru, and Venezuela." *Agronomy Abstracts*, 1981.

For long-season crops or a succession of crops, reactivity is less important. for instance, in a 4-year experiment in Columbia with grass (Brachiarai decumbens) in which a total of 14 cuttings was harvested, the yield was proportional to reactivity only for the first cutting. Thereafter all rocks increased in effectiveness, and the 4-year total yield was essentially the same for all of the six rocks tested and fully equal to the yield from TSP (Parish and coworkers, 1980). Similar results have been obtained with other crops such as beans.

When immediate response is needed, it can be obtained by partial acidulation of the ground rock with a mineral acid, such as sulfuric acid using 20%-40% of the acid that would be required for full acidulation to make superphosphate. IFDC has developed a method for partial acidulation and simultaneous granulation of ground rock that works well even with rocks of high impurity content and relatively low reactivity. This process supplies some soluble P_2O_5 for immediate effect and the remaining insoluble P_2O_5 becomes available later. Also, when sulfuric acid is used for partial acidulation, the product supplies sulfur which is often deficient in tropical soils. Figure 2 shows that in an experiment in Colombia partial acidulation of Pesca rock using sulfuric acid to the 40% level gave yields equal to TSP. In experiments in Upper Volta, ground rock (Kodjari) with 44% sulfuric acid acidulation produced yields of cotton, millet, sorghum, and ground nuts equal or superior to TSP.

1. For rocks that contain free calcium or magnesium carbonate, these carbonates should be removed by extraction with a suitable reagent before carrying out the ammonium citrate solubility test in order to obtain a correct indication of reactivity.

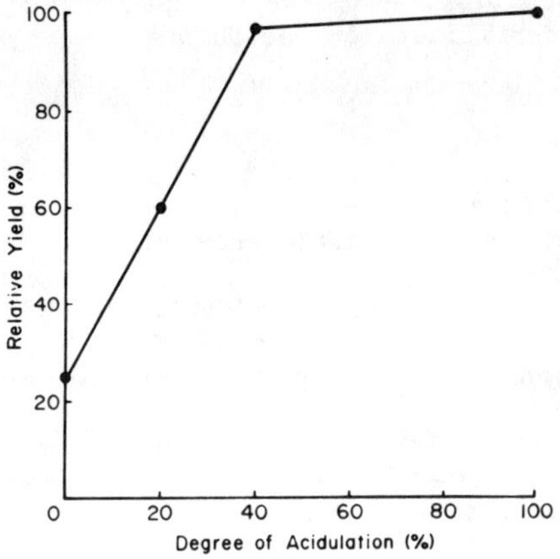

Figure 2. Response of Beans (G-4000) to Equal Amounts of Total Phosphorus (Average of Four Rates of Application) in a Field Experiment on a Volcanic Soil (Popayan, Colombia) as Influenced by the Degree of H_2SO_4 Acidulation of Pesca Phosphate Rock. (100% = TSP Rather Than SSP; However, No Sulfur Deficiency Was Present.)

Source: Hammond, L. L., and L. A. León. Unpublished IFDC data.

Energy consumption for drying is a significant cost in fertilizer manufacturing. Wherever possible, the processes are being designed to utilize chemical heat as a replacement or a substitute for natural gas, fuel oil, coal, or other types. For example, the reaction of ammonia with phosphoric acid can be used to evaporate most or all of the water that normally is done by drying. A special reactor developed by TVA called a "pipe" or "pipe-cross" reactor is being used by industry with good results, especially for MAP. For MAP the dryer is not normally used, and for DAP the energy consumption is reduced by 40% (Salladay, 1981). Several developing countries are studying the merits of this process. The reactor is also used in NPK granulation plants for producing such N-P_2O_5-K_2O grades as 12-12-12 without using a dryer (Salladay, 1977).

FERTILIZER DISTRIBUTION

The application of chemistry is also being used for improving the techniques for storage, handling, and distribution of fertilizer intermediates and products. The corrosive nature of chemical fertilizers often require special construction materials, combining metals and non-metals. For coating of mild steel, improved formulations for epoxy paints are now available. Stainless steels or rubber-lined mild steel vessels are in use for bulk transport of corrosive acids. Ceramics, able to withstand the high temperatures in rotary kilns, are commercially available. The tendency now is to move fertilizer materials in bulk as close to the farmer as possible. For example, bulk fertilizer may be shipped to a receiving port, off-loaded, and bagged immediately. The equipment for this practice is treated for protection against a salt environment.

Better materials now available for bagged fertilizers include polyethylene or polypropylene, to a great extent replacing the jute bag. Multiwall paper bags with a plastic lining are also readily available. Many developing countries import plastic pellets and convert them to continuous film for bag preparation. Certain chemical additives add strength and offer protection from ultraviolet light. In several countries fertilizers are stored outside in plastic bags. In this case polyethylene and polypropylene should contain chemicals such as various derivatives of hydroxybenzophenone, hydroxyphenylbenzotriazole, or phenyl salicylate to protect against damage by sunlight (Kirk and Othmer, 1967).

Manufacturing, storage, handling, and distribution of fertilizers, raw materials, and intermediates represent a significant part of the final product or nutrient cost. Chemical innovations such as improved chemical and physical properties, additives to improve agronomic efficiency, less energy-intensive processes, improved materials of construction, and lower losses during chemical processing will ensure more economical fertilizers in the future.

REFERENCES

Byrnes, B.H., N.K. Savant, and E.T. Craswell (1982). The Effect of a Urease Inhibitor Phenyl Phosphorodiamidate on the Efficiency of Urea Applied to Rice. Soil Sci. Soc. Am. J., In press.

Dufour, P., B. Pelletier, J.J. Predali, and G. Ranchin (1980). Beneficiation of South Florida Phosphate Rock With High Carbonate Content. In 2nd International Congress of Phosphorus Compounds, Institut Mondial du Phosphate, Paris, France, 247-800.

Flinn, J.D., and D.T. O'Brien (1982). Economic considerations in the Evaluation of Urea Fertilizers in Wetland Rice Farming. Paper presented at International Workshop/Training Course on Nitrogen Management, Fuzhou, China, April 25-May 4, 1982.

Grundt, T. and K. Christiansen (1982). Hydrogen by Water Electrolysis as Basis for Small Scale Ammonia Production. A Comparison with Hydrocarbon Based Technologies. Int. J. Hydrogen Energy, 7(3), 247-57.

Humphreys and Glasgow Consultants (1982). Nitrophosphate Fertilizer Process. Chem. Eng. World, 17(1), 29.

Ketkar, C.M. (1976). Utilization of Neem and Its By-Product. Special Report of the Modified Neem Cake Manurial Project (1969-1976). Directorate of Nonedible Oils and Soap Industry, Khadi and Village Industries Commission, Bombay, India.

Kirk-Othmer (1967). Encyclopedia of Chemical Technology, 2nd ed., 14, 273, 298, John Wiley and Sons, New York.

Kuikkola, K. (1980). Selective Flotation of Apatite From Low-Grade Phosphorus Ore Containing Calcite, Dolomite, and Phlogopite. In 2nd International Congress on Phosphorus Compounds Proceedings, Institut Mondial du Phosphate, Paris, France, 219-29.

Lehr, J.R. (1982). Phosphate Raw Materials and Fertilizers: A Look Ahead. Paper presented at TVA Ninth Shortcourse in Modern Fertilizer Technology, Muscle Shoals, Alabama, June 14-18, 1982.

Lewis, A. and R.M. Slater (1979). Reducing NH_3 Loss from Urea. British Patent 2010797A.

Livingston, O.W. and J.R. Polo (1977). Potential Small-Scale Fertilizer Processes for Developing Countries. IFDC, paper presented at seminar jointly sponsored by Council on Science and Technology of His Majesty's Government in Nepal and the Asia Society of the United States, February 27-March 3, Kathmando, Nepal.

Mudahar, M.S. and T.P Hignett (1982). Energy and Fertilizers: Policy Implications and Options for Developing Countries. IFDC Technical-Bulletin T-20, Muscle Shoals, Alabama.

Ozaki, A. (1981). Development of Alkali-Promoted Ruthenium as a Novel Catalyst for Ammonia Synthesis. Acc. Chem. Res., 14, 16-21.

Parish, D.H., L.L. Hammond, and E.T. Craswell (1980). Research on Modified Fertilizer Materials for Use in Developing Country Agriculture. IFDC Paper Series P-2, Muscle Shoals, Alabama.

Powell, R. (1968). Controlled Release Fertilizers. Noyes Development Corporation, Park Ridge, New Jersey.

Rieke, R.D. (1977). Preparation of Highly Reactive Metal Powders and Their Use in Organic and Organometallic Synthesis. Acc. Chem. Res., 10, 301-306.

Sahrawat, K.L. (1980). Control of Urease Hydrolysis and Nitrification in Soils by Chemicals — Prospects and Problems. Plant Soil, 57, 335-352.

Salladay, D.G., and E.D. Myers (1977). Use of TVA's Pipe-Cross Reactor to Conserve Energy in Fertilizer Granulation Plants. TVA Bulletin Z-111, Muscle Shoals, Alabama 35660 USA. Presented at Fifth National Conference on Energy and the Environment, Cincinnati, Ohio, November 1-3, 1977.

Salladay, d.G., F.P. Achorn, and J.L. Greenhill (1981). Recent Developments in Producing Monoammonium and Diammonium Phosphates Using the TVA Pipe-Cross Reactor. Presented at the 31st Annual Meeting of the Fertilizer Industry Round Table, Washington, DC, November 1981.

Sanchez, P.A. (1976). Properties and Management of Soils in the Tropics. P.A. Sanchez ed., 254-57, John Wiley and Sons, New York.

Stowasser, F. (1981). Phosphate Rock. Preprint from the U.S. Bureau of Mines Yearbook, Washington, D.C.

White, J.C., A.J. Fergus, and T.N. Goff (1975). Phosphoric Acid by Direct Sulfuric Acid Digestion of Florida Land-Pebble Matrix. U.S. Bureau of Mines, RI 8086, 12.

MODIFYING CROP PERFORMANCE WITH PLANT GROWTH-REGULATING CHEMICALS

Johan Bruinsma
Department of Plant Physiology, Agricultural University, Arboretumlaan 4, 6703 BD Wageningen, The Netherlands

ABSTRACT

The endogenous factors regulating crop yield are the distribution of matter and the longevity of the organs that produce and store that matter. Both factors are controlled by plant hormones so that interference with the hormonal pattern is the obvious way to manipulate crop performance. For lasting effects, the more persistent synthetic plant growth-regulating chemicals (PGR's) are usually preferred over the biolabile hormones.

With PGR's, both morphogenesis and metabolism can be influenced so that form and function are altered. Application of PGR's to cereals is an example of morphogenetic control to secure and increase grain yield, demonstrating the complexity of an apparently simple effect. Uses in fruit growing show PGR effects on metabolism improving harvest and postharvest life, and reveal the high level of organisation and co-operation required for successful application of PGR's.

KEYWORDS: apple, cereal crops, fruits (thinning, ripening, storage), grain crops, growth retardants, hormones (plant), phytohormones, plant growth regulators, postharvest physiology.

INTRODUCTION

The use that crop plants make of their environmental conditions, and the portion of their productivity that they invest in harvestable organs depend on distribution of materials within the plants as well as the relative longevity of their organs. Both the distribution of matter and the duration of organ life are largely determined by an endogenous pattern of plant hormones. Manipulation of transport and vitality in agricultural practice can enhance effects of soil and crop management and improve the quantity and quality of the yield. Such manipulation of crops can be achieved by intervening in the pattern of endogenous hormones.

THE PHYTOHORMONE PATTERN

As in animals, hormones in higher plants are produced in response to external and internal stimuli in order to adapt to environmental and internal changes.

Unlike animals, plants produce these hormones not in specific glands, but in normal tissue cells. Further, unlike animal hormones, plant hormones are not specifically directed towards a single target organ. A great many tissues contain the specific cellular receptors with which the hormones interact to exert their effects. The amount or availability of these receptors determine the sensitivity of the tissue for the corresponding hormones. Because one tissue usually contains several types of receptors, the plant hormones interfere with each other, promoting or counteracting each other's effects. In any given plant organ at any time, a shifting pattern exists of various hormones, both those locally synthesized and those derived as signals from other plant parts. The resulting vector of this hormonal pattern may induce the organ to continue or to arrest growth, to remain vital or to senesce, to absorb nutrients or to give them off. It is this hormonal state that makes seeds and buds sprout or keeps them dormant, that maintains the vegetative stage or induces flowering. Favourable soil conditions allow the root tips to produce hormones that induce the above parts to grow, whereas shortage of any nutrient has the opposite effect. Fertilizers may act as such but also by stimulating the production, by the roots, of such growth-promoting hormones.

The regulation of the growth of one plant part by the hormone-producing activity of the other, is particularly important because of the plasticity of plant growth. Table 1 presents an illustration of this.

Table 1: Yield Analysis of Winter Rye Sown at Different Densities (After Bruinsma, 1966)

Seed Density kg/ha	Number of Ears per Plant		Number of Grains per Ear		Dry Weight per Grain mg		Yield per Plant g
100	1.16	x	41	x	39	=	1.85
5	9.0	x	56	x	48	=	24.2

Normally-sown winter rye, (100 kg per ha), forms mostly one-culm plants with an ear containing less than 2 g dry weight of grains. If sown, however, at the lower seed rate of 5 kg per ha, the plants form many ears that are, on an average, 50% heavier, the total grain production being 13 times larger. Yet, both plant types develop harmoniously, the under- and above-ground parts and vegetative and reproductive parts being properly related. It is the mutual hormonal control between the different plant parts that results in such proportional development. Not only cultivated plants, but wild plants too, although constrained in their natural environment, because of this hormonal control, present examples of harmony admired by scientists and artists alike.

In the hormonal patterns at least five different groups of phytohormones interplay (Moore, 1979), three of which are derived from the mevalonate pathway (Fig. 1). The cytokinins are adenine derivatives with an isoprene unit as a side chain. The abscisins are trimers, the gibberellins tetramers of these units. Two other groups of hormones are derived from amino acids (Fig. 2): the auxins and the simple, gaseous molecule, ethylene. The latter has to be regarded as a true hormone, because of its very specific action on growth and development at very low dosages. This action can be prevented by inhibitors of the genetically controlled protein synthesis indicating that, as do other plant and animal hormones, ethylene

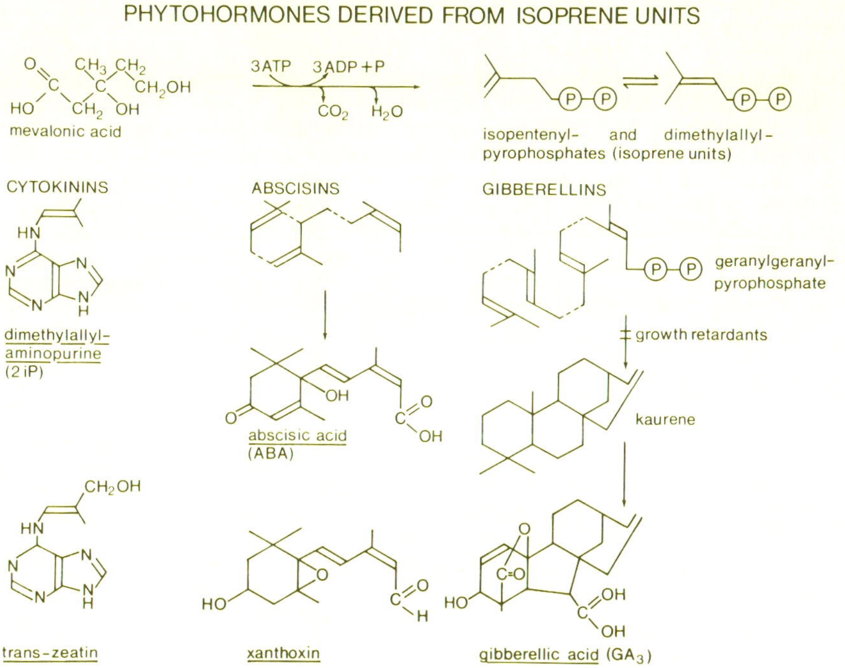

Figure 1

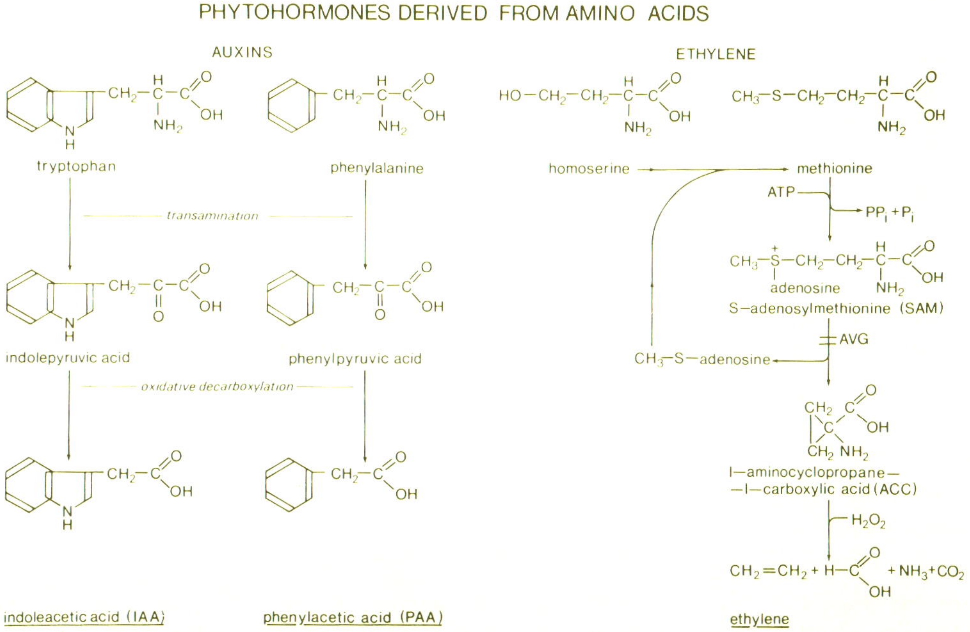

Figure 2

exerts its effect through translation of genetic information into the activity of newly produced enzymes.

It is not unlikely that new hormones will be discovered, for instance those involved in the regulation of flower induction and tuber formation (Bruinsma, 1981). Hormones that are produced at a particular site can be translocated through the vessels, also from cell to cell, and ethylene in the gas phase of the plant. At other sites they can interfere in the local hormonal patterns or be inactivated by breakdown or conjugation. As inactive conjugates they can be stored and possibly released later and redistributed as active materials. In particular, our lack of understanding of these questions of hormone transport and metabolism creates one of the main obstacles for progress in the regulation of matter distribution and organ vitality in plants.

INTERFERENCE BY SYNTHETIC PGR'S

For the application in the field, however, phytohormones themselves are less suitable, particularly because they are liable to metabolic inactivation by plant enzymes. Synthetic substances are often much more stable in the plant and consequently their effects are more persistent. This is important to their activity because the plasticity of plant growth enables the crop plant to compensate later for an earlier effect. It can, for example, outgrow an earlier growth inhibition as we will see. Synthetic PGR's are often difficult to inactivate enzymatically in the plant tissue and thus can postpone and/or reduce such compensating reactions. Alternatively, their delivery can be controlled by packaging the chemicals in time-release systems of, e.g., a membranous, matrix or laminated structure (Wilkins, 1982).

Synthetic PGR's usually interfere with the phytohormone pattern in two ways. First, their molecular structure may contain specific, configurational similarities to hormone molecules, allowing them to react with the same cellular receptors. Such PGR's can simulate the action of hormones. On the other hand, they can affect the biosynthesis, translocation, or metabolism of a hormone, thereby changing the amount of a component of a local hormonal pattern. In such cases, altered levels of hormones should occur. This can be demonstrated with the very specific, sensitive, and accurate methods available for isolation and measurement of picogram amounts of these hormonal substances (Hillman, 1978, Weiler, 1981). The hormone level, however, is not always a reliable indicator of its activity; sensitivity to, and turnover rates of hormones may be physiologically more relevant than their absolute amounts (Bruinsma, 1980). This presents another difficulty to our understanding of hormone action at present.

One problem with manipulation of crop performance by a PGR is that, because the phytohormones exert a variety of effects, PGR's are likely to do so, as well. As a consequence, undesired side-effects may occur which reduce or even cancel beneficial effects. Gibberellins, for instance, are used to prevent fruit drop in pear orchards after late night frosts, but these substances readily reduce next season's flower formation in which case the remedy may be worse than the disease.

Another problem with application of PGR's is their distribution within the treated crop plants. WHen sprayed on the foliage or applied to the soil, the penetrating molecules may be translocated to unexpected sites and exert undesired effects. If the PGR is to simulate the action of an endogenous hormone, it may arrive at sites other than where the corresponding endogenous hormone resides. Accordingly, its fate and function may be completely different, even within the same cells. For example, an exogenous PGR may penetrate the cytoplasm where it is readily metabolized, whereas the endogenous hormone may be safely stored within the vacuole (Bruinsma, 1980). Radioactive labelling may often give clues as to distribution and fate of such applied chemicals after uptake into the plant body.

Even if a PGR is sufficiently stable within the crop plant and exerts the desired effect without unwanted side-effects at the desired site, its application in practice can be hampered or even prevented by other factors such as different responses among varieties, unacceptable toxicity to producer and consumer, as well as costs of production and evaluation. In practice, PGR's are being developed only for major crops of considerable economic value.

MORPHOGENETIC EFFECTS: GROWTH RETARDANTS IN CEREALS

The peculiar feature about the grain stalk is that the heavy ear is located at the top of the elongated culm and, therefore, the stalk may lodge in rainy or stormy weather. Lodging causes severe yield losses and makes harvest difficult so that its prevention is a matter of great economic importance. Elongation of the grain stalk originates from transverse divisions of cells in the internodes and at the top of the culm, so that layers of cells on top of each other arise. Both the division and the growth of these cells are stimulated by gibberellins, so that culm height should be effectively reduced by lowering the endogenous gibberellin level. Alternatively, the level of a hormone antagonizing this gibberellin effect such as ethylene could be raised. Fortunately, a number of substances are known that either block the biosynthetic pathway of gibberellins or enhance ethylene production. Figure 3 shows some effective growth retardants active in several crops.

Many growth retardants block cyclisation of geranylgeranylpyrophosphate into kaurene (Fig. 1), but daminozide seems to increase ethylene synthesis, and ethephon serves as a substrate from which ethylene is evolved within the plant tissue. Chlormequat is used extensively in Europe to control culm length in wheat. A combination of the related compound, mepiquat, and ethephon is successfully applied to winter barley (Herbert, 1982) and also to cotton (Schott and Rittig, 1982). The amounts used are of the order of 1 kg active ingredient per ha. The time of application influences the effect on internode length; early treatment shortens and stiffens the lower internode, whereas late application reduces length of the highest and longest internodes resulting in the greatest overall shortening effect. With early application, the internodes formed later may even grow taller than in untreated plants so that early retardation is more than compensated (Fig. 4). Even then the more resistant lowest internode may have a beneficial effect.

Ample fertilization, particularly with nitrogen, enhances culm elongation and thus increases the danger of lodging. Stalk shortening with growth retardants allows a larger nitrogen fertilization resulting in a further increase of grain yield (Fig. 5).

In short-straw cultivars the resistance against lodging is genetically built in. It requires, however, a long period of breeding to obtain such varieties and, moreover, the more versatile resistance by retardant application has advantages over the fixed, genetic resistance. Because of the generally positive association between yield and plant height, there is some reluctance among breeders to develop short-straw varieties, a high biomass being a prerequisite for a high yield (Lupton, 1980, Gale and Hanson, 1982). The decision whether to apply a growth retardant, at what dosage and at what time, can be taken by the grower according to his particular conditions, offering great flexibility.

Moreover, also in these semi-dwarf varieties application of a growth

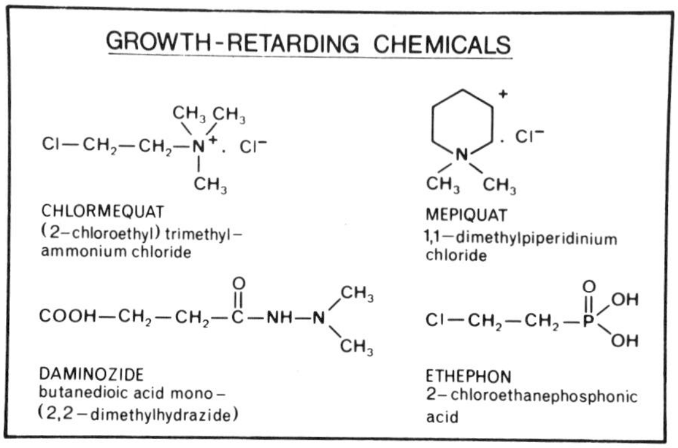

Figure 3.

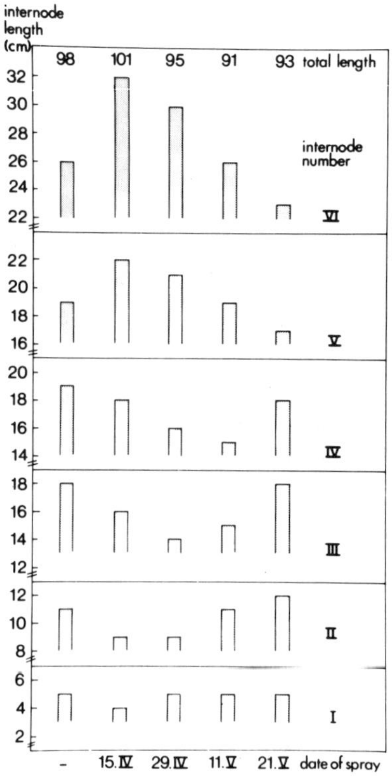

Figure 4: Effect of a practical mepiquat/ethephon spray at different dates on internode length of Ingri winter barley (unpublished results of Dr. A. Darwinkel, Research Station for Arable Farming, Lelystad, Netherlands).

retardant usually results in increased yield even in the absence of lodging (Herbert, 1982). This is a general phenomenon. In 19 trials with wheat and barley, where lodging did not significantly affect yield, an average yield increase of 12% was obtained by the use of growth retardants (Williams et al., 1982). This is because of the general impact on the morphogenesis of the crop plants by reduction of its gibberellin biosynthesis. Gibberellins promote shoot growth and generally inhibit the formation of organs for propagation, such as flowers and tubers. For this reason, growth retardants not only inhibit shoot growth but tend to enhance the numbers of, for example, pods in broad bean and tubers in seed potatoes (unpublished data, Research Station for Arable Farming, Lelystad, Netherlands). Also apical dominance of the shoot becomes less pronounced. Growth inhibition itself occurs in the aboveground parts only, so that the shoot:root ratio decreases. As a result, the relatively larger root system supplies more water and nutrients. The higher water and nitrogen content of the sprouts reflect a delay in development; the sprouts remain somewhat younger, the leaf canopy closes later and ear initiation, heading, flowering and ripening are delayed (Fig. 6).

Because of the delay in ear appearance (three days, Fig. 6), more florets complete their development. Accordingly, increased yield is not so much a matter of a larger kernel weight, although foliar senescence is also retarded, but rather of an increased kernel number. The larger number of grains results not only from more kernels per ear but also from more ears per plant (Fig. 6). Plots treated with growth retardants usually look darker than untreated plots. This is partly due to a higher chlorophyll concentration because of shortened leaves and partly to the more erect position of the leaves. As a result, less of the incident sunlight is reflected and more of it penetrates into the crop. Also, because of decreased apical dominance, more tillers survive under the better light conditions

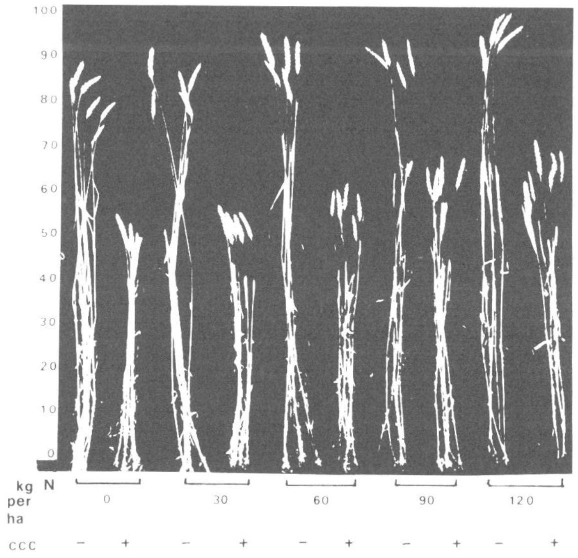

Figure 5: Effect of nitrogen fertilization and chlormequat (CCC) on culm length of Opal spring wheat (after Vos et al., 1967).

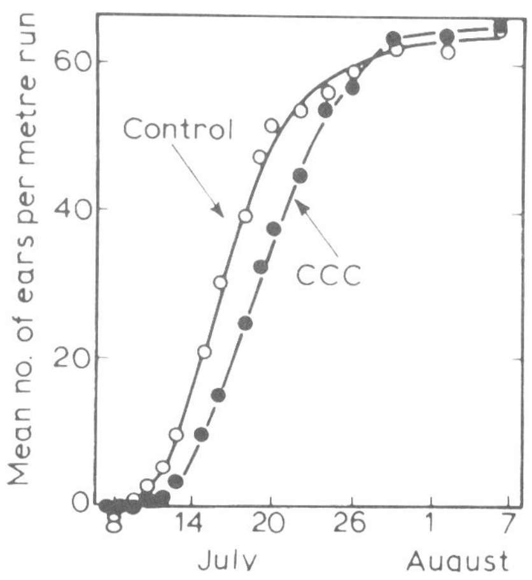

Figure 6: Effect of chlormequat (CCC) on rate of heading of Carpo spring wheat (after Bruinsma et al., 1965).

so that more ears per plant develop. More ears per plant and more kernels per ear give a larger number of grains per surface unit. The delayed foliar senescence involves a prolonged filling period for the grains so that, usually at an unchanged thousand-kernel weight, a higher yield results (Bruinsma, 1982, Hofner and Kuhn, 1982).

The application of growth retardants in cereal crops demonstrates the complexity of the mode of action of PGR's. In addition to the primary aim of culm shortening, these chemicals affect the root:shoot ratio, the change of vegetative into reproductive development, apical dominance, plant architecture, and the rates of growth and development. It is this whole complex from which increased yield ultimately results, quite apart from the resistance to lodging.

The use of growth retardants in cereals, a major application of PGR's in agriculture, is given here as an example because it illustrates the morphogenetic effects of PGR's on crop plants. Changes in morphogenesis arise from quantitative and qualitative changes in the metabolic pathways by which the plant constituents are formed and converted. Some of these metabolic effects of PGR-application will be shown in the second example.

METABOLIC EFFECTS: PGR'S IN FRUIT CULTURE

In the culture of pome-fruits, nitrogen fertilization is a main method used to control crop performance. The time of nitrogen fertilization is very important. Early application promotes strongly the vegetative growth of fruit trees, often at the cost of flower production. This is not a matter of competition between shoot growth and flower formation, rather growing shoots produce gibberellins that inhibit flower development.

On the contrary, a late nitrogen application, after cessation of shoot growth, not only enhances fruit growth but also initiation of flower primordia, probably also by the action of root-produced hormones. However, the developing seeds in the growing fruits are also a source of gibberellin production which may reduce seriously further development of the flower initials. This causes the well-known danger of biennial bearing; an on-year with heavy bearing is followed by an off-year without significant yield. The remedy is to thin the excess fruit at an early stage, before their seeds actively produce gibberellins. The remainder can then develop into large, valuable fruits without harming next season's flower formation.

Fruit-thinning by hand is far too expensive, but chemicals have been developed that when sprayed on trees cause the developing fruitlets to abscise (Fig. 7). The fruit pedicel contains a zone of cells that can be induced by the hormone, ethylene, to form cell wall-solubilizing enzymes and to secrete these enzymes in their own walls. As a result, these walls are digested so that the contacts between the cells disappear and the fruitlet drops. The same phenomenon occurs with the shedding of unfertilized flowers and with the drop of yellow leaves from deciduous trees in the fall.

The induction of ethylene in a sufficiently large number of developing fruitlets can be brought about by such scorching chemicals as DNOC or carbaryl, by synthetic auxins such as naphthylacetic amide (NAAm), and by ethephon. A side effect of ethylene, next to its thinning action, is that it antagonizes gibberellin in its inhibiting influence on flower formation. As a result of this chemical induction of ethylene synthesis the crop is properly thinned and flower formation for the following year is ensured. In the U.S. State of Washington, one of the most successful areas of apple production in the world, over 95% of the

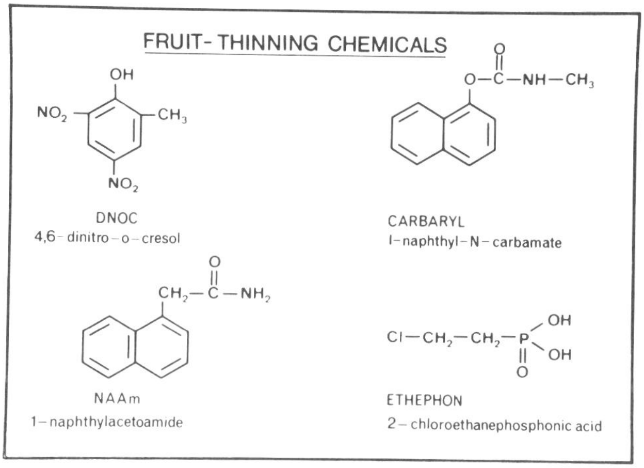

Figure 7: Fruit Thinning Chemicals.

growers use chemical thinning to improve fruit quality and yield. Biennial bearing has nearly been eliminated in Washington State apple orchards (Couey and Williams, 1982).

Another problem area in pome-fruit growing is to obtain simultaneous and regular ripening of the crop to facilitate harvesting, with optimum fruit quality for postharvest life and consumption. Ripening is a complex phenomenon, including de novo synthesis of enzymes for the development of colour and smell, conversion of starch and acids into sugars, solubilization of cell walls to soften fruits, etc. It is usually accompanied by a sudden increase in respiration and in the evolution of ethylene that is generally considered to be the ripening hormone (Bruinsma, 1982 b). Consequently, a spray with ethephon, shortly before the desired harvest, causes a rapid and regular ripening. Such a treatment increases the soluble-solids content and diminishes astringency. As a result, such a chemically advanced harvest extends the early marketing season with dessert fruit of improved quality (Couey and Williams, 1982). Apart from the risk of preharvest fruit drop, ethephon treatment reduces the postharvest shelf life of the crop. Because synthetic auxins induce ethylene evolution as well as inhibit abscission, chemicals such as 2, 4, 5-T (2, 4, 5-trichlorophenoxyacetic acid) and NAA (naphthylacetic acid) are safer PGR's to use for accelerating ripening (unpublished data, Research Station for Fruit Growing, Wilhelminadorp, The Netherlands).

In order to extend the postharvest life of fruit, either synthesis or action of the ripening-promoting hormone, ethylene, should be prevented. Both inhibitors of ethylene production and of its action are known, but these PGR's until now have not been used practically on fruits, for either toxicological or economic reasons. The vaselife of cut flowers, for example, can be considerably extended by their use. Daminozide, that otherwise stimulates ethylene synthesis in plant tissues, for some unknown reason turns out to inhibit it in pome fruits. Therefore, both in the U.S.A. and in the Netherlands a preharvest spray with daminozide is applied, not only to control the vegetative growth of the trees and to promote flower formation, but also to produce firmer, better coloured fruit that has prolonged storage life with a reduced sensitivity for such storage diseases as watercore. In combination with advanced storage conditions under controlled atmosphere remarkable results have been obtained. Daminozide-treated fruits, stored either under 2-3% oxygen and 1-5% carbon dioxide after a 15-20% carbon dioxide pretreatment during the first 10 days of storage, or under 1.0-1.5% oxygen throughout, remain firm and crisp into June and July. Together with the advanced harvest technique using ethephon a nearly year-round fresh fruit supply becomes possible. For optimum results, fertilization with nitrogen should be kept moderate, otherwise the harvest may be large but poor in quality, yielding sour and starchy apples. If these apples remain longer on the tree their taste improves but they become more liable to storage diseases (Couey and Williams, 1982). Thus, sophisticated fruit culture requires a delicate balance of optimal fertilization, PGR-application, and storage conditions. There has to be good cooperation between experienced growers and storage managers, backed up by extension service and research facilities.

CONCLUSION

The possibilities and problems of modifying crop performance by application of PGR's, chemicals that interfere with the endogenous hormonal pattern of crop plants, has been demonstrated. The examples presented show that metabolism and morphogenesis of crop plants can be carefully controlled to increase the quantity and quality of the yield and to improve harvestibility and postharvest life of the commodities. The potential of PGR's in agri- and horticulture is great, but the complexity of effects requires thorough investigation of their impact on the physiology and the morphology of the crop plants in order to ensure consistent results. For this research, the analytical tools and the biological and chemical knowledge are available, as well as the personnel. It will largely depend on the investment made as to the extent that application of PGR's contributes to the solution of man's needs for food and fodder crops.

REFERENCES

Bruinsma, J. (1966). Analysis of growth, development and yield in a spacing experiment with winter rye (Secale cereale L.). Neth. J. Agr. Sci. 14, 198-214.

Bruinsma, J. (1980). The endogenous hormonal pattern and its interference by exogenous plant growth regulators. Physiological aspects of crop productivity. International Potash Institute, Worblaufen/Bern (Switzerland), 117-124.

Brunisma, J. (1981). Present state and prospects of research into natural and synthetic plant growth-regulating substances. Aspects and prospects of plant growth regulators, Brit. Plant Growth Regulator Group, Wantage (U.K.).

Bruinsma, J. (1982a). Plant growth regulators in field crops. Chemical manipulation of crop growth and development. Butterworth Scientific, London, 3-11.

Bruinsma, J. (1982b). Hormonal regulation of senescence, aging, fading, and ripening. Postharvest physiology and crop preservation. Plenum Press, New York.

Bruinsma, J., N.M. de Vos and K. Dilz (1965). Effects of (2-chloroethyl) trimethylammonium chloride (CCC) on growth and development of cereal plants. Mededel. Landb. hogeschool Gent 30, 1990-2006.

Couey, H.M. and M.W. Williams (1982). The Washington State apple revolution. HortScience 17, 14-17.

Gale, M.O. and P.R. Hanson (1982). The plant breeding potential of genetic variation in cereal phytohormone systems. Chemical manipulation of crop growth and development. Butterworth Scientific, London, 425-449.

Herbert, C.D. (1982). Growth regulation in cereals - chance or design? Chemical manipulation of crop growth and development. Butterworth Scientific, London, 315-327.

Hillman, J.R. (1978). Isolation of plant growth substances. Cambridge Univ. Press, Cambridge (U.K.).

Hoffner, W. and H. Kuhn (1982). Effect of growth regulator combinations on ear development, assimilate translocation and yield in cereal crops. Chemical manipulation of crop growth and development. Butterworth Scientific, London, 375-390.

Lupton, F.G.H. (1980). Breeding for higher yields. Physiological aspects of crop productivity. International Potash Institute, Worblaufen/Bern (Switzerland), 27-36.

Moore, T.C. (1979). Biochemistry and physiology of plant hormones. Springer Verlag, New York, Heidelberg, Berlin.

Schott, P.E. and F.R. Rittig (1982). New findings on the biological activity of mepiquat chloride. Chemical manipulation of crop growth and development. Butterworth Scientific, London, 415-424.

Vos, N.M. de, K. Dilz and J. Bruinsma (1967). Effects of 2-chloroethyl-trimethylammonium chloride (CCC) on yield and lodging of wheat. Neth. J. Agric. Sci. 15, 50-62.

Weiler, E.W. (1981). Radioimmunoassay for mol-quantities of indole-3-acetic acid for use with highly stable (^{125}I) and (^{3}H)IAA derivatives as radiotracers. Planta 135, 319-325.

Wilkins, R.M. (1982). The potential for the controlled delivery of plant growth regulators. Chemical manipulation of crop growth and development. Butterworth Scientific, London, 111-126.

Williams, R.H., J.A. Turner and M.J. Sampson (1982). New approaches to increasing the yield capacity of cereals. Chemical manipulation of crop growth and development. Butterworth Scientific, London, 399-414.

IMPROVING THE PRODUCTIVITY OF PROBLEM RICE LANDS

F.N. Ponnamperuma
Principal Soil Chemist
The International Rice Research Institute
P.O. Box 933, Manila, Philippines

ABSTRACT

In densely populated South and Southeast Asia, where both food and arable land are scarce, about 100 million ha of lands that are climatically, physiographically, and hydrologically suited to rice cultivation lie idle largely because of soil problems such as salinity, alkalinity, strong acidity, acid sulfate soil conditions, and excess organic matter. About 100 million ha of current rice lands suffer from the same problems or deficiencies of nitrogen, phosphorus, and zinc.

Recent researches on the chemistry of submerged soils and the nutrition of the rice plant have revealed the yield-limiting factors in problem rice lands and enabled the formulation of amelioration measures. By combining soil amendments with cultivar tolerance for adverse soils, the productivity of current rice lands can be improved, and millions of ha of idle lands brought into rice production.

KEYWORDS: Acid sulfate soils, Alkalinity, Macro and micronutrient deficiencies, Peat soils, Salinity, Soil amendments, Varietal tolerance, Wetland rice.

INTRODUCTION

In densely populated South and Southeast Asia, where both food and arable land are scarce, about 100 million ha of lands that are climatically, physiographically, and hydrologically suited to rice cultivation lie idle or are cultivated with poor results largely because of soil problems. On millions of ha of current rice lands the modern improved rice varieties do not deliver their potential because of soil problems.

Five recent developments focus attention on problem rice lands: 1. the need for more food for a rapidly increasing population; 2. the scarcity of arable land; 3. soil degradation; 4. advances in the chemistry of rice soils; and 5. the advent of modern rice cultivars adapted to adverse soil conditions.

If present estimated trends persist, by the year 2000 the world population is likely to be 6.3-6.7 billion and the food need 3 billion t (Hopper, 1981). But food production growth rates are barely sufficient to keep pace with population growth especially in the poor, populous countries (Barr, 1981). Until about 1950 the increase in world food demand was met by an increase in the cultivated area. But now there is little arable land left that can be brought under crops at a reasonable cost. The pressure on the land is most severe in densely populated South and Southeast Asia (Willet, 1976). Erosion, salinization, and denudation are rendering 5 to 7 million ha per year of agricultural land unproductive (Dudal, 1978). Overcrowding, food shortages, and land scarcity are compelling developing countries to bring under crops lands lying idle largely because of soil problems.

Recent advances in the chemistry of submerged soils and the nutrition of the rice plant have revealed the growth-limiting factors in current and potential problem rice lands. Methods of diagnosing, delineating, and alleviating them have been developed, and rice cultivars with tolerance for adverse soils are now available. The main adverse soil factors are chemical, physical, and hydrological. This paper discusses the chemical problems. Chemical problems are soil toxicities, nutrient deficiencies or both.

Soil toxicities include salinity, alkalinity, strong acidity, acid sulfate soil conditions, excess organic matter, and boron toxicity. The extent and distribution of the major classes of toxic soils are in Table 1. Common deficiencies are those of the macronutrients nitrogen, phosphorus, potassium, and sulfur and those of the micronutrients, iron and zinc.

Table 1. Distribution and Extent of Problem Rice Lands in South and Southeast Asia.

Country	Problem Rice Lands (million ha)				
	Saline soils	Alkali soils	Acid sulfate soils	Peat soils	Total
Bangladesh	2.5	0.5	0.7	0.8	4.5
Burma	0.6		0.2		0.8
India	23.2	2.5	0.4		26.1
Indonesia	13.2		2.0	16.0	31.2
Kampuchea	1.3		0.2		1.5
Malaysia	4.6		0.2	2.4	7.2
Pakistan	10.5	4.0			14.5
Philippines	0.4				0.4
Thailand	1.5		0.6	0.2	2.3
Vietnam	1.0		1.0	1.5	3.5
Total	58.8	7.0	5.3	20.9	92.0

Source: International Rice Research Institute, 1980.

SALINITY

Saline soils cover 344 million ha of the earth's land surface (Massoud, 1974. Of these 230 million ha are not strongly saline and have crop production possibilities. About 27 million ha are in the humid, tropical region of South and Southeast Asia where at least one rice crop can be grown per year.

Saline soils vary widely in the kind and content of salts, salt dynamics, and in their chemical and physical properties, but they have one common feature: the electrical conductivity of the saturation extract of the soil in the root zone exceeds 4 dS m^{-1}. Some of the growth-limiting factors in saline soils other than excess salt include: deficiencies of nitrogen, phosphorus, and zinc; and toxicities of aluminum, boron, and iron (Ponnamperuma, 1982a). Rice varieties vary widely in their tolerance for salt and the accessory growth-limiting factors.

Saline tracts are converted to agricultural lands by preventing the influx of salt water, leaching out of the salts, and correcting soil toxicities and nutrient deficiencies. Advances in the knowledge of ion-exchange reactions in soils have helped soil chemists determine the amount and frequency of leaching and also avert the consequences of using low-quality irrigation water. Plant nutrition research has identified accessory growth-limiting factors and methods of alleviating them. Rice is the crop best suited to saline soils because soil submergence that is necessary for leaching out the salts benefits rice. The capital and recurrent costs of reclamation can be reduced by growing modern, salt-tolerant rice cultivars. Extensive tests in farmers' field have shown that yields of 3 to 4 t/ha per season are possible on some moderately saline tracts without costly inputs if salt-tolerant cultivars are grown (Ponnamperuma, 1982b). Use of salt-tolerant cultivars confers a comparative yield advantage of about 2 t/ha (Table 2).

Table 2. Yield Advantage Due to Soil Stress Tolerance in Modern Rices as Shown by Tests in Farmers' Fields in the Philippines (1977-1981).

Stress	Total Number			Mean Yield (t/ha)		
	Tests	Sites	Rices	Min.	Max.	Adv.
Salinity	23	14	63	1.5	3.6	2.1
Alkalinity	3	2	47	0.9	3.6	2.7
Iron toxicity	12	4	55	2.2	4.8	2.6
Peatiness	13	5	39	1.4	3.1	1.7
Al/Mn toxicity	3	1	32	2.0	3.8	1.8
P deficiency	13	2	110	1.9	4.4	2.5
Zn deficiency	25	10	91	0.8	2.9	2.1
Fe deficiency	8	3	65	0.9	2.8	1.9

Source: Ponnamperuma 1982a.

ALKALINITY

Alkali or sodic soils cover over 538 million ha. About 7 million ha are in the Indo-Gangetic plain where population pressure is high and irrigation water is available. Sodic soils contain sufficient exchangeable sodium to depress plant growth. The sodium adsorption ratio of the saturation extract exceeds 15; pH is above 8.5; and sodium carbonate or bicarbonate, calcium carbonate, and high concentrations of water-soluble silicon are present. The clay and organic matter are dispersed. The soils are sticky when wet and hard when dry. Internal drainage is very poor. Sodic soils are reclaimed for crop production by treating them with gypsum and leaching out the sodium displaced by the reaction

$$Na^{+}\text{-clay} + Ca^{2+} \rightarrow Ca^{2+}\text{-clay} + Na^{+} \quad (1)$$

Rice can be grown during reclamation if deficiencies of nitrogen, phosphorus, and zinc are corrected. The gypsum and water requirements for reclamation can be reduced by using alkali-tolerant cultivars. Such cultivars have a yield advantage of over 2 t/ha over comparable sensitive varieties (Table 2).

STRONG ACIDITY

Strongly acid soils, which include Ultisols, Oxisols, and Spodosols, cover over 2 billion ha, mostly in humid regions (Dudal, 1976). The pH values are <5.5, the cation exchange capacity and base status are low, and the degree of aluminum saturation and capacity to fix phosphate are high (Sanchez and Cochrane, 1980). Odhiambo and Sanchez (1983) discuss the problems and potential of strongly acid dryland soils for crop production in the tropics.

Knowledge of the chemistry of submerged soils and the nutrition of the rice plant have clarified the causes of poor growth of wetland rice on strongly acid soils. When a dry soil is flooded, its redox potential decreases and Fe(III) is reduced to Fe(II). Because Fe(II) compounds are much more soluble in water than their Fe(III) counterparts, large amounts of Fe^{2+} enter the solution phase. The activities of Fe^{2+} in the solution phase are governed by pE and pH according to the following equations (Ponnamperuma, 1972):

$$pE = 17.87 + pFe^{2+} - 3\ pH \quad (2)$$

and

$$pH = 1/2\ pFe^{2+} = 5.4 \quad (3)$$

According to equation (3), a unit change in pH of a reduced ferruginous soil may cause a 100-fold change in the activity of water-soluble Fe^{2+}. Nutritional studies show that >300 mg/L in the soil solution is toxic to rice. Concentrations as high as 5000 mg/L are present in extremely acid soils, such as acid sulfate soils (Ponnamperuma and others, 1973). Iron toxicity can be alleviated by liming and the use of tolerant rices. Tolerance confers a yield advantage of over 2 t/ha on unamended iron-toxic soils (Table 2).

ACID SULFATE SOIL CONDITIONS

Acid sulfate soils and potential acid sulfate soils cover about 15 million ha of flat lands in the tropics suited to rice cultivation. Of the 5 million ha in Southeast Asia less than 1 million ha are cultivated (van Breemen and Pons, 1978).

Acid sulfate soils are derived from marine sediments high in sulfidic materials such as pyrites. When submerged and anaerobic, they are nearly neutral in reaction and support halophytes. When drained, the soil material becomes aerobic and the sulfides are oxidized to sulfuric acid rendering the soil extremely acid. Acid sulfate soils are clays with a pH <4.0 and a high sulfate content. They usually have a jarosite horizon in the profile.

The growth-limiting factors in aerobic acid sulfate soils are strong acidity per se, aluminum toxicity, high electrolyte content and deficiencies of nitrogen and phosphorus. When submerged, the pH may increase to values as high as 7, and aluminum toxicity disappears according to the equation

$$pAl = 2\ pH - 4.41 \qquad (4)$$

But because of soil reduction, the concentration of water-soluble iron increases. Thus wetland rice suffers from iron toxicity on acid sulfate soils (Mai-thi-My-Nhung and Ponnamperuma, 1966; Ponnamperuma and others, 1973; Ponnamperuma and Solivas, 1981).

Acid sulfate soils can be improved for wetland rice by liming, adding manganese dioxide, and by prolonged submergence. Prolonged submergence eliminates aluminum toxicity and minimizes iron toxicity. Varieties tolerant of excess iron reduce the requirements of lime and manganese dioxide (Ponnamperuma, 1982c). Such varieties have a yield advantage of more than 2 t/ha over nontolerant rices on unamended acid sulfate soils (Table 2). On moderately acid sulfate soils with water control, yields exceeding 8 t/ha per year have been obtained by using tolerant rices and applying 50 kg N and 25 kg P per ha.

EXCESS ORGANIC MATTER (PEATINESS)

Peat soils occupy over 20 million ha of lands in South and Southeast Asia (Driessen, 1978). Because of population pressure these lands are being cleared and settlements established. Peat soils have a surface horizon at least 50 cm deep which contains 12-18% organic carbon. Unless drained, they are submerged or saturated with water for the greater part of the year. Deep or strongly acid peats are infertile and difficult to manage. Reclaiming peat land involves felling the swamp forest, draining the soil, and burning the undergrowth. Draining and burning cause subsidence of the peat due to oxidation and compaction. Subsidence can be decreased by keeping the soil anaerobic. Under anaerobic soil conditions rice is the only major food crop that can be grown.

Growth-limiting factors on peat soils have recently been identified as deficiencies of nitrogen, phosphorus, potassium, zinc, molybdenum and copper (IRRI, 1976; 1977). Figure 1 shows a marked response of rice to the combined nutrients whereas Figure 2 illustrates the role of zinc.

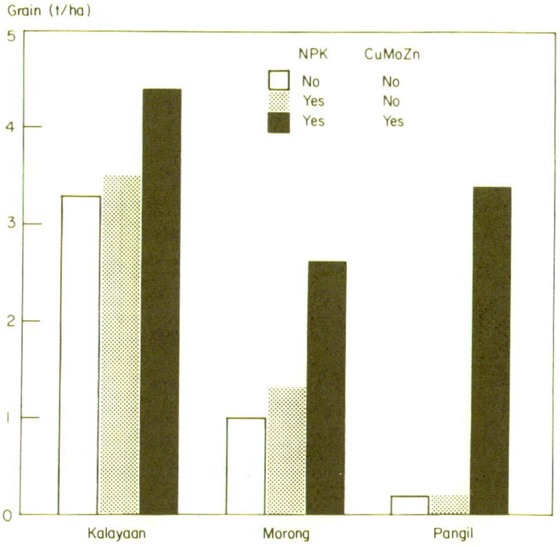

Fig. 1. Response of IR26 to N, P, K and Cu, Mo, Zn on Histosols at three locations in Laguna, Philippines, 1977 dry season.

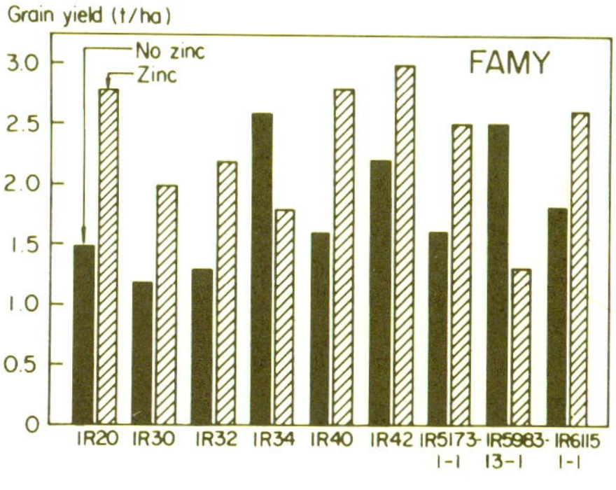

Fig. 2. Effect of Zn on the yield of some rice varieties on a peat soil at Famy, Laguna, Philippines.

Rice varieties differ in their tolerance for peat soil conditions. Such varieties have a comparative yield advantage of about 2 t/ha (Table 2). Millions of ha of shallow peat soils in the South and Southeast Asia can be brought under rice if tolerant varieties are grown with nitrogen, phosphorus, potassium, and zinc fertilizers. Two kg zinc oxide per ha costing $2.00 may spell the difference between no yield and 3 t/ha.

BORON TOXICITY

Although boron toxicity has for decades been considered a hazard on arid irrigated soils, it was only 4 years ago that the first field case of boron toxicity of rice was reported (IRRI, 1980). Boron toxicity is now recognized as a growth-limiting factor on soils irrigated with geothermal water or inundated by brackish water (IRRI, 1980). Critical soil and plant limits have recently been proposed for rice and a simple test for boron toxicity developed (IRRI, 1981). Because there is no simple soil amendment, varietal tolerance is the only solution at present. Rice cultivars with tolerance for excess boron have been identified (IRRI, 1981).

NUTRIENT DEFICIENCIES

Nitrogen

Up to 70% of the nitrogen absorbed by even a fertilized rice crop comes from the soil. Because over 75% of the rice soils of South and Southeast Asia contain <0.2% total nitrogen (the critical limit) nitrogen deficiency is the most important factor limiting rice yields. Nitrogen deficiency is corrected by applying chemical fertilizers. But one serious problem is that up to 70% of the nitrogen applied to rice fields may be lost by volatilization an ammonia or by denitrification. Deep placement of sulfur-coated urea granules in the mud has increased the efficiency of nitrogen fertilizer in wetland rice soils (De Datta and Craswell, 1982).

Initial attempts to tailor nitrogen needs of wetland rice to the nitrogen supplying power of the soil and yield target were successful (IRRI, 1981). If these results apply over a wide range of soil and climatic conditions, ammonia released by anaerobic incubation of the soil may be used as a guide to determine the nitrogen need of a rice crop. That will help farmers to add the right amount of nitrogen fertilizer where it is needed and avoid it where it is unnecessary or deleterious.

Phosphorus Deficiency

Phosphorus deficiency limits crop growth on Ultisols, Oxisols, Andepts, Vertisols, and Sulfaquents. Deficiency may be due to a low total phosphorus content or low availability due to strong fixation by soil minerals. Phosphorus deficiency can be corrected by applying phosphorus fertilizers. But strong adsorption by such soil minerals as kaolinite, halloysite, and hydrous oxides reduces the efficiency of phosphorus fertilizers to <10%. Band placement and the use of phosphorus-efficient species and cultivars are used to improve yields. Phosphorus-efficient rice cultivars have a comparative yield advantage of over 2 t/ha over comparable varieties that lack the trait (Table 2).

Sulfur Deficiency

Although sulfur is an essential element that is absorbed by crop plants in amounts comparable to that of phosphorus, it was only recently that its deficiency has been recognized as a widespread growth-limiting factor of wetland rice (Islam and Ponnamperuma, 1982). Wetland rice is more susceptible to sulfur deficiency than dryland crops because water-soluble sulfates are converted to insoluble iron, manganese, and zinc sulfides. Striking responses to sulfur applications have been reported (Wang, 1976; Blair and others, 1980). Soil and plant tests have been developed to detect sulfur-deficient soils (Islam and Ponnamperuma, 1982). Sulfur deficiency is easily corrected by applying flowers of sulfur, sulfate fertilizers, or gypsum.

Zinc Deficiency

Zinc deficiency is the most widespread micronutritional disorder of crops the world over (Lopes, 1980). In dryland crops it occurs on calcareous, alkali, and sandy soils. But in wetland rice it occurs, regardless of pH and texture, on peat soils, poorly drained soils, high-silicon soils, and soils high in chromium (IRRI, 1980). After nitrogen deficiency, it is perhaps the most important factor limiting wetland rice yields. Wetland rice is susceptible to zinc deficiency apparently because zinc is removed from the solution phase as insoluble zinc sulfide, zinc silicate, or zinc ammonium phosphate (IRRI, 1971).

Recent studies have established that the critical soil and plant limits for zinc deficiency and have shown that it can be recognized by symptoms, response to zinc application, and by simple chemical tests. Zinc deficiency is very likely if a soil has one or more of the following characteristics: a pH >6.8; an organic carbon content >3.0%, a hydrochloric-acid-extractable zinc content of <1 mg/kg, or an available silicon content exceeding 100 mg/L (IRRI, 1979; 1981). On a soil that combined these unfavorable features, in a varietal trial only 2 out of 698 rices survived (IRRI, 1979). Zinc deficiency is easily corrected by applying zinc compounds to the plant or soil. Varietal tolerance is a valuable adjunct. Varietal tolerance confers a yield advantage of about 2 t/ha (Table 2).

Iron Deficiency

Iron deficiency is a widespread micronutrient problem in calcareous and sodic dryland soils, and it is the most important nutritional factor limiting rice growth on dryland soils, regardless of pH (Ponnamperuma, 1975). Iron deficiency in aerobic soils is due to the extremely low solubility of the ferric compounds present in oxidized soils. Flooding reduces ferric compounds to their more soluble counter-parts. So iron deficiency in wetland rice is rare.

The remedies for iron deficiency are acidifying the soil, and applying iron salts or chelates to the soil or crop. But these methods are uneconomic, ineffective, and disturb nutrient balance or cause pollution. Breeding for tolerance for iron deficiency appears to be the best solution (Brown and others, 1972). Iron-efficient rice cultivars have been identified. They have a yield advantage of about 2 t/ha over varieties that are inefficient users of iron (Table 2).

REFERENCES

Barr, T.N. (1981). The world food situation and global prospects. Science 214:1087-1095.

Blair, G.J., C.P. Mamaril, and M. Ismunadji (1980). Sulfur deficiency in soils in the tropics as a constraint to food production 1980. In: International Rice Research Institute, 232-251. Priorities for alleviating soil-related constraints to food production in the tropics. Los Baños, Philippines.

Brown, J.C., J.E. Ambler, R.L. Chaney, and C.D. Foy (1972). Differential response of plant genotypes to micronutrients. In: Micronutrients in Agriculture, 389-418. Ed. J.J. Mortvedt, P.M. Giordano, W.L. Lindsay. Soil Sci. Soc. Amer. Inc., Madison, Wisconsin.

De Datta, S.K., and E.T. Craswell (1982). Nitrogen fertility and fertilizer management in wetland rice soils. In: International Rice Research Institute, 283-316. Rice research strategies for the future. Los Baños, Philippines.

Driessen, P.M. (1978). Peat soils. In: International Rice Research Institute, 763-780. Soils and Rice. Los Baños, Philippines.

Dudal, R. (1976). Inventory of the major soils of the world with special reference to mineral stress hazards. In: Proceedings of Workshop on Plant Adaptation to Mineral Stress in Problem Soils, 3-14. Beltsville, Maryland. Nov. 22-23, 1976.

Dudal, R. (1978). Land resources for agricultural development. Proceedings 11th International Soil Science Congress. Edmonton, Alberta, Canada. June 12-17, 1978.

Hopper, W.D. (1981). Recent trends in world food and population. In: Future Dimensions of World Food and Population, 35-55. Ed. R.G. Woods. Westview Press, Boulder, Colorado.

IRRI (International Rice Research Institute) (1971). Annual report for 1970. Los Baños, Philippines. 265 p.

IRRI (International Rice Research Institute) (1976). Annual report for 1975. Los Baños, Philippines. 479 p.

IRRI (International Rice Research Institute) (1977). Annual report for 1976. Los Baños, Philippines. 418 p.

IRRI (International Rice Research Institute) (1979). Annual report for 1978. Los Baños, Philippines. 478 p.

IRRI (International Rice Research Institute) (1980). Annual report for 1979. Los Baños, Philippines. 538 p.

IRRI (International Rice Research Institute) (1981). Annual report for 1980. Los Baños, Philippines. 467 p.

Islam, M.M., and F.N. Ponnamperuma (1982). Soil and plant tests for available sulfur in wetland rice soils. Plant and Soil 68:97-113.

Lopes, A.S. (1980). Micronutrients in soils of the tropics as constraints to food production. International Rice Research Institute, 277-298. Soil-related constraints to food production in the tropics. Los Baños, Philippines.

Mai-Thi-My-Nhung, and F.N. Ponnamperuma (1966). Effects of calcium carbonate, manganese dioxide, ferric hydroxide, and prolonged flooding on chemical and electrochemical changes and growth of rice in a flooded acid sulfate soil. Soil Sci. 102:29-41.

Massoud, F.I. (1974). Salinity and alkalinity as soil degradation hazards. FAO/UNDP Expert Consultation on Soil Degradation. June 10-14, 1974. FAO, Rome. 21 p.

Odhiambo, T.R., and P.A. Sanchez (1983). Physical, chemical and biological constraints on food production and on the level and efficient use of chemical inputs. Proc. International Conference on Chemistry and World Food Supplies: New Frontiers. Manila, Philippines, Dec. 6-10, 1982.

Ponnamperuma, F.N. (1972). Advances in Agronomy. 24:29-88.

Ponnamperuma, F.N. (1975). Growth-limiting factors of aerobic soils. In: International Rice Research Institute, 40-43. Major research in upland rice. Los Baños, Philippines.

Ponnamperuma, F.N. (1982a). Breeding crop plants to tolerate soil stresses. In: Plant improvement and somatic cell genetics. Eds. I.K. Vasil, W.R. Scowcroft, K.J. Frey. Academic Press, New York, 73-97.

Ponnamperuma, F.N. (1982b). Role of cultivar tolerance in increasing rice production on saline lands. Proceedings of the International Conference on Plant Improvement for Irrigated Crop Production Under Increasing Saline Conditions, Oct. 18-22, 1982, Bellagio, Italy. (In press).

Ponnamperuma, F.N. (1982c). Genotypic adaptability as a substitute for amendments on toxic and nutrient-deficient soils. In: Plant Nutrition Vol. I. Proceedings of the Ninth International Plant Nutrition Colloquium, Warwick University, England, Aug. 22-27, 1982, 467-473.

Ponnamperuma, F.N., T. Attanandana, and G. Beye (1973). Amelioration of three acid sulfate soils for lowland rice. In: Proceedings of the International Symposium on Acid Sulfate Soils, Wageningen, August 13-20, 1972, 2: 391-405.

Ponnamperuma, F.N., and J.L. Solivas (1982). Field amelioration of an acid sulfate soil for rice with manganese dioxide and lime. In: Proceedings Second International Symposium on Acid Sulphate Soils, Bangkok, Thailand, Jan. 18-25, 1981, 213-222.

Sanchez, P.A., and T.T. Cochrane (1980). Soil constraints in relation to major farming systems in tropical America. In: International Rice Research Institute. Soil related constraints to food production in the tropics. Los Baños, Philippines, 107-140.

van Breemen, N., and L.J. Pons (1978). Acid sulfate soils and rice. In: International Rice Research Institute. Soils and rice. Los Baños, Philippines, 739-761.

Wang, C.H. (1976). Sulfur fertilization of rice. In: The fertility of paddy soils. Food and Fertilizer Technology Centre for the Asia and Pacific Region, Taipei, Taiwan, 149-169.

Willet, J.W. (1976). The World Food Situation: Problems and Prospects to 1985. Oceana Publications, Inc., New York. 1136 p.

CHEMICAL APPROACHES TOWARDS INCREASING WATER AVAILABILITY TO CROPS INCLUDING MINIMUM TILLAGE SYSTEMS

R. Lal, A.S.R. Juo and B.T. Kang
International Institute of Tropical Agriculture
PMB 5320, Ibadan, Nigeria

ABSTRACT

This report describes some basic aspects of soil-water management in relation to crop production. No-tillage system with chemical weed control, the food crop/shrub inter-planted alley cropping system and the system of cover crop *in situ* mulch are promising cultural practices that improve soil moisture and nutrient availability and water use efficiency. For the easily compactable Alfisols and Inceptisols in the semi-arid tropics, however, mechanical tillage for rough seed bed preparation to increase surface retention and weed-free fallow reduce evaporation, improve root growth and hence, enable better utilization by crops of stored water in the subsoil horizons. For crop production in the fine-textured, strongly acidic Oxisols and Ultisols in the subhumid regions, sub-soil liming and phosphate application improve crop water use efficiency through increased root growth. For much of the tropical regions where soils with low CEC and low water holding capacity are predominant, inadequate nutrient supply is a key factor limiting water use efficiency by crops.

KEYWORDS: Soil-water management, no-tillage system, chemical weed control, alley cropping, soil moisture, liming, nutrients, water use efficiency, water balance run-off, soil-water storage, cover crops.

INTRODUCTION

Soils with low activity clays predominate in the low altitude tropics. These soils are characterized by low effective CEC, low plant nutrient reserve, and above all, low available water holding capacity. The kaolinitic Alfisols, Ultisols and Oxisols are predominant soils in the humid and sub-humid regions of West and Central Africa; whereas in tropical America, the oxidic Ultisols and Oxisols are wide spread. Properties and fertility limitations of soils with low activity clays have been recently described by Kang and Juo (1981).

Drought stress is one of the important constraints to crop production even during the rainy season for soils in the humid and subhumid tropics (Lal, 1979; Lawson et al., 1979; Hsiao et al., 1980; Greenland and Bhuiyan, 1982). In addition to erratic rainfall distribution, high incidence of drought stress is partly due to high evapotranspiration and direct evaporation from the soil surface. The main constraint, however, lies in the amount of available water that can be stored in the root zone and its effective utilization. With the exception of some Andisols and organic soils in Central America (Gavande, 1968; Perez - Escolar et al., 1974), the available water holding capacity of the root zone of kaolinitic and oxidic soils is rather low (Charreau, 1974; de Melo, 1974; de Silva et al., 1975; Wahab et al., 1976a,b; Lal, 1979). In order to utilise the rain water to its maximum, it is necessary therefore, to (a) increase soil-water storage, (b) decrease evaporation losses, and (c) improve water use efficiency through soil and crop management.

This report describes basic aspects of soil-water management in relation to crop production. Specific examples from some typical bench mark soils and highlights of prominent cultural practices that have proven effective under specific soil and environmental conditions are given. Research needs for improving the applicability of those techniques are emphasized.

BASIC ASPECTS OF WATER BALANCE IN RELATION TO WATER MANAGEMENT

Development of suitable cultural practices for the maximum utilization of rain water pre-supposes the understanding of different components of the hydrologic cycle, their magnitude and alterations under different systems of soil and crop management, and of the consequences of these alterations in terms of their influence on other inputs such as fertilizer use. The basic water balance equation may be represented as follows:

$$P = S + \Delta D + \Delta M + U + \int E dt$$

where P = rainfall received whose use is being optimised, S = surface runoff, ΔD = increase in surface detention, ΔM = increase in soil water storage as a net result of water flux in different directions, and U = increase in the ground water storage for layers below that for which ΔM was calculated. In regions with rolling landscape this term may also be used to denote seepage or interflow, and t = time so that $\int E dt$ is the total evaporation over the period of investigation.

If the water use efficiency is defined as the unit of economic yield produced per unit of rainfall (+ irrigation, if feasible) received, then the losses out of the root zone should be minimised. Supplementary irrigation is not feasible for grain crops in many regions. Nwa (1979) observed that it takes at least 350 mm of water to produce a satisfactory crop of 120-day maize in southwest Nigeria. Wahab et al. (1976), observed that the grain yield of sorghum on an Ultisol and an Oxisol in Puerto Rico was not improved by supplementary irrigation because the soil water was not being utilized beyond 30 cm depth in an Ultisol because of the hard and compact subsoil horizon. Improvement in soil-water extraction from a deeper horizon is a more practical solution than supplementary irrigation.

In order to improve the efficiency of rain water received, ΔM should be increased, whereas the S, E and U components should be relatively decreased. The increase in ΔM and decrease in S are achieved through soil management. Effective utilization of ΔM for improving the water use for efficiency, however, is attained through appropriate techniques of crop management.

Runoff Control and Soil-Water Storage

The rate and amount of surface runoff are governed by the infiltration capacity, antecedent soil moisture content, and the intensity and amount of rainfall received. Land relief and slope steepness has little effect on the quantity of runoff compared with the water transmission characteristics and structural stability of the surface soil horizon (Lal, 1976). In fact the soil's infiltration capacity or the equilibrium infiltration rate is the key mechanism that enables the soil to absorb and transmit water received during medium and high intensity rainstorm events. Infiltration capacity depends on structural stability of the soil and its ability to maintain the continuity of macro - or transmission pores. Coarse textured soils with low activity clays and low organic matter content are generally unstable to raindrop impact (Lal, 1981), slake on quick wetting and disperse in a fluidised soil during heavy rains. Rapid dessication on subsequent drying creates a crusted layer of low hydraulic conductance (Falayi and Lal, 1979) that encourages runoff losses of subsequent rains.

Comparison of synthetic soil conditioners with organic residue mulches have indicated that crop residue applied at 4 to 6 t/ha is more economical and effective in improving infiltration capacity than sythetic soil conditioners (De Vleeschauwer et al., 1978). The largest mean weight diameter of stable aggregates was also observed for the residue mulch treatment. In addition to improving the biological activity, mulch also protects the soil against raindrop impact, and improves the nutrient reserve.

The quantity of mulch required for maintenance of favorable infiltration capacity and structural stability depends on soil properties, rainfall characteristics, slope parameters (length and steepness), and on the rate of residue decomposition that depends on prevalent temperature regime and general ecological parameters. Field studies have indicated that both soil water "sorptivity" and "transmissivity" parameters of Philip's (1957) infiltration equation depend on the mulch rate (Table 1). The mean weight diameter of the aggregates, total porosity, and pore size distribution in favor of macropores also depend on the rate and frequency of the mulch applied. The optimum rate of mulch may be 4 to 6 t/ha although improvement in soil physical properties are observed up to a mulch rate of 12 t/ha.

Table 1: Effect of Mulch Rate on Physical Properties of an Alfisol One Year After Clearing (Lal et al., 1980)

Mulch rate (t/ha)	Soil water sorptivity	Soil Water transmis-sivity	Saturated hydraulic conductivity (cm h^{-1})	Soil bulk density (gcm^{-3})	Mean Weight diameter (mm)	Macropores (% by volume)
0	5.56	0.32	30 a*	1.32 a	1.2	26.0 a
2	7.81	0.57	45 a	1.20 b	1.4	28.0 a
4	7.50	0.67	70 b	1.17 bc	2.1	27.5 a
6	10.21	0.84	132 c	1.09 cd	2.0	29.5 a
12	15.36	1.05	129 c	1.04 d	2.4	39.5 b

* Duncan's multiple range test

One of the principal factors responsible for these improvements in soil physical properties is the enhanced activity of some species of earthworms, e.g., *Hyperiodrilus* spp. and *Eudrilus* spp. The worm casts are extremely stable and their burrowing channels are extremely effective in conducting water through the soil profile (De Vleeschauwer and Lal, 1981). The activity of Hyperiodrilus spp. is linearly related to the quantity and durability of mulch applied.

As a result of these improvements in soil physical properties the losses in water runoff decline exponentially with increase in mulch rate (Fig. 1). Repression equations relating mulch rate with runoff loss indicate a rapid decline between mulch rate of 4 to 6 t/ha (Table 2). For example, Lal (1976) reported that the mean annual runoff loss (average of 4 slopes of an Alfisol in southwest Nigeria was 393.1, 80.7, 30.1, and 12.9 mm for mulch rates of 0, 2, 4, and 6 t/ha, respectively. The average annual saving in water runoff for mulch rate of 6 t/ha was about 380 mm or 32 percent of the rainfall received.

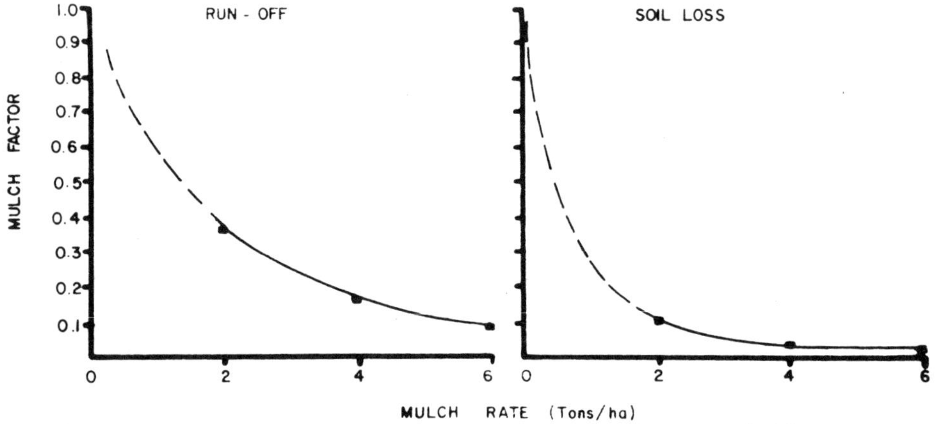

Fig. 1: Reduction in Water Runoff and Soil Loss for Different Mulch Rates (Lal, 1976).

Table 2: Mulching Effects on Water Runoff for Mulch Rates of 0 to 6 t/ha (Lal, 1976). (Y = Annual runoff loss (mm), M = Mulch rate (t/ha).)

Slope (%)	Regression Equation	Correlation Coefficient
1	$Y = 0.39\ M^{-9.73}$	0.78
5	$Y = 1.16\ M^{-0.36}$	0.80
10	$Y = 5.53\ M^{-0.27}$	0.86
15	$Y = 5.26\ M^{-0.55}$	0.75

Semi-Arid Environments

The effectiveness of mulch in improving the infiltration capacity depends on the availability of the mulch material but also on the hydrological properties of the soil profile in terms of its ability to store water or transmit it to the layers beneath. In the semi-arid region the soils of loess origin with high proportion of fine sand and silt fractions are found to be easily compactable (Nicou and Chopart, 1979), and therefore, have a limited ability to transmit water through the soil profile. Under these conditions, in addition to improving the infiltration capacity, it has been observed that improvements in surface detention component (ΔD) can play an important role in decreasing losses due to surface runoff and allowing more time for the detained water to infiltrate into the soil. The surface detention component can be drastically improved by rough and deep plowing at an appropriate time relative to the end of the rainy season.

Water management of vertisols, another important soil group of the semi-arid region, requires a careful understanding of soil-water interaction. Dispersive clays with high shrink/swell capacity, high dispersion ratio, and high surface area are extremely unstable to quick wetting and have low infiltration capacity (Collis-George and Lal, 1973). Slaking, disruption of aggregates on quick wetting, is attributed to entrapped air (Quirk and Panabokke, 1962), to the heat of wetting (Collis-George and Lal, 1971, 1973), to the drag and shearing effect of water being transmitted through the profile or hydrodynamic dispersion. For unstable slaking soils, heat of wetting controls the infiltration capacity, because experimentally it has been established that the drier the soil, and hence the more negative ΔH, the more the surface aggregates disintegrated and the lower the infiltration rate (Collis-George and Lal, 1970, 1971). In addition to the quantity of heat released, the rate of its release and dissipation are also important factors affecting slaking and, therefore, the infiltration capacity (Fig. 2). If a mechanism exists to dissipate the heat released or to minimise the quantity of heat released, then the aggregates remain stable and water is transmitted through the intra- and inter-aggregate pores. The use of crop residue mulch may maintain the surface soil at a high soil moisture potential to minimize the degradative effect of the heat of wetting.

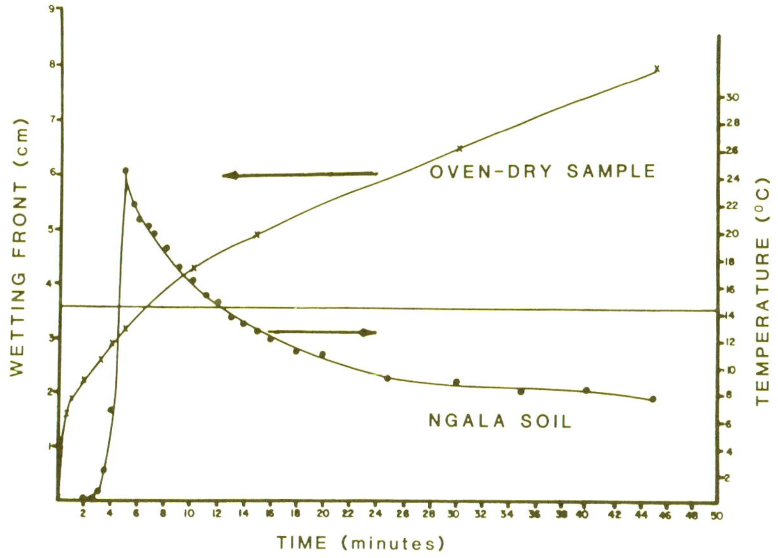

Fig. 2: Soil Temperature Profile During Infiltration into a Clayey Soil (R. Lal, unpublished).

Soil-Water Evaporation

Humid and Sub-humid Regions: The losses due to soil-water evaporation can be substantial in the tropics, and any improvement made in this aspect can significantly influence water economy and the water use efficiency. In open row crops, as much as 50 percent of the water loss in the growing season may be through direct evaporation from the soil surface. The principal source of energy for this process is the direct insulation although convection of sensible heat to and from soil may also play a minor role. The flow of vapours from soil air to atmosphere can be decreased by increase in the resistance provided by the non-turbulent layer of air in the vicinity of soil surface by creating artificial barriers such as straw mulch or polythene mulches. Residue mulching has been shown to save moisture during the first stage of evaporation (Willis, 1976). The length of the duration of the first rate evaporation stage when the soil wetness is in the vicinity of field moisture capacity increase logrithmically with the decrease in the first stage evaporation. The evaporation is also less due to low soil temperature and due to the insulating effects of the mulch layer (Ghuman and Lal, 1982). This implies that a layer of residue mulch reduces the rate of drying immediately after the rain when soil is at field moisture capacity. Consequently, the subsequent rain is more effectively utilised because the soil is already moist. The accumulative effects of slow drying during the first evaporation stage can be substantial if the rains are frequently received. The moisture profile with residue mulch is moist for a longer period during the dry season than in unmulched treatment. Moisture retention, in fact is more at high than at low mulch rates.

Semi-Arid Regions: In the semi-arid regions where annual rainfall is inadequate or poorly distributed, cropping in alternate years has proven to be more economical than cropping every year. Storing the rainwater during the fallow year in the soil profile, by decreasing its losses through runoff, evaporation, and extraction by weeds are important considerations for soil moisture conservation (Whiteman, 1975). Because of considerable losses, however, the storage efficiency usually ranges between 15 and 35 percent (Bolton, 1981). The popular belief that the protective cover of the "soil mulch" reduces losses due to evaporation is not based on flux measurements and can only be effective if the water table is present within 50 cm of the soil surface. The same or better moisture conservation can be achieved through adequate weed control with appropriate herbicides. In fact the stubble mulch of previous crop residue and dead weed growth may prevent evaporation losses more than the dust mulch. The so-called "Buckingham effect" generally refers to the discontinuity caused in capillary flow by "rapid" initial drying to exhaust the available water reserve of the upper 2 to 3 cm of soil layer. This theory, although substantiated by many field observations (Holmes et al., 1960) may not work in coarse, cloddy soil surface. On the contrary, some synthetic conditioners, if at all economical, may create hydrophobic conditions and provide discontinuity in capillary conductivity. If there are means to produce aggregate size of 0.5 to 2 mm range in the surface layer, evaporative losses are generally less.

SOIL MANAGEMENT FOR MOISTURE CONSERVATION

Humid and Sub-humid Tropics

In the humid and sub-humid tropics where the use of crop residue mulch has been shown to be advantageous in water conservation through improved infiltration and runoff control and reducing evaporation, there are different techniques available to achieve *in situ* mulch.

i) No-till Farming: This is the process of seeding through an untilled soil and crop residue mulch by opening a narrow slot or band to establish proper seed-soil contact and without mechanical primary or secondary tillage operations. The weeds are controlled by herbicides, whereby previous crop residue and dead weeds are left in the inter-row zone as mulch. The advantages for soil and water conservation are widely recognised for a wide range of conditions in the humid and sub-humid tropics. The beneficial effects in terms of moisture conservation due to reduction in water runoff through improved infiltration capacity are substantial. The mean runoff loss under maize was 5 times more in conventional system of mechanical tillage than in no-till mulched system. Less runoff in no-till than in plowed soil is partly due to high infiltration capacity caused by macro-channels created by enhanced earthworm activity (Lal, 1976b, De Vleeschauwer and Lal, 1981).

Evaporation loss of the stored water from the untilled mulched soil is generally less than from loosened plowed bare soil surface. The fine granular tilth of untilled soil surface covered with residue mulch has less evaporation than from medium to coarse textured freshly plowed surface (Holmes et al., 1960). Consequently the available moisture in the root zone of no-till soil is generally more than in plowed soil (Lal et al., 1978). This extra water may be important if the drought stress of short to medium duration (10 to 15 days of rainless period) occurred at a critical stage of crop growth. Under these conditions of soil moisture deficit, the crop is able to maintain a favourable plant-water status and sustain growth better than plowed plots where the crop suffers from severe moisture deficit. Experiments reported by Lal et al. (1978) showed that symptoms of moisture deficit in maize were observed 3 to 4 days earlier in plowed than in no-till treatments. Although leaf water potential of cowpea was not affected, that of maize showed significant differences due to tillage treatments (Fig. 3).

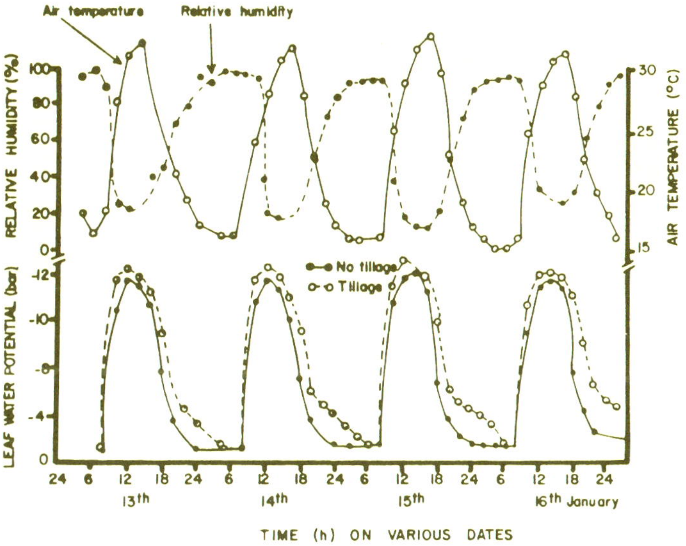

Fig. 3: Effects of Tillage Methods and Irrigation Frequency on Diurnal Changes in Leaf Water Potential of Maize (Lal, Wilson and Okigbo, 1978).

Although the leaf water potential decreased with increase in days after irrigation, the rate of decrease was slower in no-tillage than with plowing, implying that both degree and duration of soil moisture stress were affected by tillage. Consequently, the maximum water use efficiency was also obtained for the no-till system of soil management (Table 3). Although the water use efficiency was similar for the two tillage systems for some irrigation frequency the total grain yield obtained in the no-tillage method was significantly more than in plowed treatments. Results similar to these have been obtained in other regions such as in Parana, Brazil (Kemper, 1981), Puerto Rico (Chandler et al., 1966) and others. Similar studies are now being extensively conducted in Africa.

To extend the application of the no-till system to diverse soils and agro-ecological regions of the tropics, a package of cultural practices so as to enable the no-till system to be a component of the overall cropping/farming system must be developed. These cultural practices need research and development, especially for the no-till technique under diverse soils, crops, and agro-ecological regions. Successful applicability in the tropics will depend on development of appropriate cultural practices of weed control, crop rotations and crop combinations, residue management, improved implements for planting, balanced fertilizer input and adequate pest control.

No-till system works equally well for some crops for highly leached acidic Ultisols. Surface applied lime has been observed to leach readily to the subsoil horizons. In fields that received 4 tons of lime per hectare, 2.7 t/ha of applied lime was leached from the surface layer after three years (IITA, 1981), and had little effect in amending the subsoil acidity. On the other hand, however, because relatively low rates of lime sustained yields for three years or more, for these coarse-textured kaolinitic soils in the high rainfall tropics, an annual dressing of 200 to 500 kg/ha of lime should be sufficient to maintain maize and cowpea yields. At this rate lime could be regarded as a fertilizer rather than a major soil amendment. Furthermore, some crops (such as cowpea) are tolerant to high soil exchangeable Al levels (Fig. 4) and abundance of cowpea rhizobia in acid soils make this crop more attractive to smallholder farmers than other food legumes. Cowpea is also relatively drought tolerant as compared with maize and soybean.

Table 3: Influence of Tillage Methods and Irrigation Frequency on Water Use Efficiency of Maize and Cowpea (Lal et al., 1978).

	Water Use Efficiency			
	Cowpea (kg/ha/cm)		Maize (kg/ha/mm)	
Irrigation frequency (days)	No-tillage	Plowed	No-tillage	Plowed
2	20.6	20.4	7.1	6.2
4	27.5	23.9	10.8	9.5
8	29.6	24.8	10.6	6.6
12	29.1	20.0	1.9	1.1
LSD (.05)				
Tillage	8.4			
Irrigation	6.9			

ii) Cover Crops: The effectiveness of no-till farming in water conservation and in improving water use efficiency can be substantially improved if it is used in association with grass or leguminous fallow. Furthermore, no-till is less applicable on eroded, degraded and compacted soil with less residue mulch than in soils of good physical properties with adequate quantity of mulch. It is therefore, recommended that the physical, nutritional and biological properties of these soils be improved by appropriate length of planted fallows with suitable grass and legume covers. Fallowing for one or two years with *Mucuna utilis*,

Pueraria phaseoloides, Centrosema pubescens improves soil structure and infiltration capacity (Pereira et al., 1954; Kannegaieter, 1967, 1969; Bui Huu Tri, 1968; Lal et al., 1978) (Fig. 5), conserves soil water (Pereira et al., 1958) and improves crop yield through better water use efficiency. Crop yield and water use efficiency with no-tillage following a cover crop is also improved due to improved weed control, and improvement in soil organic matter content and nutritional properties. Even some of the obnoxious weeds such as Imperata cylindrica and Talinum triangulare can be effectively controlled by aggressive cover provided by these fallows.

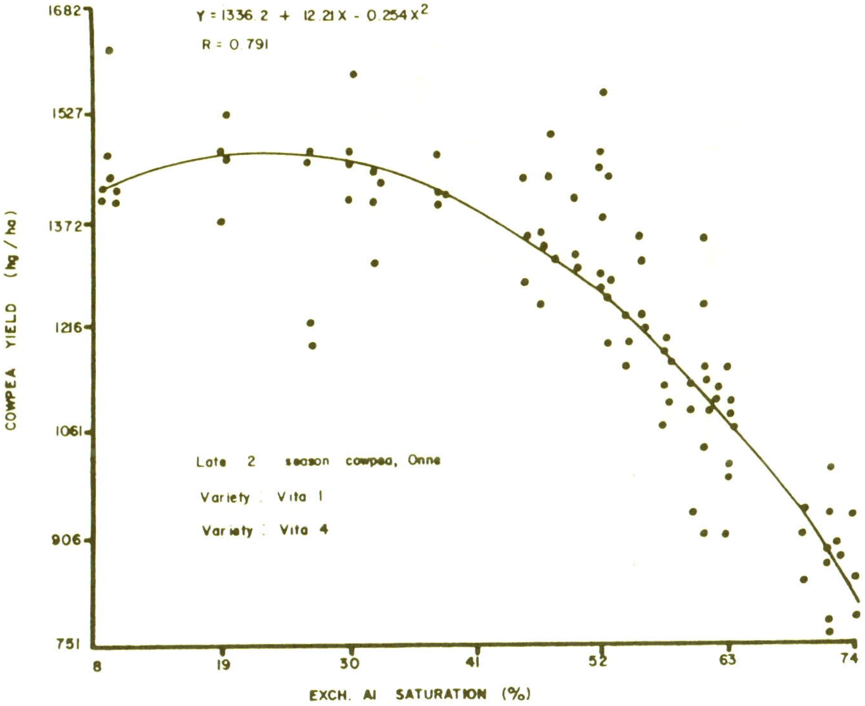

Fig. 4: Relations Between Exchangeable Aluminum Saturation in Soil and Cowpea Yield (A.S.R. Juo, unpublished).

iii) Alley Cropping: Serious problems of environmental degradation occur from expansion of intensive crop production with no or short natural bush fallow system in various upland areas of the tropics, paticularly in the humid and subhumid regions of tropical Africa. Research has concentrated on alternative crop production systems that can minimize the adverse effects of intensive continuous crop production and at the same time better enhance the productivity and continued stability of the upland areas. One solution is achieved by integrating trees and shrubs with annual food crop production in an agroforestry system (Vegara, 1982). This approach is already known and practiced by traditional farmers in many parts

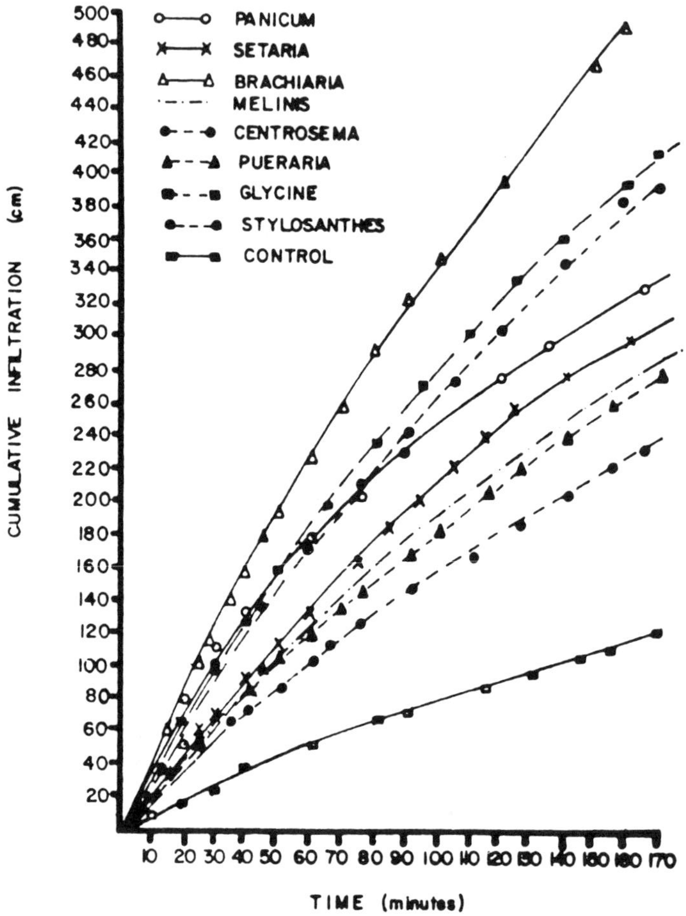

Fig. 5 Influence of Cover Crops on Infiltration Rate (Lal, Wilson and Okigbo, 1978).

of the tropics, however, its importance has only recently been realised by scientists. The effectiveness and beneficial effects of trees in reducing erosion and surface runoff, minimizing soil temperature and soil moisture loss, and increasing soil organic matter and nutrient recycling have been documented (Mongi and Huxley, 1979). The integrated food and tree/shrub fallow production system can be made more beneficial with the proper choice of the fallow species as shown wth the development of the alley cropping system. In this system, arable crops are grown in the spaces between rows of planted woody shrub or tree fallows. The growth of the fallow is controlled by periodic pruning during the cropping season to prevent shading and provide in-situ green manure (Kang et al., 1981). Promising results have been obtained in alley cropping food crops with Gliricidia and the giant type of Leucaena. This is illustrated with the results obtained from a maize-leucaena alley cropping trial conducted at the IITA site in Ibadan, Nigeria as shown in Table 4.

Table 4: Grain Yield of Maize Alley Cropped with *Leucaena* Grown on a Sandy Inceptisol at Ibadan (B.T. Kang, unpublished).

	Grain Yields (kg/ha)		
Treatments	1980	1981	1982
No N and no *Leucaena* prunings	1036	477	610
No N with *Leucaena* prunings	1918	1207	2096
80 kg N/ha with Leucaena prunings	3259	1886	2911
LSD .05	312	294	437

Large amounts of nitrogen were harvested annually with the prunings, though it is only partially effective for the associated crops during the growing season, it has the potential for reducing the dependency on chemical nitrogen fertilizers for the system. Because of the deep rooting system of the fallow trees and shrubs, the trees and shrubs can better utilize moisture from deeper soil horizons, which make the alley cropping a more efficient system in its water harvest. This point is under investigation.

(b) Semi-Arid Tropics: In the semi-arid sahel zone of west Africa with easily compactable soil, mechanical tillage operations seem to be advantageous for water conservation through improved surface detention capacity and weed control during the prolonged dry season. If soil inversion by mold-board plowing is done towards the end of the rainy season, soil hardening during the prolonged dry season is minimized (Nicou, 1976). The improved water regime observed following deep plowing is due to (a) improved infiltration, (b) reduction of evaporation, and (c) better use of stored water by crops. Charreau (1969) observed that plowing improved infiltration rate by 26% on bare soil, and 21% on cropped soil. Better use of stored water by crops is through better root system development. Charreau and Nicou (1971) observed that water uptake by rice from plowed land was 15 to 36mm more than from unplowed soil. Similar differences in water extraction were as much as 28mm for sorghum. This difference in water extraction was mainly in the deep layers through better and deep root system development by plowing. Weed control during the long fallow through the use of chemicals can further decrease water losses from sub-soil horizons and improve crop water uptake and its effectiveness.

CROP AND NUTRIENT MANAGEMENT FOR EFFECTIVE WATER UTILIZATION

Effective utilization of stored soil water determines the actual crop yield obtained and the water use efficiency. In the sub-humid and semi-arid regions, attempts should be made to utilise the limited water supply more efficiently. Some of the important management factors are discussed in this section. Other important factors include the time of planting, optimum plant population and spacing, adapted varieties, etc. In this report emphasis is given to systems and techniques that involve the judicious use of chemicals including the use of herbicides for weed control and nutrient management through balanced and judicious fertilizer use.

Nutrient Management and Chemical Amendments

Water and Nitrogen: It is well known that a balanced nutrient supply enables the crop to use available soil moisture more efficiently (Bolton, 1981). As water is directly involved in transporting soluble N from soil to the root surface by both mass flow and by diffusion, when soil moisture conditions are unfavourable, both processes are essentially stopped (Barber, 1962; Olsen and Kemper, 1968; Liao and Bartholomew, 1974). Soil management practices such as different tillage, mulching and fallow systems may require different levels of N in order to balance the moisture-nitrogen supply for maximum water use efficiency (Greb et al., 1967; Koehler and Guetinger, 1967; Oveson and Appleby, 1971; Strebel et al., 1980; Bolton, 1981). A no-tillage and crop residue mulch experiment conducted on a kaolinitic Alfisol at Ibadan (Juo, 1980) showed that higher yield of maize during an unusually dry year in 1976 (rainfall 20 percent below normal - T.L. Lawson, personal communication) in the no-till mulched plots was apparently due to better soil moisture and soluble N supplies as compared with the no-till treatment where crop residue was removed after each harvest (Fig. 6).

For much of tropical regions, where the predominant soils are characterized by low nutrient and water holding capacities, future research on improving soil productivity should give high priority to maximizing water use efficiency through balanced nutrient supply.

Root Growth and P Application: Effective utilization of sub-soil water can be economically realised through stimulating root growth by suitable methods of P application. For example, Kang and Yunusa (1977) reported that P application increased the root density at various sampling depths for an Alfisol in southwest Nigeria (Fig. 7). Irrespective of the tillage methods, there was a marked increase in root density up to 40 cm depth. P deficit soils of tropical America have been shown to respond tremendously to rate and methods of P application (Bandy and Musgrave, 1979; Gonzales-Erico et al., 1979). Lathwell (1979) reported high yield of maize with banded compared with broadcast application of P for a fine-textured Oxisol in the savanna region of central Brazil.

Liming versus Root Growth and Water Use: Crops grown in the strongly acidic Oxisols and Ultisols (i.e. pH below 5 and exchangeable Al saturation greater than 50%) in the sub-humid regions experience frequent drought stress due to unreliable rainfall distribution. Moreover, high exchangeable Al levels in the subsoil horizons restrict root penetration. Chemical amendments such as applications of lime improve soil water utilization by crops through increased root growth (Bandy and Musgrave, 1979; Gonzales-Erico et al., 1979).

Crop management studies conducted on the fine-textured Oxisols in the savanna region of Central Brazil have shown that both subsoil liming and banding placement of P fertilizer increased root proliferation and root penetration. The change in volumetric soil water content in plots receiving shallow (0 - 15 cm) and deep (0 - 30 cm) incorporation of lime for a period of 12 days without rain at tasselling stage of maize indicated more water extraction and utilization from deeper layers; for the deep lime incorporation treatment (Gonzales-Erico et al., 1979). There was greater utilization of soil water in the deep lime plots after 7 days. Plants on the shallow lime plots were severely wilted by the 6th day, while plants on the deep lime plots did not show any sign of wilting until the 9th day. The leaf water potential of deep incorporated lime treatment was about 2 bars higher than that of the shallow limed treatment.

For crops grown on the strongly leached kaolinitic acid soils in the humid tropics, however, the occurrence of severe drought stress during the growing season is less frequent than in the sub-humid and semi-arid regions. Subsoil liming under such circumstance becomes less useful. On the other hand, soil

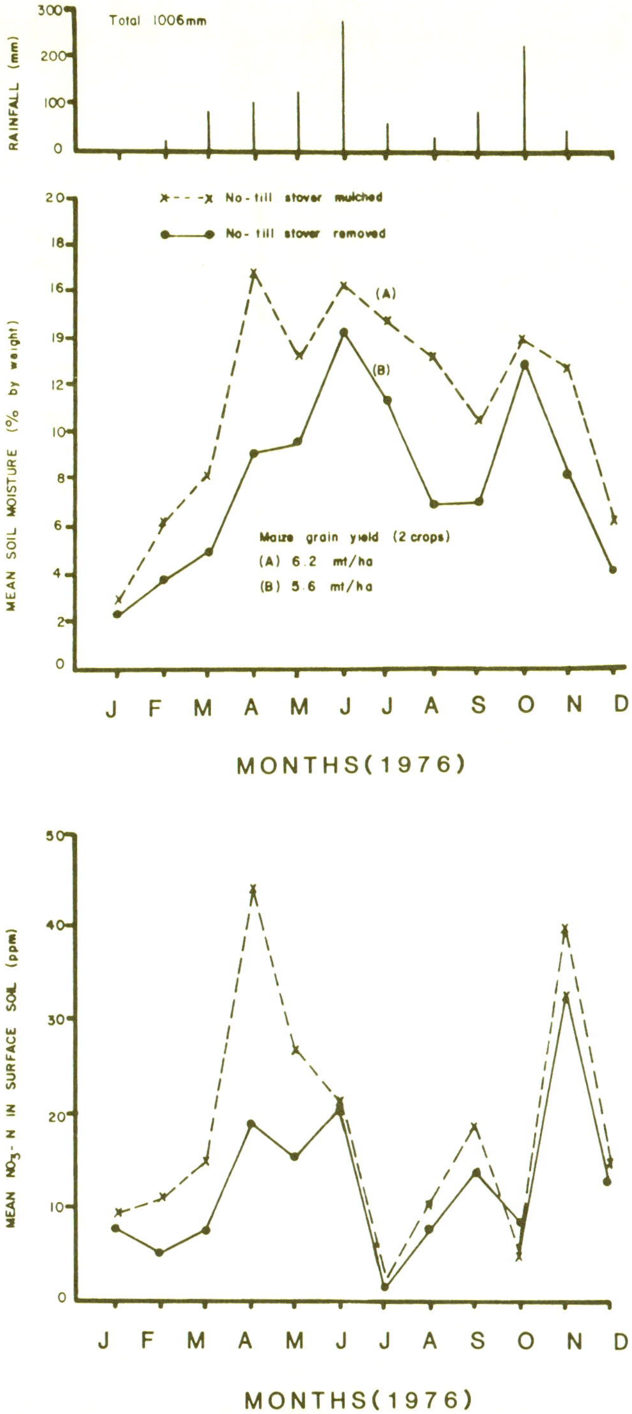

Fig. 6: Effect of Crop Residue Mulch in No-till Maize Cropping on Moisture Storage and Nitrate Levels in a Kaolinitic Alfisol in Ibadan (Juo, 1980 and Juo, unpublished data).

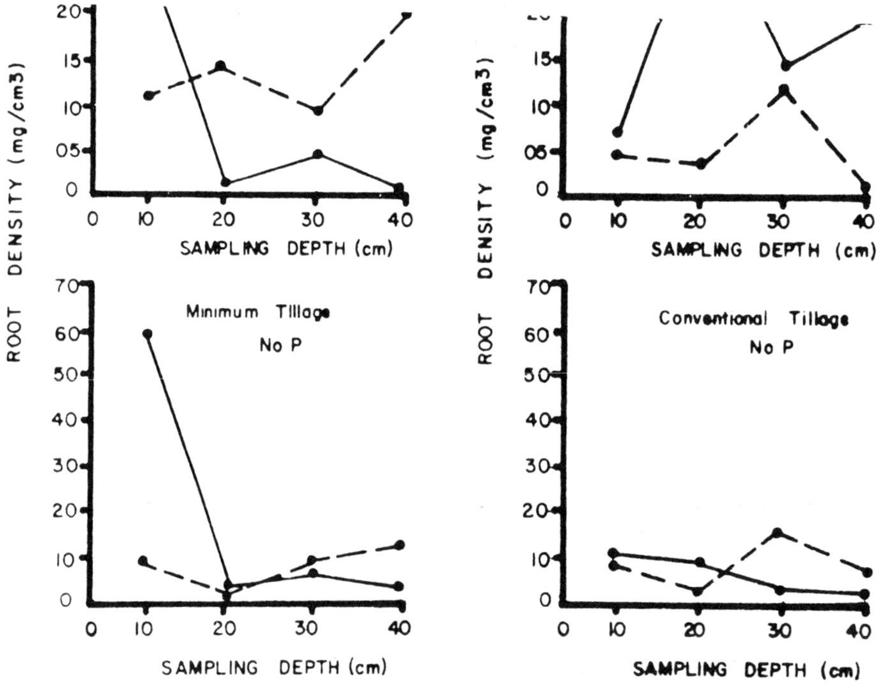

Fig. 7: Effect of Phosphate Application and Tillage Methods on Root Growth (Kang and Yunusa, 1977).

management research in the high rainfall regions gave emphasis on developing systems that minimize the disturbance of surface soil and maximize the return of plant residue as mulch. Long-term field experiments conducted at IITA's high rainfall station at Onne in Nigeria have shown that no-tillage system with annual application of relatively low rate of lime (i.e. 200 to 500 kg/ha) could maintain favourable nutrient status in the surface soil, reduce leaching losses and sustain yield of continuous cropping with maize-cowpea rotation (ITTA, 1981; Friessen, Juo and Miller, 1982).

Potassium and Plant Water Use: The uptake of K, being a diffusion-controlled process at the root/soil boundaries, is also inter-dependent with water availability. Beringer and Trolldenier (1975) indicated that K^+ ions facilitate water uptake by roots and at the same time reduce transpiration loss. It is believed that adequate supply of K ions in the soil solution stimulates water uptake via osmotic potential of root cells and xylem sap and by the involvement of K in the regulation of stomatal movement (Lauchli and Pfluger, 1978). Experimental data reported by Skogley (1976) and Mengel and Braunschweig (1972) gave strong evidence that adequate K nutrition may increase plant tolerance and avoidance of drought. With adequate K supply, plants respond almost immediately to water stress induced by hot wind by reducing their transpiration; whereas plants with moderate or severe K deficiency are unable to close their stomata effectively. This means that the same dry matter yield of maize can be obtained in soils with favourable moisture condition and lower K supply as in dry soils with high exchangeable K levels.

Crop Management

Weed Control: Competition by deep rooted weeds can drastically reduce the water use efficiency. In the humid and sub-humid tropics with forest vegetation, weeds are less of a problem during the first year following land clearing and development from primary or secondary forest than in the subsequent years. Weed problem, however, increases dramatically with continuous cropping. Furthermore, weeds compete more effectively even with quick growing and tall canopy cover such as maize when grown under low fertility conditions (Fig. 8). Experiments conducted by Kang et al., (1977) indicated that with adequate fertilization, the effects of weeds on maize yield is less. This was mainly due to the shading effect of the maize crop on weeds. The incidence of moisture stress in unweeded maize was more frequent than in weed-free maize. Five years of continuous cropping with maize indicated that weeding at low fertility improved yield by about 2 t/ha and at high fertility by 1.5 t/ha. Increase in yield by fertilizer application was 4 and 5 t/ha respectively, without and with weed control.

Stand Establishment and Plant Population: During periods of frequent drought stress, residue mulches and the no-till system may have better crop stand than bare ground surface that may have supra-optimal temperature and sub-optimal soil moisture in the seed zone. Furthermore, bare exposed soil is likely to restrict seedling emergence through crust development (Falayi and Lal, 1979). Effect of weed competition on maize grain yield was more pronounced at low plant population (26,000 plants/ha) as compared with high plant population (52,000 plants/ha). Without weeding maize grain yield with low plant population was 74% of that with high population. However, with weeding, grain yield of the low population treatment was 84% of the high population treatment (Fig. 9).

GENERAL DISCUSSION AND CONCLUSIONS

Low water holding capacity of the soil, erratic rainfall distribution, shallow effective rooting depth, and high losses by runoff and evaporation, lead to widespread occurrence of drought stress in the humid and sub-humid tropics. Although the management techniques may be different according to soils and ecologies, the basic principles towards efficient use of water are the same and include: (i) improving soil-water storage through reducing surface runoff and evaporation losses by soil management and (ii) increasing effective water utilization through appropriate crop and nutrient management.

The use of crop residue mulch, no-till farming in combination with cover crops and agro-forestry techniques, balance fertilizer application and soil amendments, and weed control through appropriate chemicals or biological means seem to have broad application through a wide range of soils and ecological environments. There are, however, basic differences that must also be considered. For example, no-tillage system with residue mulch is applicable more for the humid and sub-humid regions than for the easily compactable soils of prolonged dry seasons in the semi-arid and arid regions. Similarly, the amount of P and lime requirements for fine-textured Oxisols of tropical America for better root growth

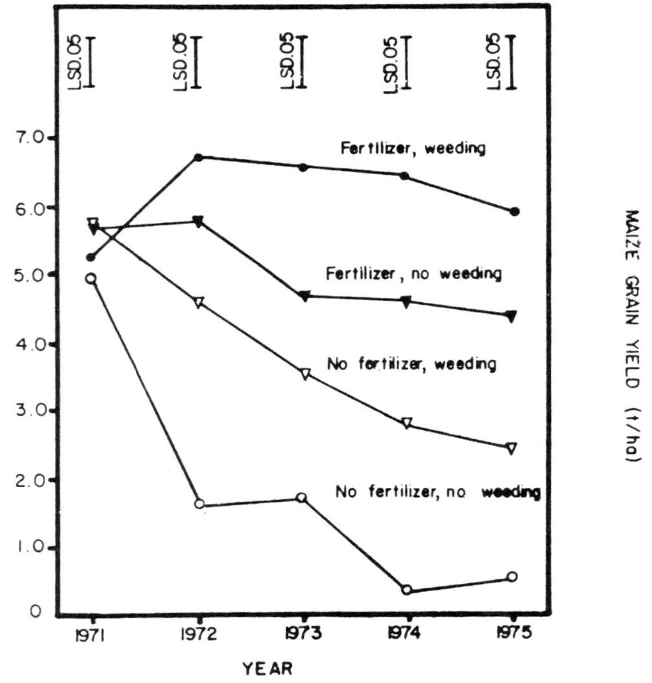

Fig. 8: Effect of Fertilizer and Weeding on Yield of Maize Grown on a Kaolinitic Alfisol at Ibadan, Nigeria (Kang, Donkoh and Moody, 1977).

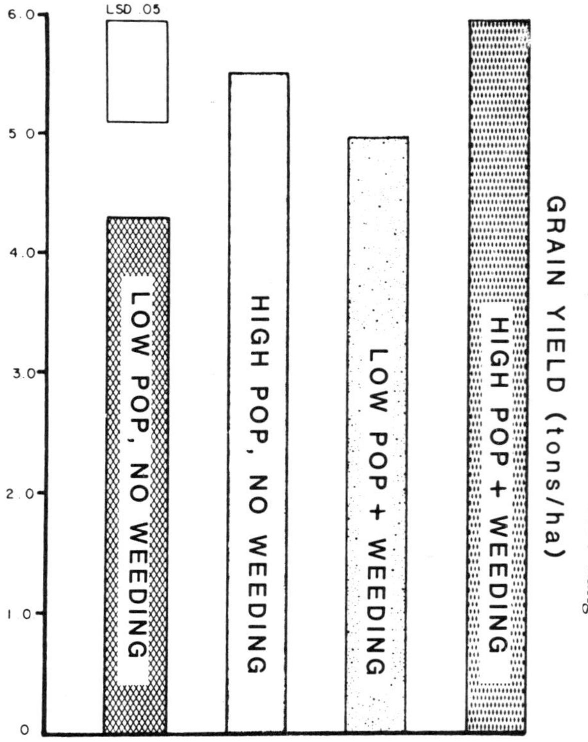

Fig. 9: Effect of Plant Population and Weeding on Maize Yield (B.T. Kang, unpublished).

and water utilization from sub-soil horizons is far larger than for the coarse-textured kaolinitic Ultisols and Oxisols of central and west Africa. These differences in soil characteristics not only influence the amount of nutrient and other soil amendment required but also their method of application.

The survey of literature presented in this report indicates serious gaps in our knowledge of water-nutrient interaction that should form the basis towards effective utilization of the scarce and non-renewable soil and water resources of the humid and sub-humid tropics. There are very few studies conducted where water and nutrient balance have been investigated simultaneously. And yet, one realises that the water and nutrient availability are interdependent. This is paticularly the case with N and K were the leaching losses occur through the processes of water movement on the one hand, and where their uptake is also a process that is governed by the water status either through mass flow or diffusion processes. Effective nutrient utilization cannot be achieved without adequate water availability and effective crop-water utilization is impossible without a balanced nutrient supply. There is an urgent need, therefore, to establish long-term experiments where the plant-nutrient-water relationships can be effectively assessed for different systems of management and for a wide range of soils and agro-ecological environments.

REFERENCES

Bandy, D.E. and R.B. Musgrave (1975). Soil management effects on water stress in corn. In 1975 Ann. Report NCSU Tropical Soils Research Program, North Carolina State Univ., Raleigh, NC, 65-88.

Barber, S.A. (1962). A diffusion and mass flow concept of soil nutrient availability. Soil Sci. 93: 39-42.

Bartholomew, W.V. (1972). Soil Nitrogen - supply processes and crop requirements. Tech. Bull. No. 6, North Carolina State Univ., Raleigh, NC.

Beringer, H. and G. Trolldenier (1978). Influence of K nitrution on the response of environmental stress. In Potassium Research - Review and Trends. Proceeding 11th Congress, Int'l Potash Inst., Bern, Switzerland.

Bolton, F.L. (1981). Optimizing the use of water and nitrogen through soil and crop management. Plant and Soil 58: 231-247.

Bui Huu Tri (1968). Dynamique de la granulation du sol sous prairies. Ann. Agr. 19(4): 415-439.

Chandler, J.V., R.C. Costas and E.g. Boneta (1966). High crop yields produced with or without tillage on three typical soils of the humid region of Puerto Rico. J. Agri. Univ., Puerto Rico, 50: 146-150.

Charreau, C. (1969). Influence des techniques culturales sur le developpement du ruissellement et de l'erosion en Casamance. L'Agron. Trop. XXIV(9): 836-842.

Charreau, C. (1974). Soils of tropical dry and dry/wet climatic areas of west Africa and their use and management. Agronomy Mimeo 74-96, Dept. of Agronomy, Cornell University, Ithaca, NY, 434.

Collis-George, N. and R. Lal (1970). Infiltration into columns of swelling soils as studied by high speed photography. Aust. J. Soil Res. 8: 195-207.

Collis-George, N. and R. Lal (1971). Infiltration and structural changes as influenced by initial moisture content. Aust. J. Soil Res. 11: 93-105.

Collis-George, N. and R. Lal (1973). The temperature profile of soil column during infiltration. Aust. J. Soil Res. 11: 93-105.

da Silva, L.F., A.C.C.P. Dias and A.A.O. de Melo (1975). Characteristics and management properties of Oxisols in Southern parts of Bahia, Revista Theobroma (Brazil), 5(4): 23-24.

de Melo, V. (1974). Agricultural aptitude of the yellow red podsolic soil. Boletin de Recursos Naturais de Sudane, 12: 77-89.

De Vleeschauwer, D., R. Lal and M. De Boodt (1978). The comparative effects of surface applications of organic mulch versus chemical soil conditioners on physical and chemical properties of the soil and on plant growth Catena 5: 337-349.

De Vleeschauwer, D. and R. Lal (1981). Properties of Worm Casts under Secondary Tropical Forest Regrowth. Soil Sci. 132: 175-181.

Falayi, O. and R. Lal (1979). Effect of aggregate size and mulching on erodibility, crusting, and crop emergence. In: R. Lal and D.J. Greenland (eds.) Soil Physical Properties and Crop Production in the Tropics, J. Wiley and Sons, Chichester, UK, 87-93.

Fox, R.H., H. Talleyrand and D.R. Bouldin (1974). Nitrogen fertilization of corn and sorghum grown in Oxisols and Ultisols in Puerto Rico. Agron. J. 66: 534-540.

Friessen, D.K., A.S.R. Juo and M.H. Miller (1982). Residual value of lime and leaching of calcium in a kaolinitic Ultisols in the high rainfall tropics. Soil Sci. Soc. Am. J. Vol. 46 (in press).

Gavande, S.A. (1968). Water retention characteristics of some Costa Rican Soils. Turrialba 18: 34-38.

Ghuman, B.S. and R. Lal (1982). Temperature regime of a tropical soil in relation to surface condition and air temperature and its Fourier analysis. Soil Sci. (in press).

Gonzalez-Erico, E., E.J. Kamprath, G.G. Naderman and W.V. Soares (1979). Effect of depth of lime incorporation on the growth of corn on an Oxisol of Central Brazil. Soil Sci. Soc. Am. J. 43: 1155-1158.

Greb, B.W., D.E. Smika and A.L. Black (1967). Effect of straw mulch rates on soil water storage during summer fallow in the Great Plains. Soil Sci. Soc. Am. Proc. 31: 556-559.

Greenland, D.J. and S.I. Bhuiyan (1982). Rice research strategies in selected areas: environment management and utilization. In: Rice Research Strategies for the Future. IRRI, Los Banos, Philippines, 239-264.

Holmes, J.W., E.L. Greacen and C.G. Gurr (1960). The evaporation of water from the bare soils with different tilths. Trans. 7th Int. Cong. Soil Sci. (Madison) I: 188-194.

Hsiao, T.C., J.C. O'Toole and V.S. Tomar (1980). Water stress as a constraint to crop production in the tropics. In: Priorities for Alleviating Soil-Related Constraints to Food Production in the Tropics. IRRI, Los Banos, Philippines, 468 pp.

IITA (1981). Management of acid soils in the high-rainfall tropics for sustained maize-cowpea production. Research Highlights, IITA, Ibadan, Nigeria, 53-55.

Juo, A.S.R. (1980). Nitrogen profile in a kaolinitic Alfisol under fallow and continuous cropping. In: T. Rosswal (ed), Nitrogen Cycling in West Africa Ecosystems. SCOPE/UNEP, Royal Swedish Academy of Sciences, Stockholm, Sweden.

Kang, B.T. and M. Yunusa (1977). Effects of tillage methods and phosphorus fertilization on maize in the humid tropics. Agron. J. 69: 291-294.

Kang, B.T., F. Donkoh and K. Moody (1977). Soil fertility management investigations on benchmark soils in the humid low altitude tropics of West Africa: Investigation on Egbeda soil series. Agron. J. 69: 651-656.

Kang, B.T., G.F. Wilson and L. Sipkens (1981). Alley cropping maize (Zea mays L.) and Leucaena (Leucaena leucocephala Lam) in southern Nigeria. Plant & Soil 63: 165-179.

Kang, B.T. and A.S.R. Juo (91981). Management of soils with low activity clays in tropical Africa. Proceedings 4th Int'l Soil Classification Workshop. June 2-12, 1981, Kingali, Rwanda.

Kannegieter, A. (1967). Zero cultivation and other methods of reclaiming Pueraria fallowed land for food crop cultivation in the forest zone of Ghana. Trop. Agriculturist (Ceylon) J. 123: 51-73.

Kannegieter, A. (1969). The combination of short term Pueraria fallow, zero cultivation and fertilizer application: its effects on fallowing maize crop. Trop. Agriculturist (Ceylon) J. 125: 1-18.

Kemper, B. (1981). Results of studies made in 1978 and 1979 to control erosion by cover crops and no-tillage techniques in Parana, Brazil. Soil and Tillage Research 1: 253-268.

Koehler, F.E. and D.L. Guetinger (1967). Moisture-nitrogen relationships for wheat production in eastern Washington. Proc. 18th Ann. Pacific Northwest Fert. Conf. Twin Falls, Idaho, 93-94.

Lal, R. (1976a). Soil erosion problems on an Alfisol in Western Nigeria and their control. IITA Monograph 1, 203 pp.

Lal, R. (1976b). No-tillage effects on soil properties under different crops in Western Nigeria. Soil Sci. Soc. Amer. J. 40: 762-768.

Lal, R. (1979). Physical characteristics of soils of the tropics: Determination and Management. In: R. Lal and D.J. Greenland (eds). Soil Physical Properties and Crop Production in the Tropics. J. Wiley and Sons, Chichester, UK, 7-44.

Lal, R. (1980). Soil erosion as constraint for crop production. Conf. Proc. Soil-related constraints to food production in the tropics. IRRI, Los Banos, Philippines, 405-424.

Lal, R. (1981). Physical and mechanical characteristics of Alfisols and Ultisols with particular reference to soils in the tropics. In B.K.G. Theng (ed) Soils with Variable Charge. New Zealand Society of Soil Science, Lower Hutt: 253-279.

Lal, R., G.F. Wilson and B.N. Okigbo (1978a). No-till farming after various grasses and leguminous cover crops. Field Crop Research 1: 71-84.

Lal, R., P.R. Maurya and S. Osei-Yeboah (1978b). Effects of no-tillage and ploughing on efficiency of wateruse in maize and cowpea. Expl. Agric. 14: 113-119.

Lal, R., D. De Vleeschauwer and R. Malafa Nganje (1980). Changes in properties of a newly cleared tropical Alfisol as affected by mulching. Soil Sci. Soc. Amer. J. 44: 827-833.

Lathwell, D.J. (ed) (1979). Phosphorus response on Oxisols and Ultisols. Cornell Int'l Agric. Bull. 33, Ithaca, NY.

Lauchli, A. and R. Pfluger (1978). Potassium transport through plant cell membranes and metabolic role of potassium in plants. In: Potash Research review and trends. Proc. 11th Cong. Int'l Potash Inst. Berne., 111-164.

Lawson, T.L., J.S. Oguntoyinbo and O. Ojo (1979). Agroclimatic conditions of West Africa. Conf. Proc. Soil & climate in relation of crop production in West Africa. Oct. 15-19, 1979, ITTA, Ibadan, Nigeria (in press).

Liao, C.F.N. and W.V. Bartholomew (1974). Relation between nitrate absorption and water transportation corn. Soil Sci. Soc. Amer. Proc. 38: 472-477.

Megel, K. and L.C. Braunchweig (1972). The effect of soil moisture on the growth of young maize plants. Soil Sci. 114: 142-148.

Mongi, H.O. and P.A. Huxley (eds) (1979). Soils research in agroforestry. ICRAF, Nairobi, Kenya.

Nicou, R. (1976). Le labour, technique d'economie de l'eau pour la zone Sahelo-Soudonienne ouest Africaine. In: Efficiency of Water and Fertilizer Use in Semi-Arid Tropics. FAO/IAEA Report 192: 75-100.

Nicou, R. and J.L. Chopart (1979). Root growth and development in sandy and sandy clay soils of Senegal. In: R. Lal and D.J. Greenland (eds). Soil Physical Properties and Crop Production in the Tropics. J. Wiley and Sons, Chichester, UK, 375-384.

Nwa, E.U. (1979). Frequency and amount of irrigation for maize in Western Nigeria. Agric. Water Management 2: 23-239.

Olson, S.R. and W.D. Kemper (1968). Movement of nutrients to plant roots. Adv. Agron. 20: 91-151.

Oveson, M.M. and A.P. Appleby (1971). Influence of tillage management in a stubble mulch fallow-winter wheat rotation with herbicide weed control. Agron. J. 63: 19-20.

Pereira, N.C., E.M. Chenery and W.R. Mills (1954). The transient effects of grasses on the structure of tropical soils. Emp. J. Expl. Agric. 22: 148-160.

Perez-Escolar, R., J. Juarez, Jr., J.A. Bonnet and M.A. Lugo-Lopez (1974). Moisture retention in the organic soils of Cano Tiburones. J. Agric. Univ., Puerto Rico, 58: 225-229.

Perez-Escolar, R. and M.A. Lugo-Lopez (1978). Effect of depth of lime incorporation on the yield of two corn hybrids grown on a typical Ultisol of Puerto Rico. Univ. of Puerto Rico, N.J. Agric. 62(3): 203-213.

Philip, J.R. (1957). Theory of infiltration. 4. Sorptivity and algebraic infiltration equation. Soil Sci. 84: 257-264.

Quirk, J.P. and C.R. Panabokke (1962). Incipient failure of soil aggregates. J. Soil Sci. 13: 60-70.

Skogley, E.O. (1976). Potassium in Montana soils. Mont. Agric. Expt. St. Res. Report 88, Montana.

Strebel, O., H. Grimme, M. Renger and H. Fleige (1980). A field study with N-15 of soil and fertilizer nitrate uptake and of water withdraw by spring wheat. Soil Sci. 130: 205-210.

Tejwani, K.G. (1979). Soil fertility status, maintenance and conservation for agroforestry systems on wasted lands in India. In: H.O. Mongi and P.A. Huxley (eds). Soils Research in Agroforestry. ICRAF, Nairobi, Kenya: 141-174.

Uehara, G. and G. Gillman (1981). The mineralogy, chemistry, and physics of tropical soils with variable charge clays. Westview Press, Boulder, CO.

Wahab, A., H. Talleyrand and M.A. Lugo-Lopez (1976a). Rooting depth, growth, and yield of corn as affected by soil water availability in an Ultisol and an Oxisol. J. Agric. Univ. Puerto Rico, 60: 316-328.

Wahab, A., H. Talleyrand and M.A. Lugo-Lopez (1976b). Rooting depth, growth, and yield of sorghum as affected by soil water availability in an Ultisol and an Oxisol. J. Agric. Univ. Puerto Rico, 60: 329-335.
Whiteman, D.T.S. (1975). Moisture conservation by fallowing in Botswana. Expl. Agric. 11: 305-314.
Willis, W.O. (1976). Soil water management and other factors that affect crop production. In: Efficiency of Water and Fertilizer Use in Semi-arid Regions. IAEA Report, 192: 1-19.
Vegara, N.T. (ed.) (1982). New directions in agroforestry: The potential of tropical legume trees. Improving agroforestry in the Asia-Pacific tropics. East-West Center, Honolulu, Hawaii.

NEW DEVELOPMENTS IN CHEMICAL CONTROL OF WEEDS

J.R. Corbett
FBC Limited
Chesterford Park Research Station
Saffron Walden, Essex CB10 1XL, England

ABSTRACT

Although over 150 herbicides are already available, there is active research worldwide by industry to improve upon them. Recent major advances include chiral wild oat herbicides, post-emergence herbicides for grass weeds, the translocated broad spectrum herbicide glyphosate, herbicide safeners, the very active chlorsulfuron, and the natural product herbicide, bialaphos. Resistance to herbicides is unlikely to become a significant problem. Herbicides technically suitable for use in developing countries exist, and suitable application machinery is available.

The new herbicides sought by industry must be aimed at the developed countries to allow recovery of the enormous costs incurred in discovery and development. Technically difficult problems for new products include tough perennial weeds, water weeds, parasitic weeds, and volunteer crops. The ideal herbicide of the future will have a broad weed spectrum, favourable crop selectivity, be non-toxic to animals, and of short environmental persistence. It is likely to be applied post-emergence and to be very active per unit weight.

No technical reasons exist to prevent the discovery of better herbicides, but industry must not be unnecessarily constrained. In particular, registration procedures should be harmonised worldwide, and patent life extended to compensate for the long development period without sales.

KEYWORDS: Herbicide, weed, development, industry, resistance, registration, patents, developing countries, technical challenge.

INTRODUCTION

This paper outlines the present range of herbicides available, and indicates the possible course of future progress in providing and using these chemical tools. The 1981 global expenditure on herbicides at the end-user level is $5.1 billion, and is estimated to rise to $7.5 billion by 1990 (Wood, Mackenzie and Co., 1982a). Since the world population is approximately 4.5 billion, the average expenditure on herbicides is now a little over $1/a per person.

PRESENT STATE OF THE ART

History of Herbicides

Current world expenditures on herbicides are best explained in terms of the history of herbicides, culminating in the present position where there is a herbicide available for most weed control situations, there being 157 herbicides listed in "The Pesticide Manual" (Worthing, 1979). (Information for this historical section was drawn from Brian (1976), Wood, Mackenzie and Co. (1982a) and Worthing (1979)).

Although salt and industrial waste chemicals had been used for hundreds of years, chemical weed killing did not start properly until it was observed in 1896 that the vine fungicide, Bordeaux mixture, would kill yellow charlock. Over the next thirty years many other inorganic herbicides were discovered and used. The next key step was the introduction of the first organic herbicide, 2-methyl-4, 6-dinitrophenol (DNOC), in 1932. Although this compound, and closely related ones, are still used as herbicides, the greatest advance so far in the development of herbicides must be the discovery of the selective hormone weedkillers early in the Second World War. Two important examples are 2,4-D and MCPA, which are both still widely used for the selective post-emergence control of broad leaved weeds in cereals. The first urea herbicide, monuron, was introduced in 1952. It acted via the soil and was used to control weeds totally in non-crop situations. More recent members of the family can be applied both pre- and post-emergence, and will control weeds in cereals, cotton, soybeans, etc.

The first thiocarbamates were introduced in 1954, and are unusual amongst herbicides in being volatile liquids, so that they must be incorporated into the soil. They are used to kill weeds in a range of crops, including corn and rice. The triazines were introduced in 1956, the most important product being atrazine, which is a soil active compound that is well tolerated by corn. The triazines have similar properties to the ureas, which they overshadowed. Recent members of the triazine family can be used to control weeds in soybeans, sugar-beet and cereals. In terms of sales value the triazines are the most important group of herbicides.

Paraquat, one of two bipyridylium compounds used as herbicides, was first reported in 1958. It became a major product since it not only kills most plant species by direct contact activity, but is rapidly inactivated by contact with soil. It can therefore be regarded as a chemical hoe, and has been used to develop the concept of direct drilling, or no-till cultivation, where crops are sown directly into paraquat treated areas, avoiding the need for mechanical ploughing. Trifluralin was introduced in 1960 and is the most important of the dinitroaniline family of herbicides. It is used pre-plant incorporated for the control of annual grasses and broad leaved weeds in cotton, soybeans and other crops. The chloroacetamide herbicides were first exemplified by propachlor, introduced in 1965 and used for the pre-emergence control of grass and some broad-leaved weeds in corn.

Recent Key Technical Developments in Herbicide Discovery

The aminopropionate class of herbicides was introduced in 1971, the first compound being benzoylprop-ethyl. Other members of the class followed, including flamprop-isopropyl, which is chiral, and can exist as a mixture of stereo-isomers. Such isomerism is common, and it is often the case that only one of the isomers shows biological activity, because the other has the wrong shape to fit to the receptor in the susceptible organism. The (R)-(-) component of flamprop-isopropyl has been introduced as a commercial herbicide having at least twice the activity of the mixture (Scott et al., 1976). This development was noteworthy since it is normally very difficult to produce a chiral isomer economically.

Diclofop-methyl, a phenoxy-phenoxy-propionic acid, was introduced for the control of wild oats and a wide range of grasses in cereals, sugar-beet and other crops in 1975 (Nestler and others, 1979), while alloxydim-sodium is used to control grasses in sugar-beet, soybeans and other broad leaved crops (Iwataki and Hirono, 1979). A recent promising introduction is fluazifop-butyl, which is chemically related to diclofop-methyl (Plowman and others, 1980).

N-phosphonomethylglycine (glyphosate) was introduced in 1971. It has a very broad spectrum of activity when used post-emergence, but is rapidly inactivated by most soils, and is essentially non-toxic to mammals, insects and bacteria (Franz, 1979). These characteristics mean that it will compete with paraquat in some markets, but it is to some extent complementary since it is readily translocated to all vital organs such as roots, rhizomes and meristems, and therefore effectively kills many deep-rooted perennial grasses, sedges and broad leaved weeds, which are not so well controlled by paraquat. The important technical characteristics of glyphosate are (i) its ease of translocation in the plant, (ii) the limited metabolism that it undergoes in the plant, and (iii) its site of action, the enzyme 5-(carboxyethenyl)-3-phosphoshikimate synthase (Amrhein et al., 1981). This enzyme is not found in animals, thus contributing to the excellent toxicity profile of the compound.

Herbicide safeners, or antidotes, are chemicals that increase the safety of a herbicide to the crop, but do not significantly alter its ability to kill weeds. The concept was introduced by Hoffmann in 1962 (Hoffmann, 1978). Safeners offer a chemical method of conferring selectivity. In 1972 naphthalic anhydride was introduced as a seed treatment which would protect corn against thiocarbamate and dithiocarbamate herbicides, and could also be used in some other crop and herbicide combinations. In 1973, N,N-diallyl-2, 2-dichloroacetamide was introduced as an antidote to thiocarbamate herbicides used on maize, and gave similar results to naphthalic anhydride, but had the advantage that it could be added to the herbicide formulation rather than applied to the seed. The next safener was introduced in 1980 when cyoxymetrinil was used to protect sorghum seed from injury by metolachlor. However, the chemically related CGA 92194 has better safety to the crop and is being developed for the same use (Rufener et al., 1982). Another safener under development for sorghum is MON-4606, a seed treatment to protect against the herbicides alachlor and acetochlor (Brinker et al., 1982).

Chlorsulfuron (DPX-4189) is a sulphonylurea herbicide with a very high level of activity. Most herbicides are used at rates of about 1 kg/ha, but chlorsulfuron will control many broad leaved weeds at 10-25 g/ha, and it shows selectivity towards cereals. Very active herbicides have been discovered before, but probably not with the same potency and selectivity shown by chlorsulfuron. Chlorsulfuron has a half-life in the soil of only 1-2 months, but, even so, lower application rates have to be used on the cereal crop if it is to be followed by a sensitive crop species than if a tolerant following crop is planned (Palm and others, 1980).

Bialaphos is a metabolite produced by some species of the fungus Streptomyces. It was discovered during an extensive screening programme to find herbicidal metabolites for agricultural use (Sekizawa and Takematsu, 1982). If bialaphos proceeds successfully to commercialisation it will be the first natural product herbicide to be sold.

There are two reasons why further highly active herbicides are likely to be sought in the future. Firstly, on environmental grounds it must be preferable to apply the minimum amount of any chemical. Secondly, if only a small amount of chemical is required it follows that more complex, and therefore more costly, molecules can be synthesized as potential herbicides.

Resistance to Herbicides

Resistance occurs when a previously susceptible species is no longer properly controlled by the herbicide. It is due to the selection by the chemical of pre-existing mutant individuals in the population that are capable of avoiding or withstanding the effects of the chemical, and is an example of man-made evolution. Resistance to fungicides and insecticides is a major problem, and products, particularly insecticides, have either been limited in their use or withdrawn because they are no longer effective (Corbett, 1979a). This has not been the case with herbicides where resistance is of limited commercial significance, the only products affected on any scale being the triazines (LeBaron and Gressel, 1982). The following factors are likely to make herbicide resistance less important than that to insecticides and fungicides: the long life cycle of weeds; their lack of mobility; the maternal inheritance of triazine resistance, so that it is not distributed by pollen; the wide range of herbicides of different modes of action that is available; crop rotations that favour the use of different herbicides; cultivation that kills resistant weeds; the diluting effect of the weed seeds in the soil seed bank; and the less competitive nature of resistant weeds compared with susceptible biotypes of the same species (LeBaron and Gressel, 1982).

It is reasonable to conclude that resistance is most likely to be a problem with persistent herbicides that create a high selection pressure. Such situations are rare, so that resistance is unlikely to become a significant problem with herbicides in general, and even if it did, there are normally alternative compounds available.

TECHNICAL OPPORTUNITIES

Exploitation of Existing Herbicides

Examining the exploitation of the existing armoury of chemical weed control agents helps to define technical reasons why these products are not used more to control weeds and therefore increase food production. In an important analysis of the effect of weed control problems on the world's food supply, Parker and Fryer (1975) concluded that there was an overall loss of 11.5% due to weeds, though they suspected that this estimate was on the low side. They divided the agricultural area of the world into three cropping systems: highly developed, intermediate and least developed, and concluded that the latter loses the highest proportion (25%) to weeds. Attention should be focussed on the intermediate and least developed classes in order to achieve the greatest increase in world food supply.

West and East Europe, North America and Japan contain the majority of the most highly developed food production areas, and some of those in the intermediate class, and they used 92% of the herbicides by value in 1981 (Wood, Mackenzie and Co., 1982b). There is relatively little use of herbicides in developing countries on small farms (Ebner, 1980), though insecticides and fungicides are used to a greater extent by peasant farmers (Hammerton, 1974).

Due to the cost of discovering and developing a new herbicide, agrochemical companies must target their new products to a large market that can afford the cost. Such markets are only found in the most highly developed systems of food production. The high level of competition between the thirty or so innovative companies in the agrochemical business worldwide ensures that the needs of the farmers in the advanced areas are met as fully as the available technology allows. Most of this paper must perforce deal with possible developments in herbicide technology aimed at this advanced market, and it is likely that new herbicides will only become available for the lesser developed parts of the world if they are developed first for use by the advanced countries. However, it is worth discussing the problems of the small farmer in the developing countries before reverting to the high technology systems of the advanced countries, if only because it is here that food shortages are already occurring and the extra production could be utilized immediately without the need for expensive transport and distribution arrangements.

In many places in the world, the size of a man's holding is governed not by the amount of land he can buy, but by how much he and his family can plant before they must start weeding (Holm, 1976; Compton, 1982). This suggests that herbicides should find a ready use. Although the problems of introducing herbicides to the small-scale farmer in the tropics are largely economic and educational (Parker, 1976), we must ask whether there are any technical problems that would explain why herbicides are so little used in these regions. Although the herbicides available were largely developed for use in North America and Europe, it seems that, by chance, there is no species of weed in the developing countries which completely defies all compounds (Parker, 1977). Neither does there seem to be a technical problem with application equipment (Ebner, 1980; Terry, 1981). One general technical difficulty, however, is the necessity of integrating herbicide use into the agricultural system employed in the particular locality (Parker, 1976; Ebner, 1980; Compton, 1982). It is clear that the factors limiting productivity in each situation must be understood before herbicides are introduced, or the result will be disappointing.

Small scale farming in the developing countries is likely to adopt a blend of old and modern practices to increase productivity (Harrison, 1982). Practices which are likely to either require, or to make particularly difficult demands upon herbicides include (i) minimum tillage, (ii) mulching, (iii) intensive cropping, and (iv) mixed cropping. Other more specific difficulties confronting the small farmer in developing countries and requiring particular attention were discussed by Parker (1976). They included the development of herbicides for (i) minor tropical crops (e.g. millets, cowpea, chickpea, pigeonpea, kenaf, jute, cassava, yams and taro), (ii) application after dry planting so that the first flush of weeds occurring with the rains is controlled and (iii) the control of parasitic weeds (Striga on cereals, Orobanche on tobacco, rape and vegetables).

Probably less than one per cent of applied herbicide reaches its site of action (Graham-Bryce, 1981). Improvement is desirable both on cost/efficacy grounds and to avoid unnecessary environmental contamination. However, it seems unlikely that there will be major breakthroughs in formulation technology in the next decade or so, but rather that there will be a steady improvement in presentation of herbicide active ingredients to the user. Efforts will continue to identify adjuvants which will improve efficacy, either by increasing contact with the plant or improving penetration. Delayed release systems are likely to be investigated more than before, and granules may be used more than in the past. Considerations of cost and, in some cases, of toxicological factors will lead to the replacement of solvent based formulations by water-based flowable (suspension concentrate) products where technically possible. Compositions for very low volume application are also likely to be sought. Since it is rare to find in one compound all the weed killing properties required, the trend towards the formulation of mixtures of two or more herbicides will continue.

With regard to application technology, most herbicides are currently applied through boom-mounted hydraulic pressure nozzles carried by a tractor or an aeroplane. Ground spraying equipment is becoming easier to use and more precise in operation due to such developments as: automatic regulation of volume; remote control of booms from the driver's cab; and swath marking systems to avoid overlap and underlap problems. There is also a trend towards the development of vehicles transmitting a low pressure to the ground so that spraying on very wet soils is possible, thus improving the timing of herbicide application.

Aside from these machinery improvements, there are developments in three areas that suggest that the dominance of the conventional hydraulic nozzle may, albeit slowly, lessen in the future. Firstly, because hydraulic nozzles cannot themselves be improved to give droplets in a more closely defined size range, attention has been paid to the use of rotary atomisers in which a spinning disc produces the spray droplets. This helps avoid drift and improve interception by eliminating very small and very large droplets. Secondly, a variety of selective application techniques is under development or in commercial use. The principle is to apply selectively a non-selective herbicide only to the plants whose destruction is desired. Most of the work has concentrated upon the weed wiping technique in which glyphosate is transferred to the plant by some sort of a wick that is kept permanently moist with herbicide solution. Thirdly, and most dramatic, is the development of a commercially realistic electrostatic spraying device, marketed under the trade name 'Electrodyn' which contains no moving parts and produces a charged spray that is attracted to plant surfaces, including the underparts of leaves (Coffee, 1979). Although this attraction to (crop) plants could be a disadvantage for herbicide application, field experience is necessary to see whether the 'Electrodyn' will be as useful in applying herbicides as it promises to be for fungicides and insecticides. The machine can only be used to apply special oil based formulations of pesticides, and cannot use those based on water.

Technical Challenges to New Herbicides

Although the herbicides discovered and developed by the agrochemical industry have revolutionised agricultural production in the developed world, there remain some problems that are more or less completely unsolved. Chemical control of perennial weeds, expecially selectively in crops, is either inadequate or lacking, so that species of the genera *Cyperus*, *Cynodon*, *Agropyron* and *Sorghum* cause considerable losses of agricultural production. No suitable chemical control method exists for aquatic weeds (e.g. *Salvinia*), but here the difficulty is not so much to achieve control of the weed but to ensure that the treated water is safe for subsequent drinking and irrigation. As already mentioned, there is no suitable chemical for the control of parasitic weeds. Finally, most crops can act as volunteer weeds in subsequent crops (Cussans, 1978). Such volunteers do not pose a problem very different from that already commonly solved by herbicides and provided effort is applied to the problem solutions should be found. However, this may not be the case where weeds are very closely related to the crop. Wild rice (comprising several species of the *Oryza* genus) sheds its seed before rice, and is a serious weed throughout much of the world (Parker and Dean, 1976). Weed beet (*Beta vulgaris*) is an annual that flowers (bolts) in the first year, and therefore produces no useful root, and it has become a significant and intractable weed of sugar-beet in Europe. It will prove very difficult to control such insidious weeds chemically. (An interesting and probably basically correct generalisation, is that, as herbicides are used more and more, the significant weeds are either perennials or they resemble the crop; wild rice and wild beet are extreme examples of the latter.)

Less than 20 of the 200,000 or so species of flowering plants have become major food crops, and none of the six crops (potato, clover, turnip, swede, oilseed rape and sugar-beet) introduced into Northern European agriculture during the last three centuries was completely new, all having been in cultivation long before their introduction into post-feudal agricultural systems (Williams, 1978). It is thus extremely difficult to introduce a new food crop, and Ebling (1982) concludes that the main approach to the world's food problem will continue to be through the improvement and extension of existing crops. Modern breeding and genetic engineering techniques will be used to improve crop strains. Of particular advantage for herbicide use is the possibility of developing herbicide resistant crop strains (Gressel, 1979). Herbicides are therefore unlikely to have to cope with totally new crops, but they will be required to control weeds infesting the improved strains of crops that will be developed. These strains may be grown in new areas of the world, possibly at the limits of the crop's geographical range. This may mean altered sensitivity to herbicides and the presence of new weeds. These challenges should not be difficult to meet with new or existing herbicides.

In contrast to the crop situation, there is no doubt whatsoever that more plant species will become weeds in the future. To deny this would be to ignore the natural plasticity of plants and the facts of evolution. Weed problems may arise by sudden genetic change, by distribution, by chemical selection, or by cultural selection, and Parker (1977) has reviewed the origin of existing weeds in an attempt to predict the future.

The ideal herbicide must be effective over a reasonable weed spectrum, possess excellent selectivity in favour of at least one major crop, be essentially non-toxic to animals (not just mammals) and have only a short period of persistence in the soil. It will probably be applied post-emergence, avoiding the prophylactic nature of pre-emergence herbicides, and their lack of effect in dry periods. Also, the ideal herbicide is likely to be highly active, thus causing minimal environmental contamination.

Calculations based on reasonable assumptions about herbicide uptake, plant volume and the number of biochemical sites that have to be inhibited per plant to cause death, suggest that it should be possible to discover herbicides that could act at an application rate of mg/ha, i.e. at least 1000 times less than the most active commercial herbicides known today. Even if this admittedly theoretical calculation proves incorrect, it suggests that g/ha should become the application rate of the future, rather than the current kg/ha. Such active herbicides may well possess a complex molecular structure, and may possibly be chiral. The development of fermentation techniques using genetic engineering may allow either the production of chiral intermediates for herbicide manufacture, or direct production of herbicides by fermentation, which, although common with pharmaceuticals, is essentially unknown for herbicides, though bialaphos, mentioned above, could be the first.

ACTIONS REQUIRED

Virtually all of the herbicides currently available have been discovered and/or developed by private industrial concerns, not by state run agencies. Since the technical achievement involved is remarkable, it would seem wise to ensure that the environment remains favourable for companies committed to discovering new herbicides and developing them for use globally. The herbicide market is already well supplied with products, and there are few gaps left on a purely technical basis. Consequently, industry is primarily intent on displacing existing products by new herbicides that are technically superior. The pesticide market as a whole is mature (Corbett, 1979b), and in this situation there is intense competition for the global market between the thirty or so chemical companies, all located in Western Europe, the U.S. or, increasingly, Japan.

It is difficult to provide an accurate estimate of the cost of discovering and developing a new herbicide since the industry does not publish such information. However, it is possible to estimate roughly the costs for pesticides in general, and there is no reason to suppose that herbicides will cost significantly more or less to discover and develop than fungicides and insecticides. Braunholtz (1979) estimated that the period from discovery to significant sales in major markets takes 7 to 8 years and costs around £10 million in research and development. His estimate excluded the costs of failed compounds and the necessary investment in manufacturing plant. Wood, Mackenzie and Co. (1982a) estimated that $5,385 million had been spent on innovative research and development of pesticides between 1972 and 1981, which period had given rise to 80 commercially justifiable products, giving a figure of $67 million per successful compound. Taking $2 = £1, the difference of $47 million between this estimate and that given by Braunholtz could be the cost of the work on failed compounds, which must, of course, be funded by the successful compounds, but could also reflect the inclusion by Wood, Mackenzie and Co. of recent (high) R and D costs that have not yet yielded successful compounds. A current figure for all R and D costs, including failures, might reasonably be £20 million per compound.

Despite the maturity and competitive nature of the market, there is currently no doubt of the overall commitment of the agrochemical industry to finding new herbicides. However, there are some unnecessary constraints imposed by governments that will inevitably reduce the amount of research and development that the industry carries out. Firstly, registration procedures that have no technical reason to be unique to a particular locality, should be totally harmonised. Local trials to establish efficacy and possibly soil persistence, etc., as well as to deal with special local environmental factors, are clearly necessary, but there is no justification for requiring tests on laboratory animals to be repeated inside a particular country. Secondly, since it normally takes at least six, and often up to ten years to put a new discovery on the market, the patent life should be extended by this period. Alternatively, the patent life should start from the time that product registration is granted.

CONCLUSION

Good chemical tools for the control of most of the world's weeds are already available. Better compounds will be produced by the agrochemical industry provided that no unreasonable constraints are put upon it by government. Although herbicides are not the sole answer to weed problems, they are a major input in most control systems, and it is essential, given the plasticity of weeds, that new compounds continue to be developed. There are no technical reasons to prevent the discovery and development of significantly better herbicides than many of those now in use.

ACKNOWLEDGEMENTS

I am pleased to acknowledge the advice received from C. Parker, Weed Research Organisation, and from many colleagues in FBC Ltd.

REFERENCES

Amrhein, N., Deus, B., Gehrke, P., Hollaender, H., Schab, J., Schulz, A. and Steinrueken, H.C. (1981). Interface of glyphosate with the shikimate pathway. Proc. Plant Growth Reg. Soc. America, 8, 99-106.

Braunholtz, J.T. (1979). Techno-economic considerations. Chem. & Ind., No. 22,

789-791.
Brian, R.C. (1976). The history and classification of herbicides. In "Herbicides - Physiology, Biochemistry, Ecology" (L.J. Audus, Ed.), 2nd edition, 1, Academic Press, 1-54.
Brinker, R., Schafer, D., Radke, R., Boeken, G., and Frazier, H. (1982). The effectiveness of MON-4606 as seed safener against Alachlor and Acetochlor in grain sorghum. Proc. British Crop Protection Conference - Weeds, 2, 469-474.
Coffee, R.A. (1979). Electrodynamic energy - a new approach to pesticide application. Proc. British Crop Protection Conference - Pests and Diseases, 3, 777-789.
Compton, J.A.F., Ed. (1982). Small farm weed control. Intermediate Technology Publication, London.
Corbett, J.R. (1979a). Resistance and selectivity: biochemical basis and practical problems. Proc. British Crop Protection Conference - Pests and Diseases, 3, 717-730.
Corbett, J.R. (1979b). Technical considerations affecting the discovery of new pesticides. Chem. & Ind., No. 22, 772-782.
Cussans, G.W. (1978). The problem of volunteer crops and some possible means of their control. Proc. British Crop Protection Conference - Weeds, 3, 915-921.
Ebling, J. (1982). Man the consumer. Chem. & Ind., No. 10, 315-321.
Ebner, L. (1980). Advances in chemical weed control in the tropics and subtropics as an integrated part of crop protection. Proc. British Crop Protection Conference - Weeds, 3, 813-822.
Franz, J.E. (1979). Glyphosate and related chemistry. In "Advances in Pesticide Science" (H. Geissbuehler, Ed.), Part 2, Pergamon Press, 139-147.
Graham-Bryce, I.J. (1981). The current status and future potential of chemical approaches to crop protection. Phil. Trans. R. Soc. Lond., B 295, 5-16.
Gressel, J. (1979). Genetic herbicide resistance; projections on appearance in weeds and breeding for it in crops. In "Plant Regulation and World Agriculture" (T.K. Scott, Ed.), Plenum Press, 85-109.
Hammerton, J.L. (1974). Problems of herbicide use in peasant farming. Adaptation of paper presented at the February 1974 annual meeting of the Weed Science Society of America held at Las Vegas, NV, USA.
Harrison, P. (1982). The new age of organic farming. New Scientist, 94, No. 1305, 427-429.
Hoffmann, O.L. (1978). Herbicide antidotes: from concept to practice. In "Chemistry and Action of Herbicide Anticotes" (F.M. Pallos and J.E. Casida, Eds.), Academic Press, 1-13.
Holm, L. (1976). The importance of weeds in world food production. Proc. British Crop Protection Conference - Weeds, 3, 754-769.
Iwataki, I. and Hirono, Y. (1979). The chemical structure and herbicidal activity of alloxydim-sodium and related compounds. In "Advances in Pesticide Science" (H. Geissbuehler, Ed.), Part 2, Pergamon Press, 235-243.
LeBaron, H.M. and Gressel, J. (1982). Eds. Herbicide resistance in plants. Wiley-Interscience, New York.
Nestler, H.J., Langelueddeke, P., Schoenowsky, H. and Schwerdtle, F. (1979). Phenoxy-phenoxy-propionic acids and derivatives as grass herbicides. In "Advances in Pesticide Science" (H. Geissbuehler, Ed.), Part 2, Pergamon Press, 248-255.
Palm, H.L., Riggleman, J.D. and Allison, D.A. (1980). Worldwide review of the new cereal herbicide - DPX 4189. Proc. British Crop Protection Conference - Weeds, 1, 1-6.
Parker, C. (1976). Weed research requirements in developing countries. Proc. British Crop Protection Conference - Weeds, 3, 801-807.
Parker, C. (1977). Prediction of new weed problems, especially in the developing world. In "Origins of Pest, Parasite, Disease and Weed Problems", (J.M. Cherrett and G.R. Sagar, Eds.), Blackwell Scientific Publications, 249-264.
Parker, C. and Dean, M.L. (1976). Control of wild rice in rice. Pest. Sci., 7,

403-416.
Parker, C. and Fryer, J.D. (1975). Weed control problems causing major reductions in world food supplies, FAO Plant Protection Bulletin, 23, No. 3/4, 83-95.
Plowman, R.E., Stonebridge, W.C. and Hawtree, J.N. (1980). Fluazifop butyl - a new selective herbicide for the control of annual and perennial grass weeds. Proc. British Crop Protection Conference - Weeds, 1, 29-37.
Rufener, J., Myffeler, A. and Peek, J.W. (1982). CGA 92194, A new safener to protect sorghum from injurious effects of Metolachlor. Proc. British Crop Protection Conference - Weeds, 2, 461-468.
Scott, R.M., Sampson, A.J. and Jordan, D. (1976). WL 43425 - A chiral form of flamprop-isopropyl. Proc. British Crop Protection Conference - Weeds, 2, 723-730.
Sekizawa, Y. and Takematsu, T. (1982). How to discover new antibiotics for herbicidal use. Abstract Xs-4, Fifth I.U.P.A.C. Conference of Pesticide Chemistry, Kyoto, Japan.
Terry, J. (1981). Adding to the weed weapon arsenal. International Agricultural Development, November/December, 14-15.
Williams, W. (1978). New crops and agricultural systems. Proc. British Crop Protection Conference - Weeds, 3, 1005-1012.
Wood, Mackenzie & Co. (1982a). Reference Volume of the Agrochemical Service. Wood, Mackenzie & Co., 62-63 Threadneedle Street, London EC2R 8HP.
Wood, Mackenzie & Co. (1982b). Update of the Agrochemical Overview Section. Wood, Mackenzie & Co., 62-63 Threadneedle Street, London EC2R 8HP.
Worthing, C.R. (Ed.) (1979). The Pesticide Manual. 6th Edition, British Crop Protection Council, Malvern, Worcestershire, UK.

CHEMICAL TECHNIQUES FOR MONITORING ANALYSIS AND AVOIDING POLLUTION OF SOIL AND WATER RESOURCES

Nikolaus J. Drescher
BASF Aktiengesellschaft, Agricultural Research Station
D-6073 Limburgerhof, FRG

ABSTRACT

In the production and protection of food, plant protection measures play an essential and growing role. The application of pesticides leads to an unavoidable contamination of soil and water, but pollution can be minimized by development of crop protection agents with high biological activity and selectivity, environmentally safe formulations and novel methods of application. Methods now exist for investigating the environmental fate and the environmental effects of crop protection agents; these produce reliable data for an assessment of environmental risk.

Modern analytical methods permit detection and determination of extremely low concentrations of active ingredients and their conversion products in numerous substrates. There is, however, a lack of simple but efficient analytical methods and instruments for performing monitoring analyses in less well equipped laboratories. A compilation of easy, tested analytical methods is recommended, together with a coordinated training program for their use in developing countries.

KEYWORDS: Pesticides; Analytical Techniques; Simplified Analytical Methods; Contamination and Pollution of Soil and Water; Environmental Fate and Environmental Effects of Pesticides; Risk Assessment; Reduction of Environmental Stress.

INTRODUCTION

Hunger in a world which has the required knowledge, the appropriate technologies and, so far, sufficient resources to nourish mankind is without doubt the greatest frustration of our time. Feeding a hungry world, now and in the future, is clearly the challenge confronting us. This means both a greater production of foodstuffs and their protection from spoilage, their distribution to those needing them and the ensuring of the power to purchase them (FAO, 1981a; FAO, 1982). Equally, to feed a hungry world means conservation of the resources necessary for food production by reducing environmental stress and improving resource management (Barney, 1981; Geissbuehler, 1982).

Ever since the pioneering work of J. von Liebig on plant nutrition in the middle of the last century, the large-scale technical production of mineral fertilizers by the Haber-Bosch process in the first decade of this century, and the beginning of modern, chemical crop protection about 40 years ago, chemistry has been an integral part of crop and food production. Even though critical voices are often the loudest, particularly in the media, the role of chemistry in both the qualitative and the quantitative improvement of our foods cannot seriously be denied (Gunn, 1976). Much of the criticism is about pollution and this term needs definition.

Chemical measures in crop protection and in crop treatment have as primary goals: control of pests, control of plant diseases, control of local plant competitors, and modification of crop performance. Accordingly, the application of crop protection agents is done in the crop itself. A number of products such as herbicides, soil fungicides, soil insecticides, and soil fumigants are also used directly on the soil, on or in surface water, for example on riverbanks, on water surfaces, or in paddy fields. The presence of the active ingredient in soil or in water is thus intentional. Some of the applied active ingredient, and in many cases, even most of it, reaches soil and water indirectly, for example by run-off from plants during treatment, wash-off by rain, drift, leaching, run-off from soil into water and precipitation of evaporated chemical by rainfall.

In contrast to the direct application, the indirect migration of crop protectants into the soil and water as a rule is an unintentional but evidently technically unavoidable effect. Thus, it is not justified to designate as pollution the mere presence of pesticides in various compartments of the environment. I would like plainly to differentiate between "contamination" and "pollution". I therefore use "contamination" (for lack of a better term) to mean the mere presence - intentionally or unintentionally - of chemicals. We should speak of "pollution" only when the presence of active ingredients or their residues involves a risk, a danger to the user or the environment, due to the inherent properties of the respective chemical.

In considering pollution, we all know that such a thing as "zero risk" does not exist. This is true for all spheres of our lives. Therefore, it must be stated that the use of pesticides may lead to a pollution of soil and water under certain circumstances. Intensive chemical and biological research, however, with developments in formulation and application technology, especially during the past two decades, have made advances in the means, methods and techniques that serve to reduce contamination of soil and water resources by pesticides, and to predict and to recognize risks that may occur as a result of contamination, and thus keep them within acceptable limits.

Contamination of soil and water (Geissbuehler et al., 1982), can be reduced by (i) products with higher biological activity, requiring lower application rates per unit area, (ii) products with less persistence in soil and water, and (iii) avoiding "useless chemistry" on farmland, e.g., by using only the biologically active isomer of the active ingredient, (iv) use of environmentally safer formulations, such as controlled-release formulations, preparations with less organic solvents or adjuvants, (v) improved application methods, such as ULV controlled droplet or electrostatic application, and (vi) ingenious inclusion of chemistry in an integrated pest management system. There is no doubt that all these factors, singly or in combination, lead to a decrease in environmental stress, and significant success in this direction has been achieved. This was made evident, for example, by a number of papers presented at the 5th International Congress of Pesticide Chemistry, Kyoto (IUPAC, 1982).

Turning to the possibilities of keeping environmental risks within acceptable limits, we first must ask how does an instance of contamination become one of pollution? This can happen in soil: if persistent active ingredients or their conversion products accumulate; if residues appear in subsequent crops; if rotational crops are damaged; if residues are washed into ground water; or if the function of beneficial soil organisms is injured. It can happen in water: if residues reach the drinking water; if fish or fish food organisms are damaged or if residues accumulate in the aquatic food chain. It is not the phenomena mentioned per se, that increases contamination toward the level of pollution. Rather, it is a question of the toxicological significance of residues and their relevance to environmental effects.

It is the state-of-the-art, that industry carefully and responsibly investigates these questions during the development of synthetic pesticides for the market. Moreover, pesticides are subjected to increasingly stringent requirements regarding the quality and quantity of data for official registration. The respective testing programs include investigations on the physical and chemical characteristics, the environmental fate and the environmental effects of the pesticides.

The data obtained from such physico-chemical investigations as determination of melting and boiling points, vapor pressure, solubilities in water and other solvents, and partition coefficients in the system octanol/water, permit initial conclusions to be drawn concerning the behavior of the active ingredient in the environment. For example, knowledge of the solubility in water and of the vapor pressure enables calculation of its rate of evaporation from water. The partition coefficient octanol/water permits a prediction of the probability of active ingredient accumulation in living organisms. Beyond this, a variety of other data is determined, but these are mainly needed to ensure safe handling of the product during production, formulation, storage, transport and use, and as such, are not directly relevant to our present topic.

The complex of questions on environmental fate may be subdivided into investigations on translocation of the product, during which the active ingredient is not altered, and on those processes in which the active ingredient is degraded. Translocation may be by adsorption/desorption, vaporization, leaching or uptake by living organisms. Degradation may be abiotic, by hydrolysis, photolysis or chemical reactions, or it may be biotic, by hydrolysis, metabolism or mineralization.

The investigative efforts in these areas typically require a processing time of a minimum of two years. Two examples suffice to illustrate this: Tests on leaching behavior generally must be done with at least three different soil types. First, model trials are conducted in the laboratory with the formulated product under extreme "weather conditions" (Figure 1). If under such conditions, i.e., rainfall of 200-500 mm in a time span of a few days, residues are found in the leachate, these tests are followed by studies under conditions more similar to those found in practice, i.e., in a lysimeter (Figure 2), or under actual field conditions. Moreover, the leaching behavior of an active compound may change after it has been in the soil a certain amount of time, adsorbed on soil particles, or altered by processes already mentioned. Leaching experiments with such "aged residues" naturally must include analysis of conversion products of the active ingredient. This again necessitates a knowledge of their chemical nature and the existence of appropriate analytical methods.

The second example concerns metabolism, in which studies can be performed rationally only with labelled compounds. Here, the active ingredient must be labelled in such a way that its conversions can be traced as long as possible, e.g., through radioactive tracers. Frequently, metabolism under aerobic and anaerobic conditions or at different pH levels is of interest. Identification of the metabolites, which often are present only in microgram amounts, demand the

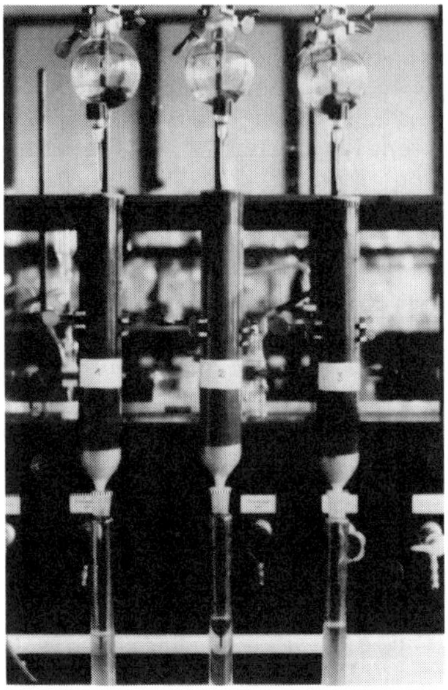

Figure 1:

Figure 2:

entire talent and often the imagination of the analyst. Unequivocal proof of their chemical identity in many cases is possible only by comparison with authentic reference compounds.

These two examples indicate the amount of work involved. The resulting data provide a well-rounded picture of the persistence, the distribution behavior, and the conversions of an active ingredient in soil and water. For an appraisal of possible risks, they must be supplemented by investigations on environmental effects.

Studies on environmental effects are mainly concerned with two topics: the influence of the product on soil function, and toxicological studies on organisms living in soil and water. Influence on soil functions may be through soil respiration, N_2-fixation or nitrification. Toxicity may be to fish, fish food organisms, or to beneficial soil organisms. Such studies, especially those on fish, are both expensive and time-consuming.

So far, we have dealt only with the investigation of isolated phenomena. In reality, an interaction of numerous factors is always involved, and they often overlap in complicated ways. There has been no lack of attempts, starting with those of Metcalf (1971), to simulate environmental conditions on a small scale using aquatic or combined aquatic/terrestrial model ecosystems. Such models provide valuable information but only a partial view of the potential risk to the environment. To be useful, the system would have to be set up on a large enough scale to maintain its own equilibrium over a prolonged period. Then it might be possible to determine whether any shifts are detectable within this system arising from the influence of crop protection agents.

The majority of investigations on the environmental fate and on the environmental effects of pesticides are possible only with assistance of highly developed, sophisticated analytical technology. During the past two or three decades, analytical techniques for the micro and ultramicro range have gone through a development that is truly amazing. In particular, it has been gas chromatography with highly sensitive and element-specific detectors, high-performance liquid chromatography, radio-gas chromatography and, more recently, combination methods like GC/MS coupling, that have introduced significant new dimensions to the analysis of traces. The detection of quantities of chemical compounds amounting to just a few nanograms, or concentrations in the range of a few μg/kg (ppb), has already become routine in many cases. Even ventures into concentrations of ng/kg (ppt) have been successful occasionally (Norstrom, 1982)). Measurement techniques with radioactive or stable isotopes have also been developed considerably and are being used increasingly for studies of the interactions in soil and aquatic ecosystems (IAEA, 1980, 1981). Moreover, automatization and electronic data processing have been introduced to the field of residue analysis and permit the analysis of larger sample series within an acceptable amount of time.

The trend to ever-increasing sensitivities will undoubtedly continue and perhaps we shall be able to identify and determine residues in substrates present in a ratio of $1:10^{15}$. This, however, is only interesting from the purely scientific point of view. For the practical purposes of monitoring analysis, methods of this sort will rarely be needed. Thus, we now have an arsenal of analytical methods at our disposal that allow the sensitive and specific determination of active ingredients and their conversion products in all relevant substrates.

Such methods often demand extremely costly analytical instruments and can be performed only by highly qualified, experience scientists. These conditions can usually be met only in specially equipped laboratories. It may well be that in our enthusiasm over the potential of modern analytics, we have neglected the development of simple but efficient methods for use in less well equipped laboratories, for example those in developing countries. This serious gap was recently studied by the IUPAC Commission on Pesticide Chemistry and an evaluation of simplified approaches to residue analysis was initiated (Batora et al., 1981; Thier, 1982). The term "simplified" must not be misinterpreted. It refers to less instrumentation at the cost of a higher limit of determination, but by no means to a reduced performance (Thier, 1982). Simplified methods should fulfil the following criteria: be sufficiently sensitive and specific; be appropriate for routine work; be simultaneously applicable to a broad range of pesticides; require inexpensive and unsophisticated equipment; work without compressed gases; work without exotic reagents; and work without large amounts of highly purified solvents.

While the present total of such methods is not large, there are indeed several that meet these criteria more or less well, namely, paper chromatography; thin layer chromatography with chemical, biological, or enzymatic detection; photometry with visible or UV detection; bioassay with Drosophila, Daphnia, mosquitoes, algae or plants. Of these listed methods, thin layer chromatography (TLC) especially deserves attention because it satisfies the above criteria almost perfectly. For example, enzymes inhibition techniques can be used to detect (Mendoza and Shields, 1970), and semi-quantitatively determine insecticidal P-esters and carbamates in the range of a few nanograms or less. For TLC determination of photosynthesis-inhibiting herbicides like ureas, acyl-aniline, triazines and benzo-thiadiazinones with chloroplasts and a redox indicator, limits of detection of a few ppb's have been achieved (Kovac and Henselova, 1977; Lawrence, 1980). Other methods, such as photometry or biotests, frequently are not sensitive enough or are inadequately specific. Through suitable combination of extraction, concentration, and separation techniques, clean-up processes, etc., with the actual method of determination, it is certainly possible that these can be improved.

While there are simple, appropriate methods on hand for monitoring analyses, what is still missing is a collection of simple, tested analysis instructions. I would like to recommend that such a manual be made, perhaps in analogy to existing ones (DFG, USDHEW), and regularly revised and expanded. This project could perhaps be put under the charge of the "CCPR Ad Hoc Working Group on Problems in Developing Countries Related to Pesticide Residues", or another international body like FAO, OECD, IUPAC or AOAC. A collection of methods alone, however, will not be enough. A coordinated training program will also be needed, especialy designed to meet the needs of developing countries, in which not only the use of such methods is taught, but also laboratory equipment, sampling techniques, sample preparation and the planning of monitoring programs. Perhaps the producers of laboratory equipment could be encouraged to develop simple but efficient analytical instruments. It should be possible, e.g., by avoiding an excess of sensitivity and operating extras, to market appropriate equipment at a reasonable price.

The assessement of environmental risk admittedly is a difficult task, as the possibilities of exact science are limited. For this reason, no generally applicable rules can be set up for risk assessment. An assessment of risk certainly does not just involve collecting a maximum array of data simply because the respective test methods are available. There is, for instance, no point in studying anaerobic degradation in soil, if it is already known that that active ingredient remains in the aerobic soil zone. Studies on rotational crops with pesticides having a half-life of a few days or weeks would be just as pointless. Requirements on the extent of investigations therefore should be flexible and follow a logical, coherent sequence. Conversely, this can also mean that additional studies may be necessary, such as life-cycle studies on fish, investigations on the effects of crop protectants on certain enzyme systems of microorganisms, run-off studies, and others (Greaves et al., 1980; OECD, 1981).

Risk assessment cannot involve simple linear extrapolations of isolated findings to the conditions found in practice, but must take into account results from studies on warm-blooded animal toxicology, use pattern, and regional and economic considerations. Certain effects of a chemical may be considered unacceptable in one country and be reason enough to restrict or avoid the use of the particular product, while another country with a different socio-economic structure with a very serious problem might well be willing to tolerate, at least temporarily, the same effects (FAO, 1981b). Risk assessment must, of course, be based on the benefits of the pesticide. Finally, risk assessment is not a process that can be completed and forgotten. It has to be revised whenever new scientific knowledge indicates this to be necessary.

CONCLUSIONS

Contamination of resources like soil and water by crop protection measures cannot be avoided in principle. We have the knowledge, experience, methods and technologies, however, to reduce this contamination and to predict with confidence what the impact to exposed resources will be. This enables us, at the present level of scientific knowledge, to make a well founded risk assessment and to take the appropriate measures to keep pollution of soil and water within acceptable limits. However, we must in addition purposefully continue along the path we have started. The development of products with higher biological activity, selectivity, and less persistence of made-to-measure formulations, and environmentally safe application methods should continue unabated. Analytical methods and equipment for the requirements of developing countries should be developed further and enhanced by a coordinated training program. Ways to study environmental fate and effects of pesticides should be further improved, with one of the primary objectives being a reduction in the enormous financial burden involved.

If we face this task responsibly and with determination, jointly with the authorities and public organizations involved, the universities and industry, chemistry will be able to meet this challenge, "The conservation of the environment for the feeding of a hungry world".

REFERENCES

Barney, G.O. (ed.) (1981). The Global 2000 Report to the President, Vol. 1. U.S. Government Printing Office, Washington, DC.

Batora, V., S.Lj. Vitorovic, H.P. Thier and M.A. Klisenko (1981). Development and Evaluation of Simplified Approaches to Residue Analysis. IUPAC Reports on Pesticides (13), Pure and Applied Chem., 53, 1039.

DFG: Deutsche Forschungsgemeinschaft (19). Ruckstandsanalytik von Pflanzenschutzmitteln. Verlag Chemie, Weinheim/New York.

FAO (1981a). Food and Agriculture Organization of the United Nations. Agriculture: Towards 2000, Rome 1981.

FAO (1981b). Second Expert Consultation on Environmental Criteria for Registrations of Pesticides. FAO Plant Production and Protection Paper No. 28, Rome 1981.

FAO (1982). Dimensions of Needs, Rome 1981.

Geissbuehler, H., P. Brenneisen and H.-P. Fischer (1982). Frontiers in Crop Production: Chemical Research Objectives. Science 217, 505.

Greaves, M.P., N.J. Poole, K.H. Domsch, G. Jagnow and W. Verstraete (1980). Recommended Tests for Assessing the Side Effects of Pesticides on the Soil Microflora. Weed Research Organization, Technical Report No. 59.

Gunn, D.L. and J.G.R. Stevens, ed. (1976). Pesticides and Human Welfare. Oxford University Press.

IAEA (1980). International Atomic Energy Agency: Agrochemical Residue-Biota Interactions in Soil and Aquatic Ecosystems. Vienna 1980.

IAEA (1981). Isotope Tracer-Aided Studies of Agrochemical-BIOTA: Interactions in Soil And water. IAEA-TECDOC-247, Vienna 1982.

IUPAC (1982). 5th International Congress of Pesticide Chemistry, Kyoto. Proceedings of the Congress. Pergamon Press Ltd. (in press).

Kovac, J. and M. Henselova (1977). Detection of Triazine Herbicides in Soil by a Hill-Reaction Inhibition Technique after Thin Layer Chromatography. J. Chromatogr. 133, 420.

Lawrence, J.F. (1980). Simple, Sensitive and Selective Thin Layer Chromatographic Technique for Detecting Some Photosynthesis Inhibiting Herbicides. J. AOAC 63, 758.

Mendoza, C.E. and J.B. Shields (1970). Sensitivity of Pig Liver Esterase in Detecting Twelve Carbamate Pesticides on Thin Layer Chromatograms. J. Chromatog. 50, 92.

Metcalf, R.L., G.K. Sangha and I.P. Kapoor (1971). Model Ecosystems for the Evaluation of Pesticide Biodegradability and Ecological Magnification. Environmental Sci. Technol. 5, 709.

Norstrom, R.J. (1982). Organochlorine Monitoring by the Canadian Wildlife Service in the Great Lakes. SECOTEX Workshop: Relevance of Laboratory and Field Tests to Predict the Fate and Effects of Chemicals in the Environment, Tokyo 1982.

OEDC (1981). Organization for Economic Cooperation and Development. Guidelines for Testing of Chemicals, OECD, Paris.

Thier, H.P. (1982). Application of Simplified Methods for the Quantification of Pesticide Residues. Proceedings of the 5th International Congress of Pesticide Chemistry, Kyoto. Pergamon Press Ltd. (in press).

USDHEW: U.S. Department of Health, Education and Welfare; Food and Drug Administration: Pesticide Analytical Manual, Vols. I-II.

INTEGRATED APPROACHES TO PEST MANAGEMENT – PRINCIPAL PESTS OF FOOD

J.C. Davies

Centre for Overseas Pest Research
Foreign and Commonwealth Office, Overseas Development Administration
College House, Wrights Lane
London W8 5SJ, United Kingdom

ABSTRACT

The background to increasing food production by proper control of pests is outlined. Factors affecting the development of workable integrated pest management strategies, mainly with reference to tropical cereals and in particular to subsistence crops in developing countries, are detailed. The current strengths and problems of adoption of an integrated pest management approach are outlined. Observations are made on the practicality of the use of pesticides on subsistence food farms. The paper gives examples of the use of integrated management of food crops and concludes with a plea for more base data and research to be carried out on integrated management on subsistence farmers' fields.

KEYWORDS: Production of staple foods, Pests and losses of food, Subsistence farmers, Pest management.

INTRODUCTION

The topic is extremely broad and could involve a semantic discussion of the four main components of the title. However it is discussed here mainly from an entomological point of view and principally from the viewpoint of cereal crops grown in developing countries as subsistence foods, to highlight both the major strengths and problems of integrated pest management on food crops and the important role the strategy must play in future agricultural policies in developing countries. Reference to definitions of integrated past management, is made in reports of the FAO/UNEP Expert Panel on Integrated Control (FAO Meeting Reports 1978, 1979, 1980) and Smith and Reynolds (1966).

BACKGROUND TO THE PROBLEMS OF FOOD - AND FOOD PRODUCTION IN DEVELOPING COUNTRIES

The dire predictions of human food shortages by 2000 AD, for Africa, S.E. Asia and parts of Latin America have been documented by FAO, UNDP, WORLD BANK and IDRC and also by the commodity based International Research Centres. Attention is drawn to the burgeoning populations of the Third World and the poor performance of the agricultural sectors in the countries concerned. Recently (Beets 1981) notes that the current consumption of rice in Indonesia is 19 million tonnes, 2.3 million of which is imported, and that there is an annual growth rate of 4% in consumption. By 2000 AD, 29 million tons will be required to feed the population and external sources of supply are declining. Already 85% of the presently cultivated land in Indonesia is devoted to rice. The World Bank (1977) has estimated that the rate of growth in demand for coarse grains, which in many countries of the Third World are the staples, will be 3.6% and thus very close to the prediction by Assiz (1976).

Unfortunately such statistics are seldom examined in depth in the light of political and technological realities. They engender a feeling of helplessness (Marnham, 1980) in a situation where there is an ability to do something significant. There is a requirement for informed and concerned comment, on both the realities of the existing state of knowledge on production and benefits of pest control in the younger nation states, and a very clear recognition of the fact that many technologies developed in the temperate latitudes are inappropriate to the tropics or need considerable adaptation. The economic, social and political factors involved in helping subsistence farmers to maximize food production now are only partly acknowledged. Such farmers produce the vast preponderance of the food on small-holdings of 0.5 ha or less. To ensure surplus food in sufficient quantities to feed adequately increasing populations calls for properly targeted and adequately funded research in crop science, socioeconomics and infrastructural (politico-economic) support within the countries concerned. Such support has to be sustained and planned for a realistic time frame. Quantum leaps in food production, such as are visualised in many projections of research benefits resulting from aid schemes, have not occurred quickly even in developed countries such as USA and UK, or developing countries, e.g. India, where very significant gains have been made in food production (Morgan 1980). Such gains have been based on broad advances and an holistic approach, which include research into fundamental principles and, particularly, educational advances. Commitment on the part of planners to stable production, with all that this entails, and investment in progressive approaches such as integrated pest management is a vital prerequisite for stability of production of food crops in developing countries. The real issues are very complex (OECD 1977, Haskell et al. 1981).

THE FOODS

The major storable and commercial foods of the world are the cereals, and the food legumes. Statistics on production exist and with some of the crops they are reasonably accurate (See Table 1). Obtaining reliable data for production from subsistence farmers will always be difficult, yet it is these farmers who produce the bulk of the coarse grains and the other major staples, e.g. bananas/plantains, yams, cassava and the various legumes, for the populations of developing countries. The crucial role of food crops such as the pulses, chickpea and pigeon pea in the Asian sub-continent in the farm level economy and diet is little appreciated. Such staple foods in the Third World have received only a minor research effort for a whole host of historical reasons, despite their pitiful yields. Most of these foods are susceptible to pests - in the broader sense - rodents, birds, insects, nematodes, weeds and plant pathogens.

Comparison of these production and yield data from developed and developing countries confirm the very considerable hectarage of food crops grown in developing countries and the great importance in their economies of subsistence food crops other than cereals. They highlight the relatively low comparative yields/ha of most food crops in the developing countries and the generally extremely low yields of legumes and groundnuts in them. They conceal the importance of energy inputs - both direct, by way of fertilisers and pesticides and indirect, by way of agricultural research, in achieving and maintaining these yields in developed countries. As was pointed out by Odum (1971), "Industrial man no longer eats potatoes made from solar energy: he now eats potatoes partly made from oil". It might also be added that he eats potatoes bred by research adequately funded and staffed in a sustained fashion over many decades.

Table 1: Production Statistics of some Major Food Crops in 1980 (Source: FAO)

CROP	AREA HARVESTED IN 1000/ha	PRODUCTION IN M TONNES	YIELD kg/ha
CEREALS	313,731 (429,795)	795,827 (774,847)	2,537 (1,803)
ROOTS AND TUBERS	13,569 (36,309)	182,763 (304,350)	13,470 (8,382)
POTATOES	13,387 (4,642)	179,944 (45,774)	13,441 (9,860)
CASSAVA	(13,926)	(122,134)	(8,770)
PULSES (DRY)	9,184 (62,840)	10,968 (36,440)	1,194 (580)
BEANS (DRY)	2,505 (23,781)	2,251 (12,413)	898 (522)
GROUNDNUTS (IN SHELL)	933 (18,842)	1,540 (17,361)	1,650 (921)
BANANAS		(38,453)	
PLANTAINS		(21,265)	

() Developing countries.

This has considerable implications within the context of our discussions here, given the price of fossil fuels and the concept of 'one world' and finite resources. These factors were emphasised by Pimental et al. (1975), who noted that only in cereals was it technologically possible to utilise existing resources and information to give the 75% increase in production required by the year 2000 AD to feed the world's population at existing levels. By the year 2100 AD they estimate that the current world production of cereals has to be increased by 330% and that of legumes by 173% to feed the enhanced population with a diet similar to that of 1975. If current projections are realistic, with the resources of land, energy and water, these increases appear too large to achieve.

When the statistics for cereals are examined several factors of relevance to the discussion of integrated pest management emerge. Rice far outstrips all other cereals in developing countries in terms of hectarage grown and production, but in fact average yields per hectare from these countries are about half those obtained from developed countries (2674 kg/ha vs. 4936 kg/ha). This possibly reflects the fact that most rice in the former is produced on small peasant holdings in the tropics with low inputs. This is on a crop where it is possible to control the growing environment to a very considerable extent from the point of view of availability of water and, given the priority status of the crop to most farmers growing it, soil preparation, fertilisation and plant population. A major 'wild card' seasonally affecting production in this and other food crops is undoubtedly pest attack. Accurate data on such losses exist for rice, but very little data which take into account the various aspects of pest loss of the type provided by Harris and Harris (1968) for sorghum, is readily available for other food crops in the tropics.

Yield differentials between developed and developing countries are much greater when other typical food subsistence crops are considered (Table 2).

Table 2: Yields of Some Major Food Crops in Developed and Developing Countries (Source: FAO)

Crop	Mean Yield (kg/ha) 1969-1971	1978	1979	1980
SORGHUM	3109 (849)	3207 (1082)	3658 (1058)	2710 (1025)
MAIZE	4041 (1595)	5124 (1735)	5595 (1815)	4768 (1826)
CHICKPEA	938 (575)	906 (560)	848 (561)	1127 (614)

() Developing Country yields.

The average yields of both sorghum and maize are about 200 kg/ha greater over the past 3 years compared to the 1969-71 period in developing countries, possibly reflecting increased research effort, particularly in Asia and Latin America, on these crops, making use of information and genetic material from developed countries and International Centres. The increase, unfortunately, probably reflects an increase in production of the cereals as a feed not a food. It also reflects big production increases in S. America and not in Africa where the crop is a staple food.

Where a crop is a subsistence food the amount of information on any aspect of it (agronomic practice, pests or any other item) in a subsistence grower situation is small. Data on pests of these food crops, and increasingly on the insect population dynamics in them, grown as monocrop, or on research stations do exist, but availability of such data is negligible in the tropical farmer's field situation. This has profound implications for the development of integrated pest management in these crops. It is impossible to transpose findings on integrated pest management derived from higher latitude developed countries, often from mechanised monocultures on large acreages, to the traditional multicrop situations of the tropics. This simple fact must be much more widely recognised by both donors and project managers concerned at the food situation in developing countries, before real progress can be made. In general, major tropical food crops, e.g., maize and sorghum, which have moved into higher latitudes are currently producing far higher yields than in their original environments giving great encouragement, since clearly the potential is there within the countries of origin. Useful steps have been taken by IBPGR and the LARC's to assemble the available germplasm of many subsistence crops and this is vitally important in the context of integrated pest management.

Data given above indicate that a plateau for the yield is being reached in developed countries and will occur unless some striking new innovation is forthcoming. The recent study (Morgan 1980), however graphically shows that the base levels of production in less developed countries have not kept pace with those in developed countries in the 1970's and there is no indication that the gap is closing in the 1980's.

THE PESTS

The biology and bionomics of pest species on food crops is usually far better understood in developed than in developing country situations, except where the pests are common to both or where related species occur. In the latter countries research emphasis was often placed on cash as opposed to food crops and the increase in available research manpower and resources, and concern for increased food production in them coincided with the age of rapid development of pesticides. These were relatively efficient in controlling insects, even in the absence of much detailed work on their biology which received limited attention owing to administrative and financial pressures. The consequent problems sowed the seeds of the "rediscovery" of the integrated control aspect, which had received some attention under various guises such as 'cultural' control earlier. This resurgence of interest has also highlighted the dearth of information, particularly on natural control elements in food crops. Review of existing information, such as by Lawani (1982) on various agronomic practices on cereal stem borers control, is needed for tropical pests of food crops to facilitate development of IPM strategies.

There is no doubt that the misuse of broad spectrum insecticides, often in increasingly large quantities has caused severe environmental problems, while in no way alleviating pest problems (Smith 1970, Conway 1971, Falcon and Smith 1973). Associated problems of insect resistance to pesticides are well documented, e.g. to DDT by Heliothis armigera on the Ord River in Australia, where a 90 fold increase in resistance has been documented, and also a five fold increase to DDT/toxaphene and endosulfan found (Wilson 1974). It must however be acknowledged that most of these problems have arisen not on food crops, but on cash crops, particularly cotton, and mainly in developed country situations. Joyce (1982), in a critical review of the use of pesticides for Heliothis management, comments that one must conclude that the strategy has been essentially one of chemotherapy which maintains the environment in a lethal state for pest species. Little note is taken of received doses or distribution of the insectide within the crop. He was essentially talking about cotton - where far more information exists than on food crops. Far too little attention has been paid to maximisation of kill of pests at an early stage before damage has begun and almost none of the effects of pesticides on early populations of pests on food crops and the parasites and predators associated with them. Nevertheless as stated by Haskell (1977) "However involved the argument of integrated control versus pesticide control becomes, the facts of the matter are that integrated control cannot be relegated to the status of a long term option given the existing finite resources of the world to feed its population". Unfortunately the emotive nature of the environmental/ecological discussions tends to obfuscate and in places completely obscure, the true pest issues.

An important issue in the consideration of management of pests is the rapidly increasing knowledge occurring in the field of plant breeding and the development of pest resistant cultivars. The literature on the subject is already extensive. Screening techniques, for ensuring maximum exposure of new plant genotypes to pest species in plant breeding programs, are being constantly improved and more readily used. Development of pest resistance in food crops have far reaching and important influence on pest management strategies. Not all of these influences are positive, as far as overall production of some food crops from large areas are concerned at the initial stages of introduction. A careful assessment of management strategies in relation to resistance breeding is required and regular reappraisal required of the pests and the damage caused to food crops.

TOWARDS AN INTEGRATED PEST MANAGEMENT APPROACH FOR FOOD CROPS

The term 'integrated control', coined by Stern et al. (1959), at that time implied selective use of pesticides. Pest management utilises any or all suitable techniques and methods, in as compact a manner as possible, to maintain pest populations below those causing economic injury. This could be a single component, e.g. spraying in an appropriate situation, or several techniques applied simultaneously or over time. Integrated pest control, (this term can be used interchangeably with integrated pest management (FAO 1980), by definition uses a multi-component approach and presupposes a knowledge of the ecosystem, crop and pest. Therein lies both the hope and the problem for the approach. A summary of the approach to the technique is given by Flint and van den Bosch (1977) who present the underlying premises for integrated pest management. However those simple premises are not well understood even by research scientists in agriculture and a continuing educative process is important.

The broad social complexities of application of the IPM approach has been detailed by Stockdale (1980), who correctly points out that pest control is but one aspect of a package of agricultural technology (and of course integrated pest control is itself a package). He also stresses that the components in the agricultural technology are themselves subject to rapid change. Almost everywhere, at present, donor agencies and world bodies are laying stress on changing of traditional systems through encouraging increased fertiliser use, commissioning irrigation projects, improving mechanisation or simply the introduction of new crops or cultivars. In these situations pest control strategies must be dynamic and perspectives realistic. The effect of the dynamic nature of the systems, pests and crops, on food production increases both the difficulty of impact assessment and the necessity for it. Impact assessment over a relatively short term is increasingly demanded by agencies, yet the long term impact of a management program on both environment and production of crops is probably more crucial. A whole range of factors outside mere pest control have to be considered, and these demand expertise beyond those just in the pest control area. Changes in pest incidence in traditional systems are not always directly created by man as exemplified by the surge in the importance of earhead caterpillars in pearl millet in the Sahel after the droughts of the early 1970's (Vercambre 1978). Constant vigilance is required in formulating integrated management strategies.

FACTORS IN FAVOUR OF AN INTEGRATED APPROACH

A factor which has tended to be overlooked in the discussion on integrated pest management is that the world's small farmers have little access to pesticides. The infrastructure for their acquisition and delivery does not exist or exists only in rudimentary form. The problems of ensuring that the chemicals and sprayers are available and maintained are formidable in the often remote areas where they are to be used. Further, the costs to farmers, government or both are large and few farmers are prepared to utilise pesticides on food crops. Reasons vary from the lack of economic incentive to sociocultural considerations which may inhibit use of even "safe" insecticides on food crops, as inadequate information delivery systems fail to educate farmers on matters of safety, non-hazard to human reproduction, or efficacy. Many existing traditional agricultural systems, nevertheless, do take cognisance of pest attacks and use of plant-derived insecticidal ash and plant extracts or oil is not unknown. Other agricultural practices which discourage pest attack are known to farmers, e.g. early sowing, distribution of crop residues etc. However, many of them, while known to farmers, are not applied and reasons for this should be sought. Mechanisms for encouraging their use involve proper application of integrated approaches which involve mundane, rather than spectacular pest management research.

Information on the basic biology of many pest species, and particularly the relationship between pests and their parasites/predators in tropical food crops, is scant. Such information is crucial in mounting wide-spread integrated control schemes, as is information on pest buildup under various agronomic situations. Such information need not be immediately available to mount a scheme as long as the general approaches are applied and a data bank built up to enable the broad strategy to be progressively refined and improved.

Much as been written on the environmental aspects that favour an integrated approach through the sensitisation of society to the short and long term 'costs' of pesticides by such writers as Carson (1962). If the purpose of such writing was to reverse the tide of pesticide use it has failed (Zwerdling 1977). However, a balanced approach to the pros and cons of insecticide usage provided by the framework of integrated control has much to recommend it, and the knowledge to use insecticides wisely often exists, but other factors impinge on utilisation of this knowledge.

FACTORS MILITATING AGAINST AN INTEGRATED APPROACH

The implementation of integrated control strategies is complex. It involves questions of economic thresholds, which may have to be determined more than once with the same pest or a series of pests during the growth period of food crops, which may be well over 100 days longer. In the instance of plague migrant pests, calculation of such thresholds is a rather academic exercise, and thoughts of interated control rather irrelevant as calculations of economic loss, particularly in the instance of small farmers really revolve round the question of no crop = no food = starvation for the family. In "normal" pest seasons a considerable amount of loss is acceptable and the question raised in these situations is development of reliable forecasting systems, as part of a management strategy. Considerable progress has been made with some pests of food crops, e.g. locusts and armyworm. However Taylor et al. (1980) comment that, in respect of migrant pests in the UK, conventional population dynamics work has been of little value as it has been concerned with temporal and not spatial change. In the tropics almost no work has been done on food crops, and much of the current work on migration unfortunately deals with movement of pests out of outbreak areas. Far more research needs to be done on movement into areas and on factors favouring the development and increase of small initial populations in a habitat. The role of forecasting in economic strategies has been discussed by Tait (1977) and Norton (1977) among others.

In situations other than those applicable to migrant pests or severe pest attack, the problems of calculation of thresholds should not be used as a means of questioning the concept of pest management and integrated control. Some loss is acceptable, and quantifiable data, especially in the tropics, are unlikely to be readily forthcoming as the resources are not available. It is doubtful, for instance, whether the problem of compensation by food crops to losses by pests (or indeed other stress factors) has ever been adequately researched in most tropical food crops, and this is very important in consideration of economic thresholds. A major question is posed as to how much more financial and staff resources should be devoted to such an exercise. In the context of integrated pest control a knowledge of the particular agroecosystem involved may be crucial to success or failure and also in assessment of the economic parameters of the strategy (Mc Carl 1981).

Integrated pest control strategies are often location and crop specific (Tarlock 1980). This can present administrators and bureaucrats with problems both of comprehension and implementation. The pest situation in an early sown, irrigated sorghum field with a uniform high yielding hybrid is very different from that pertaining in an intercropped peasant farmer crop of sorghum utilising at least two landrace cultivars. Farmers, for preference, or social reasons, may habitually grow two or three cultivars. Often such cultivars are chosen so as to maximise chance of avoiding total crop failure in situations of unreliable rainfall. Intercropping often leads to particular difficulties in a management strategy with food crops when insecticides are applied, particularly near harvest time, when it may not be advisable to utilise a persistent pesticide. In short, integrated pest control is not a uniform technology which can be recommended (or imposed) fairly on a large number of people. Flexibility is called for. As an example: should a group of farmers, who, in a large scheme, sowed their crops early and carried out suitable agronomic practices ensuring minimal pest build-up, be forced to spray to reduce the possible attack on late-sown crops of their neighbours? How does one "legislate" for early sowing? Clearly to be really efficient integrated control must be followed by most, if not all farmers in a particular agroecosystem/geographical area if possibility of success is not to be impaired. These problems have been alluded to in some publications, but seldom thoroughly discussed.

Ideally successful integrated control depends on the generation, assimilation, analysis and dissemination of a large amount of theoretical and seasonal applied information. Progress is being made in utilising IPM techniques based on these methodologies in developed countries generally with cash and high value crops and even with food crops (Teetes 1980b) - but in developing countries progress is limited (Jotwani, 1979, Kiritani, 1981) though theoretically it is in these that most is to be gained (Davies 1980).

EXAMPLES OF INTEGRATED CONTROL IN FOOD CROPS AND PROBLEMS OF ACCEPTANCE

Although integrated control is obviously well suited to food crops, given their low traditional economic value, examples of the utilisation on any great scale are few. FAO/UNEP, through sessions of its Panel of Experts, sought to encourage its use and stimulated production of guidelines on integrated control in major cereal crops including rice, maize and sorghum. These guidelines draw attention to the known difficulties of quantification of results and to the diverse issues which have to be taken into account in integrated pest control strategies, for instance, problems over pest residues in developing countries. In India use of stalks is an extremely valuable component of the total sorghum crop to small farmers. Stalks are crucially important as a summer season fodder for cattle and for milk production in urban areas and they are of high cash value at certain times of the year. Stalks are therefore held until well after sowing of the new season's crop. In Africa sorghum and millet stalks are important as building materials. It is therefore not always feasible to recommend destruction of crop residues, although these are knwon to carry aestivating larvae. Efforts are being made to carry out research into reduction of 'carry over' in some countries (Adesiyun and Ajayi 1980), but much more work is needed in determining both the importance of and methods for combatting such basic entomological problems in implementing pest management strategies. Farmers will readily accept recommendations only when they are relatively simple to use and their value is proven. It is taking time to convince farmers in developed countries of benefits from IPM, and considerable efforts in education of farmers to such benefits will be essential in developing countries. Further, the best methods of approach will differ from community to community even within a single country.

Such issues have been highlighted in a series of papers on integrated control in rice, e.g. Chang (1980) noted that in Malaysia the relative importance of rice pests changes with time, cropping pattern and agronomy, and that weather could affect the situation markedly. Currently pesticides are not widely used in Malaysia, and spot treatments are used. Revealingly in the context of our subject, he notes that only about 46% of farmers ever apply insecticides and he expresses a need for continued vigilance and work to refine the timing of insecticide applications based on insect counts, dosage, selective toxicity and method of application. Several papers by Kiritani (1979, 1981) provide strategies and tactics for good integrated pest management in rice, but in the latter paper he points out that despite repeated emphasis on the necessity of integrated control its implementation by Japanese farmers is limited. There are many obstacles - both technological and socioeconomic.

An important consideration is the energy balance involved. Udagawa (1976) observed that the energy input balance on rice in the period 1950 to 1974 increased for fertilizer, machinery, fuel and pesticides by 4, 12, 23 and 33 fold respectively, but the yield inceased only 1.5 fold in Japan. Clearly a reduction in energy input is required without a reduction in yield (Kiritani (1981)). The same author comments that of all the possible options for insect control in Japan insecticides will remain the commonest method, and of the alternatives, only varietal resistance is really practicable. A precise forecasting system is needed to enable insectidies to be applied at the time of planting for brown planthopper control in view of the increased use of machine planters. Citing the 100 fold increase in reistance to some carbamates and the 600 fold increase to malathion in Nephotettix cincticeps, an appeal is made for an orientation of control toward optimised pest management. These issues are equally important on other crops and in other countries and particularly so in the developing countries. With other food crops, such as sorghum, some progress has been made, as stated earlier, but much more effort and detailed adaptive research is required, necessitating allocation of financial and staff resources.

CONCLUSION

This brief examination of the subject shows that it would be erroneous to suggest that integrated pest control is a panacea. The method is logical but calls for concerted effort if it is to be widely applicable on food crops, and the implementation demands considerable flexibility. The plea at this time must be for more base data to be gathered on food crops in developing countries, and for more targeted and sustained research actually in the small farmer situation. Of particular importance is intensification of work on plant resistance to insects both in developing countries and developed countries. The latter could greatly assist in carrying out some of the more sophisticated analyses on bases of resistance. Little is known currently of the effects of resistance and resistance chemicals on development of pest populations or their parasites and predators. The host insect parasite/predator relationships in sprayed and unsprayed situations need to be more extensively studied in peasant farmer's fields. Results may indicate that augmentation of parasite/predator populations is possible, or as a minimum, enable indications to be obtained of methods of encouraging rapid build up of existing parasite/predator populations. To do this some of the modelling techniques developed at higher latitudes to study pest populations and migration need to be examined, in the light of the much higher pest populations which occur and the rapidity with which numbers increase on tropical food crops. Work is required on timing of spray applications, and on suitable techniques for application and insecticide formulations for tropical food crops. Actual information on received doses by insects in tropical food crops is almost non-existent.

To ensure the more rapid acceptance of Integrated Pest Management strategies, it is vital that the involvement of many disciplines - plant breeding, agronomy, pesticide chemistry, socio-economics, economics etc., is encouraged. This is crucial in the interests of formulating strategies which are in the long term, socially, environmentally, politically and economically best for the bulk of the farmers who produce the bulk of the world's food for local consumption - the small farmers of limited means operating in marginal conditions, where inputs are few and risks high.

ACKNOWLEDGEMENTS

The helpful comments of my colleagues at the Centre for Overseas Pest Research are gratefully acknowledged.

REFERENCES

Adesiyun, A.A. and Ajayi (1980). Control of Sorghum Stem Borer Busseola fusca, by Partial Burning of the Stalks. Tropical Pest Management, 26(2), 113-117.

Aziz, S. (1976). The World Food Situation Today and in the Year 2000. World Food Conference, 1976, Iowa.

Beets, W.C. (1981). Agriculture in Indonesia in AD 2000. Span (UK) Shell, 24(1), 24-27.

Carson, R. (1962). "Silent Spring". Hamish, Hamilton.

Change, Poon Min (1981). Insect Pests of Paddy in Malaysia. Proc. Symp. on Insect Pest Management in Developing Countries. Trop. Agric. Res. Series No. 14, Ibaraki, Japan.

Conway, G.R. (1971). In: The Careless Technology - Ecology and International Development. Farvan M.J. and J.P. Milton (ed.) Nat. Hist. Pres. N.Y.

Davies, J.C. (1979). Examples of Integrated Control Programs Africa. In: Elements of Integrated Control of Sorghum Pests, Plant Protection Paper 19, FAO Rome, 155-159.

Falcon, L.A. and R.F. Smith (1973). Guidelines for Integrated Control of Cotton Pests. Report AGPP Misc. 8. FAO Rome.

FAO - Reports of various sessions FAO Panel of Experts on Inegrated Pest Control. 1976-1981. FAO, Rome.

FAO (1971). Agricultural Commodity Projections 1979-1980. Vol. 1, 1971 FAO Rome.

FAO - Production Year Book, Vol. 34, FAO, Rome 1981.

Flint, A.L. and R. van den Bosch (1977). "A Source Book on Integrated Pest Management." International Center for Integrated Pest Management and Biological Control. University of California.

Harris, K.M. and Elizabeth Harris (1968). Losses of African Grain Sorghums to Pests and Diseases. PANS 14, 48-54.

Haskell, P.T. (1975). Britain's Role in International Pest Control. Proc. 8th Br. Insectic. Fungi Conf., 1059-1067.

Haskell, P.T. (1977). Integrated Control and Small Farmer Crop Protection in Developing Countries. Outlook on Agric. 9(3), 121-126.

Haskell, P.T., T. Beacock and P.J. Wortley (1981). Worldwide Socioeconomic Constraints to Crop Protection. In Kommendal. T. (Ed.) Proc. of Symposia IX International Congress of Plant Protection, Washington, 1979; vol. 1, 39-41.

Joyce, R.V. A Critical Review of the Use of Pesticides in Heliothis Management. In Proceedings of Symposium on Heliothis Management (ICRISAT), 1981, Hyderabad, India.

Jotwani, M.G. (1979). Examples of Integrated Control Programs - India. In: Elements of Integrated Control of Sorghum Pests, Plant Protection paper 19, FAO, Rome, 150-154.

Kanwar, J.S. and J.G. Ryan (1976). Recent Trends in World Sorghum and Millet Production and Some Possible Future Developments. Proceedings Symposium on Sorghum Production (CFTRI), 1976, Mysore, India.

Kiritani, K. (1979). Pest Management for Rice. Ann. Rev. Entom. 24, 279-312.

Kiritani, K. (1981). Integrated Pest Management for Rice. Proc. Symposium on Insect Pest Management in Developing Countries. Trop. Agric. Res. Series No. 14, Ibafaki, Japan.

Lawani, S.M. (1982). A Review of the Effects of Various Agronomic Practices on Cereal Stem Borer Populations. Tropical Pest Management. 28(3), 266-276.

Marnham, P. (1980). "Fantastic Invasion", Jonathan Cape.

Mc Carl, B.A. (1981). Economics of Integrated Pest Management - An Interpretive Review of the Literature. Special Report Oregon State Univ. 636.

Morgan, W.B. (1980). "Agriculture in the Third World - A Spatial Analysis." Bell and Hyman, London.

Norton, G.A. (1980). The Role of Forecasting in Crop Production Decision Making - An Economic Viewpoint. EPPO Bull. 10(2), 269-274.

Odum (1971). "Environment Power and Society." Wiley Interscience, New York.

OECD (1977). Report of the Steering Group on Pest Control under Conditions of Small Farmer Food Crop Production in Developing Countries. Development Center OECD, Paris.

Pimental, D., O. Dritschil, T. Krummel and J. Kutsman (1975). Energy and Land Constraints in Food Protein Production. Science 190, 754-761.

Ruesink, W.G. (1980). Economics of Integrated Pest Management: Discussion on Entomologists View of IPM Research Needs. Am. J. Agr. Ec. 62, 1014-1015.

Stern, V.M., R.F. Smith, R. van den Bosch and K.S. Hagen (1959). The Integrated Control Concept. Hilgardia 29, 81-101.

Smith, R.F. and C.B. Huffaker (1973). Integrated Count Strategy in VS and its, Practical Implications, LPPO Bull. 3, 31.

Smith, R.F. and H.T. Reynolds (1966). Principles, Definition and Scope of Integrated Pest Control. In: Proceedings FAO Symposium on Integrated Pest Control, FAO Rome.

Smith, R.F. (1970). Pesticides, Their Use and Limitations in Pest Management. In: Rabb and Guthrie, "Concepts of Pest Management", Carolina State University, Raleigh, NC, pp. 103-118.

Stockdale, Jerry D. (1980). Pest Management and the Social Environment, Conceptual Considerations. In Pest Control: Cultural and Environmental Aspects, Ed. Pimentel and Perkins. AAAS Selected Sympoisum 43.

Tait, E.J. (1978). The Use of Forecasting as a Method of Rationalising Pesticide Applications. Proc. 1977 British Crop Protection Conf.

Tarlock, A. Dan (1980). Legal Aspects of Integrated Pest Management. In: Pest Control: Cultural and Environmental Aspects, Ed. Pimental and Perkins. AAAS Selected Symposium 43.

Taylor, L.R., R.A. French, I.P. Wrowood, M.J. Dupuch and J. Nicklen (1980). Synoptic Monitoring for Migrant Pests in Great Britain. Rothamstead Exp. St. Report 1980, Part 2, p. 41-104.

Teetes, G.L. (1980). An Overview of Pest Management and Host Plant Resistance in US Sorghum. In Biology and Breeding for Resistance to Arthropods and Pathogens in Agricultural Plants, Ed. M.K. Harris. Texas A&M/AID.

Teetes, G.L. (1980b). Examples of Integrated Control Programs - Texas. In: Elements of Integrated Control of Sorghum Pests, Plant Production and Protection Paper 19, FAO Rome, 133-149.

Udagawa, T. (1976). Estimation of Input Energy for Rice Cultivation. Kaukyo Kagaku 5, 73-79 (Abstract).

Vercambre, B. (1978). Raghuva ssp et masalia sp. Chenilles des Chandells du mil en Zone Sahelienne. l'Agro Trop. 33, 62-69.

Wilson, A.G.L. (1974). Resistance of Heliothis armigera to Insecticides in the Ord Irrigation Area, North Western Australia. Jour. Econ. Ent. 67, 256-258.

Zwerdling, D. (1977). The Pesticide Treadmill. Environ J., Sept: 5.

PRINCIPAL DISEASES OF FOOD CROPS

G. Rangaswami
Commonwealth Technical Adviser (Agriculture)
#1, Road 23 Gulshan, Dacca-12, Bangladesh

ABSTRACT

Many diseases of food crops are prevalent in different parts of the world causing recurrent heavy losses. About 10-15% of total food production is lost due to diseases before the crops are harvested. Control measures against diseases over the past century have however yielded rich benefits. Most devastating diseases of major cereal food crops are now controllable through cultivation of disease-resistant varieties or a combination of chemical, physical, biological and cultural means. There is ample scope for improving the quantum and quality of plant protection technology presently available to farmers in developing countries, which will help further reduce the loss. The host-pathogen interactions are such that there is bound to be continuing struggle between the two. Continued and more concerted research efforts are required to improve the qualities in the host plants to withstand the onslaught of pathogens. More studies are needed on the methods to accurately assess the losses due to diseases, on the techniques for disease forecasting and to more effectively manage diseases at the time of outbreaks.

KEYWORDS: Diseases; Plant Pathogens; Food Crop Losses; Chemical Control; Resistance-Breeding.

INTRODUCTION

The importance of the world's many food crops varies with the region. Though over a thousand species of plants are known to be of direct economic value, a majority of the world's population derives its calorie needs primarily from about 15 plant species. The principal, primary or staple food crops are mostly cereals, but in some countries tuber crops, coconut and banana form staple foods. The secondary food crops are pulse grains, oilseeds, vegetables, fruits, nuts and sugarcane, sugarbeet and other sugar-yielding crops. The most important among the food crops are the five cereals, viz., rice, wheat, maize, sorghum and barley (Mangelsdorf, 1966). The world production of the major food crops is given in Table 1. About half the world population depends on rice for its staple food, and nearly 30% of human energy is derived from this one plant species. The production figures for cereals in 1980 in different parts of the world are given in Table 2. The average yields/ha of land cultivated to cereals in the developed countries are in the range of 2.0 to 3.75 t, excepting in Australia where it is about 1 t and in the USSR where it is 1.5 t. The comparable figures for most of the developing countries are less than 2 t/ha, barring China where the average yield of rice is reported to be 3.6 t/ha (Anonymous, 1981). In general, there is high correlation between the yield levels and the technology input, but this does not necessarily mean that diseases are less important in the developed countries than in the developing countries. On the other hand, with more intensive cultivation there is more disease incidence unless care is taken to control them.

Table 1: World Food Production (1980)

	Million tonnes (Mt)		Million tonnes (Mt)
Total Cereals	1,570.673	Total Vegetables	347.859
Wheat	444.543	Total Fruits	287.505
Rice	399.779	Sugarcane	730.723
Maize (Corn)	392.249	Sugarbeet	268.722
Barley	162.402	Total Oilseeds (oil equivalent)	51.075
Total Root Crops	487.113		
Potato	225.718	Total Meat	142.166
Total Pulses (grain only)	47.408	Milk	470.399
		Total Fish catch	69.700

Outlook for 1982:	Million tonnes (Mt)
Cereal Food Grains	1,470 to 1,650
Wheat	450 to 500
Rice	450 to 486
Coarse Grains	755 to 820
World carry-over stock	268

(From FAO Yearbooks and Forecasts)

Table 2: World Cereal Production (1980)

	Area ('000 ha)	Yield (Kg/ha)	Production (Million tonnes)
World	743,526	2,112	1,570.673
Africa	75,222	942	70.838
Canada	19,775	2,062	40.786
USA	71,516	3,775	269.979
Argentina	9,926	1,878	18.643
Brazil	21,060	1,573	33.126
Asia	325,270	2,038	642.488
China	102,624	2,760	283.277
India	104,509	1,386	134.879
Europe	71,084	3,642	258.912
Australia	16,043	1,028	16.497
USSR	123,975	1,475	182.811

(From FAO Yearbook)

PAST ACHIEVEMENTS

Diseases of crops are known from pre-historic days but their microbial nature came to be recognized only during the 18th century. Since then, over 30,000 pathogens have been recorded on about 3,000 plant species. Over 25,000 of the pathogens are fungi, about 600 are nematodes, more than 200 are bacteria, and at least 300 are viruses, mycoplasma and other related pathogenic entities (Rangaswami, 1979). Plant diseases are generally referred to pre-harvest ailments and hence our present consideration will be limited to losses due to diseases occurring in crop fields. However, there are close links between pre-harvest crop diseases and post-harvest losses.

Dealing with diseases of food crops, one has to take into consideration three important aspects: (1) The essential need to minimise losses so as to feed humanity - if even half of the loss of about 300 Mt of food grains due to diseases could be saved it would feed at least 750 million of the underfed and malnourished people of the world. (2) Any damage to crop plants is an economic loss to humanity as a whole and this should be minimised at any cost. (3) In order to make agriculture an increasingly economic undertaking, the productivity of arable land should be increased continuously and such an increase is possible only if the pathogens are effectively checked.

Crop Loss Due to Pathogens

On a global scale the damage caused to the economic plant species by crop pests including insects, weeds and pathogens is variously estimated at 25 to 40% of total crop production. On an average, about 10 to 15% damage to crop plants could be attributed to the diseases. As against this pre-harvest loss, the post-harvest damage to the produce takes place in transit, storage, market places and until the food is cooked or otherwise prepared and eaten. The overall loss at post-harvest stages may be anything from 5 to 25%. While the pre-harvest damage to different crops in the field vary with the varieties, virulence of the pathogen and various factors affecting the host-pathogen interactions, the post-harvest damage varies with the produce, its moisture content and the conditions under which it is stored. The more succulent and perishable commodities of fruits and vegetables often succumb to heavy damage by pathogens resulting in losses up to 50% or more. The total pre-harvest loss of cereal food grains at the 1980 level of global production of 1,570 Mt, calculated at 15% due to plant diseases, is about 277 Mt. This would amount to about US$ 40 billion. The value of loss to other plant parts of cereals and also to all other agricultural commodities would be another US$ 40 billion. To this we may add the post-harvest loss of about US$ 20 billion. By reducing this avoidable loss of US$ 100 billion, we could achieve the three objectives of feeding the unfed, improving the nutritional quality of human diet and substantially bettering the agricultural economy of the world.

Overall Assessment of the Diseases

It is known that each of the economically important crop species is affected by one or more pathogens, often scores of them on a single crop species. There are more than 100 pathogens on rice, another 100 on wheat, about 60 on maize, over 50 on sorghum and a similar number on barley (Rangaswami, 1979). Not all of them are considered major pathogens though some minor ones may occasionally become major in some locations during certain favourable seasons. The major diseases of the five important cereals are listed in Tables 3 to 7 and of the minor staple food crops in Table 8. Most of these diseases have been under study over the past several decades. The nature, etiology, factors affecting their onset and spread, and ways and means of preventing, controlling, combatting and managing the more important ones have been investigated by several thousand specialists among the

Table 3: Major Diseases Of Rice

Disease	Pathogen
*Blast	Pyricularia oryzae
*Brown Spot	Helminthosporium oryzae (Cochliobolus miyabianus)
Bakane/Foot Rot	Gibberella fujikuroi (Fusarium moniliforme)
Sheath Blight	Thanatephorus cucumeris (Rhizoctonia oryzae-sativae), (Corticium sasakii)
Leaf Scald	Rhynchosporium oryzae (Monographella albescens)
Stem Rot	Leptospheria salvinii (Sclerotium oryzae)
Sheath Rot	Rhizoctonia solani (Acrocylindrium oryzae)
Stackburn	Trichoconis padwickii
Bunt	Neovossia horrida
Leafspot	Cercospora oryzae
*Tungro	Virus:Vectors:Nephotettix virscens, N.nigropictus, Recilia dorsalis
*Bacterial Leaf Blight	Xanthomonas campestris pv. oryzae
Grassy Stunt	Virus:Vector:Brown hopper Nilaparvata lugens
Ragged Stunt	Virus (RRSV)

Table 4: Major Diseases Of Wheat

Disease	Pathogen
*Black or Stem Rust	Puccinia graminis tritici
Leaf, Brown or Orange Rust	Puccinia recondita
Yellow or Stripe Rust	Puccinia striiformis
Loose Smut	Ustilago nuda
Flag Smut	Urocystis tritici
Bunt	Tilletia caries, T.foetida, Neovassia indica
Foot Rot	Pythium graminicolum
Powdery Mildew	Erysiphe graminis var. tritici
*Leaf Blight	Alternaria triticina
*Leaf Blotch	Septoria tritici (Leptosphaeria tritici)
Leafspot	Helminthosporium sativum (Cochliobolus sativus)
Yellow Ear Rot	Anguina tritici & Corynebacterium tritici
Take-all	Ophiobolus graminis
Black Chaff	Xanthomonas translucens var. undulosum
Seedling Blight	Rhizoctonia solani & Fusarium spp.

Table 5: Major Diseases Of Maize

Disease	Pathogen
Common Smut	Ustilago maydis
Head Smut	Sphacelotheca reiliana
*Leaf Blight	Heliminthosporium turcicum (Setospheria turcica), H.carbonum (Cochliobolus carbonum), H.maydis (Cochliobolus heterosporus)
Stalk and Ear Rot	Diplodia zeae
Charcoal Rot	Macrophomina phaseoli
Top Rot	Gibberella zeae
Seedling Rot	Fusarium culmorum
Rust	Puccinia sorghi
Brown Spot	Physoderma maydis
Zonate Spot	Gloeocercospora sorghi
Downy Mildew	Sclerospora philippinensis
*Bacterial Stalk Rot	Erwinia carotovora f.sp. zeae, E.dissolvens, Pseudomonas lapsa
*Mosaic	Maize Mosaic Virus, Sugarcane Mosaic Virus, Maize Dwarf Mosaic Virus

Table 6: Major Diseases Of Sorghum

Disease	Pathogen
Seed Rot and Seedling Blight	*Pythium aphanidermatum*, *P.debaryanum*, *P.graminicolum*
Root and Stem Rot	*Rhizoctonia bataticola*
Charcoal Rot	*Macrophomina phaseoli*
Stalk Rot	*Fusarium moniliforme* (*Gibberella fujikuroi*)
*Grain Smut	*Sphacelotheca sorghi*
Loose Smut	*Sphacelotheca cruenta*
Long Smut	*Tolyposporium ehrenbergii*
Head Smut	*Sphacelotheca reiliana*
*Rust	*Puccinia purpurea*
*Downy Mildew	*Sclerospora sorghi*
Leafspot	*Cercospora sorghi*
Anthracnose	*Colletotrichum graminicolum*
*Leaf Blight	*Helminthosporium turcicum*, *H.sativum*, *H.halodes*, *Gloeocercospora sorghi*
Sooty Stripe	*Ramularia sorghi*
Rough Leafspot	*Ascochyta sorghi*
Ergot	*Sphacelia sorghi*

Table 7: Major Diseases Of Barley

Disease	Pathogen
*Stem or Black Rust	*Puccinia graminis tritici*
Leaf Rust	*Puccinia striiformis*
Covered Smut	*Ustilago hordei*
*Loose Smut	*Ustilago nuda*
Neck Blotch	*Helminthosporium teres*
*Powdery Mildew	*Erysiphe graminis*
Root and Foot Rot	*Helminthosporium sativum*
Sclerotial Rot	*Pellicularia rolfsii*
Leaf Stripe	*Helminthosporium gramineum*

plant pathologists of the world. As a result, voluminous information and valuable data have become available.

Some cereal diseases are more readily controlled than others. Blast and bacterial leaf blight of rice, and stem rust, leaf blotch and blight of wheat are controlled through resistant varieties of the respective crops. Others like tungro virus on rice and stalk rot of maize are difficult to check. Some of the diseases of sorghum, barley, oats, *Pennisetum*, *Eleusine coracana*, *Setaria italica* and a few other millets are not readily controllable through chemical means. Work on developing disease-resistant varieties is in progress.

Potato, sweet potato and colocasias are propagated vegetatively and carry the susceptibility or resistant traits in their propagules, so it is important to use healthy planting material. Much progress has been made in eliminating virus pathogens from potato tubers, before using them as seed material. Secondary field infections of these crops are generally controlled with chemicals. More damage is caused to the rhizomes and tubers by soft-rot bacteria and fungi and there are no effective remedial measures, except the prophylactic ones (Rangaswami, 1979).

Table 8: More Important Diseases Of Minor Food Crops

Crop	Disease	Pathogen
Potato:	Late Blight	*Phytophthora infestans*
	Early Blight	*Alternaria solani*
	Brown Rot	*Pseudomonas solanacearum*
	Ring Rot	*Corynebacterium sepedonicum*
	Soft Rot	*Erwinia carotovora*, *E.amylovora*, *E.atroseptica*
	Scab	*Streptomyces scabies*
	Virus	Potato viruses X and Y, Potato Leaf Roll, Mosaic and Streak
Sweet Potato:	Charcoal Rot	*Macrophomina phaseoli*
	Tuber Rot	*Sclerotium rolfsii*, *Rhizoctonia solani*, *Ceratocystis fimbriata*
	White Rust	*Albugo ipomoeae-panduranae*
Cassava:	Mosaic	Virus:Vector:*Bemisia tabaci*
	Leafspot	*Cercospora henningsii*
Yams:	Leafspot	*Cercospora colocasiae*
	Rhizome Rots	*Pythium aphanidermatum*, *Phytophthora colocasiae*, *Pellicularia rolfsii*
Coconut:	Bud Rot	*Phytophthora palmivora*
	Stem Bleeding	*Ceratocystis paradoxa*
	Wilt	*Ganoderma lucidum*
	Root Wilt/Kadang Kadang	Virus/unknown causes
Banana:	Panama Wilt	*Fusarium oxysporum* f.*cubense*
	Bacterial (Moko) Wilt	*Pseudomonas solanacearum*
	Bunchy Top	Virus:Vector:*Pentalonia nigronervosa*
	Black Tip	*Helminthosporium torulosum*
	Anthracnose	*Gloeosporium musarum*
	Leafspots	*Cercospora musae*, *Cordona musae*

Several fungi, viruses, bacteria and nematodes affect pulse crops, grown for grain or foliage. Since these crops bring poor economic returns, farmers find it uneconomical to carry out direct chemical methods of control. Disease-resistant varieties appear to be the main answer. However, it is difficult to evolve varieties resistant to all or most of the major diseases and insect pests of pulses, combining in such varieties the other desired agronomic qualities to improve economic returns.

Intensive cultivation of vegetables in localized areas, and their repeat cultivation with short rotations in the same field, lead to favourable environ for the pathogens to multiply rapidly. Combined chemical and agronomic methods along with resistant varieties are necessary to check them.

Most of the diseases of perennial fruit crops have to be controlled with chemicals, since they stand in the field for several decades after planting and replacement with disease-resistant varieties, even if available, is not easy to accomplish. In the fruit gardens, systematic measures to maintain high hygienic conditions are essential. Integrated chemical, biological and cultural methods of control must be practiced most vigorously to improve the economy of fruit farming. Since there is ready linkage between pre- and post-harvest diseases, chemical control measures with lasting effects bring better benefits. Though expensive and time-consuming, disease-resistant varieties could replace susceptible ones in disease endemic areas.

The diseases of sugarcane, sugarbeet and other high-value food crops are effectively managed by the farmers, who are financially in a better position to undertake control measures. Also, relatively more attention is paid to evolving effective control measures through financial support for the needed research. This position will continue as long as the high values of the commodities continue.

Cereal Diseases of Prime Importance

The major diseases of the five cereals with the most devastating results are marked * in Tables 3 to 7. Nine of those diseases cause the most damage. Blast and bacterial leaf blight on rice, stem rust and leaf blotch on wheat, leaf blight and bacterial stalk rot on maize, grain smut on sorghum and stem rust on barley together account for more than half of the total loss of cereal food grains in the world. These diseases should be carefully watched all over the world and adequate measures taken to prevent, check and otherwise manage them. Each one is capable of causing such severe losses in countries fully dependent on the respective crop plants that famine and starvation would occur.

Disease Control

Ever since man recognized the losses caused by plant diseases he has attempted to prevent, check and avoid the damage, and has achieved considerable success. As a result, the technology for control of plant diseases has become an important input in modern agriculture. The general principles of plant protection are: exclusion, eradication, protection and immunization. These approaches to plant disease control bring to bear: (1) direct action against the pathogen, (2) genetic modification of the host to resist the disease, and (3) alteration of the environment to make it unfavourable for the disease. Direct action includes physical, chemical and biological means of protecting the plants against the pathogen and/or eliminating or suppressing the activity of the pathogen.

Though there have been several man-made restrictions on movement of plant material from one country to another and from one region to another, the pathogens have found ways of jumping the barriers, often with the help of man himself. As a result, devastating diseases like bacterial leaf blight and tungro virus of rice have spread to many parts of the world. In the early stages of the onset of a disease the pathogen might be eradicated. However, this becomes almost impossible with major diseases of staple food crops, which are grown extensively. Therefore, direct action to protect the plants against the pathogens becomes the most important method.

The ideal method of combatting disease is to grow resistant crop varieties. Since 1900 resistance-breeding has assumed growing importance. The substantial increase in world food production has not only been possible because of improvements in the yield potential of the crop varieties but also because these varieties have genetic qualities to resist pathogens. While the crop selection practices adopted by man would place continuing pressure on the pathogens, because of the availability of wide genetic variability among the hosts and also several efficient techniques for resistance-breeding, there is ample scope for continued success in our selection process (Day, 1978; Nelson, 1979). In the coming years we may have to balance between high levels of genetic uniformity with concomitant risk of breakdown of resistance and uneconomic genetic diversity with built-in horizontal resistance. In most cases, the relative degree of resistance results from an interplay of host-pathogen, both of which possess considerable genetic variability. Under the circumstances, no one crop variety can remain continuously resistant to a pathogen. For example, over the years about 400 physiologic races of *Puccinia graminis tritici* and over one hundred races of *Pyricularia oryzae* have been identified based on their differential virulence on the cultivars of the respective plant species. Breeding for resistance, therefore, has to be a continuing process.

Chemical Control

Chemical means of protecting the crop plants against pathogens have gained importance since Millardet discovered the fungicidal effect of a mixture of copper sulphate and lime in water exactly one hundred years ago, i.e., even prior to adoption of plant breeding techniques as a means of improving crop yields. Today chemical protection is the prime means of controlling plant diseases. It will continue to be used in future as there is no other effective answer to many of the diseases of crop plants. Even where disease-resistant crop varieties have been developed, chemical control has to be a stand-by method since the resistance may break down anytime and it would take several years to provide a new resistant variety.

Several chemical formulations have come into the market for application against plant pathogens. Some of them are for protection against fungi, while a few are more specific against certain groups of fungi, bacteria and nematodes, or systemic against viruses and other pathogens. Soil-borne plant pathogens are most difficult to control. Past efforts have mostly pointed towards cultural methods of control, which are difficult to practice. Several root and seedling infections of cereals and also systemic diseases such as downy mildews of maize, sorghum, *Pennisetum*, etc. are soil-borne and have so far defied effective solution. Recent reports on the effectiveness of CGA 49104 and Oryzemate as soil fungicides to check some major pathogens of rice are very encouraging (Anonymous, 1981a). Insecticides are also in use to check the vectors which transmit certain pathogens. Some chemicals are also particularly effective for pre-treating seeds to eliminate the externally and internally borne pathogens. Systemic chemicals such as tricyclazole, benomyl and thiophanate-methyl are becoming increasingly useful against such devastating diseases as blast on rice (Anonymous, 1981a).

In the advanced countries, pesticides are in extensive use to protect economic plant species against various pests. To meet the demand, large scale industrial production is going on. The value of pesticides manufactured in 1980 was approximately US$ 10 billion, shared by the US at 35.5%, West Germany 23.3%, Switzerland 11.1%, UK 9.2%, France 8.3%, Japan 3.9% and other countries 8.8%. There are over 100 pesticides marketed in different parts of the world, for use against various pests, including plant pathogens. However, distribution and use of these pesticides in different countries on various crops vary very widely.

The level of consumption of pesticides in 1979-80 in Japan was 3,000 g/ha, in USA 1,500 g/ha and in India 330 g/ha. Often these figures, when taken as an average for the nation or region, give a misleading picture of the actual quantity used. For example, in India, of the gross cropped area of 165 Mha, only 75 Mha get plant protection coverage. Of the 75 Mha, about 55 Mha is under high-yielding varieties with rice, wheat, sorghum, maize and *Pennisetum* and the rest under cotton, sugarcane, oilseeds, vegetables, etc. Of the cereals, rice gets more plant protection coverage than the rest. On average rice receives one to two sprays mostly for controlling insect pests and to a limited extent for fungal or bacterial pathogens. In comparison, cotton gets an average of 7 to 12 sprays and grapevine 10 to 15 sprays in one fruiting season. Of the total of 392 districts in the country, 55 districts are considered agriculturally more progressive than the rest, and they use more than 50% of the fertilizers and pesticides marketed in the country. The overall picture would therefore be that only 25% of the total landspread gets any chemical spray. Of the total chemicals applied, about 50% by volume is spread over 5% of the land area (Rangaswami, 1981a).

While the use of chemicals provides an effective and profitable method of increasing crop production, there is increasing concern about the environmental pollution that may result from excessive and continuous use. While the advanced countries have every reason to be alarmed over the ill-effects of chemical pollution on humanity and on the biosphere, in most of the developing countries such a level of pollution has not been reached. It also has to be reckoned that the developing countries have come a long way over the past half a century to make farmers aware of the damage caused by pests and the benefits of chemical control. It would not be proper at this stage to intervene to prevent use of chemicals for controlling plant diseases in the developing countries. In India a recent survey had shown that only about 40% of the farmers are aware of the benefits of plant protection measures, and only about 10% adopt them. Of these not even half adopt the measures scientifically to derive the maximum benefit. Therefore, there is need for educating the farmers on proper and judicious use of plant protection chemicals, learning from the experience of the developed countries on the damages caused by continued use of higher doses of such chemicals.

EXISTING OPPORTUNITIES

While several plant disease control methods do not reach beyond experimental fields, there are many which are not adopted by the farmers because the farmers cannot economically afford investment of funds on such measures. This is especially true in the developing countries where the farm holdings are small and the farmers are poor. Since the relative loss of crops under such conditions is higher, the cost-benefit ratio and sociological benefits from such an effort would be relatively more. Therefore, steps should be taken to subsidise plant protection measures in such selected areas to make them popular.

Pre-treatment of seeds with plant protection chemicals, which is a technique relatively easier to adopt, is one which could be advocated for universal application in food crops. Popularisation of disease-resistant varieties in such areas where the relevant diseases are endemic is another important item which deserves immediate attention. In most of the developing countries only a small percentage of the total cropped area under cereal food crops is sown to the high-yielding varieties. Since most of the high-yielding varieties of rice, wheat, maize, sorghum and *Pennisetum* are known to be resistant to the more important diseases of the respective hosts, by increasing the area under such varieties crop yields can be substantially increased and disease losses reduced. Application of chemical sprays and dusts should be done in time and more scientifically to bring better results. More effective and efficient measures to transfer plant protection technology from research centres to farmers' fields should be taken. By adoption of such agronomic measures, as proper crop rotation and soil-health-care the pathogens' population could be brought below the threshold level causing economic loss to the crop, and this would bring benefits lasting for more than one cropping season. For perishable food items such as vegetables and fruits, there should be integrated pre-harvest and post-harvest plant protection measures to minimise the overall loss.

FUTURE LINE OF WORK

We have come a long way in understanding plant diseases and their causal agents, their biology and host-pathogen interactions. We have learned many ways of dealing with them - chemical, physical, biological and cultural controls, but we have not fully won the battle. Therefore, there is need for more intensive measures in the areas of resistance-breeding, developing more effective and less expensive chemical agents against the pathogens, and working out an integrated means of managing the diseases to minimise the heavy losses now incurred.

With increased intensity of crop cultivation, there will also be more of the known diseases, and also several new ones will probably appear. There is bound to be a breakdown of resistance in many of the present-day resistant varieties. We should have a second line of ready-to-release crop varieties which are equally, if not more resistant than the present ones.

Since we need to use chemicals for plant disease control, we have to find ones less harmful to the environment, knowing more about the factors favouring their breakdown in nature. More needs to be done to obtain effective chemicals with selective action on soil-borne plant pathogens affecting food crops.

Basic information on host-pathogen-environment interactions has to be collected to understand the weaker links so as to evolve more effective control measures. This is especially essential in respect of such pathogens which breed or otherwise genetically enhance their virulence capabilities to knock down even the most resistant host varieties. More needs to be done on the inter-relationships between different pathogens and between different categories of pests including insects, nematodes and other microbes on the same host. Host-pathogen interactions under physiological and environmental stresses need to be better understood.

There have been attempts, with very limited success, to forecast disease incidence so as to take advance action to minimise damage. However, disease forecasting which depends on agricultural meteorology more than on any other single factor, will become increasingly important in the coming years. Remote sensing techniques and aerial photography through satellite services should help in better understanding of the onset and spread of plant diseases. This would lead to more accurate forecasting of disease outbreak, and location-specific plant protection measures.

Before taking steps to control a disease, one has to be certain that the disease is worth controlling. Lack of accurate methods to assess crop losses continues to hamper efforts to control many diseases. The data on crop losses presently available are not always reliable, leading to hesitancy on the part both of the beneficiary and the administration to invest funds to control disease. More needs to be done to evolve accurate methods of assessing crop disease losses (Chiarappa, 1971; Horsfall and Cowling, 1978). Economics of plant protection measures, bringing out clearly the cost-benefit ratios under different sets of agronomic conditions would also help in taking adequate measures to control diseases.

In many instances the diseases are not controlled for want of human efforts in proper planning and execution. Often we resort to control measures after the pathogen had fully established itself. In many instances, even though effective methods of control are known, the administrative set up is inadequate to meet the field requirements. Due to lack of managerial skills in the production system, either the plant protection chemical is not procured and made available in the required quality and quantity and in time, or men and equipment are not available in adequate numbers to carry out the field operations. Since managerial skills to combat plant pathogens is badly lacking in most agricultural areas of the world, by improving such skills we could substantially reduce crop yield losses.

ACKNOWLEDGEMENT

The author expresses his thanks to the Secretary-General, Commonwealth Secretariat, London, U.K. for permission to participate in these proceedings.

REFERENCES

Anonymous (1981). FAO Production Yearbook: 1980, 34th Issue, FAO, Rome, pp 292.

Anonymous (1981a). International Rice Research Institute, Annual Report for 1980, Los Banos, Philippines, 467.

Chiarappa, L. (1971). Crop loss assessment methods. In FAO Manual on the Evaluation and Prevention of Losses by Pests, Diseases and Weeds, FAO, Rome.

Day, P.R. (1978). Horizontal resistance. In Plant Disease: An Advanced Treatise, Vol.II, Acad. Press, 261-265.

Mangelsdorf, P.C. (1966). Genetic potentials for increasing yields of food crops and animals. Proc. Natl. Acad. Sci., 56, 370-375.

Horsfall, J.G. and E.B. Cowling (1978). Pathometry: Disease Measurement. In Plant Disease: An Advanced Treatise, Vol.III, Acad. Press, 119-186.

Nelson, R.R. (1979). The evolution of parasitic fitness. In Plant Disease: An Advanced Treatise, Vol.IV, Acad. Press, 23-46.

Rangaswami, G. (1979). Diseases of Crop Plants in India, Prentice-Hall of India, New Delhi, 520.

Rangaswami, G. (1981). Presidential Address at the 32nd Annual General Meeting of the Indian Phytopathological Society, Feb. 1980, Hyderabad, A.P., Indian Phytopath. 33-37.

Rangaswami, G. (1981a). Plant Protection and Environment. Nilankar Prasad Memorial Lecture, at the 33rd Annual General Meeting of the Indian Phytopathological Society, Udaipur, Rajasthan, (to be published) 10.

INTERACTION AMONG PESTS, DISEASES AND WEEDS IN FARMING SYSTEMS

J.R. Finney

Imperial Chemical Industries PLC
Plant Protection Division, Jealott's Hill Research Station
Bracknell, Berkshire RG12 6EY, England

ABSTRACT

Man's crops are ravaged by pests, diseases and weeds. There are frequently interactions among them, varying widely in degrees of complexity and in the extent of influence by other factors in the agroecosystem. Only rarely can these interactions be managed within biological or integrated control systems so that they result consistently in increased agricultural productivity. More frequently the interactions are disadvantageous to productivity; here, a detailed understanding of the biology and ecology of the damaging organisms and their inter-relationships is essential for the design of effective control programmes. This paper gives examples of the many types of interaction including: biological control of one species by another, synergy or antagonism between species, alterations in competitiveness, transmission of diseases by pests, and the effect of weeds and volunteer crop plants on the survival of pests and diseases.

KEYWORDS: Agro-ecosystem; biological control; diseases; integrated control; interactions; pests; weeds.

INTRODUCTION

Pests, diseases and weeds are estimated (Anon, 1982) to reduce crop yields by about 25% in the developed world and 40% in the developing world; their effective control represents one of the most important and demanding tasks facing mankind. Although it is possible in some cases when designing control programmes to consider each individual pest, disease or weed as a problem to be tackled separately, it is always desirable and often essential, because the various components interact, to regard them as a complex requiring an integrated control approach.

This paper is concerned with the nature of these interactions among pests, diseases and weeds and attempts to highlight those aspects which are important in the context of increasing world food production. Because of the broad scope that is possible interactions involving insects, fungi, bacteria, viruses etc. which are not themselves damaging pests of crop plants have been excluded. These, however, are often essential components of biological or integrated control systems (e.g. the cultured bacterium Bacillus thuringiensis for the control of Lepidoptera, naturally occurring or introduced predatory mites for the control of red spider mites etc).

Interactions among pests, diseases and weeds are not rare events; they are frequent, particularly if the indirect effects of one damaging organism on another are taken into account. Indeed, it is possible to find examples of nearly all the 18 possible interaction combinations, positive and negative, between the 3 components. Rather than reviewing this broad subject comprehensively, this paper chooses to exemplify how a knowledge of interactions can assist in control programmes which lead to improved crop productivity. More detail on many aspects is reported in the proceedings of a recent symposium (Thresh, 1981).

TYPES OF INTERACTION AMONG PESTS, DISEASES AND WEEDS

The nature of the interactions between pest organisms are extremely diverse and no formal classification system has been developed; the interactions may be direct or indirect, single-step or multiple-step, natural or induced by a change in cultural or agrochemical practices.

Direct Interactions

The simplest types of interaction are those where one pest organism directly affects the incidence of another. A clear example of this type is where a pest acts as a vector for a disease. Many species of nematode, and *Hemiptera* (especially aphids and hoppers) transmit virus diseases to a wide range of crop plants. Frequently they are the only method of transmission and the control of the virus disease is achieved via the control of the vector as, for example, in the control of sugar beet yellows via the control of the aphid *Myzus persicae*.

In many interactions, pest, pathogen or weed species compete directly with one another in the agro-ecosystem. This is visually obvious and well understood between weed species but less so between pests or diseases. An example of direct competition between diseases on barley is provided by Sauer and Schonbeck (1976) who showed that when powdery mildew (*Erisyphe graminis*) was controlled by the specific, systemic fungicide ethirimol, it resulted in an increase in the fungal pathogen *Cochliobolus sativus* due to decreased competition on the barley leaf surface. Direct competition also exists between insect pests as illustrated by LeCato (1975). In experiments on stored maize grain involving four species of stored-product pests in a multi-factorial experiment he showed that, amongst many other interactions, the flour beetle (*Tribolium castaneum*) was the dominant species and reduced the numbers of all the others.

Direct competition between pest organisms is common and it is important that it is taken into account in control programmes to avoid 'suppressed' organisms emerging as serious problems when competitive organisms are controlled. Sometimes pest organisms are directly antagonistic or synergistic to one another. Pollet (1978) describes a case of real synergy (as defined by Wilcoscon's test) between the lepidopteran pest *Maliarpha separatella* and the fungus causing rice blast, *Pyricularia oryzae*, on the rice crop in the Ivory Coast. The fungus infects preferentially the rice already damaged by the insect, often leading to the total destruction of the rice plant.

Antagonism, as distinct from competition, is common among fungi but there does not seem to be a well documented case between specific fungal plant pathogens. An example of antagonism involving a plant pathogen, however, is provided by Munnecke et al (1981) who describes the antagonistic effect of *Trichoderma* spp on the boot-lace fungus *Armillarea mellea*, which causes serious crop losses to a wide range of bush and tree crops. A *Trichoderma* spp is also important in reducing the survival of the rice sheath blight pathogen Thanatephorus cucumeris through decomposition of rice straw (Anon, 1981). The antagonism from Trichoderma is due, at least in part, to the production of a diffusable antibiotic according to Kraft and Roquebert (1981) who studied the interaction between T.viride and Botrytis cinerea. Many soil-borne fungal pathogens, in contrast to most obligate aerial pathogens, are non-specific to hosts; they have broad host ranges or saprophytic stages and it is highly probable that many antagonisms exist.

Indirect Interactions

Indirect interactions between damaging organisms are very diverse in nature and occur frequently. Many are obvious, others more crytic and difficult to identify. An example of the most obvious type is where the activities of one organism provide micro-environments in which other organisms thrive: aphids excrete honeydew which serves as the growth medium for Cladosporium spp and other moulds which reduce the quality of many crops, including temperate cereals; Pseudocercosporella herpotrichoides causes eye-spot of cereal stem bases which induces lodging which encourages late-season weed growth which hinders harvesting resulting in reduced grain yield and quality; stem borers create micro-environments favorable for the development of many diseases, including bark canker of cocoa (Smith, 1981) and wood rot of tea (Sivapalan and Delucci, 1972); wood scab of apple (Venturia inaequalis) provides entry sites for wood canker (Nectria galligens). There are innumerable other examples.

The relationship may be much less clear and involve multiple steps. A good example is provided by Stapley (1973) from work on coconuts in the Pacific Region. Premature nutfall in this area is primarily caused by the Coreid bug, Amblypelta cocophaga which is naturally controlled by the large yellow tree-nesting ant Oecophylla smaragdina but not by the small but very numerous ant Pheidole megacephala. Pheidole, which inhabits the base of the palms and the adventitious roots under ground vegetation cover, opposes Oecophylla and prevents it occupying the palms. The balance between the ants can be substantially changed by reducing the weed cover at the palm bases by chemical or mechanical means, or by grazing, which reduces Pheidole populations and encourages Oecophylla. A change in the nature of the weed flora has also been shown to be an important factor governing the incidence of the green peach aphid (Myzus persicae) on peach trees (Tamaki, 1981). Broad-leaved weeds, but not grass weeds, act as alternate hosts for the pest (and also as hosts for many virus diseases).

The importance of weeds as sources of virus inoculum for crop plants is generally greatly under-estimated. There are very many examples: Localetti (1978) reports that three very common weeds in USA, Portulaca oleracea, Solanum sarachoides and Amaranthus retroflexus are carriers of tobacco rattle virus (TRV) and the latter two are also carriers of potato virus X. Nematodes in particular do not retain viruses over long periods of time and often become viruliferous again only by feeding on infected weed species. Fungal diseases also are often perpetuated by weeds. Volunteer 'weed' crop plants are very important carriers particularly where continuous monoculture is practised e.g. powdery mildew in temperate cereals (Hughes, 1975): sheath blight of rice is carried by many common weed species (Anon 1981), as are many soil-borne pathogens such as Pythium, Fusarium and Rhizoctonia.

Induced Interactions

When a change in a cultural or agrochemical practice is introduced to control one damaging organism it can induce the escalation of another problem. A very good example is described by Hollis (1977); the successful nematicidal control of the ring nematode Criconemoides resulted unexpectedly in reduced rice yields in plots infested with Cyperus escultentus. The control of the nematode, which infected both plant species, stimulated weed growth more than rice growth and the increased weed competition resulted in reduced rice yields. Hollis believes such interactions may be widespread in paddy rice and describes other weed species which may also be involved (Hollis, 1972). Another example, from Southern US, illustrates one of the potential problems of using broad-spectrum agrochemicals. The fungicide benomyl has been used successfully to control several diseases involved in the soybean disease complex but in some areas its use unexpectedly resulted in significant increases in populations of Lepidoptera larvae. This was traced to effects on the entomopathogen Nomuraea rileyi (Johnson et al, 1976) which was naturally controlling the Lepidopteran population.

In integrated insect control programmes, one of the major objectives is to avoid, wherever possible, effects on beneficial arthropod populations. Sivapalan and Delucchi (1972) illustrate how difficult this objective can be to achieve. Insecticidal treatments applied to control the stem borer Xyleborus fornicatus resulted in outbreaks of the hitherto relatively unimportant lepidopteran Homona coffearia; treatments applied to control it in turn favoured outbreaks of mites. Such cascade effects are experienced quite frequently yet they still remain difficult to anticipate.

Some agrochemicals have unexpected side effects in practical use. For example, the herbicides EPTC and dinoseb, which are used to control weeds in leguminous crops, alter the suceptibility of navy beans to the pathogenic fungus Fusarium solani by increasing the exudation of electrolytes, amino acids and sugars from root and hypocotyl tissue (Wyse et al, 1976).

INFLUENCE OF ENVIRONMENTAL AND MANAGEMENT FACTORS

It falls outside the scope of this paper to review the effect of external factors on pest, disease and weed interactions but it is critically important to appreciate their significance. Factors such as the degree of cultivation, irrigation, fertilizer use, crop variety, rotational practice and, of course, agrochemical use have huge effects on crop plants and their complexes of damaging and beneficial organisms. Their importance cannot be over-stressed.

INTERACTIONS DISADVANTAGEOUS TO CROP PRODUCTIVITY

Most of the interactions between pests, diseases and weed necessitate added complexity in control programmes and from this viewpoint are disadvantageous. Examples are the role of weeds as alternate hosts for viral and fungal pathogens, the generation of improved micro-environments by one damaging organism for another and the long distance transmission of viruses by insect vectors. Nevertheless, a detailed understanding of their biology and ecology using a long-term, field-orientated, multi-disciplinary approach is an essential prerequisite for the design of effective integrated control programmes.

INTERACTIONS BENEFICIAL TO CROP PRODUCTIVITY

There is a very heavy bias in the available information on interactions towards those that are unfavourable to crop productivity; the information is relatively sparse on interactions which unequivocally benefit crop productivity and these interactions almost invariably involve weeds as a component of the interaction.

Examples above have illustrated how weeds frequently act as alternative food sources or provide favourable environments for beneficial, predatory or parasitic insects and mites. These examples referred to perennial crops which, of course, are more stable agro-ecosystems than arable cropping systems which are always in an early establishment stage. Nevertheless, some weeds in arable crops have been shown to have important effects in reducing pest incidence. Altieri (1977) describes how the presence of the grass weeds Eleusine indica and Leptochloa fatiformus reduced the incidence of whitefly (Empoasca kraemeri) on beans (Phaseolus vulgaris) even when the weeds were only present as a border around plots. The effects were due to confused gustatory or olfactory function in the whitefly. In the UK, Vickerman (1976) showed that there were fewer non-specific predators of cereal aphids in weed-free fields and generated some evidence that the cereal aphids themselves were fewer in number. Very similar trends have been shown by Altieri (1980) for the incidence of Spodoptera frugiperda and its insect predators in maize.

Weeds also have a significant effect upon the visual attractiveness of a crop to pests. Flying pests are usually attracted to the yellow-green colours of spaced-out plants silhouetted against bare soil; weeds act as camouflage and lesson the contrast as illustrated by Smith (1976) for aphids in brassicae. In developed agriculture, it appears that the increasing trends towards drilling crops to a stand, to extensive monocultures and to weed-free crops are increasing the attractiveness of crop plants to pests. This contrasts with mixed cropping systems in developing agricultural systems where the aim is to maintain a continuous and varied crop canopy.

The benefits of some weeds in reducing pest attack in specific circumstances are, therefore, established; there is no doubt that it is an area warranting further study. The benefits must be set against the costs, however, of permitting weed growth in or around crops; alternative crop plants should also be considered. Weeds reduce crop yield by direct competition, they can harbour pests and disease, they can have allelopathic effects (Putnam and Duke, 1978), and they can seriously impede harvesting. Much more work is required and progress is likely to be slow because of its long-term nature and the likely variability in results depending upon environmental conditions. From an ecological viewpoint, mixed cropping systems appear favorable but these are often incompatible with the use of modern selective herbicides in particular.

Crop production is an industry and farmers employ the most cost-effective available solutions to problems provided that they are judged to be acceptably safe in the environment. Farmers tend to prefer broad-spectrum agrochemicals because they will often be cheaper than several more specific products, even though the latter will often have less effect on non-target organisms. However, if it is essential economically to control a wide range of pests, disease or weed organisms, which is very frequently the case in practice, the environmental impact of a large number of specific products may well be no less than that of fewer broad-spectrum products.

Another general trend in agricultural systems is towards increasing mechanisation and, associated with it, the removal of field boundaries to facilitate operations. Again, economics is the driving force and arguments about the loss of breeding grounds for beneficial insects, for example, have to be assessed against that background (as well as the opposing technical view that hedgerows harbour weed species which can re-infest fields and/or act as alternate hosts for pests and pathogens).

The future, undoubtedly, will see continuous progression towards integrated pest management. Manipulating the pest balance, to encourage those interactions which stabilise and reduce the incidence of damaging organisms is a key objective. Progress will be based on long-term, multi-disciplinary programmes and is likely to be slow. Most importantly the biological components of the programmes will have to be consistent in effect and economic in use if they are to be adopted widely; their importance must be assessed not in isolation but as components of the systems of which they are a part.

ACKNOWLEGEMENTS

My thanks are due to many colleagues in the Plant Protection Division of Imperial Chemical Industries PLC for helpful suggestions and especially to Mr. Richard Waterman for his searches of the technical literature.

REFERENCES

Altieri, M.A., Schoonhover, A. van, Doll, J. (1977). The ecological role of weeds in insect pest management systems; a review illustrated by bean (Phaseolus vulgaris) cropping systems. PANS, 23, 195-205.

Alteiri, M.A. (1980). The design of pest stable corn agro-ecosystems based on the manipulation of insect populations through weed management. Dissertation Abstracts International 41, 59.

Anon (1981). Control and management of rice pests. International Rice Research Institute Annual Report 1980, Los Banos, Laguna, Philippines, 153-234.

Anon (1982). GIFAP Directory 1982 - 1983. Groupement International des Associations Nationales de Fabricants de Products Agrochimiques, 1180 Brussels, Belgium.

Hollis, J.P. (1972). Nematicide - weeds interaction in rice fields. Plant Disease Reporter, 56, 420-424.

Hollis, J.P. (1977). Loss in yield of rice caused by elimination of yellow nutsedge, Cyperus esculentus. Nematologica, 23, 71-78.

Hughes, R.G. (1975). Cereals as weeds. Proc. 12th British Weed Control Conference, Brighton, 1974, 1023-1029.

Johnson, D.W., Kish, L.P., Allen, G.E. (1976). Field evaluation of selected pesticides on the natural development of the entomopathogen, Nomuraea rileyi, on the velnetbean caterpillar in soybean. Environmental Entomolgoy, 5, 964-966.

Kraft, P.L., Roquebert, M.F. (1981). Analysis of interactions between two antagonistic fungi: Trichoderma viride Pers. and Botrytis cinerea Pers. Preliminary Studies. Cryptogamia, Mycologie, 2, 137-151.

LeCato, G.L. (1975). Interactions among four species of stored product insects in corn: a multi-factorial study. Ann. Entomological Soc. Am. 68, 677-679.

Localetti, E.A. (1978). Tobacco rattle virus: weed hosts, diagnosis, characterization, purification and interaction with potato virus X. Dissertation Abstracts International, 38, 5118-5119.

Munnecke, D.E., Kolbezen, M.J., Wilbur, W.D., Ohr, H.D. (1981). Interactions involved in controlling Armillaria mellea. Plant Disease, 65, 384-389.

Pollet, A. (1978). The pests of rice in the Ivory Coast; interaction between Maliarpha separatella and Pyricularia oryzae. Zeitschrift fur Angewandte Entomologie, 85, 324-327.

Putnam, A.R., Duke, W.B. (1976). Allelopathy in agrocosystems. Annual Review of Phytopathology, 16, 431-451.

Sauer, R., Schonbeck, F. (1976). Investigations on the influence of nitrogen and ethirimol on the infection of barley by Helminthosporium sativum. Zeitschrift for Pflanzenkrankheiten and Pflanzenschutz, 83, 519-528.

Sivapalan, P., Delucchi, V. (1972). Integrated approach in tea pest control in Sri Lanka. Proc. FAP Conference, Ecology in relation to Plant Pest Control, Rome, December, 1972, 183-196.

Smith, J.G. (1976). Influence of crop background on aphids and other phytophagous insects on brussel sprouts. Ann. Appl. Biol., 83, 1-13.

Smith, E.S.C. (1981). An integrated control scheme for cocoa pests and diseases in Papua New Guinea. Tropical Pest Management, 27, 351-359.

Stapley, J.E. (1973). Insect pests of coconuts in the Pacific Region. Outlook on Agriculture, 7, (S), 211-217.

Tamaki, G. (1981). Exploiting the ecological interaction of the green peach aphid on peach trees. Technical Bulletin, Science and Education Administration, USDA, No. 1640.

Thresh, J.M. (1981). Ed. Pests, Pathogens and Vegetation. London: Pitman.

Vickerman, G.P. (1974). Some effects of grass weed control on the arthropod fauna of cereals. Proc. 12th British Weed Control Conference, 3, 929-939.

Wyse, D.L., Meggitt, W.F., Penner, D. (1976). Herbicide root rot interaction in navy bean. Weed Sci., 24, 16-21.

THE BIOCHEMICAL BASIS OF RESISTANCE IN HOST PLANTS TO INSECT PESTS

M.D. Pathak[1] and D. Dale[2]

Director, Research and Training[1], Senior Research Fellow[2]
The International Rice Research Institute
P.O. Box 933, Manila, Philippines

ABSTRACT

Resistant varieties of cultivated crops are grown throughout the world either as a primary method or as an adjunct to other insect control measures. In most cases, the host plant resistance (HPR) is of biochemical nature. However, only a few in-depth investigations on this subject have been undertaken. Lack of appropriate analytical and bioassay methods have been often the limiting factors to these studies. The phytochemicals involved in HPR mostly belong to groups like acetogenins, alkaloids, flavonoids, glycosides, isoprenoids, lignins, etc. They act as feeding deterrents, growth inhibitors, toxicants, ovipositional deterrents and repellents. Such chemicals could be of great practical significance as commercial pesticides because of their selective and non-polluting characters. Information about these chemicals also will aid the evaluation of breeding lines so as to further increase (in cultivated crops) the levels of resistance to insect pests.

KEYWORDS: Biochemical basis, Plant resistance, Nonpreference (Antixenosis), Antibiosis, Stimulant, Deterrent, Toxicant, Nutrient.

INTRODUCTION

Studies on host plant resistance (HPR) to insect pests have gained momentum during the last two decades as its potential as a practical and virtually an ideal method for pest control has come to be recognized. Consequently, much of the research on such specialized aspects as the biochemical base of HPR is of quite recent origin. Few of these studies have been of systematic or exhaustive nature and most have touched only fragments of the problem because of the complexity of factors contributing to resistance and often due to lack of close collaboration between entomologists and organic chemists. Also, the urgency of the knowledge of the chemical basis of HPR was not felt as it was not essential for breeding resistant varieties. For studies relating to the chemical basis of resistance, a detailed analysis of the various aspects of insect-host plant interaction is a prerequisite. Furthermore, the situation is more complicated because HPR may often be governed by different chemicals in different varieties.

The important role in practical pest management played by host resistance has been demonstrated for several crop plants in recent years. Resistant varieties are proving to be a practical method of pest control over large areas. For example, varieties resistant to the brown planthopper, Nilaparvata lugens, are presently occupying about 15 million ha and the level of resistance is such that no pesticide application is needed to protect the crop from this noxious pest. Furthermore, it has also been indicated that the chemicals responsible for resistance to insects in certain rice varieties have the potential of becoming commercial pesticides. The identification and characterization of these chemicals also can help in the breeding of varieties having higher concentration, and thus greater resistance. Thus, many scientists are intensively investigating the biochemical nature of host resistance.

Insect resistant plant varieties have the unique advantage of exerting insect control at all levels of pest infestations, and they do not involve any extra expense to the cultivators or cause environmental pollution problems. Furthermore, generally, varietal resistance is compatible with other methods of pest control. Thus, such resistant varieties are useful in integrated pest management programs which are receiving progressively greater attention by scientists and farmers.

Painter (1951) defined varietal resistance as "relative amount of heritable qualities possessed by the plant which influences the ultimate degree of damage done by the insect. In practical agriculture, it represents the ability of a certain variety to produce a larger crop of good quality than do ordinary varieties at the same level of insect infestations". Varietal resistance is subclassified as follows:

(a) Nonpreference (Antixenosis) - when a plant possesses characteristics that make it unattractive to insect pests for oviposition, feeding or shelter.
(b) Antibiosis - when the host plant adversely affects the bionomics of the insects feeding on it.
(c) Tolerance - when the damage to the host plant is only slight despite its supporting an insect population of a size sufficient to damage susceptible hosts severely.

The nonpreference, antibiosis and tolerance aspects of resistance result from a series of interactions between insects and plants which influence the selection of plants as hosts and the effects of plants on insect survival and multiplication. Establishment of insects on plants involves orientation, feeding, metabolic utilization of ingested food, growth, survival, egg-production, oviposition and hatching of eggs (Saxena, 1969; Saxena et al., 1974). The capacity of a plant to cause the interruption of any of these may account for its resistance to insect infestation. Such interruptions may be caused by the biochemical or biophysical characteristics of a plant. However, often such responses are due to several factors and only a few investigators have been able to identify the exact role of these plant characters in the resistance of the host plant to a particular insect.

Some insect-resistant cultivars that have been used for breeding are given in Table 1. The principles of insect resistance in crop plants and early work in this aspect have been reviewed by Painter (1951, 1958). Recent developments on HPR are reviewed by Gallun et al. (1975), Pathak and Saxena (1976), and Maxwell and Jennings (1980). This paper discusses the chemicals in plants influencing the nonpreference (antixenosis) and antibiosis types of resistance. At present, there is no evidence that the tolerance of the plants to insect attacks is influenced by any chemical.

ANTIXENOTIC CHEMICALS

Differences in the concentrations of a wide variety of chemicals in plants have been recorded to have profound effect on the host-plant selection, feeding and ovipositional behaviour of the insect pests. An abbreviated list of chemicals involved in resistance to insects is given in Table 2. These chemicals can be categorized as follows, based on their influence on insect behaviour:

Feeding Stimulants

Most of the feeding stimulants are glycosides, organic acids, flavonoid aglycones, carbonyls, phospholipids or terpenoids. They are generally 'secondary plant substances' and are not known to possess primary functions for the plant or the insects (Fraenkel, 1959). However, certain principal nutrients also stimulate the feeding activity of the insects, such as sugars and amino acids for locusts (Cook, 1977), and asparagine for the brown planthopper (Sogawa and Pathak, 1970).

Table 1: Some Insect-resistant Cultivars/lines Used For Breeding Commercial Varieties

Crop	Insect	Varieties	Nature of Resistance
Alfalfa	Pea aphid	Apex, Washoe, Dawson, Mesilla	Antibiosis
	Spotted alfalfa aphid	Mesa-Sirsa, Zia, Moapa	Antibiosis, nonpreference, tolerance
Cotton	Boll weevil	Ak Djura, NC Margin	Nonpreference, antibiosis
Maize	Corn earworm	Zapalote Chico	Antibiosis, nonpreference
	Corn rootworm	SD_{10}, B_{54}, G_{67}, B_{69}	Nonpreference
	European corn borer	B_{52}	Antibiosis
Rice	Brown planthopper	Mudgo, ASD7, IR26	Antibiosis, nonpreference
	Rice gall midge	Eswarakora, MTU15, HR42, HR63, Ptb18, Ptb21, JBS446, JBS673, Siam 29	Antibiosis
	Rice whorl maggot	IR20, CR94-13, IR40	Antibiosis
	Striped stem borer	TKM6, Taitung 16,	Antibiosis,
	Yellow stem borer	Chianan 2	nonpreference
	White-backed planthopper	N22, ARC6003, Dharia	Antibiosis, nonpreference
Sorghum	Chinch bug	'Atlas' sorgo	Tolerance, antibiosis
	Green bug	IS809, KS30, SA7536-1	Antibiosis
	Sorghum midge	AF28	Nonpreference
Wheat	Hessian fly	W38, PI 94587, Ribeiro	Antibiosis, nonpreference

Most of the studies conducted so far have shown that a complex mixture of plant chemicals rather than a single chemical factor influences the feeding activity of the insects. A classical example is the silkworm that feeds as a result of sequential production of stimuli by the mulberry leaves (Hamamura, 1970). However, in a few cases, single chemicals are involved in the resistance of the variety to the insect pests. Peng et al. (1979) attributed the resistance of rice varieties Nan You 6 and Vei You 6 to the brown planthopper to lower contents of aspartic acid, asparagine, valine, alanine and glutamic acid which act

Table 2: List of Chemicals Involved in Resistance to Insects in Crop Varieties

Host Plant	Chemical(s)	Insect Pest	Biological Effect
Alfalfa	Coumestrol	Pea aphid Spotted alfalfa aphid	Feed deterrent
	Saponins	White grubs	Growth inhibitor
	Medicagenic acid	Pea aphid Potato leafhopper	Toxicant
Corn	6-MBOA DIMBOA	European corn borer	Feeding deterrent Growth inhibitor
Cotton	Gossypol, Quercetin Rutin	Bollworm	Growth inhibitor
Crucifers	Allyl isothiocyanate	Imported Cabbageworm	Attractant
Cucurbits	Cucurbitacins	Spotted cucumber beetle	Feeding stimulant
Potato	Demissine, Tomatin Dihydro-L-solanin Solaculin Soladulcidin	Colorado potatoe beetle	Feeding deterrent
Rice	Oryzanone	Striped stem borer	Attractant
	Low asparagine contant	Brown planthopper	Feeding deterrent
	Compound A	Stem borer	Ovipositional deterrent
Soybean	Saponins, Urease	Bruchid beetle	Ovipositional stimulant
	Sapogenin	Bruchid beetle	Ovipositional stimulant
Tomato	Tomatin	Colorado potato beetle	Repellent
Wild Tomato	2-tridecanone	Tobacco hornworm	Toxicant

as phagostimulants. There were 3 to 5 times differences in the concentration of these amino acids between resistant and susceptible varieties. The concentration of 4-amino-butyric acid, which inhibited feeding by the insects was higher in resistant varieties. Dual action of the same chemical as a phagostimulant and a deterrent has been reported in the case of sinigrin (Nault and Styer, 1972). It has been shown to be a feeding stimulant in the case of cabbage aphid, Brevicoryne brassicae, but a powerful deterrent to feeding by the pea aphid, Acyrthosiphon pisum, an insect that feeds on only plants belonging to the family Leguminosae.

Feeding Deterrents

A feeding deterrent inhibits feeding by the insects. It does not kill the insects directly but they may die through starvation. Chapman (1974) made a comprehensive review of the chemicals that inhibit feeding of phytophagous insects. Phlorizin, a phenolic compound has been reported to act as a feeding deterrent for the green peach aphid, Myzus persicae, and Amorphophora agatonica (Montgomery and Arn, 1974) in apple. Yoshihara et al. (1980) recorded a higher concentration of oxalic acid in the resistant rice variety Mudgo to inhibit feed by brown planthopper.

The chemical constituents in sorghum responsible for resistance to grasshopper feeding have been extensively investigated by Woodhead and Bernays (1978). They showed that resistance in young plants was mainly due to dhurin, a cyanohydrin glucoside. Some phenolic acids on sorghum were found to occur as insoluble esters of cell wall polysaccharides. While the derivatives did not affect the feeding activity, the free acids, as a mixture, had a pronounced inhibitory effect. It was also shown that sorghum contained a few hydrolases which could convert inactive phenolic esters into active forms at such high concentrations as to reduce the feeding activity of grasshoppers (Woodhead and Cooper-Driver, 1979).

Ammonium nitrate was isolated by Akeson et al. (1969) from a resistant sweet clover species, Melilotus infesta, and was implicated as a feeding deterrent to sweet clover weevil Sitona cylindricollis. Dreyer and Jones (1981) have tested a number of naturally occurring flavonoids for their feeding deterrent activity against two aphid species, Schizaphis graminum and M. persicae. Most flavonoids showed strong deterrence at concentrations well within the range often found in plants. Three major greenbug feeding deterrents - p-hydroxy-benzaldehyde, dhurin and procyanidin - were recently isolated from sorghum leaves (Dreyer et al., 1981). It is also speculated that the concentrations of deterrent principles may vary in different parts of the plant. For example, jack pine sawflies, Neodiprion swainei and N. rugifrons feed only on the older pine needles, probably because the younger needles contain a feeding deterrent (All and Benjamin, 1975; Ikeda et al., 1977).

Ovipositional Stimulants and Deterrents

It is often reported that ovipositional stimulants act as feeding stimulants or attractants as well. Some plants in a cabbage field were found more attractive to oviposition by Pieris brassicae, because they contained higher than average amounts of volatile allyl nitriles, which attracted the cabbage butterflies (Mitchell, 1977). Stadler and Buser (1982) have assayed various fractions of carrot leaf extracts against the carrot fly and found that methyl-isoeugenol and asarone - two phenyl propanoids - stimulated oviposition.

Maxwell et al. (1969) described the discovery and utilization of an oviposition suppression factor that effectively reduced egg laying by cotton weevil, Anthonomus grandis. This unidentified factor located in Seaberry variety of Sea Island cotton, Gossypium barbadense, has been transferred through breeding into Upland varieties of commercial cotton, G. hirsutum, in which 25 to 40% reduction in oviposition has been recorded. Similarly, it was found that sapogenin in soybeans has ovipositional deterrent activity against bruchid beetle, Callasobruchus chinensis (Applebaum et al., 1965). Studies at the International Rice Research Institute have shown that plants of certain rice varieties resistant to the striped stem borer, Chilo suppressalis, contain an ovipositional deterrent factor while the susceptible varieties contain some ovipositional attractant factor. Spraying of the extracts from these plants on rice varieties dramatically altered the ovipositional behaviour of the borer moths on them.

Attractants and Repellents

Attractants mostly belong to terpenes, alcohols, esters, acids, sulphur-containing substances and phenolics. A chemical, p-methyl-acetophenone, isolated from the rice plant, was recorded to attract both the larvae and the adults of striped stem borer (Kawano et al., 1968). However, difference in its concentration in resistant and susceptible varieties was not investigated. Cucurbitacins, tetracyclic triterpenoids, are feeding attractants to the spotted cucumber beetle, Diabrotica undecimpunctata (DaCosta and Jones, 1971). A positive correlation has been shown between the concentration of the cucurbitacins and the number of beetles attracted to various cucurbit fruits (Sharma and Hall, 1973). A low cucurbitacin content in cotylendonary leaves of pumpkin appeared to impart resistance to the plants against the red pumpkin beetle, Aulacophora foveicollis

(Pal et al, 1978).

Beck (1965) has defined repellents as substances that elicit an oriented response away from their source. The property of volatility is thus inferred for the candidate chemical. Hedin et al. (1977) have listed 12 compounds as repellents: 4 hydrocarbons, 3 acids, 2 phenols, 2 terpenes and 1 alcohol. The paucity of studies on naturally occurring insect repellents is apparent.

ANTIBIOTIC FACTORS

Toxic Chemicals in Host Plants

These include alkaloids, phenolic compounds, flavonoids, terpenoids, etc. A classic example of the presence of physiological inhibitors in resistant host plants is the occurrence in corn inbreds of chemicals toxic to the first brood larvae of European corn borer, Ostrinia nubilalis. One of the potent toxicants was identified as 6-MBOA (6-methoxy-benzoxazolinone) (Smissman et al., 1957). Later, Klun (1965) found a strong correlation between the amount of 6-MBOA in 11 maize inbreds at the whorl stage of development and the field rating for resistance to the borer. Highly resistant inbred lines yielded almost 10 times more chemical than highly susceptible ones. He also suggested that a precursor of 6-MBOA may be biologically more active. Klun et al. (1967) evaluated 2-4-dihyroxy-7-2H-1, 4, benzoxazin-3-one (DIMBOA) which Virtanen (1961) and Wahlroos and Virtanen (1959) had reported as the precursor of 6-MBOA. DIMBOA was bioassayed in an artificial diet and was found to inhibit larval development and to cause larval mortality. The levels of the chemical were too low in all lines at the time of the second brood infestation and none of the corn lines was resistant to the second brood of borers. But several lines from Latin America were resistant to both first and second brood larvae; this resistance being not attributable to DIMBOA (Sullivan et al., 1975; Scriber et al., 1975). The biochemical base of resistance of these Latin American lines is apparently not yet known.

A Mexican maize strain "Zapalote Chico" was reported to have excellent resistance to corn earworm, Heliothis zea; the reasons being suggested were a silk chemical factor and long tight husks (Keaster et al., 1972; Walter, 1962; Widstrom et al., 1972). The silk factor has recently been isolated and identified by Waiss et al., (1979) as maysin (rhamnosyl-6-C-(4-ketofucosyl)-5, 7, 3', 4', tetrahydroxy flavone). The pure chemical severely retarded the growth of earworm larvae when incorporated into the insects' diet. Munakata and Okamoto (1967) reported the presence of benzoic and salicylic acids in resistant rice plants and found them to be toxic to the striped borer larvae. But Das (1976) indicated the possibility of high phenol content in Taitung-16 variety as a factor responsible for borer resistance.

The resistance of rice to gall midge, Orseolia oryzae, appears to be mainly antibiosis. The development of larvae was retarded in resistant varieties whereas it proceeded normally in susceptible varieties (Shastry et al., 1972; Wongsiri et al., 1971). Peraiah and Roy (1979) recorded more free amino acids and phenol but less sugar in the shoot tips of two resistant varieties than in a susceptible check variety tested. The foliage of the wild tomato, Lycopersicon hirsutum f. glabratum is covered with a dense vesture of glandular trichomes that physically entrap tiny arthropods in their secretions, but not larger insects (Gentile et al., 1969). However, the trichome exudate of one of the accessions was toxic to tobacco hornworm, Manduca sexta, and the corn earworm, H. zea. The toxin has been identified as 2-tridecanone, a nonalkaloid chemical (Williams et al., 1980). It was also observed that the cellular fluid of trichomes was particularly rich in the flavonol glycoside rutin, accompanied by lesser amounts of phenolics (Duffey and Isman, 1981). A breeding program is underway to incorporate the glandular trichome containing 2-tridecanone in commercial tomato varieties (Bordner, 1982).

The green peach aphid feeds on the phloem of tobacco (Nicotiania tabacum) plants and thereby avoids a powerful toxin in the xylem tissues (Guthrie et al., 1962). However, the aphid does not survive on N. gossei because the toxin in this species of tobacco exudes from leaf hairs and kills the aphid by contact (Thurston and Webster, 1962). Baker (1978) detected a trypsin-inhibitor which was toxic to the bruchid beetle grubs in the seeds of a variety of cowpea resistant to this insect. A partial characterization indicated the factor to be a mixture of several isoinhibitors, some of which were able to inhibit chymotrypsin as well as trypsin.

Saponins are triterpenoid glycosides that occur in legume seeds and plants. A correlation was reported between the toxicity of certain legume saponin fractions to the larvae of bruchids and the relative resistance of these seeds to insect damage (Applebaum et al., 1969). Saponins have also been implicated as possible resistance factors in alfalfa against white grubs (Horber, 1964), pea aphid, potato leafhopper, and certain other alfalfa insects (Horber et al., 1974; Roof et al., 1972). Hanson et al. (1973) showed that most alfalfa cultivars selected for high saponin concentration had resistance to the pea aphid. However, Pedersen et al. (1976) found that the foliage and root saponin concentrations of pea aphid-resistant and susceptible varieties were not significantly correlated with insect damage.

Three cotton gland pigments - gossypol, quercetin and rutin - when incorporated into the standard bollworm diet, decreased growth of Heliothis sp. (Lukefahr and Martin, 1966). High-gossypol cotton lines (for example, 1.7% in the bollworm-resistant variety, X-G-15, as compared to 0.4% in susceptible variety, DPL-15) have been located among dooryard cottons and appeared to be promising parents in breeding programs (Lukefahr and Houghtaling, 1969). Chan et al. (1978) showed condensed tannin in the flower buds of G. hirsutum as the antibiotic chemical to bollworm.

Certain allelochemics interfere with nutrients by blocking their availability to insects. This seems to be the case with oak leaf tannins which apparently form a complex with proteins such that the proteins are made less available to larvae of winter moth, Operophthera brumata (Feeny, 1970). Similarly, creosote resins also block insect digestive activity (Rhoades and Cates, 1976). Such interactions between diterpene acids and cholesterol have also been demonstrated (Elliger et al., 1976). The allelochemics exert their influence on the assimilation of nutrients and in the efficiency of conversion of ingested food into insect biomass. Shaver et al. (1970) found that gossypol decreased assimilation by bollworm larvae, but had no measurable effect on nutrient utilization by tobacco budworm. Erickson and Feeny (1974) demonstrated that sinigrin reduced assimilation of nutrients by the caterpillars of Black Swallowtail butterfly, Papilio polyxenes asterius, but did not reduce the efficiency with which assimilated food was converted into biomass.

Host Plant Nutrients and Insect Resistance

The role of nutritional factors in host plant resistance is little understood. Although resistance mechanisms involving host nutrition do exist, they are difficult to prove because the net effect is likely to be more quantitative and subtle than other resistance mechanisms. A few examples on the possible role of insect nutrients in HPR are reviewed below.

Quantitative differences in amino acid contents of plants can influence their resistance or susceptibility to their insect pests. Sogawa and Pathak (1970) recorded that the plants of brown planthopper-resistant varieties of rice had lower asparagine content than those of the susceptible varieties. Young planthopper females emerging from nymphs reared on the resistant variety Mudgo, which had lower asparagine titre, had underdeveloped ovaries that contained very few mature eggs, while those reared on susceptible varieties had normal ovaries with a full complement of eggs. Even though there are instances where application of excess nitrogenous fertilizers tends to lower plant resistance, it is not always observed. The asparagine content is believed to be greatly influenced by the amount of nitrogenous fertilizers applied. However, the application of high rates of N did not alter the resistance of Mudgo (Table 3).

Table 3: Effect of Different Levels of Nitrogen Fertilizer on the Reaction of Mudgo and Taichung Native 1 to the Brown Planthopper

	Insect Survival% (after 22 days)		Male:Female Ratio (after 17 days)		Insect Progeny Produced (after 37 days)	
Nitrogen (kg/ha)	Taichung Native 1	Mudgo	Taichung Native 1	Mudgo	Taichung Native 1	Mudgo
0	30	2	1:2.3	1:0.6	4775	11
50	38	0	1:1.4	1:0.7	5139	0
100	44	10	1:1.2	1:0.5	6835	19
150	54	22	1:1.4	1:1.0	8875	85
200	57	18	1:1.6	1:1.0	9363	70

Lower concentration of amino acids was recorded in the varieties of pea resistant to the pea aphid than in the susceptible varieties (Auclair et al., 1957; Auclair, 1963; Srivastava and Auclair, 1974). Sugar balance was found to be important in the resistance of certain crops. Reduction in sugar content of the plant at more critical stages of insect development may cause resistance. For example, only varieties which supplied sufficient amounts of glucose could sustain European corn borer larvae. Conversely, the rest of the varieties were resistant apparently because of the lower content of glucose. The larvae were capable of differentiating even minute changes of glucose concentration (Beck, 1957; Knapp et al., 1966). Insects depend on their food for fats and fatty acids which are the sources of energy, metabolic water and fat reserves. Grison (1958) observed that egg production rates of the Colorado potato beetle were positively correlated to the phospholipid content of potato foliage. Sterols are essential nutrients for all phytophagous insects. Sarin and Sharma (1979) analyzed β-sitosterol content of wheat grains and found that its contents were negatively correlated to the infestation of grain beetle, Tribolium castaneum. Tester (1977) has reported that soybean-resistant cultivars accumulated sterol faster, and by pod-filling stage contained 20-50% more sterol than in the susceptible cultivars.

There is ample evidence on the important role of minerals related to varietal resistance to insects. Barker and Tauber (1954) reported that pea aphid showed lower reproductive capacity on plants deficient in Ca, Mg, N, P and K. Winged forms of the green peach aphid nymphs reared on diets lacking Fe or Zn produced more alate adults, while those reared on diet devoid of P developed mainly into apterous forms (Raccah et al., 1971).

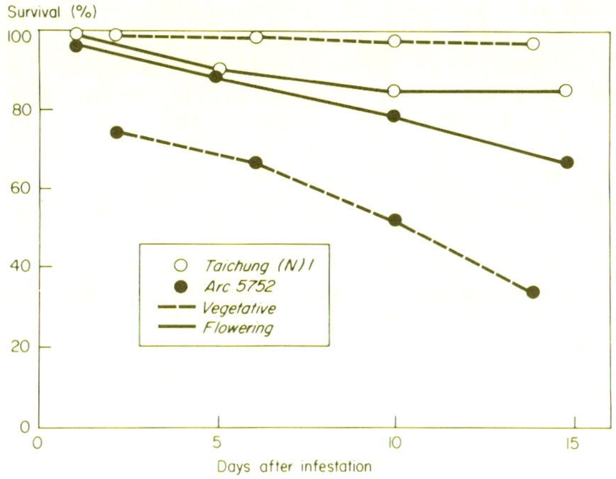

Fig. 1: Survival of first instar nymphs of whitebacked planthopper during the vegetative and flowering periods on two rice varieties.

The aphid polymorphism has very great practical significance under field conditions. It is often observed that young plants lose their resistance when they grow old. A study was carried out at IRRI for assessing varietal resistance of rice to the planthoppers and leafhoppers (Pathak, 1977). Insects caged on resistant varieties generally suffered higher mortality, had slower growth rates and were smaller in size than those reared on susceptible varieties (Figure 1). Varieties resistant at juvenile stage except ARC 5752 were also resistant at later stages.

DISCUSSION AND CONCLUSIONS

In general, the weak background of most entomologists in chemistry and the lack of formal arrangements for close working relationships between entomologists and chemists had been the main constraints in investigations on the biochemical base of insect resistance. Earlier studies in this area consisted of analysis of primary chemicals in the resistant and susceptible plants, and any differences were attributed as the factors of resistance. However, appropriate bioassay techniques have been often difficult to standardize as many insects are difficult to rear on synthetic diets or for evaluating their behavioural responses. Also, HPR is often governed by several factors such as differences in ovipositional response of the insects or survival and rate of their growth on resistant and susceptible plants. Furthermore, appropriate procedures for extraction of the plant and fractionation of the extracts are essential, as even the extracts of susceptible plants are often toxic to the insects or cause feeding deterrence. Also, certain chemicals get metabolized very fast on "wounding" the plants or during extraction, and thus plant extracts analyzed, even when complimented with bioassays, may not show the actual compound responsible for resistance. Careful planning to avoid such problems is essential.

Studies on thermal stability, polarity, interference of other phytochemicals affecting biological activity and degradation under varying conditions are some of the aspects which need primary attention. Seasonal fluctuations have also been reported to alter the concentrations of resistance chemicals. Furthermore, the concentration of the chemicals responsible for resistance frequently varies significantly in different parts of the plant tissues or at different stages of plant growth. The mode of bioassay of the chemicals on insects depends on the type of activity the compound is expected to show. As antixenotic chemicals affect the behaviour of the test insects, it is necessary that sophisticated instruments like the olfactometer, electroantennograph and electrochemical polarograph are used to locate and measure the subtle changes in insect behavioural pattern. The use of an antennograph attached to the gas-liquid chromatograph has been found very valuable for certain insects, but additional studies are required for developing the methodology for other species.

Antibiotic chemicals that can be assayed by topical application or by mixing them in synthetic diets are somewhat easier to determine. Current methodology involves fractionation of the extracts by appropriate column chromatographic methods and by bioassaying each of these fractions. The active fraction is then analyzed for its chemical constitutent. However, even these fractions frequently contain many compounds as evidenced by corresponding number of peaks in the gas chromatograms or mass-spectrometer analysis. Some of these may even be new compounds. To obtain pure samples of these compounds, and to bioassay each of these is time-consuming and expensive.

REFERENCES

Akeson, W.R., F.A. Haskins and H.J. Gorz (1969). Sweetclover weevil feeding deterrent B: Isolation and identification. Science, 163, 293-294.

All, J.N. and D.M. Benjamin (1975). Influence of needle maturity on larval feeding preference and survival of Neodiprion swainei and N. rugifrons on jack pine, Pinus banksiana. Ann. Entmol. Soc.Am., 68, 579-584.

Applebaum, S.W., B. Gestetner and V. Birk (1965). Physiological aspects of host specificity in the Bruchidae - IV. Developmental incompatibility of soybeans for Callosobruchus. J. Insect Physiol., 11, 611-616.

Applebaum, S.W., S. Marco and Y. Birk (1969). Saponins as possible factors of resistance of legume seeds to the attack of insects. J. Agr. Food Chem., 17, 618-622.

Auclair, J.L., J.B. Maltais and J.J. Cartier (1957). Factors in resistance of peas to the pea aphid, Acyrthosiphon pisum (Harr.) II. Amino Acids. Can. Entomol., 69, 457-464.

Auclair, J.L. (1963). Aphid feeding and nutrition. Ann. Rev. Entomol., 8, 439-490.

Baker, A.M.R. (1978). Protease inhibitors and the biochemical basis of insect resistance by Vigna unguiculata. Thesis, Durham University, UK, 205.

Barker, J.S. and O.E. Tauber (1954). Fecundity of the pea aphid on garden pea under various combinations of light, moisture, and nutrients. J. Econ. Entomol., 47, 113-116.

Beck, S.D. (1957). The European corn borer, Pyrasuta nubilalis (Hbn.), and its principal host plant. VI. Host plant resistance to larval development. J. Insect. Physiol., 1, 158-177.

Beck, S.D. (1965). Resistance of plants to insects. Ann. Rev. Entomol., 10, 207-232.

Bordner, J. (1982). Paper presented in the Symposium at Las Vegas meeting of the American Chemical Society (unpublished). Quoted from Science, 216, 722-723.

Chan, B.G. A.C. Waiss, Jr. and M. Lukefahr (1978). Condensed tannin, an antibiotic chemical from Gossypium hirsutum. J. Insect Physiol., 24, 113-118.

Chapman, R.F. (1974). The chemical inhibition of feeding by phytophagous insects: A review. Bull. Ent. Res., 64, 339-363.

Cook, A.G. (1977). Nutrient chemicals as phagostimulants for Locusta migratoria (L.). Ecol. Ent., 2, 113-121.

DaCosta, C.P. and C.M. Jones (1971). Cucumber beetle resistance and mite susceptibility controlled by the bitter gene in Cucumis sativus L. Science, 172, 1145-1146.
Das, Y.T. (1976). Some factors of resistance to Chilo suppressalis in rice varieties. Entomol. Exp. Appl., 20, 131-134.
Dreyer, D.L. and K.C. Jones (1981). Feeding deterrency of flavonoids and related phenolics towards Schizaphis graminum and Myzus persicae: Aphid feeding deterrents in wheat. Phytochemistry, 20, 2489-2493.
Dreyer, D.L., J.C. Reese and K.C. Jones (1981). Aphid feeding deterrents in sorghum: Bioassay, isolation and characterization. J. Chem. Ecol., 7, 273-284.
Duffey, S.S. and M.B. Isman (1981). Inhibition of insect larval growth by phenolics in glandular trichomes of tomato leaves. Experientia, 37, 574-576.
Elliger, C.A., D.F. Zinkel, B.G. Chan and A.C. Waiss, Jr. (1976). Diterpene acids as larval growth inhibitors. Experientia, 32, 1364-1366.
Erickson, J.M. and P. Feeny (1974). Sinigrin: A chemical barrier to the black swallowtail butterfly, Papilio polyxenes. Ecology, 55, 103-111.
Feeny, P. (1970). Seasonal changes in oak leaf tannins and nutrients as a cause of spring feeding by winter moth caterpillars. Ecology, 51, 565-581.
Fraenkel, G. (1959). The raison d'etre of secondary plant substances. Science, 129, 1466-1470.
Gallun, R.L., K.J. Starks and W.D. Guthrie (1975). Plant resistance to insects attacking cereals. Ann. Rev. Entomol., 20, 337-357.
Gentile, A.G., R.E. Webb and A.K. Stoner (1969). Lycopersicon and Solanum spp. resistant to the carmine and the two-spotted spider mite. J. Econ. Entomol., 62, 834-836.
Grison, P. (1958). L'influence de la plante-hôte sur la fecondite de l'insecte phytophage. Ent. Exp. Appl., 1, 73-93.
Guthrie, F.E., W.V. Campbell and R.L. Baron (1962). Feeding sites of the green peach aphid with respect to its adaptation to tobacco. Ann. Entomol. Soc. Amer., 55, 42-46.
Hamamura, Y. (1970). The substances that control the feeding behaviour and growth of the silkmoth, Bombyx mori. In Control of Insect Behaviour by Natural Products, D.L. Wood, R.M. Silverstein, and M. Nakajima, Eds., Academic Press, Inc., New York, 55-80.
Hanson, C.H., M.W. Pederson, B. Berang, M.E. Wall, and K.H. Davis (1973). The saponins in alfalfa cultivars. In "Antiquality Components of Forages", A.G. Matches, Ed., Special Publ. No. 4, Am. Soc. Agron., Madison, WI.
Hedin, P.A., L.R. Miles, A.C. Thompson and J.P. Minyard (1968). Constituents of a cotton bud - Formulation of a boll weevil feeding stimulant mixture. J. Agr. Food Chem., 16, 505-513.
Horber, E. (1964). Isolation of components from the roots of alfalfa (Medicago sativa L.) toxic to the white grubs (Melolontha vulgaris F.). Proc. 12th Int. Congr. Entomol. (London), 540-541.
Horber, E., K.T. Leath, B. Berrang, V. Marcarian and C.H. Hanson (1974). Biological activities of saponin components from Du Puits and Lahontan alfalfa. Ent. Exp. Appl., 17, 410-424.
Ikeda, T., F. Matsumura and D.M. Benjamin (1977). Chemical basis of feeding adaptation of pine sawflies Neodiprion rugifrons and Neodiprion swainei. Science, 197, 497-499.
Kawana, T., T. Saito and K. Munakata (1968). Study on an attractant of the rice stem borer, Chilo suppressalis Walker. Botyu-Kagaku, 33, 122-130.
Keaster, A.J., M.S. Zuber and R.W. Straub (1972). Status of resistance to corn earworm. Proc. North. Cent. Branch Entomol. Soc. Am., 27, 95-98.
Klun, J.A. (1965). Role of 6-methyoxybenzoxazolinone (6-MBOA) in maize inbred resistance to first-brood European corn borer, Ostrinia nubilalis. Diss Abstr., 29, 3544.

Klun, J.A., C.L. Tipton and T.A. Brindley (1967). 2, 4-dihydroxy-7-methoxy-1, 4-benzoxazine-3-one (DIMBOA), an active agent in the resistance of maize to the European corn borer. J. Econ. Entomol., 60, 1529-1533.

Knapp, J.L., P.A. Hedin and W.A. Douglas (1966). A chemical analysis of corn silk from single crosses of dent corn rated as resistant, intermediate, and susceptible to the corn earworm. J. Econ. Entomol., 59, 1062-1064.

Lukefahr, M.J. and D.F. Martin (1966). Cotton plant pigments as a source of resistance to the bollworm and tobacco budworm. J. Econ. Entomol., 59, 176-179.

Lukefahr, M.J. and J.E. Houghtaling (1969). Resistance of cotton strains with high gossypol content to Heliothis spp. J. Econ. Entomol., 62, 588-591.

Maxwell, F.G., J.N. Jenkins, W.L. Parrott and W.T. Buford (1969). Factors contributing to resistance and susceptibility of cotton and other hosts to the boll weevil, Anthonomus grandis. Entomol. Exp. Appl., 12, 801-810.

Maxwell, F.G. and P.R. Jennings (1980). Breeding plants resistant to insects. John Wiley and Sons, New York, 683.

Mitchell, N.D. (1977). Differential host selection by Pieris brassicae L. (the large white butterfly) on Brassica oleracea subsp. oleracea (the wild cabbage). Ent. Exp. Appl., 22, 208-219.

Montgomery, M.E. and H. Arn (1974). Feeding response of Aphis pomi, Myzus persicae, and Amphorophora agathonica to phlorizin. J. Insect Physiol., 20, 413-421.

Munakata, K. and D. Okamoto (1967). Varietal resistance to rice stem borer in Japan. In The Major Insects Pests of the Rice Plant, Proc. Symp. Int. Rice Res. Inst., Los Baños, Philippines, 1964. John Hopkins Press, Baltimore, 419-430.

Nault, L.R. and W.B. Styer (1972). Effects of sinigrin on host selection by aphids. Entomol. Exp. Appl., 15, 423-437.

Painter, R.H. (1951). Insect resistance in crop plants. MacMillan, New York, 520.

Painter, R.H. (1958). Resistance of plants to insects. Ann. Rev. Ent., 3, 267-290.

Pal, A.B., K. Srinivasan, G. Bharatan and M.V. Chandravadana (1978). Location of sources of resistance to the red pumpkin beetle, Raphidopalpa foveicollis Lucas amongst pumpkin germplasms. J. Entomol. Res., 2, 148-153.

Pathak, M.D. and R.C. Saxena (1976). Insect resistance in crop plants. Curr. Adv. Plant Sci., 2, 148-153.

Pathak, M.D. (1977). Defense of the rice crop against insect pests. In The Genetic Basis of Epidemics in Agriculture, P.R. Day (Ed.), Ann. N.Y. Acad. Sci., 287, 287-295.

Pedersen, M.W., E.L. Sorenson and M.J. Anderson (1976). A comparison of pea aphid-resistant and susceptible alfalfas for field performance, saponin concentration, digestibility and insect resistance. Crop Science, 15, 254-260.

Peng, Z.K., M.Y. Tang and Y.S. Chen (1979). Studies of resistance to brown planthopper in hybrid rice. Scientia Agricultura Sinica, 2, 71-77.

Peraiah, A. and J.K. Roy (1979). Studies on biochemical nature of gall midge resistance in rice. Oryza, 16, 149-150.

Raccah, B., A.S. Tahori and S.W. Applebaum (1971). Effect of nutritional factors in synthetic diet on increase of alate forms in Myzus persicae. J. Insect Physiol., 17, 1385-1390.

Rhoades, D.F. and R.G. Cates (1976). Toward a general theory of plant anti-herbivore chemistry. Recent Adv. Phytochem., 10, 168-213.

Roof, M.E., E. Horber and E.L. Sorenson (1972). Bioassay technique for the potato leafhopper Empoasca fabae (Harris). Proc. North Central Branch Meeting Ent. Soc. Amer., 27, 140-143.

Sarin, K. and A. Sharma (1979). Steroidal effect in the resistance offered by wheat varieites to Tribolium castaneum (Herbst). Experientia, 35, 1198-1199.

Saxena, K.N. (1969). Patterns of insect-plant relationships determining susceptibility or resistance of different plants to an insect. Entomol. Exp. Appl., 12, 751-766.

Saxena, K.N., J.R. Gandhi and R.C. Saxena (1974). Patterns of relationship between certain leafhoppers and plants - I. Responses to plants. Entomol. Exp. Appl., 17, 303-318.

Scriber, J.M., W.M. Tingey, V.E. Gracen and S.L. Sullivan (1975). Leaf feeding resistance to the European corn borer in genotypes of tropical (low DIMBOA) and U.S. inbred (high DIMBOA) maize. J. Econ. Entomol., 68, 823-826.

Sharma, G.C. and C.V. Hall (1973). Relative attractance of spotted cucumber beetle to fruits of fifteen species of cucurbitaceae. Environ. Ent., 2, 154-156.

Shastry, S.V.S., W.H. Freeman, D.V. Sehshu, P. Israel and J.K. Roy (1972). Host plant resistance to rice gall midge. In Rice Breeding. IRRI, Los Banos, Philippines, 353-365.

Shaver, T.N., M.J. Lukefahr and J.A. Garcia (1970). Food utilization, ingestion, and growth of larvae of the bollworm and tobacco budworm on diets containing gossypol. J. Econ. Entomol., 63, 1544-1546.

Smissman, E.E., J.B. Lapidus and S.D. Beck (1957). Isolation and synthesis of an insect resistance factor from corn plants. J. Amer. Chem. Soc., 79, 4697.

Sogawa, K. and M.D. Pathak (1970). Mechanisms of brown planthopper resistance in Mudgo variety of rice (Hemiptera: Delphacidae). App. Ent. Zool., 5, 145-158.

Srivastava, P.N. and J.L. Auclair (1974). Effect of amino acid concentration on diet uptake and performance by the pea aphid, Acyrthosiphon pisum (Homoptera: Aphididae). Can. Entomol., 106, 149-156.

Stadler, F. and H.R. Buser (1982). Oviposition stimulants for the carrot fly in the leaf surface wax of carrot leaves. Paper presented at the Vth International Symposium on Insect-Plant Relationships, Wageningen, The Netherlands.

Sullivan, S.L., V.E. Gracen and A. Ortega (1974). Resistance of exotic maize varieties to the European corn borer, Ostrinia nubilalis (Hubner). Environ. Entomol., 3, 718-720.

Tester, C.F. (1977). Constituents of soybean cultivars differing in insect resistance. Phytochemistry, 16, 1899-1901.

Thurston, R. and J.A. Webster (1962). Toxicity of Nicotiana gossei Domin to Myzus persicae (Sulzer). Entomol. Exp. Appl., 5, 233-238.

Virtanen, A.I. (1961). Some aspects of factors in the maize plant with toxic effects on insect larvae. Suomen Kemistilehti, B34, 29-31.

Wahlroos, O. and A.I. Virtanen (1959). The precursors of 6-MBOA in maize and wheat plants: Their isolation and some of their properties. Acta Chem. Scand., 13, 1906-1908.

Waiss, A.C., Jr., B.C. Chan, C.A. Elliger, B.R. Wiseman, W.W. McMillian, N.W. Widstrom, M.S. Zuber and A.J. Keaster (1979). Maysin, a flavone glycoside from corn silks with antibiotic activity toward corn earworm. J. Econ. Entomol., 72, 256-258.

Walter, E.V. (1962). Sources of earworm resistance for sweet corn. Proc. Am. Soc. Hortic. Sci., 80, 485-487.

Widstrom, N.W., B.R. Wiseman and W.W. McMillian (1972). Resistance among some maize inbreds and single crosses to fall armyworm injury. Crop Sci., 12, 290-292.

William, W.G., G.G. Kennedy, R.T. Yamamoto, J.D. Thacker and J. Bordner (1980). 2-tridecanone: A naturally occurring insecticide from the wild tomato, Lycopersicon hirsutum f. glabratum. Science, 207, 888-889.

Wongsiri, T., P. Vungislabutr and T. Hidaga (1971). Study on the ecology of the rice gall midge in Thailand. In Symp. Rice Insects, Proc. Symp. Trop. Agric. Res., Tokyo, Japan, 267-290.

Woodhead, S. and E.A. Bernays (1978). The chemical basis of resitance of Sorghum

bicolor to attack by Locusta migratoria. Entomol. Exp. Appl., 24, 123-144.
Woodhead, S. and G. Cooper-Driver (1979). Phenolic acids and resistance to insect attack in Sorghum bicolor. Biochemical Systematics and Ecology, 7, 309-310.
Yoshihara, T., K. Sogawa, M.D. Pathak, B.O. Juliano and S. Sakamura (1980). Oxalic acid as a sucking inhibitor of the brown planthopper in rice (Delphacidae: Homoptera). Entomol. Exp. Appl., 27, 149-155.

NATURALLY OCCURING PESTICIDES AND THEIR POTENTIAL

R.C. Saxena

International Centre of Insect Physiology and Ecology
P.O. Box 30772, Nairobi, Kenya
(ICIPE Scientist based at International Rice Research Institute,
P.O. Box 933, Manila, Philippines.)

ABSTRACT

Pesticides will remain indispensable to avert crop losses caused by pests in the foreseeable future. However, the problems of pesticide resistance and detrimental effects on nontarget organisms, including man, associated with the large scale use of broad-spectrum pesticides, dictate the need for effective, biodegradable pesticides with greater selectivity. There is renewed interest in naturally occurring pest control agents. This paper reviews the status and potential of pesticides of plant, insect, or microbial origin and their synthetic analogs for insect pest management.

KEYWORDS: Botanical pesticides, insect-based pesticides, microbial pesticides, naturally occurring pesticides.

INTRODUCTION

With insects, pathogens, and weeds destroying more than one-third of world crops (Cramer, 1967) and lowering crop quality, pesticide use has become intrinsic to modern crop protection technology. Without pesticides, estimated crop losses may escalate to 50%, and even higher in developing countries. World use of pesticides is over 2.3 million tons annually. About 34% is applied in North America, 45% in Europe, and 21% elsewhere, primarily in developing countries (Berry, 1979). Most pesticides developed after DDT and 2,4-D, generally have been synthetic, nonselective poisonous chemicals. They have effectively controlled some pest species, but their extensive use has led to serious social and environmental repercussions. Many cases of lethal and sublethal pesticide poisoning of humans have occurred. The poisoning of livestock, fish, wildlife, and other beneficial organisms are hidden costs in pesticide use. Pimentel et al. (1980) estimated the overall environmental and social costs at nearly US$1 billion annually. Pest resurgence associated with pesticidal destruction of natural enemies, and the development of pesticide resistance in pests have required increased doses or more powerful pesticides. That not only is uneconomical but aggravates the problem.

The demand for food and maintenance of public and animal health will not permit significant elimination of broad-spectrum, synthetic pesticides. Clearly, new pesticides will have to meet entirely different standards. They must be pest specific, nontoxic to man and beneficial organisms, bio-degradable, less prone to pest resistance, and relatively less expensive. This has led to re-examining the use of natural pesticides, which are less likely to cause ecological damage. This paper reviews the status and potential of botanical pesticides, microbial pesticides, and insect-based pesticides. Major emphasis is on insecticides because insects are the most numerous and diverse organisms.

BOTANICAL PESTICIDES

Insects have attacked plants since the Devonian period. During this prolonged interaction, many plant taxa have evolved highly sophisticated defense systems, largely a complex array of defense chemicals produced by plants themselves. Plants virtually are "nature's chemical factories", providing practically unlimited natural sources of botanical pesticides. Many of the oldest and most common pesticides, such as nicotine, pyrethrins, and rotenone, were derived from plants. The chemical or pesticide approach had its beginning in the use of botanical materials.

Renewed interest in botanical pesticides is motivated by three major objectives: 1. To encourage traditional use of simple formulations of locally-available plant materials by farmers in developing countries who cannot afford commercial pesticides. 2. To identify sources of new, botanical pesticides for commercial extraction. 3. To elucidate the chemical structure of active principals. Villagers in India traditionally mixed dried neem (Azadirachta indica) leaves in stored grains to prevent pest-caused losses. Now it is almost forgotten. Botanical pesticides extracted on a large-scale may also be used to replace or supplement the activity of existing synthetic pesticides against refractory pests. Structural elucidation of the active constituents may provide further insight into structure-activity relationships. Novel metabolites identified may serve as models for the chemical synthesis of new pesticides with more desirable properties. The greatest spin-off can be expected in the field of insect control.

Plant species reportedly screened for insecticidal properties exceeded 6,000 species by 1971. Of these, more than 2,000 were reported to exhibit measurable to considerable activity (Crosby, 1966, 1971; Jacobson, 1958, 1975). Plants that are amenable to commercial exploitation have generally received greater attention than others.

Nicotine and Other Alkaloids

Long ago, nicotine was identified as the principal toxic component of tobacco, and farmers were using ground tobacco or its water extract to kill insects. Tobacco was known as an insecticide in Europe about 1690, when the tobacco trade flourished between the American colonies and Europe. Nicotine is a fast-acting contact insecticide, which also acts as a fumigant or stomach poison. Nicotine alkaloid occurs in varying amounts in at least 18 Nicotiana sp. (family Solanaceae), but the commercially important species N. tabacum and N. rustica contain up to 6 and 18% nicotine. Some unrelated taxa such as Asclepias syriaca, Atropa belladonna, Equisetum arvense, and Lycopodium clavatum also contain nicotine. Although nicotine is highly toxic to insects, it is not as widely used as newer synthetic insecticides. Nicotine may present less of a residue problem because of its relatively high volatility and susceptibility to degradation. However, it has little advantage over synthetics because of its relatively high production costs; disagreeable smell; and insecticidal activity limited mostly to minute, soft-body insects, such as aphids. It may be possible to develop synthetic analogs of nicotine that have the required mode of action. Related alkaloids such as nornicotine and anabasine, found in a wide range of plants, also have limited insecticidal use.

Pyrethrum

Pyrethrum, derived from flowers of Chrysanthemum cinerariaefolium (family Compositae), is an outstanding example of a safe and effective natural insecticide. It has been used to control insects in the Middle East since ancient times. Pyrethrum was introduced to Europe, the USA, Japan, East Africa, and South America in the 19th century. The postwar surplus of DDT dealt a severe blow to the pyrethrum market. However, as insects developed resistance to DDT and other insecticides, and the hazards of their continued use were recognized, the pyrethrum demand rose again. Worldwide annual production of pyrethrum now averages 30,000 t. A 1972 international symposium gives an excellent review of pyrethrum (Casida, 1973).

The six insecticidal constituents in pyrethrum extract are pyrethrins I and II, cinerins I and II, and jasmolins I and II. Pyrethrin I is the most effective. Insect susceptibility to pyrethrum is attributable to the cuticular permeability and sensitivity of internal tissue "receptors" that control oxidative enzyme systems. Pyrethrins act on the central nervous system. The insecticidal activity of pyrethrins is markedly increased by sesame oil. Pyrethrum products are commonly applied as oil- or water-based sprays containing from 0.03 - 0.1% pyrethrins and 5 to 10 times as much synergists. Dusts are made by mixing pyrethrum extracts with clay or talc carriers, however, they deteriorate fast. There are five major uses of pyrethrum products: 1. household insecticides to control flies, mosquitoes, bedbugs, ants, roaches, etc., 2. livestock or cattle sprays against biting and non-biting flies, 3. sprays for mills, warehouses, and stored grains, 4. dusts or sprays against pests of vegetable and fruit crops, 5. sprays to suppress certain forest defoliators.

Although pyrethrins do not affect certain natural enemies of pests, they are nonpersistent and their effects are short-lived. Pyrethrum does not harm bees, birds, fish, and larger animals, including man. Occasionally, pyrethrum dust causes allergic reactions, but symptoms disappear when the allergen is removed. Low levels of resistance to pyrethrins have been reported in cockroaches, bedbugs, houseflies, cattle ticks, and body lice. These species, however, have high levels of resistance to organochlorines, organophosphates (OP), and carbamate insecticides. The high cost of pyrethrins and their nonpersistent character have limited their use in agriculture, but ultralow volume (ULV) applications of pyrethrum formulations permit rapid treatment of large areas effectively and economically.

Research on chemistry and structure-activity relations of pyrethrum constituents has led to the synthesis and commerical development of a number of pyrethrin-like substances called pyrethroids or synthetic pyrethroids (Elliott, 1977). Some are relatively stable and have greater insecticidal activity and lower mammalian toxicity than the natural esters. Although pyrethroids are relatively expensive to manufacture, their greater insecticidal activity, permitting fewer treatments at lower doses, gives them an advantage over conventional pesticides.

Rotenone and Rotenoids

Rotenone and rotenoids represent another class of botanical insecticides. For centuries, natives in many tropical countries used plant material from certain legumes, called toeba or tuba (Derris sp.) in the Malay archipelago, and timbo or cube (lonchocarpus sp.) in Central and South America, as fish poison or to poison arrows or spearheads. In 1848, T. Oxley suggested tuba root to control leaf-eating caterpillars. In 1877, it was mentioned that the Chinese in Singapore had a long tradition of using tuba root as an insecticide. The active ingredient isolated from Derris chinensis was called roten by natives in Formosa. Because it was a ketone, it was named rotenone. At least 10 chemically related rotenoids have been isolated. The commercially important species are D. elliptica and D. malaccensis which have an average rotenone content of 4-5%, and Lonchocarpus utilis and L. urucu containing an average of 8-10% rotenone in dried roots. A few Tephrosia sp. are a good source of rotenoids. Very low in mammalian toxicity, rotenone is extremely active as a contact and stomach poison against several insect species. Rotenone dusts, dispersible powders, and sprays have been used in crop and health protection. Crystalline rotenone is used for mothproofing. Rotenone is slow-acting, often requiring several days to kill. The toxic principals deteriorate rapidly in sunlight and air. The complicated structure and the high cost of commercial synthesis have ruled out rotenone as a competitor to the modern synthetic insecticides. Further investigations into the chemistry, structure, and the mode of action may lead to the development of simpler, more cost-effective analogs.

Other Botanicals with Pest Control Potential

Derivatives of some plants have had more temporary to restricted use in insect control, while the majority remain items of folklore, or narrow regional interest. The insecticidal activity of a few of these plants has been studied, but even after a long history of pest control potential, information on their chemistry and biological activity is scanty. This paradox can be attributed to researchers' earlier predilection for botanicals having a distinctive toxicity rather than those that subtly alter the pest's behaviour and physiology. Only recently have the potentials of behavioural and physiological aberrations been recognized. In some situations, they may be highly desirable as they minimize the risk of exposing the pests' natural enemies to poisoned food or starvation. The pest control potential of plants such as neem, chinaberry (Melia azedarach) and a few others that are more relevant to developing countries will be highlighted.

Neem "Bitters"

Centuries before synthetic insecticides became available, farmers in India protected crops with natural repellents found in the fruit and leaves of the neem tree A. indica (family Meliaceae) (Figure 1). The First International Neem Conference, in Germany in 1980, led to a useful book on the potential of natural pesticides from neem (Schmutterer et al., 1981). The Second International Neem Conference will be held in German in 1983. Although neem does not occur in any European or western country, these conferences indicate the importance of developing natural insecticides based on neem. Scientists in England, Germany, India, Israel, Philippines, and the USA are investigating the insect-repellent, antifeedant, and growth-disturbing qualities of neem oil extracted from the tree's fruit, and of neem cake from the residue. An American company manufactures neem dust and spray insecticides. They are undergoing the tier tests for toxicity required by the U.S. Environmental Protection Agency (EPA). Jotwani and Srivastava (1981a, b, c) reviewed the pesticide potential of neem.

The active principals or the "bitters" in neem have been identified as limonoids, a group of stereochemically homogeneous tetranorterpenoids. These occur predominantly in the family Meliaceae and also in Rutaceae. Azadirachtin, the best known neem and chinaberry constituent, has shown selective activity against many insect pest species affecting agriculture and health. In general, insects feed far less, grow poorly, and lay significantly fewer eggs on plants treated with neem oil, neem cake, or azadirachtin. Effects are similar for sucking insects such as the rice brown planthopper, Nilaparvata lugens, and for

Figure 1: Neem (A. indica A. Juss) is widely grown as a shade tree in Asia and Africa. It has insect-repellent, antifeedant, and medicinal properties. Approximately 14 million neem trees grow in India alone. Annual neem oil production in India is about 83,000 tons.

chewing insects such as the rice leaf-folder, Cnaphalocrocis medinalis, and the ear-cutting caterpillar, Mythimna separata (Saxena et al., 1981a, b; IRRI 1982a). Contact with or ingestion of neem oil sprays or its purified extracts or fractions disrupts insect growth (Figure 2). However emergence of parasites is not affected. Azadirachtin has been reported to act as an insect ecdysis inhibitor (Kubo and Klocke, in press), but its other effects on insect behaviour and physiology are equally important.

Neem cake mixed with urea significantly increased yields of an insect-susceptible rice selection in both wet and dry season at the International Rice Research Institute (IRRI) farm by reducing the incidence of ragged stunt, grassy stunt, and tungro virus diseases (IRRI, 1981). The reduction was probably due to the antifeedant effect on hoppers, the main virus vectors. Neem cake has also reduced the population of ostracods - crustaceans that feed actively on nitrogen-fixing bluegreen algae - thus encouraging algal growth and subsequent nitrogen fixation (IRRI, 1982b).

A neem tree becomes fully productive in about 10 years. One tree produces 30-50 kg fruits/year. Thirty kilograms of seeds yield 6 kg of oil and 24 kg neem cake. Neem oil sells in India for about US$1/kg; neem cake sells for less than US$0.50/kg. At IRRI, five applications of a 25% neem oil emulsion sprayed with an ULV applicator gave adequate protection to a rice crop against the brown planthopper. The cost of the neem oil was about US$5/ha.

Neem is relatively nontoxic to man and other vertebrates. Neem oil is used to make soaps and detergents, and neem derivatives are used in toothpaste and herbal medicines. Neem cake has been used as a feed supplement for cattle, livestock, and poultry (Ketkar, 1976). Honeybees regularly visit neem flowers to collect nectar and they commonly make their combs on neem trees. Improved formulations are necessary before a neem insecticide can be produced commercially. Sunlight degrades neem oil sprayed on plants within a week. The systemic effect of azadirachtin incorporated into soil, however, lasts more than a month.

Figure 2: Selected examples of developmental abnormalities of M. separata when 5th- or 6th-instar larvae were confined for 24 hours on rice leafcuts dipped in different solutions of methanolic neem seed extracts or partially purified fractions; in control, larvae were confined on leafcuts dipped in 250 ppm 'Citowett' solution. a-b) Larvae showing early symptoms of body shrinkage; c) larva showing pronounced body shrinkage; d) larval-pupal monstrosity with larval cuticle patches, head capsule and thoracic legs; posterior half of the body has pupal characteristics; e) pupa with vestiture of larval skin; f) pupa with malformed cuticle and appendages; g) normal pupa; h) adult with reduced, crumpled wings and deformed antennae; i) control: adult with fully developed, expanded wings.

Azadirachtin synthesis is expensive and requires a high technology. Thus, azadirachtin must be extracted from neem seeds. The labor surpluses in some countries where neem is common, however, could make that a positive attribute. The many ecotypes of neem have different concentrations of active ingredients. Standardization is essential to mass production. The first step towards commercial production will be to break neem derivatives into extracts and fractions, and then to evaluate them.

There are some plant extracts that act not only on the insect itself, but attack associated bacterial and fungal pathogens as well. The chemical and biological effects of fractions of the extracts are being studied. Yields of plant extracts of this type are low and their commercial development depends on characterizing their chemical structure to determine if they can be synthesized.

MICROBIAL PESTICIDES

Microorganisms or the chemicals they produce to suppress insects and other pests are called microbial pesticides. Being selective and biodegradable, they pose little hazard to nontarget organisms and can be a great asset in integrated pest control. Harnessing of naturally-occurring pathogens is, therefore, receiving much attention. Progress was reviewed by Burges and Hussey (1971). Later developments, including the use of inhibitors and diseases of plant pathogens as alternatives to chemical fungicides, and bactericides, were reviewed by Burges (1981). Papavizas (1981) reviewed mainly strategies of microbial pest control.

Microbial Insecticides

Insects also suffer from diseases caused by bacteria, viruses, fungi, nematodes, and protozoa. Insect populations are often suppressed and in some cases eliminated. More than 1,000 pathogens from arthropods were described by Ignoffo (1970), and the number has nearly doubled since then (NAS, 1979). Entomopathogens are associated with all major groups of insects, and a number of them have been commercialized as microbial insecticides (Table 1). Several types of pathogens may affect a particular insect pest. For example, species of *Heliothis* are susceptible to bacteria, viruses, and protozoa (Ignoffo, 1975).

Bacteria

From a large assemblage of bacteria pathogenic to insects, two are produced as microbial insecticides: *Bacillus popilliae*, the causative organism in milky diseases in scarabaeid beetle, and *B. thuringiensis*, the crystal-forming bacteria that infect a wide-range of Lepidoptera. Several urea-forming strains of *B. sphaericus* are somewhat toxic to mosquito larvae and are being considered in the World Health Organization mosquito control program. *B. popilliae* is best known for its role in the control of the Japanese beetle *Popillia japonica* in the eastern USA. After Dutky (1940) first described milky diseases in the Japanese beetle, the spores of the bacterium are still sold for grub control in lawns and golf greens under the trade name Japodemic or Doom. Japodemic is produced in living grubs, but the strong specificity of various isolates and the cultural requirements of living grubs make the production of milky disease spores expensive. Cell and tissue culture techniques may solve the problem by producing spores in culture media.

B. thuringiensis was first isolated from diseased larvae of the flour moth *Anagasta kuhniella* in 1915, but its development as a microbial insecticide was slow because of taxonomic difficulties. *B. thuringiensis* is a complex species comprising over 20 H-serotypes or subspecies (de Barjac, 1981; Dulmage, 1981). These serotypes produce spores that contain several insecticidal toxins of which β-exotoxin and δ-endotoxin are used in agriculture. The β-exotoxin is a potent insecticide with rather generalized toxicity. Its teratogenic activity in insects and possible mutagenicity has precluded its use in the USA and Canada, but it is still used in the Soviet Union.

The δ-endotoxin or the 'crystal' has valuable insecticidal properties, but the activity is limited to larvae of certain Lepidoptera, mosquitoes, chironomids, and black flies. Fast (1981) reviewed the biogenesis, chemistry, and mode of action of the crystal toxin. The crystal must be eaten to be toxic and the sickness results in gut paralysis. Genetic manipulation may amplify the toxicity

Table 1: Selected Examples of Entomopathogens Commercially Available as Microbial Insecticides.

Pathogen	Brand Name	Country
BACTERIA (*Bacillus* sp.)		
B. moritai	Rabitusu	Japan
B. popilliae	Doom Japidemic	USA
B. thuringiensis (β-exotoxin)	Bitoxibacillin, Eksotoksin, Toxobakterin	USSR
B. thuringiensis (δ-endotoxin)	Bathurin, Chemapol-Biodrma	Czechoslovakia
	Bactospeine, Sporeine	France
	Biospor	Germany
		Japan
	Dendrobacillin, Endobacterin, Insektin, Agritol, Bakthane,	USSR
	Biotrol BTB, Dipel, Parasporin, Thuricide	USA
VIRUSES		
Dendrolimus CPV	Matsukemin	Japan
Heliothis NPV	Biotrol VHZ, Virex, Viron/H	USA
Lymantria NPV	Viron-ensh	USSR
Manostra NPV	Polyvirocide	USA
Pieris NPV	Virin GKB	USSR
Prodenia NPV	Biotrol VPO, Viron/P	USA
Spodoptera NPV	Viron/s	USA
Trichplusia NPV	Biotrol VTN, Viron/T	USA
FUNGI		
Aschersonia aleyrodis	Aseronija	USSR
Beauveria bassiana	Biotrol FBB	USA
	Boverin	USSR
Metarrhizium anisopliae	Biotrol FMA	USA
NEMATODE		
Romanomermis culicivorax	Skeeter Doom	USA

and broaden the host-range of *B. thuringiensis*. It is produced by fermentation either on semisolid or liquid medium and its products are sold under a dozen names in different countries. Simple production methods for use on communes have been developed in the People's Republic of China (Hussey and Tinsley, 1981) but little progress has been made in other developing countries.

Viruses

A large number of baculoviruses, cytoplasmic polyhedrosis viruses (CPV), entomopoxyviruses, iridoviruses, densoviruses, and small RNA viruses cause disease in insects (Benz, 1981; Martignoni and Iwai, 1981; Payne and Kelly, 1981). Only a few baculoviruses, particularly the nuclear polyderosis (NPV) and one CPV, have been developed as experimental or commercial microbial insecticides. The NPVs cause destructive disease of the larval mid-gut epithelium. The NPVs have been sold as pesticides for control of Heliothis, Lymantria, Neodiprion, Trichoplusia and a few other pest species. The NPV of Heliothis was the first viral insecticide registered for use on cotton under the trade name Eclar and Biotrol-VHZ in the USA.

Although not produced commercially, a baculovirus of the rhinoceros beetle Oryctes rhinoceros has been used in pest control in Southeast Asia and the South Pacific (Bedford, 1981). It is spread by releasing virus-infected adults in the wild beetle population. Three 'nonoccluded' baculoviruses cause disease in the citrus rust mite Panonychus citri (Reed, 1981), but more research is needed on the infectious particle. A CPV that causes a flacherie disease of the pine caterpillar Dendrolimus has been registered as a microbial insecticide in Japan (Katagiri, 1981). Although less potent than NPVs in natural conditions, the CPVs are infective and kill the pest as they are sprayed.

A constraint in the production of viral insecticides is the cultural requirement of live hosts. This makes production expensive and enhances the likelihood of contamination. The possibility of in vitro production of insect viruses in cells is being explored. Natural sunlight inactivates most insect viruses. Therefore, improvements in formulations with sun-shields, e.g. carbon, carbon-based dyes, etc., are expected to increase the persistence. Safety in the use of viruses for insect control will remain a major constraint to more rapid development. Although there is no known insect virus harmful to vertebrates, agents that can reproduce in nature must be treated cautiously. Mass production and intensive application in insect control programs increase the possibility of deleterious mutations, which could survive in a natural environment and infect other species.

Fungi

Entomopathogenic fungi have received much attention lately because of their ability to decimate host populations, particularly during epizootics. Nearly 750 fungi are associated with insects (NAS, 1979). Important insect fungi are: Aschersonia, Beauveria, Coelomyces, Conidiobolus (=Entomophthora), Hirsutella, Langenidium, Metarrhizium, Nomuraea (=Spicaria), and Verticillium. Environmental safety of insect fungi has been a matter of concern, particularly possible allergic reactions during production or use in dust form and suspected production of aflatoxins. Recent data, however, have shown important insect fungi to be environmentally acceptable and noninvasive to mammals (Shadduck et al., 1982).

Beauveria bassiana is being produced commercially for control of the Colorado potato beetle in the Soviet Union. In China, B. bassiana, isolated from an endemic infection of the European cornborer Ostrinia nubilalis, has been developed for mass production (Hussey and Tinsley, 1981). In Brazil, Metarrhizium anisopliae is produced commercially for control of spittle bugs on forage and sugarcane (Ferron, 1981). In England, Verticillium lecanii is used against aphids in greenhouses (Hall, 1981). In the USA, Hirsutella thompsonii is used commercially to control the citrus rust mite Phyllocoptruta oleivora (McCoy, 1981). Control of mosquito larvae by Culicinomyces, Lagenidium, and Coelomyces has shown promise in Australia and the USA (Federici, 1981). Some insect fungi produce highly toxigenic metabolites, which rapidly kill insect hosts in bioassays (Roberts, 1981). None of the metabolites are under commercial development. Although entomogenous fungi are promising candidates for microbial insecticides, progress has been confined to temperate or subtropical fungi. Research on tropical insect fungi is urgently needed. Further taxonomic studies and improved bioassays are required for identification and selection of suitable strains and to standardize commercial production.

Genetic manipulation will contribute greatly to the use of insect fungi. Improved production and storage techniques to augment their shelf life are highly desirable. Spray formulations containing nutrients, water-retaining agents (alginates), and sun-protectants need to be tested. Appropriate spray methodology should be developed. Research on fungal toxins may lead to the discovery of chemicals with pesticide potential.

Nematodes

Many kinds of nematodes parasitize insect pests in soil or aquatic habitats and even on low-growing plants. The important parasitic nematodes are the mermithids, the sphaerulariids, and the neoaplectanids (Nickle, 1981). They attack different stages of insects belonging to Orthoptera, Coleoptera, Lepidoptera, Diptera, and other groups. The mosquito mermithid *Romanomermis culicivorax* has been reared on *Culex pipiens quinquefasciatus* larvae at a cost of \$10.10/10^6 preparasites. It has shown promise in small-to large-scale trials and, because of its wide host-range and relative safety to nontarget organisms, has been produced commercially. Other nematodes better suited to mosquito habitats and more tolerant to desiccation, salinity, and pollution are being developed. Another mermithid *Neomesomermis* has shown control potential against black flies. Mermithid parasitism of rice leafhoppers and planthoppers has been recorded in Japan and India, but no attempts have been made to utilize them for pest control. Some *Neoaplectana* sp. have wide host-ranges and have been used to control forest, soil, and plant pests. Improved culture methods and better storage techniques would aid in reducing production costs and maintaining infectivity. Environmental factors, such as desiccation, UV, and high temperatures limit field application of nematodes. The use of formulations with antidesiccants, however, increases the infection period of neoaplactenid nematodes (MacVean et al., 1982). Knowledge of host and parasite biology and their physiological interactions is expected to facilitate nematode use.

Integrated control of certain insect vectors, e.g., mosquitoes, may be more effective using nematodes and juvenile hormone compounds such as methoprene. Chicks and mice fed entomogenous nematodes and their associated bacteria have shown no disease symptoms (Poinar et al., 1982), but further tests are needed with poikilotherms.

Protozoa

Two protozoa, *Nosema* and *Vairimorpha*, have attracted some attention for use in insect control. *Nosema locustae* is pathogenic to a wide range of Orthoptera (Henry and Oma, 1981). Large-scale field trials using the microsporidian were found to be potentially useful in the integrated program of rangeland grasshopper management in the USA (Henry and Onsager, 1982). N. locustae spores must be mass produced and stockpiled because grasshopper outbreaks are extensive and require blanket treatment of infested areas. Spores are now mass produced in living hosts, but spore development in cell culture could make production more economic. Spore persistence needs to be assessed. Although *N. locustae* can be regarded as safe, it should be assessed against grasshopper parasites and predators.

Vairimorpha necatrix is highly pathogenic to a wide range of Lepidopterous pests but does not affect Hymenopterous parasites and other insects (Maddox et al., 1981). Dry spores on various substrates are sensitive to sunlight or UV radiation. Some crops, such as maize, may reduce UV degradation of spores and make UV protectants unnecessary (Lewis, 1982). Spores in baits or carriers, therefore, may be more effective in field applications. Rickettsiae also attack insects, but their pathogenicity to vertebrates, especially warm-blooded animals, makes their use unlikely.

INSECT-BASED PESTICIDES

Insect Venoms and Defense Secretions

A number of aranaeids, such as scorpions and spiders, and insects such as Hemiptera, Diptera, Coleoptera, and especially Hymenoptera, produce venoms for predation or toxic secretions for self-defense against predators. Venoms and toxic secretions are some of the most specific, most potent, and the fastest-acting insecticides. These toxicants are effective only upon injection, so they offer little promise in insect control.

Insect Growth Regulators

Insect growth regulators (IGRs) are slow-acting compounds that specifically affect growth and development of insects. Their advantage is their potential selectivity for orders, families, genera or species of insects. IGR effectiveness depends on synchrony with certain events in the insect's life in which the absence or presence of the growth regulator is critical for normal development. Four groups of IGRs based on hormonal processes are known: 1) molting hormones, 2) anti-molting hormones, 3) juvenile hormones, and 4) anti-juvenile hormones. Another group of IGR-type compounds, represented by Dimilin^{R} (diflubenzuron), interfere with cuticle synthesis in insects.

Molting Hormones

These are represented by ecdysone, ecdysterone, and other ecdysteroids. They are biochemically derived from cholesterol, still retaining its C_{27} carbon skeleton. These compounds cannot be economically synthesized. Although certain plants sources are rich in phytoecdysones, their extraction for insect control would be economical only if they are highly effective. Generally, they are effective only through injections. Also, the use of any chemical resembling steroid hormones, which play important roles in man and higher animals, would require careful testing to eliminate the possibility of side effects. The molting hormones also have a regulatory function in crustaceans. Therefore, they would be less selective. The potential for these compounds is low.

Anti-Molting Hormones

Several chemical antagonists for molting hormones have been identified, but they are still experimental. For example, certain azosterols, or even non-steroidal secondary and tertiary amines, are reported to disrupt the development of phytophagous insects, which depend on dietary sitosterol for their cholesterol requirements (Robbins et al., 1975). Such compounds may also interfere in steroid hormonal regulation in higher animals. The high cost of axosterols limits any agricultural use, but simpler antagonists may have some potential against phytophagous insects that do not have access to cholesterol in their diet.

Juvenile Hormones and Their Analogs

Insect juvenile hormone (JH) was described by V.B. Wigglesworth some 50 years ago, however, the possibility of using it to upset normal growth of insects was first suggested by C.M. Williams in 1956. Following his discovery that the abdomen of the male Cercropia moth was a rich source of JH, Williams (1967) proposed JH-active substances as powerful "third generation pesticides". The concept evoked considerable interest as can be gleaned from numerous publications (Novak, 1966; Wigglesworth, 1970; Menn and Beroza, 1972; Slama et al., 1973; Gilbert, 1975). Insect hormones, with a profound effect on insect development, do not appear to affect most other life. Their environmental impact should be negligible. Three juvenile hormones (JH1, JH2, JH3), occurring singly or in combination in different species have been identified. JH3 is a true terpene and JH1 and JH2 are homoterpenoids. The three natural JHs, differing only in the length of their side branches, may have specialized functions.

While natural JHs have never been serious candidates for use as pesticides, they have served as useful models for search and synthesis of structurally modified terpenoids with JH activity. Synthesis and evaluation of thousands of such synthetic analogs has yielded compounds with higher specific activity, greater stability, lower manufacturing costs, enhanced selectivity for several insect orders, and proprietary protection for manufacturers (Staal, 1977). The basic mode of action of JH analogs appears identical to that of exogenously applied natural JHs. All practical uses are based on their irreversible effects on metamorphosis. Other possible effects such as embryonic inhibition, chemosterilization, induced sterility through venereal dissemination of JH substances, diapause disruption, disorganization of caste system, and systemic activity in plants have not been exploited to their potential. The pest control potential of promising JH compounds is discussed here.

Control of Larvae of Mosquitoes and Aquatic Midges. Methoprene or Altosid[R] [isopropyl (2E,4E)-11-methoxy-3,7,11-trimethyl-2,4-dodecadienoate] was the first JH analog to receive the formal regulatory status from EPA. It is highly effective in controlling field populations of OP-resistant strains of floodwater mosquito *Aedes nigromaculis* larvae, which develop synchronously (Schaefer and Wilder, 1973). New microencapsulated methoprene formulations, SR10 (with charcoal) and SR10F (without charcoal), have effectively controlled larvae of *Culex tarsalis* and *Culiseta inornata*, which develop asynchronously and are refractory to other methoprene treatments (Mulla and Darwazeh, 1975). Slow-release encapsulated formulations also show excellent activity against several genera of OP-resistant aquatic midges (Mulla et al., 1974). Although larvae of several aquatic Diptera are sensitive to methoprene, it is least hazardous to nontarget organisms, such as fish, compared to most conventional larvicides.

Control of fly larvae. Methoprene has been developed for fly control in cattle manure by oral administration with food, water, or mineral supplements, and slow-release boluses. Mineral blocks containing 0.02% methoprene provided complete control of the horn fly *Haematobia irritans*, but only partial control of the face fly *Musca autumnalis*, and the stable fly *Stomoxys calcitrans*; the house fly *Musca domestica* was not affected (Chamberlain, 1975). Sustained-release boluses containing methoprene, when placed in the reticulum of cattle, inhibited horn fly development in the manure of treated cattle for several months (Miller et al., 1977, 1979). No appreciable residues were detected in treated animals (Ivey et al., 1982). An encapsulated formulation controlled house fly economically (Breeden et al., 1975). Methoprene is also effective against larvae of the sheep fly *Oestrus ovis* that are internal parasites of animals (Prasert et al., 1975).

Control of Stored Grain Insect Pests. Methoprene is ideal for insect control in stored commodities because reinfestation is not a problem and the chemical is protected from degradation by light, UV, moisture, etc. It is highly active against stored grain beetles and grain moths (Staa, 1977; Mian and Mulla, 1982). When applied directly to tobacco at a concentration of 10 ppm just before storage, methoprene prevented emergence of the adult cigarette bettle *Lasioderma serricorne* for 4 years (Manzelli, 1982). Some IRGs show selective activity against the almond moth *Ephestia cautella* without interfering with the development of its braconid parasite (Oberlander et al., 1979). The relative insensitivity of two important storage pests, the rice weevil Sitophilus oryzae and the granary weevil S. granarius, which feed and develop inside grains, limits the potential of methoprene and other JH compounds for stored grain protection.

Control of Homopterous Insects. A few IGRs are highly active against aphids, mealybugs, scales, and whiteflies. Kinoprene [isopropyl (2E,4E)-11-methoxy-3,7,11-trimethyl-2,4-dodecadienoate] combines a strong JH activity with toxicity at higher doses on several species of mealybugs and scales (Staal, 1977). It has been registered for use on foliage ornamentals in greenhouses in the USA, but more stable formulations are needed for Homoptera control in the field.

Methoprene and hydroprene or Altozar^R [ethyl (2E,4E)-11-trimethyl-2,4-dodecadienoate] were also active against aphids (Kuhr and Clere, 1973) and scales (Peleg and Gothilf, 1981). Under field conditions, however, methoprene alone may not adequately control Homoptera, but in combination with a more acute toxicant, its long-range JH activity may enhance the overall efficacy. Methoprene appears to have some compatibility in biocontrol. However, further evaluation is needed.

Control of Insect Pests of Field Crops and Forests. IGRs with JH activity have maximum scope for use against pests of field crops, provided they possess adequate chemical stability. Although their low persistence could be compensated somewhat by frequent applications and improved formulations, treatments would be costly. A new JH analog epofenonane, which combines high activity and high foliar persistence, has better prospects for use against scale insects, which lack mobility and are resistant to insecticides (Staal, 1977). Epofenonane also possesses some compatibility in biocontrol. Spray applications to orchards in the Netherlands for 3 consecutive years did not upset the relation of Panonychus ulmi spider mites and their primary predator Typhlodromus mites (van der Molen and Gruys, 1980). Again, its high cost limits use potential.

Control of Noxious Pests. Compounds with JH activity have some potential against some refractory noxious ant species. Baits impregnated with JH compounds should provide long-lasting and economic control because worker ants are likely to carry the baits into their colonies. Control of the German cockroach Blattella germanica populations was also feasible through prolonged contact with some JH compounds (Riddiford et al., 1975).

Experience with JH compounds justifies their development as insecticides, particularly in animal health, public health, and stored products. Crop protection is minimal because of the critical timing necessary to synchronize them with the narrow "windows" of pests' sensitivity, their short persistence, slow action, and pest reinfestation. The development of pest resistance to these agents is possible. Future research should aim at developing relatively more stable JH analogs, improved formulations, and application methods that take into account pest behaviour and population dynamics. Their relative safety to nontarget organisms and their compatibility with biocontrol agents should continue to be emphasized to develop ecologically safe and economical pest control programs. The high cost of JH analogs presently restricts their use primarily to developed countries.

Anti-Juvenile Hormones

Interfering with the activity of juvenile hormones became feasible with W.S. Bowers' discovery of two anti-JH compounds from extracts of the bedding plant Ageratum houstonianum. The compounds, mainly effective on Oncopeltus and Dysdercus bugs, were identified as 7 methoxy and 6,7-dimethoxy-2,2-dimethylchromene and were named as precocenes 1 and 2 (Bowers et al., 1976). Contact with precocenes induced immature bugs to skip succeeding nymphal instars and develop into tiny sterile females or males that could not inseminate females. Because of this unique feature, Bowers (1977) considered JH-antagonists as "fourth generation insecticides". Other groups of compounds, such as ethyl 4-2-(tert-butylcarbonyloxy) butoxybenzoate and piperonyl butoxide (Staal, 1977) show some characteristics of JH-antagonists against the tobacco hornworm Manduca sexta. Although the practicality of such compounds is limited because of the high and narrow dose requirements and apparent selectivity of certain species, they do pave the way toward JH-antagonists with broad-spectrum effects.

FUTURE CONSIDERATIONS AND CONCLUSION

A large array of botanical-, microbial-, and insect-based pesticides are available today to supplement and even supplant the use of persistent synthetic pesticides. The selectivity of naturally occurring pesticides makes them valuable in integrated pest control. Examples of their successful use against insect pests demonstrate their potential for development as the future generation insecticides. Their inherent nonpersistence is ecologically desirable. Thus, Brown (1971) indicated that "the best way of avoiding or at least postponing resistance is to achieve 100% control in a circumscribed area .... by nonpersistent insecticides". In many species, resistance to the slowly decaying cyclodiens and DDT has arisen more rapidly than to the more rapidly decaying organophosphates and carbamates, and there are hardly any cases of serious resistance to the very short-lived natural pyrethrins, except when frequent use has provided prolonged and consistent selection pressure. The low persistence of natural pesticides can be suitably amended by appropriate formulations for protection against too rapid degradation by sunlight, UV radiation, high temperature, moisture, or desiccation. Production methods, formulation, and method and timing of application are, therefore, more critical to the development and use of natural pesticides than of conventional pesticides.

The research and development of botanical pesticides needs to be systematically accelerated at all levels, i.e., from the direct use of locally available known pesticidal plant material by farmers in developing countries to the analysis, chemical synthesis, and development of principal constituents by the industry in developed countries. The discovery of different classes of conventional toxicants, including the much acclaimed DDT, was rather serendipitous. Presently, thousands of new synthetic compounds are ritualistically evaluated in the quest of other miracle pesticides, but the success rate is not high. On the other hand, the plant kingdom is replete with a wide variety of chemical types, which would be highly valuable in pest control. Many potent chemicals have been discovered in plants which were not even thought of possessing defense chemicals. For instance, a stemborer oviposition deterrent was isolated from a rice plant and identified at IRRI. For a long time, pest resistance was attributed to plant morphology rather than chemistry. The novel chemical is amenable to synthesis and is under evaluation. On the other hand, Margosine-O and Margosine-D insecticides, based on a neem antifeedant, are being developed in the USA.

A systematic cooperative research effort by chemists, entomologists, and others on pest-resistant plant taxa would yield new biologically active components. Such cooperation now exists between many laboratories and research institutions. A fundamental change in emphasis on properties of plant derivatives is required. Earlier researchers tested these only for insecticidal value. Subtle bioassays now show that plant chemicals can affect insects in many ways and that the susceptibility of different species may not be the same. Future pesticide technology should also consider developing blends of several active ingredients and synergists, as plants possess. Though difficult to achieve, it may ease the pesticide resistance problem. Unlike ordinary pesticides based on a single active ingredient, the possibility of a pest species developing resistance to an array of chemicals is remote. This will be a truly phytochemical approach to pest control.

The potential of microbial pesticides is considerable. Many of them are firmly established as viable commercial pesticides and their use appears to be increasing both in quantity and the range of target pests. Pathogens in use have an outstanding safety record. In contrast to chemicals, when produced according to accepted quality standards, they are virtually accident proof. Certain pathogens such as bacteria, viruses, and fungi can be mass produced in even developing countries using relatively simple technology.

The tremendous heterogeneity of insect pathogens belonging to diverse animal and plant phyla, and the variation within the species, makes identification difficult. Proper identification is essential to avoid contamination in mass production of pathogens. Also, there must be a better understanding of their pathogenicity and the nature and mode of action of their toxins. These aspects are relevant to strain selection. A greater regional and international cooperation will increase the chances of finding more effective pathogen strains. Genetic manipulation and genetic engineering will offer additional oppourtunities for developing suitable strains. In vitro culture techniques should be explored to reduce the production cost of obligate pathogens. Simultaneous use of three agents with diverse modes of action should avoid development of resistance. Patenting rights for insect pathogens need to be more realistic to attract the participation by the pesticide industry, and the use of microbial pesticides should be popularized, if possible, through government subsidy.

The insect-based pesticides will be directed against insect pests mainly as growth disturbing agents. While a number of JH-analogs have shown promise in the control of pests affecting public health, animal health, and stored products, their potential against crop pests will remain limited. Their high cost will also restrict their widespread use. Further neuroendocrinological research may lead to some more novel approaches to pest control.

The future of pest control in the 21st century looks promising as it will be supported by an increasing knowledge of chemistry, physiology, and ecology.

REFERENCES

Barjac, H. de (1981). Identification of H-serotypes of Bacillus thuringiensis. In: Microbial Control of Pests and Plant Diseases 1970-1980 (H.D. Burges, ed.). Academic Press, London, N.Y., 35-43.

Bedford, G.O. (1981). Control of the rhinoceros beetle by baculovirus. In: Microbial Control of Pests and Plant Diseases 1970-1980 (H.D. Burges, ed.). Academic Press, London, N.Y., 409-426.

Benz, G. (1981). Use of viruses for insect suppression. In: Biological Control in Crop production (G.C. Papavizas, ed.), Beltsville Symposia in Agricultural Research 5, Allanheld, Osmun Publ., London, 259-272.

Berry, J.H. (1979). Pesticides and energy utilization. In: Pesticides: Contemporary Roles in Agriculture, Health, and Environment (T.J. Sheets and D. Pimentel, eds.). Humana Press, Clifton, N.J., 57-71.

Bowers, W.S. (1977). Fourth generation insecticides. In: Pesticide Chemistry in the 20th Century (J.R. Plimmer, ed.). ACS Symposium Series 37, Washington, D.C., 271-275.

Bowers, W.S. T. Ohta, J.S. Clere and P.A. Marsella (1976). Discovery of insect antijuvenile hormone in plants. Science, 193:542-547.

Breeden, G.C., E.C. Turner, Jr. and W.L. Beane (1975). Methoprene as a feed additive for control of the housefly breeding in chicken manure. J. Econ. Entomol. 68:451-452.

Brown, A.W.A. (1971). Pest resistance to pesticides. In: Pesticides in the Environment, vol. 1 (R. White-Stevens, ed.). Marcel Dekker, N.Y., 457-552.

Burges, H.D. (1981). Microbial Control of Pest and Plant Diseases 1970-1980. Academic Press, London, N.Y., 949 pp.

Burges, H.D. and N.W. Hussey (1971). Microbial Control of Insects and Mites. Microbial Control of Pest and Plant Diseases 1970-1980. Academic Press, London, N.Y., 861 pp.

Casida, J.E. (1973). Pyrethrum: The Natural Insecticide. International Symposium on Recent Advances with Pyrethrum, the Natural Insecticide, University of Manitoba, 1972. Microbial Control of Pest and Plant Diseases 1970-1980. Academic Press, London, N.Y., 329 pp.

Chamberlain, W.F. (1975). Insect growth regulating agents for control of arthropods of medical and veterinary importance. J. Med. Entomol. 12:395-400.

Cramer, H.H. (1967). Plant protection and world crop production. Pflanzenschutznachrichten 20:1-524.

Crosby, D.G. (1966). Natural pest control agents. In: Natural Pest Control Agents, Advances in Chemistry, Series 53 (R.F. Gould, ed.). American Chem. Society, Washington, D.C., 1-16.

Crosby, D.G. (1971). Minor insecticides of plant origin. In: Naturally occurring insecticides (M. Jacobson and D.G. Crosby, eds.). Marcel Dekker, Inc., New York, 177-239.

Dulmage, H.T. (1981). Production of bacteria for biological control of insects. In: Biological Control in Crop Production (G.C. Papavizas, ed.). Beltsville Symposia in Agric. Res., May 18-21, 1980, at Beltsville, Maryland, Allanheld, Osmun Publ., London, 129-141.

Dutky, S.R. (1940). Two new spore-forming bacteria causing milky diseases of Japanese beetle larvae. J. Agric. Res. 61:57-68.

Elliott, M. (1977). Synthetic Pyrethroids. ACS Symposium Series 42. American Chem. Society, Washington, D.C., 229 pp.

Fast, P.G. (1981). The crystal toxin of Bacillus thuringiensis. In: Microbial Control of Pest and Plant Diseases 1970-1980 (H.D. Burges, ed.). Academic Press, London, N.Y., 223-244.

Federici, B.A. (1981). Mosquito control by the fungi Culicinomyces, Lagenidium, and Coelomomyces. In: Microbial Control of Pest and Plant Diseases 1970-1980 (H.D. Burges, ed.). Academic Press, London, N.Y., 555-571.

Ferron, P. (1981). Pest control by the fungi Beauveria and Metarrhizium. In: Microbial Control of Pest and Plant Diseases 1970-1980 (H.D. Burges, ed.). Academic Press, London, N.Y., 465-482.

Gilbert, L.I. (1975). The juvenile hormones. Plenum Press, New York and London. 572 pp.

Hall, R.A. (1981). The fungus Verticillium lecanii as a microbial insecticide against aphids and scales. In: Microbial Control of Pests and Plant Diseases (H.D. Burges, ed.). Academic Press, London, N.Y., 483-498.

Henry, J.E. and E.A. Oma (1981). Pest control by Nosema locustae, a pathogen of grasshoppers and crickets. In: Microbial Control of Pests and Plant Diseases (H.D. Burges, ed.). Academic Press, London, N.Y., 573-586.

Henry, J.E. and J.A. Onsager (1982). Large-scale test of control of grasshoppers on range and with Nosema locustae. J. Econ. Entomol. 75:31-35.

Hussey, N.W. and T.W. Tinsley (1981). Impressions of insect pathology in the People's Republic of China. In: Microbial Control of Pests and Plant Diseases, 1970-1980 (H.D. Burges, ed.). Academic Press, London, N.Y., 785-795.

Ignoffo, C.M. (1970). Microbial insecticies: No-Yes; now-when. In: Proc. Tall Timbers Conf. on Ecological Animal Control of Habitat Management, Tallahassee, FL, Feb. 26-28, 1970, 44-57.

Ignoffo, C.M. (1975). Entomopathogens as insecticides. In: Insecticides of the Future (M. Jacobson, ed.). Marcel Dekker, Inc., New York, 23-40.

International Rice Research Institute (IRRI) (1981). Neem cake and urea for brown planthopper management. IRRI Res. Highlights for 1980:50.

International Rice Research Institute (IRRI) (1982a). Plant derivatives for management of rice insect pests. IRRI Res. Highlights for 1981:59-61.

International Rice Research Institute (IRRI) (1982b). Neem boosts nitrogen fixation. IRRI Res. Highlights for 1981:83.

Ivey, M.C., J.A. Miller and G.W. Ivie (1982). Methoprene residues in fat of cattle treated with methoprene boluses. J. Econ. Entomol. 75:254-256.

Jacobson, M. (1958). Insecticides from plants, a review of literature, 1941-1953. Agric. Handbook 154, Agric. Res. Serv., U.S. Dept. Agric.

Jacobson, M. (1975). Insecticides from plants, a review of literature, 1954-1971. Agric. Handbook 461, Agric. Res. Serv., U.S. Dept. Agric. 138 pp.

Jotwani, M.G. and K.P. Srivastava (1981a). Neem: insecticide of the future-I-as protectant against stored grain pest. Pesticides 15(10):19-23.

Jotwani, M.G. and K.P. Srivastav (1981b). Neem: insecticide of the future-II-protection against field pests. Pesticides 15(11):40-47.

Jotwani, M.G. and K.P. Srivastava (1981c). Neem: insecticide of the future-III-chemistry, toxicology and the future strategy. Pesticides 15(12):15-19.

Katagiri, K. (1981). Pest control by cytoplasmic polyhedrosis viruses. In: Microbial Control of Pest and Plant Diseases 1970-1980 (H.D. Burges, ed.). Academic Press, London, N.Y., 433-440.

Ketkar, C.M. (1976). Utilization of neem (Azadirachta indica A. Juss) and its by-products; report of the modified neem cake manurial project, 1969-1976. Directorate of Nonedible Oils and Soap Industry, Khadi & Village Industries Commission, Bombay, India. 234 pp.

Kubo, I. and J.A. Klocke (in press). Azadirachtin, insect ecdysis inhibitor. Agric. Biol. Chem.

Kuhr, R.J. and J.S. Clere (1973). Toxic effects of synthetic juvenile hormones on several aphid species. J. Econ. Entomol. 66:1019-1022.

Lewis, L.C. (1982). Persistence of Nosema pyrausta and Vairimorpha necatrix measured by microsporidiosis in the European corn borer. J. Econ. Entomol. 75:670-674.

MacVean, C.M. J.W. Brewer and J.L. Capinera (1982). Field tests of anti-desiccants to extend the infection period of an entomogenous nematode, Neoaplectana carpocapsae, against the Colorado potato beetle. J. Econ. Entomol. 75:97-101.

Maddox, J.V. W.M. Brooks and J.R. Fuxa (1981). Vairimorpha necatrix, a pathogen of agricultural pests: potential for pest control. In: Microbial Control of Pests and Plant Diseases 1970-1980 (H.D. Burges, ed.). Academic Press, London, N.Y., 587-594.

Manzelli, M.A. (1982). Management of stored-tobacco pests, the cigarette beetle (Coleoptera:Anobiidae) and tobacco moth (Lepidoptera:Pyralidae). J. Econ. Entomol. 75:721-723.

Martignoni, M.E. and P.J. Iwai (1981). A catalogue of viral diseases of insects, mites and ticks. In: Microbial Control of Pests and Plant Diseases 1970-1980 (H.D. Burges, ed.). Academic Press, London, N.Y., 897-911.

McCoy, C.W. (1981). Pest control by the fungus Hirsutella thompsonii. In: Microbial Control of Pests and Plant Diseases 1970-1980 (H.D. Burges, ed.). Academic Press, London, N.Y., 499-512.

Menn, J.J. and M. Beroza (1972). Insect juvenile hormones: chemistry and action. In: Microbial Control of Pests and Plant Diseases 1970-1980 (H.D. Burges, ed.). Academic Press, London, N.Y., 341 pp.

Mian, L.S. and M.S. Mulla (1982). Biological activity of IGRs against four stored-product coleopterans. J. Econ. Entomol. 75:80-85.

Miller, J.A., M.L. Beadles, J.S. Palmer and M.O. Pickens (1977). Methoprene for control of the horn fly: a sustained-release bolus formulation for cattle. J. Econ. Entomol. 70:589-591.

Miller, J.A., F.P. Knapp, R.W. Miller and C.W. Pitts (1979). Sustained-release boluses containing methoprene for control of the horn fly and face fly. Southwest Entomol. 4:195-200.

Molen, J.P. van der and P. Gruys (1980). Use of insect growth regulator, epofenonane, in integrated control in orchards. In: Integrated Control of Insect Pests in the Netherlands (eds. A.K. Minks and P. Gruys, eds.). Centre for Agric. Publ. and Documen., Wageningen, 251-254.

Mulla, M.S., R.L. Norland, T. Ikeshoji and W.L. Kramer (1974). Insect growth regulators for the control of aquatic midges. J. Econ. Entomol. 67:165-170.

Mulla, M.S. and H.A. Darwazeh (1975). Activity and longevity of insect growth regulators against mosquitoes. J. Econ. Entomol. 68:791-794.

National Academy of Sciences (NAS) (1979). Microbial insect control agents. In: Microbial Processes: Promising Technologies for Developing Countries (A.K. Minks and P. Gruys, eds.). Washington, D.C., 198 pp.

Nickle, W.R. (1981). Nematodes with potential for biological control of insects and weeds. In: Biological Control in Crop Production, Beltsville Symposia in Agricultural Research 5 (G.C. Papavizas, ed.). Allanheld, Osmun Publ. London, 181-199.

Novak, V.J.A. (1966). Insect Hormones. Methuen & Co., London. 478 pp.

Oberlander, H., D. Nickle, D.L. Silhacek and D.W. Hagstrum (1979). Advances in insect growth regulator research with stored grain insects. Symp. Prevention and Control of Insects in Stored Food Products; Manhattan, KS, Apr. 9-13, 1978.

Papavizas, G.C. (1981). Biological control in crop production. Beltsville Symposia in Agricultural Research 5, Allanheld, Osmun Publ., London, 461 pp.

Payne, C.C. and D.C. Kelly (1981). Identification of insect and mite viruses. In: Microbial Control of Pests and Plant Diseases 1970-1980 (H.D. Burges, ed.). Academic Press, London, N.Y., 61-91.

Peleg, P. and S. Gothilf (1981). Effect of the insect growth regulators diflubenzuron and methoprene on scale insects. J. Econ. Entomol. 74:124-126.

Pimentel, D., D. Andow, R. Dyson-Hudson, D. Gallahan, S. Jacobson, M. Irish, S. Kroop, A. Moss, I. Schreiner, M. Shepard, T. Thompson and B. Vinzant (1980). Environmental and social costs of pesticides: a preliminary assessment. Oikos 34:125-140.

Poinar, G.O. Jr., G.M. Thomas, S.B. Presser and J.L. Hardy (1982). Inoculation of entomogenous nematodes, *Neoaplectana* and *Heterorhabditis*, and their associated bacteria, *Xenorhabdus* spp., into chicks and mice. Environ. Entomol. 11:137-138.

Prasert, V., F.W. Knapp, E.T. Lyons, and J.H. Drudge (1975). Biological effects of methoprene on the sheep bot fly. J. Econ. Entomol. 68:639-640.

Reed, D.K. (1981). Control of mites by non-occluded viruses. In: Microbial Control of Pests and Plant Diseases 1970-1980 (H.D. Burges, ed.). Academic Press, London, N.Y., 427-432.

Riddiford, L.M., A.M. Ajami, and C. Boake (1975). Effectiveness of insect growth regulators in the control of populations of the German cockroach. J. Econ. Entomol. 68:46-48.

Robbins, W.E., M.J. Thompson, J.A. Svoboda, T.J. Shortino, C.F. Cohen, S.R. Dutky and O.J. Duncan III (1975). Nonsteroidal secondary and tertiary amines: inhibitors of insect development and metamorphosis and Δ^{24}-sterol reductase system of tobacco hornworm. Lipids 10:353-359.

Roberts, D.W. (1981). Toxins of entomopathogenic fungi. In: Microbial Control of Pests and Plant Diseases 1970-1980 (H.D. Burges, ed.). Academic Press, London, N.Y., 441-464.

Saxena, R.C., N.J. Liquido and H.D. Justo (1981a). Neem seed oil, a potential antifeedant for the control of the rice brown planthopper, Nilaparvata lugens. In: Natural Pesticides from the Neem Tree (Azadirachta indica A. Juss) (H. Schmutterer, K.R.S. Ascher, and H. Rembold, eds.). Proc. 1st Int'l Neem Conf., Rottach-Egern, FRG, 16-18 June 1980, German Agency for Technical Cooperation, Eschborn, FRG, 171-188.

Saxena, R.C., G.P. Waldbauer, N.J. Liquido and B.C. Puma (1981b). Effects of neem seed oil on rice leaffolder, Cnaphalocrocis medinalis. In: Natural Pesticides from the Neem Tree (Azadirachta indica A. Juss) (H. Schmutterer, K.R.S. Ascher, and H. Rembold, eds.). Proc. 1st Int'l Neem Conf., Rottach-Egern, FRG, 16-18 June 1980, German Agency for Technical Cooperation, Eschborn, FRG, 189-203.

Schaefer, C.H. and W.H. Wilder (1973). Insect development inhibitors. 2. Effects on targe mosquito species. J. Econ. Entomol. 66:913-916.

Schmutterer, H., K.R.S. Ascher and H. Rembold (1981). Natural pesticides from the neem tree (Azadirachta indica A. Juss), Proc. 1st Int'l Neem Conf., Rottach-Egern, FRG, 16-18 June 1980, German Agency for Technical Cooperation, Eschborn, FRG, 297 pp.

Shadduck, J.A., D.W. Roberts, and S. Lause (1982). Mammalian safety tests of Metarrhizium anisopliae: preliminary tests. Environ. Entomol. 11:189-192.

Slama, K., M. Romanuk and F. Sorm (1973). Insect Hormones and Bioanalogues. Springer-Verlag, Wien, New York, 477 pp.

Staal, G.B. (1977). Insect control with insect-growth regulators based on insect hormones. In: Study Week on Natural Products and the Protection of Plants (G.B. Marini-Bettolo, ed.), Oct. 18-23, 1976, Pontificia Academia Scientarum, 353-377.

Wigglesworth, V.B. (1970). Insect hormones. W.H. Freeman & Co., San Francisco. 159 pp.

Williams, C.M. (1967). Third-generation pesticides. Sci. Am 217:13-17.

PHEROMONES AND OTHER RECENT DEVELOPMENTS IN BIOCHEMICAL PEST MANAGEMENT

P.S. Beevor, D.R. Hall and Brenda F. Nesbitt

Tropical Products Institute, Overseas Development Administration
56/62, Gray's Inn Road, London WCIX 8LU, England

ABSTRACT

The application of pheromones and other behaviour-modifying "semiochemicals" to pest management on food crops is reviewed with consideration of their advantages and disadvantages relative to conventional methods. The main scientific, commercial and organizational constraints which need to be overcome to permit full development of the potential of these chemicals as aids to food production and storage are discussed.

KEYWORDS: pheromone; semiochemical; allomone; kairomone; insect traps; population monitoring; mass trapping; mating disruption; integrated pest management.

INTRODUCTION

Pheromones are chemicals used for communication between members of the same species, and during the last fifteen years remarkable progress has been made in establishing the chemical identity of these substances, particularly in insect species. The most significant advances have been made in the identification and use of sex and aggregation pheromones, particularly those of lepidopterous and coleopterous species, but alarm, trail, dispersal and social pheromones have also been investigated. The main impetus for this work has been the potential value of pheromones in insect pest management. They are seen as a means to reduce the dependence upon conventional broad-spectrum insecticides and so to limit the effects of the latter on the environment and delay the onset of resistance in target species.

However, pheromones are only one class of chemical signals with behaviour-modifying action, and more recently attention has been directed towards other such "semiochemicals" which have interspecific activity (Figure 1.). These include attractants and deterrents for oviposition and compounds which guide parasites and predators to their hosts. Their value as pest control agents could be even greater than that of pheromones, but there are relatively few examples of chemical identification and even fewer cases of their field use, and this review describe mainly work on insect pheromones.

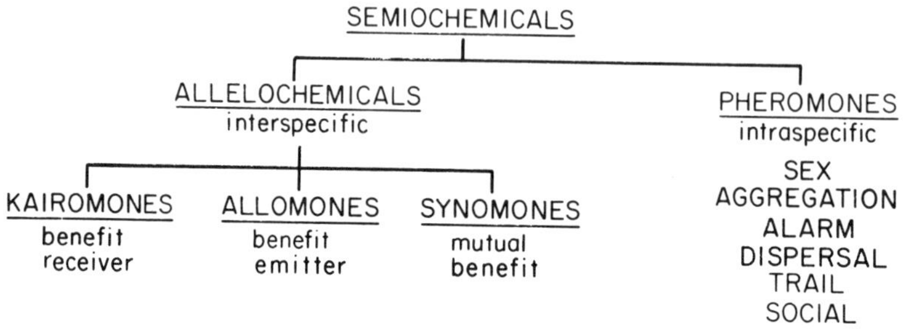

Fig. 1. TYPES OF INSECT BEHAVIOUR-MODIFYING CHEMICALS

THE USE OF PHEROMONES IN PEST MANAGEMENT

Pheromones and other semiochemicals have several characteristics which are highly desirable in insect pest control agents. They are generally extremely potent so that only small amounts of material are required to be effective. Unlike insecticides, they act by modifying behaviour and do not themselves actually kill the target species; so far they have proved to have little or no toxic effect on animal and plant life, added to which many pheromones show a high degree of biodegradability. By definition, pheromones are species-specific, and a high level of specificity is characteristic of many other semiochemicals. In practice, this specificity may not be absolute under conditions of field use of synthetic pheromones, but in general they have little adverse effect on non-target species, including predators and parasites. It is debatable whether target species will become "resistant" to control methods using pheromones. However, it might be argued that this will be relatively slow to develop because both the "transmitter" and the "receiver" have to evolve the same new communication signals simultaneously.

Most work on the field use of synthetic semiochemicals has been concerned with sex pheromones of Lepidoptera and sex and aggregation pheromones of Coleoptera. Pheromones have also been identified for certain Diptera, Hymenoptera, Orthoptera and Homoptera. The uses of pheromones fall into two main categories - population monitoring and control. The latter can be further subdivided into control by mass trapping and control by communication disruption.

Population Monitoring

Pheromone-baited traps provide a highly sensitive means of detecting the presence of insect pests with many advantages over conventional sampling methods such as light traps and scouting programmes. Pheromone traps are cheap and simple to construct and maintain, with only milligram amounts of pheromone required to keep a trap operational throughout the year. In contrast, light traps are more costly, more easily damaged and depend upon a source of electricity nearby. Scouting methods are typically tedious and time-consuming, and they depend critically upon the statistical methods used. Equally importantly, the specificity of pheromone traps eliminates the need for trained personnel to identify the species caught and counted, as are required for the other sampling methods. There are now numerous commercial suppliers of complete trap systems, and the use of pheromone traps for survey and monitoring is well-established in many countries. Some representative examples involving pests of food crops in various parts of the world are given in Table 1.

The sensitivity and specificity of pheromone traps make them ideal tools for simple detection of target species. This is particularly valuable in stored product situations and on crops with a low or zero tolerance of pest infestation. However, the quantitative correlation of trap catches with population densities is a far more complex process, and even then the populations must be related to subsequent economic damage in deciding whether control measures are necessary.

Traps baited with the sex pheromones of lepidopterous species suffer from the disadvantage that they monitor the adult moths whereas it is the larvae that do the damage. Furthermore, these traps typically attract only the male moths and not the females which are responsible for laying the eggs on the crop. The picture is further complicated in tropical areas where successive generations may overlap to a considerable extent. Nevertheless, the low cost and ease of operation by non-specialist personnel mean that pheromone traps are often the only practicable means of population monitoring. Such monitoring is a vital component of integrated pest management and essential for accurate spray timing.

Control by Mass Trapping

Pheromone traps have been used to maintain pest populations at an acceptably low level by "trapping out" both in storage situations and in the open field. Some examples of the application of this technique to pests of food crops are given in Table 2.

This method of pest control has the advantages of specificity for the target species and the use of small amounts of non-toxic chemicals in a highly restricted manner. It can only operate efficiently when populations are low and the traps can compete effectively with pheromone-producing insects. However, by its nature, a mass-trapping programme can be maintained continuously throughout the year to reduce the build-up of the pest population during the growing season.

Traps baited with sex pheromone attracting only the males must remove a very high percentage of the population if subsequent larval numbers are to be reduced, and, even then, the immigration of gravid females from outside the treated area can negate any such effect. The method has been most successful in situations where the pest population is localised by artifical barriers such as in warehouses and orchards, or by the fact that the target species does not move over great distances. Aggregation pheromones may prove more useful, and considerable success has been claimed in limiting outbreaks of bark beetles in Scandinavia using traps baited with the aggregation pheromone. In general, large numbers of traps are required - up to 100/ha in open field situations. However, since the traps can be simple and cheap to construct and relatively small amounts of pheromone are required, the cost of setting up and maintaining a mass-trapping network in countries where labour is relatively cheap may be less than that of importing insecticide and application equipment.

Control by Communication Disruption

Permeation of the atmosphere with synthetic pheromone has been shown to disrupt pheromone-based communication, although the exact biological mechanism of this is not clear and may indeed vary in different species. Application of this method to insect pest control was initially hampered by the lack of suitable slow-release formulations that could be applied over large areas, but now several effective formulations are commercially available, including hollow plastic fibres, plastic laminates and microcapsules. The microencapsulated formulations

Table 1: Examples of the Use of Pheromones For Monitoring Populations of Insect Pests in the Field And in Store.

PEST	HOST CROP	COUNTRY	REFERENCE
LEPIDOPTERA			
Laspeyresia pomonella	apples	England	Alford et al. (1979)
		Canada	Madsen (1981)
		USA	Rock et al. (1981)
Laspeyresia nigricana	peas	England	Perry et al. (1981)
Adoxyphes orana	apples, pears	Netherlands	Minks (1979)
Lobesia botrana,	grapes	France	Minks (1979)
Eupoecilia ambiguella		Israel	Gurevitz & Gothilf (1982)
Spodoptera litura	taro	Japan	Nakasuji & Kiritani (1976)
Spodoptera exempta	cereals	Kenya, Tanzania	Rose (pers. comm.)
		Malawi	Nyirenda (1982)
Ostrinia nubilalis	cereals	Europe	Minks (1979)
Chilo suppressalis	rice	Philippines	Dyck (1979)
		France	Poitout & Bues (1978)
Chilo partellus	sorghum,	India	Davies (pers. comm.)
Busseola fusca	maize	Zimbabwe	Sanders (pers. comm.)
Phthorimaea operculella	potatoes	Peru	Raman (1982)
Ephestia cautella,	stored	USA	Vick et al. (1981)
Plodia interpunctella	products	Germany	Reichmuth et al. (1980)
COLEOPTERA			
Trogoderma spp.	stored products	USA	Burkholder (1981)
HOMOPTERA			
Aonidiella aurantii	citrus	Israel	Szivos et al. (1982)

can be applied with conventional aerial or ground-based spray equipment, but the other two types of formulation have to be applied by hand or from the air with specially developed apparatus.

Considerable success has been achieved in large-scale programmes using such formulations of sex pheromones designed to disrupt mating of the target species. Much of the work has been directed against pests of cotton, such as the pink bollworm, Pectinophora gossypiella, because of the considerable commercial incentive to find a niche in the massive market for pest control agents for this crop. Attention has also been directed towards forest pests: insecticide is difficult to apply effectively on forest trees, but extensive damage can be caused by insect pests such as spruce budworm, Choristoneura spp.. The potential market is large, and government agencies are already carrying out collaborative studies with industry on the use of semiochemicals in United States, Canada and Europe. Both forests and, in many cases, cotton crops provide the opportunity for

Table 2: Examples of the Use of Pheromones For Mass Trapping of Insect Pests of Food Crops.

PEST	HOST CROP	COUNTRY	REFERENCE
Spodoptera litura	sweet potato	Japan	Tamaki (pers. comm.)
Adoxyphes spp.	tea	Japan	Negishi et al. (1980)
Laspeyresia pomonella	apples	Canada	MacLellan (1976)
Prays citri	citrus	Isreal	Sternlicht (pers. comm.)
Cryptophlebia spp.	macadamia	Malawi	La Croix (pers. comm.)
Ephestia spp.	stored	Germany	Levinson & Levinson (1979)
Plodia interpunctella	products		
Anagasta kuehniella			

treatment of large areas with pheromone formulations, thus minimising the effects of immigration of mated females into the treated area. The work on pink bollworm has shown that control with pheromone can be achieved with very low seasonal application rates down to 10 g/ha/season, which even now is competitive in cost with control by conventional insecticides.

Mating disruption has been explored as a control technique for pests of food crops, but mostly in relatively small-scale trials. Some examples are given in Table 3. However, commercial trials in California have led to registration of a hollow fibre pheromone formulation for control of the artichoke plume moth, Platyptilia carduidactyla, on artichokes in the United States.

Table 3: Examples of The Use of Pheromones For Mating Disruption of Insect Pests of Food Crops

PEST	HOST CROP	COUNTRY	REFERENCE
Ostrinia nubilalis	maize	France	Stockel & Anglade (1976)
Chilo suppressalis	rice	Philippines	Dyck (1981)
		Japan	Tatsuki & Kanno (1981)
		Korea	Lee et al. (1981)
Heliothis zea	sweet corn	USA	McLaughlin et al. (1981)
Spodoptera frugiperda	sweet corn	USA	McLaughlin et al. (1981)
Eupoecilia ambiguella	grapes	Switzerland	Arn et al. (1981)
Platyptilia carduidactyla	artichokes	USA	Haynes et al. (1981)

THE POTENTIAL USE OF SEMIOCHEMICALS AS AN AID TO FOOD PRODUCTION

Since as much as 20% of the world production of food crops is lost through insect attack and most insect species rely heavily on semiochemicals for functions vital to their survival, the potential use of these chemicals in increasing food supplies is considerable. Pheromone traps can improve monitoring systems and maximise the effects of any pest control measures taken. Two characteristics of pheromones, and probably semiochemicals in general, would seem to make them particularly suited to control of insect pests on food crops. The low mammalian toxicity of pheromones and their residues gives them an immediate advantage over the majority of insecticides. Furthermore, the negligible toxicity of pheromones to insects and the species-specificity of their behaviour-modifying effects make them entirely compatible with true biological control methods involving introduction of predators and parasites which are often used on food crops such as sugar-cane.

Despite the many factors apparently favouring development of pheromones in insect pest control, their practical application has been far more complex and has taken far longer than was originally anticipated. If techniques based on these substances are to contribute substantially to increased food production by the end of the century, large gaps in basic scientific knowledge must be filled, and economic and organizational constraints circumvented.

Scientific Constraints

More chemical work is required to ensure that all the components of a particular pheromone have been identified and are available for field work. This may involve re-examination of "known" pheromones. It is also important to examine insect material from the country where the pheromone is to be used, since geographical variations in pheromone blends are well documented. There is a great need for chemical studies on kairomones and allomones where relatively few structures are known. These substances may have advantages over pheromones for control purposes, or may be able to complement the field use of the latter.

There is an even greater need for more fundamental knowledge of the biology of insect pest species if semiochemicals are to be used successfully to manipulate their behaviour. It is vital to know the exact functions and interactions of the various chemical stimuli received by the insect in its natural environment and the responses which they elicit. Much more work on the basic biology of individual pest species is required in order to determine the most effective means of using semiochemicals alone and in conjunction with other control methods in integrated pest management programmes.

The best approach is likely to differ for each pest species and will be influenced by numerous factors. For example, the feasibility of achieving control of a pest by mating disruption with synthetic pheromone will be influenced by the nature and stability of the pheromone, the extent to which non-pheromonal communication is important in the mating process, whether the insect is mono- or poly-phagous and the distances travelled by gravid females.

Economic and Organizational Constraints

In general, agrochemical companies are understandably unwilling to carry out the long-term studies needed as a basis for the successful application of semiochemicals to pest control. It seems probable, therefore, that public funds will be needed to support this basic research into the nature, function and use of semiochemicals. Concern over the adverse environmental effects of the indiscriminate use of conventional pesticides may encourage governments to support such work. However, the potential of pheromones and other semiochemicals can never be realised without the support of the agrochemical industry as producers and the farmers and agriculturalists as users. Governments and international organizations can promote the use of semiochemicals directly in as far as they organize crop protection programmes and train entomologists and extension staff to assess the best ways to approach different pest problems. However, the agrochemical industry has been reluctant to move into products based on semiochemicals, and the problems here are much more intractable. It is significant that the majority of firms involved in producing pheromone-based systems are new to the field of crop protection.

Certain of the intrinsic factors that make semiochemicals so suitable as pest control agents in fact operate as strong deterrents of commercial development. The minute amounts of pheromone required in trapping systems are of no real interest to the larger agrochemical companies, and firms supplying traps are typically small venture companies. Even these rely on obtaining a large mark-up on complete systems of trap, dispenser and pheromone. Larger amounts of pheromone are required for control by communication disruption, but here the specificity of pheromones is the problem. In general, a different product will be required for each pest so that the market for any one product is limited. On this basis, the use of semiochemicals as control agents will tend to be restricted to major crops or crops of exceptionally high value, and to those with one or two dominant pest species or one species in a pest complex that is difficult to control with conventional insecticides.

However, exceptions to these restrictions can be foreseen. With certain pest complexes there is evidence that it may be possible to control more than one species with a single common pheromone component. Where true biological control is used against certain species in a pest complex it will be hazardous to use insecticides to control the others, but control with pheromones provides a compatible alternative. Furthermore, the above criteria could become less restrictive if key pests develop resistance to conventional insecticides and increasing consideration of environmental effects leads to greater reliance on control agents other than conventional, broad-spectrum insecticides.

Two other basic problems hindering the commercial development of semiochemicals in pest control are those of demonstrating efficacy and of obtaining patent protection. Semiochemicals act indirectly to reduce pest populations by influencing some aspect of behaviour necessary for survival and propagation of the species, and lengthy trials are generally necessary to demonstrate and effect. Furthermore, in the case of mating disruption with sex pheromones, the trials must be conducted on a large scale to minimise the effect of mated females entering the treated area from outside. Apart from being very costly, these factors introduce the further problem of obtaining untreated areas that are strictly comparable. In contrast, the obvious lethal effect of an insecticide can be demonstrated quickly in small-scale, fully replicated trials. Semiochemicals are natural products and difficult to patent per se in many countries. To date, the search for non-naturally occurring analogues of pheromones has met with little success, and it might be anticipated that this will be the case in general. Patent cover is vital for commercial development of any new product, especially ones involving new concepts such as those based on semiochemicals. Manufacturers have had to rely on patenting uses of semiochemicals or such aspects as synthetic procedures, formulations or trap designs, but these are easier to circumvent and more difficult to police than would be patents on the compounds themselves. Again semiochemicals contrast completely with conventional insecticides: the latter are typically patentable novel compounds, and the production of analogues with greater effectiveness or a certain specificity is the life-blood of the agrochemical business.

One vital factor that until recently has hindered the development of semiochemicals in pest control, but which in the future could do a great deal to promote them, is the problem of registration procedures. In most countries, semiochemicals have had to go through the same testing procedures as conventional insecticides before they can be registered for control use. Completing registration requirements is a major element in the cost of introducing a new pest control agent onto the market, and only one or two firms have risked the cost of registering a pheromone-based product when the prospect of any financial return is so uncertain. Furthermore, although many pheromones are closely related in chemical strcture, each individual compound has had to be subjected to the whole series of tests and no analogies could be assumed. Since the beginning of 1982, the U.S. Environmental Protection Agency has set much less demanding registration requirements for so-called "biorational" pest control agents which include pheromones and other semiochemicals. These changes could cut the cost of registration of such compounds to as little as one tenth of that of registering a new conventional insecticide, and consequently could provide more and profitable markets for semiochemical-based control agents.

A development of this argument is that semiochemical-based monitoring and control agents should not be seen as being in competition with conventional insecticides. The development of resistance by insect species to specific insecticides or, even worse, classes of insecticides can severly reduce the return on the enormous research, development and registration costs now required to get a new insecticide onto the market. Careful monitoring, and use of semiochemicals to maintain pest populations at a low level wherever possible, could reduce the excessive use of insecticides which encourages the appearance of resistance, and so prolong the marketable lifetime of an insecticide and increase the overall return on investment.

CONCLUSIONS

Pheromones and other semiochemicals undoubtedly have a role to play in reducing the losses of food crops due to insect pests, but realisation of their full potential will require a more flexible approach to pest management than that engendered by years of routine spraying of broad-spectrum insecticides. Fundamental changes in manufacturing and marketing strategies will be needed to ensure that these chemicals are widely available for field use. Agricultural advisers and field workers will be required who can exploit the value of semiochemicals in conjunction with other techniques in integrated pest management programmes, and who are able to select the most appropriate combination of methods for any particular situation.

REFERENCES

Alford, D.V., P.W. Carden, E.B. Dennis, H.J. Gould and J.D.R. Vernon (1979). Monitoring codling and tortrix moths in United Kingdom apple orchards using pheromone traps. Ann. Appl. Biol., 91, 165-178.

Arn, H., S. Rauscher, A. Schmid, C. Jaccard and B.A. Bierl-Leonhardt (1981). Field experiments to develop control of the grape moth, Eupoecilia ambiguella, by communication disruption. In: Management of Insect Pests with Semiochemicals: Concepts and Practice, (ed. Mitchell, E.R.). Plenum Press, New York, 327-338.

Burkholder, W.E. (1981). Biomonitoring for stored-product insects. idem 29-40.

Dyck, V.A. (1979). Sex pheromones: monitoring pest behaviour with pheromone traps. In: Annual Report, International Rice Research Institute, Manila, Philippines, 220-221.

Dyck, V.A. (1981). Stem borer control through mating disruption. In Annual Report, International Rice Research Institute, Manilla, Philippines, 60-61.

Gurevitz, E. and S. Gothilf (1982). Field studies in vineyard with sex pheromone of the European grapevine moth, Lobesia Botrana Schiff., and honeydew moth, Cryptoblabes gnidiella Mill. 3rd Israeli Meeting on Pheromone Research, Ben-Gurion University of the Negev, Beer-sheva, May 4, 1982.

Haynes, K.F., M.C. Birch, and J.A. Klun (1981). Sex pheromone offers promise for control of artichoke plume moth. California Agriculture, 13-14.
Lee, J.O., J.S. Park, H.G. Goh, J.H. Kim and J.G. Jun (1981). Field study on mating confusion of synthetic sex pheromone in the striped rice borer, Chilo suppressalis (Lepidoptera:Pyralidae). Korean J. Plant Protection, 20, 25-30.
Levinson, H.Z. and A.R. Levinson (1979). Trapping of storage pests by sex and food attractants as a tool of integrated control. In: Chemical Ecology: Odour Communication in Animals, (ed. Ritter, F.J.). Elsevier/North Holland Biomedical Press, Amsterdam, 327-341.
Maclellan, C.R. (1976). Suppression of codling moth (Lepidoptera: Olethreutidae) by sex pheromone trapping of males. Can. Ent., 108, 1037-1040.
Madsen, H.F. (1981). Monitoring codling moth populations in British Columbia apple orchards. In: Management of Insect Pests with Semiochemicals: Concepts and Practice, (ed. Mitchell, E.R.). Plenum Press, New York, 57-62.
McLaughlin, J.R., E.R. Mitchell and J.H. Cross (1981). Field and laboratory evaluation of mating disruptants of Heliothis zea and Spodoptera frugiperda in Florida. idem 243-251.
Minks, A.K. (1979). Present status of insect pheromones in agriculture and forestry. Proceedings of IOBC International Symposium on "Integrated Control in Agriculture Today". Vienna, October 1979, 127-136.
Nakasuji, F. and K. Kiritani (1976). Utilisation of pheromone traps for decision making of chemical control against Spodoptera litura. Proceedings of Symposium "Insect Pheromones and their Applications". Nagaoka and Tokyo, 1976.
Negishi, T., T. Ishiwatari, S. Asano and H. Fujikawa (1980). Mass trapping for the smaller tea tortrix control. Appl. Ent. Zool., 15, 113-114.
Nyirenda, G.K.C. (1982). The armyworm (Spodoptera exempta Walk.) pheromone trap network in Malawi. Paper presented at "Armyworm and other Migrant Pest Workshop". Arusha, Tanzania. July 5-9 1982.
Perry, J.N., E.D.M. Macaulay and B.J. Emmett (1981). Phenological and geographical relationships between catches of pea moth in sex-attractant traps. Ann. Appl. Biol., 97, 17-26.
Poitout, S. and R. Bues (1978). Le cycle evolutif de lay pyrale du riz (Chilo suppressalis Walker) dans le delta rhodanien. Utilisation due piegeage sexuel. Ann. Zool. Ecol. Anim., 10, 245-265.
Raman, K.R. (1982). Field trials with the sex pheromone of the potato tuberworm. Environ. Entomol., 11, 367-370.
Reichmuth, von C., H.-U. Schmidt, A.R. Levinson and H.Z. Levinson (1980). Das jahreszeitliche Auftreten von Speichermotten (Ephestia elutella Hbn.) in Berliner Getreideschuttbodenlagern sowie der zeitentsprechende Einsatz von Bekampfungsmassnahmen. Z. ang. Ent., 89, 104-111.
Rock, G.C., C.C. Childers and H.J. Kirk (1978). Insecticide applications based on Codlemone trap catches vs. automatic schedule treatments for codling moth in North Carolina apple orchard. J. Econ. Entomol., 71, 650-653.
Stockel, J. and P. Anglade (1976). Premiers resultats de lutte par confusion sexuelle contre Ostrinia nubilalis (Lep. Pyralidae) dans les conditions naturelles. In: Les Pheromones Sexuelles des Lepidopteres. Institut National de Recherches Agronomiques, Centre de Recherches de Bordeaux, 170-182.
Szivos, Y., I. Bar-Zakay and M. Malka (1982). Synthetic sex pheromone traps for monitoring the Californian red scale in citrus groves of the Jordan Valley. Paper presented at 3rd Israeli Meeting on Pheromone Research, Ben-Gurion University of the Negev, Beer-sheva, May 4 1982.
Tatsuki, S. and H. Kanno (1981). Disruption of sex pheromone communication in Chilo suppressalis with pheromone and analogs. In: Management of Insect Pests with Semiochemicals: Concepts and Practice, (ed. Mitchell, E.R.). Plenum Press, New York, 313-325.
Vick, K.W., J.A. Coffelt, R.W. Mankin and E.L. Soderstrom (1981). Recent developments in the use of pheromones to monitor Plodia interpunctella and Ephestia cautella. idem 19-28.

THE POTENTIAL FOR THE INTEGRATION OF PLANT RESISTANCE, AGRONOMIC, BIOLOGICAL, PHYSICAL/ MECHANICAL TECHNIQUES, AND PESTICIDES FOR PEST CONTROL IN FARMING SYSTEMS

Ida Nyoman Oka
Directorate of Food Crop Protection
Ministry of Agriculture, Jalan AUP, Pasarminggu
Jakarta, Indonesia

ABSTRACT

Intensification of food production has also increased pest damage, which acts as a deterrent. For example, plant and leafhoppers combined with the grassy stunt and ragged stunt viruses (both vectored by the brown planthopper) and the rice tungto virus (vectored by the green leafhopper), have become most destructive rice pests in South and South East Asia. Intensive use of insecticides to control the pests are extremely complex and expensive. Pest control strategies should be based on a knowledge of the complex relationships of the natural forces regulating pest populations in tropical environments. The concept of integrated pest management would be appropriate for stabilizing yields at a high level, least damaging to the environment and most profitable to individual farmers. An interdisciplinary research approach is required. For field implementation close cooperation among government authorities, extension services, field technicians, private enterprise, village leaders and farmers is essential.

KEYWORDS: pests, resistant cultivars, agronomic techniques, biological control of pests, pesticides, surveillance systems, integrated pest management.

FOOD PRODUCTION AND PEST PROBLEMS

One of the greatest challenges for humanity today is to alleviate hunger and malnutrition suffered by more than half a billion people in many developing countries. Food production, particularly in South and South Asia is being increased by intensifying the productivity per unit land area through increased inputs of chemical fertilisers and pesticides, use of high yielding varieties, improved and expanded irrigation systems, and high cropping intensity. It resulted in a marginal increase of food consumption above population growth (Table 1). Nevertheless, because of rapid population growth and increasing demands for more and better quality food in the next few decades, developing nations are challenged to continue to increase their food production.

Prolonged droughts or excessive rains and particularly inadequate pest control are the main crop yield constraints for farmers. Insect pests decreased yield of rice by an average of 0.4 ton/ha (Herdt, 1982). Increasing pest problems observed on large scale intensified rice areas also add to production instability. For example, the GLH (Green Leafhopper), the BPH (Brown Planthopper), and the WBPH

Table 1. Foodgrain consumption per capita, 1961-79 (World Dev. Reports, 1981)

Country group and region	Kilogram per capita			Average annual growth rates (percentage)	
	1961-64	1970-73	1976-79	1961-64; 1970-73	1970-73; 1977-79
World total	312.1	342.8	362.1	1.0	0.9
Developing countries	223.0	229.7	239.9	0.3	0.7
Low-income countries	207.1	202.7	202.4	-0.2	0.0
Sub-Sahara Africa	159.5	151.9	141.3	-0.5	-1.2
South Asia	215.6	211.8	213.5	-0.2	0.2
Middle-income countries	238.1	255.6	275.7	0.8	1.3
Sub-Sahara Africa	140.7	150.0	148.5	0.7	-0.2
East Asia	257.2	271.2	282.7	0.6	0.7
Latin America	235.7	244.0	249.1	0.4	0.3
S. Europe, N. Africa Middle East	390.6	441.0	495.8	1.4	2.0

(White Backed Planthopper), formerly grouped as minor rice pests have become important pests. The rice stemborers, stink bugs, leaf folders, gallmidge and rats are also becoming serious pests. The rice blast, bacterial leaf blight, sheath blight, rice tungro virus (RTV), grassy stunt virus (GSV) and ragged stunt virus (RSV), have become economically important rice disease in the intensified rice programme. Yield losses due to the important rice insect pests under field conditions vary widely depending on the insect species, varietal reactions, growth stage of the plants and the degree of damage. The BPH and the GSV and RSV (both vectored by the insect) have caused significant yield losses in most Asian countries. More than 450,000 ha of rice were damaged in 1976-1977 in Indonesia reducing yields by about 364,500 t of milled rice valued at more than $100 M. This would be enough to feed more than 3 million people for the whole year (Oka, 1979).

A combination of several factors could attribute to the increasing pest problems in the intensified areas such as (1) planting a few modern varieties over wide areas thus reducing varietal diversity and increasing the instability of community associations in favor of the pests, (2) planting varieties that do not possess resistance to all of their pests, (3) planting modern varieties with a small genetic base for resistance to a certain pest species triggering development of new races or biotypes of the pests capable of breaking down the resistance, (4) continuous and staggered planting the year around providing a permanent food source and shelter for the pests to develop, (5) increased application of nitrogeneous fertilizers enhancing the development of many pest species, and (6) overdependance on pesticide use creating pest resistance, pest resurgence, secondary pest outbreaks and other unwanted side effects.

Pest control strategies should be based on understanding the intricate interrelationships of the components of the crop ecosystem. In this context the concept of integrated pest management would be most appropriate to obtain a sustained high yield ecosystem, least damage to the environment, protection for both producers and consumers and most profit for the individual farmers. The following is a discussion of the potential for the integration of various control tactics against some important food crop pests in the farmers field. In Indonesia, this concept has been implemented with good results since 1979 to control BPH, RTV and rats attacking rice, and downy mildew on corn.

INTEGRATION OF CONTROL TACTICS

Resistant Cultivars

The use of plants resistant to pests is a control tactic that is relatively stable, cheap, non polluting and generally compatible with other tactics. Plant resistance has been extensively used by plant pathologists to combat various plant pathogens. Recently, significant advances have been made in the development of plant resistance against important rice insect pests. Six genes in rice for BPH resistance have been identified and four of them already incorporated into varieties (Bph-1, Cph-2, Cph-3 and Bph-4) which are resistant to the currently occurring biotypes in South East Asia. More than 20 million ha of rice land in Asia are now planted to BPH resistant varieties. Seven different genes for resistance to GLH have been identified and incorporated into improved varieties. Four genes for resistance to WBPH, and three genes to zig-zag-leafhopper have also been identified and are being incorporated into improved germ plasm (IRRI Quiquennial Draft Report, 1982). In addition, improved varieties with moderate resistance to the rice stem borer have been bred. Various gene sources for resistance to important rice diseases have also been identified and incorporated into improved varieties, i.e. resistance to RTV, BLB, GSV (Khush, 1980).

Employing plant resistance to cope with plant pests has certain limitations. The use of the small genetic base for resistance could trigger the development of new races (Van der Plank, 1968) or biotypes of pests capable of attacking the plant. For example, IR26, 28, 30 and 34 all possessing the Bph-1 resistance gene for BPH biotype 1 have caused rapid selection for biotype 2 of the BPH, and the varieties became highly susceptible to the insect (Oka, 1977). On the other hand, IR32, 36, 38 in which the Cph-2 resistance gene was incorporated have remained resistant to biotype-2 of BPH for more than 12-15 rice seasons in Indonesia. The reasons for this stability are not known. However, IR36 became susceptible to the RTV transmitted by the green leafhopper, *Nephotettix virescens*, after 10-12 cropping seasons in several places in Indonesia. A new biotype of the insect and/or a new race(s) of the RTV is suspected. Employing resistant plants has not been satisfactory against rice blast, *Pyricularia oryzae*, one of the most variable rice diseases. Blast disease is the number one constraint to expansion and intensification of upland rice culture. Much more information on the etiology, epidemiology, physiology and the genetics of the blast fungus is needed.

Modern resistant varieties have contributed significantly to increased yields. To arrive at a sustained high yield ecosystem, resistance should be relatively stable by broadening the genetic base for resistance by incorporating both vertical and horizontal resistance genes and by pyramiding the resistance to the particular pest biotypes. This strategy has been vigorously pursued by IRRI for some important rice pests. The wide genetic variability of the rice plants should make possible breeding of modern varieties with multiple resistance or tolerance. Several rice varieties have been released, capable of coping with a number of important pests (Table 2). In addition to conventional breeding methods, innovative breeding techniques like hybrid rice breeding, distant hybridization, mutation breeding and tissue culture is highly promising (Anon, 1980). To ensure success, cooperation among scientists of related disciplines is needed.

TABLE 2 Levels of Resistance to Important Rice Pests of Modern Rice Varieties Widely Planted in Indonesia*

No.	Varieties	BPH Biotypes 1	2	3	WBPH	GLH Isolates Bali	Lombok	S.Sulawesi	Stem borer	Gall-midge	Grassy stunt virus	Ragged stunt virus	Rice tungro virus	BLB	Blast	Sheath rot
1.	IR-5	S	S	S	S	MR	-	-	S	S	S	S	S	S	MS	MS
2.	IR 8	S	S	S	S	MS	MS	MS	S	S	S	S	S	S	S	MR
3.	IR 20	S	S	S	S	MS	MR	MS	S	S	-	-	MR	MR	S	S
4.	Pelita	S	S	S	S	-	-	-	S	S	S	S	S	MR	S	-
5.	Si Ampat (C4)	S	S	S	S	-	-	-	MR	S	-	-	MR	MS	MR	MR
6.	IR 26	R	S	R	S	MR	MR	MR	MR	S	MS	S	MR	S	S	-
7.	IR 28	R	S	R	S	MR	MR	MR	MR	S	R	S	MR	R	S	-
8.	IR 29	R	S	R	S	MR	MR	MR	MR	S	-	-	-	-	S	-
9.	IR 30	R	S	R	S	MR	MR	MR	MR	S	R	S	MR	S	S	-
10.	IR 32	R	R	S	S	MR	MR	MR	MR	S	MR	S	MR	S	S	-
11.	IR 34	R	S	R	S	MR	MR	MR	MR	S	R	S	-	S	S	S
12.	IR 36	R	R	S	S	MR	MR	MR	MR	S	R	S	S	MS	MR	MS
13.	IR 38	R	R	S	S	MS	MR	MR	MR	S	R	S	MR	MR	MS	MS
14.	IR 40	-	-	-	-	-	-	-	-	-	-	-	-	MS	S	-
15.	IR 42	R	R	S	-	MR	MR	MR	-	-	R	S	S	R	S	S
16.	Brantas	R	S	R	S	MR	MR	MR	-	S	-	-	R	S	S	-
17.	Serayu	R	S	-	-	MR	MR	MR	-	S	-	-	R	-	S	-
18.	Citarum	R	S	MR	MR	MR	MR	MR	-	S	-	-	R	S	S	-
19.	Asahan	R	S	MR	S	MS	MR	MR	-	S	R	MR	MR	S	S	-
20.	Ayung	R	MR	R	S	MR	MR	MR	S	S	R	-	S	MR	-	-
21.	Cisadane	R	MR	R	S	MR	MR	MR	-	S	MS	MS	S	MR	S	-
22.	Cimandiri	R	MR	R	-	MR	MR	MR	S	S	R	MR	S	R	MR	-
23.	Semeru	R	MR	-	-	MR	MR	MR	S	S	S	S	S	R	S	S
24.	Cipunegara	R	MR	MS	-	-	-	-	-	-	S	S	S	MR	S	S
25.	IR 50	R	R	R	S	MR	MR	MR	S	-	R	-	MR	R	S	S
26.	IR 52	R	R	R	S	MR	MR	MR	-	-	MR	S	MS	R	S	S
27.	IR 54	R	R	R	S	MR	MR	MR	-	-	-	-	MS	MR	-	-
28.	Barito	R	MR/MS	MR/MS	-	-	-	-	-	-	-	-	S	MR	-	-
29.	Krueng Aceh	R	R	R	-	-	-	-	-	-	-	-	MS	-	-	-
30.	Batang Agam	S	S	S	-	-	-	-	-	-	-	-	R	MR	-	-

* Compiled from files of the Entomology and Plant Pathology Departments of the Bogor Res. Inst. Food Corp.

To reduce pest pressures on the resistant cultivars, and therefore minimize the risks of development of new races/biotypes of the pests, the resistant cultivars should be integrated with other control methods such as agronomic techniques, biological controls and judicious use of pesticides.

Agronomic Techniques

Under tropical environment crops are subjected to continuous stresses by pests, particularly under staggered planting conditions. The existing agronomic techniques in the tropics such as synchronized planting, crop rotation, plant spacing and fertilizer management may prevent both build up of pest populations and overlapping generations of the pest. These techniques should be compatible with other control methods.

Rice in the wet season may be immediately followed by rice in the dry season (Figure 1a) or rice in the dry season may be immediately followed by rice in the wet season (Figure 1b). In many areas three rice crops per year are introduced. In all these situations the crop is under staggered planting conditions, so any pest is able to develop continuously with a constant supply of food. If a field is simultaneously prepared and synchronized planting encouraged and carried out over an entire community during April (Figure 2a) or November (Figure 2b) the following rice crop may have relatively fewer pests. Integration of resistant varieties with these agronomic techniques should further reduce pest build up. Short duration secondary food crops should be chosen to fit in the cropping systems.

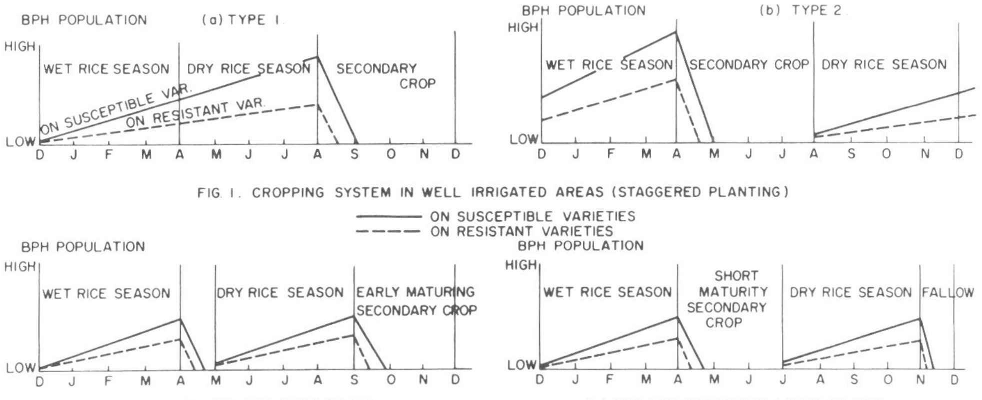

FIG. 1. CROPPING SYSTEM IN WELL IRRIGATED AREAS (STAGGERED PLANTING)

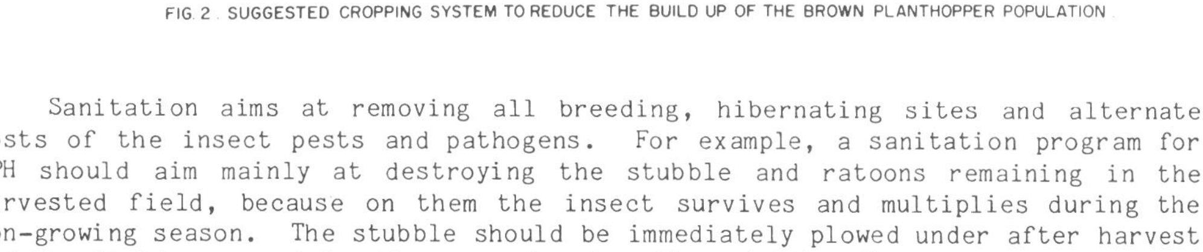

(a) WET RICE SEASON FOLLOWED BY DRY

(b) DRY RICE SEASON FOLLOWED BY WET

FIG. 2. SUGGESTED CROPPING SYSTEM TO REDUCE THE BUILD UP OF THE BROWN PLANTHOPPER POPULATION

Sanitation aims at removing all breeding, hibernating sites and alternate hosts of the insect pests and pathogens. For example, a sanitation program for BPH should aim mainly at destroying the stubble and ratoons remaining in the harvested field, because on them the insect survives and multiplies during the non-growing season. The stubble should be immediately plowed under after harvest and the field prepared for the next planting (Oka, 1979). The larvae and pupae of the rice stem borer and the GSV, RSV and RTV will also be destroyed by this practice. Burning the straw and stubble has also been suggested, but in many cases the field is still wet after harvest and the intensive schedule does not permit drying and therefore burning. Moreover, burning the stubble and straw may also destroy most of the arthropod populations that play an important role in decomposing the plant remains. Burning also wastes the available nitrogen in the plant remains. Moreover, nutrient loss by leaching is much higher after burning (IRRI, 1973).

Weed sanitation in rice fields is another requirement because it makes the microclimate less favorable for insect pests. Weeds on the ditches and fallow fields do not need to be entirely removed, because they may shelter important natural enemies for the insect pests. More research is needed on the role of the weed species in the interaction of the insect pests and their natural enemies, as well as the alternate hosts for plant disease.

Micro environments influence the development of pests. For example, blast of rice, *Pyricularia oryzae* becomes more serious under high air humidity, cool temperatures and "dry soil" conditions (Ou, 1972). In Indonesia, blast is most severe in upland rice while in irrigated rice blast has never been a problem. Close planting which results in a shaded and humid environment is unfavorable for the development of the natural enemies of the BPH, resulting in an increase in the pest population (Nishida, 1975). A dark habitat is an ideal place for the insect to congregate and multiply. Solar and ultra violet radiation act abiotically against the BPH and restrain its increase (Suenaga, 1963; Oka, 1979). Therefore, the most appropriate spacing would let enough sunshine penetrate into the basal portions of the plants to prevent pest increase, but would provide a suitable habitat for development of natural enemies. The suggested spacing is rows 15 to 20 cm apart (Kulshreshtha et al., 1974).

Draining the field at the proper time and withholding irrigation effectively controls rice water weevil, suppressing several outbreaks of BPH in Malaysia and in the Philippines (Pimentel et al., 1960; Miller and Pagaden, 1930; Dyck, 1974, in Oka, 1979). Raising the water level to control the BPH is a common practice in Indonesia (Tjoa, 1952). Sand and sawdust containing 0.25 liter kerosene for every 100 sq m was broadcast on the raised water level. The plants were then shaken to dislodge the pests. WBPH and *Scotinophora* sp. will also be affected. Raising the water level will also destroy the eggs of BPH laid in the leaf sheaths.

High inputs of chemical nitrogeneous fertilizers significantly increase certain species of rice pests (Ishii, 1945; Narayanan et al., 1973; Dyck, 1973, *in* Oka, 1979). While high nitrogeneous fertilizer is needed for high rice yields there is a balance between yields and severity of the pests at certain rate of chemical nitrogen application where the pests still can be tolerated. The Indonesian experience indicated that chemical nitrogen applied at the rate of 90 - 120 kg/ha significantly increased yields but apparently did not cause significant increases in pest populations.

Of the agronomic techniques for pest controls discussed above, the most promising are synchronized planting, crop rotation with other crops and sanitation. These approaches would break the life cycle of the pests and therefore, prevent further build up of the pests. They also will prevent development of overlapping generations of certain pest species. Monitoring of the pests will also be made easier.

Biological Control

At present knowledge on the natural enemies of major pests of rice in Southeast Asia is incomplete. In India, natural enemies of rice insect pests have been identified and some assessments made of their effectiveness. In Sri Lanka, preliminary observations on some beneficial parasitoids of insect pests have been carried out. In Malaysia, beneficial predators have been monitored (Greathead, 1982). In Indonesia, the importance of natural enemies of BPH and the yellow rice stem borer have been assessed (Eddy Mahrub et al., 1980). In Thailand, an exhaustive listing of natural enemies of rice insect pests and their quantitative benefit for biological control have been made by Yasumatsu et al. (1975). On mainland China, basic studies on biological control of rice insect pests have been made. Spiders, Trichogramma wasps, pathogen (B. thuringiensis) and ducks have been utilized in integrated control program on rice pests (Chiu, 1979). In the tropics, the role of natural enemies for pest control is important (Heinrichs et al., 1978). Therefore, within the context of IPC it is essential to find ways to maximize biological control agents. It is important to quantify the value of each of them as well as their combined action for suppressing a pest species population. For example, 79 species of natural enemies of BPH (42 species of parasites and 37 species of predators) were identified in Taiwan (Chiu, 1979). Some of them are known to have great potential to suppress the insect pest population. A small water spider, M. atrolineata is highly effective in controlling the BPH population (IRRI, 1981). The abundance of spiders, coccinellids and C. lividipennis in outbreak areas in Indonesia suggest that these are important enemies of the BPH. However, the small activity of parasitism/predation at a crucial stage may be more significant than high mortality at another (Greathead, 1982). Nishida et al., 1976 (in Chiu, 1979) reported that the combined egg parasitism rate ranged between 90 - 100% mortality of the BPH in many localities in Thailand. It is also important to identify the key species of the natural enemies under a given set of environmental conditions.

Investigation of methods of conserving, such as habitat modification and selection of specific pesticides deserves attention. Growing Digitaria sp. for example, was found to increase the activity of C. lividipennis, a predator of the BPH (Stapley, 1975, in Oka, 1979). To properly integrate pesticides with biological control, specific pesticides, proper timing and methods of application should be sought.

Physical and Mechanical Control

Heat treatment of seeds has been used to a limited extent for controlling the white tip nematode and other seedborne diseases (John et al., 1979). Various types of tillage equipment (plow and harrows) have been designed to effectively control weeds during soil preparation. On mainland China, light traps have been used along with other control tactics against some rice pests (Chiu, 1979). However, the effectiveness of light traps is questioned, because beneficial insects are also trapped. In addition, there are high costs involved. In Indonesia, the use of sulfur gas, traps and clubbing have been used for rat control and are carried out simultaneously with soil preparation by farmers.

Pesticides

Pesticides are becoming an integral part of modern agricultural technology for developing countries in increasing and maintaining high crop productivity and also in saving human lives. Therefore, pesticide use should be within the context of IPC to increase their efficiency and minimize unwanted side effects, such as pest resurgence, pest resistance to pesticides, destruction of natural enemies, beneficial insects and non-target species, persistence, residue problems and hazards. The existing pesticide regulations and their enforcement need to be strengthened to avoid adulteration and misuse of pesticides.

Selective chemical pesticides (i.e. pesticides that act against certain groups of insect pests but cause less harm to their natural enemies) should be encouraged. Insecticides such as Sumibas 75 EC and Brantasan 750 EC are effective against BPH, but have only low toxicity to a predator of BPH, the spider mite Lycosa pseudonnulata, and do not affect the other predators of BPH, Paederus tanulus, Casnodea interstialis and Coccinellids (Dandi, 1981). Pesticides should also be evaluated on selected key species of natural enemies under field conditions. Methods and time of application of pesticides need further research. Proper time of pesticide application should be based on economic thereshold levels obtained from surveillance data. For a number of rice pests economic threshold levels have been determined (Heinrichs et al., 1978).

Improved methods of pesticide application such as furrow placement or direct soil application are highly promising and economical, particularly for control of pests on irrigated rice. Monitoring of development of resistance of pests against pesticides is essential. Rotation of pesticides with different modes of action deserve investigation to avoid the development of pesticide resistance. Some of the microbial pesticides such as Thuricide (i.e. Bacillus thuringiensis) are effective for control of a number of lepidopterous pests. The potentials of fungi, viruses, and protozoa for pest control need further research. The use of pheromones to disrupt mating and the application of antifeedants to control pest populations offer other means of managing pest populations (Heinrichs et al., 1979).

Surveillance and Forecasting

Surveillance and prediction of pest outbreaks are essential for effective integrated control programs. South and South East Asian countries vary in the degree of development of surveillance systems. Some are organized and carried out by government field technicians and not by farmers (Heinrichs et al., 1979). In Indonesia, farmers' organizations also participate in surveying their own fields for pest occurrence as an integral part of their production program. The short term objective of the surveillance is mainly for the purpose of large-scale, short term forecasting of outbreaks of certain pest species. As economic injury levels for some rice pests have been developed the collected surveillance data is also being used for the issuance of pesticide control recommendations to the farmers. Thus the surveillance system is designed to become an integral part of the implementation of the integrated pest control program (IPC) for farmers.

In the future, surveillance techniques should be improved for long term forecasting purposes. The potential of behavioural chemicals such as sex attractants to monitor certain pest populations under tropical environments will receive more attention (Lewis et al., 1980). These field survey systems should remain simple and reliable based on empirical data (Rabbinge, 1979) so that they can be carried out by field technicians and the farmers themselves. But it may become unreliable when the conditions change. To aid in pest population monitoring more knowledge is needed on pest behaviour, population dynamics as affected by biological (varietal, natural enemies) and physical factors. For example, the knowledge of varietal interactions with BPH made possible the development of survey technique for BPH biotypes. Timely detection of the BPH biotypes is a prerequisite for the distribution of the appropriate resistant varieties to the biotypes existing in the region (Oka, 1980). The collected surveillance data should also be standardized for use in computers and rapid analysis.

PROBLEMS AND PROSPECTS

Countries in South and South East Asia and mainland China have been developing some forms of integrated pest control programs for farmers, such as varietal resistance x chemical control, cultural techniques x chemical control, and varietal resistance x cultural techniques x biological x chemical controls.

Varietal Resistance x Chemical Control

Modern rice varieties bred at IRRI and varieties selected from IRRI bred materials resistant to various pests are planted in about 20-30 million ha throughout the world (IRRI Quinquennial Draft Report 1982). In addition, national rice research programs periodically release pest resistant modern varieties. In this respect the resistant varieties only need chemical protection from other pests to which they are not resistant (Pathak and Dhaliwal, 1981).

Cultural Techniques x Chemical Control

To control the gallmidge, *Orseolia oryzae*, in mainland China, the emphasis is weeding of the overwintering host plant, *Leersia hexandra*, and insecticide treatments in the third and fourth generations in seedbeds and late crop rice fields (Chiu, 1980). In Indonesia, gallmidge problems can be effectively controlled by early planting for wet season rice and applying chemicals when 5 silvershoots/m^2 is detected on plants less than 40 days old. Reduction of chemical applications will give natural enemies a better chance to operate later on, such as *Platygaster oryzae*, one of the most effective larval parasites.

Combinations of plant resistance x cultural techniques x biological x chemicals are successful to control BPH, WBPH and GLH in rice centers in Indonesia. The use of multiple resistant varieties particularly to pests (Table 2) occupies almost 70 percent of the rice lands. This significantly reduces pesticide use. When needed, one or two sprays of pesticides are applied for control of WBPH and other pest species to which the varieties are susceptible. The natural enemies of the pests become less damaged. The resistant varieties are planted, synchronized and rotated with non-rice crops which prevent continuous build-up of the pests. In this way development of biotypes of the pests may be minimized. The selected sanitation programs (destruction of rice stubbles and ratoons) will also destroy both the pests and the sources of the virus diseases.

Large scale surveillance systems have been organized by government field technicians in various countries with the objective of forecasting outbreaks of certain pests and issuing of pest control recommendations. This significantly increases the efficiency of pesticide use. In addition applications of selective pesticides which are less harmful to natural enemies are encouraged.

Various problems still exist that prevent effective and successful crop protection programs. These are environmental constraints (small land holdings, water, farmers' health), politico-economic constraints (credit facilities, price policy, distribution network), socio-cultural constraints (customs, social structures) and technological constraints (communication, extension, education and training) (Haskell et al., 1979). Considering the difficult and complex interrelationships of the various constraints mentioned, any program that is based on "ad hoc" approaches will be ineffective. Instead, pest control programs must be based on a comprehensive and integrated plan. This requires an interdisciplinary approach of scientists from various disciplines. Entomologists, plant pathologists and weed specialists will have to work together with agronomists, plant breeders and others to produce a sound IPM program for both agriculture and society (Pimentel, 1982). In addition, particularly for developing nations, the following suggestions deserve attention.

1. The prevailing political will of many governments of developing nations to protect the farmers and improve their living standards should be exploited for bringing research results of integrated pest management down into farming systems. In this regard deep involvement and guidance of the government is essential. Many of the constraints have been vigorously attacked by governments, for example the

Philippines in the Masagana 99 and Indonesia in the Bimas - Inmas - Insus programs. Integrated crop protection activities should be an integral part of the total crop production system.

2. Close cooperation among government authorities, extension specialists, field technicians, researchers, private enterprises, informal influental village leaders, and farmers should be developed.

3. Massive extension programs should be organized to convince the farmers and other leaders of the advantages of integrated control methods. Regular training and visits carried out by field extension workers appears to be highly effective.

4. Farmers should be organized into production units which aid them to contribute jointly to an improved production program, including the implementation of certain pest control tactics. In Indonesia, these joint actions are made easier through the expansion and improvement of the existing farmers' irrigation organizations and the establishment of farmers' cooperatives.

5. Since the program should cover as wide an area as possible - especially with regard to crop rotation and synchronized planting, irrigation water regulations should be made for groups of farmers with land situated in one large area. Expansion and improvement of irrigation canals, dams and drainage systems are essential. Governments must put great emphasis on the development of a suitable infrastructure.

6. Legally approved pesticides and sprays should be made available to the farmers when needed. For this purpose wholesalers and particularly licenced pesticide retailers should be established at sites easily reached by the farmers. Existing pesticide regulations in developing countries should be strengthened. Emphasis should be given to pesticide distribution, packaging, storage, labelling with clear directions (in local languages) and quality control. Education for pesticide retailers and farmers through short courses, visual aids, radio and television, is essential to ensure proper handling and use of pesticides. Monitoring on the behavior of pesticides under tropical environments is also essential.

7. Research priorities in plant protection should be geared towards strengthening the integrated pest control tactics, including those discussed here. Socio-economic aspects of IPC in farmers' fields also needs further research.

8. Pilot projects on IPC that are going on in various countries in South and South East Asia need to be continued and strengthened. In this respect, a coordinated effort and exchange of information and evaluation for improvement of pilot projects has been achieved through the FAO IPC Program for rice.

9. For proper and wide scale implementation of IPC in farming systems, more trained field and research personnel are needed. IRRI's training programs on IPC play a significant role in this regard.

10. Courses in plant protection offered in universities in developing countries should be more oriented towards strengthening the IPC concept.

A tremendous challenge exists to all scientists concerned with IPC and helping the millions of farmers throughout the world. It gives scientists involved in plant protection and practitioners an opportunity to prove that the IPC concept is sound and most appropriate for achieving a sustained high crop yield and a healthy environment.

ACKNOWLEDGEMENT

The writer thanks the following specialists for reading an earlier draft of the manuscript and for their helpful suggestions: David Pimentel, Cornell University; E.A. Heinrichs, The International Rice Research Institute, B.H. Siwi, Central Research Institute for Food Crops, Indonesia.

REFERENCES

Anonymous (1980). Innovative approaches to rice breeding. Selected papers from the 1979 International Rice Research Conference. IRRI, Los Banos, Philippines, pp. 182.

Chiu, Shin-Foon (1979). Integrated Rice Insect Control in China. Proc. Symposia IX Internat. Congress of Plant Protection. Vol. II., Ed. Dr. Thor Kommendahl, Dept. of Plant Path., University of Minnesota, pp. 466-470.

Chiu, Shin-Chen (1979). Biological control of the brown planthopper. Brown planthopper: threat to rice production in Asia. International Rice Research Institute, Los Banos, Laguna, Philippines, pp. 335-355.

Chiu, Shin-Foon (Zhao Shan-huan) (1980). Integrated control of rice insect pests in China. Rice improvement in China and other Asian countries. International Rice Research Institute and Chinese Academy of Agricultural Sciences, pp. 239-250.

Dandi Sukarna (1981). Pemberian insektisida berdasar populasi wereng coklat. (Timing of insecticide application to control the brown planthopper, N. lugens based on population density). Penelitian Pertanian, Pusat Penelitian dan Pengembangan Tanaman Pangan, Bogor, 1, 1, 38-41.

Eddy Mahrub, Achmad Mukidjo, Kasumbogo Untung, Soeprapto Mangundihardjo, Sundoro dan Sri Hadiyani (1980). Laporan inventarisasi musuh alami wereng coklat dan penggerek batang padi di WKBPP Sewon, Yogyakarta (Laporan Kemajuan). Lokakarya Pengendalian Hama Terpadu, 22-27 September, 1980. 16/LPHT. 14 hal). (In Indonesian, abstract in English).

Greathead, D. (1982). The need for biological control input to the FAO Integrated Pest Control in rice programme in South and South East Asia. AGP: IPC/11/WP/7. Elevent Session of the Panel of Experts on IPC, FAO, Kuala Lumpur, Malaysia, March 1982.

Heinrichs, E.A., R.C. Saxena and S. Chelliah (1978). Development and implementation of insect pest management systems for rice in tropical Asia. Seminar on "Sensible use of pesticides", Tokyo, Japan, Nov. 29 - Dec. 3, 1979, 33.

Heinrichs, E.A., V.A. Dyck, R.C. Saxena and J.A. Litsinger (1979). Development of rice insect pest management systems in the tropics, Proc. of Symposia IX Internat. Congress of Plant Protection. Vol. II, ed. Dr. Thor Kommendahl, Dept. of Plant Pathology, University of Minnesota, pp. 463-466.

Haskell, P.T., T. Beacock and P.J. Wortley (1979). World wide Socio-economic constraints to Crop Protection. Proc. Symposia IX International Congress of Plant Protection, I: Fundamentals, pp. 39-41.

Herdt, Robert W. (1982). Changing Asian rice technology: Impact, benefits and constraints. Selected papers from IRRI Research on Consequences of new rice technology. Assembled for the Quinquennial Review, Jan., 1982. IRRI, pp. 81-129.

International Rice Research Institute (1973). Annual report for 1972, Los Banos, Philippines, pp. 246.

International Rice Research Institute (1981). Research highlights for 1980, 119.

IRRI Quinquennial Draft Report (1982). TAC Secretariat, FAO of the United Nations: Genetic evaluation and utilization, pp. 15.

John, V.T., K.C. Long and M.C. Rush (1979). Strategies for rice disease control. Proc. Symposia IX Int. Congress of Plant Protection, Washington, DC, USA. August 5-11, 1979, II, pp. 455-458.

Khush, G.S. (1980). Breeding for multiple disease and insect resistance. Rice improvement in China and other Asian countries. International Rice Research Institute and Chinese Academy of Agricultural Science, pp. 238-239.

Kulshrestha, J.L., A. Anjaneyulu and S.Y. Padmanabhan (1974). The disastrous brown planthopper attack in Kerala, Indian Farming, Dec. 1974, 3.

Lewis, W.J., K.S. Hagen, W.L. Roelofs and L.M. Schoonhoven (1980). Status and potential use of behavioral chemicals in pest management. FAO Plant Prot. Bull. 28, (4), 121-128.

Nishida, T. (1975). Causes of brown planthopper outbreaks. Rice Entomol. Newsl., 2, 38.

Oka, Ida Nyoman (1977). Occurrence of biotype 2 of the brown planthopper in Banyuwangi, East Java and Bali. Laporan Kemajuan Penelitian Seri Hama Penyakit. Centr. Res. Inst. Agric. 11, 63-67.

Oka, Ida Nyoman (1979a). Cultural control of the brown planthopper. Brown planthopper: threat to rice production in Asia. International Rice Research Institute, Los Banos, Laguna, Philippines, pp. 357-369.

Oka, Ida Nyoman (1979b). Feeding populations of people versus populations of insects: the example of Indonesia and the rice brown planthopper. Proc. Symposia IX Internat. Cong. Pl. Prot. Vol. I, Plant Protection Fundamental Aspects, pp. 23-30.

Oka, Ida Nyoman (1980). Brown planthopper biotype survey technique. Rice improvements in China and other Asian countries. International Rice Research Institute and Chinese Academy of Agricultural Sciences, IRRI, Los Banos, Laguna, Philippines, pp. 287-291.

Ou, S.H. (1972). Rice diseases. Commonwealth Mycological Institute, Kew, Surrey, England, pp. 124-125.

Pathak, M.D. and G.S. Dhaliwal (1981). Trends and strategies for rice insect problems in tropical Asia. IRRI Research Paper Series 64, 15.

Pimentel, David (1982). Perspectives of integrated pest management. Crop Protection (1982). (L), 5-2. Butterworths.

Rabbinge, R. (1979). Warning systems. Introduction Proc. of Symposia IX Internat. Congress of Plant Protection, Vol. I: Fundamentals. Ed: Dr. Thor Kommendahl, Dept. of Plant Pathology, Univ. of Minnesota, pp. 163-166.

Tjoa, T.M. (1952). Memberantas hama-hama padi di sawah dan di dalam gudang (in Indonesia). Noordhoff-Rolff N.V. Jakarta, pp. 127.

van der Plank, J.E. (1968). Disease resistance in plants. Academic Press, New York and London, pp. 206.

World Development Report, 1981. National and International Adjustment, Annex World Development Indicators, The World Bank, pp. 192.

Yasumatsu, K., T. Wongsiri, S. Navavichit and C. Tirawat, 1975. Approaches towards integrated control of rice pests. Part I. Survey of natural enemies of important rice pests in Thailand. Plant Prot. Serv. Tech. Bull., No. 24, Dept. Agric., Thailand and UNDP 9/FAO THA 68/526.

ENVIRONMENTAL ASPECTS OF PEST MANAGEMENT

David Pimentel
College of Agriculture and Life Sciences
Cornell University
Ithaca, NY 14853
USA

ABSTRACT

With pests destroying nearly half of all potential food pre- and post-harvest despite the use of pesticides and other controls, there is an essential need for pest management. This waste of food can no longer be tolerated in a world with serious food shortages and a rapidly growing human population. The benefits of pest management practices to society today are immense, amounting to several billion dollars each year. All interventions and manipulations in agriculture for pest control, however, may in turn cause some serious changes in the agroecosystem and natural ecosystem which can have environmental and social costs. The effects of pesticides on agroecosystems are examined. Biological and cultural controls are reviewed and details of some of the adverse effects presented. It is concluded that the problems with nonchemical controls have not received adequate attention, and that careful assessments of the benefits and risks of both pesticides and nonchemical alternatives are needed, if and when the controls are recommended to the farmer for use.

KEYWORDS: Pest management, agroecosystems, ecosystem, pesticides, pesticide poisoning, bioconcentration of pesticides, biological controls, hormonal chemicals.

INTRODUCTION

Throughout the world pests are destroying about 35% of all food crops before harvest (Pimentel and Pimentel, 1979). These substantial losses are primarily caused by insects, plant pathogens, and weeds. Then after the crops are harvested, an additional 10 to 20% of the crops are destroyed by insects, microorganisms, rodents, and birds. Thus, as much as 48% of the potential world food supply is being destroyed each year by pests. This occurs despite all efforts to control pests with pesticides and alternative nonchemical controls.

In the United States present estimates are that about a third of all crops is lost to pests prior to harvest, and an additional 9% is lost to pests after harvest (Pimentel, 1976). A major reason why US losses are not far different than those of the world average is that these estimated losses are based on the quality or "cosmetic standards" that exist for each particular country. The allowed and/or acceptable US quality standards are considered high compared with many developing nations, and if used in all nations, average world food losses to pests would be significantly greater than the 48% mentioned earlier.

Both pesticides and nonchemical biological and cultural controls are employed to reduce pest insect, plant pathogen, and weed losses in world crop production. About 4 Mt of pesticide are applied annually to world agricultural crops (CSM, 1978). Of this 500,000 t are used in the United States (USDA, 1980). Growth in the amount of pesticide produced and used in the United States has been rapid and continues to escalate (Figure 1). In spite of the availability of chemicals, nonchemical biological and cultural controls are the dominant methods of pest control throughout the world (Pimentel et al., 1978). At least twice as much US cropland is managed by nonchemical controls as by pesticidal controls, while elsewhere the ratio of nonchemical to pesticidal controls probably ranges from 2 to 10-fold to 1.

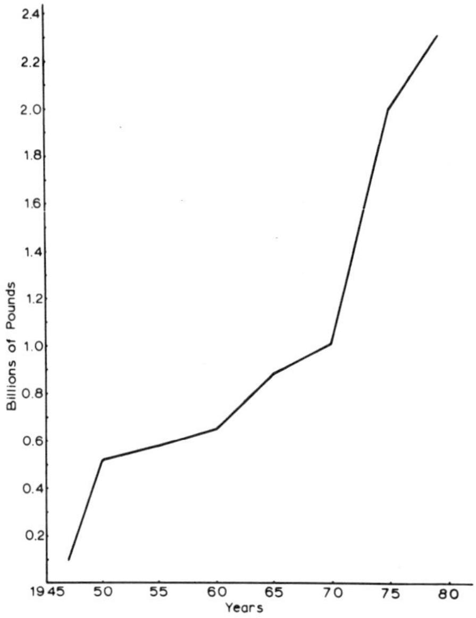

Fig. 1: Estimated amount of pesticide produced in the United States (after Pimentel et al., 1980b).

The nonchemical pest management practices include biological controls, host plant resistance, cultural controls (rotations, crop diversity, planting times, burning, etc.), soil and water management, genetic controls, and behavioral and hormonal chemicals (Pimentel et al., 1982). Although these pest management practices usually cause fewer environmental problems than pesticides, they do cause some problems.

This paper focuses on an assessment of the environmental and social costs of pest management, including both the use of pesticides and nonchemical controls. This basic information will help agriculturalists and policy makers to better understand the benefits and risks to the environment of pest management practices.

PESTICIDE CONTROLS

Pesticides play an important role in pest control and reducing pest losses. Most of the examples given here concern benefits and risks of pesticides from use in the US. Most of the US pesticide is applied to only about 1/4 of the agricultural lands (USDA, 1975). The costs of applying these pesticides are estimated to be about $2.5 billion annually, preventing an estimated loss due to pests of about $10 billion (Pimentel et al., 1978). However while every dollar invested in pest control returns about $4 in increased crop yields, it does not include the environmental and social costs that occur with the use of pesticides. These are significant and include human poisonings, domestic animal poisonings, contaminated foods, various other impacts on agricultural production, and numerous impacts on wildlife and natural ecosystems. This section assesses the environmental risks of pesticide use.

A. Human Pesticide Poisonings

Human pesticide poisonings are a major concern in producing and using pesticides in agricultural and public health programs. It has been estimated that annually in the world, there are about 500,000 reported pesticide poisonings (WHO, 1981). What proportion of these are fatalities is unknown, but it could be about 5,000 annually. In Central America (Guatemala, El Salvador, Honduras, and Nicaraugua) the estimate was about 3-4,000 pesticide poisonings annually with about 10% as fatalities (ICAITI, 1977). In the United States, the best estimate is 45,000 human pesticide poisonings annually, with about 3,000 of these serious enough to be hospitalized (Pimentel et al., 1980a). An estimated 200 fatalities occur annually due to pesticides, with about slightly more than 50 as actual accidental pesticide poisonings (EPA, 1974).

The humans that are poisoned come in contact with pesticides by various means. Workers are exposed during the production of pesticides and others are exposed when these materials are formulated, often in small formulation plants. Other important means of human exposure are during the application of the pesticides both in loading sprayers and in actual crop application, and when workers are exposed to pesticide drift during aircraft application. Also, workers may enter treated crops soon after the treatment to weed crops or harvest food and fiber products. In this situation they obtain pesticide by rubbing against the contaminated foliage with the unprotected portions of their bodies such as arms and legs (ICAITI, 1977). This was the primary means of pesticide exposure in field laborers in Central America (ICAITI, 1977).

Other means of humans obtaining semi-lethal and lethal dosages of pesticides are drinking contaminated water, breathing contaminated air, coming into contact with contaminated soil, or probably one of the major means, eating contaminated foods.

It is estimated that at least 50% of the foods sampled in US supermarkets have detectable levels of pesticides (Johnson and Manske, 1977; McEwen and Stephenson, 1979, Johnson et al., 1981). These levels, however, are said to be "no-effect" levels and therefore, are considered relatively safe. The actual contamination of the population with pesticides can be quite high, e.g. it is estimated that nearly 100% of the US population has some pesticide residue present (Kutz et al., 1977). Human milk often has detectable residues.

Several pesticides have also been found to be mutagenic in the laboratory, but no data are currently available concerning human incidence. The same applies to cancer where some pesticides have been demonstrated to be carcinogenic in laboratory animals, but where there is no clear documentation yet on the impact on humans. However, no one denies that pesticides have the potential to cause cancer. Whether or not this potential is realized in humans remains to be documented. In one epidemiological study (Clark et al., 1977) reported a significant correlation between the intensity of pesticide use by cotton and vegetable farmers and total cancer and lung cancer mortalities in this population in the southeastern United States. Other "major crops such as corn which receive less pesticide treatment were not significantly associated with cancer mortality" in the human population. The finding of this study has obviously important implications and certainly deserves further investigation. The Clark et al. study reported that cotton and vegetable farmers accounted for 1.6 - 6.7% of the total cancer variance in the sample, which is certainly significant. In estimating the fraction of the human cancer that is attributed to pesticides, Dr. Schotterfeld (1978) of the Sloan-Kettering Cancer Institute indicated that the incidence is probably less than 1%. This is relatively low and suggests that other environmental factors such as smoking are the prime cause of cancer.

B. Effects of Pesticides on Agroecosystems

When pesticides are applied in agriculture, numerous changes can occur in agroecosystems and adjoining natural ecosystems. Many of these changes can have a detrimental effect on agricultural production and quality of the environment. Some of these impacts are examined.

a. Natural enemies destroyed - Parasites and predators attack a great variety of pests and in some cases provide the primary means of control (Huffaker, 1980). When insecticides and other pesticides are employed, the poisons not only destroy the pest, but also have a severe impact on natural enemies of the target pest. In some cases the natural enemies that are destroyed are important in controlling certain other pests. When these natural enemies are eliminated, it may result in outbreaks of pests that were previously not a problem in the target crop (Pimental, 1971). When DDT, toxaphene, and several other pesticides were first used in 1945, the primary pests in cotton were the boll weevil and pink bollworm. Soon after the use of these pesticides it was noted that other pests previously not important in cotton became important. These included the cotton bollworm, tobacco budworm, cotton aphid, spider mite, and loopers. In particular, it was the cotton bollworm and tobacco budworm that became important pests, primarily due to the destruction of their natural enemies (Pimentel et al., 1977). This same pattern of outbreaks of nontarget pests occurred in the apple ecosystem. For example, when DDT and other pesticides were used in apple orchards, the European red mite, two-spotted spider mite, and apple rust mite increased to outbreak levels (Croft, 1978).

Fungicides have also been found to create pest outbreaks by reducing certain natural microorganism enemies that are normally pathogenic to insect pests. For example, the use of benomyl to reduce populations of entomopathogenic fungi resulted in increased survival of velvet bean caterpillars and cabbage loopers in soybeans, and eventually led to serious pest problems in soybeans (Ignoffo et al., 1975).

When pesticide use has destroyed natural enemies and resulted in secondary pest outbreaks in the crop, additional insecticide treatments are usually made to deal with these particular pests. In cotton, for example, it has been estimated that at least 4 to 5 additional treatments have been used to control the cotton bollworm and tobacco budworm due to the fact that the natural enemies of these two pests were destroyed when pesticides were applied to control the boll weevil (Pimentel et al., 1977).

b. Reproductive effects - Pesticides have also been found to increase reproduction in invertebrates. For example, sublethal doses of DDT, dieldrin, and parathion increased egg production by the Colorado potato bettle by 50, 33, and 65% respectively (Abdallah, 1968). DDT has also been reported to increase the reproductive rate of spider mites (Hueck et al., 1952).

c. Species diversity and food chains - Pesticides can reduce species diversity or richness, and thus alter ecosystem functions (Koeman et al., 1978). In agricultural ecosystems parasites are generally less tolerant of insecticides than their herbivore hosts. This may affect the function and structure of ecosystems. For example, treatment of cole crop plants with DDT or parathion reduced the number of attacks of herbivores, but reduced the number of parasitic and predaceous taxa even more (Pimentel, 1961a). This significantly changed the species diversity and the complex structure of this ecosystem (Figure 2). In orchards, where heavy applications of insecticides, fungicides, and herbicides were utilized, a significant reduction in species diversity resulted. This was especially noticeable with phytophagous and predaceous invertebrates (Menhinick, 1962). Some of the saprophytic types of invertebrates in the soil and litter were also reduced. When carbaryl was applied to a grassland, the species diversity of the aboveground arthropods was reduced soon after treatment (Barrett, 1968). The phytophagous arthropod species populations recovered more quickly than the predaceous species populations.

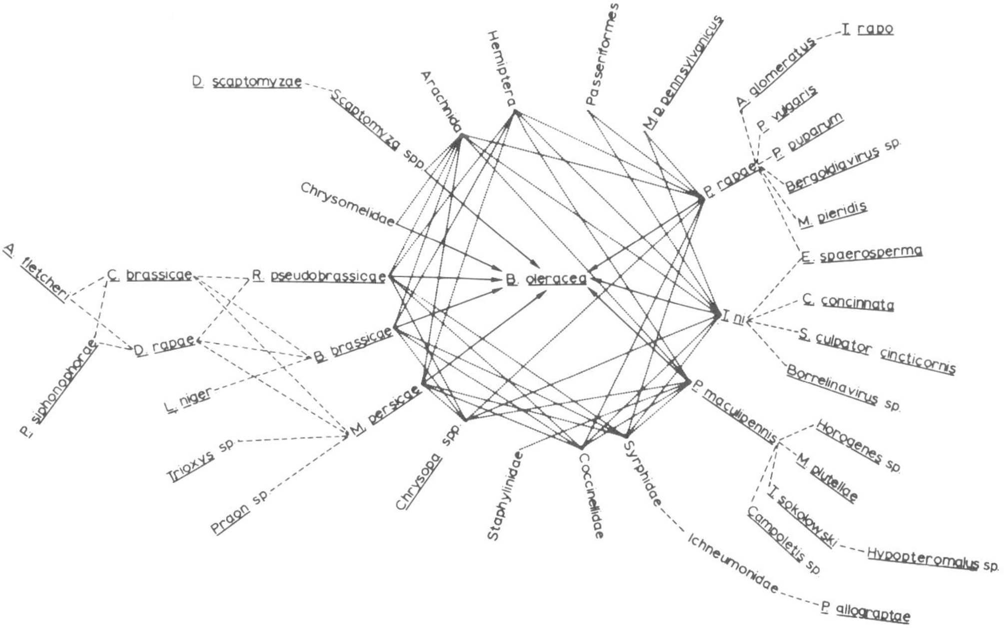

Fig. 2: Relationships between the cole crop-plant (Brassica oleracea), insect pests (——), and parasitic (-----) and predaceous (.....) enemies of the pests (Pimentel 1961c).

d. Bioconcentration of pesticides - Pesticides in the environment may be taken up and concentrated in the tissues of organisms. It is generally assumed that bioaccumulation through the trophic level of food chains is a common phenomenon (Brown, 1978). Although organisms in the higher trophic levels tend to have higher pesticide concentrations, this concept of gradual concentration of residues through the food chain is heavily dependent upon circumstantial evidence and in some cases is probably oversimplified (Pimentel and Edwards, 1982). Fish and other organisms may obtain large quantities or levels of pesticides up to 100,000 times the concentration directly from their environment without consuming prey that have been contaminated with pesticides (Benevue et al., 1972). Thus, one must be extremely careful in making estimates as to whether there is actual bioconcentration or whether the pesticides were obtained directly from the water or other environmental source.

e. Organic matter decomposition and nutrient cycling - Most dead organic matter is broken down by the activities of soil fauna and microflora which assist in the decomposition of cellulose and other organic matter. This action incorporates the material into the soil and makes available nutrients for plant growth. Clearly, pesticides that reach the soil and affect the soil microfauna directly or indirectly may alter the decomposition and nutrient cycling in agroecosystems and natural ecosystems (Thompson and Edwards, 1974).

Populations of earthworms are important decomposer organisms and can be drastically reduced by pesticides (Edwards, 1980). The value of earthworms as decomposer organisms in part is reflected by their large biomass averaging about 1000 kg/ha or more under favorable conditions (Figure 3). Changes in other flora and fauna can also affect organic matter mineralization rates and patterns of nutrient cycling. Thus pesticides may cause increases or decreases in arthropod decomposer organisms in soil, depending on the new species composition of the ecosystem. Organic matter decomposition with the new biotic association may be increased or reduced (Martin, 1976).

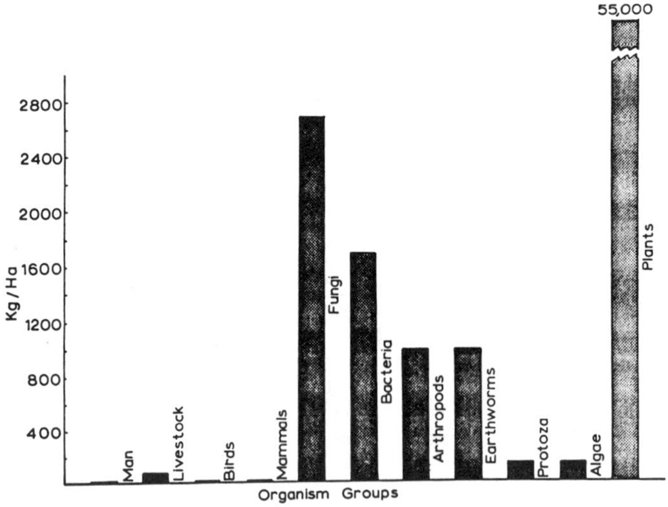

Fig. 3: The average biomass weight of man and his livestock (Pimentel et al., 1975) per acre in the United States and the estimated biomass of natural biota species groups of birds, mammals, arthropods ((Lauer et al., 1976), earthworms (Edwards and Lofty, 1977), protozoa, bacteria, algae, and fungi (Alexander, 1977) in the environment.

Pesticides also can directly alter the chemical makeup of plants. These changes are often specific for the particular plants and pesticides that interact. For example, certain organochlorine insecticides have been found to increase the amounts of some macro- and micro-element constituents of corn and beans and reduce the amounts of others (Cole et al., 1968).

Changes in plant constituents caused by pesticides can affect the insects that feed on these plants. For example, when corn plants were exposed to recommended dosages of 2,4-D in the field, the nitrogen content of the corn increased, and this in turn, resulted in nearly a three-fold increase in corn leaf aphid populations (Oka and Pimentel, 1974). The corn was also 26% more susceptible to corn borer attack. The corn borer females were about 33% larger and produced about one third more eggs than the normal female moths.

f. Pesticide resistance - The use of pesticides against pests sometimes results in populations developing resistance. The best estimate is that nearly 400 insect and mite species are now known to be resistant to some pesticides (Metcalf, 1980). A high level of pesticide resistance in crop insect and mite populations often results in additional sprays of the pesticide and/or the use of more expensive substitute pesticides, giving rise to significant environmental and social costs.

C. Honey Bees and Pollination

Honey bees and wild bees are well known for their essential role in the pollination of fruits, vegetables, forage crops, and other plants (McGregor, 1976). Thus, if pesticides destroy or reduce the populations of honey bees and other wild bees, they may have a significant impact on the agroecosystem as well as natural vegetation. In some agricultural regions where heavy use of pesticides has reduced bee populations and therefore reduced crop pollination, large numbers of colonies of honey bees have to be rented to replace the activity of the destroyed bees. In California, for example, about 700,000 colonies of honey bees must be rented annually at about $8 per colony to supplement natural pollination of almonds, alfalfa, melons, and other fruits and vegetables for seeds (Atkins, 1977).

Estimates of annual US agricultural losses due to poor pollination by destruction of bees due to pesticides range from about $80 million (Atkins, 1977) to a high of $4 billion (McGregor, 1977). Some cotton varieties well pollinated by bees can increase yield by 20 to 30% (McGregor et al., 1955). Atkins (1977) emphasized that poor pollination would reduce crop yields, but more importantly it could reduce the quality of crops such as melons and some fruits. In experiments with melons, Atkins reported that improved pollination increased yields 10%, with a 25% improvement in quality.

D. Crop Losses

Sometimes crops may be damaged as a result of pesticide treatment. This can occur when: (1) the usual dosages of pesticides are applied improperly or under unfavorable environmental conditions; (2) pesticide drifts from a treated crop to nearby susceptible crops; (3) herbicide residues prevent chemical sensitive crops from being planted in rotation or inhibit the growth of crops that are planted; and (4) excessive residues of pesticides accumulate on the crops, causing these harvested products to be rejected for use.

Drifting pesticides, herbicides in particular, are known to cause significant environmental problems. Drift occurs with any method of pesticide application, but the potential for problems is greatest when the pesticides are applied by aircraft. This method is used for about 65% of custom applied agricultural pesticides (USDA, 1976), but it has been claimed that only slightly more than half of the pesticides so applied land inside the target area (Ware et al., 1970).

BIOLOGICAL AND CULTURAL CONTROLS

Biological and cultural controls include only those management practices in which humans manipulate the environment for pest control. It does not include natural controls. Few appreciate that on nearly all US agricultural land some nonchemical control alternatives are employed. The most widespread biological and cultural control is host plant resistance aimed at controlling plant pathogens and some insect pests (Pimentel et al., 1982). Such nonchemical alternatives as host plant resistance and biological controls have been calculated to return about $30 per dollar invested in pest control (Pimentel et al., 1982). The reason for this higher return in the investment with host plant resistance and biological controls is that once the technique is developed little or no costs are associated with use. Although nonchemical pest controls employed in pest management result in fewer environmental risks than chemical pesticides, some of the biological and cultural controls can result in some adverse impacts upon the environment. Several of these impacts are examined below.

A. Natural Enemies

The deliberate use of predators and parasites, including microorganisms, for control of insect pests has proven highly successful against several insect and weed pests (Pimentel et al., 1982). The first great success in introducing predators and parasites for biological control occurred late in the 19th century when the Vedalia beetle was introduced from Australia for control of the cottony-cushion scale on citrus in California (PSAC, 1965). Presently, effective biological control is being employed for some pests on several crops such as citrus, olives, alfalfa, apples, and corn (Huffaker, 1980).

One of the potential risks of introducing insects and other invertebrates to control arthropods or weeds is that a biocontrol species may become a pest itself. For example, the lace bug, *Teleonemia scrupulosa*, a native to Mexico, was introduced to East Africa via Hawaii to control the weed, lantana. This apparently was done without adequately testing the insect's host range and later, after sesame became a major East African crop, the lace bug was observed to be attacking the sesame crop, a problem not observed on the widely-grown sesame in Mexico (Greathead, 1973).

The technology of controlling insects, weeds, and pathogens by employing microorganisms (viruses, bacteria, fungi, and protozoans) has been growing rapidly. Notable cases of success in using microorganisms are: the control of Japanese beetle larvae by the bacterium *Bacillus popilliae* (Falcon, 1971); the control of the European spruce sawfly in Canada by an accidentally introduced virus (Stairs, 1971); the control of many lepidopterous pests of crops and forests by the bacterium *Bacillus thuringiensis* (Falcon, 1971); and the control of skeleton weed on Australian wheat by the imported rust fungus *Puccinia chondrillina* (Cullen, 1976).

Some baculoviruses, such as certain nuclear polyhedrosis viruses (NPVs) of sawflies, are specific to a single sawfly species. The granulosis virus (GV) group also is not known to infect insect species outside the family of the original host (Groner, 1981). However, some baculoviruses are not as specific. For instance, the NPV that attacks *Autographa californica* can parasitize other caterpillars in at least 7 genera of 5 families (Falcon, 1978), and some NPVs attack nontarget insects from other orders (Sidor, 1960; Ignoffo, 1968). Although other insect viruses may be useful in pest control, more research is needed on their safety and specificity.

Bacillus popilliae, a relatively specific bacterium, is registered in the United States to control the Japanese beetle. It is only known to infect scarabaeid beetles (Laird, 1973). Bacillus thuringiensis (Bt) has a broader host range and infects at least 4 orders of insects (Laird, 1973). Concentrations of some Bt isolates are toxic to honey bees and earthworms (Lehnert and Cantwell, 1978).

While vertebrates have been introduced for biological control much less frequently, some like the mosquito fish Gambusia affinis, have effectively controlled aquatic breeding insect pests (van den Bosch and Messenger, 1973). Vertebrate biological control agents have caused more serious environmental problems than have invertebrate agents (Pimentel et al., 1982). The Indian mongoose was originally introduced to Jamaica and later to Puerto Rico and other Caribbean islands in 1877 to control rats in sugarcane fields (Pimentel, 1955). While the mongoose effectively controlled the Norway rat, the major rat pest in sugarcane, the competing tree rats were able to increase (Pimentel, 1955). The mongoose, which cannot climb trees, was unable to control the tree rats and the overall result was the replacement of one rat species by another. The mongoose eventually became a pest itself, and was an important reservoir and vector of rabies in Puerto Rico. It preyed on chickens, ground nesting birds, and waterfowl (Pimentel, 1955). Wolcott (1950) suggested that it also disrupted the natural balance of predators and parasites that kept the important Puerto Rican sugarcane beetles in check.

B. Host Plant Resistance

Since the beginning of agriculture, farmers have applied the technique of using resistant crops when they unknowingly planted seed from the plants that most successfully survived attacks by pests. However, the scientific selection and breeding of crops resistant to insects and diseases has developed only within the last century (Shaner, 1981). To date the greatest use of host plant resistance has been for the control of plant pathogens and varying degrees of resistance to disease have been incorporated into most major crop varieties (Shaner, 1981). There are numerous crop cultivars that resist attack to more than 25 insect pests (Sprague and Dahms, 1972).

Genetically changing the chemical or physical nature of crop plants to resist pests has some risks. Plants use many different chemical compounds as defenses against animal or microbial attack, and some of these chemicals are toxic to both humans and domestic animals. Although humans have learned to avoid or detoxify toxic plant products that they or their domestic animals consume, accidental and chronic poisoning may still occur. For example, fresh tissues of cassava will, upon injury, generate cyanide (Conn, 1979a, b), a substance toxic to humans and to many cassava pests.

The alkaloids solanine and chaconine have been associated with resistance of potatoes both to the early blight disease (Sinden et al., 1973) and to insect pests (Levinson, 1976). The accumulation of these alkaloids in potato tubers is due to such effects as irradiation by sunlight, mechanical injury, and insect or fungal attack (Jadhav and Salunkhe, 1976). The consumption of green potatoes, which contain large amounts of alkaloids like solanine, has resulted occasionally in fatal human poisonings (Jadhav and Salunkhe, 1976).

Some plants are resistant to certain pests because they are deficient in the nutrients necessary for the pests' survival or growth. For instance, the greater resistance to the brown planthopper of one variety of rice over another has been attributed to the resistant variety's lower level of asparagine (Pathak and Saxena, 1980). Asparagine, an amino acid, is a feeding stimulant to the brown planthopper development (Pathak, 1970). A deficiency of absorbic acid in a corn cultivar conferred resistance to European corn borer (Kogan, 1975).

Beneficial invertebrates may be adversely affected by crops that have been bred for resistance (Bergman and Tingey, 1979). For example, nectar can be contaminated by the alkaloids, saponins, and cyanogens that occur in plants, and it is known that honey bees can be poisoned by nectar from tobacco flowers (Barker, 1978).

Differences of morphology among crop varieties also have influenced the ability of natural enemies to locate and control pests. Two hymenopteran parasites of the imported cabbageworm, Phryxe vulgaris and Apanteles glomeratus, were more successful finding their host on collards, broccoli, and kale, which are open-leaved, than on cabbage and Brussel sprouts, which have "heads" (Pimentel, 1961b). Parasitism was 100% on the open-leaved varieties, but only 78% on Brussel sprouts, and 40% on cabbage.

The popularity of a single resistant trait in a crop can result in extensive plantings of genetically similar varieties. This genetic uniformity has been a problem in wheat. Borlaug (1965) estimated that the useful life of a variety of wheat that is resistant to stem rust averages about 5 years. New races of the rust pathogen Puccinia graminis, rapidly develop to exploit the "resistant" wheat varieties. Insect biotypes of such pests as Hessian fly, greenbug, pea aphid, corn leaf aphid, spotted alfalfa aphid, grape phylloxera, chestnut gall wasp, rubus aphid, rice gall midge, and rice brown planthopper have evolved and overcome the resistance bred into various other crops (Gallun and Khush, 1980).

D. Crop Rotations and Diversification

The corn rootworm pest complex in the United States can be effectively controlled by planting corn in rotation after such crops as soybeans and wheat. Although planting corn after a hay crop provides effective control of corn rootworms, it can result in serious attack from other insect pests such as cutworms, wireworms, seed corn maggots, white grubs, billbugs, armyworms, and chinch bugs (OTA, 1979).

Diversifying crops in the agroecosystem is related to rotations in making the environment less favorable for some pests. In some cases strips of crops can be interplanted for pest control (Dempster and Coaker, 1974). Diversifying crops to regulate pest damage creates problems in employing commercial harvesting equipment. In addition, certain mixed crop cultures can support greater numbers of endogenous pests than a monoculture (Perrin, 1977). For example, soybean looper outbreaks occurred more frequently on soybeans grown with cotton than on soybeans grown in monoculture (Glass, 1975). The reason given was that cotton produced abundant nectar and this enabled female loopers to produce more progeny.

E. Planting Times

The manipulation of planting dates is useful for controlling certain pests. However, its use is limited by various regional environmental factors (Teetes, 1981). For instance, although wheat planted late in the growing season avoids Hessian fly, it is more susceptible to winter kill than wheat planted earlier (Shaner et al., 1975). In other cases, escaping one pest by changing planting time can make the crop more susceptible to other pests (Teetes, 1981).

F. Sanitation and Burning

Crop residues often harbor some pests from one growing season to another. Destruction of pest inoculum by burning has proved highly effective for control of the rice stemrot pathogen and various blue-grass seed diseases (Hardison, 1980). Fire used in burning offers direct risks to human life and property and health risks from smoke pollution (Cooper, 1980). In addition, burning seriously reduces soil fertility, increases soil erosion, and reduces soil organic matter (Pimentel et al., 1982).

G. Soil Manipulation

Tillage is one of the oldest and most effective ways of controlling weeds, insects, and plant pathogens (Glass, 1975). Plowing and disking a seedbed can destroy 99% of the weed vegetation; breaking and tilling a corn field can destroy 98% of the overwintering pupae of the corn earworm (PSAC, 1965). However, conventional tillage, especially fall plowing, can cause heavy erosion (Pimentel et al., 1981). For example, in the US corn belt it causes an annual average rate of erosion of 22.5 t/ha, compared with only 7 t/ha if the land is not tilled and residues left on the soil surface. Soil erosion is one of the most serious threats to a sustainable US agriculture (OTA, 1982).

H. Soil Nutrient and Water Management

Fertilizer use and soil nutrition can be altered to manipulate crop pests (Huber, 1981). Some pest damage is increased by raising the levels of certain nutrients in the soil whereas damage from other pests may be increased by reducing nutrients such as nitrogen and phosphorus (Huber, 1981; Pimentel, 1982). Water management, in particular flooding crop fields, has been employed to control certain insects and weeds in rice plants (DeBach, 1974). However, irrigation can also increase crop losses to pests such as the cotton pink bollworm and apple scab (Glass, 1975). Organic agriculture often uses organic mulches to control weeds and add nutrients to the soil (Metcalf et al., 1962). These same organic mulches, however, can increase the populations of various pests such as slugs, snails, and mice. At the same time, mulches can reduce soil temperatures and result in seed rotting, poor germination, and slow crop growth.

I. Genetic Control

The most successful example of genetic pest control is the sterile insect release method (SIRM), in which large numbers of sterile insects are released into a population so the pest cannot mate effectively and propagate itself (Knipling, 1979). The screwworm control program in livestock raised in the southeast and southwest US has been a notable success (LaChance, 1979).

Potential environmental problems include the risk that pest populations will evolve some type of resistance to genetic control (Richardson et al., 1982). The development of a parthenogenetic strain in a wild population could provide the pest with insurmountable resistance. Another and probably more common form of resistance might occur if the wild population evolved associative mating in which wild females recognize and mate only with wild (fertile) males (Richadson et al., 1982). This is likely if the released insects are either genetically or behaviorally (weakened by radiation sterilization) different. Another problem is that the released "sterilized" insects are not adequately sterilized, and therefore are actually fertile, which appears to have occurred with the SIRM program in California against the Mediterranean fruit fly. Some of the Peruvian-reared and "sterilized" Mediterranean fruit flies that were released into California were apparently fertile (Marshall, 1981).

J. Behavioral and Hormonal Chemicals

Chemicals that modify insect behavior fill a vital need in pest management. Currently, the major use of these chemicals is to lure a single insect species to a trap for population monitoring (Roelofs, 1981a). Attractants are commercially available for monitoring codling moth, corn earworm, European corn borer, fruit flies, gypsy month, pink bollworm, spruce bark beetle, tobacco budworm, California red scale, San Jose scale, and the Ambrosia beetle, Trypodendron lineatum (Silverstein, 1981).

Pheromones and other behavioral chemicals that are presently used in pest control have few risks (Silverstein, 1981). Some potential, however, exists for problems. Although relatively specific, they still may have side effects on the behavior of nontarget species. Insects generally use multi-component pheromones, and some common components are used by many different species (Inscoe and Beroza, 1976). In addition, some natural enemies are attracted to their hosts' pheromones (Silverstein, 1981), an attraction that could prove lethal to natural enemies that respond to a host pheromone that is baiting an insect-killing trap. Another potentiel risk with behavioral chemicals is that resistance can develop in the insect population, rendering the technique ineffective. In laboratory tests a cabbage looper population was found to evolve tolerance to a pheromone (Shorey, 1976). "Resistance" to disruption of mating by sex pheromones could also develop through the selection of a subpopulation that used a different blend of pheromone components from that used by the majority of the species population (Roelofs, 1981b).

The right dosage of hormones applied at the right time can alter the physiology and development of insects and result in reduced insect pest populations (Morallo-Rejesus, 1980). Although juvenile hormone (JH) is relatively specific and shows promise for controlling certain insect pests, to date it has had few successes. Pests can evolve resistance to insect hormones. Brown and Brown (1974) found that Culex pipiens mosquitoes rapidly evolved resistance to the juvenile hormone methoprene. Insecticide-resistant strains of insects have demonstrated cross-resistance to juvenile hormone (Rupes et al., 1976).

CONCLUSION

Clearly the continued use of various pest controls including both nonchemical and chemical pesticides, is essential to food and fiber production in the world. In past decades, despite the use of both, pests have continued to destroy almost half of all food produced in the world. This is a waste of natural resources and food nutrients. The benefits of pest control practices to society are immense. However, as with all intervention that manipulates and controls natural species populations, some environmental and social costs result. Although pesticides are known to cause more environmental and social problems than nonchemical alternative controls, the use of nonchemical biological and cultural controls is not without risk. The problems with nonchemical controls have not received adequate attention. Careful assessments of the benefits and risks of the nonchemical alternatives are needed, if and when the controls are recommended for use.

REFERENCES

Abdullah, M.D. (1968). The effect of sublethal dosages of DDT, parathion, and dieldrin on oviposition of the Colorado potato beetle (Leptinotarsa decemlineata Say. (Coleoptera: Chrysomelidae)) Bull. Entomol. Soc. Egypt Econ. Ser. 2: 211-217.

Alexander, M. (1977). An introduction to soil microbiology. John Wiley and Sons, New York.

Atkins, E.L. (1977). Personal communication. University of California, Riverside.

Barker, R.J. (1978). Poisoning by plants. pp. 275-296 in Honey Bee Pests, Predators, and Diseases. R.A. More, ed. Cornell University Press, Ithaca, 430 pp.

Barrett, G.W. (1968). The effects of an acute insecticide stress on a semi-enclosed grassland ecosystem. Ecology 49: 1019-1035.

Benevue, A., J.W. Hylin, Y. Kawano and T.W. Kelley (1972). Pesticides in water. Organochlorine residues in water, sediment, algae and fish. Hawaii, 1970-71. Pestic. Monit. J. 6: 56-64.

Bergman, J.M. and W.M. Tingey (1979). Aspects of interaction between plant genotypes and biological control. Bull. Entomol. Soc. Am. 25: 275-279.

Borlaug, N.E. (1965). Wheat, rust, and people. Phytopathology 55: 1088-1098.

Brown, A.W.A. (1978). Ecology of Pesticides. John Wiley & Sons, New York, 525 pp.

Brown, T.M. and A.W.A. Brown (1974). Experimental induction of resistance to a juvenile hormone mimic. J. Econ. Entomol. 67: 799-801.

Clark, L.C., C.M. Shy, B.M. Most, J.W. Florin and K.M. Portier (1977). Cancer mortality and agricultural pesticide use in the southeastern United States. 8th Internatl. Sci. Mtg., Internatl. Epidemiol. Assoc., San Juan, Puerto Rico.

Cole, H., D. Mackenzie, C.B. Smith and E.L. Bergman (1968). Influence of various persistent chlorinated insecticides on the macro and micro element constituents of Zea mays and Phaseolus vulgaris growing in soil containing various amounts of these materials. Bull. Environ. Contam. Toxicol. 3: 141-153.

Conn, E.E. (1979a). Cyanide and cyanogenic glycosides, pp. 387-412 in Herbivores - Their Interactions with Secondary Plant Metabolites. G.A. Rosenthal and D.H. Janzen, eds., Academic Press, New York, 718 pp.

Conn, E.E. (1979b). Cyanogenic glycosides, pp. 21-43 in International Review of Biochemistry. A Neuberger and T.H. Jukes, eds. Biochemistry of Nutrition, IA, Vol. 27. University Park Press, Baltimore MD, 331 pp.

Cooper, J.A. (1980). Environmental impact of residential wood combustion emissions and its implication. J. Air Pollut. Control Assoc., 30: 855-861.

Croft, B.A. (1978). Potentials for research and implementation of integrated pest management on deciduous tree-fruits. pp. 101-115 in Pest Control Strategies. E.H. Smith and D. Pimentel, eds. Academic Press, New York, 334 pp.

CSM (1978). Christian Science Monitor. February 1, p. 17.

Cullen, J.M. (1976). Evaluating the success of the programme for biological control of Chondrilla juncea L. pp. 117-121, in Proc. IV Int. Symp. Biol. Control of Weeds, T.E. Freeman, ed. University of Florida, Gainesville, 299 pp.

DeBach, P. (1974). Biological Control by Natural Enemies. Cambridge University Press, London, 323 pp.

Dempster, J.P. and T.H. Coaker (1974). Diversification of crop ecosystems as a means of controlling pests. pp. 106-114 in Biology in Pest and Disease Control. D. Price-Jones and M.E. Solomon, eds., Blackwell, Oxford.

Edwards, C.A. (1980). Interactions between agricultural practice and earthworms. pp. 3-11 in Soil Biology as Related to Land Use Practices. D.L. Dindal, ed. Office of Pesticide and Toxic Substances, EPA, Washington, DC.

EPA (1974). Strategy of the environmental protection agency for controlling the adverse effects of pesticides. Environmental Protection Agency, Office of Pesticide Programs, Office of Water and Hazardous Materials, Washington, DC, 36 pp.

Falcon, L.A. (1971). Use of bacteria for microbial control of insects. pp. 67-95 in Microbial Control of Insects and Mites. H.D. Burges and N.W. Hussey, eds. Academic Press, London, 861 pp.

Falcon, L.A. (1978). Viruses as alternatives to chemical pesticides in the western hemisphere. pp. 11-23 in Viral Pesticides: Present Knowledge and Potential Effects on Public and Environmental Health. M.D. Summers and C.Y. Kawanishi, eds. Office of Health and Ecological Effects, US Environmental Protection Agency-600, 312 pp.

Gallun, R.L. and G.S. Khush (1980). Genetic factors affecting expression and stability of resistance. pp. 63-85 in Breeding Plants Resistant to Insects, F.G. Maxwell and P.R. Jennings, eds., John Wiley, New York, 683 pp.

Glass, E.H. (1975). Integrated Pest Management. Rationale, Potential, Needs, and Improvement. Entomol. Soc. Am., 141 pp.

Greathead, D.J. (1973). Progress in the biological control of Lantana camara in East Africa and discussion of problems raised ny the unexpected reaction of some of the more promising insects to Sesamum indicum., pp. 89-92 in Proc. 2nd Internatl. Symp. Biol. Contr. Weeds., P.H. Dunn, ed., Commonwealth Institute of Biological Control Misc. Publ. No. 6, 225 pp.

Groner, A. (1981). Personal communication. Institut fur Biologische Schadlingsbekampfung. Darmstadt, West Germany.

Hardison, J.R. (1980). Role of fire for disease control in grass seed production. Plant Dis. 64: 641-645.

Huber, D.M. (1981). The use of fertilizers and organic amendments in the control of plant disease, pp. 357-394 in Handbook of Pest Management in Agriculture., Vol. 1. D. Pimentel, ed. CRC Press, Boca Raton, FL, 597 pp.

Hueck, H.J., D.J. Keuner, P.J. DenBoer and E. Jaeger-Draafsel (1952). The increase of egg production of the fruit tree red spider mite (Metatetranychus ulmi Koch) under influence of DDT. Physiol. Comp. Oecol. II(4): 371-377.

Huffaker, C.B. (ed.) (1980). New Technology of Pest Control, John Wiley, New York, 500 pp.

ICAITI (1977). An Environmental and Economic Study of the Consequences of Pesticide Use in Central American Cotton Production. Final Report. Central American Research Institute for Industry, United Nations Environment Programme.

Ignoffo, C.M. (1968). Specificity of insect viruses. Bull. Entomol. Soc. Am., 14: 265-276.

Ignoffo, C.M., D.L. Hostetter, C. Garcia and R.E. Pinnell (1975). Sensitivity of the entomopathogenic fungus Nomuraea rileyi to chemical pesticides used on soybeans. Environ. Entomol. 4: 765-768.

Inscoe, M.N. and M. Beroza (1976). Insect-behavior chemicals active in field trials. pp. 145-181 in Pest Management with Insect Sex Attractants. M. Beroza, ed. American Chemical Society, Washington, DC, 192 pp.

Jadhav, S.J. and D.K. Salunkhe (1976). Formation and control of chlorophyll and glycoalkaloids in tubers of Solanum tuberosum L. and evaluation of glycoalkaloid toxicity. Adv. Food Res. 21: 307-354.

Johnson, R.D. and D.D. Manske (1977). Pesticide and other chemical residues in total diet samples (XI). Pestic. Monit. J. 11: 116-131.

Johnson, R.D., D.D. Manske, D.H. New and D.S. Podrebarac (1981). Pesticide, heavy metal, and other chemical residues in infant and toddler. Total diet samples-(II)-August 1975-July 1976. Pestic. Monit. J. 15: 39-50.

Knipling, E.F. (1979). The Basic Principles of Insect Population Suppression and Management. USDA, Agr. Handbook No. 512.

Koeman, J.H., W.M.J. Den Boer, A.F. Feith, H.H. de Iongh, P.C. Spliethoff, B.K. Na'isa and V. Spielberger (1978). Three years' observation on side effects of helicopter applications of insecticides used to exterminate Glossina species in Nigeria. Environ. Pollut. 15: 31-59.

Kogan, M. (1975). Plant resistance in pest management. pp. 103-146 in Introduction to Insect Pest Management. R.L. Metcalf and W. Luckmann, eds., John Wiley, New York, 587 pp.

Kutz, F.W., S.C. Strassman and A.R. Yobs (1977). Survey of pesticide residues and their metabolites in humanbs. pp. 523-539 in Pesticide Management and Insecticide Resistance, D.L. Watson and A.W.A. Brown, eds., Academic Press, New York.

LaChance, L.E. (1979). Genetic strategies affecting the success of the sterile insect release method. pp. 8-18 in Genetics in Relation to Insect Management, M.A. Hoy and J.J. McKelvey, Jr., eds. The Rockefeller Foundation, 179 pp.

Laird, M. (1973). Environmental impact of insect control by microorganisms. Ann. N.Y. Acad. Sci., 217: 218-226.

Lauer, G.J., D. Pimentel, A. MacBeth, B. Salwen and J. Seddon (1976). Report of the Biological Communities Task Group Report No. 8. Hudson Basic Project, Rockefeller Foundation, New York.

Lehnert, T. and G.E. Cantwell (1978). The effects of microbial insecticides on the honeybee. A Review. Am. Bee J. 118: 674-675.

Levinson, H.Z. (1976). The defensive role of alkaloids in insects and plants. Experientia (Basel) 32: 408-411.

Marshall, E. (1981). Man versus medfly: some tactical blunders. Science 213: 417-418.

Martin, N.A. (1976). Effect of four insecticides on the pasture ecosystem. V. Earthworms (Oligochaeta: Lumbricidae) and arthropods extracted by wet sieving and salt flotation. N.Z. J. Agr. Res. 19: 111-115.

McEwen, F.L. and G.R. Stephenson (1979). The Use and Significance of Pesticides in the Environment. John Wiley and Sons, New York, 538 pp.

McGregor, S.E. (1976). Insect pollination of cultivated crop plants. Agricultural Handbook No. 496. USDA, Agr. Res. Serv.

McGregor, S.E. (1977). Personal communication. United States Department of Agriculture.

McGregor, S.E., C. Rhyne, S. Worley, Jr., and F.E. Todd (1955). The role of honeybees in cotton pollination. Agron. J. 47: 23-25.

Metcalf, C.L., W.P. Flint and R.L. Metcalf (1962). Destructive and Useful Insects - Their Habits and Control. 4th ed. McGraw Hill, New York, 1087 pp.

Metcalf, R.L. (1980). Changing role of insecticides in crop protection, Ann. Rev. Entomol. 25: 219-256.

Morallo-Rejesus (1980). Juvenile hormones and insect control. Philipp. Entomol. 4: 21-54.

Oka, I.N. and D. Pimentel (1974). Corn susceptibility to corn leaf aphids and common corn smut after herbicide treatment. Environ. Entomol. 3(6): 911-915.

OTA (1979). Pest Management Strategies in Crop Protection. Vols. I, II. US Congress, Office of Technology Assessment, Washington, DC.

OTA (1982). Impacts of Technology on Productivity of the Croplands and Rangelands of the United States. Office of Technology Assessment, Washington, DC. In press.

Pathak, M.D. (1970). Genetics of plants in pest management. pp. 138-157 in Concepts of Pest Management. R.L. Rabb and F.E. Guthrie, eds., North Carolina State University, Raleigh. 242 pp.

Pathak, M.D. and R.C. Saxena (1980). Breeding approaches in rice. pp. 421-455 in Breeding Plants Resistant to Insects. F.G. Maxwell and P.R. Jennings, eds., John Wiley, New York, 683 pp.

Perrin, R.M. (1977). Pest management in multiple cropping systems. Agro-Ecosystems 3: 93-118.

Pimentel, D. (1955). Biology of the Indian mongoose in Puerto Rico. J. Mammal. 306: 62-68.

Pimentel, D. (1961a). An ecological approach to the insecticide problem. J. Econ. Entomol. 54: 108-114.

Pimentel, D. (1961b). An evaluation of insect resistance in broccoli, Brussel sprouts, cabbage, collards, and kale. J. Econ. Entomol. 54: 156-158.

Pimentel, D. (1961c). Competition and the species-per-genus structure of communities. Ann. Entomol. Soc. Am. 54: 1-69.

Pimentel, D. (1971). Ecological Effects of Pesticides on Non-Target Species. U.S. Govt. Print. Off., Washington, DC, 220 pp.

Pimentel, D. (1976). World food crisis: energy and pests. Bull. Entomol. Soc. Am., 22: 20-26.

Pimentel, D. (1982). Perspectives of integrated pest management. Crop Protect. 1: 5-26.

Pimentel, D. and M. Pimentel (1979). Food, Energy and Society. Edward Arnold (Publishers) Ltd., London, 165 pp.

Pimentel, D. and C.A. Edwards (1982). Pesticides and ecosystems. BioScience 32: 595-600.

Pimentel, D., W. Dritschilo, J. Krummel and J. Kutzman (1975). Energy and land constraints in food protein production. Science 190: 254-261.

Pimentel, D., C. Shoemaker, E.L. LaDue, R.B. Rovinsky and N.P. Russell (1977). Alternatives for reducing insecticides on cotton and corn: economic and environmental impact. Report on Grant No. R802518-02, Office of Research and Development, Environmental Protect. Agency.

Pimentel, D., J. Krummel, D. Gallahan, J. Hough, A. Merrill, I. Schreiner, P. Vittum, F. Koziol, E. Back, D. Yen, and S. Fiance (1978). Benefits and costs of pesticide use. BioScience 28: 772, 778-784.

Pimentel, D., D. Andow, R. Dyson-Hudson, D. Gallahan, S. Jacobson, M. Irish, S. Kroop, A. Moss, I. Schreiner, M. Shepard, T. Thompson and B. Vinzant (1980a). Environmental and social costs of pesticides: a preliminary assessment. Oikos 34: 127-140.

Pimentel, D., D. Andow, R. Dyson-Hudson, D. Gallahan, S. Jacobson, M. Irish, K. Kroop, A. Moss, I. Schreiner, M. Shepard, T. Thompson and B. Vinzant (1980b). Environmental and social costs, pp. 99-158 in Pest Control: Cultural and Environmental Aspects. D. Pimentel and J.H. Perkins, eds., Westview Press, Boulder, CO, 243 pp.

Pimentel, D., M.A. Moran, S. Fast, G. Weber, R. Bukantis, L. Balliett, P. Boveng, C. Cleveland, S. Hindman and M. Young (1981). Biomass energy from crop and forest residues. Science 212: 1110-1115.

Pimentel, D., C. Glenister, S. Fast and D. Gallahan (1982). Environmental risks associated with the use of biological and cultural pest controls. Final Report. NSF Grant PRA 80-00803.

SAC (1965). Restoring the Quality of our Environment. Rept. Environ. Pollut. Panel, President's Science Advisory Committee, The White House, Washington, DC, 317 pp.

Richardson, R.H., J.R. Ellison and W.W. Averhoff (1982). Autocidal control of screwworms in North America, Science 215: 361-370.

Roelofs, W.L. (1981a). Attractive and aggregating pheromones, p. 215-235 in Semiochemicals - Their Role in Pest Control. D.A. Nordlund, R.L. Jones and W.J. Lewis, eds., John Wiley, New York, 306 pp.

Roelofs, W.L. (1981b). ESA Founders' Memorial Award Lecture: Pheromones, plateaus and platitudes. Bull Entomol. Soc. Am., 27: 3-8.

Rupes, V., J. Zdarek, E. Svandova and J. Pinterova (1976). Cross-resistance to a juvenile hormone analog in wild strains of the housefly. Entomol. Exp. Appl. 19: 57-64.

Schotterfeld, D. (1978). Personal communication. Sloan-Kettering Cancer Center.

Shaner, G. (1981). Genetic resistance for control of plant disease. pp. 395-444 in Handbook of Pest Management in Agriculture, Vol 1. D. Pimentel, ed., CRC Press, Boca Raton, FL, 597 pp.

Shaner, G., R.E. Finney and F.L. Patterson (1975). Expression and effectiveness of resistance in wheat to Septoria leaf blotch. Phytopathology 65: 761-766.

Shorey, H.H. (1976). Personal communication. Dept. of Entomology, University of California, Riverside, CA.

Sidor, C. (1960). A polyhedral disease of *Chrysopa perla* L. Virology 10: 551-552.

Silverstein, R.M. (1981). Pheromones: Background and potential for use in insect pest control. Science 213: 1326-1332.

Sinden, S.L., R.W. Goth and M.J. O'Brien (1973). Effect of potato alkaloids on the growth of Alternaria solani and their possible role as resistance factors in potatoes. Phytopathology 63: 303-307.

Sprague, G.F. and R.G. Dahms (1972). Development of crop resistance to insects. J. Environ. Qual. 1: 28-34.

Stairs, G.R. (1971). Use of viruses for microbial control of insects. pp. 97-114 in Microbial Control of Insects and Mites. H.D. Burges and N.W. Hussey, eds., Academic Press, London, 861 pp.

Teetes, G.L. (1981). The environmental control of insects using planting time and plant spacing. pp. 209-221 in Handbook of Pest Management in Agriculture. Vol. 1. D. Pimentel, ed., CRC Press, Boca Raton, FL, 597 pp.

Thompson, A.R. and C.A. Edwards (1974). Effects of pesticides on nontarget invertebrates in freshwater and soil. pp. 341-386 in Pesticides in Soil and Water, W.D. Guenzi, ed., Soil Sci. Soc. Am., Madison, WI.

USDA (1975). Farmers' use of pesticides in 1971 ... extent of crop use. Econ. Res. Serv., Agr. Econ. Rep., No. 268.

USDA (1976). The Pesticide Review 1975. U.S. Dept. Agr., Agr. Stab. Conserv. Serv. 40 pp.

USDA (1980). Agricultural Statistics 1980. U.S. Govt. Print. Off., Wasington, DC.

van den Bosch, R. and P.S. Messenger (1973). Biological Control. In text Educational Publishers, New York, 180 pp.

Ware, G.W., W.P. Cahill, P.D. Gerhardt and J.M. Witt (1970). Pesticides drift IV. On-target deposits from aerial application of insecticides. J. Econ. Entomol. 63: 1982-1983.

WHO (1981). Pesticide deaths: what's the toll? Ecoforum 6: 10.

Wolcott, G.N. (1950). The sugarcane rhinocerous beetle. J. Econ. Entomol. 43: 385.

NITROGEN SOURCES AND ROUGHAGE IN RUMINANT NUTRITION

J.H.B. Roy

National Institute for Research in Dairying
Shinfield, Reading, Berks, United Kingdom

ABSTRACT

Ruminants have the ability to utilize roughage by microbial fermentation to produce valuable animal protein, provided the nutrient requirements of the micro-organisms are fulfilled. Poor quality roughages must be supplemented with nitrogen if microbial protein synthesis is to be adequate for maintenance of live weight, but preformed amino acids are not required. For growth and milk production, supplementing roughage with additional fermentable carbohydrate and further nitrogen is necessary. Basic research allows prediction of amounts and sources of nitrogen for particular productive purposes, preventing wastage of protein and the possibility of toxicity arising from excessive amounts of non-protein nitrogen, e.g. urea, in the ration. More detailed knowledge is required of how microbial protein yield can be improved and how nitrogen can be conserved in the body. Adequate indigenous field trials are needed.

KEYWORDS: Roughage; nitrogen; ruminant; energy; digestion; degradability; recycling; requirement; protein; urea.

ROUGHAGE

The term roughage has never been defined precisely and adequately. During the growth of plants, large amounts of structural cell-wall material in the form of cellulose and hemicellulose are laid down, and as the plants mature lignin is deposited within the cell wall and bonded to the cellulose and hemicellulose to give rigidity and strength to the plant. The nutritive value of this maturing material (roughage) will depend on the degree of lignification. This can be used to classify the roughage as good or poor quality, since digestibility of the cell wall organic matter (OM) in the ruminant decreases proportionately with the amount of lignification that has occurred. Poor-quality roughages also have a low concentration of readily available carbohydrates, such as sugars, starch and fructans and are deficient in N and certain minerals. Vast amounts of poor quality roughage are produced throughout the world. They include: herbage from poor range-grazing and tropical dry season standing hays; the residues, e.g. straws and stovers, remaining after removal of grain, fruits, tubers and roots from crop plants; bagasse remaining after the removal of sugar from sugar cane; and 'browse' from trees and shrubs.

There is no simple definition of a roughage, which will describe both its fibrousness, which is of physiological importance to the reticulo-rumen, if such conditions as bloat are to be avoided, and its nutritive value.

Its characteristics could be defined as plant material: a) containing >350 g crude fibre (CF)/kg dry matter (DM); b) containing >110 g lignin/kg DM (Gaillard, 1958); c) containing <7.5 MJ metabolizable energy (ME)/kg DM (Balch, 1977); d) that is resistant to digestion by proteolytic and diastatic enzymes, but can be utilised to a greater extent by microbial fermentation (Van Soest and Wine, 1967); e) that is insoluble in 'acid detergent' (ADF) or modified acid detergent (MADF) (cellulose and lignin) or 'neutral detergent' (NDF) (hemicellulose, cellulose and lignin) used in these analytical techniques; and f) that stimulates rumination; the roughage index (total time spent eating and ruminating) of Balch (1971). This index for roughage is 100-200 min/kg DM, compared with values of 5-50 min/kg DM for more concentrated diets. The various systems for partitioning the DM of a forage are shown in Figure 1 from Fonnesbeck (1968).

With cereal straws, the ME concentration varies from 5.3-7.3 MJ/kg DM, cell wall constituents represent 800-850 g/kg DM with cellulose concentration 390-430 g/kg DM, hemicellulose concentration 320-330 g/kg DM, lignin concentration 60-150 g/kg DM and ash concentration 10-20 g/kg DM. They contain <50 g crude protein (CP)/kg DM, the digestibility of the energy-yielding nutrients is <0.50, the availability of the minerals is poor and the vitamin status, particularly vitamins A and E, deficient (Wilson and Brigstocke, 1977).

Although roughage may be used for paper-making, fuel and as a substrate for industrial distillation and fermentation processes to produce single-cell proteins, with the disadvantage for some animal species of a high nucleic acid concentration, its most rewarding use is in the production of animal protein by the ruminant.

THE RUMINANT

Domesticated ruminants, cattle, buffaloes, camels, sheep and goats, and, to a lesser extent the horse, have the ability to utilise the potentially digestible OM in roughage by microbiological fermentation to produce energy, mainly as volatile fatty acids (VFA), to synthesise B vitamins and to incorporate simple nitrogen compounds into microbial protein. The ruminant thus does not require an exogenous amino acid supply to survive and competes less than do other domesticated animals for food resources that can be utilised by man. In the ruminant, about 0.9 of the fermentation occurs in the rumen and 0.1 in the large intestine (Sutton, 1976), whereas in the horse the large intestine is the main site of fermentation. Little is known about the digestion of fibre in the small intestine of animals, but there is a suggestion that at least hemicelluloses are hydrolysed, but it is not certain if the products are absorbed and used. For efficient fermentation to occur in the rumen, the microflora must be supplied with energy from fermentation of the OM and with adequate N and other elements, particularly sulphur, phosphorus, calcium and cobalt.

The voluntary food intake of ruminants declines as the density of the metabolizable energy (ME) in the DM decreases (Figure 2) (ARC, 1980), and it also

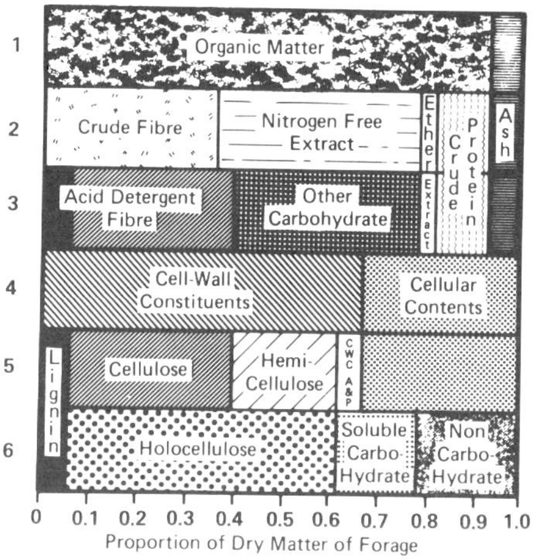

Figure 1: Various systems of partitioning the dry matter of forage: (1) organic matter; (2) Weende or proximate system; (3) Crampton and Maynard system using Van Soest's acid detergent fiber and lignin; (4) Van Soest cell-wall constitutent and cellular content; (5) partitioning of CWC into lignin, cellulose, hemicellulose, and CWC-bound protein and ash; (6) lignin, holocellulose, soluble carbohydrate, and non-carbohydrate solubles. (From Fonnesbeck, 1968.)

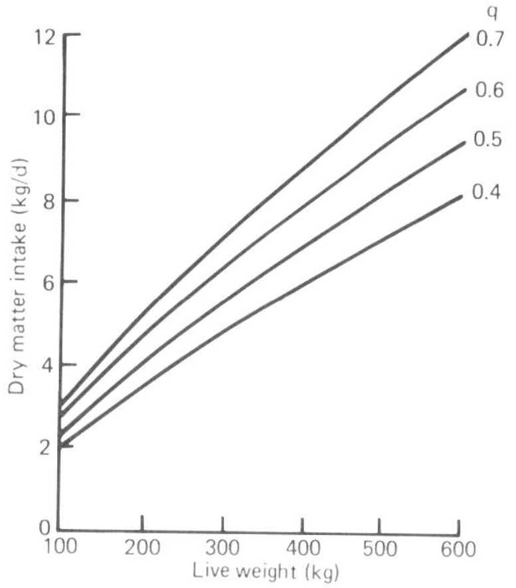

Figure 2: Voluntary dry matter intake of coarse roughage-based diets by cattle in relation to live weight and metabolizability of diet (q).

varies between species. For roughages containing 7.5 MJ ME/kg DM, equivalent to a q value (ME/gross energy) of 0.4, the maximum intake of cattle and goats is about 70 g DM/kg $W^{0.75}$, whereas the value for sheep is about 45 g/DM/$kg^{0.75}$ (ARC, 1980).

The proportion of the digested OM that is digested in the rumen, appears to be greater in cattle than in sheep (E.L. Miller, personal communication). Goats are more similar to cattle and, in comparison with sheep, appear to be able to digest more ADF (Wilson, 1977), to show an increased saliva production and rumination time, a larger rumen volume and ability to store water, an increased retention time of material in the rumen (Throckmorton, 1981) and more N recycled to the rumen (Wilkinson and Stark, 1982).

THE PROBLEM

The low density of energy in roughage means that ruminants, despite their ability to physically distend their rumens, are unable to eat sufficient DM, if the roughage is their sole feed, to satisfy their ME requirement (Figure 3) or their N requirement (Figure 4) (see below) for maintenance, let alone growth. This is partly due to the sheer bulk of the material and partly due to its slow passage out of the rumen. The rate of reduction of long to short particles, and the rate of outflow of small particles from the rumen must be improved if food intake is to be increased. As the metabolizability (q) of GE falls, the maintenance requirement of ME increases (Figure 5) (ARC, 1980). This arises because of the higher energy costs of eating, higher losses in fermentation heat and higher costs in digestion and absorption (Van Es, 1976) and possibly because of a slightly lower efficiency of utilisation of acetate than of the higher VFA's for maintenance purposes. There is insufficient rumen degradable protein (RDP = 6.25 RDN) and other elements in the roughage to supply the requirements of the microorganisms for maximal fermentation of the OM. Much of the small amount of N in the roughage is probably in the insoluble ADF fraction and unavailable to the microorganisms (Goering and Van Soest, 1970; Sutton, 1976).

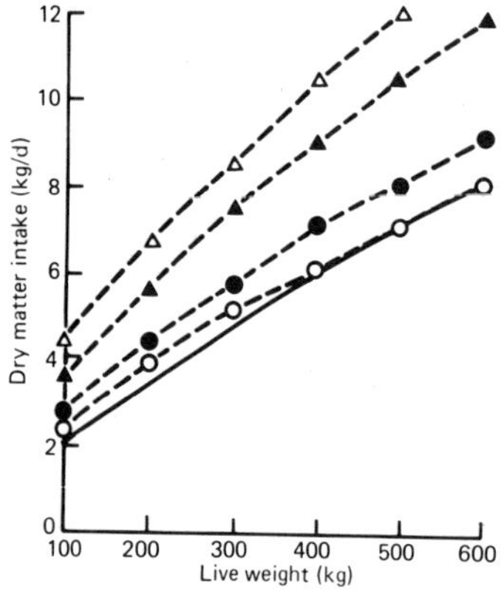

Figure 3: Dry matter intake for maintenance of body weight (O●) or 0.5 kg live-weight gain/day (△▲) for heifers of small breed size (O△) given a diet of metabolizability (q = 0.4 ≡ 7.5 MJ ME (kg DM)) in comparison with voluntary dry matter intake at q = 0.4 (——). Bulls of large breed size (●▲).

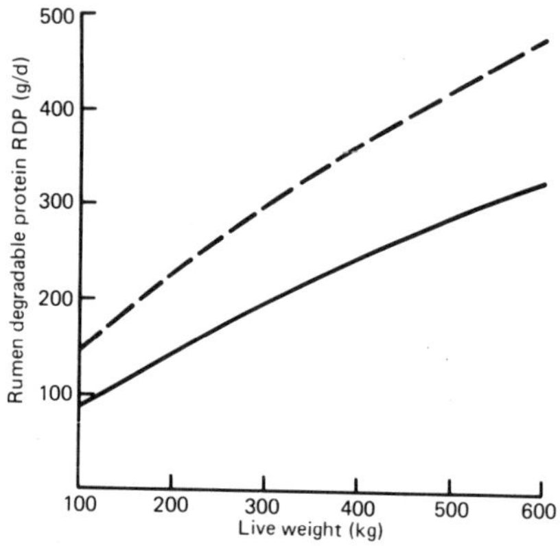

Figure 4: Maximum rumen degradable protein (RDP) intake (——) of a roughage of metabolizability (q) = 0.4 (7.5 MJ ME/kg DM) containing 40 g RDP/kg DM (50 g CP/kg DM with degradability of 0.8) by heifers and steers in comparison with their rumen degradable protein requirement for maintenance (— —).

If, by supplying the deficient nutrients, a faster rate of cellulose breakdown can be encouraged, then the particles will spend less time in the rumen and intake will increase. Moreover, the increased breakdown of cellulose will give rise to an increased digestibility of OM and thus increase the ME concentration in the roughage. This in turn will reduce the ME required for maintenance, which is inversely related to q (Figure 5), and increase the dry matter intake which is directly related to q (Figure 2).

The amount of N required for maximum forage intake is suggested as between 60 and 85 CP/kg DM, depending on the concentrate:hay ratio of the diet (Lyons et al., 1970; Andrews et al., 1972). In general, in diets consisting of low-quality roughage, N is likely to be the first limiting nutrient rather than energy. An example of this is shown in Table 1, where the proportional deficit for maintenace is 0.05 for energy but 0.35 for RDP.

Table 1: Energy and Rumen Degradable Protein Intake from a Poor Quality Roughage by a 300 kg Live Weight Steer of a Medium-Sized Breed.

	Energy	Protein (RDP)
Composition of roughage	7.5 MJME/kg DM	40 g kg DM
Max. voluntary intake (q = ME/GE = 0.4) (4.8 kg DM/d)	36 MJ/d	192 g/d
Maintenance Requirement	38 MJ/d	295 g/d
Deficit	5%	35%

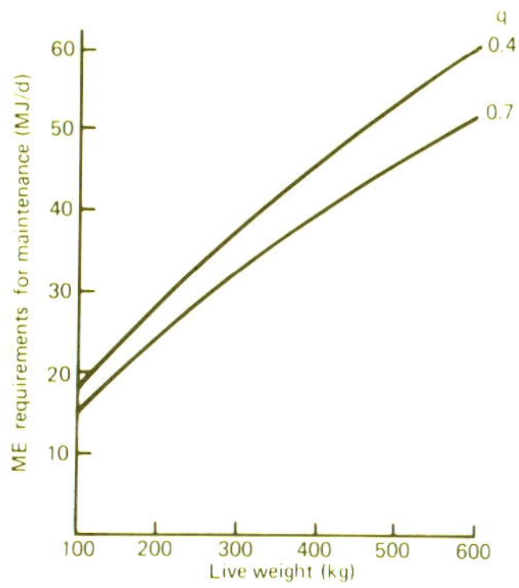

Figure 5: Metabolizable energy requirements for maintenance of heifers and steers given diets of different metabolizability of gross energy (q = 0.4 or 0.7).

The maximum production in terms of growth or milk yield that can be produced from a roughage depends on the amount of potentially digestible OM present, defined by Van Soest et al. (1980) as: potentially digestible OM (g/kg OM) = 1000 - (2.4 lignin concentration (g/kg) + ash concentration (g/kg)); and the amount of microbial protein that can be produced from that OM. Lignin concentration x 2.4 represents the indigestible OM associated with lignin (Mertens, 1973). Once the N intake deficit has been rectified, the digestible OM is likely to be the most limiting factor to increased production. Thus, the productivity of ruminants can only be increased by treatment of the roughage to improve the digestibility of the OM and/or its intake characteristics, or by supplementing the roughage with a fermentable source of carbohydrate.

The microbial protein yield from a given level of ME input will supply the tissue amino acids needs of the host animal, at that ME input, for most situations other than for the young, fast-growing animal, especially for intact male animals given diets of high metabolizability (q), e.g. the amount of microbial protein will be sufficient for cattle above 150 kg live weight for small breeds and above 250 kg live weight for large breeds, or for the lactating cow giving less than 9 kg milk/d (ARC, 1980). Perhaps this is the reason why breeds of small mature size are to be preferred if dietary N supply is limiting, and why this level of peak milk yield is quite common in developing countries.

Although factors that affect the amount of digestible OM that is digested in the rumen also affect the microbial protein yield, the digestible OM will be considered first.

FACTORS AFFECTING OM DIGESTION IN THE RUMEN

On average about 0.65 of the OM apparently digested through the whole alimentary tract is digested within the rumen (ARC, 1980). The value is probably increased when the proportion of long roughage in the diet is increased due to the longer retention time in the rumen. High-roughage diets are associated with an increased salivary flow, an increased volume of rumen contents and an increased liquid (aqueous phase) outflow rate (Hodgson and Thomas, 1975). Liquid outflow rates vary from 0.03 to 0.25/h. As well as the effects of roughage, a high liquid outflow rate from the rumen is associated with a high intake in relation to skeletal size (Van Soest et al., 1980), a high ionic concentration in the diet or induced in the rumen by salt infusion (Harrison et al., 1975, 1976, 1979), small particle size of the feed and a low environmental temperature (Kennedy et a.l, 1976; Kennedy and Milligan, 1978a). A high particulate outflow rate from the rumen is associated with feeds of small particle size, high density and high feed intake. Outflow rates of small particles are intermediate between those of long roughage and those of liquid.

An adequate supply of RDN is required by the microorganisms. Fibre digestion will be depressed when diets are only moderately deficient in RDN (McAllan and Smith, 1976; Oldham et al., 1979). In the new ARC system for estimation of the protein requirements of ruminants (ARC, 1980) the value for RDN is 30 g N from preformed amino acid or protein N, or 38 g N from urea or other source of non-protein nitrogen (NPN) per kg apparently digested OM that is digested in the rumen.

A mixture of non-structural carbohydrate (starch) and structural carbohydrate (cellulose) appears to give more efficient fermentation than either one alone (Offer et al., 1978). Tamminga (1979) has suggested that the optimum proportion of crude fibre to non-fibre extractives (NFE) is about 0.3:0.7. The digestible OM from fibre does not appear to be used as efficiently as that from starch. Readily digested sources of cellulose or hemicellulose, e.g. from sugar beet pulp and citrus pulp, avoids the low pH and depression that may occur with starch supplied by cereals. However, where cellulose digestion is very active on high-roughage diets, the rate of protein degradation is faster than when animals are fed on high concentrate diets (Mohammed and Smith, 1977; Ganev et al., 1979).

The optimum ratio between sugar and starch is not known. It has been suggested that the poor performance of ruminants on autumn-grown forages is due to lack of available carbohydrate (Van Soest et al., 1978). Free lipids in the diets are likely to reduce fibre digestion (Knight et al., 1978; Ikwuegbu and Sutton, 1981). Adequate sulphur and other micronutrients are necessary for optimum fermentation, particularly P, Ca and Co. For efficient microbial protein synthesis, it is essential that the rate of release of fermentable energy, RDN and micronutrients should be matched to each other and balanced throughout 24 h. For instance, when urea was infused continuously into the rumen, much higher microbial yields were obtained than where urea was given twice daily (Meggison et al., 1979).

PROCESSES TO INCREASE THE INTAKE AND DIGESTIBILITY OF OM OF ROUGHAGES

Grinding and pelleting is a costly process (Osbourn et al., 1976) but by reducing particle size, there is theoretically a greater surface area exposed to microbial action. The intake of DM is increased by the process, to a greater extent for sheep than for cattle, but the process tends to cause a reduction in digestibility of CF and OM due to the shorter retention time of the smaller particles in the rumen and inadequate microbial digestion (Campling et al., 1963). The increased faecal loss of energy may be compensated by the lower energy expenditure in eating and ruminating and in the lower methane loss (McDonald et al., 1966). The pelleting of all-roughage diets causes little or no change in the pattern of rumen fermentation, whereas pelleting of mixtures of concentrates and roughage lead to a narrowing of the acetic acid:propionic acid ratio (Woods and Luther, 1962).

Treatment with NaOH (Sundstol, 1981), KOH, CaO (Saadullah et al., 1981a), NH_3 (either aqueous or anhydrous), urea or even urine (Saadullah et al., 1981b) have been used to break the alkali-labile lignin-cellulose and lignin-hemicellulose bonds of straws and thus to increase the amount of digestible OM available for fermentation. KOH is expensive and CaO is less efficient than NaOH or KOH. Such treatments improve DM intake and digestibility and, in the case of the ammonia and urea treatments, provide a source of RDN. Straws from legumes are apparently resistant to alkali treatment.

In the extrusion process, kinetically-produced heat decomposes supplementary urea and the ammonia disrupts the lignin-cellulose bond. Soaking with water has been shown to increase the intakes of roughages and to improve the efficiency of utilisation of ME by changing the microbial environment and fermentation pattern so that VFA production resulted in a narrowing of the acetic acid: propionic acid ratio; it also altered the fat distribution in the animal (Chaturvedi et al., 1973; Holzer et al., 1976 a,b,c). High temperature and high pressure steam has been used (Heaney and Bender, 1970) to release organic acids derived from acetyl groups of the ligno-cellulose structure. Fungi (usually white rot fungi) degrade the ligno-cellulose bonds with minimal degradation of the cellulose.

Irradiation with gamma rays (Pigden et al., 1966; Huffman et al., 1971) or a high electron beam (Millet et al., 1970) increases the friability of fibrous material, ruptures the ligno-cellulose bonds and releases sugar from polysaccharides. Provided that the energy of incident radiation is maintained below a certain limit, there is no possibility of activating elements in the irradiated material.

SUPPLY OF NON-PROTEIN NITROGEN TO THE RUMEN

After a feed, ammonia is produced in the rumen either from the degradation of true protein or from the hydrolysis of urea or other NPN source. RDN will be limiting if the rumen ammonia is <3.5 mM, whilst a value of 5 mM would be considered reasonable (Satter and Roffler, 1975; Roffler and Satter, 1975). A value of 5.7 mM is suggested as the maximum usable amount (Buttery, 1976). The efficiency of net capture of NH_3-N by the microbes is only 0.5 to 0.7 (Salter and Smith, 1979) and is lower with diets based on urea than for those based on protein, probably because of the rapid release of the NH_3 from urea. Since fermentable energy and RDN must be available at the same time, either urea must be fed frequently, or a slow-release form, such as isobutylene diurea (Meggison et al., 1979) must be used.

Although there is no practical competitor to urea as a source of NPN, in certain circumstances it seems that preformed amino acids may be preferable to ammonia as a source of nitrogen in the substrate, especially if it consists of poor quality roughage. Extensive turnover of microbial protein (Nolan et al., 1976) in the rumen should meet the specific needs for preformed peptides, amino acids or branched chain fatty acids. However microbial protein may be deficient in methionine and cysteine (Williams and Smith, 1974). Other sources of RDN are biuret, ammonia salts and ammoniated feeds.

Spraying urea onto straw has given better results than supplementing the diet (Innes and Kay, 1978). Provision of a source of fermentable carbohydrate e.g. urea-molasses licks, helps maintain a balance between energy and ammonia release, although urea-N capture with very high intakes of molasses can be low (Oldham et al., 1977). The optimum amount of urea required for high-roughage diets is usually between 10-20 g/kg DM, but in the proposed ARC system (ARC, 1980), this information can be provided more precisely for particular situations and can prevent an excess, with its possible toxic effects, being given, i.e. RDN requirements are provided for different ruminant species, breed mature size, sex, live weight, live-weight gain, lactation yield and metabolizability of the ration (q).

The efficiency of incorporation of urea N or other NPN into microbial N is considered to be 0.80 (ARC, 1980). A similar value appears to apply for the incorporation of sulphur into microbial protein. The sulphur source should provide 0.07 g S/g N in urea, and thus for the example given in Table 2 the need for anhydrous sodium sulphate is 5 g/day. Elemental sulphur appears to be less efficiently used than sulphate and about twice the amount of sulphur should be given, i.e. 0.14 g S/g N in urea (ARC, 1980).

RECYCLING OF NITROGEN TO THE RUMEN

Of the urea N released into the blood as an end-product of N metabolism, 0.17-0.93 re-enters the alimentary tract (Oldham, 1982). Of this, as little as 0.2 enters through the rumen, the remainder entering further down the alimentary tract (Nolan and Leng, 1972; Nolan et al., 1976). Entry into the rumen is through the saliva or thorugh the rumen wall.

If there is plenty of readily available carbohydrate and a low rumen ammonia, recycling of urea through the rumen wall increases dramatically (Engelhardt et al., 1978; Kennedy and Milligan, 1978b), but if there is a high concentration of NH_3 in the rumen or a very poor quality roughage (Nolan et al., 1978), recycling of urea N across the rumen wall and in saliva secretion is depressed. Probably for this reason, endogenous urea appears to be more effectively captured than exogenous urea.

SUPPLY OF DEGRADED PROTEIN TO THE RUMEN

In the protein of feeds, there are three main fractions (Ørskov and McDonald, 1979); (a) a rapidly degraded fraction, mainly NPN, (b) an insoluble but potentially degradable fraction, (c) an insoluble and undegradable fraction. This latter is considered to be the undegradable N in ADF, which is synonymous with lignin plus cellulose, as distinct from hemicellulose. Fraction 'b', which is true protein, has been divided (Van Soest et al., 1980) into: (b_1) a water-soluble true protein, insoluble in salt buffer, rapidly degraded, a constituent of the soluble cell contents, and soluble in the NDF procedure; (b_2) true protein of intermediate degradability consisting mainly of glutelins from small cereal grains. It is insoluble in the NDF technique but soluble in ADF, i.e. it is associated with the hemicellulose, is completely available but degraded at a slower rate than fraction 'b_1' and its degradability can be markedly reduced by processing and heat treatment (Pichard and Van Soest, 1977); (b_3) true protein degraded slowly and consisting mainly of prolamin proteins, such as zein from maize, which is poorly digested inthe small intestine of cattle (Little and Mitchell, 1967).

The extent of degradation varies with the retention time of the protein in the rumen, and is the product of the amount of the protein remaining at any moment in time and the rate of microbial degradation of the material. Opinions vary as to how far the amino acid composition of undegraded protein differs from that in the protein that is fed (Smith and Mohammed, 1977; Ganev et al., 1979; Mathers et al., 1979).

Since in forages undegradable N may form a constant proportion of the DM, protein degradation will be inversely related to total N in the forage, (Wilson and Strachan, 1981). The high tannin concentration in diets may protect proteins from degradation, e.g. in sainfoin. It is suggested that the high tannin and mineral concentration in 'browse' may reduce degradation in the rumen of the goat (Throckmorton, 1981), presumably partly by protection and partly by an increased rumen outflow rate.

Providing the needs of the microorganisms for nitrogen are supplied, protein that escapes degradation provides more amino acids to be absorbed from the small intestine than does microbial protein, which contains about 0.20 of nucleic acids and other N compounds which can only be used as a source of non-specific N (ARC, 1980).

YIELD OF MICROBIAL PROTEIN

A wide variation occurs in the estimation of microbial protein yield. Part of this arises from differences in techniques (Siddons et al., 1979) and part to real differences. The highest yield occurs with a well-balanced mixture of roughage and concentrates, i.e. between 0.30 to 0.70 roughage. The slow fermentation of diets containing large amounts of cellulose and hemicellulose is likely to result in a slow uptake of NH_3 and a low microbial yield. High microbial yields are associated with high rumen liquid outflow rates (Stern and Hoover, 1979), and thus increased feed intake will increase microbial protein.

Fermentation of crops, e.g. silage, especially of high dry matter concentration, may lead to low microbial protein yields, due on the one hand to the degradation of the protein to NPN and on the other, to an increased concentration of protein unavailable to the microflora owing to the Maillard reaction (Sniffen et al., 1979). Thus microbial yield has been increased from both grass silage (Brett et al., 1979) and maize silage (Beever et al., 1982), when the diet has been supplemented with soyabean or fishmeal.

Overprotection of protein from diets by heat or formaldehyde may lead to low microbial yields as well as to poor utilisation in the small intestine. Protozoa in the rumen may improve or depress food utilisation. In calves, growth of protozoa are encouraged by high-roughage diets (Eadie et al., 1959) whereas high-concentrate diets inhibit protozoa, and the absence of protozoa may also inhibit cellulolysis in high roughage diets. Leng et al. (1981) have shown that the large biomass of large ciliates which appear to be selectively retained within the rumen has a potential detrimental effect on animal production.

TISSUE PROTEIN REQUIREMENTS

The requirements of the body tissues, including milk, of the ruminant is usually considered in terms of total amino acids. Although it is not possible as yet to consider individual essential amino acids, the requirement of total essential amino acids should be taken into account, especially for milk produciton where lysine (Hill et al., 1980) or methionine (Bergen, 1979) may be limiting and where the proportion of essential amino acids to total amino acids is probably considerably greater than in small intestinal digesta or in body tissue.

Increased intake of digestible fibre does not appear to alter the total amount of endogenous N loss, but only increases the proportion excreted in the faeces (E.R. Ørskov, Personal communication). The possibility of conserving N by reducing the amount of N excreted in the urine should be explored.

THE AGRICULTURAL RESEARCH COUNCIL SYSTEM

Like other systems advocated in France (Verite et al., 1979), USA (Burroughs, et al., 1975a,b) and Germany (Kaufmann and Hagemeister, 1973, Hagemeister et al., 1976), the ARC system (ARC, 1980), attempts to take into account many of these various complicated interacting factors that influence the protein requirements of ruminants. The principles of the ARC system are as follows:

1. For a particular ME concentration in the diet, the ME required for a particular form of production and class of stock is established.
2. The amount of OM that will be digested in the rumen is calculated, and from this the microbial protein yield and the amount of rumen degradable N (RDN) required by the microorganisms estimated. RDN requirement (g/d) has been shown to be equal to 1.25 ME (MJ). (RDP = 6.25 RDN = 7.8 ME). Conversion of RDN to microbial N is considered to have an efficiency of 1.0 for protein and amino acid and peptide N, on the assumption that when N is just limiting in the diet, any inefficiency in capture of N in the rumen is compensated by recycling of N to the rumen. For urea and other NPN sources, the efficiency of conversion is considered to be 0.8.
3. The amino acid N that can be retained in the tissues (TMN) from the microbial protein is compared with the total tissue needs of amino acid N, including that for maintenance. TMN (g/d) is estimated to be about 0.53 ME (MJ) (TMP = 6.25 TMN = 3.3 ME (MJ)).
4. (a) If the amino acid N retained in the tissue from microbial protein is > than that required by the tissues, then the requirement of the microorganisms (RDN) is the requirement of the animal.

   (b) If the amino acid N retained in the tissue from microbial protein is < than that required by tissues, then additional amino acid N must be given to the animal in a form of protein that escapes degradation in the rumen (UDN).

   UDN(g/d) = 1.91 total tissue N (TN) (g/d) - 1.0 ME (MJ)

   (UDP = 6.25 UDN = 1.91 total tissue protein (TP) (g/d) - 6.25 ME (MJ)).
5. The minimum N requirement of the animal is RND + UDN providing the diet has an optimum degradability (dg) to fulfil these requirements. If RDN only is required, this may be given in the form of urea or other NPN source.

At the present time, much research effort is being given to devise accurate, quick and cheap methods of measuring the degradabilitity of different types of feedstuffs (Orskov and McDonald, 1979; MacDonald, 1981), i.e. the proportion of RDN in total N. In Table 2, an example is given of the calculation of the RDP (urea) requirement of a heifer of 200 kg live weight from a small breed gaining 0.25 kg live weight/d on a high roughage diet with a metabolizability (q) of 0.5. In this example, no UDP is required.

Table 2: Calculation of the Requirement of Rumen Degradable Protein (RDP = 6.25 RDN) for a 200 kg Live Weight (53.18 kg $W^{0.75}$) Heifer of a Small Breed Gaining Weight at 0.25 kg/d and Given a High Roughage Diet of q(ME/GE) = 0.5

Metabolizable energy requirement (q = 0.5 $\equiv$ 9.2 MJ/kg)	34 MJ/d
Tissue protein (TP = 6.25 TN) required	
Protein retention (g/d) (153.5 g protein/kg weight gain)	38.4
Endogenous urinary protein (g/d)	43.1
Protein losses in hair and scurf (g/d)	6.3
	87.8
Tissue protein supplied by microbial protein (g/d) (TMP = 3.3 ME)	112.2

RATION COMPOSITION

	DM (g/kg)	ME (MJ/kg DM)	RDP (g/kg DM)	UDP (g/kg DM)	dg	CP (g/kg DM)
Barley straw	860	7.3	30	8	0.80	38
Rolled barley	860	13.7	86	22	0.80	108

REQUIREMENTS

	DM (kg/d) (at ME=9.2 MJ/kg)	ME (MJ/d)	RDP (g/d)	UDP (g/d)	dg	CP/kg DM
Max. DM intake = 4.3 kg/d	3.7	34	265	--	1.0	
RATION						
Barley straw	2.5	18.2	75	20		
Rolled barley	1.2	16.4	103	26		
Total	3.7	34.6	178	46	0.80	61
DEFICIT			87			
EXCESS		0.6		46		
UREA REQUIRED (kg)*	0.038		87			
TOTAL	3.7	34.6	265	46	0.87	84

* Urea required = RDP deficit/(2.3 x 1000) = 87/(2.3 x 1000)

Hence, urea addition required = 10 g/kg.

The system also demonstrates the fact, shown in Figure 6, that the microbial protein produced in breeds of small mature size will be sufficient for a higher relative performance than that produced in breeds of large mature size.

The system should be an improvement over the present 'digestible CP' system used by advisory services. It demonstrates the conditions under which urea N or other sources of NPN will be of benefit and by this means it should conserve expensive protein which could be better used by man or by other species that do not have the benefit of a fermentation vat like the rumen. However, to achieve the full potential of the system for situations where undegraded protein is required, i.e. the fast-growing young animal and a high milk yield, >9 kg/d, an accurate measurement of the total degradability of the diet is necessary.

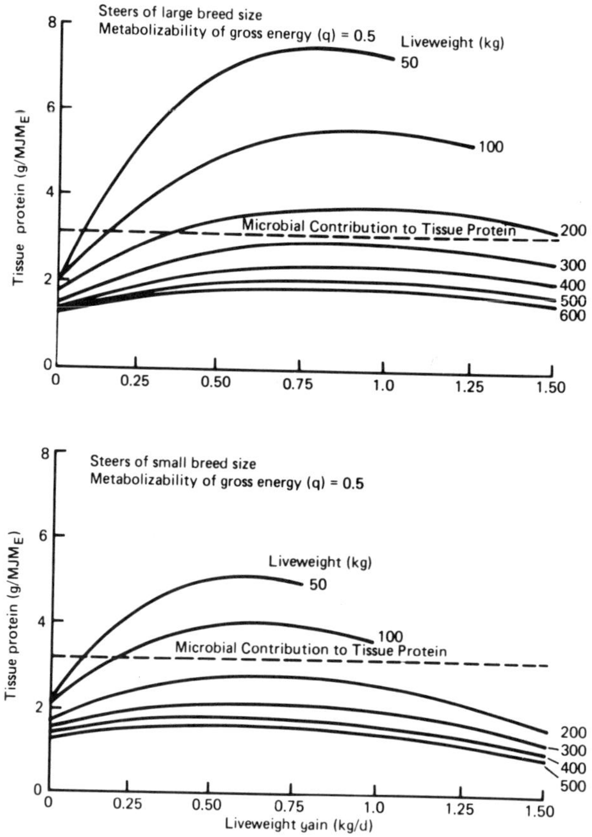

Figure 6: Comparison of the tissue protein requirement per unit of ME of steers of large and small breed size given a diet of metabolizability (q) of 0.5. The microbial contribution to tissue protein is also shown.

CONCLUSIONS

1. The world production of animal protein could be greatly increased if more of the OM in roughages could be digested by ruminants.

2. Any processing method of treating the ligno-cellulose bonds in roughages must be cheap and be able to be done near the site of production of the roughage, because of the high transport costs of a bulky material.

3. Nitrogen supply is likely to be the first limiting nutrient to the utilisation of roughage. The new systems for prediction of the requirements of proteins by ruminants, based on the N needs of the microorganisms should prevent wastage of N and toxicity problems.

4. If roughage is the sole diet during parts of the year, productivity will be low, even if adequate N is given. If higher productivity is required, additional amounts of digestible OM must be provided with concomitant increases in the nitrogen given in the diet. The raising of the quality of the diet in terms of improving the metabolizability of the ration will have a compounding effect in improving the performance of the animal since it will improve its voluntary food intake and reduce its maintenance requirement.

5. The urgent need is for more detailed knowledge of how microbial protein yield can be improved, and the exact requirements of the various organisms for specific N compounds and for other micronutrients. A hopeful aim would be to increase microbial yield so that it would support both large breeds of animal and larger outputs of milk than it does at present so that the high-performing ruminant can become even less dependent on undegraded protein.

The conservation of the N within the ruminant body is another area worthy of attention.

6. The unresolved problems in relation to nitrogen sources and roughages in ruminant diets are evident from this account of the present situation. Their solution will demand much further research, some of which is complicated and often basic in nature. It is, however, a good example of an area in which the sciences of chemistry, and particularly biochemistry, are deeply involved in the understanding of a major practical problem that is pertinent to developing and developed countries alike. Since much of the basic research has been undertaken in developed countries, it is essential that this information should not be assumed to be applicable to the use of any particular poor-quality roughage in developing countries without adequate field trials being made.

REFERENCES

Agricultural Research Council (1980). The Nutrient Requirements of Ruminant Livestock. Commonwealth Agricultural Bureaux, Farnham Royal.

Andrews, R.P., J. Escuder-Volonte, M.K. Curran and W. Holmes (1972). The influences of supplements of energy and protein on the intake and performance of cattle fed on cereal straws. Anim. Prod., 15, 167-176.

Balch, C.C. (1971). Proposal to use time spent chewing cud as an index of the extent to which diets for ruminants possess the physical property of fibrousness characteristic of roughages. Br. J. Nutr., 26, 383-392.

Balch, C.C. (1977). The potential of poor quality agricultural roughages for animal feeding. In FAO Anim. Prod. and Health Paper 4, New Feed Resources, 1-15.

Beever, D.E., B.R. Cottrill and D.F. Osbourn (1982). The effect of the quantity and composition of ruminally degradable nitrogen on microbial nitrogen synthesis and diet formulations. In Forage Protein in Ruminant Animal Production. Br. Soc. Anim. Prod. Occas. Publ. No. 6 (in press).

Bergen, W.G. (1979). Free amino acids in blood of ruminants: physiological and nutritional regulation. J. Anim. Sci., 49, 1577-1589.

Brett, P.A., M. Almond, D.G. Harrison, P. Rowlinson, J. Rooke and D.G. Armstrong (1979). An attempted evaluation of the proposed ARC protein system with reference to the lactating cow. Proc. Nutr. Soc., 38, 148A.

Burroughs, W., D.K. Nelson and D.R. Mertens (1975a). Evaluation of protein nutrition by metabolizable protein and urea formulation potential. J. Dairy Sci., 58, 611-619.

Burroughs, W., D.K. Nelson and D.R. Mertens (1975b). Protein physiology and its application in the lactating cow: the metabolizable protein feeding standard. J. Anim. Sci., 41, 933-944.

Buttery, P.J. (1976). Protein synthesis in the rumen: Its implication in the feeding of non-protein nitrogen to ruminants. In Principles of Cattle Production. (Ed. H. Swan and W.H. Broster) London, Butterworths, 121-144.

Campling, R.C., M. Freer, C.C. Balch (1963). Factors affecting the voluntary intake of food by cows. 6. A preliminary experiment with ground pelleted hay. Br. J. Nutr., 17, 263-272.

Chaturvedi, M.L., U.B. Singh and S.K. Ranjhan (1973). Effect of feeding water soaked and dry wheat straw on feed intake, digestibility of nutrients and VFA production in growing zebu and buffalo calves. J. Agric. Sci., Camb., 80, 393-397.

Eadie, J.M., P.N. Hobson and S.O. Mann (1959). A relationship between some bacteria, protozoa and diet in early weaned calves. Nature, Lond., 183, 624-625.

Englehardt, W.v., S. Hinderer and E. Wipper (1978). Factors affecting the endogenous urea-N secretion and utilization in the gastro-intestinal tract. In Ruminant Digestion and Feed Evaluation. (Eds. D.F. Osbourn, D.E. Beever and D.J. Thomson), Agricultural Research Council, London, pp. 4.1.-4.12.

Fonnesbeck, P.V. (1968). Digestion of soluble and fibrous carbohydrates of forage of horses. J. Anim. Sci., 27, 1336-1344.

Gaillard, B.D.E. (1958). A detailed summative analysis of the crude fibre and nitrogen-free extractive fractions of roughages. 11. The analysis of straw, hay, grass and marigold. J. Sci. Fd Agric., 9, 346-353.

Ganev, G., E.R. Orskov and R. Smart (1979). The effect of roughage or concentrate feeding and rumen retention time on total degradation of protein in the rumen. J. Agri. Sci., Camb., 93, 651-656.

Goering, H.K. and P.J. Van Soest (1970). Forage fibre analyses (apparatus reagents, procedures and some applications). Agric. Handbook No. 379, Agric. Res. Serv., USDA Washington, DC.

Hagemeister, H., W. Kaufmann and E. Pfeffer (1976). Factors influencing the supply of nitrogen and amino acids to the intestine of the dairy cow. In Protein Metabolism and Nutrition. 1st Int. Symp. Nottingham, 1974. (Eds., D.J.A. Cole, K.N. Boorman, P.J. Buttery, D. Lewis, R.J. Neale and H. Swan), London, Butterworths (Publ. Eur. Ass. Anim. Prod. 16), 425-439.

Harrison, D.G., D.E. Beever, D.F. Osbourn (1979). The contribution of protozoa to the protein entering the duodenum of sheep. Br. J. Nutr., 41, 521-527.

Harrison, D.G., D.E. Beever, D.J. Thomson and D.F. Osbourn (1975). Manipulation of rumen fermentation in sheep by increasing the rate of flow of water from the rumen. J. Agric. Sci., Camb., 85, 93-101.

Harrison, D.G., D.E. Beever, D.J. Thomson and D.F Osbourn (1976). Manipulation of fermentation in the rumen. J. Sci. Fd Agric., 27, 617-620.

Heaney, D.P. and F. Bender (1970). The feeding value of steamed aspen for sheep. For. Prod. J., 20, 98-.

Hill, G.M., J.A. Boling and N.W. Bradley (1980). Postruminal lysine and methionine infusion in steers fed a urea-supplemented diet adequate in sulphur. J. Dairy Sci., 6, 1242-1247.

Hodgson, J.C. and P.C. Thomson (1975). A relationship between the molar proportion of propionic acid and the clearance rate of the liquid phase in the rumen of sheep. Br. J. Nutr., 33, 447-456.

Holzer, Z., H. Tagari, D. Levy and R. Volcani (1976a). Soaking of complete fattening rations high in poor roughage. 2. The effect of moisture content and of particle size of the roughage component on the performance of male cattle. Anim. Prod., 22, 41-53.

Holzer, Z., H. Tagari, D. Levy and R. Volcani (1976b). Soaking of complete fattening rations high in poor roughage. 3. The effect of moisture content and spontaneous fermentation on the performance of male cattle. Anim. Prod., 22, 189-198.

Holzer, Z., H. Tagari, D. Levy and R. Volcani (1976c). Soaking of complete fattening rations high in poor roughage. 4. The effect on deposition and fatty acid composition of bovine fat. Anim. Prod., 23, 173-180.

Huffman, J.G., W.D. Kitts and C.R. Krishnamurti (1971). Effect of alkali treatment and gamma radiation on the chemical composition and in vitro rumen digestibility of certain species of wood. Can. J. Anim. Sci., 51, 457-464.

Ikwuegbu, D.A. and J.A. Sutton (1981). The effect of concentration of linseed oil supplementation on digestion and microbial synthesis in the stomach of sheep. Proc. Nutr. Soc., 40, 111A.

Innes, G.M. and M. Kay (1978). Supplementation of barley straw for castrated male cattle (steers). Anim. Prod., 26, 399-400.

Kaufmann, W. and H. Hagemeister (1973). Neue Aspekte zur Proteinversorgung beim-Widerkauer unter besonderer Berucksichtigung der NPN-Verwertung. Kraftfutter, 56, 608-613.

Kennedy, P.M., R.J. Christopherson and L.P. Milligan (1976). The effect of cold exposure of sheep on digestion, rumen turnover time and efficiency of microbial synthesis. Br. J. Nutr., 36, 231-242.

Kennedy, P.M. and L.P. Milligan (1978a). Effect of cold exposure on digestion, microbial synthesis and nitrogen transformation in sheep. Br. J. Nutr., 39, 105-117.

Kennedy, P.M. and L.P. Milligan (1978b). Transfer of urea from the blood to the rumen of sheep. Br. J. Nutr., 40, 149-154.

Knight, R., J.D. Sutton, A.B. McAllan and R.H. Smith (1978). The effect of dietary lipid supplementation on digestion and synthesis in the stomach of sheep. Proc. Nutr. Soc., 37, 14A.

Leng, R.A., M. Gill, T.J. Kempton, J.B. Row, J.V. Nolan, S.J. Stachiw and T.R. Preston (1981). Kinetics of large ciliatte protozoa in the rumen of cattle given sugar cane diet. Br. J. Nutr., 46, 371-384.

Little, C.O. and G.E. Mitchell, Jr. (1967). Abomasal vs. oral administration of proteins to wethers. J. Anim. Sci., 26, 411-413.

Lyons, T., P.J. Caffrey, W.J. O'Connell (1970). The effect of energy, protein and vitamin supplementation on the peformance and voluntary intake of barley straw by cattle. Anim. Prod., 12, 323-334.

Mathers, J.C., R.J. Thomas, N.A.M. Gray and I.L. Johnson (1979). The nutritive value of feed proteins which escape degradation in the rumen. Proc. Nutr. Soc., 38, 122A.

McAllan, A.B. and R.H. Smith (1976). Effect of dietary nitrogen on carbohydrate metabolism in the rumen of the young steer. Br. J. Nutr., 36, 511-522.

McDonald, P., R.A. Edwards and J.F.D. Greenhalgh (1966). In "Animal Nutrition", Edinburgh and London: Oliver and Boyd, 161.

McDonald, I. (1981). A revised model for the estimation of the protein degradability in the rumen. J. Agric. Sci., Camb., 96, 251-252.

Maggison, P.A., N.P. McMeniman and D.G. Armstrong (1979). Efficiency of microbial protein synthesis in cattle. Proc. Nutr. Soc., 38, 146A.

Mertens , D.R. (1973). Application of mathematical models to cell wall digestion and forage intake in ruminants. Ph.D. Thesis, Cornell University, Ithaca, NY.

Millet, M.A., A.J. Baker, W.C. Fiest, R.W. Mellenberger and L.D. Satter (1970). Modifying wood to increase its in vitro digestiblity. J. Anim. Sci., 31, 781-788.

Mohammed, O.E. and R.H. Smith (1977). Measurement of protein degradation in the rumen. Proc. Nutr. Soc., 36, 152A.

Nolan, J.V. and R.A. Leng (1972). Dynamic aspects of ammonia and urea metabolism in sheep. Br. J. Nutr. 27, 177-194.

Nolan, J.V., B.W. Norton and R.A. Leng (1976). Further studies of the dynamics of nitrogen metabolism in sheep. Br. J. Nutr., 35, 127-147.

Norton, B.W., R.M. Murray, K.W. Entwhistle, J.V. Nolan, F.M. Ball and R.A. Leng (1978). The nitrogen metabolism of sheep consuming Flinders grass (Iseilema spp.), Mitchell grass (Astrebla spp.) and mixed native pasture. Aust. J. Agric. Res., 29, 595-603.

Offer, N.W., R.F.E. Axford and R.A. Evans (1978). The effect of dietary energy source on nitrogen metabolism in the rumen of sheep. Br. J. Nutr., 40, 35-44.

Oldham, J.D. (1982). Patterns of nutrient utilisation - Implications for nitrogen metabolism. In Proceedings IAEA Symposium on "Isotope-aided studies on NPN and agro-industrial by-products utilisation with particular reference to developing countries". IAEA Vienna (in press).

Oldham, J.D., W.H. Broster, D.J. Napper and J.W. Siviter (1979). The effect of low-protein ration on milk yield and plasma metabolites in Friesian heifers during early lactation. Br. J. Nutr., 43, 149-162.

Oldham, J.D., P.J. Buttery, H. Swan and D. Lewis (1977). Interactions between dietary carbohydrates and nitrogen and digestion in sheep. J. Agri. Sci., Camb., 89, 467-479.

Orskov, E.R. and I. McDonald (1979). The estimation of protein degradability in the rumen from incubation measurements weighted according to rate of passage. J. Agric. Sci., Camb., 92, 499-503.

Osbourn, D.F., D.E. Beever and D.J. Thomson (1976). The influence of physical processing on intake, digestion and utilization of dried herbage. Proc. Nutr. Soc., 35, 191-200.

Pichard, G. and P.J. Van Soest (1977). Protein solubility of ruminant feeds. Proc. Cornell Nutr. Conf. 91-98.

Pigden, W.J., G.I. Prichard and D.P. Heaney (1966). Physical and chemical methods for increasing the available energy content of forages. Proc. Xth Int. Grassld Congr., Helsinki, 397-401.

Roffler, R.E. and L.D. Satter (1975). Relationship between ruminal ammonia and non-protein nitrogen utilization by ruminants. 1. Development of a model for predicting non-protein nitrogen utilization by cattle. J. Dairy Sci., 58, 1880-1888.

Saadullah, M., M. Haque and F. Dolberg (1981a). Treatment of rice straw with lime. Trop. Anim. Prod., 6, 116-120.

Saadullah, M., M. Haque and F. Dolberg (1981b). Practical methods for chemical treatment of rice straw for ruminant feeding in Bangladesh. In "Utilization of Low Quality Roughages in Africa", AUN-Agricultural Development. Report 1, Aas, Norway, 85-89.

Salter, D.N. and R.H. Smith (1979). Studies with ^{15}N of some factors that influence utilization of urea nitrogen in the steer. Proc. Jap. Soc. Anim. Nutr. Metab., 23, 143-161.

Satter, L.D. and R.E. Roffler (1975). Nitrogen requirement and utilization in dairy cattle. J. Dairy Sci., 58, 1219-1237.

Siddons, R.C., D.E. Beever, J.V. Nolan, A.B. McAllan and J.C. Macrae (1979). Estimate of microbial protein in duodenal digesta. Annals Rech. Vet., 10, 286-287.

Smith, R.H. and O.E. Mohammed (1977). Effect of degradation in the rumen on dietary protein entering the ruminant duodenum. Proc. Nutr. Soc., 36, 153A.

Sniffen, C.J., W.H. Hoover, L.L. Junkins, B.A. Crooker and C.A. MacGregor (1979). Variation in protein and soluble carbohydrate composition of various feedstuffs. Cornell Nutr. Conf. Proc., p. 119.

Stern, M.D. and W.H. Hoover (1979). Methods for determining and factors affecting rumen microbial protein synthesis: A review. J. Anim. Sci., 49, 1590-1603.

Sundstøl, F. (1981). Methods for treatment of low quality roughages. In "Utilization of Low Quality Roughages in Africa". AUN Agricultural Development, Report 1, Aas, Norway, 61-80.

Sutton, D. (1976). Energy supply from the digestive tract of cattle. In "Principles of Cattle Production". (Eds. H. Swan and H. Broster), London: Butterworths, 121-143.

Tamminga, S. (1979). Relations between different carbohydrates and microbial synthesis of protein. Institute for Livestock feeding and Nutrition, Lelystad, Netherlands, Report No. 130, 1-31.

Throckmorton, J.C. (1981). The potential for goat production. In "Recent Advances in Animal Nutrition in Australia". (Ed. D.J. Farrell), University of New England Publishing Unit, Armidale, Australia, 173-178.

Van Es, A.J.H. (1976). Factors influence the efficiency of energy utilisation by beef and dairy cattle. In "Principles of Cattle Production:. (Eds. H. Swan and W.H Broster), London: Butterworths, 237-253.

Van Soest, P.J., C.J. Sniffen, D.R. Mertens, D.G. Fox, P.H. Robinson and U. Krishnamoorthy (1980). A net protein system for cattle: The rumen submodel for nitrogen. In "Protein Requirements for Cattle" Symposium". (Ed. F.N. Owens), Oklahoma State University Division of Agriculture, 265-279.

Van Soest, P.J., D.R. Mertens and B. Deinum (1978). Preharvest factors influence quality of conserved forage. J. Anim. Sci, 47, 712-720.

Van Soest, P.J. and R.H. Wine (1967). Use of detergents in the analysis of fibrous feeds. N. determination of plant cell-wall constituents. J. Ass. Off. Anal. Chem., 56, 781-784.

Verite, R., M. Journet and R. Jarrige (1979). A new system for the protein feeding of ruminants: The PDI system. Livestock Prod. Sci., 6, 349-367.

Wilkinson, J.M. and B.A. Stark (1982). Goat production. Some possibilities for goat production in the U.K. London: Ministry of Agriculture, Fisheries and Food.

Williams, A.P. and R.H. Smith (1974). Concentrations of amino acids and urea in the plasma of the ruminating calf and estimation of the amino acid requirements. Br. J. Nutr., 32, 421-433.

Wilson, A.D. (1977). The digestibility and voluntary intake of the leaves of trees and shrubs by sheep and goats. Aust. J. Agric. Res., 28, 501-508.

Wilson, P.N. and T. Brigstocke (1977). The commercial straw process. Process Biochem., 12 (7), 17-20.

Wilson, P.N. and P.J. Strachan (1981). The contribution of undegraded proteins to the protein requirements of dairy cows. In "Recent Developments in Ruminant Nutrition". (Eds. W. Haresign and D.J.A. Cole), London: Butterworth, 228-247.

Woods, W. and R. Luther (1962). Further observations on the effect of physical preparation of the ration and volatile fatty acid production. J. Anim. Sci., 21, 809-814.

THE USE OF SUGAR CANE AND BY-PRODUCTS FOR LIVESTOCK

T.R. Preston
Tropical Veterinary Science, James Cook University
North Queensland 4811, Australia

ABSTRACT

Data are presented to support the hypothesis that the sugar cane crop offers unparalleled potential for converting natural resources in the tropics into livestock feed and fuel energy. The nutritive value of whole sugar cane is limited by the slow rate of digestion of the cell wall fibre which contributes little metabolizable energy to the animal and also reduces the efficiency of utilization of the valuable sugars through its negative effect on the rumen eco-system (due to slow rumen turnover rate). Future developments with this crop should be based on simple methods to separate juice from fibre. Sugar cane juice has supported performance rates in both ruminants and non-ruminants equal to, or exceeding, those obtained on cereal grain rations. The fibrous fraction, including any residual sugar, can be converted easily into charcoal, producer gas or an energy feed for draught animals.

KEYWORDS: Sugar cane, livestock, energy, ruminants, non-ruminants, nutrition, fractionation, separation, molasses, producer gas, urea.

INTRODUCTION

Discussion of livestock feeding in the Tropics requires consideration of energy because biomass can be worth as much for fuel as for livestock feed (Table 1). In the Tropics it is especially advantageous to integrate energy and livestock production since the climatic features of these regions favour fast-growing plant species of high yield potential for total biomass, but low nutritive value. Management of such species in order to maximise biomass yield (i.e. for energy) requires extended growing periods (Figure 1). Apparently, this practice does not require increased fertilization (Table 2). From the point of view of livestock, however, the choice of fast-growing tropical plant species, and the appropriate management practices to maximise yield, give rise to end-product material of low nutritive value due to the high content of cell wall carbohydrate.

Some livestock planners in the Tropics view energy from biomass as competitive with land use for food production (Vicente-Chandler, 1981). A more appropriate strategy is to support the concept of maximising the unit area yield of biomass and then to fractionate the end-product so that only the readily biodegradable components are directed to livestock feed usage leaving the fibrous

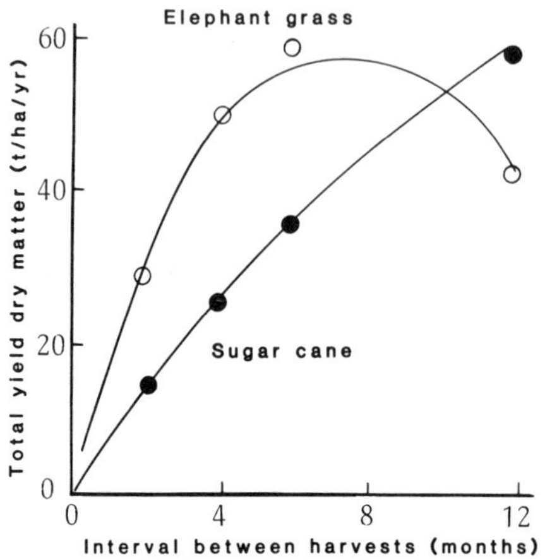

Figure 1: Effect of harvest frequency on annual yields of Elephant grass and sugar cane (Alexander et al., 1979).

Table 1: Relative Value of Biomass for Feed or Fuel.

	Energy Concentration (MJ/kg)	Price (US$/t) Exporting	Price (US$/t) Importing
For fuel:			
Oil (basis: $33/bbl)	42	175	350 (rural area)
Biomass (calculated relative to oil)	17	70	140
For feed:			
Cereal grain		120	250
Molasses		40	110

Table 2: Effect of N Fertilization on Yield (tonnes/ha) of "Energy" Cane and its Components, Harvested at 12 Months (Alexander, 1982).

N/ha/yr (kg)	Stalks	Tops	Dead Leaves Attached	Dead Leaves Detached	Total
227	246	22.8	10.3	17.5	292
454	253	33.5	16.8	22.3	326
681	247	30.8	18.0	20.3	316

residue as a source of fuel energy. The sugar cane crop is an ideal candidate for such a biomass/fractionation strategy since (1) its C4 pathway for photosynthesis confers both high-yield potential and energetic efficiency; (2) the needs of the traditional sugar industry have encouraged agronomic characteristics and practices which facilitate fractionation into leaf, juice and residual fibre; (3) its perennial growth habit and disease resistance facilitates maintenance of soil fertility and mono-cultural practices.

These characteristics of the sugar cane plant have been exploited for generations to produce crystallised sugar in systems which range from extreme rusticity, where animal and human work power still predominates (e.g. in parts of Central America, Haiti, and India), to the highly evolved computerised technology found in Hawaii, the US mainland and Australia. Alexander (1982) has extolled the same virtues of the sugar cane plant for the "energy" cane strategy that is being actively promoted in Puerto Rico.

This paper develops the hypothesis that livestock production can also benefit from the chemical attributes possessed by sugar cane.

BY-PRODUCTS OF SUGAR CANE

The two principal methods of extracting sucrose from sugar cane, and the associated crop residues and by-products, are outlined in Figures 2 and 3. In the simple technology for production of "Gur" or "Panela", there is no centrifugation and therefore no final molasses by-product.

Early interests in the utilization of sugar cane for livestock feed centred around the feeding of crop residues (the tops discarded at harvest) and the by-products (molasses from centrifugal sugar production and scums from traditional gur and panela manufacture). The main recipients of these feeds in the countries where the cane was produced were, and continue to be in some regions, the draught animals (mules and oxen) used by the sugar cane growers and processors.

Molasses

The first widespread use of molasses in production rations was in the feed compounding industry in industrialized countries which imported it as an ingredient in mixed feeds. It was valued for the palatability attributes it conferred on ground feeds by its sweetness and viscosity, the latter reducing dustiness which has economic as well as organoleptic consequences. For this usage only low concentrations (5-10%) are needed and in fact higher levels were found to make mixing more difficult. The situation in most tropical (developing) countries was quite the opposite. Molasses was exported, usually at low prices and cereal grain was imported. There was therefore strong economic pressure to develop feeding systems in which molasses played the major role.

The successful use of high levels of molasses in cattle fattening rations in Cuba (Preston and co-workers, 1967) showed growth rates exceeding 800g/d in cross-breed Zebu bulls given rations in which almost 80% of the dietary energy was derived from molasses. The ensuing research programme led to the development of large scale commercial feeding systems for cattle in both conventional feedlots (Table 3) and semi-confinement (Table 4). Molasses feeding systems were also developed for pigs, ducks, and turkeys (MacLeod et al., 1968; Perez and San Sebastian, 1970; Valarezo and Perez, 1970). For the non-ruminants there were significant advantages from using high-test molasses (the concentrated partially inverted clarified cane juice from which no sucrose has been extracted), or from adding sugar to final molasses in terms of controlling diarrhoea and in improving performance.

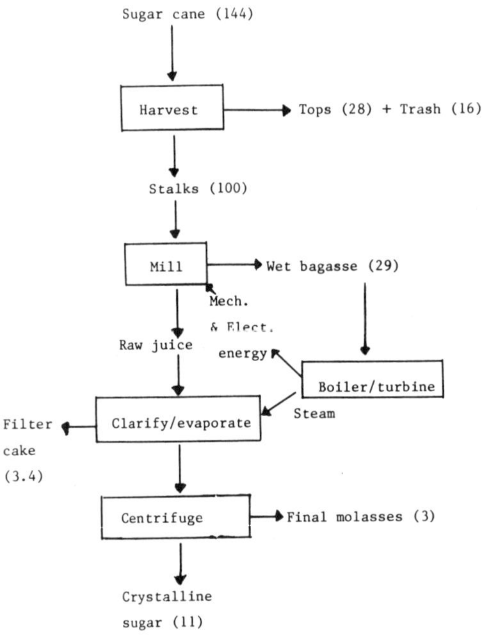

Figure 2: Residues and by-products associated with conventional "factory" production of crystalline sugar. Numbers in brackets indicate approximate quantities (fresh basis)[1] relative to sugar cane stalks = 100 (MSIRI, 1965, 1977).

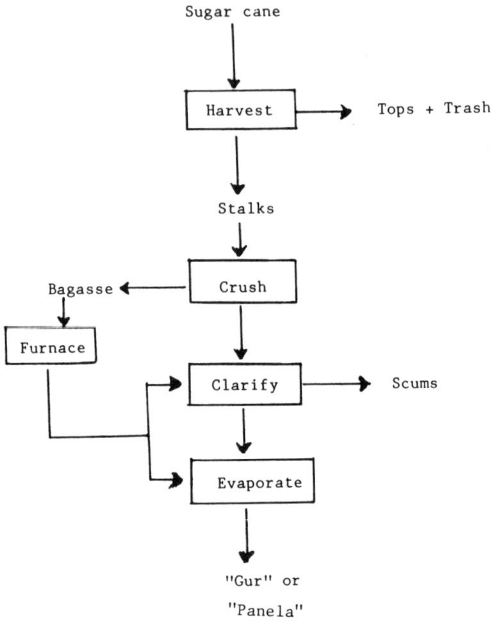

Figure 3: Residues and by-products associated with "artesan" production of Gur or Panela.

[1]Dry matter (%): Sugar cane 34; Tops 27, Trash 74, Filter cake 30, Molasses 85, Bagasse 50.

Table 3: Effects of Changing from Traditional Forage/Concentrates (1969) to a High Molasses Ration (1970) in a Commercial Feedlot in Cuba (Munoz et al., 1970)

	Forage	Molasses
Total liveweight gained daily in the feedlot (kg)	3724	8295
Daily liveweight gain (kg) per:-		
Animal	0.43	0.88
Worker	14.3	51.8
Tractor	86	420
Unit feed drymatter	0.065	0.093
Deaths, %	0.1	1.0
Emergency slaughter, %	0.4	3.0

Table 4: Input-output Data for 3500 Crossbred Zebu Bulls in 11 Commercial Units. The animals were confined for 18 hours daily where they had free access to the molasses/urea and 400g/d of a fish meal supplement. Forage was provided by restricted grazing (6hr/d) mainly on pangola or guinea grass pastures (Morciego et al., 1970).

	Best Unit	Average for all Units	Worst Unit
Liveweight gain (kg/d)	1.04	0.83	0.74
Conversion rate (kg/kg gain)			
Molasses/urea	5.9	9.1	14.7
Fish meal	0.32	0.45	0.54
Emergency slaughter (%)	-	0.44	1.33
Deaths (%)	-	0.38	1.33

The technology of high level molasses feeding for cattle has been introduced in a number of other sugar cane-producing countries and continuing research has given rise to technical and economic improvements such as the use of high quality forages to supply both protein and roughage (Tables 5 and 6) and the supplementing of poor quality roughages with poultry litter (Table 7).

Table 5: Substitution of Native Grass and Groundnut Cake by Leucaena Forage in Molasses-based Diets for Growing Bulls in Mauritius (Hulman et al., 1978)

	(Leucaena (% Liveweight/d)			Groundnut cake (g/d)	
	2	3.5	5	500	1300
Liveweight gain (g/d)	790	740	847	597	742
Feed conversion (kg/kg gain)	9.2	12.2	9.7	10.5	10.4
Molasses as % of diet	79	68	62	73	53

Table 6: Forage From Cassava or Sweet Potato as a Combined Source of Protein and Roughage in Molasses-Urea Diets for Cattle Fattening in the Dominican Republic. Additional soybean meal promoted better animal performance on sweet potato forage but not on cassava forage (from Ffoulkes and Preston, 1978).

Forage Source	Cassava		Sweet Potato	
Soybean Meal (g/d)	0	400	0	400
Liveweight gain (g/d)	853	944	570	784
Dry feed conversion (kg/kg gain)	6.28	7.19	8.28	6.74

Table 7: Poultry Litter (18% of Diet Dry Matter) Increases Cattle Performance on a Basal Diet of Molasses/urea, Wheat Bran (12% of Diet Dry matter) and Sugar Cane Tops in the Dominican Republic (from Meyreles and Preston, 1982).

	Basal Diet	Poultry Litter
Liveweight gain (g/d)	730	1010
Dry matter intake (kg/d)	6.52	8.04
Dry matter conversion	9.11	8.14

WHOLE SUGAR CANE AND DERINDED CANE STALK

An important contemperaneous development in the Caribbean was the invention by Tilby and his colleagues (Tilby 1971a, 1971b, 1976) of a new technology for processing the sugar cane stalk to give a greater range of end-product uses (Figures 4 and 5). The principal objective was to separate the rind component for use as raw material for high quality compressed boards which would substitute imported plywoods. The pith, scraped off the inside of the rind, has potential use for conventional sugar production, for alcohol fermentation and as an energy feed for livestock.

In experiments in Barbados by Donefer and his colleagues (see Pidgen, 1972), very high rates of performance were obtained in fattening cattle fed the sugar cane pith (derinded stalk) when it was adequately supplemented with chopped cane tops and a supplement providing urea and true protein (Table 8). The ground derinded stalk was also incorporated satisfactorily at quite high levels in rations for pigs and chickens (James, 1973).

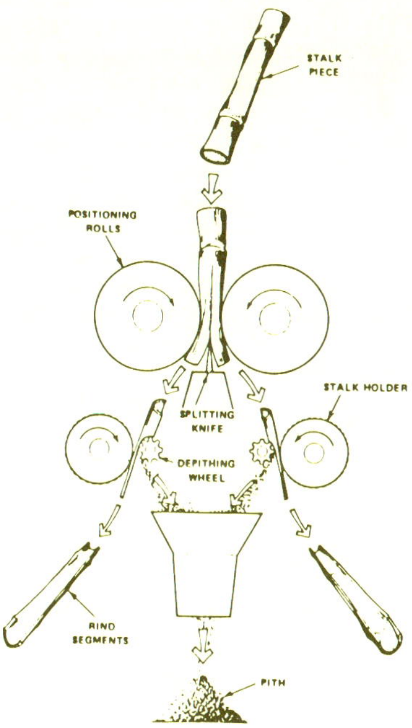

Figure 4: Simplified diagram of the Tilby separator process (from Lipinsky and Kresovich, 1982)

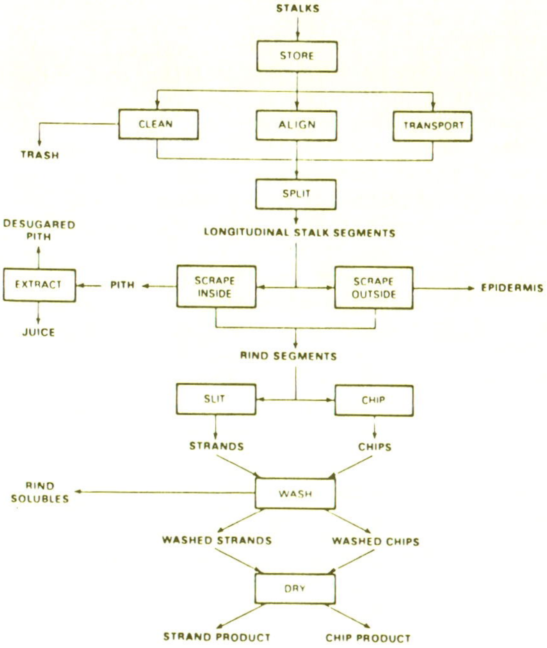

Figure 5: Tilby stalk processing operations (from Lipinsky and Kresovich (1982)

Table 8: Performance of Holstein Steers Fattened on a Basal Diet of Derinded Sugar Cane Stalk Supplemented with Cane Tops, Rapeseed Meal and Urea. Additional molasses slightly increased weight gain but made feed conversion worse: maize had a positive effect on both parameters. The trial was carried out in Barbados (from Pigden, 1972).

	BASAL DIET	ADDITIONAL ENERGY	
		MOLASSES	MAIZE
		1% liveweight/day	
Dry matter intake (kg/d)	9.6	11.4	10.8
Liveweight gain (kg/d)	0.99	1.07	1.25
Dry matter conversion	9.77	10.7	8.6

These promising developments have not yet led to commercial application. The main limitation is that the processing of the rind into compressed board requires sophisticated and expensive machinery and a minimum economic size of plant (production capacity of some 30 t/d of finished board, coming from about 400 t/d of sugar cane stalks). The by-product pith from this operation would be 300 t/d, enough to feed 15,000 head of cattle. Such a unit is not large by US feedlot standards, but is impracticable in most developing countries which is where the technology has most potential. A second factor was the realization that the derinded stalk was apparently little better than the chopped whole sugar cane plant when both were adequately supplemented (Figure 6). This is surprising in view of the higher dry matter digestibility of the derinded cane compared with the whole plant (70 vs 61%; Montpellier and Preston, 1977). The apparently poorer performance of cattle given derinded sugar cane, compared with whole cane, when only urea and minerals were given was subsequently confirmed in an experiment in the Dominican Republic (Fernandez and co-workers, 1978). Apparently in the absence of other forms of supplementation, the rind must provide some nutrients (or physical attributes) not present in the pith. The rind is certainly richer in protein and ether extract (Table 9) and presumably also in minerals and vitamins. The physical effect of the rind fibre may perhaps be the principal factor, since cattle fed the derinded stalk, even when supplemented with protein, minerals and vitamins, responded to an addition of chopped cane tops by an increased voluntary intake and liveweight gain (Pigden, 1972). However, feed conversion was worse. indicating that the beneficial factor was some characteristic which stimulated intake, but at the same time reduced the efficiency of utilization of the overall feed. A similar result was obtained in Mexico when chopped cane tops were added to a basal diet of chopped cane stalk (Ferreiro and Preston, 1976).

Table 9: Composition of Sugar Cane Tops and of the Rind and Pith Fractions Produced by the Tilby (1971) Separator Process. (Unpublished data from Division of Animal Production, Ministry of Agriculture, Mauritius).

	STALK		TOPS
	PITH	RIND	
Dry matter (%)	22	39	27
Composition of the dry matter (%)			
Protein (Nx6.25)	1.4	3.2	2.7
Ether extract	0.19	1.04	0.84
Total sugars	46.0	23.6	26.8
Fibre (non-water soluble)	45.6	69.9	56.9
Ash	1.87	3.10	5.28
Sulphur	0.19	0.25	0.40

The technology of chopping the whole sugar cane plant is simpler and much less sensitive to economies of scale than the derinding process, and became the system of choice in a number of attempts to apply sugar cane feeding of livestock under semi-commercial conditions (Preston, T.R., 1978, unpublished data; SFC, 1980, 1981).

Research has continued in order to identify the constraints associated with the use of sugar cane as a livestock feed and a number of significant discoveries have been made which have contributed to our understanding of the problems associated with the utilization of high carbohydrate - low nitrogen feeds by ruminant animals in general. The most significant findings have been that the non-sugar residual fibrous material in the sugar cane, whether in the leaf, the pith or the rind has an extremely slow rate of degradability by rumen microorganisms (Fernandez and Hovell, 1978; Santana and Hovell, 1979). It was postulated (Preston and Leng, 1980) that this results in a long residence time of fibre in the rumen which would explain the very low voluntary intake of sugar cane, whether whole or derinded, when only urea and mineral supplements are given (e.g. Fernandez and co-workers, 1979). The first step to remove this constraint appears to be the incorporation in the ration of a source of highly digestible forage, for example the foliage of sweet potato (Meyreles and co-workers, 1977; Meyreles and co-workers, 1979). Apparently, this gives rise to an improved eco-system in the rumen enabling a more efficient microbial activity, as evidenced by increased utilization of urea (Meyreles and co-workers, 1979).

The second step is to provide a highly digestible concentrate supplement containing both energy and protein which will escape, or bypass, the rumen fermentation. This apparently leads to more efficient utilization of the digestion end products, and consequently to improved feed utilization efficiency, probably by increasing the supply of amino acids and glucose precursors at the sites of metabolism (Preston and Leng, 1980). By contrast, highly digestible concentrates which do not escape the rumen fermentation because they are too soluble (e.g. molasses) lead to poorer feed utilization efficiency (Table 8). The best results are obtained when derinded sugar cane stalk is supplemented both with high quality forage and a bypass supplement (Figure 7).

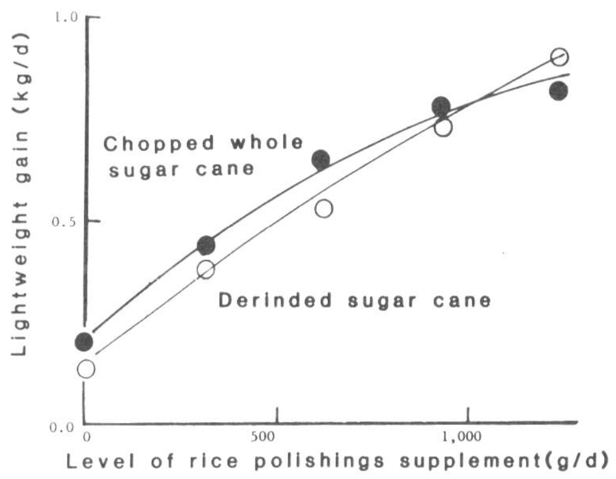

Figure 6: A large scale experiment in Mexico with 400 cattle showed little apparent difference between chopped whole sugar cane and derinded sugar cane. There was an indication that the derinded cane was inferior at low levels of supplementation but better at high levels (from Preston and coworkers, 1976).

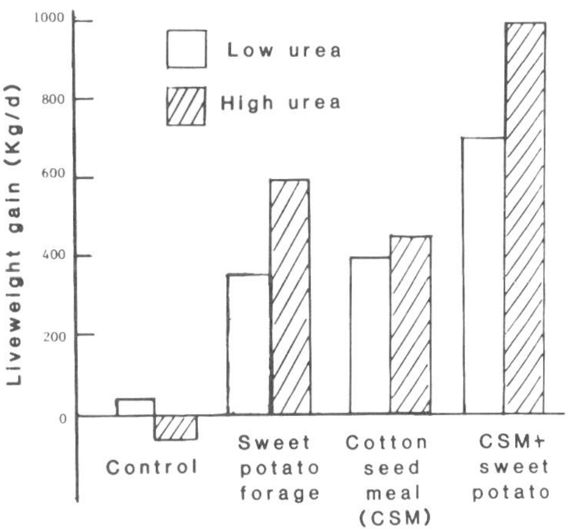

Figure 7: Supplements of sweet potato forage and cottonseed meal stimulate growth rate and response to urea in Zebu cattle given a basal diet of derinded cane stalk and urea (Meyreles and coworkers, 1979).

It can be concluded that in the absence of supplements, both chopped whole sugar cane and the derinded stalk are only maintenance feeds; that simple supplementing with urea and minerals will support some gain in weight (100-200g/d) on chopped whole cane (but not with the derinded stalk - unless cane tops are also given); and that growth rates of the order of 900-1000g/d are feasible provided that some good quality forage and a source of bypass nutrients are given in addition to the urea and minerals. In practice, the use of whole sugar cane (tops, stalk and trash) collected and chopped in the field using a strengthened maize forage harvester, has proved to be an economical solution to the problem of dry season feeding of cattle when the objective is no more than to maintain liveweight and body condition until the onset of the rains. But for productive purposes, the required levels of supplement make the system uneconomical at least in the majority of situations met within developing countries.

FRACTIONATION OF SUGAR CANE FOR LIVESTOCK AND ENERGY

Recognition of the negligible nutritive value of sugar cane fibre and of the negative effect this was having on the utilization of the valuable sugar component, especially in feeding systems designed for high productivity, led to efforts to develop processing systems in which the sugars and the fibre could be treated separately according to their individual characteristics. Fractionation in traditional sugar production requires the fibrous component to be utilized as fuel to provide the energy needed to extract the sugar-containing juice, and later to concentrate this and separate the sugar in crystalline form. Our approach (Figure 8), uses only a partial extraction of the sugar juice which drastically reduces the investment and the energy cost of milling (a single pass through a 3-roll mill is all that is used). This is justified because the economic value of sugar and fibre differs very little when the end product usage is for fuel or feed (Table 1). The hypothesis underlying this approach was that the easily extract-

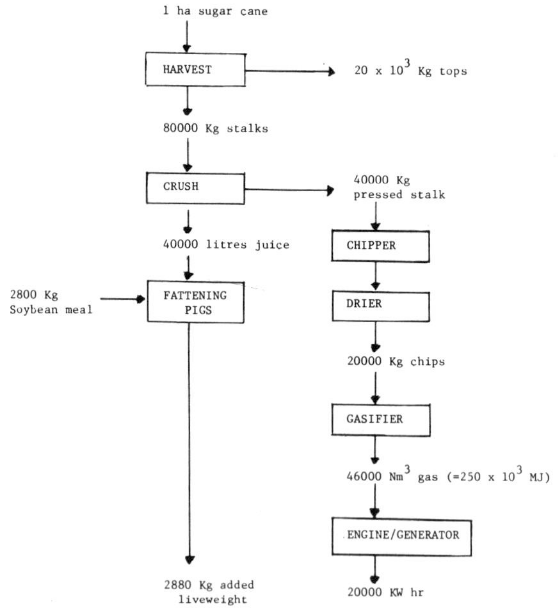

Figure 8:

Utilization of sugar cane for livestock feed and energy.

able juice would be fed to both ruminant and non-ruminant livestock while the fibrous component, including the residual sugar, could be used either as raw material to make producer gas, to substitute for gasoline and diesel oil in internal combustion engines, or as a maintenance feed for livestock, following alkali treatment to enhance its digestibility.

Sugar Cane Juice for Livestock

Considerable progress has been made in the development of animal feeding systems based on the sugar cane juice. With growing cattle the levels of performance obtained have been vastly superior to those recorded on molasses and comparable with the best that could be expected from more traditional cereal grain feeding, as practised in US feedlots (see Table 10). The difference compared with cereals has been the opportunity to save protein (which is used wastefully when cereal grain is fed to ruminant animals) because of the excellent medium for microbial protein synthesis that sugar cane juice provides. Thus it has been possible to obtain high levels of productivity, especially in terms of feed conversion, without the need for any imported protein supplements. This has been done by taking advantage of the combined protein and roughage characteristics of high quality tropical legumes such as leucaena (Table 11).

Table 10: Cattle in Mexico Grew Faster with Better Feed Conversion on Basal Diet of Sugar Cane Juice/Urea than on Molasses/Urea, Both in the Absence or Presence of a Bypass Concentrate Supplement. Initial liveweight was 260-280 kg and the trial period 84d. All animals received fresh forage (African Star grass) at the restricted rate of 3% of liveweight daily. Molasses or juice, supplemented with urea, was fed ad libitum (Sanchez and Preston, 1980).

	WITHOUT SUNFLOWER CAKE		1 kg/d OF SUNFLOWER CAKE	
	MOLASSES	CANE JUICE	MOLASSES	CANE JUICE
Liveweight gain, kg/d	0.25	0.80	0.55	1.32
Dry feed conversion (kg/kg gain)	21.5	7.42	11.8	6.44

The most appropriate application for sugar cane juice feeding, and certainly the one that offers the most attractive commercial benefits, is in the development of rations for monogastric animals such as pigs and poultry. Traditionally, these species in tropical countries have had to be fed on imported cereal grains with the associated disadvantages of competition with human food resources and/or excessive dependency on imports. The early results obtained with pig feeding (Table 12) show comparable levels of performance to what would be expected from the use of cereal grain-based diets, and better carcass merit mainly because of the higher dressing percentage. As with ruminants, there appears to be a considerable saving in dietary protein. In this case, the saving of protein arises because the required essential amino acids can be concentrated in a smaller total amount of protein, as a result of eliminating the poorly balanced proteins

Table 11: Zebu Bulls in Mexico Grow Fast and Efficiently on a Basal Diet of ad libitum Sugar Cane Juice, Preserved with Ammonia, and Supplemented with Restricted Fresh Forage (2% of liveweight) of leucaena leucocephala. Fish meal increased growth rate but not the feed conversion (Duarte and co-workers, 1982).

	FISH MEAL SUPPLEMENT (400g/d)	
	WITHOUT	WITH
Liveweight gain (g/d)	1,020	850
Dry feed conversion (kg/kg gain)	5.7	5.8

in the cereal grains.

Table 12: Pigs Fed Sugar Cane Juice in the Dominican Republic had Similar Growth Rates, but Better Feed Conversion and Carcass Yield than Controls Fed Maize and Molasses. Starting and finishing liveweights were 40 and 100 kg respectively (Mena, A., 1982. Unpublished data).

	CANE JUICE	MAIZE AND MOLASSES
Liveweight gain (g/d)	969	953
Dry feed conversion (kg/kg gain)	3.36	4.00
Dietary protein (% DM basis)	11.5	16.1
Carcass yield (%)	82.0	79.8

Experiments have just begun with poultry. The initial observations indicate interesting possibilities for simple technologies in which ducks are the preferred species because they are better adapted to utilize liquid diets. They also have the capacity to "harvest" high-protein water plants and algae, thus enabling the use of these potentially high yielding protein feeds which for conventional livestock production would require expensive harvesting and drying practices.

Utilization of the Residual Pressed Cane Stalk

The use of sugar cane juice for livestock feeding also must be accompanied by an economic means of utilizing the residual material. Options being pursued include (1) production of charcoal for rural household usage, (2) conversion into producer gas to substitute for diesel fuel and (3) development of an energy feed for livestock.

Making charcoal from pressed cane stalk is a simple process (Ffoulkes and co-workers, 1981) however, the resulting product although of high caloric value, needs to be pressed into bruiquettes in order to produce a readily saleable product. Converting pressed cane stalk into a suitable material for gasifying requires drying to about 20% moisture and chipping into particles of 10-20 mm. Preliminary tests in a wood gasifier attached to a diesel-engined lorry, indicated that the sugar-containing fibre was relatively easy to gasify and that performance was similar to that obtained with wood chips (Lindgren and Preston, 1980; unpublished observations). Some modifications to the basic gasifier design almost certainly will be required because of the lower density of the dried sugar cane chips.

Results from preliminary experiments on alkali treatment of the pressed sugar cane stalk are promising. A ration containing (dry matter basis) 44% pressed sugar cane stalk, 23% of legume forage and 24% of poultry litter, reacted for 24 hr with 5.7% NaOH and 2.8% urea, had a faster rate of digestion in the rumen than freshly harvested elephant grass (Dixon, R. and Preston, T.R., 1982, unpublished data). It is probable therefore, that the energy value of the pressed stalk will be at least comparable with that of rice straw, in which case it would be a suitable feed for draught cattle and buffalo.

CONCLUSIONS

After two decades of research it is concluded that partial fractionation is the strategy which holds most promise as a means of efficiently exploiting the sugar cane plant for livestock feeding, probably in association with energy production. What is possibly of even greater significance, compared with conventional use of sugar cane, is that this strategy is likely to have most application at the level of the small family farm. This has important socio-economic consequences as most of the alternative technologies presently being proposed for the sugar cane industry, e.g. conversion to alcohol and chemicals (Lipinsky and Kresovich, 1982), or "energy" (Alexander, 1982), still require the "plantation" approach which is anathema to the majority of the inhabitants of recently independent former colonial territories.

The strategy promises to help resolve two of the principal problems facing developing countries, i.e. food and energy. These problems are most acute in non-oil producing countries, for indigenous oil production offers the means, not only of supplying national energy needs, but through exports provides the foreign exchange with which to buy food. However even for those countries "fortunate" enough to be able to export oil, there have to be advantages from being able to supply their own rural areas with fuel without having to set up expensive infra-structures for national energy networks, or to witness the gradual demolition of their forests. Fuel saved at home, means an increasing proportion for export, or as a reserve against future needs.

The combined productivity, in energy and animal products, derivable from the integrated utilization of sugar cane is probably unequalled. Moreover, the full potential of the sugar cane crop for integrated energy and animal food production has still not been reached. It's theoretical capacity to convert energy into bio-mass was estimated by Bull and Torey (1974) to be of the order of 280 t/ha.a. Genetic progress for increasing the biomass yield of sugar cane has been con-strained by the need to maintain high sugar yields both per unit area and in terms of concentration of sucrose as a percentage of total sugars in the plant, the factor most affecting factory extraction rates. Alexander (1982) has shown that sugar cane cultivars, discarded from traditional sugar production because of low purity, have greater potential for biomass yield than the traditional "sugar" varieties. This type of plant is perfectly suitable for the "energy" cane being developed by Alexander and would be equally appropriate for the purpose described in this paper, since there are not thought to be important differences in nutritive value between sucrose and its constituent monosaccharides. The ratio of total sugars to fibre would also be of little concern provided the carbohydrates in the fibrous fraction could have a final sale value comparable with the carbohydrate in the form of sugars.

No serious limitations are envisaged in the further development of the use of sugar cane juice as a livestock feed. It is suited for most animal species, and probably only chickens are not well adapted to the use of this feed resource. This could mean that ducks should be the preferred poultry species for development in the Tropics. Further, since 40% of the world duck population is in tropical Asia, then perhaps, where simple rural technologies still predominate, natural selection has favoured ducks rather than chickens.

The shelf-life of fresh sugar cane juice is about 24 hours, by which time fermentation has proceeded to the point where the concentration of acetic acid begins to limit palatability, at least in non-ruminants. A number of compounds have proved to be effective in conserving the juice for limited periods of up to seven days. Among these are ammonia, which besides conserving the juice also provides an essential nutrient for microbial synthesis (Duarte and co-workers, 1981); and sodium benzoate (Bobadilla and Preston, 1981), which is already widely used in the food industry as a preservative. The latter is obviously more appropriate for use in non-ruminant diets.

The principal constraint to the development of the fractionation strategy, relates to the utilization of the fibrous residue in the pressed cane stalk. Development work on the design and construction of gasifiers should be accelerated, especially because of its potential for a useful and economical end-product of particular application in rural communities. Few problems seem to be associated with the use of the residual fibre to generate steam and electricity in a boiler-turbine facility, as advocated by Alexander (1982). However this usage is certainly more appropriate for large scale systems and less adapted to the small farmer except those situated within some 30km of the generating station. Conversion of the pressed cane stalk into charcoal could easily be carried out at farm and village level, although the availability of simple machines for making briquettes may provide some constraint. The processing of the pressed cane stalk with alkali to upgrade its feeding value to the point of serving as a maintenance or "working" animal diet is in its infancy and should also receive research priority.

REFERENCES

Alexander, A.G. (1982). Second generation energy cane; Concepts, costs and benefits. Symposium "Fuels and Feedstocks from Tropical Biomass II". UPR Centre for Energy and Environment Research, San Juan, Puerto Rico, 30 pp.

Alexander, A.G., W. Allison, M. Garcia, C. Rios, F.L. Chu, J. Velez-Santiago, A. Velez and R. Ramirez (1979). Production of sugar cane and tropical grasses as a renewable energy source. 2nd Annual Report to US Department of Energy, Division of Solar Technology, Biomass Energy Systems Branch. Contract No. DE-AS05-78ET20071.

Bobadilla, M. and T.R. Preston (1981). The use of sodium benzoate and ammonium hydroxide (aqueous-NH_3) as preservative for sugar cane juice. Trop. Anim. Prod. 6:345-349.

Bull, T.A. and D.A. Torrey (1974). Proceedings International Society of Sugar Cane Technologists, 15:1021-1032.

Duarte, F., A. Mena, R. Elliott and T.R. Preston (1981). A note on the utilisation of aqueous ammonia as a preservative for sugar cane juice in ruminant diets. Trop. Anim. Prod. 6:257-260.

Duarte, F., R. Elliott and T.R. Preston (1982). Sugar cane juice as cattle feed: use of leucaena as combined source of protein and roughage. Trop. Anim. Prod. 8: In press.

Fernandez, A. and F.D. DeB. Hovell (1978). The rumen degradation of rice and cane fibre in animals eating sugar cane or molasses. Trop. Anim. Prod. 3:169 Abstract.

Fernandez, A., D. Ffoulkes and T.R. Preston (1979). Sugar cane feeding: effect of derinding cane stalk and supplementation with sweet potato forage. Trop. Anim. Prod. 4:95-96 Abstract.

Ferreiro, H.M. and T.R. Preston (1976). Fattening cattle with sugar cane: the effect of different proportions of stalk and tops. Trop. Anim. Prod. 1:178-185.

Ffoulkes, D. and T.R. Preston (1978). Cassava or sweet potato forage as combined sources of protein and roughage in molasses based diets: effect of supplementation with soybean meal. Trop. Anim. Prod. 3:186-192.

Ffoulkes, D., R. Elliott and T.R. Preston (1981). Charcoal from sugar cane. Trop. Anim. Prod. 6:150-153.

Hulman, B., E. Owen and T.R. Preston (1978). Leucaena leucocephala as a combined source of true protein and roughage in fattening of cattle with high levels of molasses/urea. Trop. Anim. Prod. 3:1-8.

James, L.A. (1973). Comfith in rations for livestock. Proc. CIDA seminar on sugar cane as livestock feed. Barbados, 30-31 January, 1973.

Lipinsky, E.S. and S. Kresovich (1982). Sugar stalk crops for fuels and chemicals. Progress in Biomass Conversion. 2:89-126.

MacLeod, N.A., T.R. Preston, L.A. Lassota, M.B. Willis and M. Velazquez (1968). Molasses and sugars as sources of energy for pigs. Rev. Cubana Cienc. Agric. 2:205-210.

Meyreles, L. and T.R. Preston (1978). Sweet potato forage as cattle feed: effect on voluntary effect of different amounts added to a basal diet of chopped sugar cane stalk. Trop. Anim. Prod. 3:224-228.

Meyreles, L. and T.R. Preston (1982). The role of poultry litter in molasses/urea diets for the fattening of cattle. Trop. Anim. Prod. 7:138-141.

Meyreles, L., J.B. Rowe and T.R. Preston (1979). The effect on the performance of fattening bulls of supplementing a basal diet of derinded sugar cane stalk with urea, sweet potato forage and cottonseed meal. Trop. Anim. Prod. 4:255-262.

Morciego, S., F. Munoz and T.R. Preston (1970). Commercial fattening of bulls with molasses/urea and restricted grazing. Rev. Cubana Cienc. Agric. 4:97.

Montpellier, F.A. and T.R. Preston (1977). Digestibility of tops, rind, derinded stalk and the entire plant of sugar cane. Trop. Anim. Prod. 2:13-17.

Munoz, F., S. Morceigo and T.R. Preston (1970). Commercial fattening of bulls on molasses/urea, fish meal and restricted forage under feedlot conditions. Rev. Cubana Cienc. Agric. 4:85.

MSIRI (1965). Annual Report. Mauritius Sugar Industry Research Institute, Reduit, Mauritius.

MSIRI (1977). Annual Report. Mauritius Sugar Industry Research Institute, Reduit, Mauritius.

Perez, R. and J.R. San Sebastian (1970). Liquid molasses-based diets for ducks. Rev. Cubana Cienc. Agric. 4:205-207.

Pigden, W.J. (1972). Sugar cane as livestock feed. Report to Caribbean Development Bank, Barbados.

Preston, T.R. and R.A. Leng (1980). Utilization of tropical feeds by ruminants. In: Digestive Physiology and Metabolism in Ruminants. (Editors Y. Ruchebush and P. Thivend). MTP Press Ltd., Lancaster, pp. 621-640.

Preston, T.R., C. Carcano, F.J. Alvarez and D.G. Guttierez (1976). Rice polishings as a supplement in a sugar cane diet: effect of level of rice polishings and of processing the sugar cane by derinding or chopping. Trop.

Anim. Prod. 1:150-163.
Preston, T.R., A. Elias, M.B. Willis and T.M. Sutherland (1967). Intensive beef production from molasses and urea. Nature. 216-721.
Sanchez, M. and T.R. Preston (1980). Sugar cane juice as cattle feed: comparisons with molasses in the absence or presence of protein supplement. Trop. Anim. Prod. 5:117-124.
Santana, A. and F.D. DeB. Hovell (1979). The effect of sodium hydroxide treatment of derinded sugar cane on the rate of rumen degradation of the fibre using the in vivo nylon bag technique. Trop. Anim. Prod. 3:194-195 Abstract.
SFC (1980). Annual Report of Sugar Cane Feeds Centre. Ministry of Agriculture: Trinidad and CIDA: Ottawa.
SFC (1981). Annual Report of Sugar Cane Feeds Centre. Ministry of Agriculture: Trinidad and CIDA: Ottawa.
Tilby, S.E. (1971a). US Patent 3,567,510. March 2.
Tilby, S.E. (1971b). US Patent 3,567,511. March 2.
Tilby, S.E. (1976). US Patent 3,976,499. August 24.
Valarezo, S. and R. Perez (1970). The case of high-test molasses in liquid diets for fattening turkeys. Rev. Cubana Cienc. Agric. 4:117-120.
Vincente-Chander, J. (1981). "Biomasa, no: Alimentos, Si". Personal letter to the Spanish Language Newspaper "El Mundo". San Juan, Puerto Rico.

ADVANCES IN FODDER CONSERVATION

Ernst Zimmer
Institute of Grassland and Forage Research
Federal Research Center of Agriculture (FAL)
Braunschweig-Volkenrode
Bundesallee 50, D-3300 Braunschweig, F.R.G.

ABSTRACT

Compensation between animals' requirements and the seasonal supply of nutrients from forage, by either drying or ensiling, is the main task. Key problems include the extent of nutrient losses and land performance, and the effects of conservation on nutritive value, intake and animal performance. The present state, irrespective of climate or evolution of agriculture, can be characterized as follows: potential of grassland or forage crops is under-utilized; haymaking proves to be the least efficient technology; ensiling promises better efficiency, but control of fermentation within predictable parameters is lacking; and huge quantities of by-products and wastes could extend the feedstock basis, if used sufficiently. Fodder conservation therefore can contribute to food production if future research emphasizes the need to standardize and upgrade crop or substratum quality in terms of digestibility and suitability for conservation; and the need to control and directly influence drying or fermentation processes by means of an appropriate environment and the use of additives or active inoculants. Gaps in our knowledge of the system, and related pressing problems are discussed.

KEYWORDS: fodder conservation, haymaking, ensiling, by-products, losses, additives, intake, digestibility, inoculants.

INTRODUCTION

Compensation between the constant requirements of producing ruminants and the seasonal supply of nutrients from grassland and forage crop production and from by-products and utilizable wastes is the main task of fodder conservation. Principles of either drying (hay curing) or fermentation (ensiling) have been practised for centuries, directed at eliminating respiration and proteolysis by plant enzymes, and preventing any kind of microbial deterioration during the storage period. There are no other principles of conservation practicable for agricultural purposes. However, as a result of research in the last decades, techniques have been modified according to the specific needs (Figure 1).

Nevertheless, beside questions of technology and economics, key problems are still the biological ones: extent of nutrient losses of conservation and yield (Zimmer, 1980; Wilkinson 1981); and effects on nutritive value, intake characteristics and animal performance (Walso and Jorgensen, 1981). If we want to

Figure 1

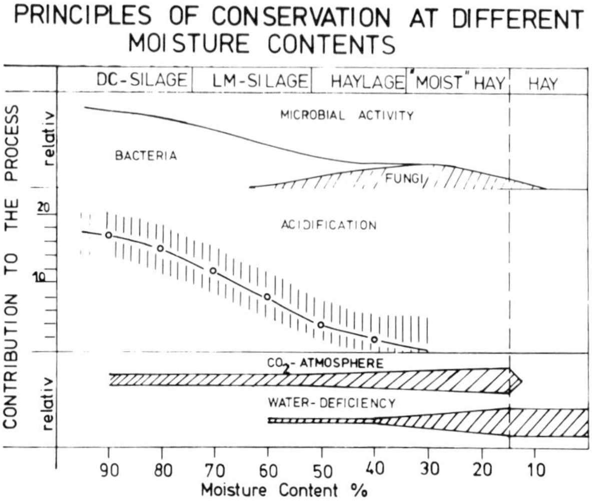

contribute substantially to food production in the future, a clearer understanding of the multifactorial system "crop - conservation process - animal" seems desirable.

THE PRESENT STATE

Crop Potential Under-Utilized

The nutritive value of ligno-cellulose type forages is determined by the degree of maturation of plants. Problems increase in the tropics, where lignification develops faster (Deinum and Dirven, 1975; Deinum et al., 1981; Hacker and Minson, 1981) compared to temperate regions with a daily decline in digestibility between 0.3 - 0.5% points (Daniel et al., 1981; Demarquilly and Jarrige, 1971, 1974).

The nutritive status in the future for any crop is unsatisfactory as a result of delay in cutting due to failures in management or technical capacity (Zimmer, 1981). A shortage of fermentable sugars from intensively fertilized grasses and legumes also appears to be a constraint for ensiling, preventing stable acidification (Zimmer, 1971; Weissbach, 1973; Wilkins, 1980). Huge quantities of by-products or wastes have a value determined by the ligno-cellulose complex remaining after processing. Their potentially valuable use for ruminants depends on upgrading these sources.

Haymaking - Least Efficient Technology

Most of the surplus forage farmers keep for either dry or cold seasons is still hay cured. Late cutting overcomes insufficient natural drying. Mature material with a crude fibre content exceeding 30%, high respiration and mechanical losses in the field, and fungal deterioration during storage often result in a hay quality suitable only for maintenance. In the tropics even minimum requirements in DCP are not reached. Irrespective of the climatic region or the system of agriculture haymaking proves to be the least efficient conservation method (Nash, 1978). This seems true despite such important advances in technical efficiency as big baling, hay stacks, and barn drying. However the existence of excellent hay qualities proves that two problems can be solved. Efficient prewilting in the field can be achieved by means of organisation, conditioning, and using desiccants. Restriction of deterioration during the last phase of drying to stable moisture conditions, e.g. during storage, can be achieved by means of known techniques: hay pods, barn drying, or the use of hay additives.

Ensiling Promises Advantages

Ensiling is of increasing importance, handling nutritious leafy grasses and legumes with less risk of weather damage. The typical "silage crops" are maize, sorghums, small grains and all kinds of moist by-products. The advantage of appropriate mechanisation is well recognized. Spontaneous fermentation by lactobacilli and its associates in principle requires a quick, lasting acidification in order to minimize proteolysis, decarboxylation of amino acids and to prevent a secondary clostridial fermentation which degrades amino acids to NH_3 and amines, forming butyric acid. It also requires complete anaerobiosis, immediate exclusion of air and prevention of gas exchange for the storage period to stop respiration, accumulation of heat oxidative losses and growth of molds and yeasts (McDonald, 1982; Zimmer, 1969). Fermentability of the crop is supported by the ensiling technique which includes conditioning and prewilting, chopping and compaction, the type of silo and the sealing method applied (Zimmer, 1977). Additives with different effects can also be used beneficially. Here desiccation in the field, substratum supplementation, selection of epiphytic microflora, direct acidification and even complete chemical preservative by strong acids or formaldehyde have been attempted (Waldo, 1978; Solsten, 1978; Lusk, 1978).

Despite advances in mechanization and technical performance, we have apparently reached a standstill in quality, unsatisfactory intake or animal performance, high losses and problems of animal health and silage quality (Zimmer 1980, 1981; Wilkins, 1980). Also there are too many failures due to poor control of fermentation within predictable parameter limits (McCullough, 1978).

The present situation can be summarized as one where technical efficiency in forage conservation is obvious in temperate, humid regions and/or developed, industrialized agricultures, but where animal and hectare performance still need futher improvement. If it can be assumed that there is surplus forage, etc., available at all in the subtropics and tropics, we still lack common, simple and appropriate methods of hay or silage making to efficiently minimize nutrient losses and deterioration.

OPPORTUNITIES FOR INCREASED FOOD PRODUCTION

Reducing Fodder Conservation Losses

Higher performance per hectare and higher nutritive value of roughage can result from development of more weather independent methods, of simple efficient technology, of better control of drying and of ensiling by using agrochemicals (Figure 2). Total unavoidable energy losses for ensiling may be less than 10% compared to the current estimate of 30-35% at the farm level. Losses reduce significantly the energy concentration (ME/TM) of the roughage. This highlights the importance of anaerobiosis and the corresponding ensiling techniques. The basic physico-chemical process of drying also could result in rather low losses (Honig, 1976) whereas those on farms are three- or fourfold the achievable.

Improving the Voluntary Intake

When applied, particularly to silage, this will help reduce concentrate

Figure 2

ENERGY LOSSES AND CAUSING FACTORS IN SILAGE MAKING
(VÖLKENRODE DATA)

PROCESS	CLASSIFIED AS...	APPROX.% LOSSES	CAUSING FACTORS
Residual respiration	unavoidable	1- 2	Plant enzymes
Lactic acid fermentation	"	2- 4	Microorganism
Effluent or field losses by wilting	mutual unavoidable	5->7 or 2->5	DM-content; Weather, technique, crop, management
Secondary fermentation	avoidable	0->5	Crop suitability,
Aerobic deterioration during storage	"	0->10	Filling time, density, silo, sealing, crop suitability;
Aerobic deterioration after unloading (heating)	"	0->15	As above, DM-content silage, unloading technique, temperature
		Total 7->40	

feeding to that of a real supplementation, and by that diminish food competition between animals and people. Factors of intake regulation, effect of protein degradation and protein value of silage, cell wall substances and digestibility, including its specificity in species and varieties, are fields of interest here. Major changes in composition during ensiling are indicated in Fig. 3. The formation of acids, increase in NPN, decrease in WSC and in digestibility are known to depress the feeding value and affect voluntary intake.

Figure 3

GENERALISED COMPARISON BETWEEN FRESH GRASS AND GRASS SILAGE

	PARAMETER OF QUALITY GRASS	SILAGE	EFFECT ON VOLUNTARY INTAKE
pH	6	3,7 - 5,2	pH very low = negative
WSC (% DM)	10 - 25	0 - 5	effect unknown, palatability?
Acids (% DM)	3	5 - 20	indirectly negative, (Acetic acid)
True protein N (%tN)	80	20 - 60	strong negative via
Ammonia N (% tN)	0	5 - 30	protein degradation
Digestibility (% DM)	75	74 - 65	negative via
Net energy (MJ/DM)	6,9	6,8 - 5,7	cellwall digestion

Source: compiled from Wilkins 1980; Honig 1981;

Efficient Fodder Conservation Systems

These will lead to better utilization of natural grasslands, avoiding the hazards of overgrazing and irreversible denaturation of the vegetation, as long as balanced stocking rates are ensured and surplus forage production actually becomes available. Adjusting technologies to local conditions, systems research to maximize biological and economic output (Corrall et al., 1982) and to reduce support energy (White, 1980; Honig and Zimmer, 1979), and the provision of better advice to farmers are necessary.

Efficiency has to be evaluated in biological terms as energy concentration, protein value, conservation losses and animal performance. Comparing the most relevant methods (Figure 4), specific differences and risks can be stated (Honig et al., 1981; Lingvall and Nilsson, 1980; Zimmer, 1980). This also appears to be relevant for tropical conditions as well, where temperature and lignification are particular hazards. However there is a lack of transferable data, indicating the need for more efforts in this particular field.

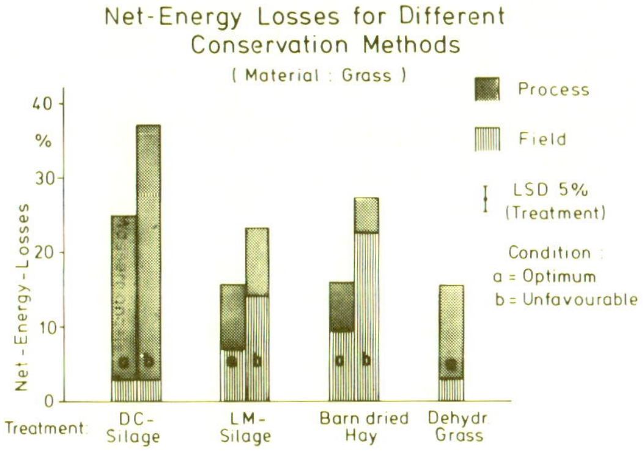

Figure 4

Careful Use of By-Products and Wastes

This can extend the feedstock basis for ruminants, but requires safe conservation of such moist, perishable materials as a precondition. Upgrading of substratum suitability and its nutritive value, control of the conservation process, and making it hygienic must all be considered.

ACTIONS NEEDED

The Impediments.

Fodder conservation, as an applied biotechnology, urgently needs an inter-disciplinary approach. Microbiology, biochemistry etc. have to answer "why it is", while agricultural sciences will develop and explain production systems, i.e. "how it can be managed". Animal performance is the final step of interest, but most of the feeding trials scattered over the world lack in full explanation. Dealing only with the relationship "crop - conservation" may help in optimizing technology, but will not give an answer to the animals' response. Labour shortage or other constraints often encourage engineers to use sophisticated techniques which may result in poorer biological output.

Too few groups in the world work by an interdisciplinary philosophy. However, a group of European laboratories is now trying to deal with a selected item "Effect of prewilting and/or additives on animal performance in different environments - why and how". Analyzing the causalities unfortunately requires better research facilities. The inexpensive field experiment, the feeding test, the comparison of equipment all make little sense. Rather, controlled laboratory scale conservation equipment, in vitro techniques and simulated models will improve understanding and add transferable data. Bearing in mind the importance of roughage as the basic feed stock for ruminants in times of need, and in comparison to the strong research impacts in mixed feeds or pasture management and ecology, some additional efforts seem advisable in this area.

FUTURE RESEARCH AND POSSIBLE ADVANCES

There is a widely-supported opinion that biological efficiency (increase in feeding value and intake, decrease in process losses) needs to be improved. This is also a precondition for technical efficiency and improved economics. Future action, therefore, is needed in the following fields (Fig. 5):

(1) better roughage evaluation: quick, cheap methods for ration formulation and as a tool for plant breeders;
(2) standardisation of crop or substratum quality by better control of agronomic measures, harvesting, and specific conditioning;
(3) upgrading the digestibility of low quality crops or other resources and their respective suitability for necessary conservation;
(4) control of drying and storage, even in unsuitable conditions, preventing fungal deterioration, but independent from support energy;
(5) better understanding of intake regulation of ruminants, overcoming existing constraints of silage feeding;
(6) better understanding of requirements, kinetics, competition and turnover of spontaneous silage microflora, aiming at selection of active strains, and further treatment for inoculants;
(7) appropriate technologies, including use of agro-chemicals, for silage making which will ensure anaerobiosis and support quick, nutrient-saving acidification.

DIRECTION OF FUTURE RESEARCH

CROP, FORAGE
- Methods of evaluation
- Upgrading low quality (conditioning, agrochemicals, cell wall research)

CONSERVATION
- "Moist hay" storage (additives, microflora, mycotoxines ?)
- Control of fermentation (silage inoculants, multipurpose additives, protein degradation, cause of losses)
- Appropriate technologies (adaptation, requirements, efficiency)

ANIMAL
- Intake and silage handicaps (regulation, constraints of silage feeding)
- Protein protection in roughage

WHOLE SYSTEM
- Efficiency of systems and economics

Figure 5

Underlining this statement and drawing attention to existing knowledge gaps, a few examples are now presented. These support the idea that advances in forage conservation will result from better step-by-step understanding of the complex system, thus affecting technology and the strategy of advisory services in disseminating knowledge.

The Unknown Value of Roughage

Efficient ruminant feeding needs supplementation with concentrates based on an exact estimate of the roughage fed. There is a demand for quick cheap methods of hay and silage evaluation for feeding purposes. Plant breeders are also looking for new methods which will help in realizing new breeding concepts. The break-through with the in-vitro technique of Tilley and Terry in the early 60's should now be followed by the proving of simplifications (using enzyme mixtures), as well as the capability of the NIR system for such heterogenous materials (Zimmer, 1982), or of the milling resistance method (Paul et al., 1981/82).

Upgrading Low Quality Forages

Digestibility of the ligno-cellulose complex of all kinds of forage and feeds is a key parameter, providing an estimate of how the ruminant uses this source of energy. Fast lignification in the tropics, late cutting due to labour and management restrictions, very mature straw, fibrous residues from leaf extraction or "predigested" wastes and manure together, all show the need for upgrading. Lower polymerized carbohydrates from the split complexes become available for silage microorganism or direct utilization by the ruminant. Treatment with either acids, alkali, ammonia, ozone or selected microorganisms (Stapleton, 1981) are already employed in some countries. Mason (1982) recently noted that "The conservation of forage as alkaline, high dry matter silages or upgraded hays and straws is becoming a practical feasibility". However available data from empirical research still cannot answer how the physico-chemical structure of the ligno-cellulose complex, varying by species and climate, prevents penetration and action of digestive enzymes or is affected by the chemical treatments (Fig. 6).

Here we need more systematic knowledge: on the substrate side about the degree and cause of the variability of the ligno-cellulose complex; of methods to quickly evaluate "structural" parameters besides the chemistry of cell walls; about microorganisms or the respective enzymes for systematic use (lignin destroying organisms are known in forestry, and basidiomycetes are already used for improving straw (Zadrazil, 1974); about appropriate conditions of preservation following such pretreatments. For example, upgrading cotton straw into a valuable roughage is a challenge in this field (Ben-Ghedalia et al., 1980).

UPGRADING LOW QUALITY FORAGES

SEKTOR	KNOWLEDGE GAPS
SUBSTRATE SUITABILITY	Lignocellulose complex - variability, evaluation, causing factors (environment, genotype)
TREATMENTS WITH	
... ALKALI, AMONIA etc.	Appropriate technologies Substrate / dosis / effect data
... MICROORGANISMS	Efficient species - requirements and effects Selection and genetic changes
CONSERVATION	Conditions - turnover, efficiency, control
FEEDING	Nutritive value, utilization

Figure 6

Field Wilting - Artifical Desiccation

Drying in the field has to be performend to different DM-levels for further conservation. Weather conditions affect both respiration and scattering losses. The efficiency of desiccants, e.g. organic acids or potassium carbonate (Wieghart et al., 1980; Tullberg et al., 1978) in support of mechanical conditioning is still an open problem, requiring better understanding of the drying process (Harris, 1980) and also the right distribution techniques.

Moist Hay, Fighting Against Fungal Deterioration.

Bulk roughage still is hay, but molding creates high losses, intake troubles, and possibly health problems. The CP content often decreases below maintenance requirements. Thus, efforts directed to preserve "moist hay" in a simple, cheap way e.g. by adding urea, ammonia or other additives will have high priority. Large quantities of nutrients could be kept for animals instead of being wasted (Kuntzel et al., 1980; Huntzel and Pahlow, 1980; Buettner et al., 1982; Westgaard, 1979).

Protein Value of Silage

An important knowledge gap in the field of conserved forage is related to its protein value. Produced with costly N-fertilizer input, or carefully using the biological N-fixation capacity of legumes, the CP of green materials is later degraded by proteolysis, deamination and decarboxylation in an uncontrolled way (McDonald and Edwards, 1976). We already know the negative effects on intake; we use protein protection to bring more amino acids to the lower intestine; and we know something about synthesis of bacterial proteins in silages (Ullrich, 1982), and about degradation and amino acid patterns (Merchen and Satter, 1982). However it is likely that low energy utilization and insufficient intake could be better explained if we had a better understanding of the whole process of protein degradation.

Controlling Fermentation Within Predictable Parameter Limits

Controlling fermentation is the biggest challenge to keeping nutrients from being wasted. In contrast to industrial biotechnology, we have to deal with heterogenous raw materials of different suitability, spontaneous fermentation by an epiphytic flora, and varying environmental conditions on the farm level.

ROLE OF ADDITIVES

EFFECT		AGENT (e.g.)	For silage (S) For hay (H)
Improving	FERMENTABILITY	Carbohydrates, Dry matter, Enzymes	S
Lowering	pH	Inorganic & Organic acids	S
Inhibiting	MICROFLORA	Formaldehyde, Acrylaldehyde, ...	S
Selecting	BACTERIA	Formic acid, Nitrite, ...	S
		NaCl	H
Restricting	FUNGI	Propionic, Butyric acid, ...	S
		Urea, Ammonia, ...	H
Upgrading	DIGESTIBILITY	Alkali, Ammonia, Enzymes	S, H
Upgrading	CRUDE PROTEIN	Urea, Ammonia	S, H

Figure 7

Major efforts have been made to affect fermentation by "additives" (McCullough, 1978; McDonald, 1982). These have included increasing the suitability of substrates by adding sugars, starch, dry matter, or mixing with other crops; lowering the pH with inorganic or organic acids; selecting epiphytic bacteria by use of bacteriostatica formic, nitrite, or benzoic acid; direct inhibition of fungal activity by propionic and acetic acids, and sulfite; and complete chemical preservation plus protein protection by formaldehyde (Figure 7). The disadvantages include the costs, technique of distribution, residue problems, and even a common uneasiness abut too much "agrochemistry". Thus use is limited to certain agricultural regions, e.g. the temperate, humid climate of northern Europe and North America. For the tropics, where silage making is expected to ensure better feed compensation for animal husbandry throughout the year, we do not yet have a general solution.

However advances in microbiology may open new and successful chances for inoculants (Moon et al., 1979; Burghardi et al., 1980). Selection of better adapted strains of lactobacilli is now possible. Obviously such strains will survive and predominate in the population in silage. They also have the ability to suppress yeasts. This will ensure aerobic stability (Pahlow, 1982). Verified activity already corresponds well to desirable levels (Fig. 8). Freeze drying enables efficient high density storage and simple handling. Existing large variations offer the chance of selecting strains for certain substrate and environmental conditions. A recent example of perserved sugar beet chips (Pahlow, 1982), ensiled at 55°C and with 22% DM, is sensitive against sporeforming bacteria, and shows a few promising adapted strains of lactobacilli. Silage microbiologists should certainly emphasize this area, as well as examining whether genetic engineering can offer a real chance to improve promising strains.

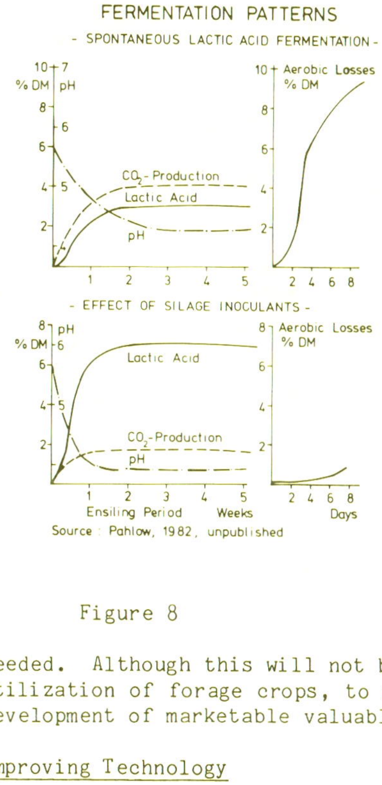

Figure 8

Processing Forage Crops for a Dual Purpose

With protein malnutrition in animal husbandry in many countries despite crop potential and success in growing legumes, Pierie's (1971) idea about extraction of leaf protein needs further development (Wilkins, 1976). Energy input for processing has been reduced. Handling and utilization of juice for pigs and poultry is being investigated. However, research on the fibrous residues as to their suitability for conservation and their intake characteristics for ruminants is still needed. Although this will not be a simple technology it may contribute to better utilization of forage crops, to more collaboration between farms, and possibly to development of marketable valuable products from green forage.

Improving Technology

Lack of technology in forage conservation by ensiling is responsible for wasting nutrients. Slow filling, low compaction, and improper sealing means delay in anaerobiosis, continuous gas exchange and temperature increase, thus promoting all kinds of microbial deterioration and a loss of highly digestible nutrients. As much as 2-3% of digestible dry matter can be mineralized. Advanced fodder conservation requires efficient provision of techniques, materials and advice to farmers.

CONCLUSIONS

Conservation of fodder from grassland, forage crops, residues and byproducts is a necessity in most climatic regions, if stable animal production is to be achieved. Large amounts of nutrients which could contribute to the world's feed and food situation are being wasted. Other large amounts are not utilized at all because of lack of special knowledge, technology or management. The "frontiers" for productive research efforts are:

(1) the complexity of the multifactorial system, crop - conservation - animal, has to be better understood and explained if there is to be successful application of biotechnology;
(2) the variability of microorganisms within this system, if it is to be controlled, promoted or used according to our needs;
(3) the constraints in producing surplus forage and the pressing problems on farms with quite different capabilities, which must be met by an appropriate, attainable technology.

REFERENCES AND RECENT RELEVANT LITERATURE

Ben-Ghedalia, D., G. Shefet and J. Miron (1980). J. Sci. Food Agric. 31, 1337-1342.

Bolsen, K.K. (1978). In: McCullough, l.c., 180-200.

Buettner, M.R. (1978). Diss. Abstr. Int., B. 39, 4134.

Buettner, M.R., V.L. Lechtenberg, K.S. Hendrix, J.M. Hertel (1982). J. Anim. Sci., 54, 173-178.

Durghardi, S.R., R.D. Goodrich and J.C. Meiske (1980). J. Anim. Sci., 50, 729-736.

Corrall, A.J., H.D. Neal and J.M. Wilkinson (1982). GRI, Techn. Rep. No. 29, January, Hurley.

Daniel, P., M.-L. Rotermund, U. Simon, M. Wermke and E. Zimmer (1981). Landbauforsch. Volkenrode, Sh. 59.

Deinum, B. and J.G.P. Dirven (1975). Neth. J. Agric. Sci., 23, 69-82.

Deinum, B., J. de Beyer, P.H. Nordfeldt, A. Kornher, O. Ostgard and G. van Bogaert (1981). Neth. J. Agric. Sci. 29, 141-150.

Demarquilly, C. and R. Jarrige (1971). Proc. 4th Gen. Meeting Europ. Grassld Fed., Lausanne, 91-106.

Demarquilly, C. and R. Jarrige (1974). Proc. 5th Gen. Meeting Europ. Grassld Fed., Uppsala, 28-41.

Giardini, A. and F. Gaspari (1981). Informatore Agrario, 37, 15433-15434, 15437-15438, 15441-15442, 15445-15446.

Hacker, J.B. and D.J. Minson (1981). Herb. Abstr. 57, 459-482.

Harris, C.E. and J.N. Tullberg (1980). Grass and Forage Sci. 35, 1-11.

Honig, H. (1976). Institute of Grassland and Forage Res., Braunschweig, Publ. No. 247.

Honig, H. and E. Zimmer (1979). Lanbauforsch. Volkenrode, Sh. 49, 71-82.

Honig, H., K. Rohr and E. Zimmer (1981). XIV Int. Grassld Congr. Lexington/USA (Proc. in press)

Kuntzel, U., Y. Leshem and G. Pahlow (1980). In: Thomas, l.c., 252-256.

Kuntzel, U. and G. Pahlow (1980). Das wirtschaftseigene Flutter 26, 39-52.

Lingvall, P. and E. Nilsson (1980). In: Thomas, l.c., 175-185.

Lusk, J.W. (1978). In: McCullough, l.c. 201-232.

Mason, V. (1982). Grass Farmer, No. 12, British Grassld. Soc., 3-7.

McCullough, M.E. (ed.) (1978). "Fermentation of Silage - A Review". Nat. Feed Ingred. Assoc., Des Moines, Iowa, USA.

McDonald, P. and R.A. Edwards (1976). Proc. Nutr. Soc. 35, 210-211.

McDonald, P. (1982). Trends in Biochem. Sci., May 15-17.

Merchen, N.R. and L.D. Satter (1982). Changes of nitrogenous compounds and sites of digestion of Alfalfa harvested at different moisture contents (in press).
Moon, N.J. and L.C. Moon (1979). J. Dairy Sci., 62, 184.
Moon, N.J., L.O. Ely, E.N. Sudweeks and M.C. McCullough (1979). J. Dairy Sci., 62, 183-184.
Nash, M.J. (1978). J. Roy. Agric. Soc. England 139, 64-73.
Pahlow, G. (1982). Unpublished data, Volkenrode.
Paul, Chr., G.-J. Schild and H. Honig (1981). Landbauforsch. Volkenrode 31, 7-10.
Paul, Chr. and V. Mika (1981). Landbauforsch. Volkenrode 31, 163-169.
Paul, Chr. and G.-J. Schild (1982). Landbauforsch. Volkenrode 32, 169-172.
Pirie, N.W. (Ed.) (1971). IBP Handbook No. 20, Blackwell, Oxford.
Stapleton, P.G. (1981). Agric. Progress 56, 15-24.
Thomas, C. (Ed.) (1980). Forage Conservation in the 80's, Brighton 1979, Occ. Symp. No. 11, British Grassld Soc., Janssen, London.
Tullberg, J.N. and D. Agnus (1978). J. Agric. Sci. Cambr. 91, 551-556.
Tullberg, J.N. and D.J. Minson (1978). J. Agric. Sci. Cambr. 91, 557-561.
Ullrich, B. (1982). Poster Paper, E.E.C. Sem. "Forage Protein Conservation and Utilization", Dublin.
Waldo, D.R. (1978). In: McCullough, l.c., 117-180.
Waldo, D.R. and N.A. Jorgensen (1981). J. Dairy Sci. 64, 1207-1229.
Weissbach, F. et al. (1973). Methoden und Tabellen zur Schatzung der Vergarbarkeit (Methods for estimation of fermentability). Academy of Agriculture, GDR, agrabuch.
Westgaard, P. (1979). NOFO, February 1979.
White, D.J. (1980). In: Thomas, l.c., 33-45.
Wieghart, M., J.W. Thomas and M.B. Tesar (1980). J. Anim. Sci. 51, 1-9.
Wilkins, R.J. (1976). In: Taylor, J.C. (Ed.). Improving the nutritional efficiency of beef production, EEC, 291-303.
Wilkins, R.J. (1980). J. Roy. Agric. Sci. 141, 127-141.
Wilkinson, J.M. (1981). Ann. Appl. Biol. 98, 365-375.
Zadrazil, F. (1974). Eur. J. Appl. Microbiol. 4, 273-281.
Zimmer, E. (1969). Proc. 3rd Gen. Meet. Europ. Grassld. Fed., Braunschweig, 113-125.
Zimmer, E. (1971). Proc. 1st Int. Silage Res. Conf., Washington D.C., 58-78.
Zimmer, E. (1977). Proc. Int. Meeting on Anim. Prod. From Temp. Grassland, Dublin, 121-125.
Zimmer, E. (1980). In: Thomas, l.c., 186-200.
Zimmer, E. (1981). Phosphorus in Agric. 35, No. 81, 3-15.
Zimmer, E. (1982). Landbauforsch. Volkenrode, Sh. 62, 91-103.

RECENT DEVELOPMENT IN FEED ADDITIVES

Zdenek Otto Muller
Chief Technical Advisor, FAO/UNDP Project PAK/80/019 –
Coordinated National Programme for Livestock Feed Resources and Nutrition
℅ FAO P.O. Box 1476, Islamabad – Pakistan

ABSTRACT

The main achievements of applied and fundamental research in feed additive science directed towards improvement of livestock performance and its economics are reviewed. The main areas covered are: vitamins, antibiotics, antimicrobial drugs, hormones, enzymes, arsenicals, lactobacilli, mold inhibitors, antioxidants, buffers, silage additives, miscellaneous feed additives such as synthetic energy and fibre sources, the use of zeolite, kaolin, taste improving substances, regulation of fly control and of insect growth via feed, and others.

The potency of new vitamin D isomers and their impact on growing poultry, egg shell quality, calcium absorption and mineralization of the bone, are reviewed in some detail. Equal attention is being paid to new findings in vitamin E-selenium research. These two constituents are evaluated in terms of their deficiency, nutrition/pathology, mutual interchangeability, interaction with several additives, effect on carcass quality and shelf-life of frozen poultry. Practical aspects of selenium toxicity modification by a polar factor contained in linseed meal and elimination of toxicity of Pisum sativum diets by vitamin E-selenium are noted. Recent findings on other feed additives and their impact on present and future food production are described. Missing links, pressing tasks and the goals envisaged for future feed additive research and development are listed with recommendations.

KEYWORDS: feed additives; vitamins; antibiotics; antimicrobials; monensin; enzymes; arsenicals; lactobacilli; mold inhibitors; antioxidants; buffers; silage additives and pigments.

INTRODUCTION

With the beginning of the antibiotic era shortly after World War II, animal science made tremendous strides. In the absence of feed additives barely half of the presently available animal products would have been on the world market. In the early 50's we witnessed the birth of the "Chicken of Tomorrow", a broiler producing more than twice the meat from the same amount of feed, hens now laying 300 or more eggs/year - more than thrice the amount of the pre-war era. The same trend occurred for pigs, beef, dairy and other areas of animal production. The evidence of past feed additives contribution over the last two decades, as illustrated in Table 1, is impressive (Muller et al., 1980). Since the early 60's scientific and technological progress has continued with unprecedented speed.

Table 1: The Impact of Feed Additives on the Performance of Meat Birds (relative values)

Feed/gain	Nature of additive[1]	Body weight
100	None[1]	100
73	Coccidiostat (C)	123
63	C+Vitamins (V)	208
57	C+V+Antibiotics (A)	222
56	C+V+A+Methionine (M)	224
55	C+V+A+M+Lysin (L)	233
54	C+V+A+M+L+Furazolidon	237
53	C+V+A+M+L+F+Enzymes[2]	241

[1] Basic diet was balanced in protein, energy and mineral elements.
[2] A mixed enzyme preparation (amylase, protease) grown on barley.

Feed additives are now accepted as an integral part of animal nutrition and play an increasingly important role in intensive livestock systems. They represent a wide interdisciplinary and coordinated endeavour on the part of nutritionists, veterinarians, biochemists, pharmacologists, bio-engineers, technologists, medicinal chemists, statisticians and many others.

The variety of feed additives is so enormous that not all of them can be included. Some of the topics are discussed in other papers in these proceedings (Amino acids, carbadox, medicated feeds) and some (trace elements, coccidiostats, nitrofurans and several other compounds) would require a separate paper. This review presents some of the most recent aspects on vitamins, antibiotics and other drugs.

VITAMINS

Fat Soluble Vitamins

Vitamin A deficiency may alter corbohydrate metabolism of the chick (Phillips and Nickels, 1977) while supplementation of vitamin A accelerates antibody production and increased phagocytosis which was proved by a significantly reduced mortality of chicks infected with E.coli (Donoghue et al., 1979). In ruminant studies additional vitamin A increased the body weight of heifers significantly but massive levels of vitamin A led to hepatic dysfunction with a resultant decrease in retinol clearance from plasma.

Most of the poultry research with vitamin D concentrated on the potency of various new isomers and on aspects related to absorption, bone mineralization processes and egg shell quality. It was found that the R isomer of 24,25-OH-CC was found to be 1.5 times higher than that of CC (McNaughton et al., 1977). As for bone mineralization it was noted that the relative potencies in decreasing order were 1-25-OH-CC, 1,24S 2,6-trans-CC and 25-OH-CC (Boris et al., 1979). In further studies it was reported that 25-OH-CC was much more effective than CC and that 5,6-trans-vitamin D_3 had 5% of the cis isomer potency (Borsje et al., 1977). The importance of the parathyroid function, presence or absence of various isomers, Ca-binding protein capacity of duodenal mucosa, Ca-resorption from the bone and Ca-absorption from the GI tract are some of the main areas influencing the potency of newly synthesized isomers and their mutual interaction and impact on the degree of CA metabolism of the chick.

Vitamin D isomer potency versus egg shell quality revealed some new findings. Oystershells improved eggshell quality better than ground limestone (McLoughlin and Soares, 1976). Furthermore it was found that Ca absorption was higher with 1_-OH-CC during the formation of the shell than CC or the 25-OH-CC molecule in the aging quail hen (Kaetzel and Soares, 1979). Similarly, markedly higher Ca-absorption was observed during shell formation as compared to the period of uterine inactivity (Bar et al., 1976). Recovery from vitamin D depletion was observed to be faster with 1 -OH-CC than CC. Recent studies have indicated that the hen can maintain vitamin D reserves and near normal Ca-absorption for around 10 days, and that hens previously exposed to vitamin D deficiency can cope better with subsequent vitamin D deficiency (Summers et al., 1981). The practical significance of extra vitamin D in rye-based diets was the increase in body weight. Interactions were also found between vitamin D_3 and Zn. Zn at 30 ppm or more increased response to vitamin D_3 but the opposite occurred with lower levels of Zn (MacAuliffe et al., 1976).

The antioxidant activity of the vitamin E-selenium complex and evidence of its other metabolic functions indicated that there are no definite answers yet regarding the role of the function of this vitamin in livestock and poultry nutrition. The interchangeability of vitamin E with Se and antioxidants had been suspected for several decades but recent research activity has extended rather than elucidated the frontiers of this research. The incidence of exudative diathesis (ED) appears to be directly related to the dietary level of Se (Usuenter and Bragg, 1977). It was also found that ascorbic acid, methylene blue, ethoxyquin and BHA were capable of preventing encephalomalacia (EM). Chicks fed 36% yeast, grown on n-paraffin, methanol or ethanol, developed ED which was prevented by the addition of extra Se or by yeast grown with added Se (Yoshida and Hoshii, 1977). In research with artificially induced ED, nicarbazin but not other coccidiostats reduced the vitamin E pathology (Barton and Dydowski, 1979). There is an overwhelming evidence that Se has the capacity to control ED even in the absence of vitamin E (Pilch and Combs, 1981). This was futher supported by deficiency studies on ducklings fed low Se-diets where vitamin E could not prevent mortality and clinical deficiency of vitamin E but Se did (Combs, 1981).

As for the interaction of vitamin E with Se it was found that ascorbic acid promoted the enteric absorption of Se but not of vitamin E. High levels of vitamin A increased the enteric absorption of Se but interfered with the absorption of vitamin E. Ethoxyquin spared the Se requirement of vitamin E-deficient chicks (Combs, 1978). Recently, Murphy et al. (1981) claimed that excessive vitamin E interfered with vitamin D utilization as well as with Ca and P metabolism. Supplemental vitamin A was found to accentuate the encephalopathology of chicks given a vitamin E-Se-deficient diet with high levels of unsaturated fatty acids. Ascorbic acid alleviated symptomatology (Dror et al., 1980). Vitamin E, ethoxyquin and extra added Se, vitamin E and DL-methionine appear also to moderate heat stress (Jensen et al., 1978). Aspirin was found to act synergistically with vitamin E in decreasing mortality from E. coli infection (Hikoff, 1981). Interaction of vitamin E-Se complex with coccidiosis was also reported. The absorption of Se is reduced during coccidial infections in the anterior of the small intestines but increased absorption of Se was found in the duodenum and anterior ileum (Pesti and Combs, 1976). Se and/or vitamin E have a significant impact in ameliorating the mortality due to coccidiosis.

Several reports suggest that vitamin E, Se and antioxidants improve the stability of carcass fat, extend the shelf-life of frozen poultry products, and that high levels of vitamin E protect substantially the onset of oxidative rancidity (Bartov and Bornstein, 1977). Additional vitamin E was found to reduce cholesterol and the proportion of oleic acid in serum triglycerides of cockerels (Klopfenstein and Clegg, 1977); and it has been suggested that linseed meal contains a heat-stable, organic polar factor that interacts with Se and counteracts the growth depression caused by excessive Se levels (Jensen and Chang, 1976). Vitamin E-alcohol was reported to be better utilized than acetate but a water soluble ester D-L-tocopheryl glycol 1000 was poorly utilized. In requirement studies it was found that 0.05 ppm Se is adequate to sustain egg production but that 0.1 ppm Se is required for optimal hatchability. Seleno-DL-methionine increased voluntary feed intake of chicks within 2-3 hours, but the selenite led to a delay in response of 3-4 hours (Combs and Scott, 1979).

Niyo and Block (1976) reported that deficiency of Se and vitamin E caused hepatic heart and skeletal damage of baby pigs which was offset by Se-vitamin E supplementation; while Malm et al. (1976) found that the progeny of gilts fed diets low in vitamin E but adequate in Se, did not develop vitamin E-Se deficiency symptoms during a 3 week lactation. A dramatic impact of vitamin E-Se deficiency was observed on high moisture maize-based diets where 90% of pigs were reported to die with multiple symptoms of vitamin E-Se pathology (Young et al., 1977). In other pig studies with 96.8% of Pisum sativum in the diet, 80% of the pigs died during the 135-day trial, but no deaths, or signs of deficiency were observed when such diets were supplemented with vitamin E, Se or both (McDowell, et al., 1977). In studies with barley diets, supplemental vitamin E improved performance of growing pigs. Similarly, maize + soybean based diets, supplemented with either 0.1 ppm Se or 11 IU vitamin E/gh, protected pigs from vitamin E-Se deficiency. As many as 65% of pigs without Se or vitamin E supplementation died (Young et al., 1978). Inorganic SE and Se from brewers grains are apparently more available to pigs than Se supplied by fish meal. Se and vitamin E were found to reduce toxicity of $CoSO_4$; arsanilic acid decreased Se excretion; and retention of Se was found to be greater with seleno-methionine compared to selenite.

The greatest demand for Se in sows appears to be immediately after parturition (Wilkinson et al., 1977). Diets of diverse ingredient composition may contribute to variations of Se retention in tissues despite the presence of similar level of Se in these diets. The optimum level of α-tocopherol acetate in the diet appeared to be 100 ppm for stability of intramuscular lipids. It was observed that although Se and vitamin E, provided by initial tissue reserves, may be adequate for pregnancy, the addition of 0.1 ppm inorganic Se and 22 IU/kg vitamin E appeared to maintain tissue levels (Piatkowski et al., 1979). Vitamin E and Se studies include those on cows (Schingoethe et al., 1978), on steers (Schingoethe et al., 1979), and on sheep (Hidiroglon et al., 1977; Whanger et al., 1978).

Water Soluble Vitamins

The growth retardation caused by feeding higher levels of Co, Se, V or Cd, and mortality from S. gallinarum were reduced by supplementing the diet with 0.1% ascorbic acid. In addition, supplemental ascorbic acid (at 0.1%) fed to birds under an artificially induced stress increased body-weight and the immune system of the chicks (Yen et al., 1977). Interactions between pyridoxin requirement and the quantity and quality of dietary protein were reported by two research groups (Kazemi and Kratzer, 1980; Bishra and Walker, 1977).

In choline studies it was found that more live pigs were born from sows fed diets with added choline and that choline was necessary to ensure progeny survival. Incidence of perosis in broilers was higher with RSM than with SBM - based diets (March and McMillan, 1980). Higher egg production was recorded with SBM than with RSM but differences were offset by choline supplementation; SBM produced larger egg yolks than RSM. The livers of the birds fed SBM contained more fat than livers of birds fed RSM but choline fortification reduced liver fat content (March, 1981).

The severity of biotin deficiency and the occurrence of fatty liver and kidney syndrome (FLKS) in chicks appeared to be affected by nutritional factors other than biotin. Rearing on wet litter apparently accelerated the incidence of foot pad dermatitis. Although biotin supplementation decreased foot pad dermatitis on dry litter, it did not on wet litter. Birds kept on litter require about 0.17 ppm biotin to control or at least reduce FLKS but on wire floors the requirement was estimated to be about 10% higher (Whitehead and Bannister, 1980). Other results indicated that 200 ppm biotin increased egg yolk biotin content and reduced foot pad dermatitis and the incidence of breast blisters of the progeny (Harms et al., 1979). A significant reduction of "flip-over" incidence and mortality was observed as a result of biotin supplementation (Hulan et al., 1980). Relevant studies on pigs include those by Brooks et al. (1977) and Bane et al. (1980).

In poultry studies with riboflavin it was suggested that chick embryos were able to replace riboflavin by 7-ethyl-8-methyl-flavin which actually inhibited riboflavin use in normal eggs. Similarly, 7,8-diethyl-flavin acted as a riboflavin antagonist for the chick embryo (Lambooy and Shaffner, 1977). It has also been established that the accelerated rate of destruction of riboflavin in a tropical environment, as compared to temperate climate, justifies a level of 5.1 ppm, i.e. much above NRC recommendations (Ogunmodide, 1977). The omission of riboflavin had the greatest effect on hatchability which, after 13 weeks of treatment, had dropped to zero level in two strains studied by Leeson et al. (1979). The availability of niacin from unroasted and roasted maize was about 30%, from soybean meal, 100% (Yen et al., 1977).

In an attempt to enhance the post-transit performance of feedlot calves, it was found that calves fed a B-vitamin supplement gained 5 kg more than the controls (Cole et al., 1982). Recently, it was reported that cows receiving niacin showed higher persistency at weeks 9 and 10 than those without added niacin. A greater stimulation of milk production was noted when niacin was added to natural protein rations than to rations containing NPN (Jimenez, 1981). Pantothenic acid deficiency was found to affect the energy utilization and body composition of chicks (Beagle and Begin, 1976). In vitamin B_{12} studies it was suggested that the level of vitamin B_{12} in a breeder's ration had a significant bearing on the vitamin B_{12} level of the chick and its tolerance to dietary fat. Chicks originating from hens fed a vitamin B_{12} deficient diet, when introduced to broiler starter containing 10% animal fat, manifested a significant growth depression. By adding 10 μg vitamin B_{12}/kg feed, the problem was overcome. More vitamin B_{12} is needed when the protein level of the ration is excessively high (Patel and McGinnis, 1980).

DIETARY ANTIBIOTICS AND OTHER ANTIMICROBIAL DRUGS

Since the discovery of the nutritional aspects of antibiotics in 1948, research on antimicrobials has been most actively pursued. Nearly 1,000 scientific papers were published in the past 5 years. The list of main antimicrobial drugs currently used illustrate this trend (Table 2).

Table 2: The Effect of Antimicrobial Drugs, Fed at Dietary Levels, on the Improvement of Animals Performance

Column 1: Improved Rate of Gain
2: Improved Feed Efficiency
3: Improved Feed Intake (appetite)
4: Decreased Mortality
5: Improved Egg Production
6: Improved Egg Fertility (hatchability)
7: Improved Egg Shell Quality
8: Improved Livability

Antimicrobial drugs	Animal specie/ class	1	2	3	4	5	6	7	8
Bacitracin group	Growing poultry	+	+	+	+	–	–	–	–
	Laying pountry	–	+	+	–	+	+	–	–
	Growing pigs	+	+	–	+	–	–	–	–
	Feedlot cattle	+	+	–	–	–	–	–	–
Bambermycins	Growing poultry	+	+	–	–	–	–	–	–
	Growing pigs	+	+	–	–	–	–	–	–
Dimitridozole	Growing turkeys	+	+	–	–	–	–	–	–
Erythromycin	Growing poultry	+	+	–	+	–	–	–	–
	Laying poultry	–	+	–	+	+	–	–	–
	Growing pigs	+	+	–	–	–	–	–	–
	Feedlot cattle	–	+	–	–	–	–	–	–
Furazolidone	Growing poultry	+	+	+	+	–	–	–	–
	Laying poultry	–	+	+	+	+	–	–	–
	Growing pigs	+	+	+	+	–	–	–	–
Impronidazole	Growing turkeys	+	+	–	+	–	–	–	–
Oleanodnmycin	Growing poultry	+	+	–	–	–	–	–	–
	Growing pigs	+	+	–	–	–	–	–	–
Penicillin	Growing poultry	+	+	–	+	–	–	–	–
	Growing pigs	+	+	–	+	–	–	–	–
Tetracyclins	Growing poultry	+	+	+	+	–	–	–	+
(CTC, OTC, TC)	Laying poultry	–	+	+	–	+	+	+	–
	Growing pigs	+	+	+	+	–	–	–	–
	Feedlot cattle	+	+	–	–	–	–	–	–
Tylosin	Growing poultry	+	+	–	+	–	–	–	–
	Laying poultry	–	+	–	+	+	–	–	–
	Growing pigs	+	+	–	+	–	–	–	–
	Beef cattle	(Reduction of incidence of liver abscesses)							
Virginiamycin	Growing poultry	+	+	–	–	–	–	–	–
	Growing pigs	+	+	–	+	–	–	–	–

Most antibiotic research with poultry has reported encouraging results with bambermycins (BM). This relatively new antibiotic, exclusively designed as a feed additive, has gained rapid momentum. In broilers, this antibiotic was widely tested with clopidol, 3-nitro, amprolium and it was found that all these substances singly or in combination improved body weight, feed efficiency and also the pigmentation of broilers. Similarly, as in the case of classical dietary antibiotics, all new drugs facilitate vitamin A transfer to eggs and BM appears to the the most effective in the pigmentation (redness and yellowness) of the egg yolk (Zotovic, 1978).

Several research groups reported that penicillin added to broiler diets containing 25-70% rye increased their growth from 24-60% respectively. Rye evidently contains antinutritional factors interfering with the utilization of lipids, particularly in diets with high concentrations of long-chain saturated fats (McGinnis and Patel, 1976). In a comprehensive test with 58 barley cultivars (Karl and 9044), a marked variation was found in their growth depressing effects. Addition of β-glucanase (150 units/kg) to barley diets improved performance on one cultivar. Penicillin (50 ppm), lincomycin (4 ppm) or M.D. bactracin (50 ppm), not single but in combination, improved growth that was not different from the maize diet (McGinnis et al., 1981). In 31 experiments with Avoparcin (5.5-22 ppm) on 35,000 birds the drug stimulated a significant improvement of performance (Penseck et al., 1981). Lambodamyzin, a product of Streptomyces griseus, was found to markedly improve protein digestibility and growth (Hennig et al., 1979). Laying birds fed bacitracin showed a lower susceptibility to low-temperature stress. Virginiamycin (VM) was found to improve shank pigmentation, egg production, feed conversion, and egg specific gravity, but reduced egg weight (Tomov et al., 1980; Janky et al., 1981).

Studies on pigs elucidated many aspects of practical significance. Higher protein (18-15%) levels in pig diets when fortified with VM resulted in better performance than lower protein (15-12%) diets (Palura et al., 1980). Comparison of Cu with VM suggested that both additives improved performance of pigs, but less when the two compounds were fed simultaneously. It was further reported that single additions of Cu, CTC or VM to a herd not exposed to antibiotics for 5 years improved performance. Cu plus an antibiotic further improved ADG but not F/G (Barber et al., 1978). Comprehensive pig studies were carried out with VM, EDDI, ASP 275, pristinamycin, efrotomycin, olaquindox, apramycin, tylosin, colistin, zinc bacitracin, indicating that all of these compounds have significant effects on improvement of pig performance though the degree may vary either due to the drug or the specific conditions under which these tests were performed.

Recent work has reviewed the effects of tiamulin, a new drug effective in the prophylaxis of swine dysentery (O'Connor et al., 1979; Taylor and Davey, 1980). Another similar drug, dichlorvos, proved to have significant effects on pig performance. Parasite-infected sows treated with dichlorvos produced more live pigs which were heavier at birth and weaning. Similarly, feeding dichlorvos for the last month of gestation produced more pigs and they were superior in postnatal performance (Young et al., 1979). Experiments on nitrovin and ronidazole (Hennig et al., 1979) have also been noted.

Many antibiotics and other antimicrobial drugs have been researched in recent years in ruminants. In studies with elfazepam it was reported that the drug increased digestibility of DM, ADF and CP in wethers (Dinius and Baile, 1977). Among numerous sulfur-containing, thiopeptin-like antibiotics, it was found that they are effective as a possible remedy for lactic acidosis. Thiopeptin given in a single-doze prevented lactic acidosis by reducing rumen lactate level required to control acute acidosis was 0.18 mg/kg BW. All members of the thiopeptin class (sulfomycin, sporangionyain, siomycin, tailomycin) have also been found to be effective (Muir and Barreto, 1979).

MONENSIN AS GROWTH PROMOTOR FOR RUMINANTS

Monensin, originally designed as an anticoccidial drug, represents a real breakthrough with far-reaching economic consequences for ruminant nutrition. The high-lights of monensin research demonstrated significant improvement in performance and ruminal VFA changes toward more propionate and less acetate and CH_4 (at least in most of the studies). There is evidence of inhibition of cellulose degradation, but not of basic changes in the population of proteolytic bacteria. Some decrease in microbial synthesis of protein and nucleic acids in the rumen, and decrease in the biological values of ingested proteins were reported (Chalupa et al., 1980). A significant improvement of carcass characteristics as a result of monensin was also observed (Potter et al., 1976b). In recent studies comparing lasalocid with monensin the former was more effective in the improvement of the feedlot cattle performance than the latter (Whetstone et al., 1981; Berger et al., 1981).

HORMONES AND HORMONALLY ACTIVE DRUGS

Despite a century-old research in the use of hormonal drugs as growth promotors for livestock, relatively little progress has been made since the introduction of DES, although some new drugs have entered the world market during the past decade. A comprehensive review on the use of drugs in animal production was recently published (Velle, 1982). In Table 3 are listed the main hormonally active substances and their effect upon livestock performance.

Oestrogens (DES, Hexoestrol and oestradiol-17 β) have been the principal targets of the research in hormonal drugs. In 17 out of 19 trials carried out over several years in the UK, body gain of steers was increased 160 g/head/day through hexeostrol implants. In other trials, oestradiol-17 β, TBA implants, Synovex-S, hexoestrol alone or with TBA resulted in a 24% increase of body gain and a 13% improvement in feed efficiency. Zeranol implants reportedly improved rate of gain of steers by 10-35% i.e. an average 150 g/head/day. Zeranol-implanted steers appear to produce carcasses with slightly higher marbling scores than controls or steers with progesterone or estradiol benzoate. Synovex-S implanted steers have a tendency to retain more DM and N and to be more efficient at converting feed gross energy to tissue gross energy. In veal calves, oestradiol-17β + progesterone increased rate of gain by 20%, and N-retention by 21%. In the surveyed literature, there appears to be a trend showing heifers to yield more than steers. Testosterone-implanted heifers exhibited a significant response though the latest findings do not hold any clearcut conclusion. In experiments with bulls, somatotropin increased rate of gain by 5-8%, with lactosomatotropin by 30% and with the mixture of dianabol and oestradiol by 34.6%. Similar effects of hormonally-active substances were observed in experiments with lambs but less pronounced responses were reported with pigs and poultry.

DIETARY ENZYMES

There has been considerable renewed interest in various enzymatic preparations. Addition of promace (mixed enzymes) was found to increase performance and carcass characteristics in pigs (Glaps et al., 1980) and hatchability in hens (Korniewicez et al., 1980). Proteases (Aspergillus and Rhizopus) were found to enhance utilization of N (Lewis and Hiller, 1980). The presence of betalucan in barley reportedly limits growth of chickens by increasing viscosity of intestinal contents Trichoderma viride, cellulase degrading enzyme, decreased the viscosity caused by beta-glucan (White et al, 1981). Bacterial diastase improved performance and carcass yield of pigs fed normal 6-row barley, and added to a huskless waxy type barley it increased backfat (Newman et al., 1980). Other researchers reported that feeding enzyme preparations to lambs resulted in intensified microbial fermentation of carbohydrate and that amylase increased growth and utilization of N in lambs (Nechipurenko et al., 1981). Research with peptic enzymes resulted in a significant increase of egg production of layers fed a rye-based diet for the first 3 weeks but by the end of 8 weeks the effect abated to a non-significant level (Patel and McGinnis, 1980).

Table 3: Hormonally-active Substances and Their Impact Upon Performance

Substance	Dose Levels	Main Uses	Body weight increase %	Improvement of feed efficiency %
Oestrogens alone:				
DES	10-20 mg/day	steers,heifers	8-14	8-12
DES	30-60 mg/day	steers	12-20	8-14
DES		veal calves	8-12	5-12
Hexoestrol	12-60 mg	steers, sheep, calves, poultry	8-30	
Zeranol (Ralgro)	12-36 mg	steers, sheep	10-35	10-16
Gestagens alone:				
Melegestrol acetate	0.25-0.50 mg/day	heifers	10-24	7-13
Androgens alone:				
TBA (Finaplix)	300 mg	heifers, culled cows	12-24	11-14
Combined preparations:				
DES+ Testosterone (Rapigain)	25 mg 120 mg	calves	11-18	6-9
DES+Methyl-testosterone (Maximin)		pigs	6-12	4-8
Hexoestrol + TBA	30-45 mg 300 mg	steers	12-30	9-15
Zeranol + TBA	36 mg 300 mg	steers		
Oestradiol-17β + (Revalor) TBA	20 mg 140 mg	bulls, steers calves, sheep	14-35	6-12
Oestradiol-17β benzoate + Testosterone	20 mg 200 mg	heifers, calves	10-20	6-15
Oestradiol-17β benzoate+Progesterone	20 mg 200 mg	steers	12-22	7-16

Source: Velle, 1982, slightly adapted.

ARSENICALS

Various arsenic-containing compounds have shown the following effects in monogastric animals: increased rate of gain of growing poultry and pigs, improved feed efficiency (poultry and pigs), improved pigmentation of egg and skin, improved bloom and feathering in poultry, prevention of coccidiosis in chicken, prevention and control of swine dysentery. Some recent studies indicated that the addition of 3-nitro-10 increased egg yolk colour, dominant wavelength and excitation purity values while decreasing luminosity values (Janky et al., 1982). In recent studies with broilers it was observed that 3-nitro alleviated the growth depression induced by high dietary Cu, reduced hepatic Cu accumulation but increased Fe, Mn and Zn content in the liver (Maurice et al., 1981). An interaction between 3-nitro, CH_3Hg, Se and S was also recently reported. It was found that arsenic had little effect on CH_3Hg toxicity but altered the ability of

Na_2SeO_3 to modify CH_3Hg toxicity (El-Begearmi et al., 1982).

LACTOBACILLI

Lactobacillus spp. were proposed as a remedy for enteric infections in the previous century by Mechnikoff and Pasteur, but the issue has remained controversial up to now. In experiments with calves receiving lactobacilli + OTC it was observed that they gain more and consume more feed than those without OTC (Morrill et al., 1977). Numerous experiments with lactobacilli in pigs did not produce a clearcut picture when compared with antibiotics (Pollmann et al., 1980a, 1980b).

The summary of research with broilers receiving L. acidophilus is tabulated below:

Trial No.	No. of birds per treatment	Body weight g per bird	%	Feed gain %
1	116,000	+ 86	+ 6.0	+ 3.0
2	31,000	+ 55	+ 3.0	+ 1.0
3	11,000	- 73	- 4.0	+ 3.0
4	11,000	+ 34	+ 2.0	+ 3.6

In all trials the average gain improvement was 25.5 g per bird and 62.5 g in F/G (Arendo, 1981).

MOLD INHIBITORS

Small particle size of the feed, dispersion of the substance in the substrate and the ability of it to penetrate into the substrate are important attributes of mold inhibitors. The antifungal activity of propionic acid in the mixed feed appared to be reduced by soybean meal, fish meal, poultry by-product meal and limestone while fat enhanced it. The activity of antifungal agents can thus be controlled by the level of antagonistic feed ingredients. The higher the moisture content of the substrate the higher concentration of the inhibitor is required (Dixon and Hamilton, 1981).

ANTIOXIDANTS

Ethoxyquin (EQ), butylated hydroxytoluene (BHT) and butylated hydroxynizol (BHA) are the main antioxidants used in feeds. Their effect is often associated with vitamin E-like activity. Lately, it was reported that some of the newer antioxidants (Ionol, Diludin, RR-123) facilitated also the absorption of dietary tocopherols and its increase in liver. In other studies it was observed that EQ and diludin enhanced the biological action of vitamin A, E, C and Se and increased the retention of Fe, Zn, Se and the vitamins (Validman et al., 1980). Investigations concerning stability of abdominal fat and broiler meat revealed that EQ, BHT and Endox-50 consistently increased α-tocopherol levels in carcass fat and the stability of abdominal fat and thigh meat increased with the combination of EQ and α-tocopherol acetate. Combination of BHT with tocopherols increased the stability of fat and meat only in birds fed diets without added fat (Bartov and Bornstein, 1981).

BUFFERS

Renewed interest has been shown in the use of buffers in ruminants fed on high concentrate diets though there are often significant variations in individual papers. The main reason seems to be that there is a variety of buffers operating within a narrow pH range. Some buffers are primarily ruminal whereas some are effective in the intestines (Wheeler, 1980). Some differ due to source and its variable physical and/or chemical properties. It also depends upon the presence of dietary stress factors which determine whether buffers are required or not. In the absence of dietary or ruminal-intestinal stress factors, the response may not exist in the diet. Studies on the buffering capacity of 35 compounds single or in certain selected combinations tested in vitro revealed that the best cation buffers in decreasing order were Na, K, Ca, Mg, Al, NH_3 and Mn with Zn having little or no buffering capacity. Acetate, bicarbonate, carbonate and dibasic phosphate were found to be the best anion buffers (Herod et al., 1978). $NaHCO_3$ as a buffer, in most experiments, exhibited increased solids-corrected milk yields, and cows fed $NaHCO_3$ adjusted to rations more quickly, consumed more feed and produced more fat-corrected milk than controls (Wheeler, 1980).

SILAGE ADDITIVES

Recent studies on the impact of formic acid and formaldehyde on digestibility of forage concluded that formaldehyde tends to depress energy digestion in the rumen but enhances the flow of total amino acids to the small intestine. Formic acid appears not to improve protein digestibility but in combination with formaldehyde there seems to be a tendency for increased nitrogen retention. In experiments with caproic and isocaproic acids a consistent inhibition of silage fermentation was observed, but caproic acid prevented deterioration of silage after exposure to air (Ohyama et al., 1977). Similarly, an ammonia-molasses-mineral solution when added to maize silage increased its stability when exposed to air. The use of cellulase in combination with other additives resulted in the reduction of cellulose content of the silage (Henderson and McDonald, 1977).

PIGMENTS

These non-nutritional additives have been researched for a considerable time because in many areas of the world where maize is not the main cereal of poultry feed the pale colour of the skin and egg yolk often creates serious marketing problems of poultry products. The major findings of recent studies offer the following conclusions: no structural changes in the xanthophylls of marigold meal were found during storage; saponification but not esterification increases the pigmentation scores in egg yolks; zeaxanthin appears to be one of the most effective broiler pigmenter producing acceptable yellow to yellow-orange colours; the availability of xanthophyll from Pro-Xan is about 1.7 times that of dehydrated lucerne and three times that of marigold meal for broilers and laying hens; the higher the level of xanthophyll in the diet the higher is the content of skin pigments which are deposited more rapidly and with less variability among the individual carcasses (Waldroup et al., 1979); β-apo-8'carotenic acid ethyl ester is two and a half times more effective than Tagetes xanthophylls (Leibetseder and Schweighardt, 1979); paprika oleoresin, combined with marigold meal and measured on egg yolk increases the visual and colorimetric response more than marigold alone (Fletcher and Halloran, 1981); availability (chick assay) of xanthophylls from four samples of corngluten meal was found to have a wide range from 47.7 to 89.1%, in eight samples of lucerne from 34.6 to 55.4% and in Coastal Bermuda grass from 18.8 to 27.9% (Middendorf et al., 1980); and xanthophylls of dehydrated turfgrass were found to be 100% available, four times the xanthophyll concentration of corngluten meal and a better protein source than lucerne meal (Robbins, 1981).

MISCELLANEOUS FEED ADDITIVES

The polyhydric alcohol, 1,3-butyleneglycol (BG) and propylene glycol (synthetic energy sources) can supply a significant portion of energy for monogastric animals. BDDO (1,3 butane-diol-1, 3-dioctanoate) was found to be equal to corn oil in chick diets and has the potential of providing rapid concentrated energy for protein anabolic processes, particularly during recovery from disease (Johnston et al., 1979).

Solka Floc - a synthetic source of fibre - was investigated in three pig studies (Moser et al., 1982). Zeolite applied on either new or reused litters reduced the incidence of foot pad burns, but incorporating 10% of zeolite in the ration significantly increased the incidence of foot pad burns. Mortality was increased when zeolite was used as a sole source of litter (Nakave et al., 1981). In experiments with turkeys it was found that addition of sand or kaolin to their diets improved energy utilization but not weight gain (Oluyemi and Harms, 1977). Sensory changes of broiler meat, flavour, aroma and moisture were reported to be improved through addition of a commercial product Enhance (Wabeck and Heath, 1982). Adding flavours to the pig starter diet increased performance but only if the baby pigs were nursed by sows that had been fed the flavour during their lactation (Campbell, 1976).

Fly control/insect growth regulators are relatively new additives. The effect of diflubenzuron (insect growth regulator) on performance of broilers and layers indicated a trend to better performance (Kubena, 1981). The effect of a synthetic pyrethroid (Decametherin) on reproduction of quails was tested. The drug produced fly-repellent action which diminished after repeated exposure (David, 1981). Methoprene applied as a feed additive, resulted in only a limited inhibition of house fly emergence perhaps due to cross resistance to methoprene followed by an induction to resistance resulting from the continuous exposure to this drug (Breeden et al., 1981). Most recently, it was observed that faeces from cattle given monensin showed 20% fewer ($P \leq 0.01$) face fly and born fly pupae and surviving pupae were smaller ($P \leq 0.01$) (Herald et al., 1982). In experiments with thiourea (NH_2CSNH_2) it was reported that inoculated house fly eggs did not develop in manure produced by the thiourea-fedhens but thiourea reduced egg production and incubation time and embryonic mortality with increasing levels of dietary thiourea (Lyons et al., 1981). Sodium taurochlorate investigated on chicks was found to improve fat retention in tallow-containing rye diets. It is speculated that bile acid metabolism may be altered in chicks fed rye (Campbell et al., 1981). Prostaglandin as a synthetase inhibitor offered some degree of protection against thermal stress in broilers (Birrenkott and Oliver, 1981). Sodium bentonite has been successfully used in dairy rations to support maintenance of milk fat test. Since loose excreta are associated with a low fat test, bentonite may be of considerable help in this respect (Harris and van Horn, 1978).

THE MISSING LINKS AND THE TASKS OF THE FUTURE

Feed additives research over the past few years has been rewarded with the development of several new drugs, improvement in the effectiveness of existing ones and the emergence of new promising additives which are ready to be transferred from laboratories and industry into the field. There are impressive opportunities for the future as the demand for livestock products is ever increasing and the productivity of livestock must keep pace. The targets are more food with less feed and less money, but safety for the consumers. The pressing tasks for the future are given in the following three sub-sections.

The General, Comprehensive Approach

- to develop a base line model for feed additives research as an integral part of nutritional research for individual species and classes of animals to disseminate and utilize the work of all scientists;
- to invent new synthetic energy sources to substitute and/or reduce dependency upon classical sources of soluble carbohydrates, in particular, grain;
- to intensify feed additive research towards substitution of maize by other cereal grains and root crops;
- to discover feed additives capable of controlling and preventing major diseases (currently prevented by vaccination) via feed;
- to intensify research in industrial production of tryptophane and threonine which together with lysine and methionine allow for a total (in pigs) or significant reduction of protein feed requirements (30-40% in poultry) which will reduce the competition with man for these resources;
- to discover new antibiotics and/or non-antibiotic antimicrobial compounds which would be effective, safe and cheap and will not be used in veterinary and/or human medicines;
- to research additives effective in modifying carbohydrate, lipid and protein metabolism resulting in animal products for consumers with lower content of fat, cholesterol, triglycerides and other unwanted constituents;
- to intensify research in detoxifying agents capable of destroying or inactivating aflatoxin, other mycotoxins, and other toxic substances in feed ingredients and in the mixed feed;
- to develop new pharmaceutical forms of vitamins which will be more effectively stored in organs and tissues for a longer period and tansferred to the progeny via placenta or, in the case of poultry, via eggs;
- to develop new hormonally active substances for control of reproductive performance which also allow for manipulation of a desired carcass composition;
- to develop new enzymes enhancing the digestibility and utilization of carbohydrates, proteins, lipids and ligno-cellulosic constituents;
- to improve reproductive performance with new additives controlling oestrus, multiple birth, size of litter;
- to discover new safe antioxidants capable of extending their action from the feed to carcass and other animal products;
- to develop new mold inhibitors which not only control but are also capable of destroying mycotoxin present in the feed;

Monogastric Species: The Main Tasks in Feed Additive Research

- to synthesize new effective coccidiostats not inhibiting growth and feed efficiency and do not interfere (in a negative manner) with other feed additives;
- to intensify research relating to new growth promotants for young poultry to further enhance their productivity and economics;
- to elucidate problems associated with egg shell quality and invent means of overcoming them particularly in tropical and substropical regions;
- to discover feed additives or their combinations effective in controlling obesity in heavy breeders and other problems associated with it such as fatty liver and kidney syndrome;
- to develop additives for the control of diseases resulting from intensive confinement systems such as hysteria, cannibalism, body, skin and leg

deformations;
- to intensify research in weaning of calf and piglet at birth or near birth with additives allowing for maximum saving on milk or, in the case of sows, for speeding up of the turnover of the herd to maximum efficiency.

Ruminants - The Main Goals of Feed Additive Research:

- to invent new additives increasing voluntary intake of low quality forage and improving its digestibility by exogenic means (the latter via ruminal manipulations);
- development of additives decreasing the toxicity of NPN compounds and supporting the maximum utilization of N to feed the least possible amount of natural proteins;
- to discover new monensin-like additives to further enhance nutrient utilization by ruminants, particularly at the low plane of nutrition with maximum level of poor quality roughages;
- to invent new additives effective in the control of mastitis via feed and ensure that no residues of the drug will be transferred into milk;
- to discover additives directly or indirectly regulating volatile fatty acid content in the rumen;
- to develop effective antibloating agents;
- to develop new buffers increasing milk fat content;
- to develop new silage and new related additives which support voluntary intake of preserved feed and maintain and/or enhance its original nutritional value.

To carry out these tasks will require a creative and motivating climate in which all concerned: research, industry and governments, will cooperate and act. Feed additive research will and must continue to advance.

REFERENCES

Arends, L.G. (1981). Influence of lactobacillus acidophilus administered via the drinking water on broiler performance. Poultry Sci. 60:1617 (Abstract).

Bane, D.P., H.D. Hilley and R.J. Meade (1980). Influence of d-biotin and housing on hoof lesions in swine. Am. Soc. Anim. Sci. Annual Meeting. Abstract No. 81, 138.

Bar, A., U. Eisner, G. Montecuccoli and S. Hurwitz (1976). Regulation of intestinal calcium absorption in the laying quail independent of kidney vitamin D hydroxylation. J. Nutr. 106:1366-1342.

Barber, R.S., R. Braude, K.G. Mitchell and R.J. Pittman (1978). The value of virginamycin (eskalin) as a feed additive for growing pigs in diets with or without a high copper supplement. Anim. Prod. 26:151.

Bartov, I. and P. Bydowski (1979). Protective effect of nicarbazin on nutritional encephalopathy in chicks. Poultry Sci. 58:597-601.

Bartov, I. and S. Bronstein (1977). Stability of abdominal fat and meat of broilers: combined effects of vitamin E, butylated hydroxytoluene and ethoxyquin. Brit. Poultry Sci. 18:59-68.

Beagle, W.S. and J.J. Begin (1976). The effect of panthothenic acid in the diet of growing chicks on energy utilization and body composition. Poultry Sci. 55:950-957.

Berger, L.L., S.C. Ricke and G.C. Fahey, Jr. (1981). Comparison of two forms and two levels of lasalocid with monensin on feedlot cattle performance. J. Anim. Sci. 53:1440-1445.

Birrenkott, G.P., Jr. and J.C. Oliver (1981). Effect of a prostaglandin synthetase inhibitor on thermally distressed broilers. Poultry Sci. 60:1595 (Abstract).

Bishra, H.N. and H.F. Walker (1977). The vitamin B_6 status of pigs given a diet containing linseed meal. Brit. J. Nutr. 37:321.

Boris, A., J.F. Turley and T. Trmal (1979). In vivo studies in chicks and rates of bone calcium mobilization by 1, 25-dihydroxycholecalciferol (calcitriol) and its congeners. J. Nutr. 109:1772-1778.

Borsje, B., A. Ileyting, J.R. Roborgh, D.B. Ross and K.W.G. Shillam (1977). Analysis of fat-soluble vitamin. XVI. Antirichitic activity of 5,6-trans-vitamin D_3 alone and in the presence of 5,6-cis-vitamin D_3 resin, using chick bioassays. JAOAC 60:1003-1009.

Breeden, G.C., E.C. Turner, Jr., W.L. Beane, R.W. Miller and L.C. Pichens (1981). The effect of methoprene as a feed additive on house fly emergence in poultry houses. Poultry Sci. 60:556-562.

Brooks, P.H., D.A. Smith and V.C.R. Irwin (1977). Biotin-supplementation of diets: the incidence of foot lesions, and the reproductive performance of sows. Vet. Rec. 101-46.

Campbell, G.L., L.D. Campbell and H.L. Classen (1981). Utilization of rye by chicks. Effects of chick microbial status, diet gamma irradiation or sodium taurocholate supplementation. Poultry Sci. 60:1633- (Abstract).

Campbell, R.G. (1976). A note on the use of a feed flavour to stimulate the feed intake of weaner pigs. Anim. Prod. 23:417.

Chalupa, W., W. Corbett and J.R. Brethour (1980). Effects of monensin and amicloral on rumen fermentation. J. Anim. Sci. 51:170.

Cohen, A., A. Bar, U. Eisner and S. Hurwitz (1978). Calcium absorption, calcium binding proetin and egg shell quality in laying hens fed hydroxylated vitamin D derivatives. Poultry Sci. 57:1646-1651.

Cole, N.A., J.B. MacLaren and D.P. Hutcheson (1982). Influence of preweaning and B-vita supplementation of the feedlot receiving diet on calves subjected to marketing and transit stress. J. Anim. Sci., 54, No. 5.

Combs, G.F., Jr. (1981). Influence of dietary vitamine E and selenium on the oxidant defence system of the chick. Poultry Sci. 60:2098-2105.

Combs, G.F., Jr. (1978). Influence of ethoxyquin on the utilization of selenium by the chick. Poultry Sci. 57:210-222.

Combs, G.F., Jr. and M.L. Scott (1979). The effects of selenium needs of laying and breeding hens. Poultry Sci. 871-884.

David, D. (1981). Laboratory evaluation of repellent properties against birds of the synthetic pyrethroid decamethrin. Poultry Sci. 60:1149-1151.

Dinius, D.A. and C.A. Baile (1977). Beef cattle response to a feed intake stimulant given alone and in combination with a propionate enhancer and an anabolic agent. J. Anim. Sci. 45:147.

Dixon, R.C. and P.B. Hamilton (1981). Effect of particle sizes of corn meal and a mold inhibitor on mold inhibition. Poultry Sci. 60:2412-2415.

Donoghue, S., D.S. Kronfield and C.F. Ramberg (1979). Plasma retinol transport and clearance in hypervitaminosis. A. J. Dairy Sci. 62:326.

Dror, Y., I. Bartov and J.J. Dubis (1980). Exacerbative effect of vitamin A on the development of nutritional encepathy in chicks. Nutr. Rpts. Int. 21:769-778.

El-Begearmi, M.M., H.E., Ganther and M.L. Sunde (1982). Dietary interaction between methylmercury, selenium, arsenic and sulfur amino acids in Japanese quail. Poultry Sci. 61:272-279.

Fletcher, D.L. and H.R. Halloran (1981). An evaluation of a commercially available marigold concentrate and paprika oleoresin on egg yolk pigmentation. Poultry Sci. 60:1846-1853.

Glaps, J., A. Korniewicz and Borowa, E. (1980). Effect of the addition of promase and nitrovin to different diets for fattening pigs. Roczniki Naukowe Zootechniki, Monografie i Rozprawy 18:239-256.

Guenter, W. and D.B. Bragg (1977). Response of broiler chicks to dietary selenium. Poultry Sci. 56:2031-2038.

Harms, R.H., R.A. Voitle, D.M. Janky and H.R. Wilson (1979). The influence of biotin supplementation on performance of broiler and breeder hens and foot pad dermatitis in the progeny. Nutr. Reports Intern. 59:561-567.

Harris, B. Jr. and H.H. Van Horn (1978). By-products roughages and other additives for optimum milk yield in sub-tropical climate. Livestock and poultry in Latin-America, 12th Am. Conf. U. Florida, Gainesvill, May 8-12, C27-C31.

Henderson, A.R. and P. McDonald (1977). The effect of cellulase preparation on the chemical changes during the ensilage of grass in laboratory silos. J. Sci. Food Agr. 24:486.

Hennig, A., H. Jeroch and G. Flachowsky (1979). Zum gegenwartigen stand des Einsatzes der Ergotropika, insbesondere der Antibiotika als Futterzusatz. Mh. Vet. Med. 34:343-351.

Herald, F., F.W. Knapp, S. Brown and N.W. Bradley (1982). Efficiency of monensin as a cattle feed additive against the face fly and horn fly. J. Anim. Sci. 54:1128.

Herod, E.L., R.M. Bechtle, E.E. Bartley and A.D. Dayton (1978). Buffering ability of several compounds in vitro and the effect of a selected buffer combination on ruminal acid production in vivo. J. Dairy Sci. 61:1114.

Hidiroglu, H., M. Ivan and K.J. Jenkins (1977). Influences of barley and oat silages for beef cows on occurrence of myopathy in their calves. J. Dairy Sci. 60:1905.

Hulan, W.H., F.G. Proudfoot and K.B. MacRae (1980). Effect of vitamins on the incidence of mortality and acute death syndrome ("flip-over") in broiler chickens. Poultry Sci. 59:927-931.

Janky, D.M., R.H. Harms and R.D. Miles (1981). Effect of virginiamycin on laying hen and broiler performance. Poultry Sci. 60:1698 (Abstract).

Janky, D.M., M.G. Dukes and R.H. Harms (1982). Effect of 3-nitro-10 egg yolk pigmentation. Poultry Sci. 61:573-574.

Jensen, L.S., R.K. Page, S.P. Wilson and T.M. Muston (1978). Muscular dystrophy in broiler breeder hens: effect of temperature. Poultry Sci. 57:729-734.

Jensen, L.S. and C.H. Chang (1976). Fractionation studies on a factor in linseed meal protecting against selenosis in chicks. Poultry Sci. 55:594-599.

Jimenez, A.A. (1981). Feedstuffs 16, March 23, 1981.

Johnston, R.K., R.L. Squibb and J.W. Frankenfeld (1979). Effect of feeding 1,3-butanediol-1, 3-dioctanoate as an energy source for chicks for catch-up growth during recovery from Newcastle disease virus. J. Nutr. 109:473-479.

Kaetzel, D.M., Jr. and J.H. Soares, Jr. (1979). Effects of cholecalciferol steroids on bone and egg shell calcification in Japanese quail. J. Nutr. 109:1601-1608.

Kazemi, R. and F.H. Kratzer (1980). Source of protein affecting the vitamin B_6 requirement of chicks. Poultry Sci. 59:95-99.

Klopfenstein, C.F. and R.E. Clegg (1977). Effects of ascorbic acid and vitamin E on serum lipids of cockerels fed Wesson Oil cholesterol diets. Poultry Sci. 56:1600-1604.

Korniewiccz, A., T. Gwara, A. Mazandowska and K. Kaczmarek (1980). Influence of the enzymic preparation "promase" on egg laying by hens and hatchability of egges. Roczniki Naukowe Zootechniki, Monografie i Rozprawy 118:257-270.

Kubena, L.F. (1981). The influence of diflubenzuron on several weight characteristics in growing male broiler and layer chickens. Poultry Sci. 60:1175-1182.

Lambooy, J.P. and C.S. Shaffner (1977). Utilization of analogues of riboflavin by the riboflavin-deficient chick embryo. J. Nutr. 1078:245-250.

Leeson, S., B.S. Reinhart and J.D. Summers (1979). Response of White Leghorn and Rhode Island Red breeder hens to dietary deficiencies of synthetic vitamins. 1. Egg production, hatchability and chick growth. Can. J. Anim. Sci. 59:561-567.

Leibetseder, J. and H. Schweighardt (1979). Comparison of Tagetes xanthrophylls (Florafil plus 50) with beta-apo-8'-carotenoic acid ethylester (Carophyll Yellow in broiler pigmentation). World's Poultry Sci. 35:236-243.

Lewis, D. and G. Hiller (1980). Biological properties in relation to chick growth of acid proteases from aspergillus SP and rhizopus SP. World's Poultry Sci. Journal 36:237-250.

Likoff, R.O., D.R. Guptill, L.M. Lawrence, C.C. McKay, C.F. Nockels and R.P. Tengerdy (198). Vitamin E and aspirin depress prostaglandins in protection of chickens against Eschedrichia coli infection. Am. J. Clinical Nutr. 34(2):245-251.

Lyons, J.J., J.M. Vandepopuliere and R.D. Hall (1981). The effects of feeding thiourea to laying hens on egg production, hatchability and the development of house fly eggs seeded on their waters. Poultry Sci. 60:1685-1686 (Abstract).

MacAuliffe, T., A. Petrasek and J.McGinnis (1976). The effect of grain component of the diet on the response of turkey poults in vitamin D_3 and penicillin. Poultry Sci. 55:183-187.

Malm, A., W.G. Pond, E.F. Walker, Jr., M. Homan, A. Aydin and D. Kirtland (1976). Effect of polyunsaturated fatty acids on vitamin E level of the sow gestation diet on reproductive performance and on level of alpha tocopherol in colostrum milk and dam and progeny blood serum. J. Anim. Sci. 42:393.

March, B.E. and C. McMillan (1980). Choline concentration and availability in rapeseed meal. Poultry Sci. 59:611-615.

March, B.E. (1981). Choline supplementation of layer diets containing soybean meal or rapeseed meal as protein supplement. Poultry Sci. 60:818-823.

Maurice, D.V., J.E. Jones and N.J. Rehrer (1981). Interaction of Roxarsone[R] and copper in broiler chicks. Poultry Sci. 60:1695 (Abstract).

McDowell, L.R., J.A. Froseth, R.C. Piper, I.A. Dyer and G.H. Kroening (1977). Tissues selenium and serum tocopherol concentrations in selenium-vitamin E deficient pigs fed peas (Pisum sativum). J. Anim. Sci. 45:1326.

McGinnis, J. and M.B. Patel (1976). The nutritive value of rye for chick growth: effect of various levels, geographical area of production and procaine penicillin supplement. Poultry Sci. 60:1687-1688 (Abstract).

McLoughlin, C.P. and J.H. Soares, Jr. (1976). A study of the effects of 25-hydroycholecalciferol and calcium source on egg shell quality. Poultry Sci. 55:1400-1410.

McNaughton, J.L., E.J. Day and B.C. Dilworth (1977). The chick's requirement for cholecalciferol. Poultry Sci. 56:511-516.

Middendorf, D.F., G.R. Childs and W.W. Cravens (1980). Variation in the biological availability of xanthophyll within and among generic sources. Poultry Sci. 59:1460-1470.

Morrill, J.L., A.D. Dayton and R. Mickelson (1977). Cultured milk and antibiotics for young calves. J. Dairy Sci. 60:1105.

Moser, R.L., E.R. Peo, Jr., B.D. Moser and A.J. Lewis (1982). Effect of grain source level of solka floc and caloric content of the diet on performance, blood and bone traits of growing-finishing swine. J. Anim. Sci. 54:1181.

Muir, L.A. and A. Barreto, Jr. (1979). Sensitivity of streptococcus bovis to various antibiotics. J. Anim. Sci. 48:468.

Muller, Z., B. Ruzicka and B. Bauer (1980). The role of chemistry and biology in modern animal nutrition. Spofa, Praha 1961 (in Czech).

Murphy, T.P., K.E. Wright and W.J. Pudelkiewicz (1981). An apparent rachitogenic effect of excessive vitamin E intake in the chick. Poultry Sci. 60:1873-1878.

Nakaue, H.S., J.K. Koelliker and L. Pierso (1981). Studies with clinoptilolite in poultry. II. Effect of feeding broilers and the direct application of clinoptilolite (Zeolite) on clean and reused broiler litter on broiler performance and house environment. Poultry Sci. 60:1221-1228.

Nechipurenko, L.I., I.V. Yavorskaya and S.N. Aitov (1981). Changes in nitrogen metabolism in lambs weaned early and given enzyme supplements. From Referativnyi Zhurn 58:9.58.129.

Newman, C.W., R.F. Eslick, J.W. Pepper and A.A. El-Negoumy (1980). Performance of pigs fed hull-less and covered barleys supplemented with or without a bacterial diastase. Nutr. Reports Intern. 22:833-837.

Niyo, Y. and R.D. Glock (1976). Effects of selenium and/or vitamin E injections on lesions in selenium-vitamin E deficient pigs. Inter. Pig. Vet. Soc. Congress, A.2 (Abstract).

O'Connor, J.J., C.O. Baughn, R.R. Pilote, W.C. Alpaugh, W.H. Linkheimer and D.C. Maplesden (1979). Tiamulin in the feed for the prevention of swine dysentery and growth promotion of growing pigs. J. Anim. Sci. 49:933.

Ogunmodede, B.K. (1977). Riboflavin requirement of starting chickens in a tropical environment. Poultry Sci. 56:231-234.

Ohyama, Y., S.I. Hara and S. Masaki (1977). The use of caproic acid to prevent aerobic deterioration of silages after opening, with special reference to the amounts and time of application. J. Sci. Food Agr. 28:369.

Oluyemi, J.A. and R.H. Harms (1977). The effects of dietary level and particle size of sand on the efficiencies of feed and of energy utilization and the rates of growth of turkey poults. Nutr. Rep. Int. 15:493-496.

Patel, M.B. and J. McGinnis (1980). The effect of vitamin B_{12} on the tolerance of chicks for high levels of dietary fat and carbohydrate. Poultry Sci. 59:2279-2286.

Patel, M.B. and J. McGinnis (1980). Effect of gamma irradiating rye or supplementing a rye-containing layer diet with penicillin or supplementing a rye-containing layer diet with pencillin or pectic enzymes on egg production. Poultry Sci. 59:2287-2289.

Pelura, J., J.L. Krider, T.R. Cline, C. Risert and L.B. Underwood (1980). Virginiamycine, protein and lysine response of young swine. J. Anim. Sci. 50:767.

Pensack, J.M., G.T. Wang and K.L. Simkins (1981). Avoparcin-growth-promoting feed antibiotics. Poultry Sci. 60:1711 (Abstract).

Pesti, G.M. and G.F. Combs, Jr. (1976). Studies on the enteric absorption of selenium in the chick using localized coccidial infections. Poultry Sci. 55:2265-2274.

Phillips, R.W. and C.F. Nickels (1977). Glucose kinetics in vitamin A deficient chickens. J. Nutr. 107:1129-1234.

Piatowski, T.L., D.C. Mahan, A.H. Cantor, A.L. Moxon, J.H. Cline and A.P. Grifo, Jr. (1979). Selenium and vitamin E in semipurified diets for gravid and nongravid gilts. J. Anim. Sci. 49:1357.

Pilch, S.M. and G.F. Combs (1981). Effects of dietary vitamin E and selenium on the mixed-function oxygenase system of male and female chicks. Comp. Biochem. Physiol. 69C:331-335.

Pollman, D.S., D.M. Danielson and E.R. Peo, Jr. (1980a). Effects of microbial feed additives on performance of starter and growing-finishing pigs. J. Anim. Sci. 51:577.

Pollman, D.S., D.M. Danielson and E.R. Peo, Jr. (1980b). Effect of Lactobacillus acidophilus on starter pigs fed a diet supplemented with lactose. J. Anim. Sci. 51:638.

Potter, E.L., A.P. Raun, C.O. Cooley, R.P. Rathmacher and L.F. Richardson (1976b). Effect of monensin on carcass composition and efficiency of converting feed to carcass. J. Anim. Sci. 43:678.

Robbins, K.R. (1981). Xanthophyll bioavailability in dehydrated turf grass. Poultry Sci. 60:254-256.

Schingoethe, D.J., J.G. Parsons, F.C. Ludens, L.V. Schaffer and H.J. Shave (1978). Vitamin E status of dairy cows fed stored feeds continuously or pastured during summer. J. Dairy Sci. 61:1582.

Schingoethe, D.J., J.G. Parsons, F.C. Ludens, L.V. Schaffer and H.J. Shave (1979). Response of lactating cows to 300 mg of supplemental vitamin E daily. J. Dairy Sci. 61:333.

Summers, J.D., H. Shen and S. Leeson (1981). Influence of a dietary deficiency of vitamin D on egg shell quality. Poultry Sci. 60: No. 7 (Abstract).

Taylor, D.J. and L.A. Davey (1980). The elimination of T. hyodysenteriae infection from pigs by medication with carbadox. 6th Intern. Pig Vet. Soc. Congress, Copenhagen, Denmark.

Tomov, T., Ya. Iliev, N. Ibrishimov, A. Kirov, L. Dyakov and V. Mentov (1980). Effect of the preparation bacitraphen premix on laying hens and broiler chickens. Veterinarnomedisinski Nauki. 17:65-71.

Val'dman, A.R., M.R. Apsite, A.B. Atlavin, I.K. Strozha, N.I. Berzin, A.Ya. Ozol and N.A. Basova (1980). Biological effects of some antioxidants on nutrition of chickens. Referativnyi Zhurnal. 58:4.58.704.

Velle, W. (1982). Hormones in animal production. FAO Anim. Prod. and Health, Paper No. 31, FAO, Rome, Italy.

Wabeck, J.C and J.L. Heath (1982). Sensory changes of meat from broilers fed the feed additive Enhance. Poultry Sci. 61:719-725.

Waldroup, P.W., T.L. Goodwin, Z.B. Johnson and K.B. Lower (1979). An estimation of the time required to develop adequate pigmentation for broilers. Nutr. Reports Intern. 20:567-572.

Whanger, P.D., P.H. Weswig and J.E. Oldfield (1978). Selenium, sulfur and nitrogen levels in ovine rumen micoorganisms. J. Anim. Sci. 46:515.

Whanger, P.D., P.H. Weswig, J.A. Schmitz and J.E. Oldfield (1977). Effects of selenium and vitamin E on blood selenium levels, tissue glutathione perioxidase activities and white muscle diseased in sheep fed purified or hay diets. J. Nutr. 107:1298.

Wheeler, W.E. (1980). Effect of limestone buffers on digestibility of complete diets and on performance of dairy cows. J. Dairy Sci. 53:803-809.

Whetstone, H.D., C.L. Davis and M.P. Bryant (1981). Effect of monensin on breakdown of protein by ruminal microorganisms in vitro. J. Anim. Sci. 53:803-809.

White, W.B., H.R. Bird, M.L. Sunde, N. Prentice, W.C Burger and J.A. Marlett (1981). The viscosity interaction of barley beta-glucan with Trichoderma viride cellulase in the chick intestine. Poultry Sci. 60:1043-1048.

Whitehead, C.C. and D.W. Bannister (1980). Biotin status, blood pyruvate carboxylase (EC 6.4.1.1) activity and performance in broilers under different conditions of bird husbandry and diet processing. Br. J. Nutr. 43:541-549.

Wilkinson, J.E., M.C. Bell, J.A. Bacon and F.B. Masincoup (1977). Effects of supplemental selenium on swine. 1. Gestation and lactation. J. Anim. Sci. 44:224.

Yen, J.T., A.H. Jensen and D.H. Baker (1977). Assessment of the availability of niacin in corn, soybeans and soybean meal. J. Anim. Sci. 45:269-278.

Yoshida, M. and H. Hoshii (1977). Preventive effect of selenum, methionine and antioxidants against encephalomalacia of chicks induced by dilauryl succinate. J. Nutr. 107:35-41.

Young, L.G., R.B. Milker, D.E. Edmeades, A. Lun, G.C. Smith and G.J. King (1977). Selenium and vitamin E supplementation of high moisture corn diets for swine reproduction. J. Anim. Sci. 45:1051.

Young, L.G., R.B. Milker, D.E. Edmeades, A. Lun, G.C. Smith and G.J. King (1978). Influence of method of corn storage and vitamin E and selenium supplementation on pig survival and reproduction. J. Anim. Sci. 47:639.

Young, R., Jr., D.K. Hass and L.J. Brown (1979). Effect of late gestation feeding of dichlorvos in non-parasitized and parasitized sows. J. Anim. Sci. 48:45.

Zotovic, M. (1978). Effect of an antibiotic (flavomycin) on weight gain, feed conversion and the occurrence of liver abscesses in fattening cattle fed concentrate diets. Arhiv. Za Poljoprivredne Nauke 31:43-60.

BIOCHEMICAL AND CHEMICAL CONTRIBUTIONS TO EFFICIENT SMALL SCALE PIG AND POULTRY SYSTEMS

A. Aumaitre

Pig and Poultry Husbandry Department
Institut National de la Recherche Agronomique
Ministre de l'Agriculture
Saint Gilles, 35590 l'Hermitage, France

ABSTRACT

Pig and poultry meat represented in 1980 60% of the total supply of Europe and North America, and more than 95% in some Asian countries. Increased production and improved performance of monogastric animals were achieved during the past 20 years thanks to more efficient control of reproduction, nutrition, environment, housing and quality of animal products.

Reproduction can be controlled in birds by light intensity and in sows by progestagens. Improved performance of highly selected animals by feeding better balanced diets has been based on intensive chemical and biochemical determination of nutrients, antinutritional or toxic factors in available and potential feed, as well as measurements of biological response of the animals. Feed supplement and growth promotors help to overcome poor environment. Control and improvement in quality of animal products is the ultimate goal of the specialist in animal production.

KEYWORDS: Biochemistry, chemistry, control, reproduction, nutrition, quality, egg, meat, pig, poultry.

INTRODUCTION

Breeding and rearing monogastric animals for meat and egg production has developed considerably in the world during the last 20 years. During this period meat consumption increased markedly in Europe (Figure 1); the proportion of pig and poultry meat in the total meat supply also increased at the expense of more costly meats. Moreover, considerable technical progress has been achieved in improving the average performance of domestic birds. The genetic improvement of specialized lines of poultry for egg and meat production was concentrated in a few large breeding companies, located in North America and Europe, which distribute parent stock throughout the world. Considerable progress has been made in the genetic potential for broiler growth rate, feed efficiency and product quality. The number of eggs per hen housed in America increased from 1958-1976 by 1.5-2 eggs per year. In addition, progress in feeding and nutrition allowed the expression of the genetic potential of the more intensively selected animals used in large as well as small herds. Similar progress have been achieved in pig production (Table 1) due to improvement in all phases of production. These

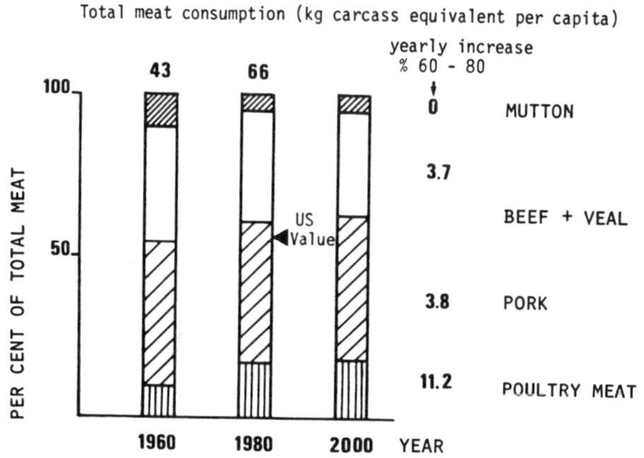

FIGURE 1 : IMPORTANCE OF POULTRY AND PORK FOR MEAT SUPPLY IN EUROPE. (ADAPTED FROM POLITIEK AND BAKKER, 1982).

improvements resulted from the tremendous increase in the knowledge of animal biology.

CONTROL OF REPRODUCTION

Feeding and housing costs of breeding and replacement stock are of extreme economic importance in the final productivity and profitability of animal production. In monogastric animals batch production was developed for improved management and hygienic conditions of intensively reared animals.

Continuous control of female reproduction is particularly necessary to designate the date of the first laying in the hen and regulate the onset of puberty in the gilt. Laying and reproductive cycles depend mainly on hormonal conditions, but environment, genetics and nutrition play indirectly an important role in their regulation. Light is one of the most important factors for hens, and social environment such as the presence of a male is a favourable factor for gilts.

Table 1: Past and Recent Progress in Pig Performance in Europe.

	Wild Animals	Intensive Production	
	1980	1960	1980
AVERAGE PERFORMANCE			
Age at puberty, d.	550	210	180
No. of young/female/year	44	18	24
Time Birth-slaughter, d.	500	200	170
Food Conversion ratio (kg/kg)	9.0	3.5	3.2

Control of Oestrus in Gilts and Sows

In recent trials, progestagens were used successfully to synchronize ovulation in gilts and to regulate delayed returns to service in weaned sows. Among them, Regumate was found to work efficiently in cyclic gilts and weaned sows and toa lesser extent in non-cyclic gilts (Figure 2). It is considerably more efficient than F.G.A. used in earlier experiment (Martinat-Botte et al., 1982).

By grouping a maximum percentage (72-85%) of gilts and sows on day 5-7 of oestrus, efficient use of artificial insemination, and subsequent batch farrowing of the sows were achieved. Field experiments (Table 2) confirmed these data and proved the favorable effect of group mating on percentage farrowing and on subsequent litter size.

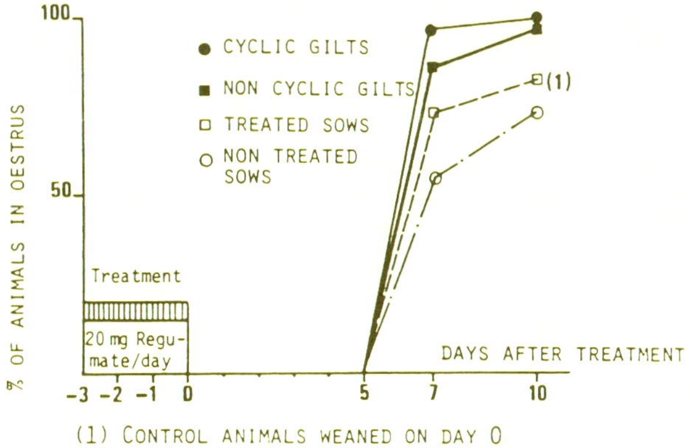

FIGURE 2 : USE OF PROGESTAGEN (RU 2267 REGUMATE) FOR OESTRUS CONTROL IN GILTS AND SOWS

Table 2: Effect of Use of Regumate for Oestrus Control in Sows on Subsequent Performance (Field Experiment)[(1)].

ANIMALS AND TREATMENTS	GILTS (PRIMIPAROUS) Control	GILTS (PRIMIPAROUS) 3 days Regumate	SOWS Control	SOWS 3 days Regumate
Percentage of animals in oestrus Day 5-7	39	72**	74	85**
Weaning[(2)] - oestrus interval d.	17.5**	6.8	6.7	6.1
Percentage of animals farrowing	65	72*	79	73
A litter size at birth	11.0	10.3	10.7	10.4

(1) Adapted from Martinat-Botte et al., 1982.
(2) Or End of Treatment - oestrus.

Control of Pregnancy and Farrowing in Sows

In a swine herd the percentage of non-pregnant sows directly affects yearly productivity and feeding costs. Numerous methods of pregnancy diagnosis have been investigated and compared on the basis of accuracy in pregnant and non-pregnant animals. Among them, vaginal biopsies or boar detection appeared to be the most efficient within 21 days post coitum (p.c.). Serum progesterone or PGF (13,14 dihydro 15 ceto PGF 2 α) between 13-15 days p.c. appeared to be valuable only in large herds (BOSC et al., 1977; Martinat-Botte et al., 1980). Farrowing control has improved management and saved labour due to intensive care of new born piglets. Injection of synthetic postaglandins (PFG 2 α) on day 112 or 113 is followed by parturition within 22.5 to 26.9 (± 8) hours (Hammond and Carlyle, 1976). But such methods should be retained exclusively for large herds.

ROLE OF CHEMISTRY AND BIOCHEMISTRY IN IMPROVING NUTRITION

Chemical analysis of feed involving physical methods for measuring energy, highly polymerised saccharides, starch, fatty acids, vitamins and trace elements. Nutritional factors concern deficiency, excess of nutrients, balance and imbalance, presence of non-nutrients, toxic and antinutritional factors. Biological factors concern availability for energy, protein, minerals and vitamins. Examples will be given of the present state of knowledge pertaining to the control of nutrition, feed resources and the efficient utilization of feed by monogastric animals.

Chemical Analysis, Energy and Protein Nutrition

Chemical Composition and DE Content: The gross energy (GE), expressed on a dry matter basis, is the main indicator of the nutritional value of a feed for monogastric animals. Equations based mostly on protein, fat, nitrogen free extract and crude fiber were initially proposed for estimating either GE, DE (digestibility energy) and even NE (net energy), (Schiemann et al., 1972). For pigs, the presence of some polysaccharides particularly crude fibre affected negatively the DE content as well as the digestive utilization of crude protein (Figure 3). Apparent digestibility of energy (ADE) or DE content of most current feeds used for pigs was measured in metabolism crates. Equations for predicting DE relative to the crude fiber content (Figure 3) show that a unique expression could not give an accurate prediction of DE content in all feeds. DE content of new or potential feeds, based only on proximate chemical analysis, in general overestimate the value of feeds. The presence of thermolabile antinutritional factors could further explain differences of 16 to 20% between the extreme values predicted by analysis and measured on animals for tropical feeds (Le Diridich et al., 1976).

Protein Nutrition and Efficiency of Feed Utilization: The main factor influencing diet efficiency in growing animals is the nutritional balance of the diet as measured by protein efficiency ratio (PER) and by the energy efficiency ratio (EER). An experiment conducted on broilers shows the important effect of a deficiency of methionine and the necessity of an accurate Energy/Protein ratio (Calet and Feudry, 1965).

Formulating feeds with minimum protein content that meets amino acid requirements, is on essential goal of nutritionists. Chemical analysis of crude protein and amino acid content of the most common feeds has been intensively employed for the past 20 years (Pion and Fauconneau, 1966; American feed tables 1974). During this period, the optimum amino acid requirements of poultry and swine have been more clearly and accurately assessed. Lysine and methionine have

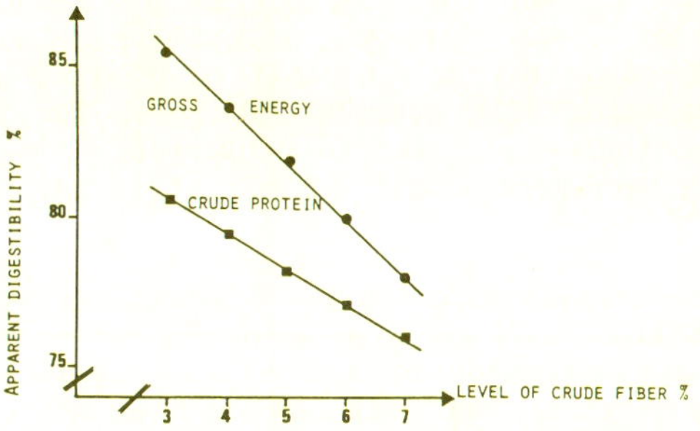

FIGURE 3 : EFFECT OF CRUDE FIBER LEVEL IN BARLEY ON APPARENT DIGESTIBILITY OF ENERGY AND NITROGEN (ADAPTED FROM HENRY AND BOURDON, 1975 ; HENRY, 1976 ; PEREZ ET AL., 1980) FOR PIGS.

ADE = 91.99 - 2.27 CF % DM		BARLEY
ADE = 96.29 - 3.04 CF % DM		WHEAT BRAN
ADE = 93.00 - 2.10 CF % DM		ALFALFA MEAL

been identified as the two main limiting amino acids. Computer computations on the basis of these requirements and according to the price of feed available in a given country permit establishment of a balanced diet at minimum cost (Fonnesbeck et al., 1976, Figure 4). The availability of synthetic amino acids has proven useful for balancing animal diets.

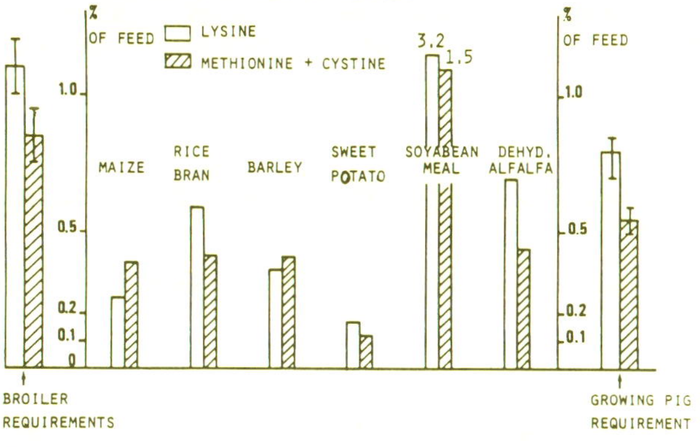

FIGURE 4 : BALANCING THE DIET OF BROILERS AND PIGS IN ESSENTIAL AMINO ACIDS

DIETS OF 13.4 MJ ME/KG, BROILERS

" " 13.4 MJ DE/KG, PIGS

Several successful techniques have been developed for reducing the protein level in the finishing diet of animals. Using a feed based on maize-soybean oil meal, supplemented with lysine for the finishing pig (50-100 kg), a formulation compatible with optimal growth performance was obtained. No addition of tryptophan was necessary, due to a satisfactory balance of dietary essential amino acids. The minimal level of undifferentiated protein after the requirements of essential amino acids had been met without diminishing growth performance and carcass grading, was 12% in castrated males and 13-14% in females, corresponding to 36 and 40-42 g crude protein per Mcal digestible energy respectively (Henry, 1977). The relation of dietary protein to optimum balance of essential amino acids must be considered within the context of dietary energy. Similarly, a reduction in the proportion of protein in the diet for finishing guinea fowls through use of dl-methionine and l-lysine hydrochloride supplementation still ensures maximum growth performance with a substantial saving of protein (Leclercq et al., 1975).

The fatness of the animals appeared to be associated more with a reduction of the average weight at the same age than with the nature and the balance of the diet. Optimum laying performance by the hen as measured by laying rate and average weight of egg vitellus was also dependent on achieving an accurate amino acid balance (Larbier et al., 1972).

These selected examples show clearly the importance of amino acid balance in the economic production of monogastric animals. Use of synthetic amino acids could help save a large amount of protein and improve the quantity of animal products available for human consumption.

Evaluation of New Feedstuffs for Monogastric Animals

Intensive research has been conducted to evaluate new or potential feedstuffs for monogastric animals, on the basis of nutrient content, level of crude fiber and presence of toxic or antinutritional factors.

Rapeseed oil meal has been recognized for its high content of protein well balanced in amino acids. However, a diet containing 17% rapeseed meal increased 10 times the thyroid weight of laying hens with a residual effect on the hatched chickens (March et al., 1972). The presence of glucosinolate (epiprogoitrin, ITC and VTO) was mainly responsible for the goitrogenic effect in poultry and pig (Clandinin and Robblee, 1976). This problem has been overcome by selection of a low erucic acid, low glucosinolate cultivar ("Zerothio"). The quality of this new cultivar was compared to normal rapeseed in a recent experiment conducted on growing pigs (Table 3).

On the basis of chemical determination, the goitrogenic effect causing liver enlargement limited the use of normal rapeseed to 10%. The first limiting factor of the "Zerothio" used at 20% in pigs is the crude fiber level; the second limiting factor is thioglucoside level. Further improvement may be possible from a new process of microwave heating, which detoxifies residual thioglucosides (Alves Pereira et al., 1981).

Field bean could be another important source of protein. Deficiency in sulphur amino acids and possibly in tryptophan was easily identified and subsequently corrected. Antinutritional factors such as antitryptic factors were identified and overcome either by cooking or by selection of zero tanin - low phenol cultivars (demonstrated in chick diet) (Guillaume and Gomez, 1977; Pastuszewska et al., 1974). Using dehulled horsebean for growing pigs increases the apparent digestibility of crude protein as well as energy, but antinutritional factors decrease metabolic utilization of nitrogen. Other protein sources such as peas, or lupines were successfully selected, produced and used as animal feeds using the same methodology (Bourdon and Perez, personal communication).

Identification and Control of Antinutritional Factors

Presence of toxic and antinutritional factors in feed: Trypsin inhibitors were identified (Chernick et al., 1948), extracted and selectively destroyed in soybean oil meal. Tests for determining the toasting degree of soybean products have been

TABLE 3 EFFECT OF LIMITING FACTORS

(RAPESEED MEAL FOR THE GROWING PIG)

(25 - 100 kg)

(Adapted from BOURDON et al., 1981)

	SOYBEAN OIL MEAL	RAPESEED OIL MEAL		
	(21) *	NORMAL (10)	ZEROTHIOGLUCOSIDE (20) -	DEHULLED
Level of ITC mg/g	-	2.2	0.8	1.3
VTO	-	5.0	1.5	2.2
(crude fiber)	(2.5)	(3.4)	(4.7)	(3.2)
Average:				
Daily gain, g	807a	775a(1)	717b(1,2)	785a
Food conversion ratio	2.9a	3.1ab(2)	3.3b(2)	2.9a
D.E. in oil meal MJ/kg D.M.	13.9	13.4	13.4	13.9
Weight of thyroid g	8.4	11.1*	9.3	12.2*
Liver kg	1.6	2.3*	1.9	2.1*

( ) level in the diet (1) Effect of ITC isothiocyanates, and VTO vinylthione oxazolidin
(2) Effect of crude fiber level

performed and their accuracy discussed (Zelter and Delort-Laval, 1971). Identification of antinutritional or toxic factors is always possible for new sources of feeds from either plant breeding or from byproducts.

Analysis of fat for predicting feed quality and grading are mostly based on simple determinations of acidity and peroxide index. However, experimental results observed on piglets and laying hens fed low grade tallows classified on the basis of chemical tests are rather inconsistent (Table 4).

TABLE 4 LIMITS OF THE CHEMICAL ANALYSES FOR PREDICTING THE NUTRITIONAL QUALITY OF FAT FOR POULTRY AND PIGS
(Adapted from LECLERQ and JOUANDETT, 1969 and AUMAITRE, 1969)

Grade of industrial tallow	1st Quality "Human"	Solvent extracted "Poultry"	Pressure extracted "Soap"
Acidity (% oleic acid)	0.3	6.4	26.2
Peroxyde index	1.0	2.5	2.7
Cru de protein	0	0.4	0.9
Performance of laying hen(1)			
Laying rate%	73	74	73
Total eggs (kg)	13.5	13.6	13.6
Performance of piglets(2)			
Daily gain g	400	265**	286**
Food conversion ratio	1.7	1.9*	1.8*
A digestibility of fatty acids (%)	85	79**	76**

(1) 5-10% in the diet: 11 months
(2) 8% in the diet: 8-25 kg of live weight
* ** significant effect vs control

These fats should be more accurately characterized, particularly with respect to protein content and rate of lipolysis (Aumaitre and Rerat, 1968), as an indication of the presence of antinutritional factors. Sensitivity of monogastric animals to such factors seems to depend on the species and the physiological stage.

Estrogenism was identified and described in immature pigs, but development of this reproductive disturbance was only recently associated with ingestion of large amounts of mouldy grain. The factor responsible for this syndrome was identified in maize as zearalenone and quantified in feed and tissues (Etienne and Jemmali, 1982). The consequences of ingesting a diet containing contaminated maize at 3.6 to 4.3 ppm of zearalenone were studied in gilts. When fed to non-pregnant gilts, zearalenone induced a pseudopregnancy; the weight of the uterus and of the foetus significantly decreased (Table 5).

Further Tests to be Developed

As a complement to visual assessment of colour, smell and general aspects (mostly useful for maize) (Hill, 1976), several chemical methods for predicting quality of feeds for pigs and poultry have been developed. Some methods are not very accurate, in particular, protein solubility, amylolytic activity or level of alcohol soluble carbohydrates as a means of testing drying conditions of grains. Recent detection and quantitative determination of toxins such as zearalenone and aflatoxins are of further interest as indicators of storage conditions (Anonymous, 1971; Hill, 1976). Moreover specific methods for testing the biological value of soya protein which includes toasting are: protein solubility in alkali (pH: 7.7), modified binding test of cresol red and corrected tryptic activity. Further chemical tests are necessary for a more comprehensive determination of limiting antinutritional or toxic factors in new protein sources (Aumaitre et al., 1982).

TABLE 5 EFFECT OF FEEDING GILTS A DIET CONTAINING ZEARALENONE PRIOR OR AFTER MATING ON THEIR REPRODUCTIVE PERFORMANCE. (ETIENNE and JEMMALI, 1982)

BASIC FEED IN THE DIET	NORMAL MAIZE	CONTAMINATED MAIZE (1)	
		Puberty to 2nd oestrus	During Pregnancy
Percentage of animals			
Pregnant 80d. p.mating	72	56**	74
Not pregnant, not cyclic	0	6*	22**
Embryonic mortality %	29	37*	36*
Total foetus/litter	10.5	8.7	9.1
Foetal mean weight 80 d. of pregnancy, g	423	393	323**

(1) Diet containing 4 mg zearalenone/kg

QUALITY OF ANIMAL PRODUCTS

Quality is not a simple concept when it concerns human food. In particular it is not sufficiently correlated with nutritional value and balance and no simple unique biochemical or chemical criteria gives a definitive and single rating of animal products.

In the 1960's, spectacular improvement in growth rate of animals and decrease in the age at slaughter were associated with a so-called "decrease in quality of meat". Consumer attitude and pressure as well as an improvement in the research methodology and knowledge have generated new progress in expressing quality of animal/products. Some examples of factors influencing the quality of meat and eggs will be given. They mainly concern modification of the chemical composition, the presence of natural components in the meat of male animals and the possible consequences of nutritional balance or imbalance in the diet of animals.

Frontiers for the Improvement of Quality

Consequences of using a large amount of yeast grown on alkane were measured by lipid composition of the egg and of the fatty tissues in the laying hen (Leclercq, 1976). The presence of a large amount of odd-numbered fatty acids (15 and 17) was measured in yeast and subsequent to its ingestion, an increase in incorporation in the lipids of egg and adipose tissue was demonstrated. Apparently no effect on the quality of products was observed. The long term effects, especially toxicity of ingesting a diet rich in odd-numbered fatty acids on the health of human consumers should be assessed.

Consumers' Attitude Towards Boar Meat

Research is needed to define more precisely and objectively the nature of fresh pig meat quality. Abnormal flavours originating from feedstuffs too rich in fish byproducts were observed in the past. The production of meat from boar instead of from castrated males could decrease markedly production costs by saving 30 kg of food and improving the amount of lean meat by 4 kg/animal. Nevertheless consumer objection to possible "boar taint" considerably limit development of this technique. A recent survey noted a difference in preference between British consumers who accepted boar meat and French ones who were reluctant (Kempster and Cuthbertson, 1982).

Androstenone storage level in fat is one component responsible for the boar taint (Figure 5). Thus judgement of boar meat quality requires a quick and accurate method for detecting high androstenone levels. Above certain limits of androstenone (1 μg/g fat) this meat cannot be recommended for roasting. Further information on this defect and on the favourable effect of technological treatment such as cooking or drying are expected in the future (Figure 5). Such results may assure development of a more profitable production technique.

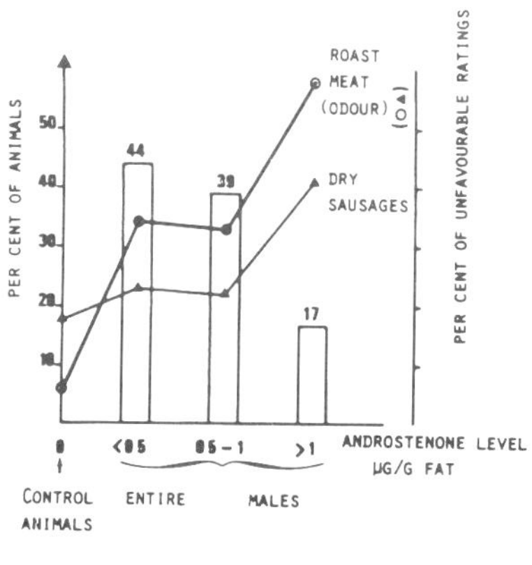

FIGURE 5 : FRENCH CONSUMMERS ATTITUDE TOWARDS BOAR MEAT (54 LARGE WHITE BOARS OF 100 KG LIVE WEIGHT), ADAPTED FROM BONNEAU AND DESMOULIN, 1980 ; AND DESMOULIN ET AL., 1982.

Nutritional Balance of Animal Feeds and Quality of Products

Numerous experiments were performed to ascertain the effect of nutritional balance on product quality. We will give two examples.

Chlorine deficiency of several feedstuffs used currently in poultry nutrition such as maize, wheat and bran, yeast, groundnut and soyabean oil meals is well recognized. A lack of sodium chloride supplement in the feed of the laying hen decreases the average weight of egg from 60 to 50g after 1 week. But it increases the frequency of blood splasches in eggs up to 30% and decreases egg quality of 4 haugh units after 3 weeks of deficiency (Amin et al., 1970). The economic consequences of accidental deficiency could be serious.

Choline is a vitamin particularly known for its essential role in the liver fat metabolism - a deficiency preventing fat accumulation. A diet supplemented with choline was used to cram geese. Significant improvement in the weight of the liver and in product quality was observed subsequent to this supplement (Blum and Leclercq, 1970).

The nutritional balance of micronutrients plays an important role in improving performance and profitability and also in controlling the quality of animal products. Mineral imbalance can damage the quality of animal products, particularly with respect to reproductive ability. Hatchability of guinea fowl eggs was lowered progressively by a deficiency in the diet of available phosphorus (Figure 6). As a consequence of accidental deficiency, profitability of production could be lowered through loss of a high proportion of selected fertile eggs (Sauveur, 1979).

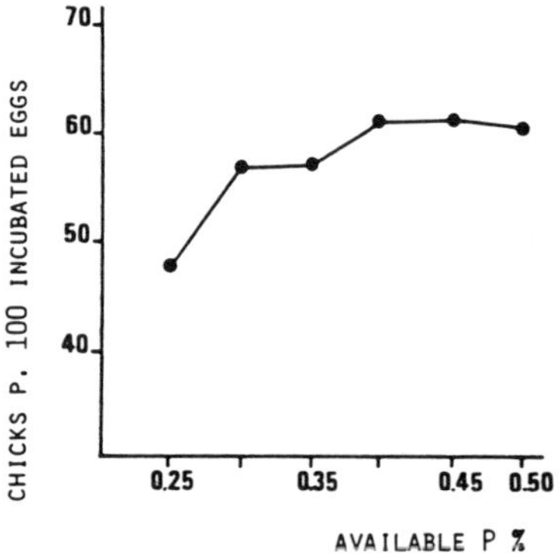

FIGURE 6 : EFFECTS OF THE LEVEL OF AVAILABLE PHOSPHORUS IN THE DIET OF GUINEA-FOWL BREEDERS ON EGG HATCHABILITY EXPRESSED AS CHICKS P.100 INCUBATED EGGS.

CONSEQUENCES OF INTENSIFICATION OF ANIMAL PRODUCTION SYSTEMS

Monogastric animals have been selected for high performance in association with intensive production systems over the last 20 years. Progress in labour saving, in housing conditions and in control of environment were suggested as a means of further improvement. Most workers, however are convinced that high performance requires more accurately, controlled and balanced climatic and nutritional conditions. Some examples of the extreme sensitivity of monogastric animals to unbalanced conditions are given.

Housing, Management and Possible Nutritional Deficiencies

Highly productive monogastric animals raised indoor require a diet balanced for all nutrients including vitamins. Recent work shows an improvement in the survival rate of piglets born from mothers receiving a diet supplemented in vitamin A. A lack of vitamin A storage in the liver of newborn piglet as well as in the milk of sows was observed in animals deprived of a vitamin supply (Salmon-Legagneur and Cabonis, 1969). On the contrary, animals raised on pasture or receiving continuously forage fail to exhibit such a deficiency. Similarly, a folic acid deficiency (3.4 vs 16.4 mg/kg) in the diet of hens had no direct effect either on growth rate and laying rate or on the percentage of hatched chicks. It was followed, however by a high mortality rate of the progeny (16 to 86% of the chicken) (Lillie et al., 1950). These two examples show that intensively raised small pig or poultry herds of highly selected animals, to be profitable, require a continuous supply of trace elements and vitamins.

Piglets born and suckled under intensive conditions, particularly on concrete floors, develop anaemia due to iron deficiency of sow's milk. Young animals supplied with fresh soil never exhibit anaemia (Ullrey et al., 1959). But suckling animals deprived of iron (Figure 7) or to a lesser extent animals supplied orally with iron sulphate between birth and five weeks develop severe anaemia. Iron dextran (fervatol) injection was recognized for a long time as an efficient solution to anaemia prevention. Thus, growth retardation (- 40%), animals more sensitive to digestive diseases and poor and inconsistent performance can be prevented by an appropriate supply of iron. Further deficiencies already demonstrated in sows milk for Mn, Co, etc., have not up to now been efficiently prevented.

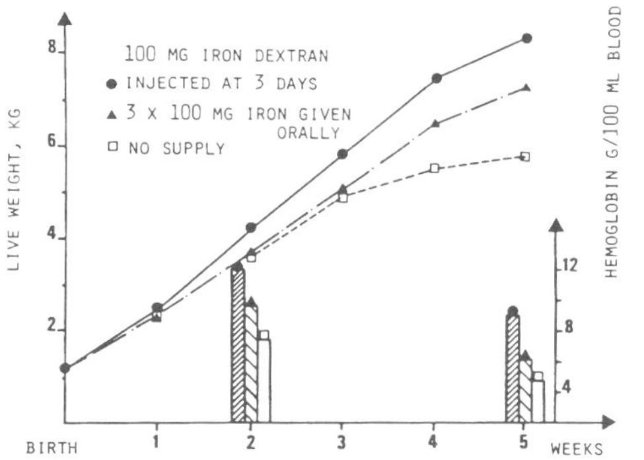

FIGURE 7 : CONSEQUENCES OF IRON DEPRIVATION ON THE DEVELOPMENT OF ANAEMIA AND PERFORMANCE OF SUCKLING PIGLET (ADAPTED FROM LE DIVIDICH ET AL., 1970).

Sensitivity of Highly Selected Chicken to Mineral Interaction

Selection for a high growth rate in poultry is sometimes associated with appearance of a cartilage abnormality such as tibial dyschondroplasia. The gross appearance of the lesions shows some similarities with rickets but histological examination fail to confirm. Unknown growth disorders have appeared as ossification defects, in connection with complex relationships between minerals. Strict interrelationships between Na^+, K^+ and Cl^- levels in the diets of broiler chicks were demonstrated by Sauveur and Mongin, 1978. The importance of these relations appears when new cultivars, new feeds or industrial byproducts are proposed as feed components for highly selected broilers.

Housing Conditions and Efficiency of Growth Promotants

Growth promotants, originally antibiotics, have been considered irreplaceable for stimulating growth of animals and for saving feed in large industrial herds. In pigs their effect on prevention and control of gastrointestinal disorders has been proven during weaning stress or after sudden changes of housing conditions even in small units. The livestock producer of today is just as concerned about human health and nutrition and food quality as any consumer group. The present legislation of FDA, EEC etc., stipulates the use of non-antibiotic antibacterial agents stringently tested for their efficiency, quick elimination from body fluids and non-residues in meat.

Under intensive housing and rearing conditions quinoxalin derivatives were shown to be efficient for pig and poultry (Figure 8). Olaquindox supplements demonstrated greater efficiency in weaned piglets raised on deep litter than in caged piglets, stimulating growth and decreasing food conversion ratio, even with batch production. The growth rate of chicken (1-5 weeks) and finishing pig (30 - 100 kg) raised under optimum housing conditions was improved by 3-4%. A further saving of 4% in the amount of feed required per unit of gain is of great economic importance.

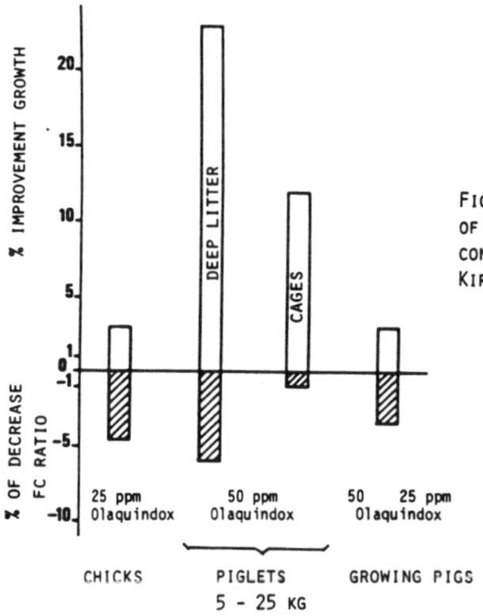

FIGURE 8 : EFFICIENCY OF GROWTH PROMOTANT ON PERFORMANCE OF GROWING CHICKS, PIGLETS AND PIGS. EFFECT OF HOUSING CONDITIONS (ADAPTED FROM BOURDON AND BAR, 1979 ; ROTH AND KIRCHGESSNER, 1977 ; SCHNEIDER ET AL., 1976).

REFERENCES

Alves Pereira, J., Kirleis, A.W and Cline, T.R. (1981). J. Anim. Sci., 53, 1278-125.
American Feed Tables Feed Comp. (1974). University of Florida, 53 pp.
Amin, M., Mongin, P. and Sauveur, B. (1970). Ann. Zootech. 19, 367-383.
Anonymous (1971). Maize: Drying treatments and quality, 1971. Ann. Zootech. 20, H.S. 603-715.
Aumaitre, A. and Rerat, A. (1968). Proc. Sec. World Conf. Anim. Prod.

Beltsville, MD, ed., 381.
Aumaitre, A. (1969). Jour. Rech. Proc. France I.N.R.A. I.T.P. Paris, 1, 125-130.
Aumaitre, A., Sybesma, W., Vanderhaegen, J., Whittemore, C.T. and Zivkovic, S. (1982). Livest. Prod. Sci., 9, 127-161.
Blum, J.C. and Leclercq, B. (1970). Ann. Zootech. 19, 347-351.
Bonneau, M., Desmoulin, B., Frouin, A. and Bidard, J.P. (1982). Ann. Tech. Agri. 29, 69-73.
Bosc, M.J., Martinat-Botte, F. and Nicolle, A. (1977). Jour. Rech. Porc. France I.N.R.A. I.T.P. Paris, 9, 33-37.
Bourdon, D. and Bar, L. (1979). Rech. Med. Vet. 155, 873-885.
Bourdon, D., J.M. Perez and J.J. Baudet (1981). Jour. Rech. Porc. France, I.N.R.A. I.T.P. ed. Paris 13, 163-178.
Calet, C. and Fendry, M. (1965). Ann. Zootech., 14, 351-359.
Chernick, S.S., Lepkovsky, S. and Chaikoff, L.L. (1948). Amer. J. Physiol. 155, 33.
Clandinin, D.P. and Robblee, A.R. (1976). Agri. Bull. University Alberta, 7-10.
Desmoulin, B., Bonneau, M., Frouin, A. and Bidard, J.P. (1982). Livest. Prod. Sci. (In press).
Fonnesbeck, P.V., Harris, L.E., Kearl, L.C. (1976). First Inter. Symp. Feed composition. Utah Univ. ed. 799 pp.
Etienne, M. and Jemmali, J. (1982). J. Anim. Sci., 55, 1-10.
Guillaume, J. and Gomez, J. (1977). Anim. Biol. Anim. Bioch. Biophys. 17, 549-552.
Hammond, D. and Carlyle, W.W.H. (1976). 8th Inter. Cong. Anim. Repr., Krakow, 103 Abs.
Henry, Y. and Bourdon, D. (1975). Jour. Rech. Porc. France I.N.R.A. I.T.P. Paris, 7, 71-80.
Henry, Y.M. (1976). First Inter. Symp. Feed composition, Fonnesbeck. Harris-Kearl, ed. Logan, 1, 270-281.
Henry, Y. (1977). Jour. Rech. Porc. France. I.N.R.A. I.T.P. ed. Paris 9, 259-264.
Hill, L.D. (1974). Corn quality in world market. Danville ed. IL, 270.
Kempster, A.J. and Cuthbertson, A. (1982). E.A.A.P. Meeting Leningrad Pig Commission 3 (1), 11 pg.
Larbier, M., Blum, J.C. and Guillaume, J. (1972). Ann. Biol. Anim. Bioch. Biophys. 12, 125-138.
Leclercq, B. and Jouandet, C. (1969). Ann. Zootech. 18, 331-336.
Leclercq, B., Larbier, M. and Blum, J.C. (1975). Ann. Zootech. 24, 229-23.
Leclercq, B. (1976). Ann. Zootech. 16, 155-158.
Le Dividich, J., Aumaitre, A., Salmon-Legagneur, E. and Rousseau, P. (1970). Rec. Med. Vet. 146, 237-250.
Le Dividich, J., Aumaitre, A and Seve, B. (1976). First Inter. Symp. Feed composition. Fonnesbeck-Harris-Kearl, ed. Logan, 1, 125-132.
Lillie, R.J., Combs, G.F. and Briggs, M. (1960). Poult. Sci. 29, 115-129.
March, B.E., Biely, J. and Soong, R. (1972). Poult. Sci. 51, 1589-1596.
Martinat-Botte, F., Bariteau, F., Mauleon, P., Scheid, J.P. and Signoret, J.P. (1982). Jour. Rech. Porc. France I.N.R.A. I.T.P. Paris, 14, 75-80.
Martinat-Botte, F., Gauthier, J., Depres, P. and Terqui, M. (1980). Jour. Rech. Porc. France I.N.R.A. I.T.P. Paris, 12, 167-170.
Pastuszewska, B., Duee, P.H., Henry, Y., Bourdon, D. and Jung, J. (1974). Ann. Zootech. 23, 537-554.
Perez, J.M., Ramoelintsalama, B. and Bourdon, D. (1980). Jour. Rech. Porc. France I.N.R.A. I.T.P. Paris, 12, 273-284.

Pion, R. and Fauconneau, G. (1966). Amino acides, peptides, proteines. Cahier no 6. A.E.C. ed. 155-175.
Politiek, R.D. and Bakker, J.J. (1982). Livestock Production in Europe. Lives. Prod. Sci. 9, 335 pp.
Roth, F.X. and Kirschgessner, M. (1977). Zuchtungskde 49, 66-74.
Salmon-Legagneur, E. and Cabanis, J. (1969). Jour. Rech. Porc. France I.N.R.A. I.T.P. Paris, 1, 185-189.
Sauveur, B. and Mongin, P. (1978). Ann. Biol. Anim. Bioch. Biophys. 18, 87-98.
Sauveur, B. (1979). Ann. Zootech. 28, 109-119.
Schiemann, R., Nehring, K., Hoffmann, L., Jentsch, W. and Chudy, A. (1972). Energetische Futterbewertung und Energienormen. V.E.B. Deutscher Landwirtschaftverlag, Berlin, 34 p.
Schneider, D., Bronsch, K. and Richter, L. (1976). Zeits. Tierphysiol. Tierernahrg. Futtermittelk. 36, 241--249.
Ullrey, D.E., Miller, E.R., West, D.R., Schmidt, D.A., Seerley, R.W., Hoefer, J.A. and Luecke, R.W. (1959). J. Anim. Sci. 18, 256-262.
Van den Eynden, G. (1978). Proc. World Poult. Congr., 64-74.
Zelter, S.Z. and Delort-Lavel, J. (1971). Ann. Zootech. 20, 17-29.

THE ROLE OF CHEMISTRY IN AQUACULTURE SYSTEMS: PROBLEMS IN BREEDING AND FEEDING FISHES

Rafael D. Guerrero III
Aquatic Biosystems
Philippines

ABSTRACT

The contribution of aquaculture to world food supply will be more significant in the future. Constraints to aquaculture development include the shortage of fish seed and the high cost of feeds. The application of hypophysation techniques in the induced spawning of cultured fishes has provided a useful and effective means of increasing fish seed artificially. More studies, however, are needed to develop a standard low cost hormone or chemical which would give consistent results.

Chemicals also play an important role as diluents and cryoprotectants for the preservation of fish gametes and embryos. A better understanding of the pheromone-like substances detected in fishes can enhance artificial breeding. The use of synthetic sex steroids and other chemicals has tremendous potential for production of monosex and sterile fish.

The nutritional requirements of only a few cultured fishes have been studied. Further research on the development of efficient and economical feeds is essential for intensive culture systems. Substitutes for fish meal as a source of animal protein in fish diets need to be found. Various compounds and products of lipid oxidation present in feedstuffs are known to harm fish or cause "off flavour".

KEYWORDS: Fish breeding, feeding, hypophysation, induced spawning, sex hormones, fish nutrition.

INTRODUCTION

Aquaculture is the farming of aquatic organisms primarily for human food. Production from aquaculture development in 1985 is expected to reach 12 million tons (Pillay, 1979). The contribution of aquaculture to food supply in the future can exceed that of marine fisheries with improvement in production systems (Webber and Riodan, 1979). With the increasing trend towards intensive and semi-intensive aquaculture in many countries, several constraints have been identified. Among these constraints are the shortage of fish seed and the high cost of feeds (Reay, 1979; Neal and Smith, 1982). Increased fish production is predicated on selective breeding, improved water quality and optimal feeding (Hastings, 1979). The supply of quality seed should be adequate to meet the needs of fish farmers. Equally important is the manufacture of efficient and economical feeds which are essential for intensive culture.

This paper presents a review of aquaculture systems in fish breeding and feeding with emphasis on chemistry-related problems.

PROBLEMS IN FISH BREEDING

Induced Breeding of Fishes

There are more than 300 different species of finfish cultivated by man. Only a few species, however, can be spawned naturally in captivity (Davy and Chouinard, 1981). Techniques for induced spawning of cultured fishes using gonadotrophin hormones are now widely practiced in hatcheries throughout the world. Induced spawning of Chinese carps and the Indian major carps has been successful with the application of hypophysation or the injection of pituitary gland extract (Harvey and Hoar, 1979). A semi-purified gonadotrophin from the chinook salmon (SG-G100) has been used to induce ovulation in the Indian catfish (Sundararaj et al., 1972), the grey mullet (Kuo and Nash, 1975) and milkfish (Chaudhuri and Juario, 1977). Human chorionic gonadotrophin (HCG) has also been widely used in spawning cultured fish (Pickford and Atz, 1967; Yamazaki, 1965).

The problems in the application of hypophysation are manifold. The dosage to be injected is difficult to standardize because the activity of the extract depends on the age, sex, maturity of donor, method of collection and preservation of the pituitary gland (Jalabert et al., 1977) as well as phylogenetic relationship between the donor and recipient. Pituitary extracts contain a mixture of hormones that make assessment of specific activity problematic. Moreover, collection of pituitary glands involves the sacrifice of potential breeders and is time-consuming (Harvey and Hoar, 1979). There is therefore a need to develop a natural or synthetic hormone with standard potency and low cost to meet the increasing world-wide demand for commercial production of fish seed (Chaudhuri and Tripathi, 1979).

Aside from hypophysation and purified gonadotrophin, the use of hypothalamus releasing hormones (e.g. LH-RH), sex steroids and other chemicals such as clomiphene citrate (an antiestrogen) has shown promise for influencing gonadal maturation in fish (Harvey and Hoar, 1979). In China, the synthetic analogue of LH-RH has been found to be effective for inducing ovulation in grass carp (Davy and Chouinard, 1981). The cost of the analogue, however, is prohibitive. The administration of progesterone for induced ovulation of goldfish has been reported by Khoo (1974). Jalabert et al., (1977) demonstrated the effectiveness of 17α-20β Pg in the induction of oocyte maturation in goldfish, rainbow trout and northern pike.

The action of chemicals such as clomiphene citrate is believed to be on the feedback control of gonadotrophin secretion (Pandey and Stacey, 1975). The chemical competes with endogenous gonadal steroid for binding sites in the pituitary and hypothalamus thereby stimulating gonadotrophin secretion and inducing ovulation. With variable results obtained in the application of clomiphene citrate attributed to a lack of knowledge on the effective dosage and the responsive stage of gonad maturation, further research on the chemical is recommended by Harvey and Hoar (1979).

Pheromones in Fish Breeding

Pheromones are chemical substances transferred between members of an animal population for control of behaviour and development. Pheromone-like substances have been described in freshwater fishes (Solomon, 1977). In the cichlid, Aequidens portalegrensis, spawning frequency is believed to be influenced by a pheronome (Folder, 1971). There is a dearth of knowledge, however, concerning such chemicals. A better understanding of the origin, chemical nature and activity of pheromones in fishes may improve breeding techniques.

Preservation of Fish Gametes

The preservation of fish sperm and eggs is advantageous in the breeding of the cultured fish which spawns only during certain seasons. It also allows the crossing of genetic strains for selective breeding (Holtz et al., 1979). Much work has been devoted to the cryopreservation of fish sperm. The process essentially involves the use of a diluent ("extender") to keep the sperm alive but inactive prior to freezing and a cryoprotectant to minimize damage during freezing and thawing (Harvey and Hoar, 1979). In the preservation of trout sperm, Holtz et a., (1979) found dimethylsulfoxide (DMSO) to be a better additive than glycerol. Hara et al., (1982) recommended the use of milkfish serum as diluent for Chanos chanos sperm. Successful cryopreservation of unfertilized salmonid (Atlantic salmon, rainbow trout and brook trout) eggs and zygotes has been reported by Zell (1978). The eggs and zygotes were frozen to $-20^{\circ}C$ in liquid nitrogen for 5 minutes using Hank's solution as medium. No cryoprotectant was added. Future work on the preservation of gametes and embryos of other cultured fishes will enhance breeding programs and propagation of quality seed.

Sex Control in Fishes

Artificial sex control in fishes has been achieved with the use of snythetic hormones or steriods (Yamamoto, 1969; Schreck, 1974; Guerrero, 1979). Hormone treatment to elicit sex reversal or sterilization in cultured fishes has tremendous potential for commercial application.

In aquaculture, the raising of monosex or sterile fish offers many advantages. Monosex male culture of tilapia, for instance, limits reproduction of the prolific species and increases the yield of harvestable fish (Guerrero, 1975). Introduction of monosex grass carp in the United States without the capability of breeding in open waters has been proposed (Shelton and Jensen, 1979). The shortage of breeders for spawning as in the case of the grouper, Epinephelus tauvina, can be alleviated with the administration of male sex steroid (Davy and Chouicard, 1981). The problem of precocious maturation of males in salmonid culture may be solved by hormone sterilization of juveniles (Goetz et al., 1979).

For production of all-male tilapia, the synthetic sex steriods 17α-methyltestosterone and 17α-ethyltestosterone are orally administered to sexually undifferentiated fry at a dosage of 30-60 mg/kg diet for a period of 3-4 weeks (Guerrero, 1975; Shelton and Hopkins, 1978). High percentages of female Tipapia aurea were produced by Hopkins (1979) using oral administration of ethynylestradiol (with cypropterone acetate) at 100 mg/kg diet for 35 days. Goetz et al. (1979) produced all-female stocks of rainbow trout and Atlantic salmon with oral administration of 17α-oestradiol at a concentration of 20 mg/kg diet for 21-30 days following first feeding. The same workers also succeeded in inducing sex reversal of genetic females of rainbow trout and salmon with oral treatment of 17 α-methyltestosterone at 3 mg/kg diet for 90 days. Treatment with 17α-methyltestosterone by immersion and oral administration at a dosage of 30 mg/kg diet for 120 days was found effective for sterilization.

In Singapore, work on the sex reversal of the protogynous hermaphroditic, Epinephelus tauvina, is in progress (Davy and Chouinard, 1981). Experiments have shown that oral administration of methyltestosterone can accelerate female-to-male transformation for the production of functional males for breeding. Duration of methyltestosterone treatment was considered critical for influencing sex reversal. The levels of steroids during sex transformation need to be assessed; the age or size when the species is amenable to treatment has to be determined.

Methallibure and cypropterone acetate have been applied in fishes as chemosterilants. Methallibure acts as a hypophysectomizing agent that blocks the production and release of gonadotrophins (Pandey and Leatherland, 1970). Cypropterone acetate is an anti-androgen that competes with the uptake of testosterone by the testes (Schreck, 1973). Dadzie (1972) found degeneration of all cells in the testes of adult Tilapia aurea and T. mossambica with treatment of methallibure. The chemical may also be used to synchronize spawning of fish (Hyder, 1972). One week after cessation of methallibure treatment, spawning of fish occurs indicating that the effect of the chemical is temporary and that the compound is excreted from the body (Hoar et al., 1967; Dadzie, 1975).

The mode of action of cypropterone acetate is believed to be its interferrence with the biosynthesis of testosterone (Sorcini et al., 1971). When administered to rainbow trout (Salmo gairdnerii) it was found to block the uptake of testosterone (Schreck, 1973). Attempts to produce all-females by feeding the chemical to the Japanese medaka (Oryzias latipes) and Tilapia aurea have been unsuccessful (Smith, 1976; Hopkins, 1979).

PROBLEMS IN FISH FEEDING

The development of efficient and economical feeds is essential for intensive fish culture systems. While extensive studies have been conducted on the nutritional requirements of salmonids, catfish, eels and sea bream, relatively little is known about the nutrition of most cultured fishes. Fish are regarded as efficient converters of feed. They require higher levels of dietary protein than land animals (Spinelli, 1980). According to Hastings (1979), the protein requirement of fish is related to the energy requirement at a given temperature, size and age, stocking density, and environmental stress (i.e., low dissolved oxygen, presence of toxicants, etc.). The efficiency of feeding is very much dependent on water quality of the fish's environment. With high rates of feeding, the buildup of organic matter and ammonia, and increase in oxygen demand can be detrimental to fish (Allison et al., 1979).

Specifically, more systematic research is recommended by Halver (1979) for: (a) specific nutrient requirement values, (b) digestibility coefficients, and (c) sparing effect of one nutrient upon another as practical, effective, efficient fish rations are formulated from the ingredients available from agricultural and aquacultural industries.

Studies on the replacement of fish meal in aquaculture diets have shown promising results. The use of single cell protein (SCP) and krill meal is believed to be commercially feasible (Spinelli, 1980). SCP is produced from microorganisms such as yeast, bacteria, fungi and algae. It has a well-balanced amino acid profile and is rich in certain vitamins and minerals. Tests on feeding of rainbow trout have shown that yeast SCP can substitute up to about 25-40% of the fish meal in standard diets. Krill meal consists of the zooplankton (minute crustaceans) harvested in the Artic and Antartic regions. It contains about 55% protein. An estimated volume of 300 million tonnes can be harvested yearly. In feeding trials with rainbow trout, growth of fish fed with krill as the main source of protein was lower than that of the fish fed the control diet. The poor performance of the krill is attributed to its high ash content. A milling and screening process has been sugggested to improve the quality of krill meal.

Certain compounds occurring in feedstuffs are known or suspected of causing physiological abnormalities and retarding fish growth. These compounds include glycosides, phytates, mycotoxins, cyclopropenoid fatty acids, trypsin inhibitors, mimosine, glucosinolates, haemoglutinins, plant phenolics, oxidized and polymerized lipids, histamine and putrescine, and nitrosamines.

The problem of lipid oxidation in the storage of feedstuffs is a major one. Feedstuffs with highly unsaturated lipids are susceptible to oxidation. Oxidation of fats to aldehydes, ketones and acids may be toxic to fish and may accelerate the loss of certain vitamins and amino acids. The products of lipid oxidation also contribute to "off flavour" of fish flesh (Hastings, 1979). To inhibit lipid oxidation, anti-oxidants such as ethoxyquin butylated hydroxytoluene and alpha tocopherol are added to feedstuffs.

REFERENCES

Allison, R., R.O. Smitherman and J. Cabrero (1979). Effects of high density culture and form of feed on reproduction and yield of Tilapia aurea, pp. 1-10. In: T.V.R. Pillay and Wm. A. Dill, eds. "Advances in Aquaculture." FAO, Rome, Italy.

Chaudhuri, H. and J.V. Juario (1977). Use of hormones in breeding cultivated warm-water fishes with special reference to milkfish, Chanos chanos (Forskal). Fish. Res. J. Philipp. 2(2):1-6.

Chaudhuri, H. and S.D. Tripathi (1979). Problems of warmwater fish seed production, pp. 127-134. In: T.V.R. Pillay and Wm. A. Dill, eds. loc. cit.

Dadzie, S. (1972). Inhibition of pituitary gonadotropic activity in cichlid fishes of the genus Tilapia by a dithiocarbamoyllhydrazine derivative (I.C.I. 33, 828). Ghana J. Sci. 12: 41-53.

Dadzie, S. (1975). A preliminary report on the use of methallibure in tilapia culture. African J. Hydrobiology and Fisheries 4 (special issue) 3: 127-140.

Davy, F.B. and A. Chouinard (1981). Induced fish breeding in Southeast Asia. Report of a Workshop held in Singapore, 25-28 November, 1980. Ottawa, Ontario. IDRC. pp. 48.

Goetz, F.W., E.M. Donaldson, G.A. Hunter and H.M. Dye (1979). Effects of estradiol-17β and 17α-methyltestosterone on gonadal differentiation in the cohoe salmon, Oncorhynchus kisutch. Aquaculture, 17: 267-278.

Guerrero, R.D. (1975). Use of androgens for the production of all-male Tilapia aurea (Steindachner). Trans. Amer. Fish. Soc. 104(2): 342-348.

Guerrero, R.D. (1979). Use of hormonal steroids for artificial sex reversal of tilapia. Proc. Indian Natn. Sci. Acad. 45(5): 512-514.

Halver, J.E. (1979). The nutritional requirements of cultivated warmwater and coldwater fish species, pp. 574-580. In: T.V.R. Pillay and Wm. A. Dill, eds. loc. cit.

Hara, S., J.J. Canto, Jr. and J.M.E. Almendras (1982). A comparative study of various extenders for milkfish Chanos chanos (Forskal) sperm preservation. Aquaculture, 28: 339-346.

Harvey, B.J. and W.S. Hoar (1979). The theory and practice of induced breeding in fish. International Development Research Centre, Ottawa, Canada, 48.

Hastings, W.H. (1979). Fish nutrition and fish feed manufacture, pp. 568-574. In: T.V.R. Pillay and Wm. A. Dill, eds. loc. cit.

Hoar, W.S., J. Wiebe and E.H. Wai (1967). Inhibition of the pituitary gonadotropic activity of fishes by a dithiocarbamoylhydrazine derivative (I.C.I. 33, 828). Gen. Comp. Endocrinol. 8: 101-109.

Holtz, W., S. Buyukhatipoglu, J. Stoss, B. Oldigs and H.J. Langholz (1979). Preservation of trout spermatozoa for varying periods, pp. 141-143. In: T.V.R. Pillay and Wm. A. Dill, eds. loc. cit.

Hopkins, K.D. (1977). Sex reversal of genotypic male Sarotherodon aureus (Cichlidae). M.S. Thesis, Auburn University, Auburn, Ala, 38.

Hopkins, K.D. (1979). Estrogen sex-reversal of Tilapia aurea. Aquaculture, 18: 263-368.

Jalabert, B., B. Breton, E. Brzuska, A. Fostier and J. Wieniawski (1977). A new tool for induced spawning: the use of the 17α-hydroxy-20β-dihydroprogesterone to spawn carp at low temperature. Aquaculture, 10: 353-364.

Khoo, J.H. (1974). Steroidogenesis and the role of steroids in the endocrine control of oogenesis and vitellegenesis in the goldfish, Carrasius auratus. Ph.D. Thesis, Univ. of B.C., Vancouver, BC, Canada, 126.

Kuo, C-M. and C.E. Nash (1975). Recent progress on the control of ovarian development and induced spawning of the grey mullet (Mugil cephalus L.). Aquaculture, 5: 19-29.

Neal, R.A. and I.R. Smith (1982). Key problem areas in world aquacultural development. ICLARM Newsletter, 5(1): 3-5.

Pandey, S. and J.F. Leatherland (1970). Comparison of the effects of methallibure and thiourea on the testes, thydroid and adenohypophysis of the adult and juvenile guppy Poecilia reticulata (Peters). Can. J. Zool. 48: 445-450.

Pandey, S. and N. Stacey. (1975). Antiestrogenic action of clomiphene citrate in goldfish, Can. J. Zool., 53: 102-103.

Pickford, G.E. and J.W. Atz (1957). The physiology of the pituitary gland of fishes. New York Zoological Society, New York.

Pillay, T.V.R. (1979) The state of aquaculture 1976, pp. 1-10. In: T.V.R. Pillay and Wm. A. Dill, eds. loc. cit.

Polder, J. (1971). On gonads and reproductive behaviour in the cichlid fish Aequidens portalegrensis (Hensel). Neth. J. Zool. 21: 265-365.

Reay, P.J. (1979). "Aquaculture." Edward Arnold (Publishers) Limited, London, pp. 60.

Schreck, C.B. (1974). Hormonal treatment and sex manipulation in fishes, pp. 84-106. In: C.B. Schreck, ed. "Control of Sex in Fishes", Extension Division, Virginla Polytechnical Inst. and State Univ., Blacksburg, Virginia.

Shelton, W.L. and G.L. Jensen (1979). Production of reproductively limited grass carp for biological control of aquatic weeds. Water Resources Research Institute, Auburn Univ., Auburn, Alabama, pp. 174.

Shelton, W.L. and K.D. Hopkins (1978). Use of hormones to produce monosex tilapia for aquaculture, 10-33. In: R.O. Smitherman, W.L. Shelton, J.H. Grover, eds. Culture of Exotic Fishes Symposium Proceedings. Fish Culture Section, American Fisheries Society, Auburn, Alabama.

Smith, H.T. (1976). Various aspects of reproductive control in freshwater fishes. M.S. Thesis. Virginia Polytechnic Inst. and State Univ., Virginia, 70.

Solomon, D.J. (1977). A review of chemical communication in freshwater fish. J. Fish Biol., 11: 363-376.

Spinelli, J. (1980). Unconventional feed ingredients for fish feed, 187-214. In: Fish Feed Technology. United Nations Development Programme, Rome, Italy.

Sorcini, G., F. Sciarra, D. Silverio and F.R. Fraioli (1971). Further studies on plasma androgens and gonadotrophins after cypropterone acetate (SH 714). Foliа Endocr. Roma. 24: 196.

Sundararaj, B.I., T.C. Anand and E.M. Donaldson (1972). Effects of partially purified salmon gondotropin an ovarian maintenance, vitellogenesis, and ovulation in the hypopysectomized catfish, Heteropneustes fossilis (Bloch). Gen. Comp. Endocrinol. (US), 18: 104-114.

Weber, H.H. and P.F. Riordan (1979). Problems of large-scale vertically-integrated aquaculture, pp. 27-34. In: T.V.R. Pillay and Wm. A. Dill, eds. loc. cit.

Yamamoto, T. (1969). Sex differentiation, 117-175. In: W.S. Hoar and D.J. Randall, eds. "Fish Physiology." Vol. III. Academic Press, New York.

Yamazaki, F. 1965. Endocrinological studies on the reproduction of the female goldfish Carassius auratus L., with special reference to the function of the pituitary gland. Mem. Fac. Fish. Hokkaido Univ. (Japan) 13: 1-64.

Zell, S.R. (1978). Cryopreservation of gametes and embryos of salmonid fishes. Ann. Biol. Anim. Biochem. Biophys., 18: 1089-1099.

CHEMISTRY IN THE CONTROL OF RUMINANT ANIMAL DISEASES AND REDUCTION OF PHYSICAL AND BIOLOGICAL STRESS

Robert H. Dunlop
Dean, College of Veterinary Medicine
University of Minnesota, St. Paul, Minnesota 55108, USA

ABSTRACT

Chemistry and biochemistry have emerged as the most important disciplines related to veterinary medicine in the potential for solving the problems of animal disease. Biochemistry is essential to gaining an understanding of the basic veterinary biology of the productive performance and disease processes of ruminants. It plays an expanding role in developing new technologies for the manipulation of performance or the control of diseases. Analytical chemistry has helped to characterize naturally-occurring antibiotics and pharmaceuticals, as well as to monitor the pharmacokinetics of drugs and metabolics in the body compartment. Synthetic chemistry, linked to drug design, has generated new classes of pharmaceutical chemicals with many options for derivatives. It has also allowed the development of many modifications of the nucleus of complex antibiotics obtained from microorganism. It serves as the spearhead of the development of products to treat or prevent animal diseases and to enhance the efficiency of livestock production. The important advances and the chemotherapy of ruminant diseases are assessed, along with the development of products to prevent metabolic disorders, relieve stress or enhance productivity.

KEYWORDS: Ruminants, diseases, pharmacology, stress, biotechnology, chemotherapy, antibiotics, anthelmintics, inflammation, metabolic diseases, growth promotants, veterinary biology, veterinary medicine, drugs, vaccines.

INTRODUCTION

A major thrust of this paper will be to review and update the achievements and contributions of chemistry in gaining control of animal diseases. The following sections cover the drugs developed to treat or prevent infectious diseases by the class of pathogenic agent. Chemotherapeutic drugs have become the major pharmaceutical agents used in animal diseases. Drugs to manipulate the physiology or pathophysiological changes have found less application although there are notable exceptions such as the antibiotics used to modify digestion and some endocrine products used to modify growth or reproductive functions.

Because the chemotherapeutic drugs have found such wide application and since many of them are used in human medicine, concerns have been expressed. One is that use of antibiotics in animals, particularly at low levels in feed to improve productive performance, might favor the development of resistant strains of pathogenic bacteria. Another is the potentially hazardous effects of drug residues in animal products for human consumption.

CHEMOTHERAPY

Antibacterial Chemotherapy

Inhibitors of Folinic Acid Synthesis: Pathogenic organisms that are unable to absorb preformed folic acid are susceptible to chemicals that impair their ability to synthesize folic and folinic acids. The discovery of the anti-bacterical action of sulfanilamide and the subsequent demonstration that it acts via inhibition of the incorporation of para-aminobenzoic acid (PABA) into dihydropterioc acid, the precursor of folic acid, stands as one of the great milestones of chemotherapy. Synthetic modifications of the para-aminobenzene sulfonamide nucleus yielded a large number of congeners, some of which have remained important in veterinary therapeutics. Important variables are solubility, pK_a value, rate of absorption and bioavailability of active drugs. Molecular manipulation has affected the distribution kinetics of the compounds and this has allowed some selection to achieve specific therapeutic goals. However, the action of sulfonamides is bacteriostatic, not bacteriocidal.

A second means of inhibiting folinic acid synthesis can be achieved by inhibiting folate reductase thereby preventing the final step of conversion of the dihydrofolate (folic acid) to tetrahydrofolate and folinic acid. Since this enzyme is required by both microbial and animal cells, a preferential inhibitor that would selectively inhibit the microbial cell was required. Trimethoprim was found to inhibit some microbial folate reductases at 10^{-5} times the concentration required to inhibit the mammalian enzyme. Although not very effective when given alone, a combination of trimethoprim with a sulfonamide gives a much greater efficacy that the sum of the effects of the two agents given alone. This exquisite example of synergism between two drugs has been exploited in therapeutics (Figure 1).

Penicillins: Separation and identification of the nucleus of the penicillin molecule, 6-amino-penicillanic acid, (6APA) was achieved in 1957. This compound has a double ring structure comprised of a β-lactam ring connected to a thiazolidine ring. Chemical characterization of this nucleus led to one of the most interesting series of synthetic manipulations in pharmacological history. The pharmaceutical industry has made a remarkable contribution to human and veterinary medicine via these new semi-synthetic penicillins (Figure 2). The potency, duration of action, acid-resistance, breadth of anitibacterial spectrum and penicillinase-resistance were modified over the two decades following the isolation of 6-APA.

The original acid-labile, penicillinase-susceptible penicillin was penicillin G or benzyl penicillin. Acid-resistance and improved oral efficacy was achieved by isolating another natural form, penicillin V. Increased duration of action was attained by complexing the penicillin G with either procaine or benzathine, reducing solubility. Modification of the microbially produced penicillin nucleus began in earnest in the late 1950's. The original penicillins were only effective against certain gram positive bacteria. One of the most significant synthetic achievements was the imparting of efficacy against many gram negative bacteria by using amidases to remove the side chain then adding an aminobenzyl side chain to 6-APA to yield ampicillin. This compound slowly penetrates the cell wall of gram negative bacteria, unlike the naturally-occurring forms. Adding a hydroxyl to the benzyl ring of ampicillin in the para position results in amoxycillin which

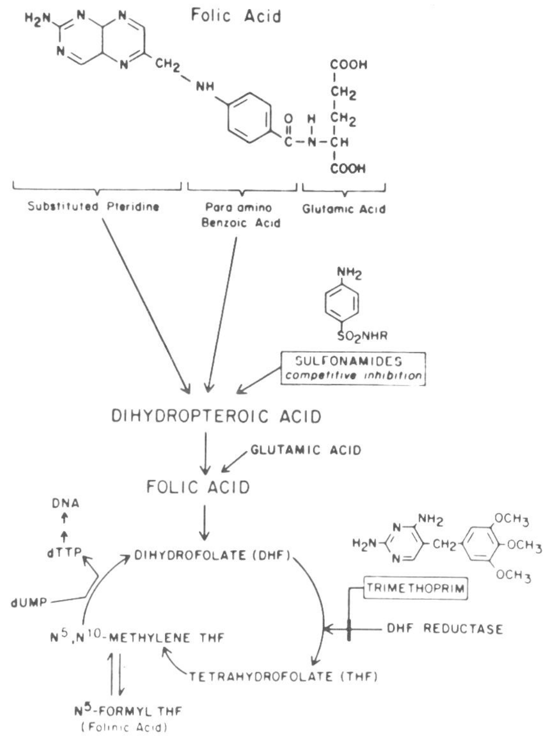

Figure 1: Sites of action of sulfonamides and pyrimidines in the synthetic pathway for folinic acid; dUMP is deoxyuridylate, which is methylated to deoxythymidylate, dTTP.

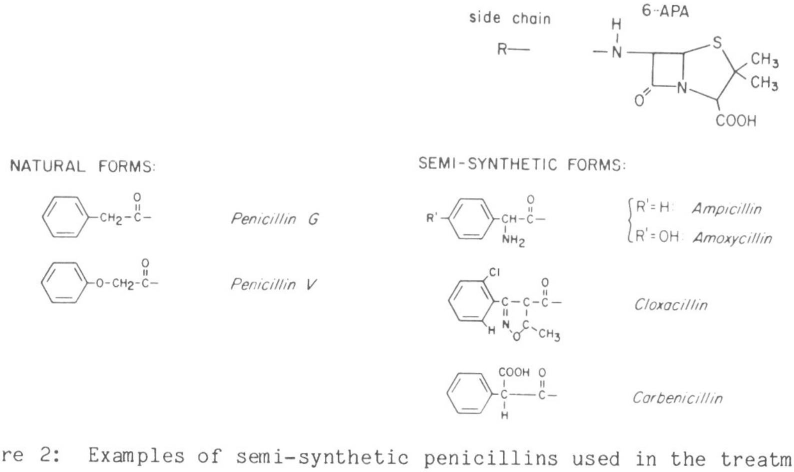

Figure 2: Examples of semi-synthetic penicillins used in the treatment or prevention of ruminant diseases. In addition the procaine and benzathine salts of penicillin G are used to provide insoluble repository forms for IM use.

penetrates into gram negative bacteria more readily and is more effective as a result (Osborne et al., 1978). Another group of derivatives was found to be effective against the notorious Pseudomonas aeruginosa. However, one of these, carbenicillin, when tested against mastitis caused by this organism was not fully effective and resistant strains soon appeared (Ziv et al., 1969). The isoxazole penicillins were found to have activity against penicillinase-producing organisms. One of these, sodium cloxacillin, has been developed to treat mastitis in lactating cows. The benzathine derivative of cloxacillin has low solubility and protracted persistence when introduced to the udder of a dry cow. Such dry cow therapy has become one of the most important tools in controlling mastitis on a herd basis (Brander et al., 1975).

The latest development in the penicillin saga is the discovery of inhibitors of penicillinases (β-lactamases). The idea is to find one that is effective in protecting penicillinase-susceptible forms of penicillin when administered concurrently. The compound clavulanic acid, derived from a streptomyces species, has been found to have this property and is currently under investigation (Figure 3). In combination with amoxycillin it appears to be very promising in improving efficacy against penicillinase-resistant staphylococci and enterobacteria. Allergic reactions to penicillin may involve a penicilloyl-protein complex. The cephalosporins are a related group of substances that are derived from cephalosporin C obtained from Cephalosporium fungus. Similar to the penicillins, a common nucleus of 7-amino-cephalosporanic acid has been modified to yield many semi-synthetic derivatives.

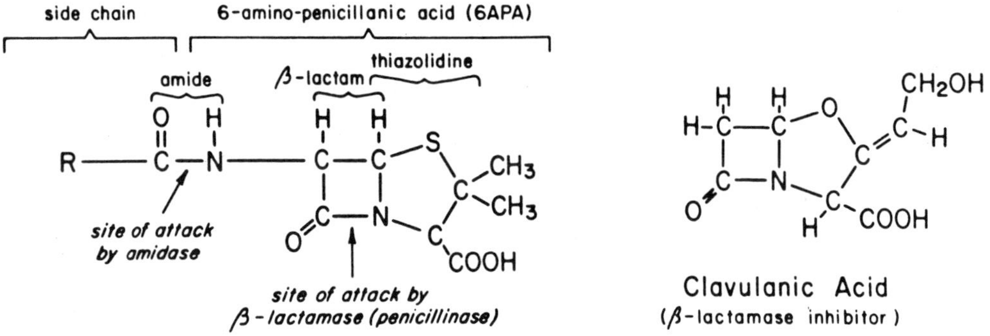

Figure 3: The structure of the essential nucleus of penicillin, 6APA indicating the site of action of β-lactamase and the structure of the irreversible inhibitor clavulanic acid.

Other Antibacterial Antibiotics and Chemotherapeutics: After the first wave of chemotherapy attributable to the sulfonamides and penicillins, many great searches were undertaken to seek new naturally-occurring antibiotics. The main sources proved to be soil organisms. The last forty years has witnessed the advent of several major groups:

1. Aminoglycosides, or aminocyclitols (streptomycin, dihydro-streptomycin, neomycin, kanamycin, gentamycin, spectinomycin, paromomycin, tobramycin and amikacin).
2. Tetracyclines (chlortetracycline, oxytetracycline, tetracycline. Other derivatives have been developed.)
3. Chloramphenicol
4. Polypeptide Antibiotics (bacitracin, colistin, polymyxin B)
5. Macrolide antibiotics and lincosamides (erythromycin, tylosin, spiramycin;

lincomycin, lindamycin).
6. Miscellaneous Chemotherapeutics (nitrofurans, naldixic acid, and carbadox).

Strategy of Antibacterial Chemotherapy: Infections in young animals are one of the most important causes of losses, often involving enteric or respiratory infections, or both. Predisposing factors of environmental and nutritional stress are frequently contributory to the onset of such diseases, wtih lack of adequate maternal passive immunoglobulin support via colostrum in the neonatal period a common occurrence. Determining the microbial components of the disease complex may be difficult. Both viruses and bacteria may be involved and often more than one species of each. Thus the problem is multifactorial both in the stress components that reduce host resistance and the microbial pathogens or opportunists that proliferate. Control of such complex syndromes requires much more than cultures of aerobic bacteria coupled with antibiotic sensitivity tests. The whole managerial situation must be evaluated and a corrective strategy developed. Chemotherapy may have a place in this but it will be only one component. Selection and usage of the appropriate class of antibiotic requires consideration of the antibiotic sensitivity of pathogenic organisms that have been isolated, knowledge of the clinical pharmacology of the drugs and clinical judgement based on experience and the severity of the illness. Often preliminary decisions must be made before laboratory results are available. Since not all effects of chemotherapeutics are beneficial, there are situations where they may be contraindicated.

Similar reasoning applies to enteric disorders that occur post-weaning, the respiratory disease complex that occurs after cattle are shipped or kept in confined housing, and septicemia.

Virus Diseases, Cancer and Immunological Disorders

Although it is too early to review progress in application of pharmaceuticals to the treatment of these problems, it is important to take note of their importance. They remain as one of the areas where a great challenge to scientific exploration remains. The lack of specific antiviral chemotherapy for the whole range of virus diseases leaves the veterinarian with only supportive therapy and treatment of secondary bacterial infections to fall back on. Considerable advances have been made in selective chemical manipulation of nucleic acid systems that may lead to effective antiviral agents. The first prototypes that block viral replication are starting to appear, such as amantadine (Oxford, 1982) and rimantadine for prophylaxis of influenza A and methisazone for smallpox. Another development has been that of nucleoside analogues such as ribavirin and idoxuridine. Development of aerosol therapy to distribute appropriate agents deep into the respiratory tree is being investigated as a possible way to enchance efficacy against respiratory infections (Knight et al., 1981). Acyclovir or acycloguanosine shows selective inhibition of DNA polymerase of some viruses, especially herpesviruses. Deeper understanding of the molecular biology of viral attack on the host cells and of the latter's responses may be a prerequisite to mastery of the consequences of virus infections. The complexity of the host's interferon system of anti-viral proteins alone has proven to be much greater than at first expected. However the application of biotechnology to this complex field has greatly increased the momentum of the effort to attain immunological and pharmaceutical solutions to some of the outstanding unresolved disease problems of ruminants.

Faster progress can be expected in the application of chemical and biological ingenuity to the development of safer and more effective vaccines against microbial infections of livestock using recombinant DNA technology (Bachrach, 1982). The first of these to be attempted was foot and mouth disease virus. The viral protein VP_3 has been prepared by genetic manipulation of Eschericia coli plasmids. Smaller peptide fragments of this protein have been synthesized chemically (Bittle et al., 1982) and shown to have immunogenicity.

Antifungals

Dermatomycosis is a major disease of cattle having widespread distribution with very high morbidity albeit with low mortality. Repeated treatment with a wide range of empirical keratolytic and disinfectant-type products applied topically after scrubbing off the crusts used to be the rule. More specific effects were obtained subsequently with fatty acids and their salts and more selective antifungal disinfectants such as captan and tolnaftate. The antibiotic, griseofulvin, from Penicillium griseofulvin was the first effective oral antifungal agent against dermatophytes. However it is expensive and not cleared for use of food animals. Pimaricin (natamycin) sprays have been used to sterilize lesions and arrest the spread of infection. The newer imidazole-based antifungals such as imazalil (Desplenter, 1980; de Keyser, 1981), clotrimazole, econazole and miconazole show promise for topical administration. The development of ketoconazole which can be used for both dermal and systemic mycoses merits further investigation (Niemegeers et al., 1981). Inhibition of synthesis of ergosterol and other structural elements required for fungal cell wall integrity may be one mechanism of action of these imadazole derivatives. Very little information is available on the treatment of gastro-intestinal and systemic fungal infections in ruminants but these new agents that can be used orally or parenterally warrant further study.

Antinematodal Drugs

Helminthic parasitic diseases have been more amenable to efficacy studies in chemotherapy than the microbial diseases because most of the adult forms are readily visible. However, after it was realized that controlling early larval forms is essential if the detrimental effects of helminths are to be avoided, more sophisticated studies became necessary. Current research is being directed to this end as well as to the breadth of the anthelmintic spectrum. Manipulation of distribution kinetics by modifications of formulations has shown promising results, particularly the use of slow-release forms to minimize the risk of reinfestation. Correct timing of dosing to meet the seasonal changes in challenges from the environment is also a critical factor in successful anthelmintic chemoprophylaxis.

No aspect of veterinary pharmacology has led to such a diverse variety of chemical forms as the search for selectively toxic agents against helminth parasites. The early work with phenothiazine was followed by the development of methyridine and selected organophosphorus inhibitors of cholinesterase such as haloxon, coumaphos, trichlorphon and naphthalophos. Several new classes of compounds have been found to have selective action against nematodes (Armour and Bogan, 1982).

The tetrahydropyrimidines, pyrantel and morantel, stimulate cholinergic sites on nematode ganglionic and muscel cells causing depolarization. Binding morantel tartrate to ion-exchange resin slows release and improves safety. Another very interesting group consists of the imidazothiazoles, the most important members being tetramisole and its levo-isomer, levamisole, for ruminants. These compounds also kill worms by paralysis, this being due to prolonged ganglionic stimulation. they are effective against both gastro-intestinal and pulmonary nematodes and may be given orally or parenterally. Levamisole was also found to have the interesting property of modulating the responsiveness of the immune system. This finding has led to the creation of a new class of "immunomodulator" drugs.

Benzimidazoles: Benzimidazoles emerged in the 1960's and 1970's as very safe and effective broad-spectrum anti-nematodal agents (Prichard, 1978). Since thiabendazole was launched in 1961, several congeners have been introduced (Figure 4). Despite low aqueous solubility in most cases, the benzimidazoles tend to be quite readily absorbed, uniformly distributed throughout the body, metabolized in the liver and excreted via the kidney. Their toxicity for the host is remarkably low except that several of the derivatives show embryotoxicity or teratogenic activity and should be avoided in early pregnancy. They have a remarkably broad spectrum of activity against lungworms, tapeworms and liver fluke although they may not be the drugs of choice. The presence of a side chain at the 5-position on the benzene ring slows metabolism and prolongs the presence of the drug at effective levels in the plasma, while deferring the time of the peak concentration. The effect on the worms appears to be attributable to inhibition of fumarate reductase leading to impaired anaerobic energy metabolism that is essential to survival at some stages of development. Thiophanate undergoes cyclization to yield benzimidazole carbamate which is believed to be the active form so that it can be classed in this group.

Figure 4: Benzimidazole anthelmintics.

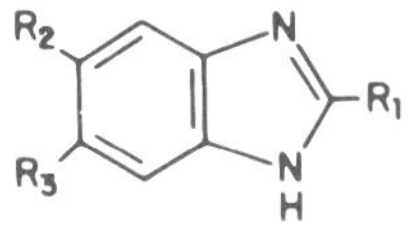

Substituted Benzimidazoles

	R_1	R_2	R_3
Thia-bendazole		H	H
Par- "	$-NH\text{-}CO\text{-}OCH_3$	H	$-C_4H_9$
Cam- "	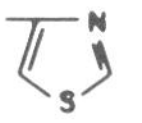	$-NH\text{-}CO\text{-}O\text{-}CH{<}^{CH_3}_{CH_3}$	H
Me- "	$-NH\text{-}CO\text{-}OCH_3$	H	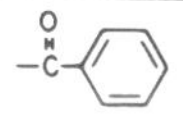
Oxi- "	$-NH\text{-}CO\text{-}OCH_3$	$-OCH_2\text{-}CH_2\text{-}CH_3$	H
Fen- "	$-NH\text{-}CO\text{-}OCH_3$	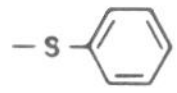	H
Al- "	$-NH\text{-}CO\text{-}OCH_3$	$-S\text{-}CH_2\text{-}CH_2\text{-}CH_3$	H
Oxfendazole	$-NH\text{-}CO\text{-}OCH_3$	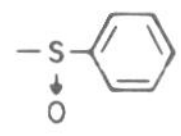	H

Avermectins: Avermectins are macrocyclic lactones formed by Streptomyces avermitilis that have remarkable activity against parasitic nematodes, arthropods, and selected protozoa (Campbell, 1981). The active ingredient of the commercial product is the chemically modified 22,23-dihydroavermectin B_{1a} or "ivermectin" which has a secondary butyl group attached to the 25-carbon of the complex structural nucleus. It is administered subcutaneously, although it is also effective orally (Armour and others, 1982). It has been proposed that this compound has the unique property of promoting the release of gamma-aminobutyric acid (GABA), the inhibitory transmitter of interneurons in the nematode parasite's nervous system, and enhancing its binding at post-ganglionic receptors. This allows an influx of chloride ions causing paralysis of electrochemical excitability and failure to respond to neural stimuli. In the case of arthropods, the action occurs at the inhibitory neuro-muscular junction instead of a ganglion but the end result is similar, failure of the muscle to respond to neural stimuli.

Lack of efficacy against trematodes and cestodes may be explained by their lack of a GABA inhibitory system. The drug is not active against adult heartworms in dogs which may be a fortunate circumstance although it is effective in blocking the development of immature forms. These new agents show great promise as they have efficacy against lungworms, against gastrointestinal nematodes already resistant to established anthelmintics, and versus inhibited fourth stage larvae. The completely new property of efficacy against ectoparasites has considerable potential. The name, distilled from anti-vermes and anti-ectoparasites, belies this new dimension.

Antitrematodal Medication: Fasciola hepatica causes major economic losses in sheep and cattle. Hepatic damage is most acute when the immature flukes migrate through the liver. After 2 months they move to the bile ducts where they attain maturity and cause chronic inflammation. Many drugs have been discovered that destroy adult flukes but even these may become inaccessible due to proliferation of connective tissue, in heavy infections. However, few agents are really effective against the very young migrating stages. Carbon tetrachloride and hexachloroethane have been used for a long time in many sheep raising countries, although they are toxic and are primarily effective against adult forms. Hexachloroparaxylene has similar efficacy and is less toxic. A wide range of bisphenolic compounds have been developed that are active against flukes. The first of these, hexachlorophene, is similar in action to the simpler hydrocarbons and very effective against adults. Bithionol sulfoxide has a similar action and has been used in conjunction with hexachlorophene. A dinitro compound related to hexachlorophene, niclofolan, has greater efficacy against immature forms and is made in oral and injectable formulations. Nitroxynil or 4-hydroxy-3-iodo-5-nitro-benzonitrile is not a biphenyl but is also used by subcutaneous injection and has greater although not complete efficacy against immature flukes at well tolerated dose levels. A major class of drugs that has been found to have trematocidal activity is the halogenated salicylanilides. This includes the bromsalans (di- and tri-bromsalans), brotianide, oxyclozanide, closantel (de Keyser, 1982) and rafoxanide. These agents have broader therapeutic ratios and greater activity against immature flukes. A related iodinated compound, clioxanide, may be less satisfactory against larval forms. The benzimidazole compound, albendazole, has some efficacy against liver flukes and has been found to give protection when used prophylactically. A newer benzimidazole, fasinex, is reported to have high activity against both the parenchymal and bile duct stages of the parasite. A very interesting trematocidal drug is diamfenetidine which is by far the most active agent to date against immature flukes are being metabolized in the liver to an amine form. It holds promise of being an effective prophylactic against fluke infection (Baldwin, 1982).

Anticestodal Drugs: Since the early day of compounds containing copper, arsenic, or nicotine several milder and more effective cestocides have been developed against the moniezia and thysanosoma tapeworms of ruminants. Several benzimidazoles give good control of cestodes. The substituted salicylanilide, niclosamide, is very effective, disintegrating the tapeworm thereby destroying the evidence of its efficacy. Resorantel is a new hydroxybenzanilide cestocide that also removes the rumen fluks or paramphistomes. It is 4'-bromo- -resorcylanilide. Bithionol is used against cestodes and rumen flukes. A less toxic form, the sulfoxide, has been developed; it also has greater trematocidal activity. The isoquinoline derivative, praziquantel, is efficacious against the cestodes of ruminants by either the oral or subcataneous route.

Protozoal Diseases: Ruminant species are susceptible to several major blood-borne protozoal diseases including trypanosomiasis, theileriosis, babesiosis and anaplasmosis. This is a very complex field as several species of each have been described. Enteric coccidiosis can also become a significant problem particularly in intensive systems of management. The complex urea derivatives, amicarbilide isethionate and quinuronium sulfate, are partially effective against various species of Babesia. Transient systemic toxic reactions and irritation at the injection site are commonly observed, while a carrier state may persist. The introduction of imidocarb dipropionate has allowed complete elimination of the parasite (McHardy, 1982). However it must survive questions about potential toxic hazards. This compound is also active against anaplasma but lacks prophylactic effect. Gloxazone or hetoxal is another new product being tested in anaplasmosis. New long-acting dosage forms for tetracycline antibiotics may obviate the need for the continuous medication with these agents previously resorted to in this condition (Roby et al., 1978; Magonigle et al., 1978).

The dreaded killing disease of cattle, theileriosis or East Coast Fever, eluded the quest for therapeutic agents for a long time. Tetracyclines given early in the course of the disease appeared to be beneficial. Recently, progress has been made with the finding that the quinazolinone coccidiostat, halofuginone, has efficacy given promising results against T. parva. Two related compounds have been reported to be effective - menoctone or 2-hydroxy-3-(8-cyclohexyloctyl)-1,4-naphthoquinone (McHardy and Rae, 1981) and parvaquone or 2-hydroxy-3-cyclohexyl-1,4-naphthoquinone (Morgan and McHardy, 1982).

Trypanosomiasis renders large regions of the planet unfit for livestock production. Despite massive efforts to find selectively toxic agents against trypanosomes, only modest success has been attained. The main groups of drugs that have been developed include: the diamidines, of which diaminazene aceturate is the major one; the quinapyramines, notably quinapyramine sulfate (therapeutic) and chloride (prophylactic), and the aminophenanthridium compounds, such as homidium bromide or chloride, prothidium and isometamidium. The complex high molecular weight compound, suramin, has a narrow margin of safety and is long acting.

Although pharmaceutical chemists have developed considerable fire-power against hemoprotozoal infections, the goals of sure prophylaxis and truly selective toxicity are still far off. Resistance develops to drugs, reducing their efficacy. The possible hazard of drug residues in products for human use is another concern. A careful strategy is required to achieve immunity and host resistance while avoiding disease. Some drugs are "too effective" in that their biocidal action does not allow adequate immunization to occur and an animal may be vulnerable to reinfection after treatment. Biotechnology is being deployed to produce biologics that may be more effective in generating immunity without disease. Since these diseases involve invertebrate vectors in their transmission, control of the intermediate hosts is a strategic possibility. The development of chemicals to control ticks and biting flies is another field to which chemical and biological ingenuity has been applied. Regular spraying or dipping is one approach. Risks to the environment are substantial if major vector eradication programs are undertaken.

Intestinal coccidiosis due to Eimeria species can cause heavy losses in intensively managed ruminants (Radostits and Stockdale, 1980). Sulfamethazine has been used effectively to control bovine coccidiosis but efficacy may be even better if it is supplemented with chlortetracycline. The anti-thiamine coccidiostate, amprolium, is also effective. The ionophore antibiotic, monensin, may be particularly valuable as a preventative in exposed animals.

Ectoparasiticides (Insecticides and acaricides): External parasites can cause heavy losses in livestock, either directly (e.g. sheep scab; cattle grub; screwworm) or indirectly as vectors of diseases (e.g. tsetse flies for trypanosomiasis; ticks for theileriosis, babesiosis etc.). A vast array of compounds has been developed and tested for control of such pests. The older persistent organochlorine (OC) insecticides (DDT, dieldrin, benzene hexachloride, toxaphene, methoxychlor, lindane etc.) have largely been displaced because of environmental hazards. Their replacements have been the more toxic but less persistent organophosphorus esters (OPs) and carbamates. The insect growth regulator, methoprene, is used to control insects that breed in feces. Improved formulations of pyrethrins are still in use for knock down of stable flies, face flies and horn flies. A synthetic pyrethroid, permethrin, with similar properties has been developed (Bailie and Morgan, 1980; Morgan and Bailie, 1980). Fly treatments can be applied as sprays, dips, via self-treatment devices or impregnated ear tags. Stirofos and methoprene feed additives prevent fly development in manure. A variety of OP and OC insecticides have been used in controlling lice. The acaricides, diazinon and propetamphos (Kirkwood and Quick, 1982), can control sheep scab, as can gammahexachlorocyclohexane (Benzene hexachloride or BHC). A variety of OP, OC and carbamate compounds have been applied against ticks by strategic dipping but ticks are notorious for developing resistance by increased ability to detoxify the OPs. Strains of cattle tick resistant to older OPs may be susceptible to the newer acaricidal products such as promacyl (a carbamate), chlormethiuron (a thiourea), phosmet (an OP), chlordimeform and amitraz (formamidines).

MANIPULATION OF PHYSIOLOGICAL REGULATORY PROCESSES

Stress: Cannon focussed on the sympathetic system in conceptualizing an "Emergency Syndrome" (1932). Selye was struck by the similarity of the clinical signs he observed as complications following a variety of primary maladies. These signs included gastrointestinal dysfunction, loss of body weight, and mental depression. He developed a new concept of a "general adaptation syndrome" or nonspecific "stress syndrome", involving the three phases of (1) an alarm reaction, (2) heightened resistance and (3) the alternative conclusions of remission or exhaustion. The latter stage involved atrophic degenerative changes in the thymus gland, the lymphoid tissues, and the stomach along with congestion of the adrenal cortex (Selye, 1936). Selye recognized that the body's reactive processes could have harmful effects and that these could be partially offset by the corticoid hormones secreted by the adrenal cortex in response to increased production of adrenocorticotrophic hormone from the anterior pituitary gland (1950). Selye also recognized that life involves stress and that survival required either adequate adaptive responses to stress or avoidance of excessive levels of it (1970), identifying a new class of catatoxic steroids that induce reactions which inactivate toxic substances.

Early workers studied the significance of environmental stresses on livestock, particularly the consequences of excessive rates of gain or loss of heat. Simulated unfavorable climatic conditions in environmental chambers were used to develop the desired levels of thermal stress. Environmental and physiological interactions influencing resistance to infectious disease were reviewed by Webster (1969). Tropically adapted animals have a critical temperature of between +20°C and +30°C, below which the thermal stress is too severe and metabolic rate increases rapidly (Scholander et al., 1950). Polar-adapted mammals, on the other hand, have critical temperatures ranging down to -40°C (Arctic Fox) partly because of superior insulation and adaptive physiological and behavioral mechanisms. More recent reviews have covered the impact of heat stress (Fuquay, 1981) and cold stress (Young, 1981) on animal production as well as the ways such stresses can be reduced by modifying housing and management (Hahn, 1981; Stott, 1981).

Stephens (1980) defined stressors as stimuli perceived by an animal to be at a rate significantly deviant from normal or to be unusually prolonged or intense. this definition should be enlarged to encompass nutritional stresses which do not fit readily under the heading stimuli. His review draws on the great achievements of metabolic, physiological and behavioral research that have helped to characterize the phenomenon of stress. The advances in knowledge of the interactions between neurological and endocrine systems in response to stress are particularly important. Starting with reflex autonomic responses releasing catecholamines from the adrenal medulla and progresing via release of ACTH from the anterior pituitary evoking release of corticoids from the adrenal cortex, the physiological responses to stress are complex. The border area between normality and reduced productivity or feed efficiency is one field where extra research is required. Another is the "breaking point" phenomenon when excessive stress leads to definite abnormality or "disease".

The Bruce County Beef Project in Ontario has illustrated how little is really understood in this field. Discriminant analysis was applied to identify variables that were best correlated with differences in mortality due to respiratory disease in groups of cattle introduced to the feedlot. The three critical factors that correlated with the higher risk of mortality were feeding corn silage early in the feeding period, mixing and shipping of cattle from different sites of origin, and the use of respiratory vaccines shortly after arrival (Martin et al., 1980 a,b).

Ruminant Growth Promotants: Oral antibiotics in young lambs and calves led to improved weight gain and feed efficiency (Pritchard et al., 1955). After the animals become dependent on rumen microbial fermentation the antibiotic effect is reduced and may become detrimental under some conditions. Concerns about the possibility that such usage leads to spread of plasmid-mediated resistant strains of pathogenic bacteria have led to constraints on their use.

Another approach to the manipulation of growth has been via chemicals with hormonal effects stimulating weight gain. This anabolic stimulus was achieved with stilbene compounds notably diethylstibestrol (DES) or hexestrol. These agents appear to act by increasing growth hormone secretion (Davis et al., 1978). However the risk of carcinogenesis has been deemed unacceptable and these implants have been disallowed in the U.S.A. Subsequently the use of anbolic steriod hormones (estradiol, testosterone) or related compounds (trenbolone acetate) has increased.

The natural hormones appear to be more acceptable and attention has been focussed on developing an implant capable of sustained delivery of predictable doses of the hormone. One example is the use of a silicone rubber implant to deliver estradiol-17 β. Estradiol and trenbolone implants have been found to reduce thyroxine levels in sheep plasma, leading to a suggestion that the resulting hypothyroid status might increase feed efficiency and growth rate (Donaldson et al., 1981). Other theories of how estrogen type growth promotants may act involve increases in secretion of growth hormone, prolactin and insulin. The compound, zeranol (a resorcylic acid lactone), may have a similar action stimulating the pituitary gland. Recent isolation and synthesis of growth hormone releasing hormone that is found in the hypothalamus will no doubt lead to studies of its value as a growth promotant if its effects are specific. Similarly immunological inhibition of the growth arresting hormone, somatostatin, is being investigated. Adrogenic steroids act directly on skeletal muscle by either increasing protein synthesis or blocking glucocorticoid promotion of protein degradation.

The development of new polyether antibiotics as feed additives for ruminants has led to their widespread use. Monensin, a product of *Streptomyces cinnamonensis*, forms ionophore complexes with monovalent cations that may change membrane permeability to ions. It has selective inhibitory activity against rumen protozoa and some strains of rumen bacteria leading to a modification of the pattern of rumen fermentation. The net effect is an increase in propionate production and a decrease in methanogenesis. Lasalocid is a related compound that has a lower toxicity. Avoparcin and salinomycin have similar effects.

Prevention of Rumen Dysfunctions: Polyether antibiotics may be of value in preventing an important disease of ruminants that occurs on diets containing readily-fermentable carbohydrates (Dennis et al., 1981). This is known as D-lactic acidosis because of the development of severe lactic acidosis involving D-lactate accumulation (Dunlop, 1972). Lasalocid and thiopeptin are also effective in suppressing the lactic acid fermentation in the rumen (Nagaraja et al., 1982). Similarly, monensin may be effective in preventing the 3-methylindole (3MI) formation in the rumen by lactobacilli. 3-MI is believed to be responsible for the pulmonary lesions in fog fever (Carlson et al., 1982).

Polioencephalomalacia (PEM) or cerebrocortical necrosis is a unique disease of ruminants, characterized by neurological signs and massive cerebral necrosis. It occurs as a result of the appearance of thiaminase I in the rumen contents. Under certain, but as yet unspecified, dietary conditions there is sufficient enzyme present to destroy all the ingested or synthesized thiamine in the rumen. This transferase also leads to the formation of a series of new pyrimidine derivatives, some of which may have antithiamine activity. One of these, delta-pyrrolinium, has been isolated from cases of PEM but its role in the disease may not be essential (Edwin et al., 1982). Use of a purified thiamine-free milk substitute diet in newborn lambs led to reproduction of the disease within 5-6 weeks (Thornber et al., 1979). In this model it was shown that the neurological signs appears after brain thiamine and thiamine diphosphate (TDP) levels fell below 20% of control levels (Thornber et al., 1980). Unique changes appear on the electroencephalogram in the terminal stages (Dunlop et al., 1981) and these can be reversed by early parenteral thiamine administration. Prevention of further losses during an outbreak can be achieved by a large subcutaneous dose of thiamine every three weeks.

Metabolic Diseases: Failure to adapt to physiological stresses relating to nutrition, metabolism, production or the physical and biological environment leads to a variety of specific syndromes categorized as metabolic diseases. Examples include milk fever (hypocalcemia), grass tetany (hypomagnesemia), ketosis (acetonemia), bloat, various trace mineral deficiencies (e.g. selenium, copper, cobalt) or excesses (e.g. selenium, copper), fatty liver syndrome, displaced abomasum, and locomotor disturbances in growing animals.

As knowledge of the regulation of calcium metabolism has progressed, strategies for minimizing the risk of milk fever or parturient hypocalcemia have been developed. Probably the most effective is to utilize metabolites or analogues of vitamin D during the week preceding calving. Both the natural metabolite, 25-OH-vitamin D_3 in a few mg per day or a single oral dose (Hove and Kristianson, 1982) and the synthetic analogue 1-α- hydroxy vitamin D_3 are effective if the time of parturition can be predicted. The latter compound, in a single dose of 500 mg some time in the last week of gestation will prevent up to 75% of cases. The use of glucocorticoids to treat ketosis by promoting gluconeogenesis is well established as a back up to intravenous glucose.

New techniques to prevent nutritional deficiencies and bloat involve the development of continuous release devices or materials for oral administration to be repository in the rumen. Special devices for delivery of anti-bloat detergents (poloxalene or pluronics), magnesium salts or iodine have been invented to prevent legume bloat, hypomagnesemia and I deficiency respectively. Copper oxide needles for slow release of copper and dense "bullets" for prolonged release of magnesium are available. Controlled release glass implants are under development for continuous supply of trace minerals such as copper, cobalt and selenium (Allen et al., 1982).

Prostaglandins in Reproduction: Mastery of the reproductive process in livestock can have enormous economic implications. Consequently a major effort has been invested in physiological and pharmacological research of reproductive regulatory systems. Great progress was made after the discovery of the gonadal steroid hormones, the pituitary gonadotrophins that promote their synthesis, and of the hypothalamic releasing factors that stimulate the anterior pituitary gland. However, a greater stimulus to pharamaceutical manipulation of reproduction has been the discovery of the prostaglandins and their subsequent chemical modification (Figure 5).

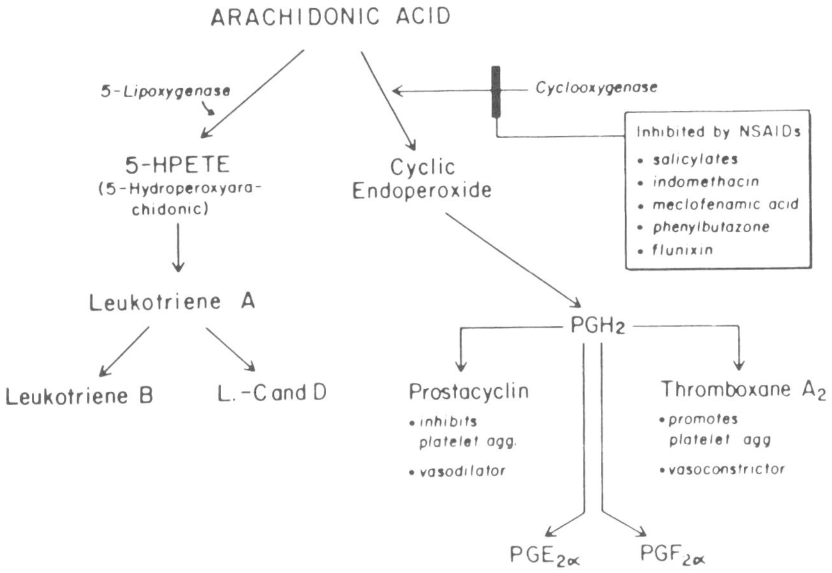

Figure 5a: Synthesis of some prostaglandins and leukotrienes (SRS-A), indicating site of action of non-steriod anti-inflammatory drugs (NSAID).

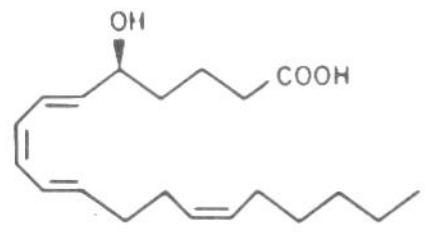

Leukotriene B

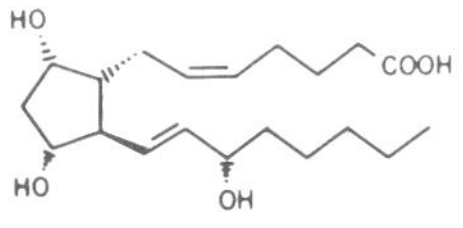

Prostaglandin $F_{2\alpha}$

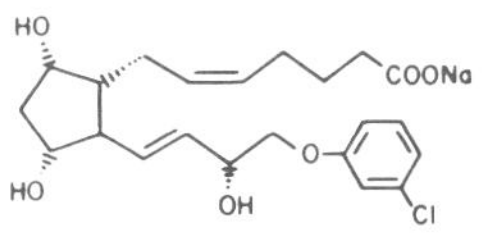

Cloprostenol Sodium

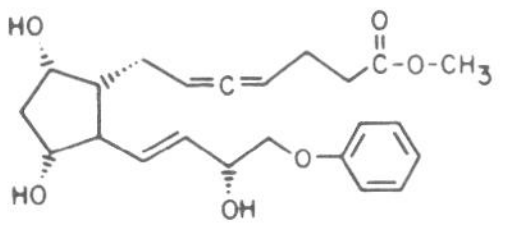

Fenprostalene

Figure 5b: The structures of two chemically modified forms used to manipulate bovine reproduction.

The natural uterine luteolytic factor, prostaglandin $F_{2\alpha}$, leads to the demise of the corpus luteum and termination of its endocrine secretion of progesterone. Thus this substance and its chemical analogues can be used to determine the timing of estrus, thereby allowing a group of females to be synchronized to come into estrus with a period of a couple of days. This allows much more efficient use of artificial insemination and embryo transfer procedures. The prostaglandins are useful in treating several reproductive problems including hydrops allantois (Harker, 1981), undetected estrus, some types of ovarian cysts, pyometritis, mummified fetus, and unwanted pregnancy (Seguin, 1980). They can be used to induce parturition (Johnson and Jackson, 1982), but there may be a high incidence of dystocia and retained placenta. Examples of prostaglandins modified chemically for greater efficacy are cloprostenol and fenprostalene. The latter has a prolonged action because it is resistant to attack by the 13,14-reductase. An implant involving an osmotic minipump has been developed to control dosage (Herschler and Reid, 1982).

Modification of Inflammatory Responses: A review of the psychological and behavioral aspects of stress has not been attempted. Studies of sedatives and tranquilizers to control stress have found more application in poultry and swine that in ruminants (Dantzer, 1977). A study of the neuroendocrine and metabolic changes in lambs during round-up, shipping, washing, yarding and slaughter following electrical stunning (Kilgour, 1978) showed a massive adrenergic response at slaughter and various changes along the way. The significance of such findings for animal welfare and predisposition to disease remains controversial.

Inflammation and hypersensitivity are two major areas of investigation where much has been learned but the puzzle is not yet fully assembled or solved. Various components can be identified including the metabolic and vascular changes of fever. Underlying factors include pyrogens. The responses involve autonomic cholinergic and adrenergic mediators. These can now be blocked pharmacologically. More recently the roles of the vasodilator, histamine, and the vasoconstrictor, serotonin, and other local factors have been reexamined. Anaphylactic reactions to drugs occur and can lead to lethal shock. Recommendations for emergency treatment include intramuscular epinephrine followed by intravenous H_1-antihistamine (Editorial, 1981).

New blocking agents against proposed H_2 and S_2 receptors have been developed. The former (e.g. cimetidine) provide protection against gastro-dudenal ulcers while the latter (e.g. ketanserin) relieve hypertension. The S_2-blockers may find application in therapy of conditions in ruminants that involve vascular and edematous changes such as laminites and wound healing by opposing venoconstriction (Ooms and Awouters, 1982).

Some prostaglandins of the E series are potent vasodilators and often act in concert with bradykinin leading to pain. Unstable intermediates of prostaglandin synthesis have potent transient actions (Figure 5). These include thromboxane A and prostacyclin which inhibits platelet aggregation (Vane et al., 1982). The leukotrienes are related compounds that make up the slow-reacting substance of anaphylaxis (SRS-A). They are spasmogens for airway smooth muscle, are chemotactic for leukocytes and cause exudation of plasma.

Currently many potent derivatives of the adrenal corticosteroid nucleus are marketed for use in inflammatory conditions and metabolic disorders but they lead to suppression of natural adrenal corticoid secretion (Toutain et al., 1982). The rationale for their use is not clearly established and firm evidence of efficacy in infectious states is lacking even though they may provide a degree of symptomatic relief (Eyre, 1980). A similar situation prevails with respect to non-steroid anti-inflammatory agents (Figure 5). Until the interactions and time courses of all mediators and cellular changes can be defined and correlated with biological events, it is unlikely that empirical use of anti-inflammatory drugs will be successful. They may impede the body's autoregulatory attempts to restore homeostasis.

REFERENCES

Allen, W.M., B.F. Sansom, C.F. Drake P.R. and Moore (1982). Recent developments in the treatment of metabolic diseases. Pharmacologie et Toxicologie Verterinaries, INRA Publ., Paris. Colloques de l'I.N.R.A., 8, 287-288.

Armour, J., K. Bairden, and T.M. Preston (1982). Anthelmintic efficacy of invermectin against naturally occurring gastrointesinal nematodes of sheep. Vet. Rec., 111, 80-81.

Armour, J. and J. Bogan (1982). Anthelmintics for ruminants. Brit. Vet. J., 138, 371-382.

Bachrach, J.L. (1982). Recombinant DNA technology for the preparation of subunit vaccines. J. Amer. Vet. Med. Assn., 18, 992-999.

Bailie, H.D. and D.W.T. Morgan (1980). Field trials to assess the efficacy of permethrin for the control of flies on cattle. Vet. Rec., 106, 124-127.

Baldwin, T.M.A. (1982). The effects of diamphenethide on the morphology of Fasciola hepatica. Parasitol, 84, 1. (Abstract)

Bittle, J.L., R.A. Houghten, H. Alexander, T.M. Shinnick, J.G. Sucliffe, R.A. Lerner, D.J. Rowlands and F. Brown (1982). Protection against foot-and-mouth disease by immunization with a chemically synthesized peptide predicted from the viral nucleotide sequence. Nature, 298, 30-33.

Brander, G.C., J.H. Watkins and R.P. Gard (1975). A large scale mastitis control programme in Somerset. Vet. Rec., 97, 300-304.

Campbell, W.C. (1981). An introduction to the Avermectins. N.Z. Vet. J., 29, 174-178.

Cannon, W.B. (1932). The wisdom of the body. New York: W.W. Norton Co.

Carlson, J.R., M.R. Nocerini, R.G. Breeze, B.M. Olcott and W.W. Laegreid (1982). The role of metabolism in the pathogenesis of 3-methylindole-induced pneumotoxicity in ruminants. Proc. XII World Congr. Dis. Cattle. Amsterdam: International Congrescentrum RAI, Vol. 1, pp. 27-34.

Dantzer, R. (1977). New aspects of the use of tranquillizers in animal husbandry, with particular reference to pigs. Vet. Sci. Comm., 1, 161-169.

de Keyser, H. (1981). Activity of repeated administration of enilconazole (R23979) against Trichophyton verrucosum in cattle: a clinical trial. (in Dutch) Tijdschr. f. Diergeneesk., 106, 799-805.

de Keyser, H. (1982). The endo and ectoparasitic use of closantel in cattle. Proc. XII World Congress on Diseases of Cattle. Amsterdam: RAI, vol. II, 1159-1162.

Dennis, S.M., T.G. Nagaraja and E.E. Bartley (1981). Effects of lasalocid or monensin on lactate-producing or -using rumen bacteria. J. Anim. Sci., 52, 418-426.

Desplenter, L. (1980). New possibilities in the treatment of dermatomycoses and of Aspergillus fumigatus. In: Trends in Veterinary Pharmacology and Toxicology. A.S.J.P.A.M. van Miert, J. Frens and F.W. van der Kreek, editors. Amsterdam: Elsevier, pp. 266-274.

Donaldson, I.A., I.C. Hart and R.J. Heitzman (1981). Growth hormone, insulin, prolactin and total thyroxine in the plasma of sheep implanted with the anabolic steriod trenbolone acetate alone or with estradiol. Res. Vet. Sci., 30, 7-13.

Dunlop, R.H. (1972). Pathogenesis of ruminant lactice acidosis. Adv. Vet. Sci., 16, 259-302.

Dunlop, R.H., L. Bueno and Y. Ruckebusch (1981). Slow wave spindle bursts in bovine polioencephalomalacia. In: Metabolic Disorders in Farm Animals. Gieseck, D., Dirksen, G. and Stangassinger, M., Edistors. Munchen: Tierarztliche Facultat, pp. 16-19.

Editorial (1981). Treatment of anaphylactic shock. Brit. Med. J., 282, 1011-1012.

Edwin E.E., L.M. Markson and R. Jackman (1982). The aetiology of cerbrocortical necrosis: the role of thiamine deficiency and of deltapyrrolinium. Brit. Vet. J., 138, 337-349.

Eyre, P. (1980). Pharmacological aspects of hypersensitivity in domestic animals: a review. Vet. Res. Comm., 4, 83-98.

Fuquay, J.W. (1981). Heat stress as it affects animal production. J. Amin. Sci., 52, 164-174.

Hahn, G.L. (1981). Housing and management to reduce climatic impacts on livestock. J. Anim. Aci., 52, 175-186.

Harker, D.B. (1981). Treatment of a case of bovine hydrops allantois. Brit. Vet. J., 137, 575-577.

Herschler, R.C. and J.F.S. Reid (1982). New prostglandins: present studies and future. Pharmacologie et Toxicologie Veterinaires. INRA Publ., Paris. Les Colloques de l'I.N.R.A., 8, 401-402.

Hove, K. and T. Kristiansen (1982). Prevention of parturient hypocalcemia: effect of a single oral dose of 1,25-dihydroxyvitamin D_3. J. Dairy Sci., 65, 1934-1940.

Johnson, C.T. and P.S. Jackson (1982). Induction of parturition in cattle with cloprostenol. Brit. Vet. J., 138, 212-224.

Kilgour, R. (1978). Minimizing stress on animals during handling. Proc. 1st World Congress on Animal Ethology. Madrid: Editorial Garsi, pp. 303-322.

Kirkwood, A.C. and M.P. Quick (1982). Propetamphos for the control of sheep scab. Vet. Rec., 111, 367.

Knight, V., H.W. McClung, S.Z. Wilson, B.K. Waters, J.M. Quarles, R.W. Cameron, S. Greggs, J.M. Zerwas and R.B. Couch (1981). Ribavrin small particle aerosol treatment of influenza. Lancet ii: 945-949.

Magonigle, R.A., J.E. Simpson and F.W. Frank (1978). Efficacy of a new oxytetracycline formulation against clinical anaplasmosis. Amer. J. Vet. Res. 39, 1407-1410.

Martin, S.W., A.H. Meek, D.G. Davis, R.G. Thomson, J.A. Johnson, A. Lopez, L. Stevens, R.A. Curtis, J.F. Prescott, S. Rosendal, M. Savan, A.J. Zubaidy and M.R. Bolton (1980 a). Factors associated with mortality in feedlot cattle: the Bruce County beef cattle project. Can. J. Comp. Med., 44, 1-10.

Martin. S.W., A.H. Meek, T.E. Feltmate, R.A. Curtis, J.A. Johnson and D.G. Davis (1980b). Factors related to sickness and death in feedlot calves. Proc. 13th Ann. Convention Amer. Assn. Bov. Practitioners, pp. 32-36.

McHardy, N. (1982). The activity of imidocarb against tick-transmitted parasitic infections. Pharmacologie et Toxicologie Veterinaires, INRA Publ., Paris. Les Colloques de l'I.N.R.A., 8, 427-428.

McHardy, N. and D.G. Rae (1981). Treatment of stabilate-induced East Coast Fever with menoctone. Trop. Anim. Hlth. Prod., 13, 227-239.

Morgan, D.W.T. and H.D. Bailie (1980). A field trial to determine the effect of fly control on milk yields in dairy cattle in the U.K. Vet Rec., 106, 121-123.

Morgan, D.W.T. and N. McHardy (1982). Comparison of the antitheilerial effect of Wellcome 993C and halofuginone. Res. Vet. Sci., 32, 84-88.

Nagaraja, T.G., T.B. Avery, E.E. Bartley, S.K. Roof and A.D. Dayton (1982). Effect of lasalocid, monensin or thiopeptin on lactic acidosis in cattle. J. Anim. Sci., 54, 649-658.

Niemegeers, C.J., J.C. Levron, F. Awouters and P.A. Janssen (1981). Inhibition and induction of microsomal enzymes in the rat. A comparative study of four antimycotics: Miconazole, Econazole, Clotrimazole and Ketoconazole. Arch. Int. Pharmacodyn., 251, 26-38.

Ooms, L. and F. Awouters (1982). Serotonin and serotonin 2 antagonists: pharmacology and veterinary applications. Pharmacologie et Toxicologie Veterinaires, INRA Publ., Paris. Les Colloques de l'I.N.R.A., 8, 433-436.

Osborne, A.D., A.H.K. Nazer and C. Shimeld (1978). Treatment of experimental calf salmonellosis with amoxycillin. Vet. Rec., 103, 233-237.

Oxford, J.S. (1982). New biochemical and clinical approaches to the control of epidemic influenza virus. J. Antimicrobial Chemotherapy, 10, 161-163.

Prichard, R.K. (1978). In the epidemiology and control of gastro-intestinal parasites of sheep in Australia. Ed. A.D. Donald, W.H. Southcott and J.K. Dineen. Australia, CSIRO, pp. 75-107.

Pritchard, G.I., J.A. Newlander and W.H. Riddell (1955). Aureomycin effects - growth and digestibility studies with identical twin calves. J. Anim. Sci., 14, 336-339.

Radostits, O.M. and P.H.G. Stockdale (1980). A brief review of bovine coccidiosis in Wastern Canada. Can. Vet. J., 21, 227-230.

Roby, T.O., J.E. Simpson and T.E. Amerault (1978). Elimination of the carrier state of bovine anaplasmosis with a long-acting oxytetracycline. Amer. J. Vet. Res., 39, 1115-1116.

Scholander, P.F., R. Hock, V. Waters, F. Johnson and L. Irving (1950). Heat regulation in some arctic and tropical mammals and birds. Biol. Bull., 99: 237-258.

Sequin, B.E. (1980). Role of prostaglandins in bovine reproduction. J. Amer. Vet. Med. Assn., 176, 1178-1181.

Selye, H. (1936). Thymus and adrenals in the response of the organism to injuries. Brit. J. Exp. Pth., 17, 234-248.

Selye, H. (1950). The physiology and pathology of exposure to stress. Montreal: Acta, Inc.

Selye, H. (1970). Adaptive hormones: retrospect and prospect. Perpect. Biol. Med., 13, 343-363.

Stephens, D.B. (1980). Stress and its measurement in domestic animals: a review of behavioral and physiological studies under field and laboratory conditions. Adv. Vet. Sci., 24, 179-210.

Stott, G.H. (1981). What is animal stress and how is it measured? J. Anim. Sci., 52, 150-153.

Thornber, E.J., R.H. Dunlop, J.M. Gawthrone and C.R. Huxtable (1979). Polioencephalomalacia induced by experimental thiamin deficiency in lambs. Res. Vet. Sci., 26, 378-380.

Thornber, E.J., R.H. Dunlop and J.M. Gawthrone (1980). Thiamin deficiency in the lamb: changes in thiamin phosphate esters in the brain. J. Neurochem., 35, 713-717.

Vane, J.R., S. Bunting and S. Moncada (1982). Prostacyclin in physiology and pathophysiology. Int. Rev. Exp. Path., 23, 161-207.

Toutain, P.L., R.A. Brandon, M. Alvinerie, R. Garcia-Villar and Y. Ruckebusch (1982). Dexamethason in cattle: pharmacokinetics and action on the adrenal gland. J. Vet. Pharmacol. Therap., 5, 33-43.

Webster, A.J.F. (1970). Environmental and physiological interactions influencing resistance to infectious disease. R.H. Dunlop and H.W. Moon, editors: Resistance to Infectious Disease. Saskatoon: Modern Press, pp. 61-84.

Young, B.A. (1981). Cold stress as it affects animal production. J. Anim. Sci., 52, 154-163.

Ziv, G., A. Rosenzuaig, R. Risenberg-Tirer, Y. Danieli and I. Miller (1969). Carbenicillin milk levels following intramammary treatment of bovine pseudomonas mastitis. Refuah Vet., 26, 152-158.

CHEMISTRY AND THE CONTROL OF ANIMAL TRYPANOSOMIASIS AND THEILERIOSIS

A.R. Gray
International Laboratory for Research on Animal Diseases
P.O. Box 30709, Nairobi, Kenya

ABSTRACT

Trypanosomiasis and theileriosis are widespread vector-borne diseases of livestock caused by several species of parasitic protozoa. Control measures for trypanosomiasis include tsetse control, mainly by insecticides, and chemotherapy. Control of theileriosis is based on acaricidal applications to livestock to destroy ticks, vaccination against Theileria annulata and on a limited scale against T.parva, and chemotherapy. The efficacy of existing control measures is variable and the diseases still exact a heavy toll on livestock numbers and productivity. In Africa, better methods for the control of tsetse populations and trypanosomiasis would release an area of approximately 7 million square kilometers for increased food production by mixed agriculture and livestock farming. In Africa and Asia, improved control of trypanosomiasis and theileriosis would greatly reduce livestock losses and allow improved meat and milk production. Chemical and biochemical research could contribute to the achievement of disease control objectives by the development of new classes of biodegradable insecticides and acaricides, less toxic and structurally new therapeutic drugs for treatment of infections, the identification and production of parasite antigens for vaccination and, possibly, the identification and multiplication of disease resistant varieties of livestock.

KEYWORDS: Trypanosomiasis; Theileriosis; Vector-control; Ticks; Tsetse Flies; Disease; Chemotherapy; Vaccination.

INTRODUCTION

The impact of disease in various forms on the health and well being of mankind is a familiar matter of concern to all of us. Infectious diseases of livestock contribute to this problem by their effects on economic and social development, and by limiting food production. Many viral and bacterial diseases of livestock have been brought under control by vaccination, chemotherapy and other forms of preventive medicine, but serious parasitic diseases are still prevalent in many parts of the world. The two disease complexes comprising the topic of this paper, the trypanosomiases and theilerioses, belong to this last category.

Trypanosomiasis is casued by several species of blood-dwelling pathogenic trypanosomes which affect cattle, sheep, goats and pigs, and to some extent, horses and camels. Pathogenic trypanosomes have a world-wide distribution, but they assume major importance in tropical Africa where they are transmitted from wildlife reservoir hosts to domesticated livestock by various species of tsetse flies (Glossinidae). Elsewhere, pathogenic trypanosomes are transmitted from host to host by other species of blood-sucking flies. The pathogenic trypanosomes present in Africa undergo a cycle of development usually lasting 1-3 weeks in the tsetse fly, but direct transmission by other species of fly may also occur. In domestic animals, trypanosomes develop in the blood and tissues, causing anaemia, unthriftiness, reproductive disorders and death.

Theileriosis is also a widespread complex of diseases caused by protozoan parasites. The most important species are Theileria annulata, which affects cattle in wide areas of North Africa and Asia, and Theileria pava, also in cattle, which is confined to Central and East Africa. Other less pathogenic species of Theileria occur in the same regions and also in more temperate areas. All Theileria species are transmitted by hard ticks (Ixodidae) in which they undergo cyclical development culminating in the production of infective sporozoites. Cattle are the main victims of theileriosis, but sheep and goats are affected in certain areas. In these hosts, the parasites invade the lymphocytes and erythrocytes where they develop as schizonts and piroplasms. The form the disease takes varies from chronic anaemia, unthriftiness, reduced productivity and death in the case of T.annulata infection, to an acute, lymphoproliferative syndrome and death, or greatly reduced productivity, in the case of T.parva.

CURRENT DISEASE CONTROL PRACTICES

The methods currently available for the control of trypanosomiasis include vector control, to break the disease transmission cycle, and chemotherapy to deal with the parasites in infected hosts. In tropical Africa, a variety of methods have been applied to reduce tsetse flies. These include vegetation clearance, human settlement and agricultural development to create an adverse environment for the flies, and game destruction to reduce their food sources and the size of the natural reservoir of infection (Jordan, 1974). Various forms of tsetse trap and the sterile insect technique have also been used, but often only on an experimental basis (Jordan, 1978).

Reduction of tsetse flies can also be achieved by the use of insecticides. Large areas have been freed from tsetse populations by the careful application of residual insecticides, such as dieldrin and to vegetation providing the habitat. In recent years repeated aerial applications of endosulphan have been used to control tsetse populations with less general adverse ecological impact. Trials have been conducted with new synthetic pyrethroid insecticides and new methods of formulation which may provide tsetse control in response to very low levels of toxic compound.

Vector control has been less effective for the control of infections with T.evansi which is transmitted in extensive areas by a variety of arthropod species. Trypanocidal drugs have been in use for many years to treat trypanosome infections, thus reducing the reservoir of infection in domestic livestock and losses from the diseases. Suramin has been widely used since 1925 to treat infections with T.evansi and other brucei group trypanosomes, while quinapyramine, homidium, diminazene and iosmetamidium salts have been successively introduced for the treatment of infections with T.brucei, T.vivax and T.congolense at various times during the past 35 years. With certain notable exceptions, chemoprophylaxis, as opposed to treatment, has been limited in effect. Failures in this area have been attributed to the rapid development of drug resistance in trypanosomes, logistical difficulties in maintaining regular drug treatment schedules and, occasionally, excessively high trypanosomal challenges.

In the case of theileriosis, vector control varies in importance in relation to the prevailing form of the disease. In general, control of ticks in the environment is difficult, but control of animal movement and pasture burning in the case of Rhipicephalus appendiculatus, the vector of T.parva, and heat sterilisation and acaricidal treatment of animal barns to reduce Hyalomma spp. populations carrying T.annulata, can each be used occasionally. More usually, tick control is achieved by treating infected cattle with acaricide applied in dips or spray races. Regular once or twice weekly acaricide applications are often necessary to achieve effective tick control and the method is demanding in terms of capital investment, acaricide costs and supervision. A succession of acaricides has been introduced commercially over the years including arsenical, organochlorine and organophosphate compounds. Many compounds have been discarded because of the development of resistant populations of ticks, mainly species of Boophilus, and there is concern about development of further acaricide-resistant tick strains. Interest is currently focussed on cyclic amidines and synthetic pyrethroids for future use.

Two drugs have recently become available for the treatment of clinical theileriosis. Halofuginone has been found to have a substantial therapeutic effect on both T.annulata and T.parva and should be beneficial in closely supervised situations. The identification of theilericidal activity in menoctone by McHardy et al (1976) led to the development of experimental analogues, one of which, identified as 993C (McHardy et al, 1980), should soon be marketed for the treatment of T.parva infections in Africa under the name Parvaquone. These drugs appear to be active predominantly against schizont forms of the parasites and provide a useful addition to available disease control measures (Dolan, 1981). Vaccination procedures against both T.annulata and T.parva are also possible, but they are usually practised on a relatively small scale or experimentally. Vaccines prepared from the schizont stage of T.annulata have been used with considerable success to immunize cattle, notably in Israel and Iran (Pipano, 1977). Vaccines could probably be used more widely if the necessary technical knowledge and expertise was extended. Immunization against T.parva infection can be achieved by initiating infections with sporozoites and concurrent treatment with oxytetracycline (Cunningham, 1977). This method is subject to considerable limitations at present, notably the need for sporozoite production, the occurrence of antigenically different parasite stocks which make up the T.parva disease complex, and the insensitivity of certain T.parva stocks to treatment with oxytetracycline. Its use is therefore mainly experimental.

OPPORTUNITIES FOR INCREASED FOOD PRODUCTION

In considering the opportunities for increasing food production by better control of trypanosomiasis and theileriosis, one situation stands out. Approximately one third of the land mass of the African continent is infested to variable degrees by one or more species of the tsetse fly. These flies, carrying pathogenic trypanosomes, occupy extensive tracts of light woodland and savannah with good rainfall, preventing their development for the production of livestock for use in mixed farming enterprises and for meat production.

The present cattle population of Africa, located mainly in tsetse-free areas, is approximately 160 million. It has been estimated that the region could support a further 120 million cattle and many more sheep and goats of tsetse flies and trypanosomiasis could be brought under control. Of course, suitable precautions would be needed for proper management of the livestock population and protection of the environment (FAO, 1974). The U.N. Food and Agriculture Organization was charged at the World Food Conference in Rome in 1974 with the organization and co-ordination of activities associated with this difficult task. Elsewhere in the world, infections caused by Trypanosoma evansi and T.vivax also affect food animal production. In some areas, these parasites cause relatively mild forms of disease, but occasionally T.evansi infections in cattle and buffalo cause severe losses. Better control methods for these forms of trypanosomiasis would also contribute to improved livestock productivity.

In the case of theileriosis, there are also several situations where improved disease control could contribute significantly to improved food production. Infections with T.annulata tend to occur in the severest forms in highly productive livestock introduced directly, or by cross breeding projects, into areas where the disease is endemic. Efforts to provide more milk for human consumption, by crossing local cattle with highly productive European varieties, have resulted in dairy cattle populations which are very susceptible to the parasite, for example in Israel, Iran and India (Henson and Campbell, 1977). Bovine theileriosis caused by T.parva, also causes heavy losses among livestock, and greatly limits the realisation of the full food-production potential of an extensive region of Eastern and Central Africa. Actual cattle mortality figures caused by East Coast fever in the whole region are not available, but in Kenya with a cattle population of eleven million some 50-70,000 animals over 1 year old succumb to the disease annually (Duffus, 1977). In addition to such direct losses, the disease causes unthriftiness and greatly reduced growth rates in a high proportion of the animals which survive an attack.

DEFICIENCIES IN EXISTING CONTROL METHODS

There are many deficiencies in the existing disease control methods. For example, in the case of trypanosomiasis, current methods of tsetse control are applicable in only a small part of the total infested area, which covers some 10 million square kilometres. In some cases, tsetse control methods are only effective on the fringes of infested areas where the insects survive near the extreme range of their physiological tolerances. Some tsetse species inhabit areas where there is, as yet, no known method of control. High rainfall and dense vegetation limit the feasibility and efficacy of insecticidal application (Jordan, 1974). Also, the use of insecticides in tsetse control operations can result in pollution of the environment with associated, unwanted ecological effects (Koeman et al, 1971). The use of newer, highly active, more biodegradable insecticides, which have recently undergone field trials, could reduce this criticism (FAO, 1981) but the problem of comprehensive applicability in all tsetse habitats and in all ecological zones still exists.

In the case of chemotherapy for trypanosomiasis, the capacity of the parasites to develop resistance to each new compound within a few years of its introduction for field use is a familiar problem. Strains of trypanosomes of the major pathogenic species resistant to quinapyramine, phenanthridine and diamidine drugs are commonly encountered. As a long term measure, the prospects of trypano-somiasis control by chemotherapy are therefore poor. At best, this approach should be regarded as a supportive measure. In tropical Africa, trypanosomiasis is maintained in a sylvatic cycle involving tsetse flies and wildlife. The problems of tsetse control have been mentioned; the problems of eliminating the reservoir of the parasite from wildlife cannot even be approached by chemotherapy. However, in areas where the tsetse vector population has been eliminated by insecticidal application, or other means, chemotherapy has a part to play in removing parasites from domestic lifestock and decreasing the possibility of mechanical transmission by other arthropod species. The continued existence, however, of infections of domestic livestock with T.evansi in extensive areas of southern America, northern Africa and Asia, and T.vivax infections in areas of central and southern America, in the absence of significant reservoir host populations or vectors in which the organisms can undergo cyclical development, must leave some doubt about the ultimate efficacy of chemotherapy as a control measure for this disease.

In the case of theileriosis, it is important to distinguish between different forms of the disease when considering deficiencies in presently available control measures. For T.annulata, both tick control measures and effective vaccination procedures are available (Pipano, 1977). The incidence of the disease tends to be seasonal thus reducing the need for acaricidal applications, but the development of acaricide resistance is still a problem, because of the need for tick control to prevent other tickborne diseases. Vaccination against T.annulata in different countries means that, for maximum efficacy, vaccines must be prepared from locally isolated organisms. Lack of appropriate administrative infrastructure often results in a failure to mount effective disease prevention campaigns. Also, extended tick control campaigns can result in the development of a livestock population which is highly vulnerable to a variety of tickborne diseases, since there has been no opportunity for the animals to develop immunity following exposure to ticks. A break in vector control or acaricide application resulting from mechanical breakdowns can result in livestock losses (Norval, 1981).

CHEMICAL AND BIOCHEMICAL APPROACHES TO IMPROVED DISEASE CONTROL

Insecticides and Acaricides

Dissatisfaction with current insecticides for tsetse control has resulted in increased interest in "third generation" insecticides, based on hormonal and growth regulatory factors of the insects (Jordan, 1978). Such compounds, their analogues or antagonists, might be used as very specific agents against the Genus Glossina provided that means can be found of applying them to the target insects. This might be achieved by insect attractants, such as host odour constituents or pheromones in the case of tsetse which are highly mobile (Hargrove and Vale, 1979), but this approach might have less to offer in the case of ticks (Rechav and Whitehead, 1981).

Improved Therapeutic Compounds

Chemotherapy has much to offer in any concerted effort at disease control. It is important that efforts should continue to find new drugs active against parasitic trypanosomes and better ways to use existing drugs. For example, trypanocidal activity in vivo has been conferred on the antibiotic daunorubicin by coupling it to a carrier molecule. Alone, the antibiotic is only active in vitro (Williamson et al., 1981).

Success in establishing the three major species of African pathogenic trypanosomes in culture systems has also opened up opportunities for drug screening in vitro which should be of major assistance in drug development (ILRAD, 1981). Such culture systems are also proving useful in basic research on the biochemical pathways by which essential cell components are synthesized. This work could lead to the design of parasite - pathway specific metabolic inhibitors for use as therapeutic agents (Williamson, 1980). Carbohydrate catabolism (Clarkson and Brohn, 1976) and lipid synthesis involving threonine are receiving attention in this respect (Newton, 1980). The synthesis and attachment of trypanosome surface glycoprotein, which forms a protective coat on the organisms, is under examination. Unique features of the glycosylation of trypanosomal coat protein are particularly attractive (Rovis and Dube, 1981).

In theileriosis, effective therapeutic drugs have been lacking for so long that it came as a major surprise when two compounds of different classes were found to exhibit a useful degree of schizonticidal activity. Anti-Theileria activity was identified in halofuginone in infected animals by chance, but a parasitised lymphoblastoid cell culture provided the in vitro drug screening system which led to the discovery of the activity of menoctone (Dolan, 1981). Both halofuginone and a menoctone derivative (Parvaquone) have been used experimentally to treat cases of infection with T.annulata and T.parva in cattle and these drugs may soon be released for field use. However, halofuginone is toxic and the safety margin is low. Considerable diagnostic skill and care in treatment will be required if these drugs are to be used effectively and better drugs are still required. Theilericidal activity has also been found in a third class of compounds, the ionophore coccidiostats, using the same in vitro screening system (McHardy and Rae, 1982). The availability of lymphoblastoid cell cultures infected with Theileria should prove very useful in the search for further and safer therapeutic compounds for treatment of these diseases (Brown, 1980).

Vaccines Based on Parasite Antigens

In the absence of measures for the complete elimination of the vectors of trypanosomiasis and theileriosis, or of the parasites from the reservoir hosts, the development of vaccines to protect livestock must be seriously considered (Holmes, 1980).

In the case of trypanosomiasis, chemistry is already contributing substantially to vaccine research. The immunogenic antigens of trypanosomes are associated mainly with the parasite surface and have been found to be glycoproteins (Cross, 1975). Unfortunately each trypanosome stock can produce many different surface glycoproteins, so protective antibodies produced by infected hosts have only limited effects (Gray and Luckins, 1976). However, it is conceivable that work on the amino acid sequence of trypanosome proteins will reveal regions of homology among stocks of particular trypanosomal species which could be used as a basis for vaccine antigen production. Work on structural proteins, which show antigenic similarities between trypanosome species, also offers a possible approach to vaccine production (Shapiro and Murray, 1982), if means can be found to render such proteins immunogenic.

In theileriosis, vaccine development is at an exciting stage. Theilerial schizont vaccines have been available for some time for protection against T.annulata, but the development of a similar vaccine against T.parva has not been so successful, probably because of incompatibility problems between cells used in the vaccine and those of recipient hosts (Cunningham, 1977). Vaccines against the sporozoite stages of T.parva offer more promise for the prevention of East Coast fever. Chemical knowledge of the protective antigens of T.parva does not yet exist, but research in this field is gathering momentum and there has been speculation that it might be possible to produce, by genetic manipulation techniques, hybrid parasites incorporating in one organism the essential antigenic features currently displayed by the different stocks of parasites encountered in the wild (Irvin and Boarer, 1980).

Vaccines Based on Vector Antigens

There is evidence that animals exposed to haematophagous insects and ticks produce antibodies which can affect ectoparasite survival (Willadsen, 1980). In the case of tsetse flies, there seems little merit in considering this approach as a disease control measure since many species of fly are catholic in their feeding behaviour and are highly mobile. Contact between immunized hosts and the vectors is likely to be sporadic. However, animals exposed repeatedly to feeding ticks are known to develop resistance to tick attachment and the presence of experienced hosts can reduce the efficacy of tick feeding and the size of tick populations (Chiera et al., 1981). Recent research has suggested that antibodies produced by tick-infested hosts also have significant effects on tick egg laying and hatchability (Cunningham, 1981). The identification of the antigens produced by ticks, their chemical characterization and production in vitro, could contribute to the development of a new method of tick control and an indirect improvement in theileriosis control.

Exploitation of Natural Resistance

Natural resistance to trypanosomiasis and theileriosis is displayed in various forms in domestic livestock. These vary in trypanosomiasis from examples of absolute host restriction, such as the failure of *T. vivax* to establish infection in pigs, to the race resistance exhibited by trypanotolerant West African Shorthorn cattle to the more common species of pathogenic trypanosmoes affecting ruminants (Murray et al., 1979). In theileriosis, both humped stock of Zebu (*Bos indicus*) origin, and humpless, *Bos taurus*, animals are highly susceptible to infection. However, these two major cattle types are known to differ in the degree to which they support feeding tick populations (Hewetson, 1981; Bonsma, 1981).

Such differences in susceptibility undoubtedly have a genetic basis and chemistry might well play a role in the identification of the factors concerned. If resistance can be attributed to specific protein molecules or secretions, it could well be exploited by genetic micromanipulation and embryo transfer technology to produce livestock with appropriate genes.

CONCLUSION

In cases of diseases like trypanosomiasis and theileriosis, which each result from interactions between arthropod vectors, protozoan parasites and mammalian hosts, it is unlikely that an attack on any one component of the complex will provide a wholly effective measure of disease control. This can best be achieved by an integrated approach against all of the components of each disease complex, by whatever means become available. Thus, both trypanosomiasis and theileriosis may eventually be controlled by a combination of insecticide and acaricide applications to destroy the vector, chemotherapy for active cases and vaccination using parasite antigens. Adequate attention must also be given to environmental and livestock management, and to possibilities of genetic manipulation of the host immune response and the use of resistant livestock. In anticipation of success in disease control, plans for proper land use and animal management should also be prepared.

REFERENCES

Bonsma, J.D. (1981). Breeding tick-repellent cattle. In Tick Biology and Control. Eds. G.B. Whitehead and J.D. Gibson. Rhodes University, Grahamstown, South Africa, 67-77.

Brown, C.G.D. (1980). In vitro cultivation of *Theileria*. In: The in vitro Cultivation of the Pathogens of Tropical Diseases. Eds. D.S. Rowe and H. Hirumi. Schwabe and Co., Basel, 127-143.

Chiera, J.W., R.M. Newson and M.P. Cunningham (1981). The cumulative effects of host resistance on *Rhipicephalus appendiculatus*. In Annual Report of the

International Centre of Insect Physiology and Ecology for 1980. Nairobi, 37-38.
Clarkson, A.B. and F.H. Brohn (1976). Trypanosomiasis: an approach to chemotherapy by inhibition of carbohydrate catabolism. Science, 194, 204-206.
Cross, G.A.M. (1975). Identification, purification and properties of clone-specific glycoprotein antigens constituting the surface coat of Trypanosoma brucei. Parasitology, 71, 393-417.
Cunningham, M.P. (1977). Immunization of cattle against Theileria parva. In Theileriosis. Eds. J.B. Henson and M. Campbell. International Development Research Centre, Ottawa, 66-67.
Cunningham, M.P. (1981). Biological control of ticks with particular reference to Rhipicephatus appendiculatus. In: Advances in the Control of Theileriosis. Eds. A.D. Irvin, M.P. Cunningham and A.S. Young. Martinus Nijhoff, The Hague/Boston/London, 160-164.
Dolan, T.T. (1981). Progress in the chemotherapy of theileriosis. In: Advances in the Control of Theileriosis. Eds. A.D. Irvin, M.P. Cunningham and A.S. Young. Martinus Nijhoff, The Hague/Boston/London, 160-164.
Duffus, W.P.H. (1977). Theileriosis in Kenya. In Theileriosis. Eds. J.B. Hensen and M. Campbell. International Development Research Center, Ottawa, 28-32.
FAO (1974). Programme for the Control of African animal trypanosomiasis. AGA/TRYP/74/1, (Mimeographed).
FAO (1981). Report of FAO Panel of experts on the ecological and technical aspects of the programme for the control of African animal trypanosomiasis and related development. Food and Agriculture Organization, Rome. 1-5 June, 1981.
Gray, A.R. and A.G. Luckins (1976). Antigenic variation in salivarian trypanosomes. In: Biology of the Kinetoplastida, Vol. 11. Eds. W.H.R. Lumsden and D.A. Evans. Academic Press London/New York/San Francisco, 493-542.
Hargrove, J.W. and G.A. Vale (1979). Aspects of the feasibility of employing odour-baited traps for controlling tsetse flies (Diptera: Glossinidae). Bull. ent. Res., 69, 283-290.
Henson, J.B. and M. Campbell (1977). Theileriosis. International Development Research Centre, Ottawa, 112.
Hewetson, R.W. (1981). Results of selection for cattle tick resistance in cattle. In Tick Biology and Control. Eds. G.B. Whitehead and J.D. Gibson. Rhodes University, Grahamstown, South Africa, 91-93.
Holmes, P.H. (1980). Vaccination against trypanosomes. In: Vaccines against parasites. Eds. A.E.R. Taylor and R. Muller. Blackwell, Oxford, 75-105.
ILRAD (1981). Annual Report of the International Laboratory for Research on Animal Diseases, Nairobi, Kenya, 28.
Irvin, A.D. and C.D.H. Boarer (1980). Some implications of a sexual cycle in Theileria. Parasitology, 80, 571-579.
Jordan, A.M. (1974). Recent developments in the ecology and methods of control of tsetse flies (Glossina spp) (Dipt. Glossinidae) - a review. Bull.ent.Res., 63, 361-399.
Jordan, A.M. (1978). Recent developments in techniques for tsetse control. In Medical Entomology Centenary, Symposium Proceedings Trans. R. Soc. Trop. Med. Hyg., 76-84.
Koeman, J.H., H.D. Rijksen, M. Smies, B.K. Naisa and K.J.R. Maclennan (1971). Fanunal changes in a swamp habitat in Nigeria sprayed with insecticide to control Glossina. Netherlands J. Zool., 21, 443-463.
McHardy, N. and D.G. Rae (1982). Anti-theilerial activity of the coccidiostat monensin. Trop Anim. Hlth Prod.; 14, 13-14.
McHardy, N., A.J.B. Haigh and T.T. Dolan (19176). Chemotherapy of Theileria parva infection. Nature, 261, 698-699.
McHardy, N., A.T. Hudson and D.G. Rae (1980). Therapy of Theileria parva:

correlation of results in culture and in cattle. In: The in vitro Cultivation of the Pathogens of Tropical Diseases. Schwabe and Co., Basel, 149-152.

Murray, M., S.K. Moloo, K. Hirumi, F. Dar, C.A. Hinson and H. Hirumi (1982). Reduction of trypanosome infection in Tsetse flies by antibodies raised against procyclic forms of Trypanosoma congolense. Proceedings of Fifth International Congress of Parasitology. 7-14 August 1982, Toronto, Canada.

Murray, M., W.I. Morriso, P.K. Murray, D.J. Clifford and J.C.M. Trail (1979). A review: trypanotolerance. Wld. Anim. Rev., 31- 2-12.

Newton, B.A. (1980). Chemotherapy of trypanosomiasis: current views on mechanism of action of anti-trypanosomal drugs. In: Report of the Expert Consultation on Research on Trypanosomiases. Food and Agriculture Organization, Rome. Appendix XV, 111-119.

Norval, R.A.I. (1981). A reassessment of the role of dipping in the control of tick-borne disease in Zimbabwe. In: Tick Biology and Control. Eds. G.B. Whitehead and J.D. Gibson. Rhodes University, Grahamstown, South Africa, 87-90.

Pipano, E. (1977). Basic principles of Theileria annulata control. In Theileriosis. Eds. J.B. Henson and M. Campbell. International Development Research Center, Ottawa, 55-65.

Rechav, Y. and G.B. Whitehead (1981). Can Amblyomma pheromones be used for control. In: Tick Biology and Control. Eds. G.B. Whitehead and J.D. Gibson. Rhodes University, Grahamstown, South Africa, 209-210.

Rovis, L. and D.K. Dube (1981). Studies on the biosynthesis of the variant surface glycoprotein of Trypanosoma brucei: sequence of glycosylation. Mol. Bioch. Parasit., 4, 77-93.

Shapiro, S.Z. and M. Murray (1982). African trypanosome antigens recognized during the course of infection in N'dama and Zebu cattle. Inf. Immun., 35, 410-416.

Willadsen, P. (1980). Immunity to ticks. Adv. Parasit, 18, 293-313.

Williamson, J. (1980). Present situation of research for new trypanocidal drugs. In: Report of the Expert Consultation on Research on Trypanosomiases. Food and Agriculture Organization, Rome. Appendix XIII, 90-96.

Williamson, J., T.J.S. Finnigan, M.A. Hardman and J.R. Brown (1981). Trypanocidal activity of daunorubicin and related compounds. Nature, 292, 466-467.

SYNTHETIC FOODS: TECHNICAL, CULTURAL AND LEGAL ISSUES

Sanford A. Miller
Director, Bureau of Foods
Food and Drug Administration
Department of Health and Human Services
Washington, DC 20204 USA

ABSTRACT

Contemporary science and technology is being increasingly applied to the problem of meeting the world's current and future food needs. The scope of these developments imply an urgent need for assurance of safety and nutritional quality of new products, for an understanding of the dynamics of food selection including cultural factors, and for progress in regulatory activities. These issues are reviewed in detail with an emphasis on the role of the scientist requiring flexibility and adaptability.

KEYWORDS: nutritional quality, toxicological safety, food selection, cultural factors, regulatory acts, testing programs, dietary choice.

INTRODUCTION

Food has always been a limiting factor in the survival of man. Faced by the hazards of a fundamentally hostile environment, humankind has managed to survive only through a series of adaptations resulting in an exceptional flexibility to utilize a wide range of materials as substrate for its metabolic machinery. Man alone has the ability to add further to his food resources by manipulation, but nevertheless has always been faced by the spectre of famine. Relatively minor changes in climate and environment has resulted in deaths of millions due primarily to an inability to modify food production rapidly enough to meet the challenge. Today, this problem of "time" has become even more acute as changes are occurring with such rapidity and have political and social implications of such magnitude that the possibility of ultimate disaster has become a continuous problem both in individual lives and for national and international political structures. The need for action to prevent starvation or, at least the physical and mental enervation of malnutrition, is an essential component of modern politics. Part of the solution to these problems must lie in increased utilization of available resources by application of contemporary science and technology. The solutions might include the modification of food materials for direct utilization by man; the replacement of biological processes for the conversion of food to man by chemical systems; the provision of nutrients through direct synthesis; and the use of these nutrients in the formulation of new products having nutrient compositions closely resembling nutritional optimums (and therefore increasing the metabolic efficiency of such products). However, it is also clear that the development of this technology is only part of the solution. A better understanding of the dynamics of food needs and wants must also be included. With this understanding, technology can provide the foundation for organoleptic equivalence in the exchange of such new foods for more traditional products. These properties of foods are by far the most important stimuli for making "food", food. Moreover, these developments in technology can permit high acceptability for a variety of non-traditional nutrient sources. Finally, this concept of organoleptic equivalence of new nutrient sources could permit, for the first time, the construction of foods to the specification of human need rather than dependence upon the vagaries of natural products. Nevertheless, as with all such advances, there are potential problems of equal magnitude.

For the first time, there exists the possibility that the human dietary can be restricted to a relatively few formulations of varied organoleptic structure, thus limiting the traditional safety inherent in the enormous variety of the traditional dietary. Equally, for the first time, a significant, profound change can be made in the diet of a major part of the human population within a very short period of time. Since better marketing practices and efficient distribution systems can result in a broad exposure to new substances or foods before historical experience with their use is obtained, there is a need for assurances of safety and nutritional quality for such products far beyond any required in the past. Moreover, because of our increasing knowledge and the fundamental role of these products, the level of biological function required to be investigated in these evaluations is orders of magnitude more specific than have been previously considered. Finally, as a result of political, ecnomic, and social and historical forces, society has developed a series of regulatory processes designed to protect itself from ultimate disaster through inappropriate modification of the food supply which must be considered in the development of any new food product. In these evaluations, it is important to recognize, that, to a significant extent, such regulatory activities designed to be protective of human health often are antithetical to this goal by interfering with the innovative process. Similarly, it must be remembered that the political imperatives which lead to this regulation are often based upon considerations other than that associated with the welfare of the individual, but rather directed towards the maintenance of the political body.

It is clear then that the ultimate utilization of the most important concepts developed in contemporary chemical research will depend not only on our ability to exploit advances in modern chemistry but also on our ability to define, understand and resolve three broad questions: nutritional and toxicological safety, the dynamics of food selection and regulatory and political impact.

SAFETY AND NUTRITIONAL QUALITY

The issues concerned with nutritional quality and toxicological safety are the most susceptible to scientific inquiry and to the development of a technical solution to the problem. The extent to which problems in nutritional utilization need to be solved is a function of the nutritional uses to which these products are to be put. In the case of a product designed to replace a traditional nutrient resource, the problem becomes important only when the intake of the new food becomes a significant part of the total daily intake. Thus, if a meat substitute is consumed as a snack food in addition to the major diet, it must meet different standards of nutritional equivalence to meat than if it is to represent the major share of the total animal protein intake. The possible absence in these products of substances for which requirements have not been determined is also of significance. Finally, the problem of meeting the needs of different age groups living in varying states of health becomes of importance when the product is a major source of nutrients. Although some attempt has been made to consider the needs of the infant and child in considerations of nutritive quality and safety, little attention has been paid to the elderly or to the effects of disease, stress or prior nutritional state.

Problems associated with the development of products designed for nutritional superiority in relation to natural products can mostly be defined with two questions: (1) what represents a more optimal pattern of nutrients? (2) are we developing other problems of unknown dimensions by modifying a pattern of long standing? For those substitutes that are to be used as a principal part of the diet, the problem of possible nutritional imbalance must also be considered, not only in terms of interactions within the food, but even more importantly, as a resuult of changing dietary patterns overall, and thus possibly modifying interactions among foods.

For some new products, particularly vegetable protein mixtures, there is a large number of nutritional studies available. They are, however, almost entirely concerned with the problems of protein quality and utilization rather than those of product safety or, for that matter, total product utilization. This arises from the fact that the principal reason for the development of such products has been to expand protein resources. In general, these studies demonstrate that a mixture of high protein quality can be derived either as a single source supplemented with the most limiting amino acids or through use of a mixture of proteins. For example, there are several studies, both on animals and man, that clearly demonstrate the high nutritional quality of textured soy protein supplemented with methionine. More recently, it has been indicated that the use of egg albumin as a binder in such products also helps improve their biological value. Thus, it seems clear that the protein quality of such a mixture is assured, or at least capable of being assured. In fact, the data suggest that even in long term exposures, these products appear capable of maintaining maximal growth of weaning animals and maintenance of adults when included in a normal diet otherwise complete. Thus, in my opinion, any study of wholesomeness of an analog does not have to concern itself primarily with protein quality. It does, however, have to concern itself with several other questions of nutritional importance to which existing studies have not really addressed themselves.

It is clear, for example, that no study has examined the ability of these products to act as a major or primary food source. In contrast to its current use as a simple source of protein, the major role of vegetable protein and other similar products in the future will be in the development of food products which, as described earlier, may be a source of several nutrients, not only protein. With these products, the subtle effects of a marginal deficiency may not be revealed in testing under standard protocols. For example, research has shown that marginal intakes of methionine, lipotropes, and Vitamin B-12 during early life results in a significant decrease in the ability of the animal to resist infection later in life even though no other gross effect was observed in the animals studied. Thus, products expected to have a broad impact on the food habits and patterns of an entire society require special, more sensitive and precise determinants of overall nutritional quality than those products having more limited use. Such tests must include exposure of infants and the aged, measurements of response to stress and determination of the status of the organism for several micro-elements, as well as evaluation of several other dynamic, functional and interactive activities.

The problems of toxicology and safety of foods as compared to food additives are much more difficult to resolve. They are complicated by the fact that they include not only questions of science, but also of social forces such as the legitimate demands of consumers for greater assurances of safety resulting in increasing resistance to the inclusion of various chemical and nontraditional substances in the food supply. This social phenomenon occurs in the developing world as well as in industrialized societies and results in new products that must not only perform as claimed, but must do so with the minimal, highly rationalized, additive mixture. Traditional techniques for the evaluation of safety which involve studies having as their endpoint only survival or histopathological effects, are usually not sufficiently sensitive to provide information appropriate for estimating the safety of potentially widely used substances, such as newly formulated food products. For most such products, the potential hazards involved are not those leading to death, but rather concern for the effect of such substances on the long term capacity to perform. Over the years, a highly structured series of studies have been developed to address some of these questions. These include not only lifetime exposure of the animal to the test substances, but also special tests for reproductive function, for teratogenic effect, for carcinogenesis and for mutagenesis. More recently, these requirements have been expanded to include such functional, dynamic parameters as resistance to infection, behavior, etc. Nevertheless, in spite of the increasing complexity and scope of these test protocols, they do not include investigation of several areas essential in making hazard assessments of food and nutrient substitutes. To date, most such studies begin with healthy, weaning animals and rarely directly involve the neonate or aged animal, nor do they examine the effect of stress or of ill health on response to the test product. However, most importantly, these protocols are designed to study individual additive substances, i.e., substances that are added to the diet in relatively small amounts. Because they represent such a small proportion of the diet, it is possible to investigate them under grossly exaggerated conditions to assure the most conservative estimate of human hazard. This approach and these protocols designed for the study of substances present in the diet in small quantities are not appropriate for the study of foods.

One of the best examples of the complexity of such issues is illustrated by the studies to evaluate the safety of irradiated foods. Most of these studies cannot be interpreted since they involve feeding to animals such exaggerated amounts of a food that metabolic derangements are common. In its evaluation of irradiated foods, the Food and Drug Administration came to the conclusion that the evaluation of the safety of these products could not be made solely on the results of animal tests since there was no way to provide for a sufficient margin of safety in the protocols. Their approach was to utilize as much of the total knowledge about the process including data concerned with the changes occurring in the chemical composition and identity of the components as a result of food irradiation. Based on these considerations and the data available from the animal testing, the agency has proposed a regulatory approach to the use of this process. Thus, it seems clear that any sort of protocol designed to deal with the problem of the safety of new, nontraditional foods and food analogs must take into account the relative insensitivity of animal testing protocols for such products and provide, in addition, an opportunity for human experience and for the use of the total body of knowledge concerning the new product. Some of these issues were addressed in the reports of a UN protein and calorie advisory group in the series of PAG guidelines on the testing of novel sources of protein. These proposed protocols involved the development of chemical information, preclinical testing in animal models, as welll as clinical testing at some point after the animal studies are completed.

In addition to the question of the potential toxic effects of novel nutrient sources, concern must also be expressed about the presence of naturally occurring toxic substances in products derived from new food sources. Toxic substances, which even in small amounts may pose a threat to animals and humans, occur in many widely used foods unless the foods are specially processed or restricted in their use. The wide variety of such substances which may be natural components of animal or plant food include glycosides saponins, favism and lathyrism (as factors in legumes), goitrogens, and a variety of anti-vitamins. Contamination of natural products with molds and other microbiological entities may also result in the reproduction of potent toxins and carcinogens such as the large family of mycotoxins. Some food sources may contain natural substances in excess of that considered to be safe for man. For example, normally humans consume nucleic acids in many foods. However, in non-photosynthetic single-call protein sources, nucleic acid concentrations may be high enough to cause a long-term hazard.

Many of the problems associated with the evaluation of nutritional quality and toxicological safety of novel sources of nutrients are amenable to scientific inquiry. Modern experimental techniques can be used to examine the effect of such substances on resistance to disease, on immune competence or on behaviour either during development or later on in life. Protocols with which direct measurements of nutritional utilization can be made in man have also been developed. Most importantly, increased concentration on the utilization of metabolic and biochemical measurements of performance make it less difficult to translate data to man. Nevertheless, it must be remembered that these testing programs are extensive and expensive. Their consideration has to occur early in the process of the exploitation of any new chemical technology, and their cost must be considered a routine component of the economic base for deciding the value of any new technology.

DYNAMICS OF FOOD SELECTION

The resolution of the above questions of safety and nutritional quality of new and unusual components of the food supply leads to a series of more difficult and complex questions. Having decided that a particular food product is safe and will function effectively as a nutrient source, how then do you assure that it can function by having people eat it? It has become very evident to nutritionists in recent years that people eat for reasons satisfying to themselves, not to the nutritionist, planner or technologiest. Palatability and emotional satisfaction are of primary importance to the consumer; nutritional benefits are of secondary concern. Thus, the issue of how groups of people resolve the complex task of choosing what to eat is of great concern to the proper exploitation of new technology. The omnivore is faced with a particularly difficult problem in such a food selection. The world is full of things that might be eaten. Some are nutritious, some are poisonous, and a great majority are of no biological significance. As a highly developed and reasonably successful omnivore, man has a special problem in resolving this complex issue. While there is surely a biolgical base that guides our food selection, it represents, at best, only a minimal component of the process. A review of the data on human food selection can lead only to the conclusion that the prime determinant of what people eat is their culture and, more specifically, their cuisine. Most people do not eat most foods not because they have tried and rejected them, but rather because they never had the opportunity to eat them. They are unavailable and/or not part of their cuisine. Thus, to a significant extent, the bulk of these choices are made on the basis of familiarity. New foods are approached with a suspicion described by Rozin as neophobia. Thus, the universe of potentially eatable foods in any culture is usually highly restricted and determined in large part by availability and by the cuisine of the particular culture.

Humans are virtually alone in the world in having such a body of rules or limitations about what to eat and how to eat it. They characteristically do something to natural foods such as mix or cook them before ingestion. Nevertheless, the process by which they modify their food supply is also highly regulated by the rules of their society. The role of custom, culture and availability of foods and fuels and status on food selection has been reviewed by many workers. For example, food is used to maintain interpersonal relations, to cope with emotional stress and to reward or punish and, perhaps, most importantly, to maintain status. Often, in a particular culture, choices are limited by economic factors and by the limitations of the particular market. If either factor changes so that the buyer has increased economic power or the market is expanded, the procurer of food in a particular family may choose other additional foods for the status that accrue to the entire family. The prestige so conferred may be so important as to influence a food procurer to reject a nutritionally superior food in favor of an inferior one if the latter can add to the family's prestige. The switch from brown rice to polished rice and the resulting dietary deficiency among affluent Asians is an outstanding example of this kind of behavior. Nevertheless, most food selection patterns are conservative and are based upon an important survival characteristic. In this way, the experience of one's ancestors with nutritious or poisonous foods and with the neophobic-neophyllic conflicts surrounding eating can be easily passed along.

Rozin has classified cuisines into 4 basic components. The first component is that of primary and secondary foods: the principal sources of calories and most of the nutrients in the diet. These are often limited as in the heavy reliance on corn and beans in the Mexican diet or on rice in the diet of many East Asian groups. The second component is the manner of food preparation. This again, is typically restricted for any given cultural group. The third component is that of the flavorings added to the basic foods during their preparation. It has been pointed out that major world cuisines repeatedly added a characteristic flavoring of flavor principle of their foods. For example, the combination of tomato and chili constitutes a Mexican flavor principle, while soy sauce and ginger root pervades Chinese cuisines. The fourth important component of cuisine is the mixed bag of additional rules about eating that relate to the social conditions of eating such as ritual or festival foods and food taboos. In particular, when food selection is strongly modulated by ethnic or religious concerns, the problems of modifying the selection process become extraordinarily difficult. On the other hand, modification of food selection can often be used as an ideological marker. In particular, a modification in lifestyle or philosophy, can result in rather profound changes in food selection and cuisine.

In spite of the complex and apprently nonquantifiable nature of the food selection process, an understanding of the role of cuisine itself must be an important component of the process of modifying food selection. Cuisines play an important function in the mediation of the constant conflict between neophobia and neophilia in the ingestion of foods. By defining a body of familiar foods and particularly by placing a familiar testing flavor combination on such foods, much of the psychological danger and stress is taken out of eating. Moreover, the development of such definitions provides a possible insight into mechanisms for the incorporation of new foods into the dietary. Thus, a cuisine can function as a means of circumscribing acceptable foods, reducing the variability in food selection within a given culture and rejecting new foods. On the other hand, proper exploitation of the components of cuisine can be used to introduce new food substances within the context of an apparently limited universe of acceptable food items. For example, as we have indicated, it is widely known that food habits are extremely difficult to change and that, in particular, immigrant groups are most reluctant to abandon their traditional food practices. Long after they have lost linguistic and other ethnic markers of their native country, the immigrants retain strong preference for their native cuisine. Nevertheless, work done in several countries on the pressures resulting to change food habits of a variety of immigrant groups have suggested that changes do occur and that when they do so, they occur with great rapidity. The inertia and conservativism of food habits appear to be significantly less rigid in relation to the pressures for changing these habits in children. Under these conditions, ethnic habits are maintained mostly in terms of cooking styles and the use of herbs, spices and condiments, i.e., flavor principles of the diets. The primary and secondary foods of the dietary, however, are expanded to include the components of the new culture in which the immigrants find themselves. It is unclear what causes these changes. These factors include not only the increased availability of new foods and better communication, i.e., advertising but, perhaps, more importantly, the pressures for status transmitted largely through children, which result from the desire of the children to conform to the mores of their peers at school.

The introduction of any new food product into a fundamentally conservative culture thus will require attention to a variety of factors. The data strongly suggest that there are two particular areas in the process of food selection most amenable to modification. At least one critical point for intervention is in the market, through the radio or televesion, and generally in the schools as a traditional advertising campaign based on a basic desire to attain status and the satisfaction of other emotional needs. Food choices in modern society increasingly appear to be influenced by advertising which creates no new needs, but merely echos man's tendency to multiply and differentiate his necessities. Any program to introduce new foods must recognize that most advertising campaigns are designed to meet the desires of the public. More importantly, for most cultures in the developing world, a second point of modification may be in the development of food products which can be added to the primary and secondary food supply and yet not be intrusive enough to modify the basic favoring component in the diet which is important in defining appropriate and acceptable foods in any culture. New food products must be provided in such a form that they do not require any new methods of preparation and can accept traditional flavoring principles readily. The development of Incaparina in Guatemala is an exceptional example of such an approach where the product was made and designed to represent a drink traditional in the culture, and as a result found great acceptability among the population. The problems of modifying food selection processes are dependent upon our understanding the complex and often highly emotional nature of the role of culture and custom in this process. While much can be done in a empirical sense, the development of a theoretical basis for such changes will prove to be one of the most important contributions to the ultimate exploitation of contemporary technology in resolving this most basic problem of world food needs.

REGULATORY AND POLITICAL ISSUES

The third set of questions deals with regulatory and political issues and are equally complicated and difficult to define. In terms of regulatory activities, not only are there potential barriers to the acceptance of new technologies in national regulatory structures, but differences in such structures among countries constitute a barrier of equal or greater dimensions. Regulatory activities in any country are principally derived from the basic concept that it is the function of government to protect the health and welfare of its people. In each case, regulations are derived from a perceived need to address an issue that individuals are unable to address for themselves. In the United States, the development of technology, which had led to a national food industry beginning in the last century, resulted in the passage of the first Federal Food and Drug Act in 1906 when it became clear that technology had made it difficult for the consumer to make judgements of safety and quality. By 1938, new technical advances led to a public perception of increasing problems in recognizing quality and safety in food, and another set of laws were passed which further expanded the role of government in making determinations of safety and quality. By 1958, this concern again reached a peak and Congress responded a third time with the passage of the food safety amendments of that year. Food safety legislation and the regulations based on tis legislation are designed to control those segments of the population who are unable or unwilling to follow the basic mores and conventions of their culture. Thus, regulations tend to be restrictive, specific, and, unfortunately in a society that believes in equal justice under the law, applies to all segments of society without restriction to those whose action produce the legislative response. In addition, the long delays involved in the implementation of such regulatory activites are usually the result of the statutory need and the philosophical desire to provide a broad spectrum of views on the validity and value of proposed regulation. In the United States this desire results in a mandated series of steps which often include a public hearing to develop thse views on a particular issue. There is always a conflict between the desire for rapidity of action and the need for developing the largest, most diverse body of opinion on the validity and desirability of a particular regulatory action. It is interesting to note that many of the people who complain most bitterly about delay in the process are the very ones who use those delays to prevent an action with which they do not agree.

Differences in national laws generally arise from varying concepts of the role of government in incurring consumer protection against fraud and assuring the safety of the products. This role is often modified by the equally important desire of government to promote local agricultural and industrial interests against those products derived from areas outside of the national structure. Further, cultural and religious predilections play a major role in accounting for and justifying differences in legislative and regulatory approaches to consumer protection. However, with the development of international bodies such as the Codex Alimentarius and the joint expert committees of the WHO/FAO, there has occurred over the last 25 years an increasing tendency towards uniformity in the development and application of many of these regulations. Nevertheless, the process remains extremely complicated and slow largely as a result of the desire, particularly in the international program, of assuring that national interests are considered in the process. In every case, however, the final decision concerning the utilization of a new technology will be based first and foremost on the ability of its proponents to demonstrate that it is safe and its use does not present any long term hazard. As in each of the areas discussed the development of good science is an important requisite for rational regulatory decisions.

In view of the stringent provisions of the US Food, Drug, and Cosmetic Act, and the elaborate structure for enforcing regulations encompassed under this act, decisions of the FAS are recognized and adopted by much of the world. The basic thrust of US law is designed to maintain the level of safety and quality of the food supply at the very highest levels. To a certain extent, it is unfortunate that US law provides the basis for action in much of the world. US law is predicated on an affluent society, having a luxus food economy. Nevertheless, the US position, based as it has been for 75 years on the development of an appropriate scientific base to make regulatory decisions, can provide an important benchmark in the determination of national food safety policy. This regulatory policy of the United States has generally kept pace and evolved with changes in science. In 1958, the agency operated under a stringent absolute ban on "carcinogens" in the food supply. Through increasing knowledge and better understanding of the carcinogenic process and the development of information that suggests that such substances are widely distributed throughout our environment as a result of both natural and made-made effects, the agency has modified this position considerably and is attempting to regulate, to the greatest extent possible, on the basis of some understanding of the hazards such substances represent to human health rather than as a reflex to a legal stimulus. More importantly, as a result of changes in science, the agency is proposing changes in the basic Food, Drug, and Cosmetic Act to account for these advances in science and to allow for more judgement in making such regulatory decisions.

Perhaps the most significant barrier in the adoption of any new technology is that concerned with the political issues. One of the most essential conditions for acceptance of any new technology is that it be politically acceptable. Even if the need is great and other conditions are favorable, efforts to introduce new technology will abort if political considerations are not favorable for its acceptance. New ideas may be politically unacceptable for many different reasons. High among these is the perception by those in power of the adequacy of already existing solutions. If political forces can rationalize that traditional approaches to these problems will resolve a particular problem, then those solutions will be selected because new ideas tend to threaten the classical structure of a society. Thus, there is a tendency for political bodies to attempt to either define the problem in such a way as to suggest that traditional solutions will function well or to restate the classical solution so that it appears to resolve the issue. On the other hand, new technologies may be proposed as a device to postpone dealing with more fundamental problems such as more just distribution of resources and land reform. The new technology is ultimately rejected and abandoned because there was no real interest in it in the first place. For the reasons given earlier, new technologies also are often rejected because they might be considered to increase dependence on foreign sources of nutrients and foods in general. Or, new technolgies might be considered harmful to a useful practice such as breast-feeding or new technologies might be considered an alien idea and represent a threat to religious or traditional ethnic practices.

Generally speaking, politicians do not like to face up to the problems of food balance, particularly when the resolution of these problems requires long term modification of existing economic and social structures. They tend to argue that food problems are temporary, brought about by a combination of natural circumstances and the world political and economic situation and as a result, sooner or later, will go away. For example, in 1974, the US Department of Agriculture concluded that "the factors which have given rise to the present world food situation are largely transitory and can be corrected by intelligent policies." Other officials have stated "what we face now is the prospect of a population famine where the population simply outruns the increasing food supplies." The result of this kind of thinking is a belief that there are simple solutions to these problems and little research is required for their resolution. Another way in which political forces block utilization of new technology is to change the issue by changing the rules. In recent years, there has been a significant downward change in the scientific consensus on protein requirements for human beings. This has been reflected in downgrading the protein aspect of the food problem and has led to the notion that what is needed is "more of the same kind of food". While this debate and the issues raised by scientific inquiry are important and essential for the ultimate resolution of these problems, the existence of the debate has tended to obscure the food energy protein trade-off and the dynamics of the interrelationship between protein supply and total food supply. It, therefore, provides an opportunity for political leadership to delay or fail to face up to the political decisions that are required while "the scientific issues are being resolved."

The important point of these discussions is not that political entities are callous about or unsympathetic to the problems faced by the populations which they represent, but rather the tendency is for political structures to take the most conservative and traditional approaches to the resolution of their problems. It is difficult for a national government to admit that continued malnutrition in their population might result in a significant decrease in intellectual capacity. As a result, they are unwilling to enter into or promote the use any technology which is predicated upon this fact. It is clear, then, that the ultimate exploitation of new technology to improve the health and welfare of any population is dependent in large measure upon the development of an appropriate political strategy to ensure the adoption of the new technology will not cause serious, social and political disruption. Similarly, as we saw in our discussion of cultural and psychobiological problems associated with the use of new technology, the most useful approach to the problem would have to entail the use of those technologies which maintain fundamental, widely accepted components of the food supply in place and modify those components which are difficult for the consumer to recognize as being changed. Science, therefore, plays its most important role in the beginning of the process. The development of the technology and the determination of its safety and nutritional quality and the recognition of those factors involved in food selection and acceptance are among the earliest requirements. However, when this work is done, then the role of the political scientists, economist and strategist must now predominate. The issues are no longer scientific, but social and economic and, at that point, the best role the scientist can play is to be willing to be flexible in the development of their technology and to always keep in mind that the ultimate goal is to provide for the welfare of the entire human population.

In 1978, Svetlana Kharatyan quoted Marcellin Berthelot, a founder of organic chemistry, in the conclusion of an article on "Microbes as Foods for Humans". In 1897, Berthelot offered the following prediction in his book, Science et Morale as follows: "People often speak about the future of the human society and I should like to foresee it as it will be in the year 2000, naturally, for the point of view of the chemist. Then there will be neither shepherds nor ploughmen; food products will be produced chemically. Once cheap energy is generated, it will become possible to synthesize food products from carbon isolated from carbon dioxide, hydrogen obtained from water, nitrogen and hydrogen extracted from the atmosphere. Nitrogen compounds, synthetic fats, starch and sugar - all these substances will be produced in great quantities by our plants; manufacture of artificial food will be independent of the season, rain, drought, frost; moreover, foodstuffs will contain no pathogenic microbes - the primary cause of epidemics and enemies of the human life. Chemistry will produce a revolution, the signficance of which cannot be understood today. There will be no difference between productive and non-productive regions. But one should not think that in the worldwide power of chemistry there will be no art, beauty and charm of human life. If the soil is not used for farming it will be again covered by grasses, forests, flowers. It will turn into a large garden irrigated by underground waters where people will live enjoying an abundance of foodstuffs and joys of the legendary gold age". These Scientific Proceedings consider the most recent developments in chemistry and chemical technology as they apply to the resolution of worldwide food problems, and provide another step towards the achievement of Bethelot's prediction. We are closer today than we ever have been. It is essential if this prediction of a new golden age is to be fulfilled, that we understand all of the factors that are involved in its achievement.

NEW PROTEIN SOURCES OF FOOD AND FEED

Yasuji Minoda
Department of Agricultural Chemistry, The University of Tokyo
Bunkyo-ku, Tokyo 113, Japan

ABSTRACT

Research for new microbial protein sources of food and feed focussed on yeasts from palm oil and hydrogen bacteria. An isolated strain, identified with Torulopsis candida, effectively assimilated crude and refined palm oil. The addition of surfactant, vigorous agitation and high oxygen supply accelerated yeast growth, optimally at $30^{\circ}C$ and a pH of 3.0 when the cell yield on 3% of palm oil was 95%. Another strain, which could grow at about $40^{\circ}C$, was isolated and identified with Candida tropicalis. Thermophilic hydrogen bacterium, Pseudomonas hydrogenothermophilia showed very rapid growth (μ = 0.73 hr^{-1}) at $52^{\circ}C$ on simple inorganic medium supplied by a mixture of H_2, O_2 and CO_2. The most suitable medium and the best gas ratio for cell growth were determined. For each kilogram of bacterial cells, 5.4 m^3 of hydrogen was consumed and the carbon dioxide intake ratio was 65%. Protein levels of T.candida and P.hydrogenothermophilia were 46.8 and 75.0% respectively and the amino acid composition of both proteins was well balanced. These organisms are promising strains for SCP production.

KEYWORDS: Single cell protein, yeasts, autotrophic microorganisms, hydrogen bacteria, culture media, cell yield, protein content, amino acid composition.

INTRODUCTION

Growing concern for the food needs of the world's exploding population has led to examination of a variety of potential food resources. A critical shortage of protein, present and future, has stimulated agricultural and industrial research and development in many countries aimed at meeting this special need. Among these, single cell protein (SCP), a novel source of protein of microbial origin, has enabled industrial production of proteins for both humans and animals.

Microbial cells contain 40-80% protein of good quality and can be mass-produced in a short time, largely avoiding any adverse weather influences. This is a protein productivity far larger than that of animal or plant origin. SCP as a class covers all proteins obtained from many kinds of microorganisms (bacteria, algae, yeasts, fungi) cultivated on various substrates (carbohydrates, agricultural by-products, hydrocarbons, alcohols and others). Thus, there is excellent opportunity to combine judiciously microorganisms and raw materials for the preparation of SCP. Molasses and sulfite waste liquor have been used for production of food and fodder yeasts. The technology for massive production of SCP by fermentation from fossil sources, including normal paraffins, methanol and ethanol, has been developed and has led to large-scale production in several countries. However, the energy crisis has driven up the price of these raw materials and they lose economic advantage. Consumers, in some countries, moveover, have not yet accepted SCPs from these unconventional resources although their safety has been confirmed through precise and long-term examination. SCP processes using various kinds of waste materials from agriculture, forestry and live-stock industries have also been developed. These processes, however, present problems of scale because supply and handling of these raw materials is not easy. For these reasons search for new microbial protein sources that are safe, acceptable and inexpensive has been carried out.

Palm oil is a high-yielding tropical product whose production has increased steadily over recent years, currently being in excess of 3 million t/a. Cell yield from palm oil is expected to be twice as high as that from carbohydrates. Moveover, since the cells are coated in an edible product, SCP from palm oil might be readily accepted by consumers. For tropical countries it could be a good source of protein as well as of vitamins.

In view of the depletion of our natural resources, the SCP of autotrophic microorganisms which feed on the infinitely renewable resource of carbon dioxide assumes urgent importance. There are two major types of autotrophic microorganisms: photolithotrophs and chemolithotrophs. The former include algae and photosynthetic bacteria, and already, algae such as *Scenedesmus*, *Chlorella* and *Spirulina* are being commercially produced as SCPs. The photolithotrophs in general, however, seem to be uneconomical for industrial production of SCP, which leaves the chemolithotrophs. They generally use the energy provided by oxidation of inorganic materials, assimilating carbon dioxide. Within the autotroph class, hydrogen bacteria are thought to be the most promising, because a supply of cheap hydrogen can be expected in the future. Organic products are produced effectively in conventional large-scale fermentors by these organisms. Cell production of hydrogen bacteria has been examined by only a few investigators (Repaske, 1962, and Schlegel and Lafferty, 1971). From the industrial point of view, thermophilic hydrogen bacteria are promising. Because the risk of contamination is reduced, cooling water costs can be reduced. Gases are supplied more efficiently at the higher temperatures. A strain with high growth rate has been obtained; a high yield of cells is expected.

SCP FROM PALM OIL

Many yeast strains capable of assimilating palm oil were isolated from various sources such as flowers, plants and soils. The isolation medium contained 2% palm oil as a sole carbon source, 0.4% NH_4NO_3, 0.47% KH_2PO_4, 0.03% $Na_2HPO_4 \cdot 12H_2O$, 0.1% $MgSO_4 \cdot 7H_2O$, 0.001% $FeSO_4 \cdot 7H_2O$, 0.001% $CaCl_2 \cdot 2H_2O$, 0.001% $MnSO_4 \cdot 4H_2O$, 0.01% yeast extract. To control bacterial growth 0.002% chloramphenicol was added. More than 200 strains were isolated from 240 samples. Among these the strain Y-128 assimilated crude and refined palm oil most effectively. The yeast has a round shape with a size of 2.5 - 5.6 μ. It showed multilateral budding, no pseudomycelium was formed and spores were absent. From these morphological and other physiological properties, the strain was identified with *Torulopsis candida*.

In shaking culture of this strain, corn steep liquor as a natural nutrient was suitable for yeast growth, and ammonium sulfate was the most suitable nitrogen source. Addition of a surfactant, sucrose fatty acid ester in a concentration of 0.05 to 0.1%, was effective for yeast cell growth because of its nontoxic properties. Culture conditions to produce SCP from palm oil were studied by using a jar fermentor. Optimal conditions are given in Table 1. Yeast growth depended largely on the culture temperature with an optimum at 30°C. The optimum pH of 3.0 as shown in figure 1 is a remarkably acidic condition for yeast growth and is thought to protect against bacterial contamination. At a concentration of 3%, the palm oil was effectively converted to cell mass. Increase in agitation speed and oxygen supply during cultivation increased cell production.

Table 1: Optimal Culture Conditions Of *T.candida*

Palm oil	30.0 g/L	$(NH_4)_2SO_4$	7.0 g/L
KH_2PO_4	7.0 g/L	$NaH_2PO_4 \cdot 7H_2O$	1.0 g/L
$MgSO_4 \cdot 7H_2O$	1.0 g/L	$CaCl_2 \cdot 2H_2O$	0.01 g/L
$FeSO_4 \cdot 7H_2O$	0.01 g/L	Corn steep liquor	0.2 g/L
Sucrose fatty acid ester P-1670		0.5 g/L	

pH 3.0; aeration 1.5 vvm; temperature 30°C; agitation 1500 rpm.

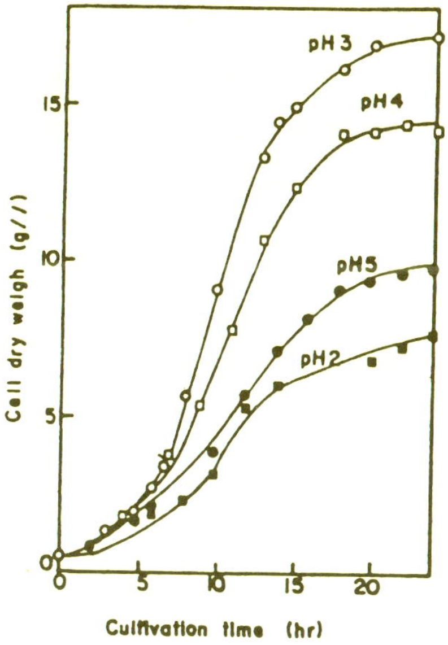

Fig. 1 Effect of pH on yeast growth at 3% palm oil concentration Aeration rate: 1.0 vvm; agitation speed: 1000 rpm.

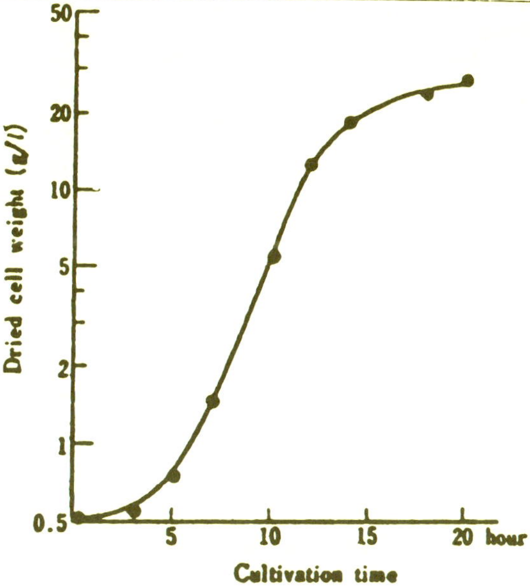

Fig. 2 Time course of yeast growth under optimum conditions. pH controlled at 3.0 with 5% NH_4OH. Aeration rate: 1.5 vvm; agitation speed increased from 700 to 1500 rpm.

Figure 2 shows the time course of the yeast growth under the optimal conditions. Palm oil (3%) was emulsified using sucrose fatty acid ester. Specific growth rate at logarithmic growth was 0.53 hr^{-1} and the doubling time was about 1.3 hr. More than 90% of the palm oil was consumed by the yeast, for an overall yield of 95% based on the consumed oil in the stationary phase.

As palm oil does not melt completely below 40°C, the liquidity of palm oil is not sufficient at 30°C. Therefore the higher cultivating temperature appeared to be more favorable. Screening of the thermophilic yeasts which grew at 45°C on palm oil was carried out. Several strains were obtained from soils and flowers and a strain was selected, because it showed good growth in liquid palm oil medium at about 40°C. It was an ellipsoidal shaped yeast with a size of (5-8μ) x (5-10μ). It showed multipolar budding and made pseudomycelium, but no spores. Assimilation of KNO_3 was negative. It was thus identified with Candida tropicalis. Optimum growth temperature was 38°C where it had very high specific growth rate of 0.95 hr^{-1}.

HYDROGEN BACTERIA, A NEW SOURCE OF SCP

The author and his co-workers have isolated hydrogen bacteria, one mesophile (Kodama et al 1975), two thermophiles (Goto et al 1977, 1978) and several extreme thermophiles (Kawasumi et al 1980). One of the thermophiles showed very rapid growth and seems to offer good advantages for SCP production. It is a rod-shaped bacterium, motile with polar lophotrichous flagella, and a GC content of 62.3-62.5 mole %. It was identified as a new strain belonging to genus Pseudomonas and named as Pseudomonas hydrogenothermophila.

The bacterium rejected all of the sugars which were tested, but propagated well on organic acids, such as acetic, succinic etc. Growth however was much better under autotrophic conditions. It can be cultivated autotrophically in simple inorganic medium under a supply of mixed gas of H_2, O_2 and CO_2. Ammonium sulfate was the best nitrogen source and addition of Ca^{2++} showed good effect. Positive correlation existed between the initial partial pressure of CO_2 and the specific growth rate, with the rate falling when the pressure was below 0.09 atm. There was very little effect on growth rates when $P(O_2)$ was in the range of 0.09 to 0.33 atm., although the $P(O_2)$ must be kept low during the initial stage of cultivation. The most suitable culture medium is as follows: $(NH_4)_2SO_4$, 5g/L; KH_2PO_4, 5.66g/L; K_2HPO_4, 2.38g/L; $MgSO_4 \cdot 7H_2O$, 1g/L; $CaCl_2$, 30mg/L; NaCl, 0.5g/L; $FeSO_4 \cdot 7H_2O$, 0.1g/L; pH=7.0; $H_2:O_2:CO_2$ = 400:50:60 (OD < 5), 400:100:60 (OD > 5).

The relationship between culture temperature and growth rate is shown in Figure 3. The highest rate of growth occurs at 52°C, $\mu = 0.73\ hr^{-1}$, and the doubling time is a little less than an hour, which is not inferior to that of the current heterotrophic bacteria in common use. We believe that it is the highest value obtained for all carbon-fixing organisms, including plants, algae and bacteria. The growth curve for jar cultures at 52°C is shown in Figure 4. In this experiment $P(O_2)$ in the initial ventilating gas ratio was reduced to H_2:Air:CO_2 = 65:25:10. At the point indicated by the arrow, the ratio was changed to $H_2:O_2:CO_2$ = 7:3:3. After a lag of two hours, the culture entered logarithmic growth and continued to a marginal concentration of (X_c) = 2.3 g/L. At sixteen hours, the concentration reached 6.8 g/L. Our current studies of continuous cultures have shown a production efficiency of DX = 3.14 g/L/hr, which we believe can be improved.

The material balance obtained for a gas ratio of H :O :CO = 68:23:9 (the best ratio for cell yield on closed flask culture) was

$$23.8\ H_2 + 3.5\ CO_2 + 11.6\ O_2 + 0.84\ NH_4^+ = C_{3.45}H_{6.41}O_{2.12}N_{0.84} + \text{other products.}$$

For each kilogram of bacterial cells, 5.4 m^3 of hydrogen was consumed and the carbon dioxide intake ratio was 65%. The conversion efficiency from hydrogen energy to cell energy was 36%. The heats of combustion of hydrogen and bacterial cells were 236.4 kJ/mol and 2029 kJ/100g, respectively.

The composition of yeast cells from palm oil is shown in Table 2. A rather high content of carbohydrate and ash were detected. Table 3 shows the results of amino acid analysis. The amino acid composition of T.candida was well balanced. In the case of C.tropicalis the content of sulfur-containing amino acids was low.

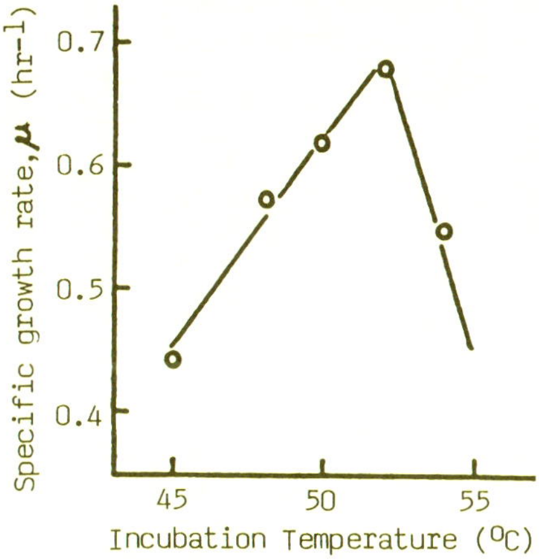

Fig. 3 Effect of incubation temperature on cell growth in the optimal medium.

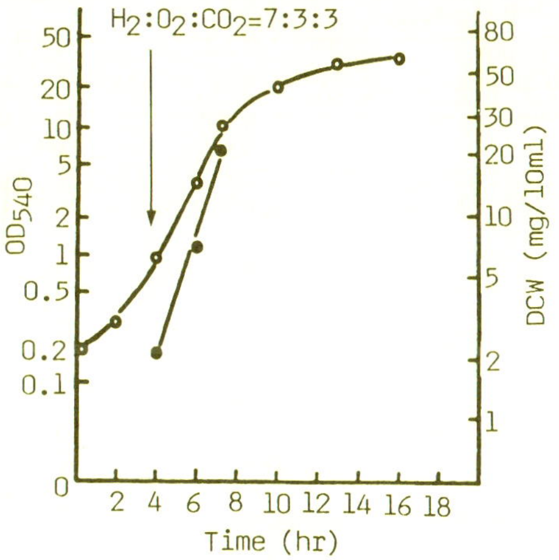

Fig. 4 Growth curve of hydrogen bacterium in a jar fermentor at 52°C and agitation speed 1000 rpm in the optimal medium.

o-o, Optical density at 540 nm;
●-●, Dried cell weight.

Table 2: Compositions of Yeast Cells

Component	Torulopsis candida	Candida tropicalis
Protein (Lowry)	46.8%	34.8%
Crude fat	1.4	7.7
Nucleic acid		3.05
Carbohydrate	39.2	41.6
Ash	10.5	6.6
Vitamin B_2		62.1 mg/kg
Niacin		543.0 mg/kg

The protein content of the hydrogen bacterium, like most bacteria was higher than that of yeasts. The amino acid composition was excellent and the methionine content was high in contrast to most SCPs.

Table 3: Amino Acid Compositions of Yeasts and Bacterial Protein g/100g Protein

Amino acid	FAO Reference	Torulopsis candida		Candida tropicalis		P.hydrogeno-thermophilia
Asp		8.33		8.94		9.21
Thr	4.0	4.69		6.37		5.29
Ser		5.03		5.75		3.79
Glu		9.20		15.63		12.75
Pro		2.86		3.49		0.50
Gly		3.86		4.46		6.63
Ala		4.95		6.18		9.21
Val	5.0	4.86		5.72		8.13
Cys	2.0 (Cys + Met)	1.61	4.17 (Cys + Met)	0.46	1.74 (Cys + Met)	
Met		2.56		1.28		2.82
Ile	4.0	4.25		5.11		5.34
Leu	7.0	6.47		7.19		9.45
Tyr		3.47		2.83		3.45
Phe		4.21		4.31		5.12
Lys	5.5	5.99		8.42		6.39
His		1.95		2.25		2.84
Arg		3.91		5.35		7.99
Crude Protein %		46.8		34.8		75.0

DISCUSSION

The yeast identified with Torulopsis candida, we believe, is a promising strain for SCP production. The low optimum pH for growth is especially favorable, because of the problem of contamination. A high cell yield and favorable amino acid composition were observed. At 30°C, the optimum temperature for this strain, palm oil was poorly emulsified and moreover, the oil adhered to the wall of the fermentor preventing the yeast from utilizing it completely. The addition of surfactant, an increase in rate of agitation speed and a high oxygen supply were effective to some degree. However, cell growth in a medium containing more than 3% palm oil was not satisfactory. To solve this problem yeast strains which grow at a temperature higher than 40 C were screened. The C.tropicalis thus obtained had a high cell productivity without addition of surfactant, and contained a large amount of riboflavin and niacin. However, the protein content of the strain was low and moreover the safety of C.tropicalis as a source of SCP may be suspect.

The excellent characteristics of Ps.hydrogenothermophilia are encouraging and may land to industrial production of SCP from carbon dioxide and hydrogen. The use of this hydrogen bacterium as SCP remains under development. Production technology requires feed evaluation tests, safety tests, and a bioengineering investigation. There is a special problem because of the danger of an explosion of the mixed gases. Safety procedures therefore will have to be established. The current cost of hydrogen is uneconomical, and commerical success depends on the future direction of hydrogen source development and on protein needs.

REFERENCES

Goto, E., T. Kodama and Y. Minoda (1977). Isolation and culture conditions of thermophilic hydrogen bacteria. Agric. Biol. Chem. 41, 685-690.

Goto, E., T. Kodama and Y. Minoda (1978). Growth and taxonomy of thermophilic hydrogen bacteria. Agric. Biol. Chem. 42, 1305-1308.

Kawasumi, T., Y. Igarashi, T. Kodama and Y. Minoda (1980). Isolation of strictly thermophilic and obligately autotrophic hydrogen bacteria. Agric. Biol. Chem. 44, 1985-1986.

Kodama, T., Y. Igarashi and Y. Minoda (1975). Isolation and culture conditions of a bacterium grown on hydrogen and carbon dioxide. Agric. Biol. Chem. 39, 77-82.

Repaske, R. (1962). Nutritional requirements for *Hydrogenomonas eutropha*. J. Bacteriol. 83, 418-422.

Schlegel, H.G. and R.M. Lafferty (1971). Novel energy and carbon sources. Production of biomass from hydrogen and carbon dioxide. Adv. in Biochem. Engineering. 1, 143-168.

CONVERSION OF PLANT AND ANIMAL WASTE TO FOOD

Antoni Rutkowski
Professor of Food Technology
Agricultural University
Grochowska 272
03-847 Warsaw, Poland

ABSTRACT

Factors such as the slow increase in agricultural productivity, growing costs, and enlarged world population suggest that present plant and animal waste in the future will be intensely converted into food. At present only a limited volume of waste may be utilized directly in food production. Waste should be partially convertable by animals into meat, milk or eggs. This form of conversion however is limited to local production. Research on application of physical and chemical methods of waste utilization to food grade products is already under way. Top attention in many laboratories in the world is given to bioconversion of waste. Processing of waste to food products demands that close attention be paid to the high quality and nutritional properties of the products.

KEYWORDS: algae, bacteria, biomass, byproduct, conversion, pulp, single cell protein, waste, whey, yeast.

INTRODUCTION

"The War on Waste" is currently regarded as one of the leading methods of increasing food resources. The notion "plant and animal waste" varies for different climatic zones and different countries especially since the distinction between "waste" and "by-product" is fluid (Figure 1).

There are two possible ways of converting waste to food: (1) via animals, and (2) via some industrial process (van der Wal, 1979). The conversion may take place at the farm in cooperation with a processing plant, directly at a processing plant where utilization of wastes is carried on by a separate department for by-products, or at an industrial plant specially designed for that purpose.

Most food industry processing plants are located in rural areas which facilitates simple and direct utilization of waste via feeding of domestic stock, i.e. leading to production of milk, meat and eggs. There are, however, unavoidable losses in the bioconversion in the animal organism (see Figure 2).

A considerable proportion of waste is not utilized, however, because most processes using waste are seasonal in character with up to 75% of the waste being accumulated within 2-3 months. This results in direct or indirect reduction of feeding potential due to losses in nutrients as well as a quantitative and qualitative rise in volume of wastes burdening the environment.

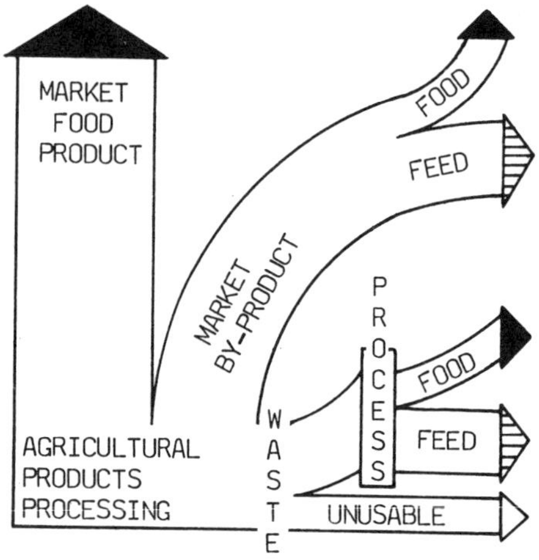

Figure 1: By-products and Wastes from Processing Agricultural Products.

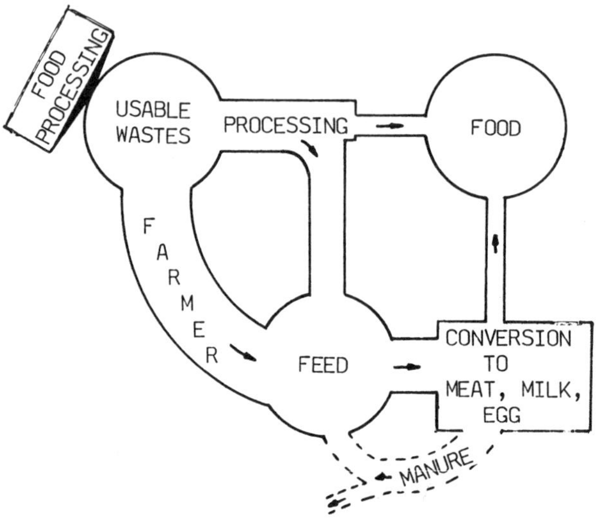

Figure 2: Pathway for Usable Wastes

DIRECT UTILIZATION OF WASTES AS FEED

Many food industry wastes are not acceptable as a component of human food even though their nutritional value is considerable. The reason stems, firstly, from the negative organoleptic factors such as flavour, smell or texture, and, secondly, from the excessive content of ballast substances (cellulose), and antinutrition factors such as phenols, saponines, glucosides, etc. Although some of these can be removed physically, chemically or biologically, thus improving waste quality, as a rule such processes are uneconomical. Hence, the most effective manner of utilizing wastes is to convert them into animal feed. While converting the organic substrates of waste into meat, milk or eggs, the animal organism also functions as a natural filter of the antinutrition factors. Food industry wastes used as feed include the following:

- wet wastes: whey from cheese processing, effluents of starch plants (potato, maize, etc.), pulp from fruit juice factories, etc.
- dry wastes: molasses from sugar factories, oilseed cake and meal from oilseed processing, residues from grain milling, i.e. bran, etc.

Wet waste is characterized by its high content of water and perishability, and thus, its utilization as animal feed should take place almost immediately after rendering. That is feasible only when the distance to the processing plant is short enough that the cost of transporting the waste is substantially lower than its value as fuel. Direct use of wet wastes in animal feeding thus requires small- and medium-size processing plants located in a region of intensive animal husbandry. At large plants, or those situated in urban areas, only a minor proportion of wastes can be used for animal feeds.

Conversion of wastes on the farmer's premises is usually limited to their conservation and storage for feed to be used by his own stock. This applies in particular to seasonal wet wastes which may be used over a period exceeding the seasonal nature of production. The most frequently used technology of silo storage does not demand costly equipment, high qualifications of the farmer or any additional sources of energy. Using silos, however, one must take into account relatively high losses in nutrients and up to 20% of organic matter, which are not balanced by an increment in bacterial biomass. The silo technique is commonly used to conserve potato pulp, beet pulp, and fruit and vegetable pulps. The positive aspects of silo methods include fermentational deactivation of many antinutrition factors as, for instance, glucosides of cruciferae plants. Coffee pulp is a further example, in which residual caffeine and polyphenols are reduced by ensiling and water washing (IDRIC, 1981).

Utilization of dry wastes as feed poses no significant problems as long as their composition meets requirements of quality. Dry wastes normally contain relatively large proportions of ligno-cellulose, hardly fit for utilization even by ruminants. Hence, studies in this field focus on raising the digestibility and feed value through physical, chemical and microbial processing. Alkali treatment increases the digestibility of ligno-cellulosic materials such as sugarcane, bagasse and the stems and straw from oilseed and cereal crops. Sodium hydroxide, most often recommended, is expensive and hazardous, and is manufactured in relatively few countries. Consequently, University of Alexandria investigators are exploring the effectiveness of alkaline solutions derived from the ashes of burned wood available in most village communities. Also being studied are several novel approaches to ensilaging of agricultural materials, including straw mixed with urea and covered with citrus pulp to absorb the ammonia generated (IDRIC, 1981).

DIRECT UTILIZATION OF WASTES AS FOOD

Direct utilization of wastes as human food would be an ideal solution yet our capabilities in this regard are quite limited. This group of wastes includes those animal products which could be fully used as food, but for diverse reasons are not acceptable to consumers (Table 1).

Table 1: Percentage of Slaughter By-products and Wastes

Cattle	50 - 60%
Pigs	20 - 30%
Poultry	30 - 40%

Basically, division between edible and inedible animal products is determined by the purchasing power of the consumer, his food habits, religion and customs. In particular, this applies to slaughterhouse wastes, i.e., lungs, intestines and other giblets, which, regarded as waste in one country, are sought after as dishes or components of meat products in many other countries. Although attempts at introducing them to markets in forms of special kinds of processed food should not be given up, their full use depends mainly on proper preservation and transfer to those markets where they are acceptable. In this group of wastes, blood deserves a special position. Even though the nutritional value of blood is not especially high, its application in food products grows continually. Blood has been used for many years in Europe, in the manufacture of traditional butcher's products such as black pudding and blood sausage.

Large reserves of meat remain attached to bones of slaughtered animals at industrial slaughterhouses, and of fish after filleting. Mechanical deboning methods have already been used making it possible to turn this kind of meat (MDM) into an edible variety. Its nutritional value may indeed correspond to that of regular meat (Table 2) but any wider application is hampered by the pink colour and presence of minute bone crumbs (poultry).

Table 2: MDM - Composition Red Meat
(Federal Reg. 42, No. 194,, 54437-54442, 1977)

Protein	14%
Fat	30%
Calcium	0.75%
PER	2.5%

MDM is used up to 20% in such products as frankfurters, fresh pork sausage, luncheon meat, chili con carne, etc., but FDA does not permit using MDM in the following: hamburger, ground beef, fabricated steaks (Madlansacay, 1979). Results of current research create an even broader prospect for utilization of MDM in meat- or fish-processed foods. There is also a similar reserve of food in the form of small fish which are not worth eviscerating. Using a meat-and-bone separator, investigators obtained almost 70% flesh from mixed catches of small whole fish. The mince was mixed with salt (10%) and after 30 minutes pressed into cakes and dried (Hansen, 1979).

An important problem consists in proper utilization of collagen-containing raw materials (e.g. skin or hide residual tissues). Bone meal has, for a long time, been used in manufacturing of edible gelatin serving as an extender, binder, moisturizer and texturizer of food. Hide collagen is effectively used in manufacture of edible sausage casings which replace animal intestines.

Until recently, soybean meal was used exclusively for animal feed. Following some minor quality-increasing operations, it now serves as a component (25-50%) of inexpensive high protein nutrients and as an additive to bread (6-12%). It can be used for production of higher quality edible products such as protein concentrates, isolates and their derivatives.

CONVERSION OF PLANT AND ANIMAL WASTE TO FOOD

The direct utilization of waste for human food remains an ideal, limited potentially to a few materials. In most cases, the effectiveness of waste conversion into feedstuffs or food does not depend on raw material costs which nominally are either insignificant or zero value (Figure 3). The cost-intensive expenditures are for preservation and particularly transportation of wet wastes. Thus, while the cost of conversion may seem to be low when we want to obtain a food grade product of proper hygienic and nutritional standards, it may become unacceptably high. The production cost of final products from waste can often be higher than their market value.

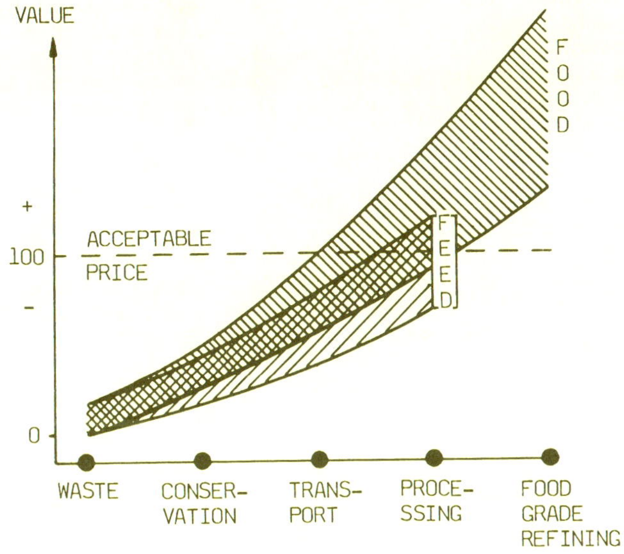

Figure 3: Conversion Costs for Processing Wastes

The most effective utilization of wastes is obtained by direct conversion in medium and large processing plants. As environmental protection regulations grow more stringent, waste conversion of byproducts is more economically feasible and offers an opportunity of arriving at a total wasteless factory - the ideal of the modern processing of agricultural products.

Conversion of wastes in separate industrial units may lower energy outlays per product item, create better qualitative and quantitative utilization of wastes and offer a more sophisticated control of process and product. In such factories, the share of the processing cost per product unit is indeed higher. This limits investment in those factories to those that convert wastes into products of high market value, e.g. pharmaceuticals, photographic gelatine, etc.

Methods used in the conversion of waste to food in general divide into physical and chemical processes and biological processes.

WASTE CONVERSION BY PHYSICAL METHODS

As mentioned above, the basic difficulty in utilizing fluid and wet wastes from food industries is their perishability plus the cost of transporting large volumes. Removal of excess water is achieved by such operations as mechanical separation of mass by decantation, centrifuging, pressing, etc., thermal evaporation of water from the mass, and isolation of valuable components from the mass.

Mechanical and thermal methods are used in primitive technologies in which, for example, pressing is performed with simple machines and human physical force, and also in industrial technologies with the use of decanters, continuously operating presses and various kinds of driers. Mechanical dehydration is indeed the least expensive, however, it does not reduce water content to a level safe for storing and it does not solve the problem of sewage wastes. This happens, for example, while pressing sugar beet as well as fruit and vegetable pulps. Drying is a technologically simple process, but its effectiveness is determined primarily by the cost of energy for a drying device. For example, drying of wastes by drum drying, in spite of its relative economy, usually verges on the border of economic feasibility on account of the low value of the final product.

A considerable problem is posed by fluid wastes from dairy plants which consist of 1-4% of processed milk. These are relatively "pure" wastes containing valuable feedstuff components. In particular, much attention is paid to whey because the price of food continues to grow and stringent environmental regulations are enforced (Imbs, 1978). The simplest technique for its utilization is to obtain whey powder by condensing in evaporators and subsequent spray-drying. Production and food use of whey powder in EEC countries has risen dramatically (Table 3). Similarly, in the USA, in 1955, only 25% of cheese whey was converted in food but by 1975 the figure reached 75% (Muller, 1977). In that country, whey powder is used in the manufacture of bread, ice-cream, meat products and processed cheese. Whey, deionized by electrodialysis or ion exchange, is used in baby foods (Coton, 1977).

Table 3: World Dry Whey Production (FAO Production Yearbook V. 34, 1980)

	Annual Average (1000 tonnes)	
	1969-71	1978-80
NC-America	298.5	370.0
Europe	217.3	614.8
Asia	1.3	8.0
Oceania	1.4	7.3
Total	518.5	1000.0

Attempts at utilizing raw whey that contains 6-8% solid components (80% lactose, 15% proteins) as a substrate for fermentation proved it is an uneconomical process unless there is full use of the protein (Rogers, 1978). This can be done by ultra-filtration techniques and reverse osmosis to separate whey into its two main components: whey protein and lactose. Whey proteins obtained in this manner are highly nutritious, finding application as protein components of diet mixtures as well as baby foods. Demand for pure lactose, unfortunately, is limited. One possibility is to convert it enzymatically to a galactose-glucose syrup, a quality sweetener. A lactose-based non-caloric sweetener, Lacticol, is manufactured in Japan. It is not absorbed by the digestive system and has found use in a number of low calorie products such as soft drinks, fruit syrups, ice cream, cakes, preserves, and condensed milk (Coton, 1976).

Experiments have also been carried out to recover feed components from liquid waste through ion exchangers which consist of cross-linked regenerated cellulose modified by the introduction of anion or cation exchange groups. Both on a laboratory and pilot-plant scale, encouraging results were obtained in recovery of protein from whey, slaughterhouse residues including gelatin, fish residue, and meat and vegetable plant sewage (Kapsiotis, 1977). It was observed, for example, that between 2% and 5% of total carcass protein is lost in the effluent from abattoirs and poultry plants. This can be recovered together with simultaneous purification of the effluents. This process becomes economical, it should be stressed, only when production volume reaches 20 x 10^3 birds/day, corresponding to ca 400 x 10^3 liters of effluents per day (Grant, 1976). As is the case of reverse osmosis and ultrafiltration, this method will take many years before it can be applied on a global scale.

A major potential source of food is rice bran. Its production could give annually about 4 x 10^6 t of proteins and 5 x 10^6 t of edible oils. Several rice bran products are produced, i.e. stabilized-, defatted-, dephytinized-, defibered-bran, protein concentrates and isolates, etc. Only a minimum part, however, goes to human nutrition, because many technical and economic problems remain to be solved before extensive use of bran and bran fractions in formulated foods is achieved (Barber, 1979).

Great hopes have been pinned on the processing of by-products and wastes from the edible oil industries into food products. Lecithin is isolated from waste from oil degumming, and is put on the market in a number of categories depending on the degree of refining and chemical modification such as hydrogenation, hydroxylations, etc. The most desirable is soya lecithin which is used in reducing the viscosity of chocolate, as an emulsifier of margarine and mayonnaises, as a facilitating agent for mixing of components, and for extending the freshness time of breads, noodles, and cakes. Waste after deodorization and soapstock from which sitosterols and tocopherols may be obtained by molecular distillation methods is another processing possibility.

Rich sources of edible materials are found also in wastes accumulating during the processing of fruit and vegetables, e.g. skins, peels, cores, stalks, trimmings, etc. They are used in a simple manner as fodder. Chances for their application in human nutrition depend on local economic and technological circumstances, and as usual, the decisive factor is the market value of the final product (Zamorra, 1979). Here are some examples:

- Wastes and pulp resulting from tomato juice manufacturing account for 10% to 20% of input. After drying and pulverizing, they can be used for production of sauces, soups, etc.
- Wastes accumulating from pressing apples (20% to 30%) as well as lemons and oranges, following lixiviation serve as a source of low- and high-methylated pectins by means of extraction. The pectins are high quality gelling materials used for manufacturing jellies and gels.
- Calcium citrate, a waste resulting from neutralization of citrus fruit and pineapple pulp may be used in production of citric acid.
- Pineapple core extract is used for producing bromeoline, a protein-digesting enzyme used as a component of chill-proof beer, a texturizer in baking and meat tenderizer.
- Fruit pulp, e.g. of apples, black currants, raspberries, etc., when subjected to distillation with steam, provides condensates from which aromatic substances can be extracted. Aromatic substances are mixtures of substances whose arrangement is significantly affected by climatic conditions. It is difficult when launching technical-scale production to obtain consistently a product corresponding to the natural aroma of the fruit. Similar remarks concern taste-and-flavour extracts of garlic, onions, leeks, mushroom waste, etc. (Spicer, 1974).

WASTE CONVERSION BY BIOSYNTHESIS

Conversion of waste with biological methods follows two basic patterns (Johnston and Greaves, 1969):

(1) rendering of biomass products (BMP) whereby the waste is subjected to endo- and exoenzymatic processes. The microbes used in the process remain an organic residue in the mixture,

(2) biosynthesis of food components exuded from biomass after termination of the process. When the microbes are harvested and separated from the substrate we refer to the product as Single Cell Protein (SCP).

BMP (commonly obtained by ensilaging) is used very rarely in producing food. One of the very few and interesting examples is the preparation of fish sauce, a popular South-East Asian food. The waste serving this purpose is fish byproducts, small whole fresh fish and shrimp, mixed with salt at a ratio of 3:1 and subjected to the enzyme actions of fish and bacteria. After 4-6 months, an autolysate is available which is a mixture of soluble proteins, peptides and aminoacids (Hansen, 1979). This paste-like product (bagoong) or a separated clear sauce (patis) provides approximately 7% of the daily protein consumption in the Philippines (Arroyo, 1979), Table 4. Products of a similar kind are obtained through fermentation of pressed copra pulp (Bongkrek) and of peanut outjam (Rogers, 1978).

Table 4: Composition of Fish Sauce (Arroyo, 1974)

	Bagoong	Patis (regular)
Total solids	40%	32%
Protein (N x 6.25)	12.5%	6%
NaCl	20-25%	.
Spec. gravity	.	1.21-1.22%

Waste containing organic components and paticularly carbohydrates is used as raw material for biosynthesis of various organic compounds (Fig. 4).

These are used mainly in the production of:

- food yeast (SCP) by fermentation of lactose from deproteined whey or of molasses saccharose,
- protein biomass in symbiotic fermentation of starch-industry effluents where starch is first decomposed into fermenting sugars that serve as substrate for the generation of yeast biomass,
- various edible products such as acetic, lactic, citric acids, aminoacids such as lysine and methionine, oils, etc.

SCP biosynthesis has shown considerable promise. One hundred grams of a carbohydrate substrate yield approximately 35 g bacteria and 25 g yeasts or fungi protein. In effect, this is the most broadly and relatively well-controlled process of utilization of wastes. Rendering of edible SCP from food industry waste has recently acquired a new impetus as an alternative to large-volume production of SCP from crude oil paraffins which turned out to be unreliable due to technological, economic and health factors.

One of the first major applications of yeast in human nutrition was the

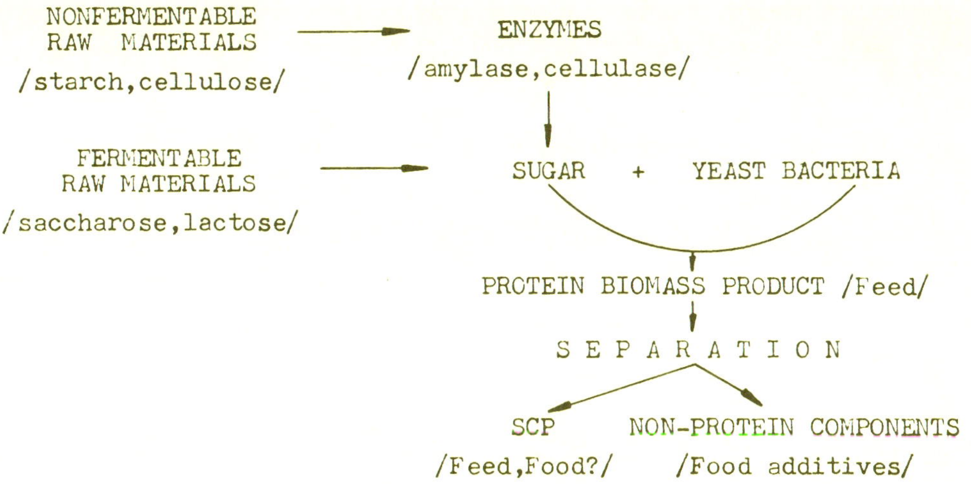

Figure 4: Conversion of Carbohydrate Waste to Fodder and Food.

manufacturing of "Torula Yeast". An example of production of edible yeast protein is the so-called Symba process based on symbiotic culture of two strains of yeast Endomycopsis fibuliger (starch) and Candida utilis (sugar). This process has been applied to wastes developed in the course of processing tapioca, sweet potatoes, vegetables and by obtaining starch products from wheat, rice, maize, potatoes and cassava. A Symba-type plant with a yeast output of 1-2 x 10^3 t per year operates in Eslov, Sweden (Skogman, 1976). Useful raw materials for food protein biomass include distilling pulps, whey, dairy and starch industry effluents, and those from fruit and vegetable processing.

We also have considerable experience in using de-proteined whey as a substrate for yeast biomass cultivation. Even though the combined process of whey conversion is the best method of its utilization, it calls for substantial investment, and large quantities of whey necessary for optimal scale (Szczepanik, 1977). The first industrial plants (1971-1974) processed 100-200 x 10^3 litres of whey per day (Horton, 1974). According to Coton (1976), the process becomes economical only when an installation capable of handling about 500 x 10^3 litres of whey per day is in operation and when the resulting proteins can be used for human nutrition. The possibility of running such a vast factory is rather limited.

The largest, specific, and at the same time relatively little utilized group of raw materials consists of cellulose-containing wastes. Resources of this type are vast (Table 5).

At Natic Laboratories USDD, a method has been developed to use Trichoderma Viride to convert cellulose into glucose, which is subsequently fermented into alcohol (Mark, 1979). Another study reports on a method in which fermentation of 100 kg of coniferous wood waste yielded 22.1% alcohol and 4-5 kg dried yeast containing ca 50% protein. In spite of these and other aspects, chances for utilization of this kind of raw materials are still very poor. Cellulose woody tissue, and to a lesser degree, lignin of one-year plants are resistant to chemical or enzymatic hydrolysis. It is difficult to work out a method to hydrolyse cellulose for fermenting sugars such as glucose that would be economically competitive with other standard fermentation processes.

Table 5: Annual World Production of Carbohydrates in Some Processing Waste (Worgan, 1976)

Sugar cane bagasse	83.0×10^6 t
Wheat bran	57.3×10^6 t
Maize cobs	30.1×10^6 t
Molasses	9.3×10^6 t

Utilization of cellulose- and hemicellulose-rich wastes such as brans and other seed coat materials, maize cobs, press cakes, vegetable and fruit residues such as peels by means of higher-order fungi such as Basidiomycete-Sporotrichum pulverulentum is being researched. Results are encouraging and the mycelium obtained may be acceptable as a component of mushroom soups, burgers or pizzas (Van Hofsten, 1976). Thai scientists have isolated fast-growing fungi that increase the digestability and microbial protein content of sugarcane bagasse, rice straw, and other lignocellulosic wastes. Fungi have been selected that display a high lignolytic and cellulytic activity at tropical temperatures and that generate neither toxic materials nor present any hazard to human and animal health (IDRIC, 1981). Last year the Rank-Hovis-McDoughal Co. in the U.K received regulatory approval for producing and marketing an edible fusarium protein product called Mycoprotein.

In discussing biosynthesis of protein mass, it is not possible to omit cultures of the algae Chlorella, Scenedesmus and Spirulina which, on several occasions, have already electrified the world as a potential source of large masses of biomass. Unfortunately, because of the high cost of energy (light, heat) the process is not applicable in the moderate climate zones. It seems, however, to be a realistic proposal to breed these algae in regions of high sun exposure and an average annual temperature of 25°C. Research in Singapore and in Bangkok focusing on Scenedesmus acutus deserves mention. Obtaining full-value edible protein from algae is difficult since it requires removing the thick cellular membrane, as well as chlorophyl, carotene, unsaturated lipids and other compounds that lower consumer acceptance of algae-containing foods. Feeding of raw algae biomass to animals results in unpleasant off-flavours and over-colouring of products.

Utilization of protein biomass in human nutrition requires complete removal of substances that reduce food value, thus providing a product with suitable organoleptic features. The basic method for obtaining protein from biomass depends on enzymatic decomposition of cellular membranes and on release of protein. Alternatively, physical breakup of the membranes in alkali (pH 10) provides a protein extract which after dialysis is subsequently dried or condensed to the consistency of paste. Positive results were also obtained by decomposing cellular membranes with urea solution. Presently, those methods require considerable improvement in technique, particularly in the mode of separation of protein from the cell, and removal of substances such as nucleic acids, fatty acids (yeast), chlorophyll (algae), etc. These are rather costly operations. To improve the quality of yeast sufficiently for feeding purposes increases the cost of production by some 50% to 80%. This is even higher in the production of food grade SCP.

Currently, bioconversion of wastes produces, in most cases, a SCP product unacceptable to consumers. Most of the protein preparations have a typical off-flavour and look like a formless powder. Providing it with fibrous structure by thermoplastic treatment or spinning increases production costs considerably. The nutritional value of SCP is low and its PER is 1.5 - 1.8. Raising it to a higher level may be possible through the economically feasible process of the Fujimaki plastein reaction. It extends the polypeptide chain through condensation or transpeptisation with simultaneous building in of L-methionine. The resulting product is almost white, flavourless (slightly meaty), and not very sensitive to pH and temperatures up to 100°C. Biosynthesis of lipids has no significance at the present moment. In principle, plant oils and animal fats are abundant. The cost of natural production is significantly lower than that of biosynthesis (Ratledge, 1976).

CONCLUSIONS

1. Increasing costs of agricultural production and the lag behind human food requirements, in addition to the increasing stringency of environmental protection, compel us to utilize wastes more effectively in the manufacture of feedstuffs for animals and foodstuffs for man.

2. At present, only a limited volume of wastes may be utilized directly in food production, and effective use of some of them (mainly animal wastes) is constrained by food habits, religious traditions, and the like.

3. Mankind has at its disposal, methods that make it possible to produce good quality food products from wastes. Unfortunately, in most cases, these are so costly that they exceed economic feasibility even when raw materials are available at no cost. However, the difficult world food situation in the forthcoming century, as well as improvements in processing methods, will make fuller conversion of plant and animal waste to food both desirable and feasible.

REFERENCES

Arroyo, P.T. (1974). The Science of Philippine Foods, Abaniko, Quezon City.

Arroyo, P.T. (1979). Management of Wastes and Byproducts of Fish Processing. Phil. J. Food Sci. Techn., 3, No. 2, 74-79.

Barber, S. (1979). Rice Bran as a Potential Source of Food. In: Proc. V Congress FST, Kodaha-Elsevier, Tokyo-London, 112-121.

Coton, S.G. (1976). Recovery of Dairy Waste. In: Food from Waste, Applied Sci. Publ., London, 221-231.

Coton, S.G. (1977). Characteristics of Products Obtained by Membrane Process when Applied to Milk Products. Int. Dairy Fed., Doc. 55.

Grant, R.A. (1976). Protein Recovery from Meat and Fish Processing Plants. In: Food from Waste, Applied Sci. Publ., London, 205-218.

Hansen, P. (1979). Fish Protein from Underutilized Species. In: Proc. V Congress FST, Kodaha-Elsevier, Tokyo-London, 58-65.

Horton, B.S. (1974). Advances in Preconcentration and Dehydration of Foods. Applied Sci. Publ., London.

IDRIC (1981). A Decade of Learning. IDRIC 170e, Ottawa.

Imbs, B. (1978). The State and Perspectives of World Milk Production, Consumption and Processing. Pol., CZSR, Warszawa.

Johnston, B.F. and Greaves, J.P. (1969). Manual on Food and Nutrition Policy. FAO Rome, 1969.

Kapsiotis, G.D. (1977). Food from Residues and Nutritional Considerations. UNDP/FAO Seminar, Rome 18-21.I.1977.
Madlansacay, P.L. (1979). Management of Wastes and Byproducts of Meat Processing. Phil. J. Food Sci. Techn., 3, No. 2, 81-86.
Mark, E.M. (1979). The World Food Problem and Meeting the Challenge. In: Proc. V Congress FST, Kodaha-Elsevier, Tokyo-London, 3-15.
Muller, L. (1977). The Future for Dairy Proteins. Int. Dairy Fed., Doc. 9.
Ratledge, C. (1976). Microbial Production of Oils and Fats. In: Food from Waste, Applied Sci. Publ., London, 98-112.
Rogers, P.L. (1978). Single Cell Protein from Food and Agricultural Wastes. Food Technol. in Australia, No. 3, 109-115.
Rutkowski, A. (1979). Technological Bearings of Non-Conventional Proteins and Fats Use in Human Nutrition. Pol., Ossolineum, Wroclaw, 41-43.
Skogman, H. (1976). Production of Symba-Yeast from Potato Wastes. In: Food from Waste, Applied Sci., Publ., London, 167-176.
Spicer, A. (1974). In: Advances in Preconcentration and Dehydration of Foods, Applied Sci. Publ., London.
Szczepanik, E.F. (1977). Socio-Economic Aspects of Agricultural and Agroindustrial Residue Utilization. Doc. UNEP/FAO/ISS 4/03, Rome.
van der Wal, P. (1979). SCP for Animal Feeding. In: Proc. V. Congress FST, Kodaha-Elsevier, Tokyo-London, 47-57.
von Hofsten, B. (1976). Cultivation of Thermotolerant Basidiomycete on Various Carbohydrates. In: Food from Waste, Applied Sci. Publ., London, 156-165.
Worgan, J.R. (1976). Wastes from Crop Plants as Raw Materials for Conversion by Fungi to Food or Livestock. In Food from Waste, Applied Sci. Publ., London.
Zamorra, E.L. (1979). Utilization of Fruit and Vegetable Wastes. Philippines, J. Food Sci. Techn., 3, No. 2, 68-73.

IMPROVING THE SUPPLY, QUALITY AND UTILITY OF CARBOHYDRATES

R. Rajalakshmi
Biochemistry Department
M.S. University of Baroda
Baroda, India

ABSTRACT

Plant foods are rich in carbohydrates and hence are the most important component of human diets from the agro-economic point of view as they involve the least pressure in soil, water and production costs. Carbohydrates are also the most suitable from the physiological viewpoint as they are fully digested and assimilated, involving the least stress on metabolism for elimination. Regardless of the diets consumed, the obligatory metabolic requirement for carbohydrate is at least 4-5 times as much as for protein. The major carbohydrates consumed are starch and sucrose from plants and lactose from milk. Starch is the most abundant and most suitable because of its physiological characteristics and presence in staple foods which also provide most of the other nutrients. The value of carbohydrate sources can be enhanced by the right choices, combinations and procedures which exploit favorable characteristics and minimize or eliminate adverse ones. The supply of these foods can be increased by a greater emphasis on feeding low on the food chain, limiting the production of animal feeds to the extent permitted by plant products not edible for man, identifying new suitable sources, safeguarding the quality and availability of required food production resources (soil, water, air and labor), and ensuring a more equitable distribution of land and its produce. Studies exemplifying these concepts are presented.

KEYWORDS: Carbohydrate sources, carbohydrate supply, carbohydrate quality, carbohydrate utility, synthesis of carbohydrates, dietary substitution, starch, lactose, sucrose, lysine, toxic factors.

INTRODUCTION

Both cellular synthesis and the maintenance of homeostasis require energy which is derived mainly from the oxidation of carbohydrates. Although proteins and fats also serve as 'fuels' they have to be first converted to one or other of the intermediates of carbohydrate metabolism. The energy released during oxidation is trapped in the form of high energy phosphate bonds by the formation of ATP from ADP and AMP.

It is well known that animals get the substrates needed for metabolism from the foods they eat which may be of microbial, plant or animal origin. But in the ultimate analysis the substrates needed by living organisms are either synthesized as in plants and certain microorganisms both of which are called 'autotrophs', or obtained from such 'autotrophs' which use either solar radiation (photoautotrophs) or energy derived from thermo-chemical currents (chemoautotrophs), the former being dominant in terrestrial life. Both these classes of reactions can be represented by the general equation $CO_2 + 2H_2O$ light/enzyme > $(CHOH) + H_2O + 2D$ where D represents oxygen, sulphur etc. as appropriate (Cartwright, 1972).

Carbohydrate is the major substrate oxidized by the body although the reserves of carbohydrates in the body are limited, being of the order of 0.2 kg in adult man as compared to about 10 kg each of protein and fat. This is the case even when one subsists on a high protein, high fat, low carbohydrate diet. Physiologically speaking, it is the most economic fuel as it involves the least stress on digestion, absorption and elimination of waste products. This is obvious when we consider that the physiological fuel value of carbohydrate is the same whether it is utilized by the body or burnt in a calorimeter. The two values (kcal/g) differ in the case of fats (9.0 and 9.4) because of incomplete digestion, and in the case of protein (4.0 and 5.65), because of both incomplete absorption and the need for eliminating the nitrogen derived from amino acids in the form of nitrogenous carbon compounds such as urea (Davidson et al., 1975).

Although carbohydrate is not a dietary essential, it is a metabolic requirement for the brain, which uses only glucose as fuel in the adult except during adaptation to prolonged starvation, and also for red blood cells. In the absence of a dietary supply, it has to be synthesized from glucogenic amino acids of protein and the glycerol moiety of fats. Such synthesis imposes an extra load on metabolism. People switched to high fat-high protein diets, as well as starving individuals, are prone to develop ketosis (Cahill and Owen, 1968).

Agro-economic considerations also suggest that carbohydrate rich foods derived from plants are the most economical ones as the synthesis of carbohydrate by plants and algae constitutes the first step in the food chain, i.e. in the transfer of energy (and protein) through ecosystems. Out of a total of 1.3×10^{23} kcal of solar energy available per year on the earth's surface, the energy trapped by plants is only 1×10^{21} calories, that derived by herbivores from plants, 5×10^{20}, and by carnivores eating such herbivores, 1×10^{20}, and secondary carnivores, 3×10^{19} (Cole, 1958).

Energy yields per acre in India (kcal $\times 10^5$) are of the order of 50-100 for roots and tubers, which in some cases store a large part of the energy trapped by plants, 25-50 for leafy vegetables, 10-20 for other vegetables, 9-12 for grains and less than 2.5 and 0.5 respectively for milk and animal products. The corresponding figures for protein (kg per acre) are 100-200, 200-400, 50-100, 40-60, 10 and 5 (Rajalakshmi and Ramakrishnan, 1969 a,b). An essentially similar pattern is found in the West (Schuphan, 1961). Plants also require much less water than animals for producing equivalent amounts of protein and energy (Altschul, 1965), a vital consideration as fresh water supplies form only about 4% of the earth's water reservoir. Of this, 80% is present in the form of ice and snow mainly in the arctic cap, and more than 99% of the remainder as subterranean water (Pirie, 1976). It is important to conserve this scant supply of fresh water by minimizing losses due to evaporation, drainage into the seas, seepage through the earth, and loss of rain water likely to result from deforestation (Fiennes, 1972, Pirie, 1976). A greater dependence on foods of plant origin should help to reduce the amount of water needed for food production.

The development of agriculture freed man (or at least the more privileged sections) from the perpetual need for hunting and gathering but also led to a society based on division of labour. This division of labour led to gross disparities between the life styles of different classes of people. These disparities increased with the advent of industrialization, and even more with the advent of colonialism and establishment of more efficient modes of international trade and commerce which have often resulted in decreased supplies of essential foods for the local poor (Lappe and Collins, 1977). They have led to more or less monotonous diets, particularly among the poorer sections, because of the loss of skills that hunting-gathering societies had in eking out a diverse variety of foods from their environment. Such skills still exist in primitive societies which are found to maintain a better state of nutrition than rural societies (Ganlin and Komner, 1977). A typical example is Kerala where local availability of rice, coconut and fish to the poor has diminished because of the large scale switch in cultivation from rice to coconut and fish through improved facilities for storage and transport. Cassava, which compares poorly with rice, has been introduced to fill part of the resulting gap in food supplies (Prathapkumar et al., 1982).

Other developments include change from crop rotation to intensive agriculture, failure to replenish soil from which organic matter is removed, and efforts to compensate for this by using chemical fertilisers. Used injudiciously, fertilizers can interfere with soil texture and biology and the ecosystem and thus facilitate the growth of some pests, and the use of pesticides with their possible serious consequences for mankind and the ecosystem.

CARBODHYDRATE SOURCES, QUALITY AND UTILITY

The predominant dependence of man on carbohydrates, which account for as much as 65-80% or more of food energy in poor diets, takes us to a consideration of carbohydrate sources, advantages, disadvantages and the possibility of minimizing the degree of dependence on them (Rajalakshmi, 1974). Quantitatively starch, lactose and sucrose account for the bulk of utilizable carbohydrates in human diets but their proportions vary in different diets, with starch predominating in poor diets, and the other two in the diets of the affluent west. For instance, the dietary amount (g) derived from starch, sugar and lactose are of the order of 400, 5-10 and 2.5 in tribal Gujarat, 250-300, 15-20 and 3-6 in rural Baroda, 250, 50 and 25-30 in urban Baroda and of the order of 200, 100-150 and 25 in Western diets (Rajalakshmi, 1976; Rajalakshmi and Ramakrishnan, 1977).

While sucrose is a readily utilizable carbohydrate and a convenient source for increasing the energy value, palatability and carbohydrate content of diets, it is taken largely in the form of refined sugar and as such is devoid of nutrients essential for its metabolism. Even the unrefined form contains only minerals and is practically devoid of protein and vitamins. Its excess consumption is associated with dental cares, hypercholesterolemia and diabetes. It also compares poorly with starch in maintaining beneficial intestinal microflora (Bender and Damji, 1972). Although the sparing action of carbohydrates on proteins is well known, this capacity seems to be greater in the case of starch. In several studies rats fed purified diets deficient in protein or devoid of a specific vitamin or mineral are found to develop the deficiency symptoms much faster with sugar as carbohydrate source than with starch. The observation is probably accounted for by the fact that protein digestion takes about the same time as starch digestion so that the local availability of glucose for oxidation facilitates the active transport of amino acids as well as other nutrients. The differences in vitamin status could be due to those in intestinal microflora on the basis of rat studies in this laboratory (Varsh Shah, unpublished).

Although sucrose is a disaccharide, the response of blood sugar levels to sucrose is faster than that to glucose (Crane, 1968) presumably because of its rapid hydrolysis to glucose and fructose and the simultaneous absorption of both by the two different mechanisms of active and facilitated transport (Wohl and Goodhart, 1964). On the other hand, the response of blood sugar levels to starch is relatively slow. In studies carried out in this laboratory, blood sugar levels in the fasting state and 30, 60, 90 and 120 minutes after the administration of a standard dose of carbohydrate were found to be 72, 135, 91, 67 and 68 in the case of glucose, 63, 86, 69, 64, and 61 in the case of wheat and rice, and 60, 90, 75, 58 and 63 in the case of coarse grains such as kodri which have a high fibre content resulting in slower digestion of starch (Vaguala, unpublished). Also, whereas starch is consumed in the form of food grains or roots and tubers which provide not only carbohydrates but also protein, minerals and vitamins, refined sugar provides no other nutrients.

Starch sources also contain fibre which has both beneficial and deleterious effects. The former include the provison of satiety, regulation of intestinal motility, inhibition of the absorption of cholesterol and its precursors, water holding capacity and also perhaps the promotion of favourable intestinal microflora. However, different substances which are grouped in this category such as cellulose, hemicellulose, pectin, algal polysaccharides, gums and mucilages may have different effects (Spiller and Shipley, 1977). Mammals do not synthesize cellulase but synthesize cellobiase whose functional significance is unknown but which could have a role in the further degradation of cellobiose released from cellulose by intestinal microflora. This would facilitate the supply of glucose for certain organisms and also contribute to the digestibility of partially digested cellulose. Not all components of fibre are totally non-nutritive. For instance, about 20% of fibre in cooked cereals and tubers and more than 25% in vegetables and fruits may be utilized ultimately (Kay and Truswell, 1977 a,b; Kay, 1978). Cellulose is also convertible to edible foods such as mushrooms by microbial organisms.

The other common carbohydrate is lactose. Much is said about the lactose intolerance of non-whites despite the fact that lactose is the first carbohydrate consumed by mammalian infants with the exception of those of the sea-lion whose milk is devoid of carbohydrate (Pilson and Kelly, 1962). In adult men and women (Indians) in this laboratory who were initially given 50 g of lactose and the amounts then gradually reduced or increased by 25 g every week according to whether they showed clinical 'tolerance' or 'intolerance', the minimum level tolerated was found to be 30 g as compared to 15 g in a cup of milk. The maximum was 175 g or more. One individual was found to tolerate 175 g both clinically and chemically (Jarullah, unpublished). Clinical or chemical intolerance to lactose does not signify intolerance of milk. This may be because lactose intolerance is less evident in the presence of protein and fat. Even those unable to tolerate milk are able to take it along with cereal products or in the fermented form which has been advocated for infants with lactose intolerance (Hanafy, 1980).

Much of the so called lactose intolerance develops in the postweaning period. The decline in lactose activity during the suckling period is a well known phenomenon in many species. In man it must be partly attributed to the high prevalence of gastrointestinal disorders in this age group as lactose is the first disaccharidase affected in such disorders which are rampant in 'non-white' areas. Sucrose is less affected and maltase and amylase not much affected, again a point in favour of starch (Crane, 1968). The only food source of lactose, namely milk, is a food of choice because of its easy digestibility and excellent nutritive value, especially with regard to calcium, vitamins and protein. However is in short supply in poor areas and its distribution is far from equitable. In India, the urban population, forming 20-25% of the total, consumed 35-40% of the milk produced primarily in rural areas. Mean consumption per capita per day is 100 g but varies from less than 50 g to about 800 g or more. Another dimension to the problem is the lack of equitable distribution even within the family, with children getting less than adults with the advent of tea and coffee, cheese and ice-cream. This is a reversal of the traditional practice according to which the youngest child in the family got the highest priority for milk supplies. Similarly buttermilk, formerly a by-product of home-scale butter-making, is no longer freely available to the poor in rural areas with the advent of dairies which sell skim milk powder to be used as such, or for products consumed by the upper class such as chocolates and malted beverages. Dairy animals which live on grains do not compete directly with man for food grains. They contribute to the enrichment of the soil in many ways and convert plant proteins to milk proteins more efficiently (3:1) than meat animals (typically, 5:1, 10:1). Thus there is a strong case for improved production of milk and ensuring the right priorities in its distribution.

In the ultimate analysis, starchy foods such as roots, tubers and other vegetables, as well as cereals, millets, and legumes are the most common sources of carbohydrates which provide the bulk of calories and other nutrients for the poor. Among foods in the first category, potatoes and wild yams are more satisfactory with regard to protein which accounts for 6.6% and 9% of total calories as compared to 6, 4.1, 4, 1.5, 1.2, 0.2, and 0.2% for elephant yams, plantains (a species of bananas), sweet potatoes, cassava, sugar cane juice, sago and arrowroot respectively. Traditionally, such foods were consumed along with other foods such as cereals or millets, legumes, meat and/or fish. Where agro-economic developments or cultural restraints (as in the case of young children during episodes of illness) have limited the use of the latter, severe malnutrition has resulted, particularly in young children in areas where such foods are given to them preferentially. Deficiency symptoms are occasionally found in agricultural labourers before harvest when the food supplies are scarce and the energy requirements are high. However, where such starchy foods are judiciously used along with other food sources, after the removal of toxic factors where necessary (e.g. cyanogens in cassava), they can provide an easily digestible source of calories in the poor man's diet. We have found that root vegetables can be incorporated in dishes based on cereals and legumes so as to provide about 15-20% of total calories without jeopardizing either their palatability, which is often enhanced, or nutritive value (Rajalakshmi and Ramakrishnan, 1969a).

Cereals and millets, however, are the most predominant sources of carbohydrates and account, in the poor Indian diet, for 77 percent of food energy and 77, 36, 88, 63 and 31 per cent of other nutrients such as protein, calcium, iron, riboflavine and vitamin A (as carotene) respectively (Rajalakshmi, 1974, 1981). They vary in protein content from 6% in rice and kodri (Paspalum scorbiculatum) to 12-13% in wheat and maize but the differences in nutritive value are less evident in mixed diets. For instance, weight gain per week in weanling rats when fed only the staple as protein source varied from less than 2 g for kodri, 2-3 g for maize and jowar, 5-6 g for wheat and 6-7 g for bajra and barley. When supplemented with legumes and leafy vegetables, the corresponding values were 9-10 g in the first three cases and 11-12 g for wheat and 10% casein diet. Rats fed diets consumed by the low and high income groups in Baroda gained 8 g and 11-12 g per week. Rats fed the upper class diets compare well with those fed a 10% casein diet. The growth of pre-school and school children is very similar in different parts of India such as Kerala, Tribal Gujarat, Tamil Nadu, rural and urban Baroda, and Varanasi when matched for economic status (Rajalakshmi, 1976). A similar pattern is evident from the studies compiled by Patwardhan (1961) and the Indian Council of Medical Research (Copalan et al., 1971). Thus, economic status, by determining the intake of critical supplementary foods, seems to determine the overall nutritive value of diets based on different staples rather than the staple itself. Deficiency diseases such as beriberi and pellagra, associated with the consumption respectively of rice and maize, were and are found only among the poor, except in alcoholics. Incidentally they are seldom found in population groups consuming parboiled rice as in Tamil Nadu (Patwardhan, 1961) or whole maize as in tribal Gujarat (Rajalakshmi, 1974, 1981) or Latin America (Davidson et al., 1975). The prevalence of pellagra in southern USA and other regions early in this century could have been due to the consumption of maize grits rather than whole maize. The amounts of nutrients provided by the two differ, being 70 and 106 mg per 1000 calories for tryptophan and 1.8 and 5.0 mg for niacin (Horwitt et al., 1956). Thus, the introduction of modern techniques for the processing of grains may be fraught with dangers if due precautions are not taken. The prevalence of beriberi in the west following the advent of refined flour and its practical disappearance, except in alcoholics, after the enrichment of flour is an example of the triumph of technology, even though such solutions may have no validity for other cultures. For instance, to advocate lysine-enriched bread in India where wheat is consumed in the form of chapaties borders on absurdity. Baking bread requires more fuel which is already a scarce commodity for the poor. Production of bread is less than 1-2 g per capita per day (Rajalakshmi, 1974, 1981). It also involves the preferential use of refined flour because of its better acceptability and keeping quality. Refinement of flour results in the loss of 20-40% of edible portions, appreciable losses of protein as well as of critical amino acids such as lysine, methionine and tryptophan, because the germinal portions removed during refinement are richer in albumin and globulins containing more of these amino acids, whereas the left-over refined flour contains more of prolamin and glutelin deficient in these amino acids (Naik and Das, 1972). Rats fed whole wheat flour gained 6.0 g per week as compared to 2 g gained by those fed refined flour with standard vitamin and mineral mixtures (Rajalakshmi, 1976).

The high fibre content of grains such as kodri and ragi is an advantage in some respects but results in certain disadvantages such as decreased digestibility, decreased availability of minerals and unsuitability for children when cooked as such. However, certain tribals are found to subsist on whole paddy (rice with husk intact) ground into flour and baked like chapaties and do very well on this diet (Sohonie, 1972). In spite of its high fibre content which limits the value of ragi to infants and young children, it can provide excellent weaning food when processed according to tradition. The grain is steeped in water, allowed to sprout, wet-ground, tied in a muslin cloth and hung up so that the milky fluid filtering through can be collected in a large platter, dried, stored and used for making a product similar to corn-mash. A more general procedure is to allow the grain to sprout, dry in the shade, roast lightly, grind into flour, and sieve off only the coarse flour (Rajalakshmi, 1974, 1981). Wheat processed in this way has a malty flavour highly acceptable to children on the basis of long term feeding trials in which this was used along with bengal gram with or without added groundnut. The increase in vitamins during sprouting more than compensates for the small losses during roasting. Grains such as kodri with a high fibre content are well accepted and tolerated even by young children if incorporated in fermented foods such as dhokla and idli. We have also found puffed rice and parched grains suitable for use in infant and child feeding. The procedure used for parching in this country differs from that used for breakfast foods such as corn flakes and involves a brief exposure (2-3 minutes) of low moisture grains to a temperature of 230-240 C in country ovens with one side open. The grains are put in long-handled pans containing pre-heated sand, inserted in the oven with stirring, almost immediately removed and the sand sieved off. The vitamin losses in this procedure are less than 5% but some lysine losses are suspected on the basis of biological evaluation with rats (Rajalakshmi, 1974, 1976, 1981; Rajalakshmi and Ramakrishnan, 1977).

Lysine, however, is not a serious problem in diets containing adequate quantities of legumes, milk, meat, fish and or eggs all of which contain 0.4-0.5 g of lysine per g nitrogen. A diet providing cereals and legumes in ratio 4:1 compares with a 10% casein diet in protein value. Supplementation of wheat with isonitrogenous quantities of bengal gram, milk powder or fish flour is found to result in comparable improvements in nutritive value. Thus, the nutritive value of roots and tubers as well as cereals and millets can be considerably enhanced by the addition of legumes in appropriate amounts (Rajalakshmi, 1976; Rajalakshmi and Ramakrishnan, 1977).

The most common legumes used in India and adjacent regions are bengal gram, red gram, green gram and black gram, generally used after dehusking and decortication. Such dehusking is not associated with loss of minerals or vitamins. The resulting 'dals' have good keeping quality and are readily cooked and digested. Bengal gram dal after soaking in water and roasting has been used with sprouted and roasted wheat in breakfast gruels in feeding trials mentioned earlier. These as well as soybean have been incorporated in fermented foods such as idli and dhokla after sprouting and/or wet grinding where appropriate. Such foods are found to be well-accepted and tolerated by children (Rajalakshmi and Ramakrishnan, 1977).

Studies in this laboratory show that toxic factors in legumes such as trypsin inhibitors and hemagglutinins are hydrolysed at least partially during fermentation. The preparation of fermented foods in this country involves the making of a batter without addition of any inoculum and the fermentation is effected by the microorganisms in the raw materials used. The procedure used, namely, fermenting the batter at room temperature for 10-15 hours, is fairly simple compared to the involved procedures used for Tempeh and other similar products. Both germination and fermentation reduce cooking time and fuel requirements appreciably. They also result in other changes such as the partial breakdown of starch and increases in amino-nitrogen (indicating the partial digestion of proteins), in inorganic phosphorus (due to hydrolysis of phytate), and in the proportion of ionizable iron and vitamins such as riboflavin, niacin, pyridoxine, ascorbic acid total and alpha tocopherol. Increases in thiamine which are variable also occur under some conditions. The effects are additive when both procedures are combined i.e. sprouting followed by wet grinding and fermentation (Ramakrishnan, 1979; Rajalakshmi and Ramakrishnan, 1977). Fermentation, under some conditions, results in the synthesis of vitamin B_{12} which is absent in plant foods (Parekh and Varsha Shah, unpublished). The organisms involved in these changes in various fermented foods have been identified (Ramakrishnan, 1979).

Phytate in cereals and legumes interferes with the utilization of divalent conditions. Its adverse effects seem to diminish in the order, calcium, iron and zinc, probably because of differences in their affinity for phytate and/or in the amounts consumed (400-500 mg, 15-30 mg and 10-15 mg). The effects also appear to vary with habituation to high-phytate diets (Walker, 1951) and the form in which phytates are present. Sodium phytate has greater adverse effects than complex phytates present in wheat (Davidson et al., 1975). Those which are not readily released from cellulose in the intestine may have a smaller effect because of their decreased ability to chelate with ions in the bound form. The mammalian intestine secretes both acid and alkaline phytases which are capable of hydrolysing phytate (Ramakrishnan and Bhandari, 1979). Other factors are also likely to be involved in the etiology of zinc deficiency associated with whole wheat diets in the Middle East (Reinhold, 1976, 1980) as such prevalence is not seen in Northern India where the diets are similar. The amounts of zinc in the diet and the relative proportions of divalent cations which compete for chelation with phytate and vitamin D status (which influences intestinal alkaline phosphatase activity) may all be involved as factors.

The legumes Lathyrus sativus consumed in central India along with varying proportions of wheat, a scarce commodity following a poor harvest, is also found to contain a neurotoxin (B-N-Oxalyl-L- , B-diamino propionic acid) and its prolonged consumption results in lower limb paralysis (Patwardhan, 1961). Rats whose intestinal microflora are capable of breaking down this neurotoxin are able to grow as well on this legume as on bengal gram (Baroda studies, unpublished). Microorganisms capable of breaking down this neurotoxin have been identified in this laboratory (Ramchand, unpublished) although the best approach is to ban the cultivation of this legume altogether because of its tragic consequences.

Aflatoxin contamination of groundnut and its cake is found to be prevented by slightly roasting or drying the grain before oil extraction (Rampal, 1965; Chandrasekhara, 1972). Similar efforts to control moisture, humidity and storage conditions should prevent such contamination found occasionally in food grains such as parboiled rice.

CARBOHYDRATE SUPPLY

On the question of supplies, urgent steps are needed to increase production, minimize losses before and after harvest and ensure a more equitable distribution of the foods produced. Two thirds of the world's population live in poor areas where calorie deficits are of the order of 300-500 kilocalories per capita per day. Bridging this gap is well within the realm of possibility by increasing the production of foods requiring less land and feeding low on the food chain (Rajalakshmi and Ramakrishnan, 1969a; Pirie, 1976; Lappe and Collins, 1977). Foods advantageous from the point of view of yield, resistance to adverse environmental conditions, good digestibility and acceptability, cookability and nutritive value should be identified from the different categories mentioned above, their genetic potential improved and their cultivation and consumption popularized. Foods such as those consumed by tribes who have not yet lost all their traditional skills as hunter-gatherers should not be ignored. Those consumed by non-human primates in their natural habitats are also worth examination (Ganlin and Komnes, 1977).

With increasing population and increasing requirement of land for purposes other than food production, the cultivable land available per capita is unlikely to increase. Hence an increase in productivity can only be achieved by conserving soil texture and microorganisms, exploiting nitrogen fixing bacteria and returning to the soil all that has been removed from it instead of burning it up as fuel (e.g. cowdung, dry leaves) or dumping valuable organic matter in the rivers and oceans which also affects marine life (Fiennes, 1972; Pirie, 1976; Lappe and Collins, 1977). For instance, the cost of chemical fertilizers supplying isonitrogenous quantities of cowdung is much more than that of kerosene providing an equivalent amount of fuel (Rajalakshmi, 1969), without taking into account the water, manpower and energy requirements for the production of fertilizers and the problems posed for the ecosystem. Such shortcuts for increasing cultivation as the widespread and indiscriminate use of chemical fertilizers and pesticides without organic manure may ultimately be counter-productive. A greater dependence on foods of plant origin should help to reduce the amount of the scant supplies of fresh water needed for food production. International trade and commerce should also be reoriented consistent with the philosophy of 'Food First' and for every one (Lappe and Collins, 1977).

Finally, agriculture which is largely based on man-power should be geared to a more efficient and just exploitation of human labour. An underpaid agricultural labourer who is unemployed for 4-6 months of the year, who is left high and dry after a bad harvest, and whose diet even normally provides not much more than sufficient calories for basic metabolism, is unlikely to be efficient in agricultural production. Both enlightened self-interest and humane considerations show the need for giving him a better deal and enabling him to develop himself and the land he cultivates.

ACKNOWLEDGEMENTS

The author thanks Harjit Singh, Nikita Shah, Suresh Siddhanti and Sunita Zalani for their clerical and bibliographical help. The Baroda studies referred to in this paper were carried out under a variety of research projects including those supported by the Ford Foundation, UNICEF/FAO/WHO, CARE, USDA and the University Grants Commission.

REFERENCES

Altechul, A.M. (1965). Proteins, Their Chemistry and Politics. Basic Book Inc. Publishers, New York.

Bender, A.E. and K.B. Damji (1972). Some effects of dietary sucrose. World Rev. Nutr. Diet., 15, 106-141.

Cahill, G.F. and O.E. Owen (1968). Some observations on carbohydrate metabolism in man. In: Carbohydrate Metabolism. Vol. I, (F. Dickens, P.J. Randle and W.H. Whelan Eds.) Academic Press, London, New York.

Cartwright, T.E. (1972). The ways of acquiring and utilizing energy. In: Biology of Nutrition (R.N.T.W. Fiennes Ed.) Pergamon Press, Oxford, pp. 205-248.

Chandrasekara, M.R. (1972). Development of processed foods based on oil seed meals. In: Perspectives in Nutrition (R. Rajalakshmi, Ed.), University of Baroda, India, p. 80.

Cole, L.C. (1958). The ecosphere. Sci. Amer. Reprint No. 144.

Carne, R.K. (1968). Digestion and absorption of carbohydrates. In: Carbohydrate Metabolism. loc. cit.

Davidson, S., R. Passmore, J.F. Brock and A.S. Truswell (1975). Human Nutrition and Dietetics. Churchill Livingston.

Fiennes, R.N.T.W. (1972). Biology of Nutrition. Pergamon Press Ltd., Oxford.

Ganlin, J.C. and M. Komner (1977). In: Nutrition and the Brain (R.J. Wurtman and J.J. Wurtman) Raven Press, New York.

Gopalan, C., S.C. Balasubramanian, B.V. Ramasastri and Rao K. Viswanwara (1971). Diet Atlas of India. National Institute of Nutrition, ICMR, Hyderabad, India.

Hanafy, N.N. (1980). Maternal Nutrition and Lactation Performance in Egypt. In: Maternal Nutrition During Pregnancy and Lactation. (H. Aehi and R. Whitehead, Ed.), Hans Huber Publishers, Bern. p. 222-232.

Horwitt, Harvey, Rothwell, Cutler and Haffron (1956). c.f.: Modern Nutrition in Health and Disease. 3rd edition (N.G. Wohl and R.S. Goodhart, Eds.). Lee and Febiger, Philadelphia, pp. 200.

Kay, R.M. and A.S. Truswell (1977a). Effect of citrus pectin on blood lipids and fecal steriod excretion in man. Am. J. Clin Nutr., 37, 171-175.

Kay, R.M. and A.S. Truswell (1977b). The effect of wheat fibre on plasma lipids and fecal steroid excretion in man. Brit. J. Nutr. 30, 227-235.

Kay, R.M. (1978). Kinds and Functions of Plant Fibre. In: Nutrition in Transition (P.L. White and Nancy Selvey, Eds.) American Medical Association. Monree, U.S.A.

Lappe, F.M. and J. Collins (1977). Food First. Houghton Mifflin Co., Boston.

Naik, M.S. and N.B. Das (1972). Amino acid imbalance in cereal grain proteins. In: Perspectives in Nutrition (R. Rajalakshmi, Ed.) University of Baroda, India.

Pilson, M.E.Q. and A.L. Kelly (1982). Composition of milk from Zalophus californianus, the Californian Sea-lion. Science, 135, 104-105.
Pirie, N.W. (1976). Food Resources Conventional and Novel. 2nd Ed. Penguin Books Ltd., Middlesex, England.
Pratapkumar, S., R. Vijayaraghavan, Sail Suresh, N.V. Rao, R. Rajalakshmi, C.V. Ramakrishnan (1982). Comparative profiles of severely malnourished children in different categories in coastal Kerala and interior Tamil Nadu, Baroda J. Nutr., Vol. 9 (in press).
Rajalakshmi, R. (1969). Applied Nutrition, 1st edition, published by the author.
Rajalakshmi, R. and Ramakrishnan, C.V. (1969a). Dietary and nutrient allowances for Indians. Pl. Fds. Hum. Nutr., 1, 163-192.
Rajalakshmi, R. and C.V. Ramakrishnan (1969b). Horticulture in relation to nutritional requirements. Pl. Fds. Hum. Nutr., 1, 79-92.
Rajalakshmi, R. (1974). Applied Nutrition. 2nd edition published by Oxford and IBH Publishing Co., New Delhi.
Rajalakshmi, R. (1976). Monograph on Pre-school child malnutrition - Pattern, Prevalence and Prevention. University of Baroda, Baroda, India. Also Baroda J. Nutr. 3, 2.
Rajalakshmi, R. and C.V. Ramakrishnan (1977). Formulation and evaluation of meals based on locally available foods for young children. Wld. Rev. Nutr. Diet., 30, 274.
Rajalakshmi, R. (1981). Applied Nutrition. 3rd Edition. Published by Oxford and IBH Publishing Co., New Delhi.
Ramakrishnan, C.V. (1979). Studies in Indian fermented foods. Baroda J. Nutr., Vol. 6.
Ramakrishnan, C.V. and S.D. Bhandari (1979). Differential developmental pattern of alkaline and acid phytase and phosphatase activities in rat intestine - Experientia, 35, 994.
Rampal Veena (1965). The effect of different supplementations on the biological value of groundnut cake. M.Sc. Dissertation, M.S. University, Baroda, India.
Reinhold, J.G., B. Farari, P. Abadi and F. Ismail Reigi (1976). J. Nutr., 106, 493.
Reinhold, J.G., Parichehr Abadi, Bahran Faraji, Faramars Ismail Beigi and Robert M. Russel (1980). Depletion of calcium, magnesium, zinc and phosphorus associated with consumption of humans of whole meal wheaten bread rich in fibre and phytate. Baroda J. Nutr., 7, 55-62.
Schuphan, W. (1961). Zur qualitat das Nahrungs Pflanzen, urban and Schwarzenberg, Munchen.
Sohonie, K. (1972). Nutritive value of paddy flour and rice mill residue. In: Perspective in Nutrition (R. Rajalakshmi, Ed.) University of Baroda, India.
Spiller, G.A. and H.A. Shipley (1977). Perspectives in Dietary Fibre in Human Nutrition. Wld. Rev. Nutr. Diet., 27, pp. 125-131.
Walker, A.R.P. (1951). Cereals, Phytic Acid and Calcification. Lancet ii: 244.
Wohl, N.G. and P.S. Goodhart (1964). Modern Nutrition in Health and Disease. 3rd Edition, Lea and Febiger, Philadelphia.

IMPROVING THE SUPPLY, QUALITY AND UTILITY OF FATS

Joseph Moolayil
Technical Manager, Lever Brothers (M), Sdn Bhd
Kuala Lumpur, Malaysia

ABSTRACT

Oils and fats, besides being the most concentrated energy source, also supply essential fatty acids, carry fat-soluble vitamins, control blood lipids, build cells and add palatability and variety to our food. Among the numerous oils, soybean oil, sunflower oil, palm oil, butter and tallow currently constitute nearly 60% of the world's production of oils and fats, and nearly 80% of its trade. By the year 1990, soybean and palm alone will account for 40% of the world's oil supply, and thus become the main sources of any future increased supply.

From a nutritional point of view, a dietary intake of at least 30% of energy as fat is recommended. In many developing countries, fat consumption provides even less than 10%. In order to enable the under-nourished millions to achieve at least the minimum fat intake, about 20 million tons more oil are needed anually. Increased supply from additional acreage, improved irrigation, and better manure application have their limitations; increased supply from better yielding varieties and strains, improved pollination, etc., are preferred. Relatively untapped but high potential sources of additional oil are available from many tree seeds in Asia, Africa and South America. A very high quantity of rice bran oil would also be available, if available bran is properly collected and promptly extracted.

Improving the quality and utility of fats necessitates the application of correct technology from their origin to their eventual consumption. Influencing the fatty acids formed, whether in plants or animals, by genetic engineering, and applying appropriate technology in extraction, storage, transportation and processing are some of the essential steps in maintaining quality and nutritional value. While at one time all the oils exported were in the crude form, currently, substantial quantities are exported after refining. Modifications to current processing and handling techniques have to be made to obtain optimum results from this new pattern.

KEYWORDS: fats, fat requirement, fat availability, fat modification, oils, edible oils, soybean oil, palm oil, coconut oil, rapeseed oil, rice bran oil, oils of tree origin, oil quality deterioration, processing fats.

INTRODUCTION

Fats are used predominantly for human consumption and are the most concentrated source of energy, providing about 9 Calories/g compared to 4 Calories/g for protein and an equivalent amount for carbohydrates. Although the main function of fats is considered to be the provider of energy, they have other vital uses. Some of the most important of these are: (1) improving the flavour of foods and increasing palatability; (2) being the source of fat-soluble vitamins like A, D and E; (3) being the source of essential fatty acids, which cannot be made in our body; (4) controlling blood lipids; (5) being structural components of membranes; (6) participating in the storage and transportation of metabolic fuels; (7) being involved in tissue immunity; (8) delaying digestion of food and thus preventing premature sensation of hunger; and (9) being of special significance to mothers when breast-feeding their babies. Without sufficient fat content in mother's milk, the calorie requirement of the infant cannot be adequately fulfilled. There are probably other deep-rooted reasons for the natural craving of the human organism for fatty foods.

FAT REQUIREMENTS: PRESENT AND FUTURE

The energy requirements of a person varies with age, sex, the type of activity, etc. For calculation purposes, energy requirements of an average person may be taken as 2300 Calories per day. From a nutritional point of view, a level of between 20% and 30% energy input by fat is recommended. Based on 25% energy input from fat, the recommended fat intake per person per day is 64 g or about 23.3 kg per year.

In developed countries, about 50% of the fat intake is obtained from invisible fat due to the large proportion of fatty foods consumed. In the developing countries, where the food consists mainly of vegetables, rice, tubers, etc., the daily intake of invisible fat is extremely low. On a world-wide basis, invisible fat intake is less than 20%. Allowing for an invisible fat intake of 20%, the average fat intake of visible fat per person per day should be at least 51 g or 19 kg per year.

World population projections up to the year 2000 are given in Table 1 and can be used to show that the yearly fat requirement, to provide 19 kg per person per year, is 91 Mt in 1982, 105 Mt in 1990, and 125 Mt in 2000. Current and future world production of fat is given in Table 2.

Table 1: World Population

	Population (million persons)		Net Growth	% Growth in 25 years
	1975	2000	(in millions)	(from 75-2000)
DEVELOPING COUNTRIES	2997	5080	2083	69.5
DEVELOPED COUNTRIES	1093	1270	177	16.2
WORLD	4090	6350	2260	55.0

Table 2: Forecast of World Production of Major Oils and Fats. (Source: Palm Oil Research Institute of Malaysia, (provisional data unpublished.))

OIL/FAT	YEAR				
	1981*	1985	1990	1995	2000
Soybean oil	12.17	17.09	22.88	29.99	38.54
Palm oil	5.06	7.07	10.59	15.39	21.77
Tallow	6.05	6.73	7.60	8.50	9.42
Butter	4.75	5.21	5.54	6.88	6.24
Lard	3.83	4.00	4.00	4.00	4.00
Sunflower seed oil	4.76	5.39	6.04	6.72	7.40
Rapeseed oil	3.86	4.49	5.66	7.03	8.59
Coconut oil	3.13	3.50	3.75	3.95	4.09
Cottonseed oil	3.20	3.31	3.39	3.44	3.47
Groundnut oil	2.80	3.08	3.04	2.99	2.95
Olive oil	1.97	1.79	1.91	2.02	2.12
Fish oil	1.14	1.05	1.05	1.05	1.05
Palm kernel oil	0.71	0.78	0.93	1.10	1.27
TOTAL	53.43	63.49	76.38	92.06	110.91

* Actual

Comparing the world's fat requirement to furnish minimum visible fat needs and the actual production, we are faced with the reality that we are not producing enough. In 1982, while we would have required 91 Mt, we have produced only 55 Mt. While we will require 105 Mt in 1990, we may produce only 76 Mt. If the fat available in 1982 is divided equally among the world's population, availability is about 11.5 kg per person per year, or about 60% of the recommended fat intake. Producing about 60% of the recommended fat intake would not be too bad a situation if all the fat produced in the world was shared equally by all the inhabitants. The situation is far from this. Developed countries in 1980 produced about 40% of the fat and consumed nearly 42%. In 1990, production and consumption by developed countries are expected to be about 35% and 38% respectively. On the other hand, production and consumption by developing countries in 1980 were 34% and 33% respectively, while in 1990, they are expected to be 38% and 39% respectively. More than 75% of the world's population lives in developing countries and they consume about 34% of the fat produced. On the other hand, 25% of the population living in developed and centrally planned economies consume 66% of the fat produced. That is, while the fat availability of an average person in developing countries is 5.1 kg per year, it is 31.3 kg per year in developed countries.

Table 3 gives per capita availability of fats for consumption in India over the last 10 years. From the table, the following conclusions can be made: (1) Per capita availability of edible fat has not improved significantly in India over the last 10 years. (2) Although on a world-wide basis, availability per person per year is 11.5 kg, for the average Indian, the availability is about 5.0 kg. (3) Fat availability to an average Indian is about 42% of that available to a person on a world-wide basis. (4) Fat availability to an average Indian is only about 26% of the recommended fat intake.

Table 3: Per Capita Availability of Edible Oil and Vanaspati in India. (Source: Economic Survey 1981-82, Government of India.)

Year	Kg/Year
1970-71	4.5
1975-76	4.3
1976-77	4.1
1977-78	4.7
1978-79	4.8
1979-80	4.7
1980-81	5.2

Table 4: Daily Availability of Fats and Oils for Consumption in Some Countries (per capita).

	AVAILABILITY OF FATS (kg/year)
DEVELOPING COUNTRIES	
Thailand	1.8
Afghanistan	2.2
Nepal	3.3
Sri Lanka	3.6
Indonesia	3.6
Philippines	4.0
India	4.7
Burma	5.5
Pakistan	6.9
Malaysia (Peninsular)	10.2
DEVELOPED COUNTRIES	
Sweden	19.7
Switzerland	22.3
United Kingdom	23.7
United States of America	24.8
France	26.6

ENSURING MINIMUM FAT INTAKE FOR ALL

Even if it is considered possible to increase fat production by the necessary quantity, the problem of ensuring recommended fat intake by all would not be solved. As can be seen from Table 5, while imports of edible fats of developing countries, other than Malaysia, Brazil and Argentina, increased from 1.9 Mt in 1969/70 to 6.8 Mt in 1979/80, exports decreased from 2.9 Mt in 1969/70 to 2.6 Mt in 1979/80.

Table 5: World Trade in Total Edible Oils. (Source: Paper presented by Mr. Thomas in his Leverhulme lecture at the SCI Annual Meeting in Liverpool on 16 July 1982, published by Chemistry and Industry, 17 July 1982.

	EXPORTS (Mt)		IMPORTS (Mt)	
	1979/80	1969/70	1979/80	1969/70
Developed countries	8.977	4.038	8.349	6.095
Malaysia	2.624	0.454	0.081	0.026
Brazil	0.906	0.128	0.202	0.010
Argentina	1.054	0.128	0.004	0.003
TOTAL	13.6	4.7	8.6	6.1
Other developing countries	2.6	2.9	6.8	1.9

There has been a tremendous growth in the production and, consequently, exports of fats from high income countries, mainly North America. By contrast, with the exception of the above-mentioned three developing countries, increase in production rate in many developing countries is not even sufficient for meeting the increase in population growth. A number of developing countries, formerly exporters of vegetable oils and fats (e.g., India, Ghana and Nigeria), have over the years, become importers of these commodities (van der Hoven, 1981). The extent of this rising tide of oil imports can be illustrated by an Indian example. In 1970, it imported 80,000 t of vegetable oils at a cost of $30 million US. By 1980, its imports of vegetable oils had risen to 1.35 Mt at a cost of more than US $700 M. The immensity of this increased burden can be better appreciated if one bears in mind that, during the same period, that country's import bill for petroleum also rose from $164 million in 1970, to $7000 million in 1980. This diversion of resources will inevitably slow down the rate of economic development of such countries (Thomas, 1982). The country cannot afford to import all the fat necessary to meet the minimum recommended intake of fat. Most of the developing countries are in the same situation, while some are in worse situations. Therefore, in order to meet fat intake requirements of the needy people, it is not only necessary to increase world production of fats, but also to ensure that the increase is made in those developing countries where the population is under-nourished.

MAJOR FAT SOURCES OF THE WORLD AND IMPROVING THEIR SUPPLY

In the past, increase in the production of fats has been largely due to increase in the area cultivated. Taking the major oil crops, about 70% increase in oil production during the 1970's was due to larger areas being cultivated with only 30% of the increase coming from higher yields. However, there is only a finite amount of land which has so far not been brought into cultivation, and much of this land is marginal and difficult to exploit agriculturally. In the future, we will have to rely more and more on improved yields to increase production of oils and fats (Thomas, 1982). Although sources of edible fat are quite numerous, most of the world depends for its fat supply on a very few. In 1981/82, soybean, rapeseed, sunflower, cottonseed, coconut, palm, butter, tallow and lard constituted 85% of the world's fat supply. Among these, production/demand for the last three seemed to have stagnated and it is not possible to expect vast improvement in production in the future. Therefore, it is clear that the world will depend for the increase in future fat supply on soybean, rapeseed, sunflower, cottonseed, coconut and palm. Another potential source of increased supply of fat is from conventional sources.

Soybean Oil

Soybean is by far the largest of the six vegetable sources of fat, which together provide 60% of the total fat output, constituting about 25% of the world fat supply. Soybean will continue to be, by a great margin, the largest single source of fat supply even at the turn of the century. The United States is the main producer followed by Brazil and now Argentina. The world, especially the fat-hungry millions, owes a lot to the soybean for being their main source of fat in times of scarcity. However, the main consideration in growing soybean, which contains 18% oil, is the market for meal from which it derives about 60% of its value. It is an annual crop and, therefore, production can be easily controlled, unlike the perennial crops, oil palm and coconut, which remain as steady suppliers of fat through thick and thin. The main consideration in controlling production will be the economic returns from meal, although the returns from oil are also taken into consideration.

Palm Oil

The second largest source of vegetable fat supply is the oil palm, a tree which produces two distinctly different fats in abundant quantities. The growth of this source has been spectacular in recent years. Although increase in production has been mainly due to increase in the area planted, recently, yield per unit area also has increased tremendously. When commercial cultivation started in Malaysia, oil yield ranged from 1.25 t/ha to 3.1 t/ha. Currently, oil yield ranging from 5.5 to 6.6 t/ha is being achieved. Individual field yields of oil of over 8 t/ha are becoming quite common.

The recent introduction of the pollination weevils, the *Elaedabius camerunicus*, has eliminated the substantial cost of artificial pollination, and additionally improved upon the yield per unit area by a more successful overall pollination (Nielsen, 1982). Increase in yield by this alone will be more than 10%. Techniques for vegetative propagation of oil palm via tissue culture are now being developed and improved. The first clones produced from selected ortets have recently been planted in the field. Most of the ortets have yielded over 60 kg of oil per year, equivalent to over 8 t/ha at a density of 138 palms/ha. In several instances, these high yields of the progeny trials in which the ortets are situated, come largely from high oil to bunch, so it is likely that the clones will be similarly high yielding. Thus, a 30% increase in oil yield from clones seems quite feasible, and it has been estimated that yields of 12 t/ha might ultimately be achieved (Corley et al., 1981). Possibilities also exist for modifying the chemical composition of palm oil. By cross-breeding the common variety, Elaeis quineensis, with a related species, the South American variety, Elaeis olifera, it has been found that palm oil with higher content of unsaturated fatty acids can be produced.

The oil palm has responded positively to man's research ambitions not only to increase the oil yield tremendously, but also to altering the oil's chemical composition. No other major oil bearing crop has responded as well. Producing from its third year continuously until it is thirty, photosynthesizing 365 days a year under tropical sun and rain, oil palm is the most prolific synthesizer of fat known to man. For comparison, while the soybean produces about 340 kg of oil from one hectare, the oil palm currently produces 20 times as much and in the future, it is going to be even higher. The oil palm appears to have the best potential to solve the fat deficiency in developing countries situated in the tropics where suitable land for its cultivation is available. Its growing, processing and application techniques are widely known and can be put into practical use within the shortest time possible. Increases in supply due to genetic engineering and other means, however attractive in the long term, take years to be implemented widely at the grass root level, and therefore cannot be considered a solution to the present acute shortage of fat in developing countries, at least for the immediate future.

Coconut Oil

The coconut is another source of fat which has great potential for increased production, although its production has not increased significantly in recent years and no dramatic increase is forecast for the near future. Average yield per hectare per year is about 1.0 t. Increased yield of about 2.5 t/ha from higher yielding hybrids has been found to be possible. Well grown hybrids have given 5.0 t/ha. Ironically, its durability - productive life is said to be 100 years - inhibits replacement of low yielding varieties with high yielding ones. This is unfortunate because the coconut palm produces a relatively high yield of oil from a unit of land, and also is a plant, every portion of which is useful and economically remunerative to the producer. Further, coconut oil, having been used extensively from time immemorial by the tropical housewife, is much more appreciated than any other imported oil. Although no breakthough [in tissue culture and weevil pollination] has been found yet for coconut, it is reasonable to assume that such success may be too far off because of the success in the case of its brother, oil palm.

Rapeseed Oil

The production of rapeseed oil is expected to more than double in the next twenty years. The oil was introduced to the Canadian farmers as an alternative crop to wheat. A setback was suffered when the long chain fatty acid component of rapeseed oil, erucic acid, was shown to have possible nutritional hazards. However, within a few years, an exceedingly efficient research effort by plant breeders succeeded in producing varieties almost free of erucic acid in sufficient quantity to enable the Canadian farmer to change his product. As a result, the growth in rapeseed has been encouraged by the EEC agricultural policy, so that countries in which it was not an economical crop are now significant producers (Berger, 1980).

Sunflower Seed Oil

The major grower of sunflower seed has been the U.S.S.R., and since harvests fluctuate rather widely, the surplus has been variable. In recent years, sunflower has become popular because of its polyunsaturated acid content and is now more widely grown. The U.S.A. has become a significant exporter in recent years, contributing much to the overall growth in production (Berger, 1980). Cottonseed oil and groundnut oil tonnages are not expected to increase significantly over the next 20 years.

Non-conventional Sources

Rice Bran Oil: World production of rice currently has an oil equivalent (using an 80% factor) of 5.5 Mt. The main producers of rice are China, India, Indonesia, Thailand, South America, and Africa. In those areas, there is an acute shortage of fat. Therefore, it would have been ideal if at least half of what is available in those countries could be extracted and used for edible purposes. Even 20% would go a long way. In India, the current potential for rice bran oil is 0.48 Mt. The actual production of rice bran oil has been much less, of the order of 120,000 t, and of this, edible grade rice bran oil has been only about 10,000 t. Recent recommendations from India (National Seminar, 1981) should be noted: "The matter regarding fuller exploitation of this important source of rice bran oil is under active consideration of the Ministry of Civil Supplies. A high level Inter-Ministerial Committee has been constituted by this Ministry to look into the various aspects relating to increased production of rice bran oil in general, and edible grade rice bran oil in particular. Government is also taking various measures to encourage stabilization of rice bran so as to enable production of edible grade rice bran oil in increased proportions. For example, oil usage policy in Vanaspati has been formulated in such a way so as to give encouragement for increased production of indigenous oils including rice bran oil."

Oils of Tree Origin: Many trees in the tropical forest bear seeds which contain both edible oil and meal. No country other than India appears to have made serious efforts to collect these. In India, there are several oilseeds of tree origin which are exploited at present. These include *mahua*, *neem*, *sal*, *kusum*, *karanj* and mango kernels. A beginning has also been made in exploiting other sources like *babul* (National Seminar, 1981). The potential for salseed, which has an oil content of 13% and a meal content of 83%, is 6 Mt; and that for mango kernel, with an oil content of 9% and meal content of 90%, is said to be 2 Mt. The fat from these two sources are important because they can be used as cocoa butter substitute and extender, and thus command a very high unit price.

Setting up of autonomous bodies, one each for rice bran and for oils of tree origin with representations from all concerned, has been recommended to improve the use of these two important sources of oil for edible purposes. Such bodies would look into all problems connected with procurement of raw materials, extraction, quality control, and adequate remuneration.

Genetic Manipulation of Oil Bearing Crops

Existing methods of breeding and tissue culture utilize variations among plants already in nature. These variations can be increased by using various agents such as γ-irradiation to increase the rate at which genetic changes or 'mutations' occur. In the near future, it may also be possible to redesign the genetic blueprint of plants in highly controlled and specific ways by the techniques of genetic engineering to transfer individual genes. Although the feasibility of each step has been demonstrated, these techniques are still in th early stages of development. It now remains to bring them together to produce specific designed changes. The possibility of introducing the 'nitrogen fixing' properties of legumes into other species, notably grain crops, and thereby achieving enormous savings in the use of chemical fertilizer, has attracted much interest. It may also be possible to improve the amino acid balance of plant proteins or modify the fatty acid composition of fats and oils far more quickly using genetic engineering than by classical methods of plant breeding (Thomas, 1982).

IMPROVING QUALITY OF FATS

Quality deterioration of fats starts at the time the oil bearing seed or fruit matures and continues until it is finally consumed. Some of the fats, if left on their own, deteriorate faster than others. However, adequate precautions through harvesting, handling, extraction, storage, processing, packing and eventual supply to the customer can minimize the degree of deterioration. Chemistry plays a very important role in preserving the quality of fats. Quality deterioration of the fat before extraction is in some cases very severe. In some cases, like copra, groundnut, etc., after these have been adequately dried, oil deterioration is less dramatic. When insufficient drying is coupled with bad storage conditions, severe quality deterioration in oil occurs. Mold growth also sets in. In the well known case of groundnut, due to mold growth, toxic materials are present in the nut, and after the oil has been extracted, both the fat and the meal are contaminated. Serious diseases are caused to those who consume both the meal and fat, if consumed without refining. However, refining effectively removes toxic materials present in crude oil. In extreme cases, the meal becomes unsuitable even as animal feed. Although the well known case is groundnut, spoilage occurs in many other cases also.

Oxidation Due to Contact with Pro-Oxidant Metals/Alloys

Quality deterioration due to contact with pro-oxidant metals, notably copper and its alloys, brass and bronze, can start at the extraction stage and through storage, transportation, pumping, processing and packing. Brass couplings, valves/cocks, draining facilities, etc., should not be used.

Storage Tanks

Storage tanks should be constructed with facilities aimed for minimizing quality deterioration, for energy conservation and for easy cleaning. Size of the tank itself is very important. When liquid crude oil is stored, capacities of a few thousand tons per tank is possible, but for refined oil, smaller tanks are preferable. This allows easy and less costly segregation of the oil, if some of the refined oil goes off specification. Smaller tanks also allow more frequent emptying and are preferable for storing high melting fats. In general, refined oil storage tanks should have a maximum capacity of one day's output. As the material of construction is usually mild steel, it should be suitably coated. Cylindrical tanks with conical bottoms are preferred.

Heating Facilities

With fats of high melting point, it is necessary to heat the oil in order to make it homogenous, especially before charging in or charging out and for sampling. As oxidation is greatly accelerated at higher temperatures, heating should be the minimum necessary and carried out for minimal quality deterioration. Hot water heating is preferred to direct steam heating. If steam is used, it should only be of the minimum pressure necessary. Heating should be gradual and controlled and local over-heating should be avoided. Suitable side entering stirrers facilitate quicker and more uniform heat distribution. The temperature of the fat may be allowed to fall in the period between charging and discharging, such as during a long voyage, but it should not be allowed to fall too low, making the fat very hard. If this happens, when the oil is heated, local over-heating is bound to happen, especially if no stirring is possible. Heating should be such that the maximum temperature rise in 24 hours is 5°C, and the difference between different parts of the tank not exceed 5°C. Thermostatically controlled heating is preferable to manual control, and insulation of the tank minimizes heat loss and saves energy.

Aeration

Aeration of the oil, especially at higher temperatures, causes oxidation and thus causes quality deterioration. Even the obvious ways to minimize aeration are not always practised. Nitrogen sparging/blanketing will minimize the oxygen level in contact with the oil.

Contamination of Fat

Contamination of fat, sometimes with poisonous materials, occurs during transfer of oil, especially when it is shipped from one country to another. The pipe lines used for pumping, lorry tankers used for inland transport, storage tanks, ship's tanks, etc., are the sources of contamination from non-edible products previously contained in them. Previous use for non-toxic cargoes should be a requirement before the fat is pumped in. Inspection for cleanliness and fitness for transporting/containing edible fat should be certified.

Toxic Materials Present in the Fat

Many seed fats contain toxic materials at the time of their extraction. These could be derived from the residue of the weed killers and insectides used to protect the oil bearing plants from weeds and insect attack. International law should be promulgated to ensure that such chemicals are either non-toxic to animals or they degrade within a short time.

Addition of Preservatives/Anti-Oxidants to Fats

Quality deterioration of fats can be retarded to a certain extent by the addition of suitable preservatives and or anti-oxidants. The addition should be done as soon as the fat is extracted. Additional chemical research is necessary to find the most effective but non-toxic preservative/anti-oxidant for each type of fat. Further efforts are needed so that the use of these preservatives/anti-oxidants will truly ensure preserving the quality of the fats. International agreement would be the best level of government action to cover the usage of these materials.

IRMPOVING UTILITY OF FATS

Man has managed to greatly enhance the utility and versatility of fats in many ways, and chemistry has played a major role. Some of these improvements are: (1) Removal of free fatty acids, pro-oxidant metals, phosphatides, etc., by degumming and refining, resulting in enhanced shelf life for the fat. (2) Removal of colour and flavour by bleaching and deodorization which enables the fat to be subsequently flavoured and coloured to suit specific requirements. (3) Modifying chemical and physical properties by fractionation, hydrogenation, and inter-estification, enabling end products to be made to suit every conceivable functional and consistency requirement. (4) Blending and processing of fat to suit desires in appearance, taste and smell.

Modification to the present techniques and the invention of new techniques will further enhance the utility of fats. The effect of each of these in improving the utility of fats and their future use is a major topic, and only two recent developments with important implications for the future are given here.

Processing/Modifying Fats at Its Origin

Until the emergence of palm oil as one of the major sources of fat, most of the export of fats was in the crude and unmodified form. Now about 90% of palm oil exported from Malaysia is refined/modified which has further improved its utility. Two examples may be given.

Palm oil is reddish brown in colour as it is extracted from the fruit. The colour can be removed either by adsorption to a bleaching medium or destruction by heat. Bleachability is best when the oil is fresh and at that time it can be bleached almost colourless. When palm oil is shipped in the crude form for a long distance or a length of time, its bleachability gradually decreases. Depending on the conditions of storage, handling, heating, proximity to pro-oxidant metals, etc., the oil may not now be bleachable to such a good colour and hence cannot be used for products for which light colour is very important. Thus, if the oil is bleached at its point of origin, the cost is less and the final result better.

Fractionation is one of the modifications carried out to palm oil before exporting. By fractionation, palm oil is separated into two or more components. If it is separated into three fractions, they are known as palm liquid fraction or olein, palm mid-fraction, and palm solid fraction or stearine. Characteristics of each fraction can be varied to suit a customer's requirements. The mid-fraction, depending on conditions at which fractionation is carried out, is rich in triglycerides of structure similar to those in cocoa-butter. These mid-fractions are now available directly to confectionery manufacturers. This is another example of improving the utility of fats.

Biochemical Principles in Fat Modification

Until quite recently, mostly chemical and to a lesser extent, physical technologies were used in modifying fats. One such modification is interesterification in which a chemical catalyst is used. The catalysts used are sodium metal, sodium methoxide or sodium ethoxide. The catalyst, under suitable conditions, brings about acyl migration between glyceride molecules in such a way that the end product consists of glyceride mixtures in which the fatty acid components are randomly distributed among triglyceride molecules. Recently, it has been found that it is also possible to catalyze interestification using a group of enzymes known as lipases. Enzymes have several advantages over chemical catalysts in many processes since most are highly specific in the reaction which they catalyze and have a high catalytic power; most are active in conditions of pH, temperature and pressure which are mild in comparison to those required by chemical catalysts; and they are more rapid and efficient.

If a non-specific lipase is used to catalyze the interesterification of a triglyceride mixture, the triglycerides produced are similar to those produced by chemical interesterification. Many lipases, however, are more specific in the reactions they will catalyze. One group, for instances, catalyzes the releases of fatty acids specifically from the 1- and 3- positions of the triglyceride. In that case, acyl migration is confined to the 1- and 3- positions and a mixture of triglycerides which is unattainable by chemical interesterification is produced (Thomas, 1982).

The ability to produce novel triglyceride mixtures using specific lipases is of interest to the oils and fats industry because some of these mixtures have properties which make them commercially very valuable. This is illustrated by the following examples: 1,3-specific lipase catalyzed interesterification of 1,3-dipalmitoyl-2-monooleine (POP), which is the major triglyceride of the mid-fraction of the palm oil, with either stearic acid or tristearine gives products enriched in the valuable 1(3)-palmitoyol-3(1)-stearoyl-2-mono-oleine (POSt) and 1,3-distearoly-2-monooleine (StOSt). POSt and StOSt are the main components of cocoa-butter, and therefore it is possible by the interesterification reaction to produce a valuable cocoa butter equivalent from cheap starting materials (Macrae).

SUMMARY AND RECOMMENDATIONS

Fat has been an integral part of man's diet from time immemorial. Besides being the most concentrated energy source, fat has both nutritional and functional roles in our diet. Based on visible fat intake per person per year of 19 kg, on a global basis, only some 60% of the recommended fat intake is produced now. The serious problem is that the fat is not shared equally. For about three-quarters of the world's population living in developing countries, the average availability per person per annum is 5.1 kg, while the availability of fat for the average person in developed countries is 31.3 kg per year.

Developing countries, with few exceptions, are not producing sufficient fat and their imports have increased compared to 10 years ago, adding to their financial burden. Therefore, in order to meet the recommended fat intake of the needy people, it is not only important to increase the world's production of fat, but also to ensure that the increase takes place in those developing countries where the population is under-nourished. Although numerous fats are produced, most of the people in the world get their fat supply from very few of them. Soybean is by far the largest, and its share is going to be higher at the end of the century. Within the last few years, palm oil has attained second place. Oil palm produces much more fat per unit area than any other plant known to man. For those developing countries situated in the tropics, oil palm seems to be the best way to increase the fat supply. The potential for rice bran oil production is also very high. Fat from trees grown in the forest also has a high potential. Developing these two sources is particularly important because of their potential in countries where need is the greatest. Although performance in the past does not call for optimism, concerted effort will improve these two sources. Genetic engineering may revolutionize fat production in a way that is not predictable at present.

Improving the quality of fat necessitates attention to detail from the point of fats' origin in the plant or animal until eventual consumption. International standards/agreements in the construction of storage tanks, transporting, heating and pumping procedures for fats aimed at minimizing quality deterioriation is recommended. Further research into effective but safe preservative/anti-oxidants and international agreement on their usage will also minimize quality deterioration and should be undertaken.

Various processing/modifications being carried out to fat has improved its utility. Traditionally, processing/modifications are being carried out by the consuming countries. Process/modifications being carried out to palm oil at its origin has further enhanced its utility. Further research is necessary in ascertaining optimum processing/modifications of fats at their country of origin and at their point of eventual consumption. Application of this technique to other fats also should be investigated. Recent development in the lipase catalyzed interesterification further enhances utility of fats. Further work in this field coupled with wider application of this technology aimed at improving utility of fats should be explored.

REFERENCES

Berger, K.G. (1980). Oils and fat supplies - temperate and tropical. 10th Anniversary Seminar of University Pertanian Malaysia, Serdang, Malaysia.

Corley, __, Wong, __, Wooi, __ and Jones, __ (1981). Early results from the first oil palm cone trial. International Symposium on "The Oil Palm in Agriculture in the Eighties", Kuala Lumpur, Malaysia (1981).

Macrae, A.R. Lipase catalyzed interesterification of oils and fats. Am. Oil Chem. Society.

National Seminar (1981). Proceedings National Seminar on Problems and Prospects of Non-Conventional Oils. Calcutta, India, Nov. 13, 1981. Pub. by East India Oil Millers Assoc., Oil Technologists Assoc. of India (East Zone).

Nielsen, B. Bek (1982). Forecasting consumption and long term costing of palm oil. Address to visitors to Unite Plantation, Telok Anson, March 1982.

Thomas, T. (1982). Chemistry and biology - An interface. Chemistry and Industry, 17 June 1982.

van den Hoven, H.F. (1982). Some aspects of the World Food Problem. 57th World Congress organized by the International Association of Seed Crushers, San Francisco, 10 June 1982.

RUMEN-PROTECTED AMINO ACIDS

Heribert Offermanns and Herbert Tanner
Degussa AG, Weissfrauenstrasse 9, D-6000 Frankfurt, FRG

ABSTRACT

Addition of the limiting amino acids methionine and lysine to animal feeds allows over 5 million tons of protein annually to be saved or freed for other uses. To date, however, these savings have only been realized for monogastric animals. Based on our present knowledge, it should also be possible to provide polygastric animals with synthetic amino acids using a protected form. In contrast to poultry and swine, the feeds for sheep and cattle cannot be directly supplemented with synthetic amino acids since free amino acids are metabolized by the microorganisms in the rumen. At present, amino acids may be protected from microbiological catabolism by both physical and chemical methods. While results achieved thus far, especially with protected methionine, are quite promising, these developments are still in the beginning stages.

KEYWORDS: Nutrition, Limiting Amino Acids, Protein Quality, Supplementation, Ruminants, Microbial Synthesis, Rumen-Protected Amino Acids, Protein Supply.

PROTEIN AND AMINO ACID NUTRITION

Supplying the necessary human protein requirements means supplying the necessary amino acids. Today, about 20 amino acids have been identified as end products of protein digestion. Eight of these amino acids are essential for humans. Our organism is not capable of synthesizing them, by modification of other amino acids, for example, and thus must ingest these essential amino acids as a part of our nutritional requirements, predominantly as vegetable or animal proteins. The organism is capable of synthesizing the two semi-essential amino acids, but only to a limited extent. Non-essential amino acids, in contrast, can be easily produced from other building blocks. The distribution of the 20 amino acids in proteins varies considerably. The quality of a nutrient protein depends primarily upon how closely its amino acid content corresponds to the amino acid requirements of man. This is especially true of the essential amino acids. When the requirement for one of the essential amino acids is not met, the utilization of the others and thus the value of the entire protein is limited. The amino acids which are the most nutritionally and physiologically important are the limiting ones and it is no simple matter to exactly balance our amino acid supply with our requirements.

If we take lysine as an example and compare the human requirement in various stage of life with the lysine content of various foodstuffs, we can clearly see that the corresponding values agree well at some times and less well at other times (Figure 1). The lysine requirement of a school child can only be met by milk protein, if we assume that a given foodstuff is the single source of nutrition. In practice of course, this does not hold. As far as posible, proteins with complementary amino acid distributions are combined. In general, imbalances can be equalized in this manner. The optimization of diets may be represented in a simplified manner by the classical example of Liebig's barrel (Figure 2). The contents of the barrel, representing the nutritional value of a given protein or protein mixture, is limited by the height of the lowest stave, which corresponds to the quantity, relative to requirement, of the least abundant amino acid. By lengthening the stave - that is, supplementing the limiting amino acid, the contents of the barrel - that is, the nutritional value of the protein or protein combination, is increased.

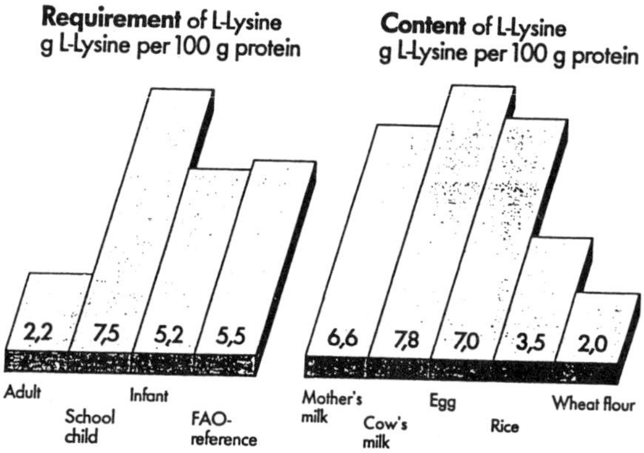

Fig. 1: L-Lysine: Requirement and Content

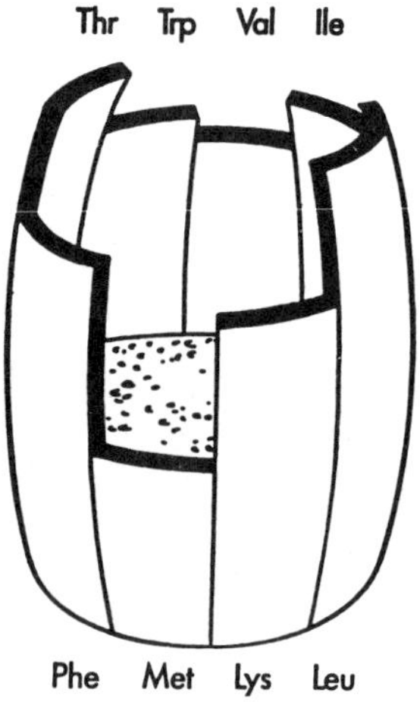

Fig. 2: Liebig's Barrel

The supplementing of nutrient proteins with amino acids has so far only been tested in exceptional cases in the human diet. The enrichment of rice with lysine and threonine and of soy protein with methionine have not yet attained widespread application. For animal nutrition, on the other hand, the addition of synthetic amino acids to feeds is the state of the art. Approximately 200 Mt/a of feed mix for fowl and swine are prepared worldwide in this manner. The feed industry utilizes annually 100,000 t synthetic methionine and 40,000 t synthetic lysine for this purpose. The addition of these limiting amino acids has allowed the average protein content of feed mix to be lowered to less than 20%. This means that <40 Mt/a of protein are sufficient to adequately nourish our agricultural livestock. Supplementing with methionine and lysine has improved protein utilization by about 20%, which corresponds to a savings of over 5 Mt of protein. Supplying the major part of the methionine, which is the primary limiting amino acid in poultry feed, by addition of methionine-rich fishmeal, as was common in the past, would today require catching 30 Mt/a of fish and processing them to fishmeal. Synthetic methionine thus makes a certain contribution to maintaining the ecological balance on earth.

The monogastric animals have a digestive system which essentially consists of a stomach and a small intestine. The nutrient proteins are enzymatically hydrolyzed to amino acids, from which the body's own proteins are synthesized after resorption.

ENZYMATIC CAPABILITIES OF RUMEN MICROORGANISMS

The situation is different for the case of the polygastric animals, the ruminants. In contrast to monogastric animals, sheep and cattle have a multiple stomach system. The first and largest of these stomachs, the rumen, is a fermentation vessel which is populated by microorganisms capable of altering practically all feed components. Therefore, we must pay particular attention to the fate of the substances contained in feed, especially the feed proteins (Figure 3).

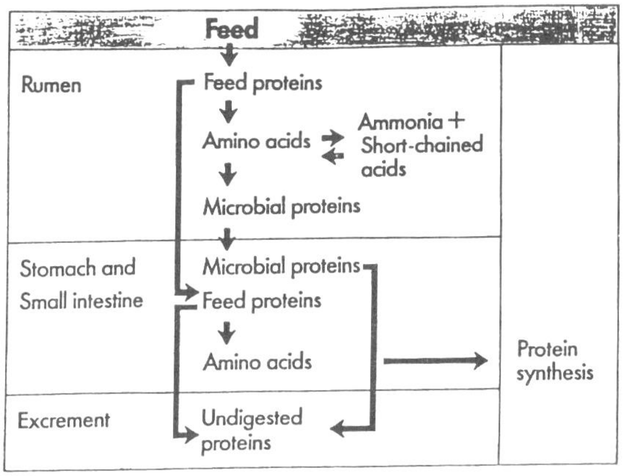

Fig. 3: Metabolism of Ruminants.

Today we know that a portion of the feed protein passes through the rumen system of the animal without alteration. According to Kaufmann and Lupping (1979), this portion averages about 30%. This feed protein which arrives unchanged in the abomasum is handled by the same procedure as for monogastric animals. Most of the feed protein, in contrast, is degraded in the rumen system by the action of microbial enzymes. Thus, about 70% of the feed which is consumed by ruminants is initially utilized to nourish their microorganisms. A ruminant, therefore, lives predominantly on protein from microorganisms. Microbial proteins can only be formed, however, after the feed proteins have been broken down, by way of short-chain peptides, into amino acids, short-chain fatty acids and ammonia (Hatfield, 1970). These are the only building blocks which the rumen microorganisms use to build their particular proteins.

This capability of ruminants was impressively demonstrated by Virtanen (1966). For many years, a large herd of milk cows was provided with urea and ammonium salts as the only source of nitrogen-containing compounds in their feed. They were also fed carbohydrates and minerals to provide for a balanced diet. The cows gave approximately 4,000 L/a of milk per animal during the experiment. They remained healthy and bore calves. Nothing about the animals was unusual, except for the fact that they had not consumed a single gram of amino acids for protein during the entire time span of the experiment. This proof that ruminants do not require protein in their feed is quite important; it means that the ruminants need not be competitors for protein with humans and monogastric animals. Grass, hay, silage and other green vegetable matter which is unsuitable for nourishing monogastric life forms can provide the ruminants a sufficient feed basis. In essence, Loosli (1949) had pointed this out when he determined that there are no essential amino acids for ruminants, or more precisely, for their microbial systems.

Scientists agree that the addition of synthetic amino acids to ruminant diets can have no effect, since the amino acids are decomposed in the rumen. Lewis (1971) and Chalupa (1976) have demonstrated this. This does not exclude the possibility, however, that methionine (Remond, 1979) for example, or methionine-hydroxy-analogue (Chandler, 1976) could have a positive effect on the microbial growth. The key to ruminant nutrition is the microbial protein. If we are to improve the protein utilization of sheep and cattle, we must consider the composition of these microbial proteins.

A simplified presentation of such composition is shown in Figure 4 (Tamminga, 1975). It clearly indicates, with regard to the sulfur-containing amino acids methionine and cystine, that the microbial proteins compare unfavorably with those of milk and wool. The microbial protein synthesis thus leads to a methionine deficiency. It should therefore be possible to improve the performance of ruminants by the use of sulfur-containing amino acids - and methionine is capable of fully replacing cystine biochemically - provided the methionine is brought intact through the rumen so that it is available to the monogastric part of the animal's digestive tract.

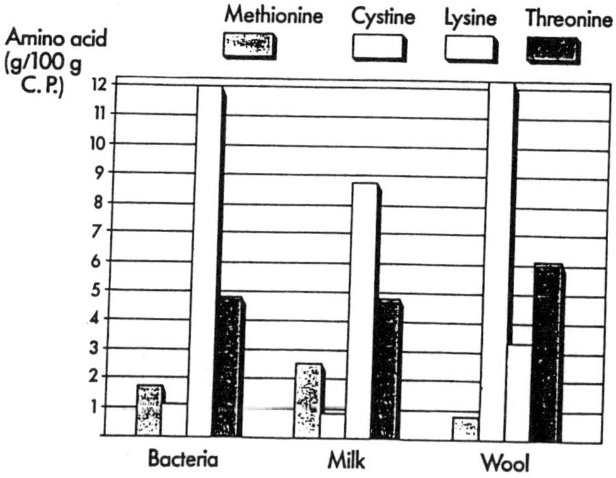

Fig. 4: Amino Acid Composition of Ruminant Protein.

Reis and Schinkel (1963) confirmed this in a brilliant experiment. They hypothesized that sheep, which produce wool with the extremely high content of sulfur-containing amino acids of 12%, must have a large requirement of methionine and cystine. Since feeding the additional methionine brought no result, they infused synthetic cystine or methionine in aqueous solution into the small intestine of the sheep. The rumen and its microbial system, which would have decomposed the amino acids, was thus avoided. The infused cystine or methionine could be resorbed from the small intestine, analogous to monogastric animals, and used for synthetic purposes such as wool production. They observed an increase in wool production of 16 - 130% under these conditions (Figure 5). As expected the largest effect was found for the infusion of 2 g cysteine per animal per day of 2.46 g methionine per animal per day. The infusion of a high-value protein, which likewise contains the limiting amino acid, also brought about a very large increase in wool production.

Experiments of this type in which synthetic amino acids singly, in various combinations, or even in the form of intact proteins such as casein were infused as aqueous solution into the abomasum small intestine system of the ruminants, were repeated quite often. The principle of these infusion experiments was explained by Lewis (1971). When the primary limiting amino acid, usually methionine, is infused into the small intestine of a ruminant, the level of this amino acid in the blood after resorption does not increase, in spite of increasing dosages. This is because the additional amino acid is immediately withdrawn from the amino acid pool and used for intensive protein synthesis which is now possible. Only at the point at which this amino acid is no longer limiting does its content in the blood increase with further infusion. When, on the other hand, a non-limiting amino acid such as lysine is infused into the duodenum, it is unable to effect increased protein synthesis. Therefore, the concentration of this amino acid in the blood increases immediately from the beginning of the infusion.

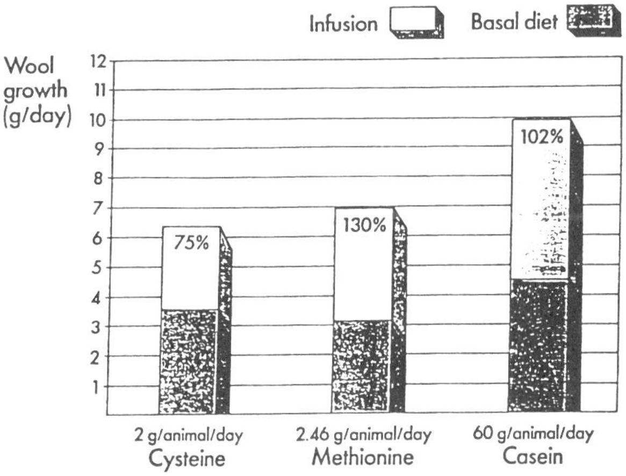

Fig. 5: Increased Wool Production by Cysteine and Methionine.

Unfortunately, infusion experiments are not a practical method for providing ruminants with supplementary amino acids on a large scale. In principle, however, there are two practical possibilities. Either the decomposition of proteins and amino acids in the rumen can be prevented by modifying the properties of the microorganisms which are responsible for their degradation, or the feed proteins and supplemental amino acids can be protected from attack by the microorganisms. The first possibility has been pursued by the Smithkline Corp. (1973). It is necessary to either block or inhibit the microbial enzyme systems which decompose the amino acids. The difficulty of this approach lies in the fact that hundreds of different types of bacteria and protozoa live in the rumen and each of them possesses a large number of enzymes. Nonetheless, the inhibition of microbial deaminases, and thus a reduction in the decomposition of amino acids, has been achieved using diaryliodonium salts. The second possibility is discussed in detail in the following section.

RUMEN-PROTECTED AMINO ACIDS

Attempts were made at first to protect suitably the feed proteins from decomposition in the rumen. The reasons were clear: proteins were fed to ruminants in large quantities anyway, the protection of certain proteins looked relatively simple at first glance, and no one knew with any certainty which amino acids under which conditions were limiting for which ruminants. The fundamental investigations concerning the protection of certain proteins were carried out by Chalmers (1954). Protection was achieved by heat treating. Later, Zelter and Leroy (1966) used tannic acid to protect feed protein from decomposition in the rumen. A certain effect was achieved by this method. Reproducible results, however, could not be obtained.

One interesting development of rumen-protected proteins can be traced back to the work of Ferguson (1967) in Australia and the first use of formaldehyde to protect natural feed protein. The coating of proteins with suitable polymers was also developed. The results which were obtained for proteins protected in this manner in simple in vitro and in vivo experiments were impressive. Once again, however, great difficulties arose in putting these findings into practice. Only recently, with improved methods, has it become possible to control the reaction of formaldehyde with proteins, especially soy and rape protein, so that sufficient protection in the rumen and sufficient "deprotection" in the abomasum/small intestine is obtained. In spite of this progress, it is still difficult to obtain reproducible results in view of the numerous cross-linking possibilities open to formaldehyde in the presence of amino groups. Protected proteins are on the market today under the names "Protane" from Ucanor (1977) in France, and "Milkipro" and Trouw/BP (1978) in Germany. They are providing initial success in practice.

Parallel to the investigations of the protection of proteins, an increasing knowledge of the limiting amino acids for ruminants has been gained. This was achieved for the most part in infusion experiments in which the rumen was completely avoided. Today we may assume that methionine and lysine are, for ruminants also, the amino acids which usually limit performance. It is thus necessary to protect these amino acids from decomposition in the rumen and still have them subsequently available to the stomach/small intestine as free amino acids. Two particular possibilities exist. The first is more of a physical solution to the problem of protecting the amino acids from decomposition in the rumen. Particles of amino acids - crystals of a certain size, for example, or agglomerates of crystals with the proper dimensions - can either be coated with or imbedded in a suitable material. The second possibility consists of chemical alteration of the amino acid molecules and thus involves genuine chemical reactions.

In the coating process amino acid crystals, either singly or as agglomerates, are covered with a material which is stable to the microorganisms in the rumen. A particle protected in this manner can pass unchanged through the rumen. In the abomasum/small intestine, on the other hand, the amino acids must be released. This is the prime requirement for a coating material. Often, the pH-dependent solubility or swelling capacity of the coating material is used to effect this. Examples of suitable coating materials are the cellulose-propionate-morpholinobutyrate polymers developed by Eastman Kodak (1972, 1978), or the polymethyl methacrylates described by Astra Chemical (1980). Feeding experiments with sheep which were conducted in the US by Komarek (1978) using coated methionine particles developed by Eastman Kodak resulted in increases in wool production up to 42% and in nitrogen balance of 13%. The coating materials require special attention. Generally, they have no nutritional value but must be completely non-toxic, and, for economic reasons, must be thin. If they are too thin, however, the amino acid may begin to diffuse through the wall while the particles are still in the rumen. If, on the other hand, the coating is too thick, the product is not only expensive but also may not set its amino acids free in the digestive tract after passing the rumen.

Imbedding amino acids in materials which remain unchanged in the rumen but lose their protective properties in the abomasum/small intestine is also carried out using amino acid crystals or crystal agglomerates. The protection at the surface of such products is only partial, since some amino acids particles naturally are exposed at the surface and are thus subject to attack by the rumen microorganisms. Recent examples of this type of rumen-protected amino acids are the protected methionine Ketionine from Rumen Kjemi (1972) and the protected methionine Loprotin from Lohmann (1978). The rumen-resistant imbedding material consists essentially of long-chained fatty acids such as stearic and palmitic acid. Earlier, hydrogenated fats were used by Labatt (1968) for this purpose. In feeding experiments using Ketionine, Daugaard (1978) found an improvement of about 5% in the quantity of fat-corrected milk and the milk fat content. Similar results have been obtained using Loprotin, under practical as well as experimental conditions.

The chemical protection of amino acids, in contrast to the physical methods, involves amino acid molecules. The requirement is still the same: the protecting derivative must be stable in the rumen but must subsequently release the amino acid in the abomasum/small intestine. There are numerous, varied possibilities to find the particular products which suit the three criteria: protection in the rumen, "deprotection" in the abomasum/small intestine, and economic feasibility. Of the numerous products which have been tested, N-hydroxymethyl-methionine-calcium should be mentioned. In this compound, developed by Degussa (1973), the amino group of the methionine is protected by the hydroxymethyl group, which is pH-sensitive. The carboxyl group is further stabilized by formation of the calcium salt. The protecting group provides sufficient stability against attack by the microorganisms in the rumen, but it instantly cleaved to the free amino acid under the acidic conditions of the abomasum. In subsequent digestive processes, this methionine is resorbed and used in protein synthesis exactly the same as that from the bacterial protein or the remaining feed protein. This methionine derivative, which is marketed under the name "Mepron", has shown good results in feeding studies under suitable conditions.

The milk production of dairy cattle, for example, could be improved with "Mepron" (Kaufmann, 1979). The improvement was essentially an increase in milk quantity rather than a change in the protein or fat content of the milk. In the investigations, addition of about 15 g protected methionine resulted in an average increase in milk production of 1.9 l/day or 7.5% (Figure 6).

FUTURE OUTLOOK

When Hatfield (1979), noted that the use of amino acids for cattle was no longer speculation, it seemed to be a somewhat premature and quite daring statement. However, for those involved with human and animal nutrition and the production of food and feedstuffs, the present developments can only mean that we must begin to realize that the rumen-protected amino acids are one possibility which can contribute to the protein supply of mankind.

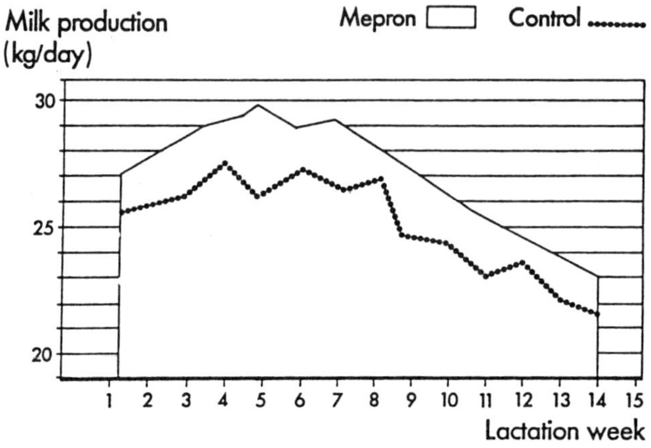

Fig. 6: Increase in Milk Production with Protected Methionine (Mepron).

It is not enough merely to develop effective products. We must also find a way to feed these products to the ruminants. How are Australian wool sheep to be supplied with protected methionine in the vast expanses of that land? What possibilities exist to supply rumen-protected amino acids to cattle which are nourished solely from grazing? There are still many questions to be answered.

When we consider that the quantity of feed consumed by ruminants is much larger than that which is fed to poultry and swine, then the opportunities simply must be good. In the future the ruminants will perhaps live more predominantly from plant products which cannot be utilized by monogastric life forms. We have succeeded in increasing the protein efficiency in poultry and swine nutrition by almost 20% through the addition of synthetic amino acids. Over 5 million tons of proteins have thus been freed for other uses. Is this not encouragement enough for us to develop rumen-protected amino acids for ruminants?

REFERENCES

Astra Chemical Products AB, Sodertalje, Sweden (1978, 1980). US-Appl. 892 869, US-Pat. 4 234 565.

Chalmers, M.J., Cuthbertson, D.P., Synge, R.L.M. (1954). Ruminal Ammonia Formation in Relation to the Protein Requirement of Sheep. Duodenal Administration and Heat Processing in Factors Influencing Fate of Casein Supplements, J. Agr. Sci., 44, 254.

Chalupa, W. (1976). Degradation of Amino Acids by the Mixed Rumen Microbial Population. J. Anim. Sci., 43, 828-834.

Chandler, P.T., Brown, C.A., Johnston, R.P. Jr., McLeod, G.K., McCarthy, R.D., Moss, B.R., Rakes, A.H. and Sattler, L.D. (1976). Protein and Methionine Hydroxy Analogue for Lactating Cows. Dairy Sci., 59, 1897-1909.

Daugaard, J. (1978). Investigation on Methionine Supplement to Lactating Cows. Thesis, Copenhagen, Denmark.

Degussa AG, D-6000 Frankfurt (1973). DOS 2 307 836.

Eastman Kodak Co., Rochester, NY, USA (1977, 1979). US-Appl. 830 281, DOS 2 838 298.

Ferguson, K.A., Hemsly, J.A. and Reis, P.J. (1967). Nutrition and Wool Growth. Austr. J. Sci., 30, 215.
Hatfield, E.E. (1979). Amino Acids for Ruminants-Myth or Merit. Feed Management, March, 40-41.
Hatfield, E.E. (1970). Proc. Am. Soc. Exp. Biol. 29, 44-50.
Kaufmann, W and Lupping, W. (1979). Zum Einfluss von Protected Protein und HMM-Ca auf die Leistung von Milchkuhen. Z. Tierphysiol., Tierern. u. Futtermittelkde. 41 (4), 202-217.
Komarek, R.J. and Jandzinski, P.A. (1978). Increased Wool Growth and Nitrogen Balance with Dietary Rumen-Protected Methionine. Abstract 70th ASAS-Meeting, East Lansing, MI.
Kuther, K. (1980). Moglichkeiten und Wirksamkeit verbesserter Methionin-Versorgung bei Rindern, Vortrag d. Wissenschaftlichen Lohmann-Tagung, Lohmann-Tierernahrung GmbH, D-Cuxhaven.
John Labatt Ltd., London, Ontario, Canada (1969), DT-PS 1 692 452.
Lewis, D. (1969). Protein Metabolism in the Rumen. Beef Cattle Science Handbook, 6, Agriservice Foundation.
Lewis, D. and Swan, H. (1971). The Role of Intestinal Flora in Animal Nutrition. Symp. Soc. Gen. Microbiol. No. XXI, 149-175.
Lohmann Tierernahrung GmbH, P.O. Box 446, D-2190, Cuxhaven 1.
Loosli, J.K., Williams, H.H., Thomas, W.E., Ferris, F.H. and Maynard, L.A. (1949). Synthesis of Amino Acids in the Rumen. Sci. 110, 144.
Milkivit-Werke, A., Trouw GmbH, Gempfinger Str. 15, D-8859 Burgheim.
Reis, P.J. and Schinkel, P.G. (1963). Some Effects of Sulphur Containing Amino Acids on the Growth and Composition of Wool. Aust. J. Biol. Sci., 16, 1, 218-230.
Remond, B., Champredon, C., Decaen, C., Pion, R. and Journet, M. (1971). Influence d'un apport de DL-methionine des vaches au debut de la lactation sur la production laitiere et la composition du sang. Anim. Biol. Anim. Bioch.-Biophys. 11, 455-469.
Rumen Kjemi A/S, Peter Mollervei 1, Oslo 5, Norway (1972), DT-PS 2 212 568.
Smithkline Corp., Philadelphia, PA, USA (1973), DOS 2 413 222.
Tamminga, S. (1975). Observation on Protein Digestion in the Digestive Tract of Dairy Cows. Z. Tierphysiol., Tierern. u. Futtermittelkde, 35, 337-346.
Ucanor, 6, Rue Roquemonts, Caen, France.
Virtanen, A.I. (1966). Milk Production of Cows on Protein-Free Feed. Sci. 153, 1603.
Zelter, S. and Leroy, F. (1966). Schutz der Nahrungsproteine gegen mikrobielle Desaminierung im Pansen. Z. Tierphysiol., Tierern. u. Futtermittelkde. 22, 39.

IMPROVING THE SUPPLY, QUALITY, AND UTILITY OF PROTEINS

Soichi Arai
Department of Agricultural Chemistry, The University of Tokyo Bunkyo-Tokyo 113, Japan

ABSTRACT

Food proteins inevitably undergo deterioration preharvest and postharvest. The deterioration may jeopardize efficient supply of proteins to any section of the world and their usefulness for human consumption. A solution to this problem requires basically the use of up-to-date methods for minimizing the factors involved in the deterioration and, at the same time, for maximizing desirable protein properties. The sophisticated methodology currently available for this purpose, emphasizes the potential of novel enzymic modification of proteins. The subtlety and versatility of protein structures permit their intentional modification in the form of covalent attachment of suitable amino acid esters with the aid of a protease action. Examples are given of recent work on improving the nutritive value of soy protein by attachment of L-methionine ethyl ester, and of endowing hydrophilic proteins (e.g., gelatin) with a greater surface action by attachment of lipophilic L-leucine n-alkyl ester. The utility of enzymically modified proteins and future research strategies are discussed.

KEYWORDS: enzymic modification; flavor; functional property; gelatin; lipid; methionine supplementation; pigment; plastein reaction; protease; soy protein.

INTRODUCTION

As the requirement for high quality foodstuffs increases, it becomes imperative to improve the supply, quality, and utility of conventional as well as non-conventional proteins. The less affluent sections of the world critically need protein foodstuffs. Any of the currently available proteins, however, has limitations for use as human food. All are deficient in one or more of the essential amino acids. Many suffer from lack of ideal functional properties, and several contain undesirable chemical components. These deficits can be totally or partly minimized by use of modern methods of modifying the properties of food proteins. Although proteins are fundamentally important as nutrients, they possess subtle and versatile chemical and physical properties that make them key functional ingredients of food. As the world's need for high quality proteins becomes more acute; more and more of this need can be met only by intentional modification of the properties of proteins in order to increase their utility for human consumption. This is one of the main reasons why studies on the intentional modification of food proteins has been stressed in recent years.

Every protein originates in a living organism and it is likely that the protein undergoes deteriorative changes while the organism goes through its own life cycle. Preharvest life chronologically consists of: genesis, growth, maturation, and reproduction. A variety of environmental factors may affect this process in whole or in part. These include ambient temperature, light, humidity, water activity, composition of the atmosphere, nutrient concentration, etc. Sometimes exogenous enzymes, invading microorganisms, and even air-borne or water-borne pollutants may affect the growth cycle. Failure to control properly such factors eventually has an undesirable effect on the quality of the proteins harvested from a particular organism.

In addition, many deteriorative processes occur postharvest. Most crops and food products after harvest require sophisticated treatment prior to being consumed. More often than not, proteins as food constituents deteriorate to a greater or lesser extent from the time of harvest to the time of consumption. When exposed to a high or low temperature, proteins often are denatured. The use of an extremely high temperature may produce potent carcinogens. Adding alkali can cause the formation of crosslinked lysinoalanine. Oxidation of amino acids occurs commonly, which is particularly serious when photosensitization leading to the formation of singlet oxygen takes place. Even food constituents themselves can act as deteriorative factors; carbonyls, including reducing sugars, induce the browning reaction involving proteins; oxidized lipids cause changes in amino acid residues through a free radical mechanism.

Although a great many methods are available for controlling the deterioration that may occur, their primary purpose focuses only on lengthening the time from harvest to consumption. The longer the time, the farther the protein can be transported from its site of harvest to the consumer. This factor is extremely important in supplying efficiently protein foods to people living in all parts of the world.

There is another purpose which may fundamentally be more important. It is to maximize the protein quality in terms of nutritional and functional properties. In particular, the latter has long received significant industrial attention because a world-wide demand for functional protein products has been continuously increasing. Modern food chemistry has disclosed that, in any particular food, its constituents at a molecular level have multilateral interactions which affect its favorable qualities. Protein molecules play a central role in building up food products with a desirable functionality. Among the currently available proteins, however, only a limited number are utilized; the rest remain neglected because of poor functionality. With the objective of maximizing the utility of these proteins, it is imperative to improve those aspects of their structure that determine their basic functionality.

Minimization of undesirable properties and maximization of the desirable properties should be attainable through chemical and enzymic modification. Usually, chemical modification aims at improving structure by covalent attachment of suitable groups. For example, acylation in order to mask sensitive amino groups is a typical case. Divalent reagents often are used to cross-link protein molecules. Enzymic modification also can be used to attach covalently specific groups to protein molecules. The use of proteases is rather a conventional way of modification, whereby protein molecules undergo peptide bond hydrolysis. An extensive literature on chemical and enzymic modification of food proteins exists. It cannot be reviewed completely here. Recent reviews should be consulted (Feeney, 1977, 1980; Pour-El, 1979; Whitaker, 1982; Fox et al., 1982).

In the present paper, discussion concentrates on enzymic modification of food proteins. Special emphasis is placed on novel approaches to removing undesirable factors from proteins and to confering desirable nutritional as well as functional properties on low quality proteins.

Minimization of Undesirable Factors

Several different types of chemical compounds act as ligands, binding to proteins particularly when they get denatured. This results in physiological and organoleptic problems. Two independent ways are available to remove protein-bound ligands enzymically. A less specific process is based on the degradation action of proteases. Nonspecifically, certain non-protease enzymes are capable of removing the protein-bound ligand without degrading the protein. Figure 1 illustrates schematically the mode of action of these two processes. Either or both have been successfully applied to the removal of, for example, pigments, lipids, and flavors.

Fig. 1: Schematic representation of two processes for enzymatic removal of a protein-bound ligand.

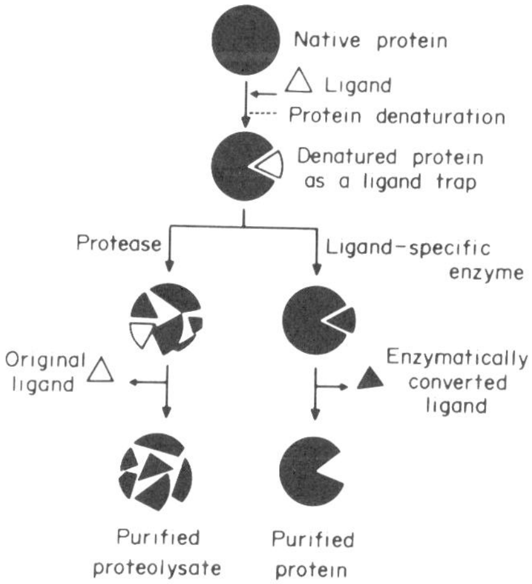

Naturally occurring pigments, as well as added food colours, are often bound by proteins. Photosynthetic pigments may obstruct the preparation of pure leaf protein. A chlorophyll-lipoprotein complex present in an algal chloroplast fraction is often a nuisance in that it resists digestion. There is a distinct possibility that some fluorescent compounds with photosensitizing and mutagenic activity remain in foods in a protein-bound state. Besides these, closely related compounds such as chlorogenic acid and other ubiquitous phenols occur naturally, readily interacting with proteins and affecting their quality. Several of them, though colorless per se, undergo enzymic oxidation to pigments. When such pigmentation takes place, significant deterioration in protein quality results. Arai et al. (1976) have succeeded in removing protein-bound pigments from single-cell proteins by treatment with a protease and in obtaining a colorless hydrolysate from, for example, a Spirulina protein (Figure 2). Details of these studies were described in a recent reveiw (Arai, 1980).

Lipids that consist of unsaturated fatty acids are precursors of a series of aliphatic carbonyls with objectionable flavors. These lipids often associate with proteins primarily through hydrophobic bonding. A serious problem arises when such lipids undergo peroxidation. Free radicals originating from lipid peroxidation could transfer to several amino acid residues. As a result the protein molecules undergo radical-induced polymerization, with loss of their original properties including solubility and nutritive value (Schaich and Karel, 1976). Proteases are effective in removing lipids from proteins (Arai, 1980).

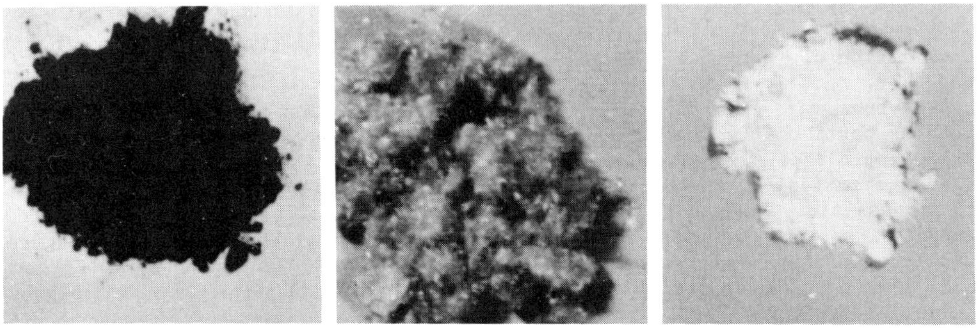

Fig. 2: Pictures of a Spirulina mass (left), an extracted protein (middle), and its hydrolysate (right).

Volatile carbonyls, when bound to proteins, behave as if they were non-volatile compounds and can be retained for a long period of time during storage (Franzen and Kinsella, 1974). Such interactions quickly cause organoleptic problems, due to chemical deterioration of the proteins. Soy protein, for example, when it is heated, interacts with 1-hexanal, a principal beany odorant. Interaction of this odorant with heat-denatured protein is lessened by partial hydrolysis with selected proteases (Arai et al., 1970). A recent study (Chiba et al., 1979) has disclosed that use of aldehyde dehydrogenase isolated from bovine liver is effective in minimizing 1-hexanal and other aldehydes present in soy protein and flour. The main mechanism involved in this enzymic process is believed to be conversion of these aldehydes to alkanoic acids which have lower binding affinities toward soy protein.

Maximization of Desirable Properties

Modification of food proteins through partial hydrolysis by proteases has long been practiced (Pintauro, 1979). The plastein reaction, a unique protease-catalyzed process leading to formation of a plastein product from a protein hydrolysate, has been extensively studied with the specific objective of improving the nutrtional and functional properties of proteins (Fujimaki et al., 1977). A great advantage of this reaction may be its capability of covalently incorporating any amino acid (ester form). Details have been reviewed (Schwimmer, 1981; Whitaker, 1982; Fox et al., 1982).

Recently, Yamashita et al. (1979a) developed a modification of the classical plastein reaction which permits use of a protein as the substrate. Since this process is designed to incorporate efficiently amino acid esters, it is necessary to make the reaction alkaline. At a moderately alkaline pH, most added amino acid esters act as effective nucleophiles, attacking the acyl-enzyme intermediate and, consequently, are incorporated. Papain, although its activity for peptide bond hydrolysis is maximized at a weakly acidic pH, is capable of catalyzing efficiently this incorporation at pH 9. A schematic representation of the new process is shown in Figure 3.

Application to Improvement of Nutritional Properties

The new enzymic process (Figure 3) was first applied to soy protein isolate (SPI) with the objective of enhancing its limiting amino acid, methionine. An

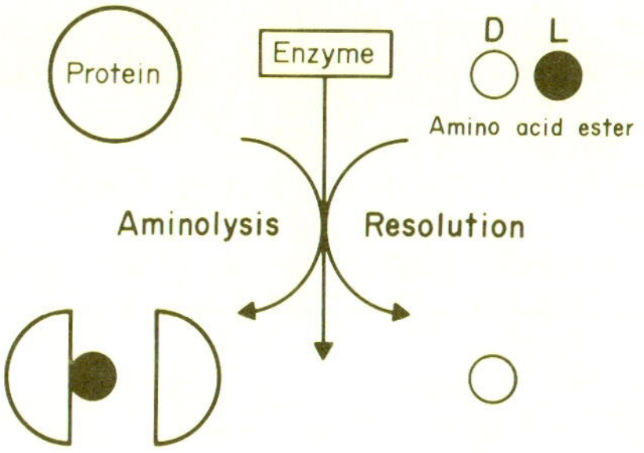

Fig. 3: Schematic representation of a new process for producing an enzymatically modified protein (EMP) with covalent incorporation of L-amino acid ester.

enzymically modified protein (EMP) with covalently attached L-methionine at a sufficient level was obtained in high yield (Yamashita et al., 1979b). A further study (Arai, 1981) made it possible to produce, on large scale, a similar EMP whose methionine level was enhanced up to approximately 11 percent (weight basis) (Figure 4). The product, in spite of such a high methionine content, was free of any objectionable flavor. Chemically, this product was defined as a peptide mixture having a molecular weight distribution in the range of 1,000 - 8,000 daltons.

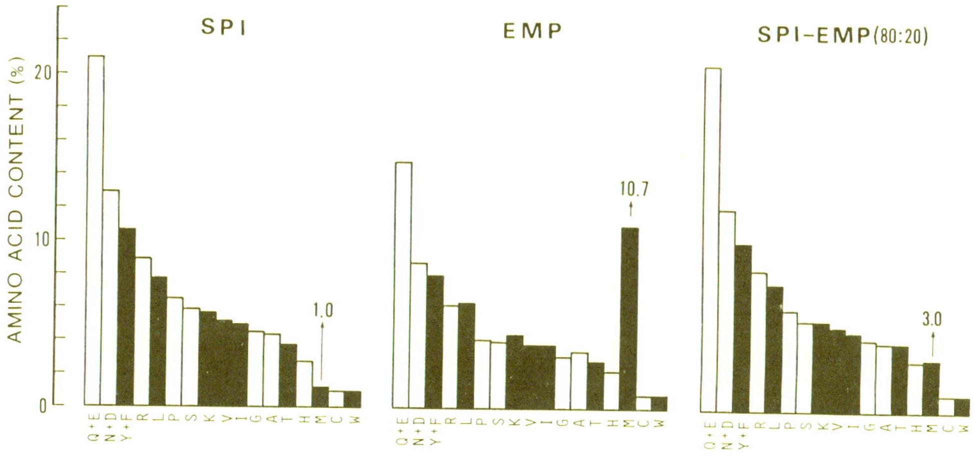

Fig. 4: Amino acid patterns of soy protein isolate (SPI), enzymatically modified protein (EMP), and an SPI-EMP (80:20) formula.

The EMP product was fed to rats in order to investigate the small-intestine contents. Four hours after feeding, a major part of the methionine remained in an oligopeptide form. Interestingly, however, efficient entry of free methionine into the portal vein was observed during the same period (Arai, 1981). This observation in view of modern concepts of peptide transport may not be contradictory (Matthews, 1977). Another experiment was carried out using the EMP product as a methionine supplement to SPI. Formulation of SPI and the EMP product in a ratio of 80:20 both theoretically and experimentally, gives a formula whose methionine level approximates 3% (Figure 4). It should be noted that in this case the resulting formula compares well to SPI in terms of the relative levels of amino acids other than methionine. Different ratios of formulation would afford formulas of different methionine levels, with all other amino acid remaining unchanged. To find out the nutritionally optimal level of covalently attached methionine, feeding tests with rats were conducted. The optimal nutritive value measured by protein efficiency ratio (PER), occurred at about 2.5 per cent. This result may contribute to our understanding of the indispensable minimal level of methionine in soy protein.

The two experiments above indicate that the covalently attached methionine occurring in the EMP product is bioavailable and that this product can be used as a new type material for improving the nutritional quality of soy protein and, possibly, other low-quality proteins. Also, similar EMP products may be obtainable from proteins other than SPI.

Application to Improvement of Functional Properties

The new enzymic process (Figure 3) was applied in a series of studies to imparting a surface active function to proteins lacking this property. Surfactancy is a highly useful property in food processing. In a model experiment, our group (Arai and Watanabe, 1980; Toiguchi et al., 1982) used succinylated α_{s1}-casein (hydrophilic substrate), L-leucine n-dodecyl ester (lipophilic nucleophile) and papain (EC 3.4.4.10). The reaction afforded a 20,000-dalton product with potent oil-emulsifying properties. A 13-NMR study and many others demonstrated that formation of the 20,000-dalton product resulted from degradation of the substrate at the peptide bond between Phe-145 and Tyr-146. This product was defined as a macropeptide bearing a leucine dodecyl moiety at the C-terminal residue, Phe-145 (Figure 5). Formation of an amphiphilic structure (Figure 5) is probably responsible for the emulsifying function.

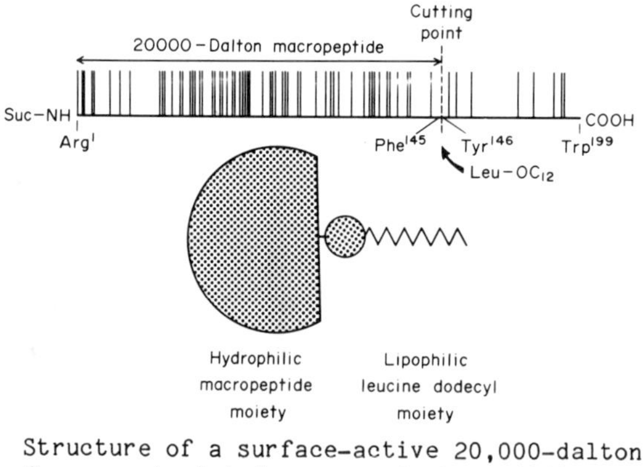

Fig. 5: Structure of a surface-active 20,000-dalton macropeptide formed from succinylated α_{s1}-casein by modification with papain in the presence of L-leucine n-dodecyl ester. All the ionizable or hydrophilic amino acid residues of this protein are marked.

In a further attempt to apply the new enzymic process to the preparation of functional EMP products from various sources, we selected gelatin as a hydrophile and attempted to attach a homologous series of L-leucine n-alkyl esters as lipophiles. Chemical analyses showed that the products had an average molecular weight of approximately 7,500 daltons, each bearing the leucine alkyl esters at a level of 1.1 - 1.5 mol/7,500 g. Regardless of the similarities between the products, their functionality varied depending on increased lipophilicity. In particular, EMP-6 and EMP-12, (referring to the products attaching leucine hexyl ester and leucine dodecyl ester, respectively), possessed characteristic surface functions; the former was characterized by a highly whippable property and the latter by a great ability to emulsify oil finely (Watanabe et al., 1981a). These products effectively improved quality when applied to colloidal foods of many kinds (Watanabe et al., 1981b).

A concentrated emulsion simulating mayonnaise could be prepared by using EMP-12 (Shimada et al., 1982). A gel was formed when a dispersion of EMP-12 at a high concentration was emulsified with a sufficient volume of oil. EMP-12 also has the property of giving a highly viscous emulsion. The increased viscosity may reflect interaction of this surfactant with the surrounding water molecules. Watanabe et al. (1982), using a pulsed NMR technique, verified that a larger amount of water interacted with EMP-12 after emulsification. Also, a significant portion of the oil molecules interacted with EMP-12 as a result of emulsification. This observation was based on an ESR study of measuring the average rotational correlation time (τ_c) of the spin probe, 2-(10-carboxydecyl)- 2-hexyl-4, 4-dimethyl-3-oxazolodinyloxy methyl ester, which had been dissolved in the oil used.

From the data presented above, and that of other studies, we conclude that the EMP products, EMP-6 and EMP-12, are bifunctional in the sense that each possesses both a surface activity and a viscosity-adding property. These data are discussed at greater length in a review (Watanabe and Aria, 1982).

Further Research Needs

The creation of protein through the growth of animals, plants and microorganisms and the direct recovery of proteins from wastes in the meat, fishery, dairy and vegetable processing industries are of primary importance. It is equally important to modify these proteins in order to lessen their original undesirable properties and to enhance their desirable properties.

The safety and wholesomeness of such modified proteins should, of course, be confirmed. In the event that a compound is attached covalently to a protein by chemical or enzymic modification, it should be selected from those items legally permitted for food use. Preferably, the attached compound should be liberated in the digestive tract prior to being absorbed and entering the body. Based on these criteria, we optimistically expect that EMP-12 and related products can be used as food ingredients or as additives. Final confirmation of their safety, however, awaits further experiment. Regardless of this limitation, the basic and applied studies on intentional protein modification should be extended. Chemical and biochemical research along these lines will undoubtedly contribute to the improvement of the supply, quality, and utility of proteins for human consumption.

REFERENCES

Arai, S., M. Noguchi, M. Yamashita, H. Kato, and M. Fujimaki (1970). Studies on flavor components in soybean: some evidence for occurrence of protein-flavor binding. Agric. Biol. Chem. 34, 1569-1573.

Arai, S., M. Yamashita, and M. Fujimaki (1976). Enzymatic modification for improving nutritional qualities and acceptability of proteins extracted from photosynthetic microorganisms, Spirulina maxima and Rhodopseudomonas capsulatus. J. Nutr. Sci. Vitaminol. 22, 447-456.

Arai, S. (1980). In "Chemical Deterioration of Proteins" (ACS Symposium Series 123), American Chemical Society, Washington, D.C., 195-209.
Arai, S. and M. Watanabe (1980). Modification of succinylated α_{s1}-casein with papain: covalent attachment of L-norleucine dodecyl ester and its consequence. Agric. Biol. Chem. 44, 1979-1981.
Arai, S. (1981). Improvement of the essential amino acid pattern of soy protein isolate by enzymatic modification: procedures for the implementation and evaluation of the products. Nutr. Sci. Soy Protein (Japan) 2, 23-26.
Chiba, H., N. Takahashi, N. Kitabatake, and R. Sasaki (1979). Enzymatic improvement of food flavor. III. Oxidation of the soybean protein-bound aldehyde by aldehyde dehydrogenase. Agric. Biol. Chem. 43, 1891-1897.
Feeney, R.E. (1977). In "Food Proteins: Improvement through Chemical and Enzymatic Modification" (Adv. Chem. Series 160), American Chemical Society, Washington, D.C., 3-36.
Feeney, R.E. (1980). In "Chemical Deterioration of Proteins" (ACS Symposium Series 123), American Chemical Society, Washington, D.C., 1-47.
Fox, P.F., P.A. Morrissey, and D.M. Mulvihill (1982). In "Developments in Food Proteins-1", Applied Science Publishers, London and New Jersey, 1-60.
Franzen, K.L. and J.E. Kinsella (1974). Parameters affecting the binding of volatile flavor compounds in model systems. I. Proteins. J. Agric. Food Chem. 22, 675-678.
Fujimaki, M., M. Yamashita, and S. Arai (1977). In "Food Proteins: Improvement through Chemical and Enzymatic Modification" (Adv. Chem. Series 160), American Chemical Society, Washington, D.C., 156-184.
Matthews, D.M. (1977). In "Peptide Transport and Hydrolysis" (Ciba Foundation Symposium 50), Elsevier, Amsterdam-Oxford-New York, 5-14.
Pintauro, N.D. (1979). In "Food Processing Enzymes: Recent Developments", Noyes Data Corporation, Park Ridge, New Jersey, 181-193.
Pour-El, A. (1979). In "Functionality and Protein Structure" (ACS Symposium Series 92), American Chemical Society, Washington, D.C., 4-7.
Schaich, K.M. and M. Karel (1976). Free radical reactions of peroxidizing lipids with amino acids and proteins: an ESR study. Lipids 11, 392-400.
Schwimmer, S. (1981). In "Source Book of Food Enzymology", the AVI Publishing Company, Inc., Westport, Connecticut, 618-621.
Shimada, A., E. Yazawa, and S. Arai (1982). Preparation of proteinaceous surfactants by enzymatic modification and evaluation of their functional properties in a concentrated emulsion system. Agric. Biol. Chem. 46, 173-182.
Toiguchi, S., S. Maeda, M. Watanabe, and S. Arai (1982). Purification and characterization of a surface-active macropeptide produced from succinylated α_{s1}-casin by modification with papain. Agric. Biol. Chem., in press.
Watanabe, M., A. Shimada, and S. Arai (1981a). Enzymatic modification of protein functionality: implantation of potent amphiphilicity to succinylated proteins by covalent attachment of leucine alkyl esters. Agric. Biol. Chem. 45, 1621-1625.
Watanabe, M. and A. Shimada, E. Yazawa, T. Kato and S. Arai (1981b). Proteinaceous surfactants produced from gelatin by enzymatic modification: application to preparation of food items. J. Food Sci. 46, 1738-1740.
Watanabe, M. and S. Arai (1982). In "Modification of Proteins: Food, Nutritional, and Pharmacological Aspects" (Adv. Chem. Series 198), American Chemical Society, Washington, D.C., 199-221.
Watanabe, M., N. Fujii, and S. Arai (1982). Characterization of foam- and emulsion-stabilizing functions of enzymatically modified proteins with surfactancy. Agric. Biol. Chem. 46, 1587-1592.
Whitaker, J.R. (1982). In "Modification of Proteins: Food, Nutritional, and Pharmacological Aspects" (Adv. Chem. Series 198), American Chemical Society, Washington, D.C., 57-87.

Yamashita, M., S. Arai, Y. Imaizumi, Y. Amano, and M. Fujimaki (1979a). A one-step process for incorporation of L-methionine into soy protein by treatment with papain, J. Agric. Food Chem. 27, 52-56.
Yamishita, M., S. Arai, Y. Amano, and M. Fujimaki (1979b). A novel one-step process for enzymatic incorporation of amino acids into proteins: application to soy protein and flour for enhancing their methionine levels. Agric. Biol. Chem. 43, 1065-1068.

EXPANDING AND IMPROVING THE FOOD SUPPLY WITH ENZYME SYSTEMS

Bernard Wolnak
Bernard Wolnak Associates
360 N. Michigan Avenue (Ste. 706)
Chicago, Illinois, 60601, U.S.A.

ABSTRACT

The possibility of producing and using single enzyme systems to increase the food supply in the developing countries seems remote. However, multienzyme systems, namely fermentations, offer a much more promising situation. Many of the major components required for the development and operation of a successful fermentation process are present in the developing countries. Such factors as the nutrients, microorganisms, and manpower are or can be made available. The major components of equipment and capital can be obtained, particularly since such projects can be structured to return all invested capital and to make a profit. In addition, the products obtained from a fermentation plant, e.g., antibiotics, amino acids, and vitamins could materially assist the food supply in the developing countries. These concepts are discussed in detail.

KEYWORDS: Enzymes, food, feed, food supply, fermentation, aerobic, anaerobic.

INTRODUCTION

Enzymes are defined as organic catalysts or more precisely: "An enzyme is a biologically derived, water soluble, protein of comparably high molecular weight which acts as a catalyst". As with any chemical reaction susceptible to catalytic change, the following equation represents an enzyme catalyzed reaction:

$$A \underset{}{\overset{(E)_1}{\rightleftarrows}} B$$

where A is the substrate, B is the product or products of the reaction, and $(E)_1$ is the enzyme or enzyme system acting as the catalyst. The reaction is reversible and the enzyme, functioning as a true catalyst merely overcomes the energy barrier allowing the reaction to proceed more completely to produce the desired product, B. The basic facts of enzyme reactions, the classes of enzymes and their advantages and disadvantages are well known.

The degree of catalytic activity shown or possessed by any one enzyme varies widely from enzyme to enzyme. Some such as catalase are very active. One molecule of catalase will catalyze the decomposition of 3.5 M molecules/min. Some enzymes are comparatively less active, e.g. one molecule of amyloglucosidase catalyzing the production of 10,000 to 100,000 molecules of glucose/min. While reaction rate increases with increasing temperature, in the case of enzymic-catalyzed reactions, an increase in temperature may be self-defeating since the enzyme is a protein and may be inactivated or destroyed. Most enzymes have their optimum temperature range between 25-40°C and a few in the 55-65°C range. These optimum temperatures may change when the enzyme is supported or bound. Ionic substances present in the solution may either inhibit the activity of an enzyme or may stabilize the enzyme. In general the CN^- ion seems to be the worst inhibitor, while the Ca^{++}, Co^{++} and Zn^{++} ions seem to stabilize many enzymes; in some cases these ions are necessary for maximum activity.

Enzymes, as indicated previously, are peptides of comparatively large molecular weight, ranging from a value of 14,000 to a million daltons or higher. Direct chemical synthesis of enzymes therefore does not appear to be a likely prospect in the near future. Some work in recent years has indicated that smaller organic molecules possessing active site centers similar to those found in enzymes, do have significant levels of catalytic activity. However, the levels of activity of these synthetic molecules are orders of magnitude less than those found in the analogous enzyme, and these molecules tend to be less stable than the enzyme itself. Since this work is advancing slowly, it may well be two decades or more before it reaches the point of being competitive with the natural enzymes.

The three future souces of enzymes for use in the food industry are microbial, animal, and plant. Prior to the 1950's most of the important enzymes were derived from plant or animal tissues. With a growing capability in fermenation more and more enzymes are being produced microbially, and currently most of the important food enzymes are produced that way. We believe this trend will continue so that by the end of the decade very few enzymes will be produced from animal or plant tissue.

UTILIZATION OF ENZYMES

The worldwide markets for industrial enzymes are today estimated at $360 M, about with 1/2 in the US and the remainder distributed worldwide. This is not an insignificant amount but disappointing for an industry which began at the turn of the century. With the tremendous amount of basic scientific information available, there appears to be difficulties in transferring the technology from the academic sector to industry in the industrially developed countries. Some of the important reasons are probably (1) a serious lack of communication between academia and industry, (2) a lack of personnel interested in and trained in both the theoretical and industrial aspects of enzymology, (3) the concept taught in academia that enzymes in foods are to be destroyed rather than that they may be useful, (4) a serious lack of communication between the potential industrial user and the producer of enzymes.

While commercial enzyme production and use may be at a low level, the value of the products obtained by the use of enzymes is very large. In the U.S. the bulk of the enzymes produced are the amylase group, used to convert starch derived from corn (maize) to various sweeteners including corn syrups, dextrose, and high-fructose corn syrup. The other two major enzymes were rennin for cheese production and papain for chill-proofing beer and tenderizing meats. The amylases plus these latter protease enzymes accounted for more than 80% of total enzyme usage in 1980.

ENZYMES IN THE DEVELOPING COUNTRIES

Based on the economics of the developing countries and the enzyme industry as it is presently constituted in the developed countries, we are hard pressed to develop a rationale for operating an enzyme facility in a developing country. This conclusion is particularly true for fermentation-produced enzymes. The technology required is troublesome. There are problems in the application and utilization of the enzymes. There is the problem of obtaining skilled manpower. There is a lack of an obvious major area of utilization of enzymes. When these considerations are coupled with the lack of infrastructure and finances, the possibility of developing an enzyme-based program which will significantly improve the food supply of the developing countries is discouraging.

FERMENTATIONS

In fermentation processes organisms grow in an aqueous solution of nutrients. During the growth process, the organism metabolically produces the desired product; in some organisms this metabolite remains internally in the microorganism; more often the metabolite passes through the cell wall of the microorganism into the surrounding aqueous medium. Following the fermentation it is necessary to separate the cellular mass from the solution and recover by chemical or physical means the desired product. The aqueous solution of nutrients (media) is usually a complex misture of a carbon source (molasses, glucose, etc.), a nitrogen source (urea, ammonium salts, etc.) and various minerals. Other substances may be needed to maximize the production of the desired compound. Process variables such as pH, temperature, agitation, air supply, and concentration of the nutrients must be controlled to obtain the best yield of the desired product. The process is capital intensive and sensitive to raw material costs. Since the average concentration of product in the fermentation broth is on the average of about 20 g/L. Large volumes of water are involved. The fermentation reactions may be exothermic and, therefore, may require a supply of cooling water. However, many products, verying from simple molecules such as citric acid to complex ones as the tetracyclines, can best be produced by fermentation.

The three major types of fermentation used worldwide are tray or Koji type fermentations, anaerobic submerged fermentations, and aerobic submerged fermentations. The first is the oldest type and is still used extensively in Japan and in Europe. Anaerobic submerged fermentation forms the basis of production of such products as ethanol or lactic acid. It produces the largest volume of products today and dominates the fermentation industry.

A large number of fermentation processes impinge directly and indirectly upon the food and animal feed supply. These are fermented foods, food or feed ingredients, and food or feed additives. While there are some 200 fermentation products produced commercially today, some of the more important products are listed in Table 1.

An important advantage of a fermentation process accrues from the fact that many products can be produced by fermentation techniques which cannot be produced easily by any other means, e.g., organic syntheses. In a few cases synthetic processes prevail, e.g., for riboflavin, ascorbic acid, chloramphenicol, but these are the exception rather than the rule. Fermentation products are used by three major industries, viz., the pharmaceutical, food, and animal feed industries. For the pharmaceutical industries the principal products are the antibiotics, vitamins, steroids, and amino acids, and are usually produced in large amounts, in highly purified forms, and sold to the health-care industry in finished "dosage" form. For the production of these pharmaceutically based products, a very high

Table 1: Important Fermentation Products

Foods	Food or Feed Ingredients	Food or Feed Additives: Amino Acids	Food or Feed Additives: Food Acids
Sauerkruat	Yeasts	Lysine	Citric
Pickles	Single Cell Protein	Tryptophane	Lactic
Cheese	Soy Sauce		Gluconic
Olives	Vinegar (Acetic Acid)		Fumaric
	Mushrooms		Propionic
	Cocoa		Acetic
	Tea		

Vitamins	Thickeners	Antibiotics
Riboflavin		Penicillin
B-12		Tetracycline
		Tylosin
		Erythromycin
		Bacitracin

level of technology is required along with a comparatively high investment and high processing costs.

In the food area a number of important food additives, e.g., food acids, vitamins, etc. are produced by fermentation. These additives in total also represent a large industry and production volume, but seem, at this time, to offer little to the problem of increasing the food supply in the developing countries.

However fermentation is an important step in the preparation of a number of valuable food condiments produced in the large part in the developing countries, e.g., cocoa, tea, and to a lesser extent, coffee. The fermentation involved in producing these products is usually uncontrolled and crude, and leads to high levels of spoilage and loss with attendant poor economics. More needs to be done to apply available technology to improve these industrial operations.

Information on the food supply in the developing countries indicates that deficiencies of amino acids, and/or protein now exist. The major sources of these vital nutrients in the developed countries are dairy products, meat, and fish. In regard to the first two, and particularly meat, multi-enzyme systems (fermentations) can make a valuable contribution. They provide a strong incentive to expand upon the small base of "cottage" industries already present in developing countries and to build an expanded animinal-raising industry. The inclusion of fermentation processing could be a most economically valuable adjunct. The basic reasons for this conclusion are the following. In raising aminals in the developing countries, the diet and the health of the animals in many cases is not much better than those of the persons growing the aminals, and for very much the same reasons. From the dietary standpoint certain amounts of starches and sugars are available but protein is not available to the animal. From the health standpoint, the animal is susceptible to a variety of bacterial infections, which, if they do not kill the animal, do greatly interfere with its rate of growth and production of meat. Fermentation can assist in the nutrition and maintaining the health of the animal by producing amino acids, vitamins and antibiotics.

In the fermentation industry antibiotics represent the major product in terms of monetary value. The technology began to be developed during World War II and has expanded continually since that time accompanied by a large increase in the size and breadth of the antibiotics markets. About fifty antibiotics valued at nearly 10 billion dollars at the manufacturer's level are now produced worldwide, which means that the fermentation industry is now very much dominated by the pharmaceutical industry.

In the developing countries to obtain a position in fermentation production that is technologically oriented and capital intensive it is useful to examine the production of an antibiotic (tetracycline, bacitracin, or penicillin) and the steps and ingredients necessary. The media or nutrients required are usually carbohydrate-based with small amounts of organic nitrogen required; in a few cases inorganic nitrogen can be used. The amount of carbohydrate necessary should be readily available in most countries. The specific microorganism for some fermentations may present a problem but for many antibiotics it may be purchased or may be developed in a microbiological laboratory. The technology for the "mature" antibiotics can be purchased along with the specific microorganism. With the exception of the fermenters, the equipment is straightforward chemical equipment. Capital may be a problem but highly purified antibiotics need not be prepared. When antibiotics are used in animal feeds in the developed countries, they are added as impure products. In fact, after the fermentation is complete, the total fermentation mass is removed from the fermenter, and dried by conventional means to give the desired antibiotic additive product. On this basis capital requirements are drastically reduced as compared to producing pharmaceutical grade antibiotics. The requirement for scientific personnel may be a problem, but training of a cadre supported by technicians could solve the problem.

Thus, the application of fermentation processing in the developing countries could significantly increase the supply of food in these countries.

REVIEW OF STORAGE AND PROCESSING

J. Solms
Laboratory of Food Chemistry and Technology
M. Bachmann
Laboratory of Dairy Science
Department of Food Science, Swiss Federal Institute of Technology
8092 Zurich, Switzerland

ABSTRACT

The present state of food processing and storage is discussed with special reference to heat processing, irradiation, freezing preservation, dehydration, intermediate-moisture foods, cooking estrusion, chemical-biological preservation, and fabricated foods. The most typical chemical reactions occurring in processing and storage are also mentioned. A subsequent discussion is devoted to the aspects of storage and processing in developing countries.

KEYWORDS: heat processing, irradiation, freezing preservation, dehydration, intermediate-moisture food, cooking extrusion, heat sterilization, amino acid degradation, freezing, lipid oxidation.

THE CHALLENGE WE FACE

Our present food processes are the result of a scientific and technical development of approximately 80 years in regions with a rich food supply. This has resulted in a shift in the preparation of food from the kitchen to a highly processed and diversified "prosperity" food supply, adapted to industrialized countries and their needs. A great need exists, however, for more effective and more suitable methods of food processing and preservation for developing countries.

It is now widely accepted that a simple transfer of the present food technology of industrialized countries to economically less developed regions of our world will not be suitable. As Slater (1979) has pointed out, a detailed systems approach, including the careful assessment of the available resources in order to transfer adaptable technologies, is necessary. In contrast to industrialized countries, developing countries are short of the most elementary inputs for industrialization. In many cases they lack well-trained manpower and advanced know-how, transport and communication facilities, and well-organized infrastructure of energy supply, water supply and services. Another important difference is the extremely low purchasing power of up to 80% of the population of developing countries with a great proportion living in urban centres. Planning the food supply of tomorrow's population is not possible without studying the demographic development of these cities. Increasing proportions of the world's population will not live near the production sites of food but far away in huge and badly organized cities (Singh et al., 1982). Between 1980 and 2000 urbanization will proceed at a fast pace in the developing countries, with an increase of 180% projected.

It can be expected that these trends in population development will lead to a corresponding trend in the development of food processing techniques. Reviewing the present food technology situation in industrialized countries as compared with developing countries, we found a set of characteristic parameters, as presented in Table 1 (Bachmann, 1979).

Table 1: Some Characteristics of Food Technology of Industrial Countries Compared with the Respective Requirements of the Third World

Characteristics of Industrial food technology	Food Technology as required by Third World countries
Centralized	Decentralized, labour intensive factories in rural areas
Mainly machinery, labour saving	Creation of many working opportunities
Employment of few, but highly qualified workers	Employment of many poorly or nonqualified workers
Capital intensive	Requiring minimum or no foreign exchange
Energy intensive	Energy saving
Use of oil and electricity	Use of locally available, decentralized energy sources
Incomplete use of raw material, much waste	Maximum use of raw material, little waste
High degree of refining, high purity of products	Making use of every available food value
Great convenience, high costs	As inexpensive as possible
Expensive, one-way packaging	Economical, return packaging
Internationally and nationally standardized products	Making use of local products, tastes and production methods

In this review the present state of food processing and storage is discussed, followed by a more specific consideration of the requirements of developing countries.

THE PRESENT STATE - TECHNOLOGICAL ASPECTS

General Aspects

Limited availability and limited stability of food raw materials are a major reason for processing. Processing and preservation of food materials can be accomplished by physical and by chemical-biological means. Physical treatments comprise heating, irradiation, controlled reduction of the temperature (chilling, freezing), controlled reduction in the product's water content (concentration, dehydration, freeze drying), and the use of protective packages. Chemical-biological treatments involve a change in chemical composition - often also in physical properties - of the food material in order to prolong stability. These changes can be achieved by addition of substances such as sugars, salts, organic acids, antimicrobials, or by biological processes such as alcohol and acid fermentation. Finally the production of fabricated foods can be considered to be a special operation of its own. In this operation traditional or non-traditional raw materials are transformed by new processes into new types of food, often imitation foods.

Heat Processing

Heat treatment of foods, especially canning, is perhaps the best studied process and a well established operation. It consists of product pretreatment, followed by sterilization and storage. There is a wealth of information available concerning the sterilization process, its temperature-time relationships and its effects on other food components, namely, nutrient retention, enzyme inactivation, pigment retention etc. (Lopez, 1975; Ball and Olson, 1975; Stumbo, 1973; Lamb et al., 1982; Downey, 1978). Recent advances deal with the development of fully automatic batch-type equipment and continuous-flow hydrostatic sterilizers.

The necessary heating and cooling treatment may also be carried out in a continuous-flow heat-exchanger with subsequent aseptic filling of the product into the package. Aseptic packaging has resulted in a number of advanced systems capable of handling aseptic filling of milk, fruit juices, and other liquid beverages into presterilized flexible or rigid containers (Emch, 1978; Hersom and Shore, 1981). It permits preparation of pasteurized and UHT-treated beverages of high quality in low-cost containers and is extensively applied in Europe and in the Far East.

In sterile processing the retort-pouch is a newcomer (Lampi, 1977). It has better geometrical proportions giving better heat-transfer conditions. For certain products shorter process times are possible compared with ordinary cans. Moreover the retort-pouch is lower in cost and needs less energy for its production. This gives the retort-pouch a considerable potential in the future. Retort-pouches are alreay in wide use in Europe and in Japan. A most interesting application is sterile low-pH foods in retort-pouches, which combine the advantages of low pH and retort-pouch packaging during processing. A large variety of this type of product has recently appeared on the Swiss market. Other new devices and processes like tray-packs and flame sterilization of cans are at present under study or in limited application.

Irradiation

Sterilization or pasteurization to inhibit the development of spoilage can also be accomplished by ionizing radiation treatments (USAEC, 1965; Urbain, 1978; Diehl, 1977). Foods so processed are referred to as "irradiated foods". The effectiveness of irradiation in many applications is well established. However, since progress in application has been slow for various reasons, a more detailed discussion will not be given here.

Freezing Preservation

Freezing preservation is another well established technology (Van Arsdel et al., 1969; Fennema, 1982; I.I.R., 1982). It consists of pre-freezing treatment, freezing, frozen storage, and thawing. It is a method of long-term food preservation, and is generally regarded as superior to heat sterilization or dehydration when judged on the basis of retention of most quality criteria. As a typical physical processing operation this system can be described with a time-temprature curve. However, this curve is not restricted to factory operations as for canning, but extends over the total life-time of the product.

Recent developments are concerned with important improvements such as shorter freezing times, the preparation of free-flowing products, and cryogenic freezing. The application of freezing preservation is suitable for regions with ample refrigerated distribution channels and availability of home refrigerators. Newer data suggest that freezing preservation is also a process which is energetically favorable compared with other preservation methods.

Dehydration

Although drying is a rather energy-intensive process, due to its versatility it is a most important operation (Van Arsdel et al., 1973; Spicer, 1974; Downey, 1978). Drying of foods can be accomplished by direct or indirect sun-drying, by drying in simple kiln dryers and tray dryers, in continuous spray dryers or drum dryers, all at atmospheric pressure, and in rather sophisticated vacuum or freeze dryers. Dehydration in all its forms is a process of considerable importance, applied in the preparation of sugar, starch, coffee, milk products, flour mixes, breakfast foods, potato products, snacks, fruits and vegetables. Many dried products can be stored for a long time in simple and low-cost packages.

This process implies wide changes of all product parameters including concentration relationships in the product. Due to these variations, drying processes are difficult to describe theoretically, but they offer many practical possibilities in application. Recent practical developments have mainly been oriented towards advanced processing in drying. Much less practical work has been done with simpler drying technology, although it seems possible to improve quality of dehydrated products by fairly simple and thus inexpensive methods. To illustrate this point, investigations on the improved stability and quality of salt-treated sulfited air-dried carrots and potatoes may be noted (Escher et al., 1978). Extensive studies have also been made on the relationships between water content and stability of dehydrated foods leading to the concept of water activity and its determination by sorption isotherms (Heiss, 1968; Gal, 1967; Karel, 1973).

Intermediate-moisture Food

Intermediate-moisture foods are a special application of dehydration processes. They are characterised by a water activity of 0.6 to 0.8, which is low enough to prevent the growth of microorganisms. However, it is high enough to convey a soft texture to the product, permitting direct consumption. Several dried fruits belong to this category. They are prepared either by a simple dehydration process or by admixing suitable compounds like sucrose and glycerol to a wet product to further depress the water activity. Antimicrobial agents can also be used. The production of intermediate-moisture foods is often fairly simple and inexpensive, and the product is stable, ready to eat and does not require expensive packaging. Intermediate-moisture foods are still awaiting introduction for human consumption on a larger scale (Davies et al., 1976; Karel, 1973). Possible potentials, however, are illustrated by traditional baked Swiss specialities called Birnenweggen and Birnenbrot. Both are typical intermediate-moisture type foods of very high acceptability based on dehydrated pears.

Cooking Extrusion

The cooking-extrusion process combines heating and extrusion of food products followed by a dehydration step to create a cooked and shaped dry food product. In a typical cooking-extrusion process the food product can reach quite high temperatures of up to 180°C under a corresponding pressure; but the residence time is very short, from 5 to 10 seconds. Physical and chemical changes characterize the extruded material. Typically a rigid often porous structure is imposed upon the product. Chemical changes occuring under cooking-extrusion are similar to ordinary heating processes. However, due to the very high temperature and very short processing time the reaction parameters are different from other processes (Harper, 1981). The extrusion process is carried out advantageously with moist starchy and/or proteinaceous foods. Cooking-extrusion has a high productivity at a relatively low cost and is probably rather energy efficient. However, it produces a restricted type of new food products which need an introductory time for acceptance (Harper, 1981).

Chemical-biological Treatments

By combining food-grade chemicals with suitable food raw materials, eventually in combination with a preservation or fermentation process, a variety of products can be obtained. In several processes the added compounds are an important ingredient of the food product, e.g. sugar in jams and jellies, and salt in pickled products. In other cases the chemicals are additives with special and specific activities and are applied in combination with other treatments. Examples are SO_2, nitrite, antimicrobial and antioxidant agents.

Conservation by chemical treatment is restricted to special applications. Most applications, however, are simple and economical and do not require advanced technology. Throughout the world one can find fermentation processes that have been developed as a means of preserving foods against spoilage. During fermentation, bacteria, yeast and/or molds multiply in the food material which acts as a culture medium. They consume a certain amount, mostly of the carbohydrate component, transform the composition of the culture medium itself in the process, and enrich it with the products of their metabolism. This often conveys to the product additional stability (Reed, 1982; Prave et al., 1 · Steinkraus, 1980/81; Farr, 1980/81; Pederson, 1979).

Fermentation alters the sensory, physical and nutritional characteristics of the starting material, and can help to improve diets. Examples are the use of yeasts to prepare beer, wine and bread, the use of bacteria to produce acidified milks and cheeses, vinegar, sausages and pickled vegetables. Fungi are used to produce products such as Tempeh and Sake. Taking into account the great number of possible fermented food products, only a relatively small number of products are produced under controlled fermentation conditions with advanced equipment in centralized production units. Many fermented foods can also still be produced successfully in smaller decentralized food-processing installations. This technology, also termed intermediate technology, has in fact been adopted in Switzerland with considerable success for the production of Swiss cheese. This example of a functioning intermediate technology can certainly be of interest to countries with a decentralized infrastructure. There are 2000 cheese factories in Switzerland!

Fabricated Foods

For the preparation of fabricated foods conventional or non-traditional raw materials in more-or-less pure form are mixed in novel ways and subjected to appreciable changes in chemical and physical composition by new types of processes. Examples are restructured meats from plant proteins, non-dairy whiteners, and snack foods based on starchy or proteinaceous plant materials. These fabricated foods make possible the introduction of new raw materials and their transformation into new food products of high acceptance. A number of interesting developments have potential (Inglett, 1975).

Introduction of engineered foods with pure substances merits, however, careful consideration. Pure substances often represent a category of food materials containing so-called "empty calories". In industrialised countries, like Switzerland, it is quite usual that people consume 50 to 60% of their daily energy requirement in the form of empty calories. It means that the micronutrients such as vitamins and minerals would be derived from the remaining 50% (Brubacher, 1978). If the fabricated foods replace meat, for example, and do not contain the necessary micronutrients in correct composition, an excess of empty calories in the daily diet, with all its consequences, could be manifest upon consumption of these foods.

THE PRESENT STATE - CHEMICAL ASPECTS

Microbiological and Chemical Aspects of Heat Sterilization

In heat sterilization most effects are due to a controlled treatment at a constant concentration. It is possible to use conventional reaction kinetic parameters and their temperature dependence to describe thermal resistance of various food constituents. In general the data are presented in time-temperature sterilization curves. Data from such curves are presented in Table 2. In this table z is the increment in temperature necessary to change the reaction rate by one order of magnitude, and E_a is the Arrhenius activation energy. The D-value is defined as the time at a constant temperature for the concentration of the constituent in question to decrease by 90%.

Table 2: Thermal Resistance Of Various Food Constituents (Lund, 1977)

Constituent	z (^{o}F)	E_a (kcal/mole)	D_{121} (min)
Vitamins	45-55	20-30	100-1000
Color, texture, flavor	45-80	10-30	5-500
Enzymes	12-100	12-100	1-10
Vegetable cells	8-12	100-120	0.002-0.02
Spores	12-22	53-83	0.1-5.0

Of special interest are the differences for these values for vitamins, color, texture, flavor and enzymes on one hand and for vegetative cells and spores on the other. These differences furnish the theoretical basis for sterile high quality products, and explain the beneficial effects of high-temperature-short-time sterilization processes.

Amino Acid Degradation Reactions

There are probably three groups of reaction sequences where proteins are involved with a negative influence on nutritive quality. Proteins in the presence of large amounts of sugars undergo the classical Maillard or non-enzymatic browning reaction. Proteins in more-or-less pure form undergo, at elevated temperatures, several interactions with formation of inter-molecular peptide-bridges and isopeptides. Alkali pre-treated proteins undergo various cross-link reactions, where lysinoalanine is of special importance as a reaction product in foods.

The Mallard reaction and its advanced reaction sequences, the Strecker degradation, are related to the so-called caramelization reaction and can form interesting food pigments and flavor compounds. Therefore the protein degradation reactions not only cause amino acid destruction with negative influence on nutritive quality, but also form compounds which affect food quality in a positive way (Adrian, 1982; Mauron, 1979; Hurrell & Carpenter, 1977).

Considerable work has been done on chemical reactions, especially on amino acid degradation, occurring during dehydration. An important aspect is the loss of available lysine upon dehydration. The content of total lysine and available lysine of a food product measured after various drying processes is an indicatorof the dehydration conditions (Escher et al., 1978).

For evaluating storage stability after dehydration ascorbic acid is often considered as an index of quality of the product. It is one of the most readily destroyed water-soluble vitamins and its retention can also serve as an indicator. The effect of temperature and humidity during storage on the stability of ascorbic acid in drum-dried baby-foods is presented as an example in Table 3. The data demonstrate the importance of temperature and humidity during storage.

Table 3: Ascorbic Acid Halflife (months) in Drum-dried Baby-foods, Stored at Different Temperatures and Moisture Contents (Escher et al., 1978)

Drum-dried baby-foods	10.7% H_2O	7.0% H_2O	5.0% H_2O
45°C	0.41	0.47	0.79
37°C	0.52	0.86	1.47
26°C	0.68	1.56	3.39

Lipid Oxidation

It is generally recognized that in many instances oxidation reactions of lipids limit the stability of processed foods (Witting et al., 1982; Lundberg, 1962; Labuza, 1971; Solms, 1973). Lipid autoxidation is a rather complex free-radical reaction, and takes place initially in three steps: free-radical initiation, propagation, and termination. Fatty acid hydroperoxides formed during this reaction sequence are then decomposed by a series of dismutation reactions. Several of the products formed are radicals and undergo further radical reactions. Others, such as hydroxy acids, keto acids and aldehydes are commonly found in oxidized lipid systems. Many of the short chain volatile compounds are responsible for the off - flavors and odors characteristic of stale or rancid foods. However, many of these compounds are also found in normal food flavors as important constituents. There are physical and chemical methods to retard the reaction effectively. The use of antioxidants is one such method.

In addition to chemical autoxidation, enzymatic oxidation catalysed by lipoxidase is a most important reaction in many raw and processed foods. The reaction is dependent on specific reaction conditions, but finally gives similar reaction products to those from chemical autoxidation. An interesting nonvolatile oxidation product described recently is one consisting of bitter-tasting hydroxy acids (Baur et al., 1977).

STORAGE AND PROCESSING IN DEVELOPING COUNTRIES

There are aspects of storage and processing which must be matched to the requirements of developing countries. The underlying principles of food processing and of chemical reactions do not require any reconsideration. However, the different economic and social parameters are of great importance. Selected aspects of storage and processing in developing countries, based on practical experience, as well as a few typical case studies are presented here.

Heat treatment of food is one of the most important and most common processes in developing countries. However, there are two major drawbacks: the price and scarcity of fuel in many regions, and the frequent necessity of rapid cooling after heat treatment. In the pasteurization processes, for instance, heating does not present a major problem. However, rapid subsequent cooling down to temperatures below 10°C is problematic. This operation always requires a cooling installation. Most cooling facilities in rural areas of developing countries, however, are unreliable and expensive. Therefore heating processes have to be designed without subsequent cooling to low temperatures or cold storage. A practical example for milk processing, which is easy to handle, is sterilization with subsequent cooling, storing, and marketing at ambient temperatures. The use of ambient temperature usually does not pose any problems, and can be achieved with the available local water supply. In regions without any or with an unreliable supply of electricity the application of continuous processing operations is questionable. This is also often true with diesel-operated power stations. It is, for instance, not advisable to use continuous-flow heat-exchangers. However, any batch-operated installation, e.g. by using hand-operated stirrers, can be used with many fewer problems. Aseptic filling under conditions of rural food processing plants is also not recommended. Most reliable are heating processes, e.g. sterilization processes, batchwise in wholesale or retail packages. The Dairy Science Laboratory of our Department is developing a special flame sterilization process for milk in wholesale containers suitable for rural dairy plants in the Third World.

In these low-cost and simple operations the price of the packaging material is of special importance. Every difference in price, e.g. cans vs. pouches, will be of great importance. Products where the packaging material is more expensive than the food inside, a well-known phenomenon in our country, are certainly not recommended.

Cold storage and freezing preservation of food will certainly find limited application in Third World countries. The difficulties are reliable refrigeration facilities and ample refrigerated distribution channels. Average retailers and consumers in developing countries will not have refrigerators for a rather long time to come. Therefore, freezing preservation in developing countries will be restricted to special situations. Freezing preservation as a unique technique adapts a climate to a food. However, that is the reason why it poses problems in developing countries with hot climates.

Dehydration is one of the most important techniques in developing countries. The process can be applied in many variations, and the final product is in most cases fully adapted to local climate and market conditions. For small processing plants there exist improved low-cost solar drying systems which do not require scarce and expensive fuel (Gomez, 1981). Controlled dehydration is also possible by using as fuel methane gas which can be generated locally. Dried food is low in weight and thus easily transportable, and it can be stored and marketed at ambient temperatures. The same applies to intermediate-moisture foods whose development should be promoted throughout the Third World. However, introduction of dried foods on a larger scale requires special attention to nutritional aspects.

Cooking-extrusion processes should be taken into consideration mainly in more advanced developing regions, i.e. in the so-called "threshold countries". The process needs relatively sophisticated and expensive equipment. However, the output of even a medium-size extruder is high. Most extruded products are marketable at ambient temperatures.

An excellent technique for food processing under conditions found in developing countries are chemical-biological treatments. Salting, pickling or smoking as well as alcoholic or lactic-acid fermentation are traditional well-tested preserving methods for food. One advantage of handling such food in warm climates is the possibility of storing and marketing without cooling. In many instances fermented foods can be consumed directly or with little cooking. Miso is an excellent example (Steinkraus, 1981). Rice and soybeans are transformed into consumable foods by simple processing operations. Miso is preserved chemically by its high sodium-chloride content and biologically by lactic acid fermentation. It is ready for consumption after mixing with water and warming. Compared with a normal process, soybeans require cooking for several hours in order to destroy the trypsin inhibitors, the substances causing flatulence, and to make the product organdeptically acceptable. The same can be said for other beans, pulses, and lentils. Therefore, the development of similar transformation-preservation processes for these commodities, as described above, is highly recommended. Another practical example of biological treatments is the production of sour milk, which has been introduced recently in a school milk program in an African country (Bachmann, 1981). The milk is processed in a small, centralized processing plant. The product is stable, without cooling, in open milk cans for several days until consumption. The plant has a capacity of not more than 3,000 kg/day and does not require expensive installations or packaging materials.

Where should food processing take place in developing countries? Under conditions of a tropical or subtropical climate, perishable raw material has to be processed as soon as possible. Therefore, processing of the raw material at the site of production is mandatory. This is especially important for milk, juicy fruits, and vegetables. Moreover it provides working opportunities for the local population and thereby slows down the migration of the rural population to the cities. The rural location of processing plants demands new approaches to energy provision, water supply, and effluent treatment. All these problems, however, are easier to deal with, if rural food processing is organized in smaller units. Smaller factories are easier to manage and they allow for a better utilization of unconventional energy sources from man, animals, wind, biogas, etc. Location and size of the processing plants, as well as the limitation in value which can be added to processed food, determine to a great extent the technology to be applied. In most cases, the technology will be rather simple, will make use of low-cost human labour, and will not draw heavily on limited financial resources. The result is often a unique approach with modified and/or adapted processes. It will even be necessary to create tailor-made food-processing installations. As an example of what has been achieved, a small dairy factory is shown in Figure 1. It shows a cross-section of the factory, where milk is being transported from one processing site to the other without any pumps, just by gravity. Another appropriate approach which is under study in our Department concerns the purification of effluents from rural food-processing installations with the help of water hyacinths and other green plants, combined with the production of biomass for anaerobic digestion.

FINAL REMARKS

Food storage and processing has reached a high degree of perfection in industrialized countries and related regions. It is now widely accepted that a

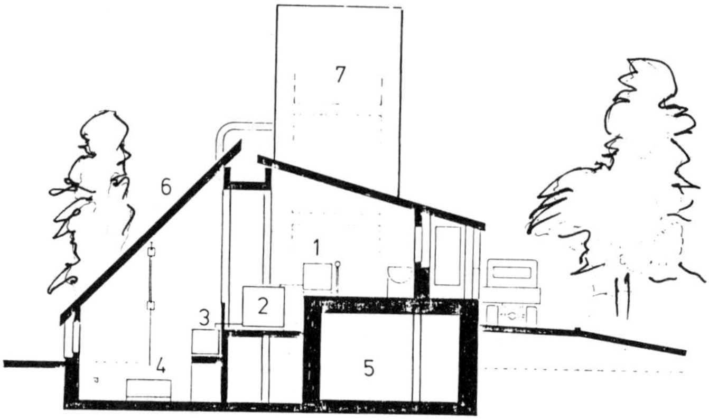

Fig. 1: "Cascade - type" Dairy Factory

1, reception basin; 2, pasteurizer; 3, balance tank; 4, cheese vat; 5, cheese ripening room; 6, roof with integrated solar collector; 7, water tower.

simple transfer of this technology to economically less developed regions is not suitable. The solution of food problems of the developing countries must be based on their own resources. As discussed by Slater (1979), food technology in industrialized countries has finished a period of evolution and has reached a more static phase of optimization. Contrary to this, food technology of developing countries will be, or certainly is already, in full evolution. Summing up the most important points in the evolution, the following are the most important trends for processing and storage in Third World countries: (1) production of low-cost and safe food for an urban population with limited purchasing power; (2) use of locally available raw materials and energy; (3) employing marginally trained labour; (4) use of equipment, tools, and packaging materials of local origin, if possible.

These trends are different from the classical trends of development, which have already been outlined in Table 1. However, the present review and the examples cited have shown that food technology will be able to adjust to these developments. Food technology is an instrument of change.

REFERENCES

Adrian, J. (1982). The Maillard Reaction. In: M. Rechcigl, Handbook of Nutritive Value of Processed Food, I. CRC Press, Boca Raton, 529.

Bachmann, M. (1979). Does the Third World Need a Third Way for Processing. In: P. Linko, Food Process Engineering. Applied Science Publ., London, 926.

Bachmann, M. (1982). How to Approach Food Technological Problems in Developing Countries, Lebensm. Wiss. Technol. 14, 348.

Ball, C.O. and F.E.W. Olson (1975). Sterilization in Food Technology. McGraw-Hill Co., New York.

Baur, C., W. Grosch, H. Wieser and H. Jugel (1977). Enzymic Oxidation of Linoleic Acid: Formation of Bitter-Tasting Fatty Acids. Z. Blebensm. Unters. Forsch. 164, 171.

Brubacher, G. (1978). Importance of Meat Products for the Supply of Micronutrients in Human Nutrition. In: W.K. Downey, ed., Food Quality and Nutrition. Applied Science Publ., London, 463.
Davies, R., G.G. Birch and K.J. Parker, eds. (1976). Intermediate Moisture Foods. Applied Science Publ., London.
Diehl, J.F. (1977). Lebensmittelbestrahlung. Kertechnik 19 (11), 494.
Downey, W.K., ed. (1978). Food Quality and Nutrition, Research Priorities for Thermal Processing. Applied Science Publ., London.
Emch, F. (1978). Aseptische Verfahren in der Lebensmittelindustrie. Lebensm. Technol. 11 (1), 5.
Escher, F. and B. Blanc (1978). Quality and Nutritional Aspects of Food Dehydration. In: W.K. Downey, ed., Food Quality and Nutrition. Applied Science Publ., London, 297.
Farr, D. (1980/81). Traditional Fermented Foods: A Great Potential for the Future. Nestle Research News, 31.
Fennema, O. (1982). Effect of Freezing on Nutritive Value of Food. In: M. Rechicigl, Handbook of Nutritive Value of Process Food, I. CRC Press, Boca Raton, 31.
Gàl, S. (1967). Die Methodik der Wasserdampf-Sorptionsmessungen. Springer-Verlag, Berlin.
Gomez, M.I. (1978). Carotene Content of Some Green Leafy Vegetables of Kenya and Effects of Dehydration and Storage on Carotene Retention. J. Plant Foods, 3, 231.
Harper, M.J. (1981). Extrusion of Foods. CRC Press, Boca Raton.
Heiss, R. (1968), Haltbarkeit und Sorptionsverhalten wasserarmer Lebensmittel. Springer Verlag, Berlin.
Hersom, A. and D.T. Shore (1981). Aseptic Processing of Foods Comprising Sauce and Solids. Food Technol. 35 (5), 53.
Hurrell, R.F. and K.J. Carpenter (1977). Maillard Reaction in Foods. In: T. Høyem and O. Kvale, eds., Physical, Chemical and Biological Changes in Food Caused by Thermal Processing. Applied Science Publ., London, 168.
Inglett, G.E. (1975). Fabricated Foods. AVI Publ. Co., Westport.
Intern. Institute of Refrigeration (1972), Recommendations for Processing and Handling of Frozen Foods. I.I.R. Paris.
Karel, M. (1973). Recent Research and Development in the Field of Low-Moisture and Intermediate Moisture Foods. CRC Crit. Review Food Technol. 3, 329.
Labuza, T.P. (1971). Kinetics of Lipid Oxidation in Foods. CRC Crit. Review Food Technol. 2, 355.
Lamb, F.C., R.P. Farrow and E.R. Elins (1982). Effect of Canning on Nutritive Value of Food. In: M. Rechcigl, Handbook of Nutritive Value of Processed Food, I. CRC Press, Boca Raton, 11.
Lampi, R.A. (1977). Flexible Packaging for Thermoprocessed Foods. Adv. Food Res. 23, 305.
Lopez, A. (1975). A Complete Course in Canning. The Canning Trade, Baltimore.
Lund, D.B. (1977), Maximizing Nutrient Retention. Food Technol. 31, (2) 71.
Lundberg, W.O., ed. (1962). Autoxidation and Antioxidants. J. Wiley and Sons, London, 1961.
Mauron, J. (1979). Die Proteinwertigkeit der Nahrung. In: B. Blanc and H. Bickel, eds., Der Energie- und Proteinbedarf des Menschen. D. Steinkopff Verlag, Darmstadt, 10.
Pederson, C.S. (1979). Microbiology of Food Fermentations. AVI Publ. Co., Westport.
Prave, P., U. Faust, W. Sittig and D.A. Sukatsch (1982). Handbuch der Biotechnologie. Akademische Verlagsgesellschaft, Wiesbaden.
Reed, G. (1982). Prescott and Dunn's Industrial Microbiology. AVI Publ. Co., Westport.
Singh, J. and A. Mattis (1982). The City. Development 24, (2), 3.

Slater, L.E. (1979). In: P. Linko, ed., Food Process Engineering. Applied Science Publ., London, 17.
Solms, J., ed. (1973). Functional Properties of Fats in Foods. Forster Publ. Ltd., Zurich.
Spicer, A., ed. (1974). Advances in Preconcentration and Drying and Foods. Applied Science Publ., London.
Steinkraus, K.H. (1980/81). The Indigenous Fermented Foods. Nestle Research News, 23.
Stumbo, C.R. (1973). Thermobacteriology in Food Processing. Academic Press, New York.
Urbain, W.M. (1978). Food Irradiation. Adv. Food Research 24, 155.
U.S. Atomic Energy Commission (1965). Radiation Preservation of Food. Publ. 1273, Nat. Academy of Sciences, Washington.
Van Arsdel, W.B., M.J. Copley and R.L. Olson, eds. (1969). Quality and Stability of Frozen Foods. Wiley-Interscience, New York and London.
Van Arsdel, W.B., M.J. Copley and A.I. Morgan, eds. (1973). Food Dehydration. AVI Publ. Co., Westport.
Witting, L.A. and P.S. Dimick (1982). Effects of Processing on Food Lipids. In: M. Rechcigl, Handbook of Nutritive Value of Processed Food, I. CRC Press, Boca Raton, 403.

FISH PROCESSING

Torger Børresen and Terje Strøm
Institute of Fishery Technology Research
P.O. Box 677, N-9001 Tromsø, Norway

ABSTRACT

In recent years the total annual aquatic catch has been relatively constant at 70 M tonnes. Production can be increased by a more efficient utilization of conventional resources, increased aquaculture, and by exploitation of unconventional resources. Increased fish production requires improvement of existing preservation and processing technology. Examples of improved methods of preservation are chilled storage of fresh fish in a mixture of ice and seawater, chilling under controlled atmosphere containing CO_2, and acid preservation. The lower organic acids like acetic or propionic acid may be used for processing of mince from small, pelagic fishes, and in processes where fish material is autolysed by endogenous enzymes present in the fish. Fish mince should find a wider application in reformed and dry products. Fermented fish products should be further developed, particularly through the use of lactic acid bacteria. For a successful exploitation of unconventional resources, new technology has to be developed. The new principles of biotechnology and enzyme technology may provide important tools for this purpose.

KEYWORDS: Fish processing, chilled storage, acid preservation, minced fish, fermentation, autolysis, fish silage.

INTRODUCTION

The total aquatic catches have been relatively constant at 70 M tonnes in recent years. The developing countries account for roughly 50% of this. Recent data from FAO indicate that most of the increase in the world catch is occurring in developing countries, while the catches in the developed countries have shown little change (FAO, 1981a). On the other hand, it is thought that the change in fish consumption shows an opposite trend, with a decrease in developing countries and an increase in the developed ones. (FAO, 1981a). As people in developing countries depend very much on fish as a protein source in their diet, the present trend in consumption should be watched carefully. Steps should also be taken to initiate enterprises that could work towards a change in that trend.

The future demand for fish, and anticipated fish production have been treated in an excellent review by Krone (1979) According to this review there will be a demand of 110 M tonnes of food fish in the year 2000. In addition there may be a demand of 20 to 25 M tonnes of fish for fish meal production if the feed markets are maintained. To meet this total demand, annual fish production will have to be increased by 60-65 M tonnes. According to Krone (1979) the production of conventional resources could be increased by more intensive inland fisheries, increased aquaculture production and reduction of post harvest losses. Taken together, this could be enough to meet the food fish demand, but would probably not be sufficient to meet the total demand. Unconventional resources will therefore also have to be exploited. Among these are the Antarctic krill, cephalopods (oceanic squid) and mesopelagic fish. The exploitation of these resources will however meet great technological challenges, for which research will be urgently needed.

FAO statistics (FAO, 1980) show that of the total catch, 70% is used for human consumption, and 30% for other purposes. Of the fish used for human consumption in 1980, 29.3% was consumed fresh, 31.0% frozen, 20.2% canned and 19.5% processed by various curing methods such as drying, salting, smoking, etc. However, the pattern of utilization in specific regions may differ very much from this average world total. It has been estimated that in the Latin American countries 50% of the food fish are utilized as fresh, 25% as frozen, 17% as canned and only 8% as salted, dried, or cured in other ways (Slavin, 1980).

A high percentage of fresh fish consumption may indicate that the fish is mainly eaten by people living relatively close to the sea. It could be that an increased production of fish preserved by freezing, canning or various curing methods would facilitate and increase fish consumption by making the products available to larger population groups. However, freezing and canning represent energy intensive technologies. With rising energy process these methods are probably only above to contribute marginally toward an increased use of fish. Curing methods like drying, salting and smoking are examples of simple, inexpensive processing methods, and, consequently, are promising principles for preservation of fish in the future.

FISH PRESERVATION

One of the unique features of fish is that it is a highly nutritious muscle food. On the other hand it spoils quickly. Preservation is therefore essential. The simplest way of prolonging the storage life of fresh fish is to chill it (FAO, 1975). Crushed ice is very efficient, and the fish may be kept on ice for one to two weeks as long as the temperature is kept close to $0^{\circ}C$ (Hansen, 1980). The shelf life may also be extended beyond this if a combination of a controlled atmosphere containing CO_2 and ice is applied (Regenstein & Regenstein, 1981).

An important aspect that should be emphasized is that chilling seems to be much more effective in tropical regions than in cold regions (Jones and Disney, 1977). The reason for this is apparently that psychrotrophic spoilage microorganisms are virtually absent in the warm waters. Under such conditions it is extremely important that uncontaminated, fresh water is used for ice production. Otherwise, psychrophilic spoilage bacteria may be enriched in the ice plant, such that when the fish are chilled with ice, the spoilage bacteria will be introduced together with the ice.

When fish are chilled, it is important to obtain good contact between the fish and the cooling medium. Two methods of chilling that have proved to be very efficient are: the Chilled Sea Water (CSW) and the Refrigerated Sea Water (RSW) systems. In the first, the fish are chilled in a mixture of ice and seawater in containers of various sizes (Hansen, 1977). In the second, the fish are chilled in tanks with circulated, refrigerated seawater (Merritt, 1969).

FREEZING AND CANNING

The main quality problems are protein denaturation and lipid rancidification. Both problems are under intensive study. Most of the studies on denaturation of white, salt water species have been focused on the effects of formaldehyde produced from trimethylaminoxide during frozen storage (Parkin and Hultin, 1981).

The quality changes associated with freezing are considered to have a negative influence on the acceptability of the fish. This is natural as long as the thawed product will resemble and have the image of fresh fish. With whole fish and fish fillets this will probably also be the prevailing situation in the future. For comminuted and formed products the situation is different. Controlled freezing or reformed products has been demonstrated to lead to desired textural changes. In Japan such methods are successfully applied in manufacturing of seafood products (Noguchi et al., 1978). This technique however, is an example of a highly sophisticated technology mainly applicable in the idustrialized countries.

Canning has been used for many years as a method of preservation of fish, and has the advantage that the product may be stored for very long times at ambient temperatures. In a report by FAO (Lanier, 1981) considering the use of canned small pelagic fish it is concluded that the marketing prospects seem to be good. It should however be noted that most canning plants today are highly mechanized, and the traditional tin used is expensive. However, the canning operation may also be performed on a small scale without sophisticated equipment, and if a cheaper packaging material could be found, canning may find a wider application in the future. Retort pouch technology also may provide a solution to the packaging problem.

IMPROVEMENT OF CONVENTIONAL CURING METHODS

Before discussing conventional curing methods like drying, salting, smoking, etc., the potential of using acid as an aid in preservation should be considered. Acids have mainly been used to lower pH in pickled products, and for the manufacture of fish silage (Raa and Gildberg, 1982). The main objective in these processes has been to prevent the growth of spoilage microorganisms and to add taste and flavour (pickled products). Effects of acids that have not been considered in the past are, for example, their ability to decrease water binding capacity, and other, more specific effects that will be discussed below.

The simplest, and possibly the oldest way of preserving fish is to dry it. Today, fish are either dried in artificial driers, or in the open air, exposed to sun and wind (Waterman, 1976). In order to avoid contamination and to speed up the drying process during sun drying, the fish has to be placed on racks above the ground. A serious problem occurring, particularly in tropical regions, is the infestation of flies. One way of reducing the problem is to treat the surface of the fish with pyrethrum, which is an insecticide (Losnegard and Bakken, 1973). Another method is to increase the temperature during drying such that the fly maggots cannot survive. Polyethylene tent driers have been constructed for this purpose (Doe et al., 1977). The temperature should be 45-50°C inside the tents. This will prevent fly infestation without too serious heat denaturation of the fish. The air circulation in the tents is regulated through adjustments of vents at top and bottom. It may however be difficult to provide sufficient air flow by such adjustments. The temperature inside will then be too high, and the fish will be "cooked".

Another approach to the fly problem that has been tested, and found satisfactory, is the use of organic acid. If the fish is dipped in acid, or acid is sprayed on prior to drying, it will act as a fly repellant. The acid will evaporate together with the water during drying, such that virtually no acid is left in the product. Any of the lower organic acids may be used but propionic acid has proved to be most effective (J. Raa, personal communication).

Lowering the pH to about 4 will weaken the connective tissue. This principle can be successfully applied in detaching tough scales from the fish skin. The fish must then be soaked in acid for some time, before the scales are scraped off. Acetic acid may be preferred for this treatment. After descaling the fish may be further processed by drying, salting, etc.

USE OF ACIDS IN NEW PROCESSING TECHNOLOGY

If moderate heat is applied in combination with acid, the fish skin may be completely removed. This is done in a process for the production of fish mince from small, pelagic fishes (Eide et al., 1982). The process, Figure 1, was developed, and is particularly suitable for the processing of mince from capelin, which is a small arctic, pelagic fish, ca. 20-30 g. The fish are first cut transversely in a set of rotating knives, water and acid are added, and the mixture is stirred at 35-40°C for 45. min. During this time the viscera and the fish skin are removed, and the lipid will be washed off the fish pieces. Fish mince is obtained after bone separation.

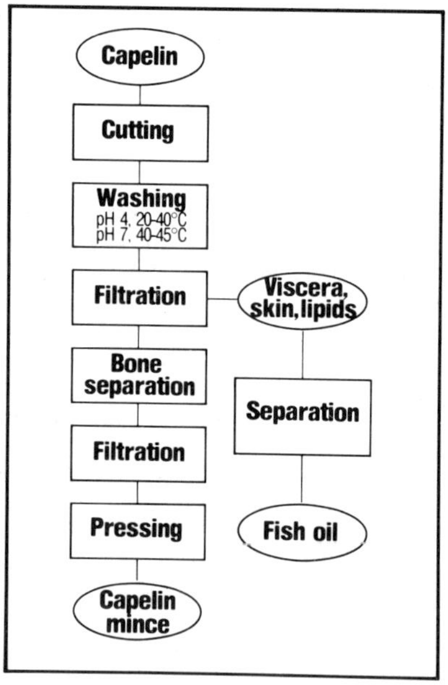

Figure 1

The mince process may be done on a small scale, requiring manual labor, or it may be adapted to large scale production with a high degree of mechanization and automation. When the fish contains roe, this may be removed by washing before acid or heat is introduced. The roe may then be purified in a separate process. Water is easily removed from the acid mince, as the water binding capacity is low. The mince is therefore suitable for further processing into dried products. However, the acid treatment in combination with elevated temperature will lead to a marked decrease of the functional properties. The mince is therefore not suitable for surimi production, in which it is required that the muscle proteins have good gelling properties. On the other hand the nutritional properties are still very good, and the mince may be used in several seafood products together with other ingredients prossessing adequate functional properties.

The process just described for fish mince production is an example of how acid treatment may lead to new technology within fish processing. Another example is the autolysis of fish material with acid present. As long as the temperature is kept low, the acid will stabilize the material as in normal fish silage. If however the temperature is raised to 30-35°C, acid proteases will be activated, and autolysis will occur. Since the viscera contains the highest level of proteolytic enzymes, fish offal containing guts will hydrolyze rapidly (Raa and Gildberg, 1982). When autolysis is complete, the offal is turned into a liquid with high content of protein - N, which is well suited as an animal feed (Strom et al., 1980). Application of this process may help in reducing the post harvest losses by turning fish offal that is ordinarily discarded into valuable feedstuff.

The acid autolysis principle may further be applied for hydrolysis of liver from the white fishes. After digestion of the liver tissue, the liver oil is easily recovered by conventional separation methods. This liver oil will be of very high quality, for example with a high level of vitamin B_{12}, in contrast to ordinary liver oil which is obtained after steaming the liver at high temperature.

MINCED FISH

Minced fish is generally produced by mechanical separation of bones and skin from the fish flesh in machines called bone separators or mechanical deboners (Grantham, 1981). The mince may be recovered from frames and cuts remaining after filleting, or it may be obtained by passing split, small fishes through the deboners. In the Western world most of the mince is frozen in blocks, and processed into products. In Japan the mince is used for surimi-based products (Suzuki, 1981). Minced fish may be salted and dried (Del Vallee, 1974). A tough, fibrous product is obtained, and, when mixed with starches, the mince may be processed in the form of crispy flakes which produce a puffed appearance when fried (Hansen, 1981).

Fish mince may be artificially dried at a high speed in roller driers (Herborg et al., 1974). This equipment consists of two horizontally mounted, rotating drums. The slit opening between the drums is less than one mm. wide, and as the mince is pressed through this slit, it adheres to the heated (150-160°C) surfaces of the drums. As the drums are rotated slowly, the mince dries and can be scraped off before a full cycle of rotation. The temperature in the mince does not exceed 80-85°C, and drying is completed within 20-40 sec. FPC-B is a dried protein concentrate that, in contrast to the FPC-A, has a flavour and taste much like ordinary fish meal. As the FPC-B has a distinct flavour characteristic it may be easier to incorporate in the diet than the FPC-A, which is virtually tasteless. The FPC-A is a valuable food protein, but it lacks functional properties, is expensive and cannot compete with other commercially available protein concentrates (Pariser et al, 1978). A specially textured dry fish product called marinbeef is produced in Japan, but is also fairly expensive (Suzuki, 1981).

FERMENTED FISH PRODUCTS

Fermented foods have been prepared both in the East and West for centuries (Hesseltine, 1965). The present trend shows that the consumption of fermented foods in the West is increasing, while it remains constant in the East (Hesseltine, 1981). The increased consumption of fermented foods in the West is among other things, due to the introduction of products originally known in the Orient.

In spite of a long tradition of eating fermented foods, very little is known about the biochemical and microbiological changes occurring (Whitaker, 1978). The basic principle of fermentation is to establish conditions that will prevent growth of spoilage microorganisms, and stimulate growth of favourable microorganisms. The microbes will add flavour and taste to the product, possibly digest toxic constituents, and produce valuable nutrients. Fermented fish may be prepared as dried products, as fish paste, or as fish sauce. An excellent review of fermented fish products has been written by Amano (1962). These products may be processed from whole fish or squid, viscera, roe, milt, etc. Cured-fermented fish products may be prepared by fermenting fish meat in rice. Katsuobushi is an example of a dry, fermented fish product. It is prepared by evisceration and cooking of the fish, followed by fermentation until it is dry (Tanikawa, 1971).

Fish paste is also produced from eviscerated fish. Salt is added, and it is fermented for about 3 months. The texture is then very soft and hydrolytic enzymes in the fish have produced a characteristic flavour and aroma (Van Veen, 1965).

Fish sauce is widely distributed in the Orient, and is used as a condiment providing valuable nutrients to carbohydrate - rich diets mainly based on rice (Saisithi et al., 1966). Fish sauce in prepared by mixing the fish with salt in a volume ratio of 3:1. After fermentation is concrete vats (about 40°C) for at least 6 months, the resulting liquid is filtered, and further ripened in the sun for 1-3 months in earthenware containers. The final sauce is dark-brown and has a high salt content. It has a distinctive flavour and aroma resulting from the combined action of protein hydrolysis and growth of halophilic bacteria.

Fermentation with lactic acid bacteria has been used for several foods, and this principle could represent a very important method for preservation of fish in the future (Raa and Gildberg, 1982). In addition to the fish a carbohydrate source and a starter culture is needed. It is possible to use such very cheap carbohydrate sources as molasses, and as the fermentation is proceeding, samples may be removed and used as starter cultures for later fermentation batches. Preservation with lactic acid will prevent the growth of spoilage bacteria by the low pH established, and by antimicrobial compounds produced by the lactic acid bacteria. The taste of lactic acid fermented fish products should be acceptable both in the West and in the East, and the bacteria will provide essential nutrients in the diet.

Traditional fermented foods in the East are often based on mycelial yeasts. Such products could possibly also be introduced in the West, although most Western cultures have a deep-seated prejudice against moldy products. Taste preferences should not be underestimated. It is therefore imperative, in any product development, that the taste preferences of the target population groups be carefully investigated (Kreutzer, 1974).

EXPLOITATION OF UNCONVENTIONAL RESOURCES

It was stated above that the total demand for fish products could be met if unconventional resources were exploited. Some approaches toward utilization of these resources have already been tried, e.g., the production of protein concentrates from krill (Suzuki, 1981). It is believed that for future successful exploitation of the unconventional resources untraditional processing technology will have to be developed. In order to achieve this goal the new developments in biotechnology and enzyme technology may provide very important tools. In addition, some organisms, e.g. krill and squid, seems to contain an array of unfamiliar enzymes and enzyme inhibitors which have very interesting properties. An approach towards the utilization of enzymes present in krill has been proposed by Ellingsen and Mohr (1979).

CONCLUSION

It seems possible to meet the projected demand for fish products in the year 2000, if 1) the conventional resources are utilized better, and 2) research efforts are directed toward the development of new technology for exploitation of unconventional resources. It is thought that a more efficient utilization of the conventional resources can be achieved if the production is turned towards an increased use of conventional curing methods like drying, salting, smoking etc. In addition the principles of acid preservation and procesing should be further developed. Finally, an increased use of fermented fish products looks promising, particularly through the use of lactic acid bacteria. Fish fermentation technology should thus be further developed.

ACKNOWLEDGEMENT

The authors wish to thank Professor Jan Raa for helpful suggestions and inspiring discussions during the preparation of the manuscript.

REFERENCES

Amano, K. (1962). The influence of fermentation on the nutritive value of fish with special reference to fermented fish products of south-East Asia. In Fish in Nutrition (Heen, E. and Kreutzer, R., eds.). Fishing News (Books) Ltd., London, England, pp. 180-200.

Del Valle, F.R. (1974). A quick-salting process for fish. In fishery products (Kreutzer, R., ed.). Fishing News (Books) Ltd., London, England, pp. 304-308.

Doe, P.E., M. Ahmed, M. Muslemuddin and K. Sachithananthan (1977). A polyethylene tent drier for improved sun drying of fish. Food Tech. Australia, 29, 437-441.

Eide, O., T. Borresen and T. Strom (1982). Minced fish production from capelin (Mallotus villosus). A new method for gutting, skinning and removal of fat from small fatty fish species. J. Food Sci., 47, 347-349 and 354.

Ellingsen, T and V. Mohr (1979). A new process for the utilization of Antarctic krill. Process Biochem. 14, 14-19.

FAO (1975). Ice in Fisheries. FAO Fish. Rep. No. 59, Rev. 1, pp. 1-57.

FAO (1980). Yearbook of fishery statistics. Fishery Commodities, vol. 51, FAO, Rome, pp. 22-23.

FAO (1981a). Commodity review and outlook, 1980-1981, FAO Economic and Social Development Series, No. 20, Rome, pp. 92-97.

FAO (1981b). The prevention of losses in cured fish, FAO Fish Tech. Pap. 219, 1-87.
Grantham, G.J. (1981). Minced fish technology: A review. FAO Fish. Tech. Pap. No. 219, pp. 1-87.
Hansen, P. (1977). Improved ice chilling of trawl catches in boxes and containers. Scadinavian Refrigeration 6, 262-266.
Hansen, P. (1980). Fish preservation methods In: Advances in Fish Science and Technology (Connell, J.J., ed.) Fishing News Books Ltd., Farnham, England, pp. 28-34.
Hansen, P. (1981). Fraktionering of foraedling af industrifisk. Arsberetning 1981, fiskeriministeriets Forsogslaboratorium, Lyngby, Denmark, pp. 1-9.
Herborg, L., F. Vilien, A. Brunn and B. Eggum (1974). Roller-dried fish protein. In: Fishery Products (Kreutzer, R., ed.). Fishing New (Books) Ltd., London, England, pp. 309-311.
Hesseltine, C.W. (1965). A millenium of fungi, food, and fermentation. Mycologia 57, 149-197.
Hesseltine, C.W. (1981). Future of fermented foods. Process Biochem. 16, 2-6 and 13.
Kreutzer, R. (1974). Fish and its place in culture. In: Fishery Products (Kreutzer, R., ed.). loc. cit. pp. 22-47.
Krone, W. (1979). Fish as food. Present contribution and potential. Food Policy 4, 259-268.
Lanier, B.V. (1981). The World supply and demand picture for canned small pelagic fish. FAO Fish Tech. Paper No. 220, pp. 1-111.
Losnegard, N. and K. Bakken (1973). Pyrethrum treatment for the protection of drying fish from blow-fly infestation. Norwegian Fisheries Research Institute, Bergen, Norway, pp. 1-10.
Merritt, J.H. (1969). Refrigeration on Fishing Vessels. Fishing News (Books) Ltd., London, England, pp. 71-86.
Noguchi, A., N. Sannomiya, K. Umeda and S. Kimura 91978). Freezing process for production of textural protein food materials from krills. J. Agr. Chemc. Soc. (Japan) 52, 454-552.
Pariser, E.R., M.B. Wallerstein, C.J. Corkery and N.L. Brown 91978). Fish protein concentrate: Panacea for protein manlnutrition? The MIT Press, Cambridge, Massachusetts, USA, pp. 1-285.
Parkin, K.L. and H.O. Hultin (1981). A membrane fraction from red hake muscle catalyzing the conversion of TMAO to dimethylamine and formaldehyde. In: Advances in Technology in the Chilling, Freezing, Processing, Storage and Transport of Fish, Especially Underutilized Species (Annon., ed.). International Institute of Refrigeration, Paris, France, pp. 475-482.
Raa, J. and A. Gildberg (1982). Fish silage: A review. CRC, Crit. Rev. Food Sci. Nutr. 16, 383-419.
Regenstein, J.M. and Regenstein, C.E. (1981). The shelf-life extension of fresh fish. In: Advances in Technology in the Chilling, Freezing, Processing, Storage and Transport of Fish, loc. cit. pp. 357-364.
Saisithi, P., B. Kasemsarn, J. Liston and A.M. Dollar (1966). Microbiology and chemistry of fermented fish. J. Food Sci. 31, 105-110.
Slavin, J.W. (1980). Overview of mince fish utilization in developing countries. In: Third National Technical Seminar on Mechanical Recovery and Utilization of Fish Flesh (Martin, R.E., ed.). National Fisheries Institute, Washington, DC, USA, pp. 451-456.
Strøm, T., A. Gildberg, B. Stormo and J. Raa (1980). Fish silage: Why not use propionic and formic acid? In: Advances in Fish Science and Technology, loc. cit. pp. 352-355.
Suzuki, T. (1981). Fish and krill protein: Processing Technology. Applied Science Publ. Ltd., London, England, pp. 1-251.
Tanikawa, E. (1971). Marine products in Japan, Koseisha-Koseikaku Co., Tokyo, Japan, pp. 1-498.

Van Veen, A.G. (1965). Fermented and dried seafood products in Southeast Asia. In: Fish as Food (Borgstrom, G., ed.) Vol III, Academic Press, New York, USA, pp. 227-250.
Watermann, J.J. (1976). The production of dried fish. FAO Fish Tech. Pap. No. 160, pp. 1-52.
Whitaker, J.R. (1978). Biochemical changes occurring during the fermentation of high-protein foods. Food Technol. 32, 175-180.

PROGRESS IN PRESERVATION OF FOOD THROUGH FERMENTATION

Keith H. Steinkraus
Professor of Microbiology and Food Science
Institute of Food Science
Cornell University
Geneva/Ithaca, New York 14456 USA

ABSTRACT

Fermentation as a means of processing and preserving food was discovered many centuries ago, long before the sciences of microbiology and chemistry. Most fermentations were originally spontaneous, for example, the souring of milk, production of cheese, production of wine from fruit juice, and production of vinegar from wines. Fermentation offers low-cost methods for preserving fresh vegetables by lactic acid fermentation, leavening bread, increasing the protein content of high starch substrates, and fortifying foods with important vitamins and essential amino acids. Fermentation provides the means whereby meat-like flavors and textures can be introduced into cereal grain/legume substrates decreasing the need for the inefficient cycling of vegetable proteins through animals. Fermentation is responsible for the wide diversity of flavors and textures in our foods, and often decreases cooking time and fuel requirements. It is a major means of processing foods at low cost to the developing world, and is likely to play an ever-increasingly important role in the future as world population increases.

KEYWORDS: Fermentation; fermented foods; low cost foods; food preservation.

INTRODUCTION

World population, presently about 4 billion, is expected to reach 6 billion by the year 2000 and finally level off at about 8 to 12 billion in the 21st century. It is estimated that between 450 million and 1 billion people are presently underfed and undernourished (National Research Council, 1977). The Green Revolution has resulted in vast increases in the world-wide productivity of rice and wheat that has enabled mankind to continue to feed its burgeoning population, but it has not relieved the millions of hungry and malnourished in the developing world. In the developing world, about 180 kg/a of cereal grains are available per person and they are consumed by humans (Brown, 1974). In the US about 900 kg/a of cereal grains are available per person. Of this amount approximately 90 kg are consumed directly in the form of bread, cereals, etc. The rest is used for production of alcoholic beverages or fed to animals where the efficiency of conversion of vegetable protein to animal protein ranges from 31% (milk) to 27% (eggs) to 18% (chicken broilers) to 9% (pork) to 6% (beef) (Pimentel et al., 1975). If Americans became vegetarians, the cereal grains/legumes presently fed to animals could feed another 800,000,000 people.

There also are enormous losses of food in the developing countries resulting from spillage, contamination, insects, birds, rodents, and microbial spoilage during storage. The National Research Council, U.S.A. (1978) estimated that a minimum of 107 million t of food were lost in 1976. The food lost could have fed approximately 168 million people. We must look for alternate ways of increasing the food supply or modifying the distribution of cereal grains, legumes, and other foods between animals and man.

In the Western World the major methods of preserving food include canning, freezing, and dehydration with heat all of which are too expensive for most of the developing world. In the developing world, major methods of preserving food include sundrying, salting, and acid, alcoholic, and fungal fermentation. The indigenous food fermentations developed many centuries ago, are generally low-cost, important in the nutrition of the poor, yield aromas and flavors highly acceptable to millions of people, and often reduce fuel requirements and cooking times. Some offer the means to raise the protein content of high starch substrates, provide low-cost means of raising the vitamin content of the diet, produce meat-like flavors in vegetable protein substrates, and provide a means of introducing a meat-like texture into vegetable protein substrates (Steinkraus, 1982).

FERMENTATION AS A MEANS OF EXPANDING THE FOOD SUPPLY

Fermented Foods Involving an Acid Fermentation

Most food spoilage and most disease organisms cannot survive in an acid environment. Thus foods that undergo an acid fermentation generally can remain wholesome for a long period of time. There are approximately 176,000,000 kg of sauerkraut (lactic acid fermentation) produced per year in the United States (Dr. John Stamer, personal communication). The cabbage is shredded and mixed with 2.25% w/w of table salt (NaCl). The salt extracts liquid from the cabbage which is fermented anaerobically by a sequence of organisms, *Leuconostoc mesenteroides*, followed by *Lactobacillus brevis* and then completed by *Lactobacillus plantarum* (Pederson and Albury, 1969). Acidity reaches 1.7-2.3% (as lactate) sufficient to preserve the sauerkraut as long as it remains anaerobic.

Pickling of cucumbers and other vegetables is widely practiced in the world today. Although a variety of methods are used, placing the cucumbers in 5% salt brine is a satisfactory method. The cucumbers absorb salt until there is an equilibrium between the salt in the cucumbers and the brine (about 3% salt in the brine). Acidity reaches 0.6-1.0% (as lactic) with a pH of 3.4-3.6 in about 2 weeks, depending upon the temperature (Pederson, 1979).

Korean kimchi, made from Chinese cabbage, radishes and other vegetables, is closely related to sauerkraut and pickle fermentations. The vegetables are placed in 5-7% salt w/w for 12 hours or in 15% salt brine for 3-7 hours, then rinsed and drained. Optimum salt concentration is about 3% and fermentation runs over one day at 30 C, 2-3 days at 20 C or 30-60 days at 5 C. Optimum acidity of kimchi is 0.4-0.8% (as lactic). Organisms involved include *Leuconostoc mesenteroides*, *Streptococcus faecalis*, *Lactobacillus brevis*, *Pedicoccus cerevisiae* and *Lactobacillus plantarum* (Kim and Whang, 1959; Kim and Chun, 1966).

The Malaysians pickle a variety of vegetables including cucumbers, ginger, onion, leeks, chillies, bamboo shoots, mustard leaves, green mangoes, papaya, nutmeg and limes by similar processes (Merican, 1977). The Egyptians pickle carrots, turnips, cucumbers, cauliflower, green and black olives, onions and hot and sweet peppers by a variety of salting and brining processes (Morcos, 1977; Mahmoud, 1977; Mahmoud et al., 1972). Brining with lactic acid fermentation is a useful and wholesome method of preserving a wide variety of fresh vegetables and fruits.

Lactate fermentation is very important in the processing of many cereal grain/legume foods, particularly in Africa. Middle Eastern kishk, Greek trahanas and Turkish trahanas are prepared by boiling sour milk with parboiled wheat, constantly stirring until it reaches a very high viscosity due to loss of water (sometimes vegetables are added). The milk/wheat mixture is formed into balls or flat cookies and sun-dried. The final product can be stored for several years retaining its excellent nitrutive value. It is used primarily as a base for soups (Morcos et al., 1973; Economidou and Steinkraus, 1977).

Nigerian ogi and Kenyan uji are yogurt-like products made by natural lactate fermentation of maize, sorghum or millet flour or mixtures. Soybean is sometimes mixed with the cereals to increase the protein and nutritive value (Akinrele and Edwards, 1971; Okafor, 1981; Banigo et al., 1974). Generally, the grains are steeped for 1 to 3 days, ground wet, slurried, fermented for 1 or 2 days, and then boiled at a solid content of about 8% to yield a porridge. The pH should be between 3.6-3.7 for optimum flavor.

Mahewu (Magou) is a traditional sour maize beverage popular among the Bantu people of South Africa. A maize porridge of about 8% to 10% solids is produced by boiling. It is then mixed with a small amount of wheat flour which serves as inoculum for the acid fermentation (Schweigart and de Wit, 1960). Final acidity is 0.4 to 0.5% (as lactic) with a pH of about 3.5. This process has been commercialized and mahewu is now produced in 3785 l tanks using pure culture inocula. The fermented slurry is then enriched with added soybean meal and spray dried (Hesseltine, 1979). Nigerian gari is made by an acid fermentation of grated cassava pulp followed by a semi-dextrinization of the starch and drying (Onyekwere and Akinrele, 1977; Okafor, 1977).

Milk contains the microorganisms necessary for a lactic acid fermentation and spontaneously sours if held at a favorable temperature. Sour milk has an excellent keeping quality and, if the curd is separated, leads to a variety of cheeses which also have excellent keeping quality. In 1979, there were an estimated 214,589,000 cows in the world which produced 419,464,000 t of milk (FAO, 1980). In addition, 25,879,000 t of buffalo milk, 7,291,000 t of sheep milk and 7,236,000 t of goat milk were produced in 1979. The total milk supply amounted to 315 g of milk each day for each of the world's 4 billion people. The US alone produced 56,074,000 t of milk amounting to 668 g/day of milk per capita. These huge quantities of milk could not be consumed entirely in the fresh state. Some means of preservation is essential. Some of the fresh milk was fermented to yogurt. World-wide, 11,012,350 t of cheese were produced amounting to 7.5 g/day of cheese per person. Cheese fermentation involving curdling of the milk followed by lactic acid fermentation is a striking example of how fermentation is used to preserve the nutrients in one important food - milk. There are hundreds of varieties of cheese, most involve a lactic acid fermentation (*Lactobacillus*, *Leuconostoc*, *Streptocuccus*). Roquefort or Blue Cheese and Cambembert cheese also utilize the molds *Penicillium roqueforti* and *Penicillium camemberti* respectively. Generally acid fermentation requires no heat at any stage of its preparation. Thus, it is fuel efficient. All these indigenous fermentations continue to offer excellent means of preservation in a world where fuel is becoming increasingly scarce.

Leavened Bread Without the Use of Wheat

In some parts of the world, wheat and rye, the basis for flours used to make leavened breads, are not readily available. Thus, there is considerable interest in processes that permit leavened breads to be made using other grains, such as rice. The Indians have just such an indigenous food process (Mukerjee et al., 1965; Steinkraus et al., 1967). Rice and a legume, such as black gram dhal or dehulled soybeans are soaked separately during the day. In the evening they are drained and ground to a slurry. A small amount of salt is added for flavor. The rather thick batter is incubated overnight in a warm place. During the night, the essential microorganism, Leuconostoc mesenteroides, produces lactic acid and CO_2 basically the same as it does in a sauerkraut fermentation. The substrate becomes acid and leavened by the CO_2. Gums present in the legume and/or dextrans produced by L. mesenteroides retain the gas so that the product called idli, when steamed, is essentially a leavened, sour-dough bread. This process could be applied in many parts of the world where wheat or rye flour is unavailable. The acidity of the idli also makes the product resistant to deterioration or invasion by food spoilage or food poisoning organisms.

Fish Sauces and Pastes (van Veen, 1965)

It is estimated that fish provides about 17% of the world's animal protein intake (National Research Council, 1978). About 60% of the fish is consumed fresh. The remainder is dried, salted or fermented to produce sauces and pastes. Most of the fish sauce/paste fermentation occurs in Asia (Vietnam, Thailand, Burma, Philippines). Processing methods are relatively simple. Small fresh fish are layered with crude sea salt in large wooden vats. Liquid extracted from the fish by the salt drains from the bottom of the vat into wooden containers and is recirculated by pouring it into the top of the vat. This continues for about 3 months during which time, most of the fish tissue liquifies and separates as a salty (25-30% salt) mixture of amino acids and peptides - a meat flavored condiment very important in the diets of the Southeast Asians.

Fish pastes are produced by grinding small fish with salt and incubating the mixture anaerobically in large vats. Under such conditions, hydrolysis is less and the fish tissue remains a salty paste which is consumed following frying as a side dish along with rice. Larger surplus fish can be fermented to sauces or pastes following gutting and mixing with salt. The fish sauce/paste fermentations offer a method of preserving much of the surplus perishable fish that cannot be dried. Microorganisms play a minor role in the fish sauce/paste fermentations. Hydrolysis is the result of proteases and lipases within the fish tissues - primarily the gut. In recent years the fish sauce/paste fermentations have been accelerated by the addition of either vegetable proteases such as papain or bromelein or by the addition of fungal enzymes produced on cereal grain/legume substrates (kojis) that will be described later. Thus fermentation time can be decreased to 1 or 2 weeks in place of the months required by natural fermentation. Fish sauces/pastes are wholesome, nutritious, and have an attractive flavor. The minimum salt content required for long-term preservation is 12% salt/fresh weight of fish (Tan et al., 1967). The relatively high salt content limits the amount of fish that can be consumed; but the fermentation preserves much of the nutritive value of surplus fish that might otherwise spoil. The fish sauces/pastes are known and widely utilized in Southeast Asia, but remain largely unknown in much of the rest of the world. Their use could be expanded wherever there are surplus fish that could be preserved.

Fermented Foods Involving an Alcoholic Fermentation

Fruit juices such as grape or apple, sugar cane juice, and the sap collected from damaged palm flowers are all attractive foods in the diets of the human race (Steinkraus, 1979). Without pasteurization, canning, bottling, vacuum concentration or freezing, most juices cannot be kept in their fresh form for any length of time. However, nature has made yeasts a part of the environment wherever fruits are grown. The prepared juices, unless pasteurized by heat, contain yeasts which immediately start to ferment the sugars producing ethanol which also serves as a preservative. If the wines are kept anaerobic, they can be preserved in wholesome condition for long periods of time. Wines exposed to air undergo a secondary aerobic fermentation by Acetobacter microorganisms producing acetic acid (vinegar) - a staple among man's condiments and also a very useful, preservative pickling agent for vegetables.

Primitive Beers

Kaffir (Sorghum) Beer (Steinkraus, 1979)

Kaffir beer, an alcoholic beverage with a pleasantly sour taste and the consistency of a thin gruel, is the traditional beverage of the Bantu people of South Africa. Alcohol content may vary from 1 to 8% v/v. Kaffir beer is generally made from kaffir corn (Sorghum caffrorum) malt and unmalted kaffir corn meal. Maize or finger millet (Eleusine coracana) may be substituted for pat or all of the kaffir corn depending upon the relative cost (Schwartz, 1956). Even cassava and plantains may be used (Platt, 1964). The kaffir corn grain is steeped for 6 to 36 hours. It is then drained and placed in layers and germinated with periodic moistening for 4 to 6 days. Germination continues until the plumule is 2.5 cm or longer in length. It is then sun-dried. Over 80% of the malt is sold for home brewing of kaffir beer.

The essential steps in brewing are mashing, souring, boiling, conversion, straining, and alcoholic fermentation. Mashing is carried out in hot (50 C) water. Proportions of malted to unmalted grains vary, but 1:4 is satisfactory. Four 1 of water are added for about every 2 kg of grain. Souring begins immediately because of the presence of lactobacilli. A temperature of 50 C favors development of Lactobacillus delbrueckii. Souring is complete in from 6 to 15 hours. Water is added and the mixture is boiled. It is then cooled to 40-60 C and additional malt is added. Conversion proceeds for 2 hours and then the mash is cooled to 25-30 C.

Yeasts present on the malt are responsible for the natural fermentation. However, Saccharomyces cerevisiae isolated from kaffir beer is frequenty inoculated in the more modern breweries. Kaffir beer is ready for consumption in 4 to 8 hours. It is drunk while still actively fermenting. This state may continue for up to 40 hours. Ethanol content is generally from 2 to 4% w/v. The beer also contains from 0.3 to 0.6% lactic acid and 4 to 10% solids. Production of acetic acid by Acetobacter is the principal cause of spoilage. van der Walt (1956) has described the microbiology of kaffir beer production. Other studies on the process include those by Platt and Webb (1946), Platt and Webb (1948), Aucamp et al. (1961), Horn and Schwartz (1961), von Holdt and Brand (1960a,b), O'Donovan and Novellie (1966), and Novellie (1959, 1960a,b, 1962a,b,c, 1966).

Platt (1964) described village processing of kaffir beer. Brewing is a woman's responsibility. Every girl learns the procedure before she marries. Consumption each day of 2.8 l of kaffir beer, which requires approximately 425 gm of grain, is not unusual for a working man. How the consumption of the beer diet compares in nutrient intake with a diet in which the grains are consumed directly is shown in Table 1 (Platt, 1964). It is seen that caloric content of the two diets is quite similar. Only 37 calories are lost in the diet containing the beer. However, most notable is the doubling of riboflavin and near doubling of nicotinic acid content of the diet containing beer because of synthesis of vitamins during malting and foermentation. Pellagra, which is relatively common in those subsisting on maize diets, is never noted in those consuming usual amounts of kaffir beer (Platt, 1964).

Table 1: Comparison of Diet With and Without Maize Beer (from Platt, 1964)

Food item	Amount of food eaten: Diet without beer	Amount of food eaten: Diet with Kaffir beer
Maize, whole meal	350 gm	13.5 gm
Maize, 60% extraction	350 gm	137.5 gm
Maize beer	----	2.8 l (5 pints)
Vegetables	130 gm	130 gm
Sweet potatoe	470 gm	470 gm
Kidney beans	30 gm	30 gm
Calories	3016	2979
Vitamin B_1	2.00 mg	1.95 mg
Riboflavin	1.13 mg	2.32 mg
Nicotinic acid	11.70 mg	20.30 mg

Table 2: Effect of Germination on the Vitamin Content of Kaffir Corn and Maize[a]

Sample	Moisture (%)	Vitamin content (micro g/gm)(dry weight basis): Thiamin	Riboflavin	Nicotinic acid
Kaffir corn				
Grain	11.6	3.34	1.29	35.1
Malt	10.7	1.73	2.41	34.1
Maize				
Grain	12.1	3.50	1.6	19.6
Malt	8.6	2.32	2.02	29.1

[1]Adapted from Goldberg and Thorp (1946).

The effect of germination on vitamin content of kaffir corn and maize is resented in Table 2. Thiamine decreases, riboflavin shows a substantial ncrease, nicotinic acid remains nearly constant in kaffir corn, but increased by bout 50% in maize. Thus, the microorganisms involved in the fermentation must be contributing additional thiamine, riboflavin, and nicotinic acid to the product.

Aucamp et al. (1961) report the results of chemical analyses of municipal kaffir beers (Table 3). According to these figures, the content of ascorbic acid in kaffir beer is negligible. However, consumption of about 2 l/day of beer which is rather common, would supply more than the minimum daily requirements (Harris, 1959) of thiamine, riboflavin, and nicotinic acid as shown in Table 4. The vitamins supplied, however, would be less than the allowances recommended for American adult males by the Food and Nutrition Board, National Research Council (1974). Nevertheless, kaffir beer nutritionally is a very important part of the South African village diet.

Table 3: Composition of Municipal Kaffir Beers (from Aucamp et al., 1961)

	Range	Mean	Number of analyses
pH	3.2-3.7	3.4	10
Alcohol (%w/v)	1.8-3.9	3.0	17
Solids (%w/v)			
Total	3.0-8.0	5.4	17
Insoluble	2.3-6.1	3.7	6
Nitrogen (%w/v)			
Total	0.059-0.137	0.093	16
Soluble	0.010-0.017	0.014	9
Thiamin (g/100 ml)	20-230	93	21
Riboflavin (g/100 ml)	27-170	56	21
Nicotinic acid (g/100 ml)	130-660	315	21
Ascorbic acid (mg/100 ml)	0.01-0.15	0.04	7

Table 4: Amounts of B Vitamins Supplied by Two Liters of Kaffir Beer[a]

	Thiamin (mg)	Riboflavin (mg)	Nicotinic acid (mg)
Minimum	0.45	0.61	2.9
Maximum	5.2	3.9	15.0
Mean	2.1	1.3	7.2
Minimum daily requirement[b]	1.0	1.2	10.0
Recommended allowance[c]	1.5	1.8	20.0

[a]Calculated from values in Table 3. [b] United States Food and Drug Administration (Harris, 1959). [c]Adult males; Food and Nutrition Board, National Research Council (1974). Production of Sweet/Sour Alcoholic Pastes from High Starch Substrates.

The Orientals have their rice wines, which were originally produced in indigenous fermentations. In the Japanese sake process, amylolytic enzymes required for saccharification of the starch are generated by growing selected amylolytic molds on a steamed rice substrate (to produce a koji)(Kodama and Yoshizawa, 1977). In the case of Western beers, malted barley generally supplies the amylases. The Indonesians produce a sweet/sour alcoholic paste from rice in an indigenous fermentation involving yeast-like molds and mold-like yeasts that have the ability both to hydrolyze starch and to produce ethanol from the released sugars (Cronk et al., 1977; Ko, 1972). The products called tape ketan (from rice) or tape ketella (from cassava) are highly acceptable in flavor and aroma. They contain up to 6 or 8% ethanol which, along with acids produced by the microorganisms, preserve the product and keep it wholesome. In addition, lysine, the first limiting amino acid in rice, is selectively synthesized improving the protein quality. Thiamin is increased 300% making tape an essential source of this vitamin for people subsisting on polished rice.

Using rice as a substrate, the per cent protein approximately doubles during the tape fermentation due mainly to conversion of a portion of the starch to sugars, acid and ethanol reducing the total quantity of starch in the substrate. Millions of the world's poor and economically deprived people try to subsist on cassava as a staple. Cassava has a desirable flavor, but generally contains less than 1 or 2% protein. Thus, it is impossible to obtain one's daily protein requirements by consuming cassava as a sole protein source. However, when the tape fermentation is applied to cassava, the protein content can be raised as high as 6% significantly improving its nutritive value as a source of protein.

Modern Methods of Introducing Meat Textures into Vegetable Substrates

Modern food science and technology have taught us how to isolate and spin soybean proteins into fibers that provide a meat-like texture which when fortified with fats and meat-flavors serve as meat analogs or substitutes for animal meats in the diet. Meat-like nuggets can be produced by extruding formulations of soybean protein concentrates and/or gluten with appropriate fats and flavors through small orifices at high temperatures and pressures (Odell, 1968; Smith and Circle, 1972; Wanderstock, 1968). Already a number of meat analogs such as lean bacon flavored bits and imitation hamburger have been commercialized. If these meat-analogs replaced meats in the diet, huge quantities of cereals/legumes presently fed to animals would be released for human food. This could improve the overall nutrition of large masses of the human race particularly among the presently economically deprived.

An alternate method for producing meat analogs is a process developed by Rank, Hovis, MacDougall Research Institute in England (Spicer, 1971a, 1971b). By this process, a selected edible mold is grown on low-cost carbohydrates in large sophisticated tank fermenters. The mold is recovered by filtration and the fibrous mycelium which provides a meat-like texture is formulated with fats and meat flavors yielding a type of meat analog, competitive in price with meat. This is modern, highly developed, capital intensive applied microbiology/food technology.

The production of microbial protein (Single-cell protein - SCP) by growing selected edible, non-toxic microorganisms on suitable substrates continuously in huge computer-controlled fermenters is one of the scientific marvels of this century. The processes were developed initially using petroleum as the substrate which must be available in large quantities at low cost in order to make SCP competitive in cost with soybean and fish meal as ingredients for animal feeds. Optimum size for such a factory is a capacity of 100,000 t dry SCP per year. Capital investment is approximately $75,000,000 to $100,000,000. If the price of petroleum had not risen so drastically, it was anticipated that SCP would provide 3% of the total world protein needs by now (Wells, 1975). If SCP production had been expanded to the level originally expected, SCP would have been fed to animals thus releasing quantities of grain and legumes for feeding the human population.

Eventually microbial cells will be fractionated much as soybeans are fractionated today. The proteins will be concentrated, isolated and spun into fibers for use in meat analogs as is soybean protein. In fact, all the necessary steps have been developed in the laboratory. However, the price of petroleum is presently too high and SCP cannot compete with the soybean at present prices.

Indigenous Methods for Introducing Meat-Like Textures into Vegetable Substrates

Although modern food science/technology developments involve highly sophisticated, relatively expensive technology, the Indonesians centuries ago developed low-cost fermentation processes by which meat-like textures are introduced into vegetable protein substrates, particularly soybean, by use of selected molds belongong to genus *Rhizopus* (van Veen and Schaefer, 1950; Steinkraus et al., 1961; Hesseltine et al., 1963; van Veen and Steinkraus, 1970). The product called tempeh kedelee when made from soybean, tempeh bongkrek when made from coconut or ontjum (oncom) when made from peanut press-cakes is an example of low cost technology applicable at the village level the world over, wherever protein-rich meat analogs are needed. Soybeans are soaked, dehulled by rubbing them between the hands or stamping with the feet, partially cooked, cooled, inoculated with the mold previously grown in relatively pure culture on leaves and sun-dried, wrapping the inoculated bean cotyledons in packets about the size of the adult human hand using wilted banana leaves and allowing the beans to ferment in a warm place for 36 to 48 hours. The cotyledons become a firm cake through overgrowth with the fibrous mold mycelium. The cake can be thinly sliced and deep-fried or cut into chunks and used in soups. Protein content on a dry solids basis is above 40%. Used in soups, the tempeh has a texture and flavor resembling veal. Soybeans generally require from 5 to 6 hours boiling to soften them sufficiently for human consumption. Allowing approximately 1 hour boiling for initial partial cooking, the tempeh requires only 10 minutes boiling in a soup before consumption. Thus, tempeh is one of the world's original quick cooking foods. While the soybeans are soaking in the first step of tempeh fermentation, the beans undergo an acid fermentation which reduces the pH of the beans to between 4.5 and 5.0 satisfactory for subsequent mold growth but unfavorable to the development of microorganisms that might spoil the tempeh.

Most of the world's population are vegetarian for economic reasons. The affluent meat-eaters get their daily vitamin B_{12} requirements from meats or milk. There is always a hazard that vegetarians will not get sufficient vitamin B_{12}. Interestingly enough, it has been found that commercial tempehs invariably contain vitamin B_{12} activity due to a specific bacterium that accompanies the mold during the fermentation (Liem et al., 1977). Digestibility of the soybean is improved by the tempeh fermentation. Also, it has been found that riboflavin doubles and niacin increases nearly 7 times (Steinkraus et al., 1961). Increase in vitamin content is very important in the diets of the economically deprived.

It is obvious that the meat-like texture for both the Rank, Hovis, MacDougall meat analogs and for Indonesian tempeh comes from mold mycelium. In the Rank, Hovis, MacDougall process the substrate is separated from the fungal mycelium which then undergoes extensive, relatively expensive formulation to make a human food. In the tempeh process, the fungal mycelium is consumed with the substrate and no further formulation is required except cooking for a short time. Indonesian tempeh offers the world a process of making low-cost, protein-rich meat analogs with an excellent nutritional quality. Tempeh has been adopted by

American vegetarians and is now being produced commercially in the United States. It is anticipated that it will become at least as highly accepted as yogurt in the American diet.

Utilization of Food Processing and Agricultural Waste to Produce High-Quality Foods

The Indonesians have developed methods for converting food processing by-products, such as peanut and coconut press-cakes, which the Western World has traditionally fed to animals, to nutritious human foods called ontjom and bongkrek. They have done this by using the basic tempeh process. The press-cakes are hydrated, coarse ground, steamed, cooled and inoculated with either the tempeh mold or Neurospora intermedia. The mold grows over the particles, knitting them into tight cakes that can be sliced or cut into chunks and used in soups (van Veen and Steinkraus, 1970). These products are low-cost, protein-rich meat analogs. The basic changes are similar to those that occur during tempeh fermentation. In addition, it has been found that the content of aflatoxin, always present in peanut press-cake, is reduced (van Veen et al., 1968). The strains of Neurospora intermedia also contain cellulases that reduce the natural fibre content of the peanut or coconut press-cake.

These indigenous fermented food processes offer a unique opportunity for increasing the quantity and quality of protein in areas of the world where the staple food is largely comprised of starch. They not only contribute to Western food science, but are also suitable, at the village level, for low-cost production of foods with acceptable flavor, texture, and nutritive values.

The human race seems to have a great desire for meat-like flavors. Thus, it was one of the great advances in food science and technology when the Orientals over a thousand years ago discovered how to produce soybean sauces and pastes. Soy sauce and Japanese miso processes are examples of indigenous science and technology that have been already considerably commercialized. However, these products can be and are still manufactured at the cottage or village level in parts of Asia. In the soy sauce/miso process in its most primitive form, soybeans are soaked, cooked, cooled, made into a ball tied together with rice straw and hung under the rafters (Shurtleff and Aoyagi, 1976). After about 30 days, the ball is overgrown with molds belonging to the Aspergillus oryzae group. Subsequent study has shown that these molds are rich in amylases, proteases and other enzymes (Yokutsuka, 1960; Shibasaki and Hesseltine, 1962). The mold covered ball containing microbial enzymes (a type of koji) is crushed and mixed with salt and water. Proteases from the fungus hydrolyze the proteins in the soybeans freeing amino acids and peptides with a meat-like flavor. The liquid portion is essentially a soy sauce. The past-like residue is essentially a miso. Both have a meat-flavor. Miso is used as the basis for a nutritious breakfast soup eaten traditionally daily in Japan. Soy sauce is used as a condiment for flavoring soups, bean curd, and many other Chinese and Japanese dishes. It also is widely used in the US. When one considers that most of the world lives on bland diets containing relatively small amounts of meat, the importance of meat flavors derived from vegetable protein takes on enhanced importance for the world of today and the future. In addition, soy sauce and miso are important sources of amino acids/peptides in the diets of the consumers.

Mushroom Production on Ligno-Cellulose

It is possible to use cellulose or ligno-cellulosic wastes such as waste paper, cotton waste, straw, wheat, or rice bran and go directly to a food. This idea has already been developed to a high degree in Asia in the production of mushrooms such as Volvariella volvacea, the padi mushroom, and Pleurotus ostreatus, the oyster mushroom, on cellulosic and ligno-cellulosic wastes (Chang, 1972, 1977; Eger et al., 1976). Mushrooms contain 2 to 5 percent protein on a fresh weight basis, but from 30 to 47 percent on a dry weight basis (Kurtzman, 1975). As much as 1.25 kg of fresh mushrooms can be produced on 1 kg of straw. In Hong Kong there is an estimated 30,000 t of cotton waste per year. This could serve as a substrate for producing approximately an equal weight of fresh mushrooms. The padi mushroom is grown by many farmers in Asia, using rice straw as a substrate. Thus, the Asians have demonstrated to the world a practical way to ferment ligno-cellulosic wastes into highly acceptable food for man. They are literally growing a type of microbial protein (SCP) directly on cellulosic waste as a nutritious, delicious food.

The padi and oyster mushrooms can be grown under rather simple conditions. Paper or cotton substrates are shredded. Straw can be trimmed, coarse ground, or used directly. Five percent wheat or rice bran and 5 percent $CaCO_3$ are added, along with sufficient water to raise the moisture content to about 60 per cent. This requires that approximately 1,500 ml of water be added per kg of ligno-cellulosic waste. The substrate is steamed for 30 minutes. Alternatively, the substrate can be composted in heaps where microbial activity results and the temperature rises to about 55 C. The substrate is cooled and inoculated with the mushroom spawn. The spawn is the desired mushroom species grown on soaked, sterilized wheat, corn, or rice straw. Approximately 160 gm of spawn are added to each kg of starting (dry weight) substrate. Within a few weeks, under tropical temperatures and humidities, several flushes of fresh mushrooms are produced (Eger et al., 1976). The developing countries in Asia are already expanding their own use of mushrooms in the diet. Taiwan is producing canned mushrooms for export. In 1977 Americans consumed 163,000 t of mushrooms, 22 percent of which were imported (Hayes, 1978).

Most food fermentations were originally indigenous and restricted in their development and use to certain geographical areas. For example, originally meat-flavored soy sauces (Japanese shoyu) were known only in Asia, but today very nearly every American kitchen has its bottle of soy sauce and fermented soy sauce is being manufactured in the United States. Soy sauce is less well known in Europe but its use is gradually spreading. As the Japanese have increased their consumption of meat, their use of miso has leveled off, but American vegetarians have discovered the rich diversity of miso and consumption of miso in the United States is increasing. Thirty years ago Indonesian tempeh, was known only to a very few scientists in the United States. It is now produced by at least 33 factories in the United States (Shurtleff and Aoyagi, 1979) and is a staple meat substitute in the diets of American vegettarians coast to coast.

Recently efforts have been made to collect information on indigenous fermented foods on a world-wide basis. In 1977, an International Symposium on the Indigenous Fermented Foods was convened in Bangkok, Thailand as part of the Congress Global Impacts of Applied Microbiology (GIAM), and the proceedings will be published as a Handbook of the Indigenous Fermented Foods (Keith H. Steinkraus, Editor-in-Chief), Marcel Dekker Inc. publishers. The use of fermented foods have developed an increasing prestige. More research is being done to improve the fermentations and expand their use.

Not all the potential contributions that indigenous food fermentations can make to feeding humans in the future have been reviewed. However, as science delves into the underlying principles behind other indigenous food fermentations, it is likely that the supply of low-cost, nutritious foods for feeding the world of the future will further expand.

REFERENCES

Akinrele, I.A. and C.C.A. Edwards (1971). An assessment of the nutritive value of a maize-soya mixture, Soy-Ogi, as a weaning food in Nigeria. Brit. J. Nutr. 26: 177-185.

Aucamp, M.C., J.T. Grieff, L. Novellie, B. Papendick, H.M. Schwartz, and A.G. Steer (1961). Kaffircorn malting and brewing studies. VIII. Nutritive value of some kaffircorn products. J. Sci. Food Agric. 12: 449-456.

Banigo, E.O.I., J.M. de Man adn C.L. Duitschaever (1974). Utilization of high-lysine corn for the manufacture of ogi using a new, improved processing system. Cereal Chem. 51: 559-572.

Brown, L.R. (1974). By Bread Alone. Praeger Publishers, p. 8.

Chang, S-T. (1972). The Chinese Mushroom (Volvariella volvacea) - Morphology, Cytology, Genetics, Nutrition and Cultivation. The Chinese University of Hong Kong Press.

Chang, S-T. (1977). Cultivation of the Straw Mushroom (Volvariella volvacea). Unesco/UNEP/ICRO/CSCHK/CUHK Regional Training Course on Cultivation of Edible Fungi (Mushrooms) Laboratory Manual. The Chinese University of Hong Kong Press.

Cronk, T.C., K.H. Steinkraus, L.R. Hackler, and L.R. Mattick (1977). Indonesian tape ketan fermentation. Appl. Environ. Microbiol. 33: 1067-1073.

Economidou, P. and K.H. Steinkraus (1977). Studies on Greek trahanas. Symposium on Indigenous Fermented Foods (SIFF). Bangkok, Thailand. Nov. 21-27.

Eger, G., G. Eden and E. Wissig (1976). Pleurotus ostreatus - Breeding potential of a new cultivated mushroom. Theoret. Appl. Genet. 47: 155-163.

Food & Agricultural Organization of FAO United Nationa (FAO) (1980). 1979 FAO production yearbook. Vol. 33, FAO Statistic Series, 28. Food & Agric. Organization of the United States. Rome, Italy.

Food and Nutrition Board, National Research Council (1974). Recommended Dietary Allowances, 8th ed., Natl. Acad. Sci., Washington, DC.

Goldberg, L. and J.M. Thorp (1946). A survey of vitamins in African foodstuffs. VI. Thiamin, riboflavin and nicotinic acid in sprouted and fermented cereal products. S. Afr. J. Med. Sci., 11: 177-185.

Harris, R.S. (1959). Supplementation of foods with vitamins. Agric. Food Chem. 7(2): 88-102.

Hayes, W.A. (1978). Edible mushroom. In: L.R. Beuchat (ed.) Food and Beverage Mycology, 301-333. The AVI Publishing Co., Westport, CT.

Hesseltine, C.W. (1979). Some important fermented foods of Mid-Asia, the Middle-East and Africa. J. Am. Oil Chem. Society. 56: 367-374.

Hesseltine, C.W., M. Smith, D. Bradle and K.S. Djien (1963). Investigations of Tempeh, an Indonesian Soybean Food. Devel. Ind. Microbiol. 4: 275-287.

Horn, P.J. and H.M. Schwartz (1961). Kaffircorn malting and brewing studies. IX. Amino acid composition of kaffircorn grain and malt. J. Sci. Food Agric. 12: 457-459.

Kim, H.S. and J.K. Chun (1966). Studies on the dynamic changes of bacteria during the kimchi fermentation. J. Nuclear Sci. (Korea). 6: 112-118.

Kim, H.S. and K.C. Whang (1959). Microbiological studies on kimchi. I. Isolation and Identification of anaerobic bacteria. Bull. Sci. Res. Inst. (Korea). 4: 56-62.

Ko, S.D. (1972). Tape fermentation. J. Appl. Microbiol. 23: 976-978.

Kodama, K. and K. Yoshizawa (1977). Sake. In: Economic Microbiology. Edited by Rose, A.H. Academic Press, New York, pp. 423-475.

Kurtzman, R.H. (1975). Mushrooms as a source of food protein. In: M. Friedman (ed.) Nutrition and Clinical Nutrition. I. Protein Nutritional Quality of Foods and Feeds. Part 2. pp. 305-317. Marcel Dekker, Inc., New York.

Liem, I.T.H., K.H. Steinkraus and T.C. Cronk (1977). Production of vitamin B-12 in tempeh - a fermented soybean food. Appl. and Envir. Micro., 34: 773-776.

Mahmoud, S.A.Z. (1977). Chemical and microbiological composition of Egyptian kishk. Symposium on Indigenous Fermented Foods (SIFF), Bangkok, Thailand, Nov. 21-27.

Mahmoud, S.A.Z., S.M. Taha, Y.Z. Ishak, M. Abdel-Halim and M.N.S. El-Nakhal (1972). Cucumber pickling in Egypt. II. Old brine method. J. Microbiol. (Egypt), 7: 65-78.

Merican, Z. (1977). Malaysian pickles. Symposium on Indigenous Fermented Foods (SIFF), Bangkok, Thailand, Nov. 21-27.

Morcos, S.R. (1977). Egyptian fermented foods. Symposium on Indigenous Fermented Foods (SIFF), Bangkok, Thailand, Nov. 21-27.

Morcos, S.R., S.M. Hegazi and S.T. El-Damhougy (1973). Fermented foods in common use in Egypt. I. The nitritive value of kishk. J. Sci. Food & Agr. 24: 1153-1156.

Mukherjee, S.K., M.N. Albury, C.S. Pederson, A.G. van Veen and K.H. Steinkraus (1965). Role of Leuconostoc mesenteroides in leavening the batter of idli, a fermented food of India. Appl. Microbiol. 13: 227-231.

National Research Council, Commission on International Relations (1977). World Food and Nutrition Study: The Potential Contributions of Research. National Academy of Sciences, Wasington, DC, page 1.

National Research Council, Commission on International Relations (1978). Postharvest Food Losses in Developing Countries. National Academy of Sciences, Washington, DC, page 7.

Novellie, L. (1959). Kaffircorn malting and brewing studies. III. Determination of amylases in Kaffircorn malts. J. Sci. Food Agric. 10: 441-449.

Novellie, L. (1906a). Kaffircorn malting and brewing studies. IV. The extraction and nature of the insoluble amylases of kaffircorn malts. J. Sci. Food. Agric. 11: 408-421.

Novellie, L. (1960b). Kaffircorn malting and brewing studies. V. Occurrence of beta-amylase in kaffircorn malts. J. Sci. Food Agric. 11: 457-471.

Novellie, L. (1962a). Kaffircorn malting and brewing studies. XI. Effect of malting conditions on the diastatic power of kaffircorn malt. J. Sci. Food Agric. 13: 115-120.

Novellie, L. (1962b). Kaffircorn malting and brewing studies. XII. Effect of malting conditions on malting losses and total amylase activity. J. Sci. Food Agric. 13: 121-124.

Novellie, L. (1962c). Kaffircorn malting and brewing studies. XIII. Variation of diastatic power with variety, season, maturity and age of grain. J. Sci. Food Agric. 13: 124-126.

Novellie, L. (1966). Kaffircorn malting and brewing studies. XIV. Mashing with kaffircorn malt. Factors affecting sugar production. J. Sci. Food Agric. 17: 354-365.

Odell, A.D. (1968). Incorporation of protein isolates into meat analogues. In: Single-Cell Protein, edited by Mateles, R.I. and Tannenbaum, S.R., M.I.T. Press, Cambridge, MA.

O'Donovan, M.B. and L. Novellie (1966). Kaffircorn malting and brewing studies. XV. The fusel oils of kaffir beer. J. Sci. Food Agric. 17: 362-365.

Okafor, N. (1977). Micro-organisms associated with cassava fermentation for gari production. J. Appl. Bacteriol. 42: 279-284.

Okafor, N. (1981). A scheme for the improvement of fermented foods of Africa South of the Sahara. In: Global Impacts of Applied Microbiology. Editors: S.O. Emejuaiwe, O. Ogunbi and S.O. Sanni, Academic Press.

Onyekwere, O.O. and I.A. Akinrele (1977). Nigerian gari. Symposium on Indigenous Fermented Foods (SIFF), Bangkok, Thailand, Nov. 21-27.

Pederson, C.S. (1979). Microbiology of Food Fermentations. Second Edition, AVI Publishing Co.

Pederson, C.S. and M.N. Albury (1969). The sauerkraut fermentation. New York State Agr. Exp. Stat. Bull., 824, 84 pp.

Pimentel, D., W. Dritschilo, J. Krummel and J. Kutzman (1975). Energy and land constraints in food protein production. Science, 190: 754-761.
Platt, B.S. (1964). Biological ennoblement: Improvement of the nutritive value of foods and dietary regimens by biological agencies. Food Technol., 18: 662-670.
Platt, B.S. and R.A. Webb (1946). Fermentation and human nutrition. Proc. Nutr. Soc., 4: 132-140.
Platt, B.S. and R.A. Webb (1948). Microbiological protein and human nutrition. Chem. Ind. (London), Feb. 7, 88-90.
Schwartz, H.M. (1956). Kaffircorn malting and brewing studies. I. The kaffir beer brewing industry in South Africa. J. Sci. Food Agric., 7: 101-105.
Schweigart, F. and J.P. Wit (1960). Favourite beverage of the Banta. Preparing and drying of mahewu and its nutritional value. Food Indust. S. Africa, 12: 25-27.
Shibasaki, K. and C.W. Hesseltine (1962). Miso fermentation. Economic Botany, 16: 180-195.
Shurtleff, W. and A. Aoyagi (1976). The Book of Miso. Autumn Press, Kanagawa-ken, Japan.
Shurtleff, W. and A. Aoyagi (1979). The Book of Tempeh. Harper and Row Publishers.
Smith, A.K. and S.J. Circle (1972). Protein Products as Food Ingredients. In: Soybeans: Chemistry and Technology I. Proteins. Edited by Smith, A.K. and Circle, S.J. AVI Publishing Co., Westport, Connecticut, p. 365.
Spicer, A. (1971a). Synthetic proteins for human and animal consumption. Vet Record. 89: 482-487.
Spicer, A. (1971b). Protein production by microfungi. Tropical Sci. XIII: 239-250.
Steinkraus, K.H., Yap Bwee Hwa, J.P. Van Buren, M.I. Provvidenti and D.B. Hand (1960). Studies on tempeh - An Indonesian fermented soybean food. Food Res. 25: 777-788.
Steinkraus, K.H., D.B. Hand, J.P. Van Buren and L.R. Hackler (1961). Pilot plant studies on tempeh. In: Proc. Conf. on Soybean Products for Protein in Human Foods. NRRL. USDA. Peoria, IL. pp. 75-84.
Steinkraus, K.H., A.G. van Vee and D.B. Thiebeau (1967). Studies on idli - an Indian fermented black gram-rice food. Food Technol. 21(6), 110-113.
Steinkraus, K.H. (1979). Nutritionally significant indigenous foods involving an alcoholic fermentation. In: Fermented Food Beverages in Nutrition. Editors: C.F. Gastineau, W.J. Darby and T.B. Turner. Academic Press. Chap. 3, pp. 35-59.
Steinkraus, K.H. (1982). The indigenous fermented foods. Nestle Research News 1980/81, pages 23-28. Nestle Products Technical Assistance Co. Ltd., 1814 La Tour de Peilz; Switzerland.
Tan, T., A. van Veen, D. Graham and K. Steinkraus (1967). Studies of the manufacture of low salt fish paste. Philippine Agric., LT 51: 626-635.
van der Walt, J.P. (1956). Kaffircorn malting and brewing studies. II. Studies on the microbiology of kaffir beer. J. Sci. Food Agric. 7: 105-113.
van Veen, A.G. (1965). Fermented and dried seafood products in Southeast Asia. In: Fish as Food, vol. 3. Edited by Gorgstrom, G., Academic Press, New York, 227-239.
van Veen, A.G. and G. Schaefer (1950). The influence of the tempeh fungus on the soya bean. Doc. Neer. Indones. Morbis Trop., 2: 270-281.
van Veen, A.G., D.C.W. Graham and K.H. Steinkraus (1968). Fermented Peanut Press Cake, Cereal Sci. Today, 13: 96.
van Veen, A.G. and K.H. Steinkraus (1970). Nutritive value and wholesomeness of fermented foods. Agric. Food Chem., 18: 576-578.
von Holdt, M.M. and J.C. Brand (1960a). Kaffircorn malting and brewing studies. VI. Starch content of kaffir beer brewing materials. J. Sci. Food Agric., 11: 463-467.

von Holdt, M.M. and J.C. Brand (1960b). Kaffircorn malting and brewing studies. VII. Changes in the carbohydrates of kaffircorn during malting. J. Sci. Food Agric., 11: 467-471.

Wanderstock, J.J. (1968). Food analogs. Cornell H.R.A. Quarterly, August, 29-33.

Wells, J. (1975). Analysis of potential markets for single cell proteins. Paper presented at Symposium on SIngle Cell Protein, American Chemical Society Meeting, Philadelphia, PA.

Yokotsuka, T. (1960). Aroma and flavor of Japanese soy sauce. In: Advances in Food Research, 20: 75-134, Academic Press, New York.

AMINO ACID PRODUCTION AND USE TO IMPROVE NUTRITION OF FOODS AND FEEDS

Takekazu Akashi
Managing Director, Central Research Laboratories
Ajinomoto Co., Inc., Kawasaki 210, Kanagawa, Japan

ABSTRACT

The practical application and industrial production of amino acids are reviewed and future prospects explored. Emphasis is given to the relationships of amino acids to Asian food specialties and their flavoring properties. Monosodium glutamate, which induces the umami (tastiness) sensation, demonstrates a physiological effect of food protein digestion on human nutrition. Amino acids can also significantly increase the nutritional value of grain, a major source of protein. Methinonine and lysine have, for two decades, been used as supplements. Tryptophan and threonine will probably follow. Amino acids as fish attractants are important in fish culture. New amino acid applications for direct human use as well as animal use require more study. Adequate storage systems for foods with amino acids are required. The amino acid industry continues to improve conventional production processes through cost reduction and utilization of unused resources. Unique process combinations of fermentation with enzymation or organic synthesis should be developed.

KEYWORDS: amino acid; amino acid industry; amino acid production; feed fortification; fish culture; food flavor; food fortification; protein shortage; recombinant DNA; umami.

INTRODUCTION

Amino acids, during the late 1800s, were identified as the basic building units of protein, a component of all living organisms. Industry has now succeeded in manufacturing amino acids from material other than natural protein. As a consequence it is expected that crystalline amino acids will help satisfy world needs in the food and feed sector. This paper reviews the present situation and future aspects of the practical application of amino acids from the standpoint of industrial production of amino acids. Selected historical aspects of amino acids, with an emphasis on Asian topics, are presented. Finally, the contribution of amino acids to the world's future food needs and current research efforts required are mentioned.

PRESENT APPLICATIONS OF AMINO ACIDS

The desire for flavorful food and a healthy life is a basic wish and traditional world need of mankind. From ancient times, people have looked for a variety of flavoring agents, such as spices and seasonings. For instance, Southeast Asia has customarily used fish sauce, and for many centuries East Asia has used soy sauce. While the former utilizes small fish as the raw material, the latter uses soybean and wheat. The main flavoring components of both sauces, however, are composed of amino acids derived by hydrolysis of proteins. This tradition led Professor Kikunae Ikeda of the University of Tokyo into research on tangle stock flavor. He discovered in 1908 that monosodium glutamate was a key component of its flavor and might be an excellent flavor enhancer for various foods. Ikeda's discovery is one of the monumental works of chemical research as applied to world needs. Thereafter monosodium glutamate, which contributes directly to the desire of mankind for flavorful food, became the first amino acid commercially produced. It also became the parent of our current amino acid industry.

Amino acids participate in natural food flavors. A recent report indicates that the Chinese dish, abalone, is distinctly flavorful by virtue of its content of glycine and glutamate. Similarly in the dish, shrimp, glycine plays a remarkable role. Amino acids obviously have contributed to food flavoring, especially in Asian foods, for a very long time. We use the term umami for flavor enhancing by monosodium glutamate (MSG). Our study in beagle dogs showed that oral stimulation of MSG causes an increased flow of pancreatic juices. This suggests that the taste sensation of umami mediates a message to the gastrointestinal tract to prepare for food protein metabolism. This observation shows how MSG contributes to human nutrition from the physiological point of view, although it is not a direct nutrient.

World food protein supply in 1977 was 96 million t. Plant protein was the major source, twice as much as animal protein. Protein supply varies geographically in different localities (Table 1), some areas being short of food protein, especially animal protein. Plant protein is inferior to animal protein in terms of its essential amino acid pattern. As a typical example of studies in rats, Figure 1 shows the effect on weight gain of feeding amino acids. Fed only plain corn, the rats lose weight. By supplementing the corn with both lysine and tryptophan the rate of weight gain increases. Table 2 shows amino acid supplementations of grain. The Protein Efficiency Ratio (PER) is used to assess nutritional value. Addition of just 0.2% lysine to wheat flour doubles the PER. Both lysine and threonine addition gives a PER close to the value for casein. Thus, small amounts of amino acids can raise significantly the nutritional value of grain, and ultimately can save food resources.

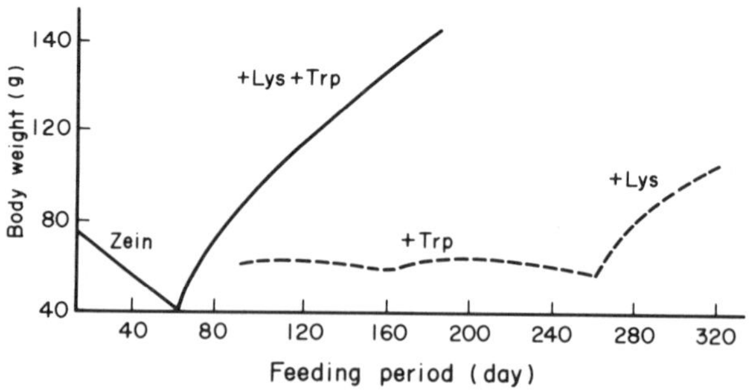

Fig. 1. Effect of Amino Acid Supplementation

Table 1 Protein Supply

(g/person/day)

	Animal	Vegetable	Total
The World	25	45	70
Developed countries	56	43	99
Developing countries	13	46	58
Asia	8	43	50
Africa	11	44	55
APEC*	14	50	64
Near east	15	60	75

* Asian planned economy countries

FAO Information (1977)

Table 2 Amino Acid Supplementation to Grain

Protein source	Amino acid		PER
			Unfortified
Rice	L-Lys DL-Thr	0.2 % 0.2 %	2.6 (1.5)
Wheat	L-Lys	0.2 %	1.6 (0.7)
	L-Lys DL-Thr	0.4 % 0.3 %	2.7 (0.7)
Corn	L-Lys	0.4 %	1.1 (0.9)
	L-Lys DL-Trp	0.4 % 0.07 %	2.6 (0.9)
Casein			2.5

Howe and Jansen : Am. J. Clin. Nutr. (1965 and 1967)

Amino acid supplementation, in the feed industry, has already put the results of research to practical use. Methionine for 20 years has been used to fortify soybean meal, and lysine for 15 years has been used to fortify feed grains. World demand for both amino acids continues to increase every year because of the demand in the feed industry. Furthermore, this year has seen the opening of a new age with a third amino acid, tryptophan, being used by Japanese manufacturers to fortify baby pig pre-starter feed. Threonine will undoubtedly follow these other amino acids in future application.

NEW APPLICATIONS OF AMINO ACIDS

New Food Technologies

With respect to development of new protein resources, Southeast Asian governments have strongly supported projects for improving the PER of the winged bean, a local product. Professor Yoshida of Nagoya University showed in rat feeding studies, that addition of methionine to winged beans raised their nutritional value nearly to that of egg protein. An important technology in amino acid application has been launched by Professors Fujimaki and Arai, University of Tokyo. Details are presented in another paper in these proceedings. Ordinarily, supplementation requires simple mixing of protein and amino acids. The plastein reaction modifies the original protein by incorporation of amino acids into the primary protein structure. The resulting product offers better physical texture, nutritional value and flavor characteristics. The reaction will have future practical applications.

Substances containing the peptide bond of amino acids have attracted a good deal of attention because of their nutritional value, medicinal effect or other physiological activity. Aspartame (L-Aspartyl-L-Phenylalanine methyl ester) has a taste profile close to that of sugar, but is 200 times sweeter. The taste specificity of this material was unexpectedly found in 1965. Aspartame has already been approved as a low-caloric sweetener of high safety in about twenty countries including the Philippines, the United States, Canada and many European countries. Although sugar will continue to play its leading role as a good quality sweetener, it has been implicated in the United States and European countries, as contributing to obesity and to dental caries by virtue of excessive intake. Aspartame represents a new way in which an amino acid relates directly to a healthy diet. Amino acids in clinical nutrition will not be discussed, although their value is widely recognized in the area of parenteral solution, elemental diet, defined formula diet and milk substitutes.

New Feed Technologies

Recent research has disclosed that amino acid supplementation may bring benefits beyond the saving of feed grains. For instance, laying hens raised in the tropical and subtropical zones of Southeast Asia have produced heavier eggs by feeding them a diet enriched with crystalline lysine rather than with proteinous lysine such as soybean meal. This observation indicates that by formulating feeds in this way it is possible to approach directly the requirement levels of each amino acids for the animal, thus exceeding the marginal efficiency of any natural protein mixture and eliminating unusable surpluses. It is also expected that tryptophan supplementation may improve antibody formation. The availability of the newer amino acids could be particularly useful in formulating feeds from high lysine corn or other corns developed through gene engineering.

Fish culture continues to increase both in variety and in volume. Research on an artificial diet for fish culture has been conducted since 1970, especially in the United States and Japan. In Japan yellowtail is produced at a rate of 180,000 t/a. The traditional feeding of yellowtail is limited to fresh fish. Professor Takeda of Kohchi University however has achieved good growth of yellowtail fries using an artificial diet in which the amino acids, alanine and proline, are supplemented together with disodium inosinate. The supplements work synergistically stimulating strongly the appetite of young yellowfish. Amino acid as a fish attractant in artificial diets opens a new field of amino acid application.

Two other valuable technologies should be mentioned. One is related to new protein resources which includes efficient utilization of animal waste. The other is a total storage system of foods together with crystalline amino acids. These technologies are described in other papers in these proceedings.

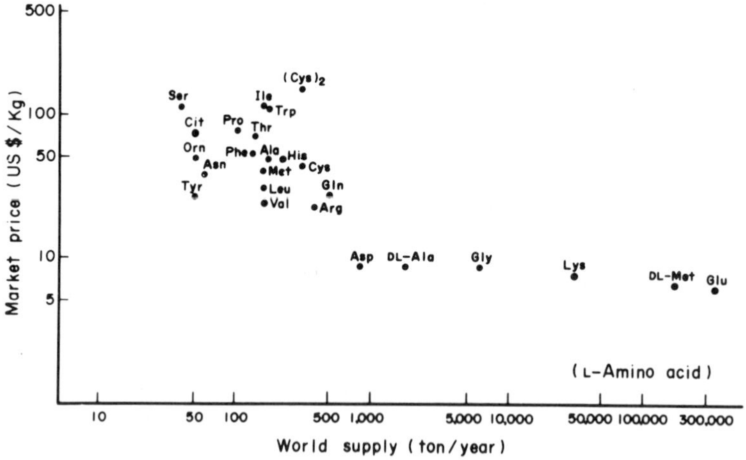

Fig. 2. Relationship between Supply and Current Market Price of Amino Acid (1980)

INDUSTRIAL PRODUCTION OF AMINO ACIDS

Table 3 shows the world supply of amino acids (1980), classified according to production method. More than 95% is accounted for by three amino acids, L-glutamic acid (monosodium salt) as a flavor enhancer, DL-methionine and L-lysine as feed additives. Figure 2 shows the relationship between supply and current market price of amino acids. A greater demand induces a lower cost, and lower costs encourage greater demand. As a consequence, the amino acid industry constantly strives to improve the technology of production in an effort to reduce costs.

Table 3 World Supply of Amino Acid (1980)

Fermentation	(ton/year)	Enzymatic synthesis	(ton/year)
L-Glu	340,000	L-Asp	450
L-Lys	34,000	L-Met	150
L-Arg	500	L-Ala	130
L-Gln	500	**Chemical synthesis**	
L-His	200	DL-Met	120,000
L-Thr	160	Gly	6,000
L-Ile	150	DL-Ala	2,000
L-Val	150	L-DOPA	200
L-Pro	100	L-Trp	200
L-Cit	50	L-Phe	150
L-Orn	50	**Extraction**	
L-Ser	50	L-Cys, L-$(Cys)_2$	700
		L-Leu	150
		L-Asn	50
		L-Hyp	50
		L-Tyr	50

Bacterial fermentation, in effect, is a one step reaction in aqueous solution under conditions close to room temperature and atmospheric pressure. Organic synthesis however, consists of a multi-step process under a wide range of conditions from high to low temperature and/or pressure. The reactions moreover, take place in water or in non-aqueous solvents (Table 4). Importantly, both approaches contrast with each other with respect to the stereo-specificity of the final product. Whether to use fermentation or organic synthesis has to be decided for each amino acid after due consideration of production costs and raw material availability. Imaginative combinations of both fermentation and organic synthesis may be feasible in some cases.

Table 4 Comparison between Fermentation and Organic Synthesis in Amino Acid Production

		Fermentation	Synthesis
Raw material		Agr. products	Chem. resources
Reaction	Temp.	Room temp.	Low ~ High
	Pressure	Low	Low ~ High
	Solvent	Water	Water, Org. solvent
Product	Concentration	Dil.	Conc.
	D,L-Form	L	DL
Process	System	Batchwise	Batch or Flow
	Step	Single	Multiple
Plant and Equipment	Investment	Small	Large
	Scale merit	Small	Large
	Applicability	General	Special

Recombinant DNA and other advanced biological techniques have been introduced into the field of amino acid fermentation. The Ajinomoto Company, for example, has recently reported two successes with respect to threonine fermentation. One is a DNA recombinant, Eschrichia coli hybrid, that significantly increases the yield of threonine. The other concerns cell fusion technology applied to commercially used bacteria. For the future it is expected that organic synthesis will be carried out either in space or in deep sea where conditions are quite different from those at ground levels.

CONCLUSION

In conclusion, Table 5 shows our list of challenging objectives and R&D strategies for the amino acid industry. The first category of objectives is development of new production technologies that take into account cost reduction and utilization of unused resources. Unique combinations of fermentation with enzymation or organic synthesis should be investigated. Frontier methods should be vigorously introduced. The second category of objectives is development of new application technology, including such subjects as feed fortification, food fortification, and food preservation. Having good food and a healthy life are basic desires of mankind. Progress towards meeting the challenging objectives listed above will contribute towards meeting these basic desires.

Table 5 Challenging Objectives of Amino Acid Industry for World Food Supplies

Production technology	Application technology
(1) Improvement of conventional process	(1) Feed fortification
Cost reduction	Newcomer ------------ Trp, Thr
Resources saving	New field ----------- Fish diet
Unused resources utilization	(2) Food fortification
(2) Development of process combination	Food flavoring ------ *"Umami"* and New sweetener
Fermentation × Enzymation × Synthesis	Supplementation ------ Value addition to resources
(3) Introduction of frontier technology	Incorporation -------- Protein modification
Recombinant DNA method	(3) Basic strategy for world food shortage
Space / deep sea engineering	Total storing system --- Food + Amino acid

Clinical nutrition ----- Elemental diet
Intravenous solution

STUDIES ON HEAT TREATMENT OF WHEAT GRAINS IN CHINA

Ruizheng Liu, Yiqiang Xiong*, and Cheng Wang
Cereal and Oil Chemistry Institute of China, Beijing, China
*(Present Address: *Department of Grain Science and Industry Kansas State University, Manhattan, KS 66506 USA)*

ABSTRACT

Heat treatment of wheat as a traditional method to control insects in stored grain has been practiced in China for more than fifteen centuries. To improve this old method, a series of studies have been conducted over the past 25 years, and the results are reviewed in this paper. For effective control of insects, wheat grains are first subjected to sun-curing to raise the grain temperature to 50-52°C, and the grain moisture is reduced to about 12%. The grain temperature is maintained at 50-52°C for at least two hours, and the warmed grains are then moved into a semi-airtight bin where the grain temperature is retained above 40°C for a week or so. Under these conditions, the chemical composition of wheat does not change nor does its baking quality. For grain which is to be used as seed, however, the above procedure only applies to newly-harvested dormant seeds. For newly harvested seed grains having passed through dormancy, the grain moisture content should be below 10% when heat treatment is practiced. For old seed harvested in a previous year, heat treatment is not recommended. Enzymatic studies as well as germination tests revealed that the aging process was accelerated by heat treatment, and the effect of heat treatment on the aging process was influenced by the grain moisture content and the level of interseed carbon dioxide. Heat treatments of grains other than wheat are also discussed briefly.

KEYWORDS: wheat, heat treatment, enzymatic studies, germination tests, aging process (wheat), moisture content (grain).

INTRODUCTION

Semi-airtight storage of sun-heated wheat as a safe storage method, especially as an effective insect control method, has been practiced in China for more than 1,500 years. In the 6th century, this method was outlined in a Chinese agricultural encyclopedia, -- Chi Min Yao Shu -- (Shik, 1962), as: "Pitting heat-corn: sun to dryness, and bury while still quite warm".

At the present time the general application of the method begins with spreading freshly harvested wheat grains in a thin layer on the ground under the bright sun until the grain temperature increases to above 50°C, and the grain moisture decreases below 12%. Then the grains are piled up to maintain the grain temperature for several hours. The sun-cured grains are moved into a storage bin and covered with some insulation material (Fig. 1). In a well insulating storage bin, the high grain temperature can be retained for several days before gradually dropping to ambient in two months or so. Thereafter care is given to avoid insect reinfestation.

Fig. 1: A type of insulating storage bin in the countryside of China, which is insulating not only to heat, but also to ground-water and rodents. 1 - insulating materials; 2 - grains; 3 - inner wall; 4 - dry sand.

To develop this old technique, experiments on heat treatment of wheat have been carried out by our group and other researchers in China, and information from chemical, biological and entomological investigations has accumulated. This paper is a brief review of these studies.

HEAT TREATMENT STUDIES

Determination of Effective Temperature for Insect Control in Commercial Wheat Storage

Several storage experiments during the 1950's involving thousands of tons of wheat were conducted in Jiangsu, Hebei, Henan and Shandong provinces (Grain Ministry, 1958). The results showed that insect control depended on the grain temperature, especially the initial temperature at which the grain was put into the bin. For example, insects were completely controlled when the initial grain temperature was 44-47°C, but when the initial temperature was below 43°C, not all insects were killed. Based on these findings, it was recommended with the traditional sun-drying system, that grain temperature on the drying floor should reach 50-52°C and remain there for more than 2 hours, and that the initial grain temperature in the bin should be above 46°C and the grain temperature should be kept above 40°C during the first week or so.

Effect of Heat Treatment on Nutritional Value and Baking Quality of Wheat

A chemical and physiological study of heat treated wheat was conducted by Chao (1956). In his study, wheat grains already past dormancy and with 12.7% moisture were used. After four hours of sun-curing when the grain temperature had reached about 45°C, part of the grain (about 90 tons) was put into storage immediately, and the rest was cooled down to ambient conditions (25-30°C) before storage. A bin of wheat without heat treatment was used as control. Chemical analyses of ether extracts, crude protein, soluble nitrogen, vitamin B_1, fat acidity value, total sugars and gluten were made at intervals during a six-month period. Tests of seed viability and baking quality were made at the end of the study. The results indicated that, except for seed viability, there was no significant difference in quality between the heat-treated grain and the control.

To further check the commercial test, a laboratory experiment under precisely controlled conditions was also carried out by Chao (1956). The experimental material was newly harvested dormant wheat with moisture of about 12%. Grain samples were subjected to heat treatment at 46-48°C for 8 or 18 days and then were stored at room temperature. The results of chemical analyses are shown in Table 1. There was no significant change in vitamin B_1, total sugars or fat acidity value between untreated grains and the 8-day heat-treated grains. A decrease in vitamin B_1 was found when the heat treatment was extended for 18 days. In commercial storage, the high grain temperature is usually maintained for not more than one week. The author concluded that under practical conditions heat treatment of wheat affects neither the nutritional value nor the baking quality.

Table 1: Effect of Pre-storage Heat Treatments (46-48°C on Vitamin B_1, Total Sugars, Fat Acidity Value and Seed Germination of Wheat Stored at Room Temperature

Heat Treatment	Weeks After Harvest	Moisture Content (%)	Vitamin B_1 (g/g)	Total Sugars (%)	Fat Acidity Value (mgKOH/100gDM)	Germination (%)
None	1	11.74	4.58	2.31	23.35	1.3
None	12	11.60	4.56	2.72	30.40	90.3
46-48°C 8 days	12	9.86	4.55	2.63	25.93	98.6
46-48°C 18 days	12	10.68	4.00	2.85	32.35	98.0

Recently we sent some samples to the Grain Market Research Center of USDA, which included two varieties with different baking characteristics and two harvest years (1981 and 1982). The primary results indicate that although there were significant differences in baking quality between both varieties and harvest years there was no detectable differences among treatments. The mixograph and baking results are shown in Table 2. The variety difference is apparent. For both varieties, the "1982 harvest" gave shorter mixing times and smaller loaf volumes indicating poorer quality, probably due to the unusual rainy weather at harvest time. However, there was no significant difference among heat treated and untreated samples, confirming that heat treatment of wheat doesn't affect the baking quality.

Table 2: Protein Content, Mixogram, and Bread-Making Data for Straight-Grade Fours Milled from Samples of Zhong Yin 779 and Jing 771 wheats that were Harvested in the Peoples Republic of China in 1981 and 1982 and that Received Heat Treatments Under Aerated (HTAR) and Air Tight (HTAT) Conditions. a/ b/

Treatment	Protein (%)	Dough Absorption Bake (%)	Dough Absorption Mixogram (%)	Dough Mix Time Bake (%)	Dough Mix Time Mixogram (%)	Loaf Volume (cc)	Bread Crumb Grain c/
Zhong Yin 779 - 1981 Crop							
Control	13.4	59.4	61.4	2	1-7/8	68.0	Q
HTAR	13.6	59.6	61.6	2-1/4	2	67.3	Q
HTAT	13.5	59.5	61.5	2-1/8	2	67.5	Q
Zhong Yin 779 - 1982 Crop							
Control	12.7	57.0	58.5	2-1/4	1-7/8	60.5	U
HTAR	12.7	57.0	58.5	2-1/4	2	61.0	U
HTAT	12.6	57.4	58.4	2-1/8	1-7/8	62.0	U
Jing 771 - 1981 Crop							
Control	16.2	62.9	64.9	2-1/8	2-1/8	96.0	S open
HTAR	16.3	63.0	65.0	2-1/4	2-1/4	95.3	S open
HTAT	16.4	63.1	65.1	2-1/4	2-1/4	96.3	S open
Jing 771 - 1982 Crop							
Control	15.7	60.2	62.2	1-3/4	1-3/4	88.5	S open
HTAR	15.6	60.1	62.1	1-3/4	1-3/4	89.5	S open
HTAT	15.6	60.1	62.1	1-7/8	1-3/4	89.5	S open
CS-80	12.9	61.0	64.0	4		88.0	S

a/ Chemical data expressed on a 14% moisture basis.
b/ The condition of treatments are the same as in Table 3.
c/ S = satisfactory; Q = questionable; U = unsatisfactory.

Effect of Heat Treatment on Viability of Wheat Seed

Various experiments indicated that heat treatment accelerated the process of passing through the post-harvest dormancy period, as shown by the higher germination rate of treated seeds. In a planting experiment (Grain Bureau of Shandong Province, 1960) there was no difference in germination, growth or yield between heat-treated and untreated seed when newly harvested dormant grains received the heat treatment. On the other hand, for wheat with a high germination rate, i.e., having passed the dormancy period, contradictory results were reported. Generally, seed germination was significantly decreased following heat treatment. However, some experiments gave different results: no decrease in germination was noted although post-harvest dormancy was broken prior to heat treatment. This finding was also recorded by Chao (1956). A bulk of wheat already past dormancy was subjected to heat treatment and subsequently stored in a semi-airtight bin. A sample of the grain was taken from the bin after the high temperature period and was stored in the laboratory. After three months the germination of wheat from the semi-airtight bin dropped from an initial 76% to 6%, while that of the laboratory stored grain dropped to 49%. Chao postulated that the composition of the interseed atmosphere caused this difference. He also suggested that moisture content of grain could be involved in the effect of heat treatment on the viability of wheat which had passed dormancy.

PHYSIOLOGICAL EFFECTS OF HEAT TREATMENT

The objectives of our work (Liu and Xiong, 1962; Liu and Wong, 1981) were to determine how the grain moisture content and interseed atmosphere were involved in the physiological effects of heat treatment on wheat seeds that have passed dormancy.

Six varieties of newly harvested seeds [Nanda-2419 (white seed coat), Shite-14 (white), Nonda-183 (white), Bima-1 (white), Zaoyang (red) and Jinmai-2148 (red)] and three varieties of seeds harvested in the previous year [Zaoyang, Baiyupi (white) and Jinmai-2148] were used to observe the effects of heat treatment on new and old seeds.

Dormancy of the newly harvested seed was broken prior to beginning the experiment. Seeds were sun-dried to about 12% moisture and then stored at room temperature until germination reached at least 90%. The time required varied from about 6 to 11 weeks. Seeds from two lots (Nonda-183 and Zaoyang) were further dried to about 10% moisture to compare the effect of heat treatment on seeds with different moisture contents.

A sample of seed from each lot was placed in a cloth bag ("heat treatment aerated" HTAR) and also in a stoppered glass bottle ("Heat treatment airtight" HTAT) to investigate the effect of interseed atmosphere. Thermostatically controlled chambers were used for heat treatment, and relative humidity of chamber air was adjusted to limit loss of seed moisture content to 0.2-0.4%. The chamber temperature was 48-50°C for 1 hour and then 42-45°C for 2 weeks. After heat treatment samples were stored in open bottles at room temperature. Untreated check samples were stored in open bottles at room temperature.

Germination tests were made before heat treatment and at intervals after treatment on 400 seeds of each samples allowed to germinate at 21-25°C. Average seedling lengths were measured for some samples. Seed germination was counted on the third day ("germination vigor") and again on the seventh day ("germination rate"). Seedling lengths were measured on day three. Fat acidity value by titration of extract, and peroxidase activity (Honold and Stahmann, 1968) were determined for several samples.

The germination vigor and germination rate for the test materials are shown in Table 3. The germination rate of HTAR-treated seeds was generally higher than that with HTAT, although in a few cases both treatments gave about the same rate, suggesting a difference in response to high temperature among wheat varieties.

When the CO_2 contents of interseed atmosphere of samples with moisture levels of 9% and 12% were measured, the CO_2 with HTAR was equal to that of atmospheric air, i.e. 0.03%, but with HTAT treatment the CO_2 of interseed air of samples at 9% and 12% moisture increased to 0.3% and 1.4% respectively. This suggested that high CO_2 concentration might accelerate the aging process of heat treated seeds.

Table 3: Germination Vigor and Germination Rate of Heat-treated, Post Dormant Wheat Seed

Sample No.	Variety	Age[a] (days)	Moisture Content (%)	Treatment	Germination Vigor[b]/Germination Rate[c] (%) Days after treatment: 1	30	60	90	450
1	Nanda-2419	5	12.8	CK[d]	93/99	97/99	98/99		
				HTAR[d]	91/98	95/99	96/99		
				HTAT[d]	92/98	90/96	71/86		
2	Shite-14	3	12.8	CK	96/99	97/100	95/99		
				HTAR	83/95	94/97	91/96		
				HTAT	56/87	78/92	49/69		
3	Bima-1	37	12.8	CK	97/98	99/100	99/99		
				HTAR	60/91	91/95	67/88		
				HTAT	22/81	57/76	26/39		
4	Jinmai-2148	17	11.9	CK	87/88	84/86			
				HTAR	86/86	81/85			
				HTAT	84/85	75/78			
5	Jinmai-2148	365+	11.9	CK	78/79	81/82			
				HTAR	72/74	69/70			
				HTAT	51/53	34/37			
6	Zaoyung	35	12.0	CK	74/95			84/92	
				HTAR	78/95			73/90	
				HTAT	69/93			54/79	
7	Nonda-183	70	12.0	CK	81/94			95/97	
				HTAR	83/95			88/95	
				HTAT	44/74				
8	Buiyupi	365+	12.0	CK	96/99			63/71	
				HTAR	81/92			29/37	
				HTAT	25/42			1/1	
9	Nonda-183	43	9.8	CK	95/96			90/95	87/96
				HTAR	96/97			92/96	78/94
				HTAT	96/98			90/95	83/92
10	Zaoyang	8	9.8	CK	91/93			89/98	64/91
				HTAR	92/95			81/95	75/93
				HTAT	86/90			80/94	75/94
11	Zaoyang	365+	9.8	CK	38/46			27/41	24/42
				HTAR	37/42			25/48	19/35
				HTAT	35/39			15/32	----

[a]Days from end of dormancy to the beginning of heat treatment.
[b]Germinated seeds on day 3
[c]Germinated seeds on day 7
[d]CK - Check; HTAR - Heat treatment aeration; HTAT - Heat Treatment airtight

Grain moisture content also affected the aging process of heat treated seeds. Wheat seeds at 12% moisture content and above generally aged with heat treatment, as shown by low germination vigor and germination rate (Table 3). On the other hand, when wheat seed moisture content was less than 10%, the detrimental effect of heat treatment was slight, as, for example, in samples 7 and 9.

The detrimental effect of heat being closely related to grain moisture content was further substantiated when a check was made of all available published and unpublished data on heat treatment of wheat in China. All the reports showing that heat treatment did not affect germination had involved grain with moisture content lower than 10%. All the studies indicating that heat treatment decreased germination rate had been conducted on wheat with moisture content in the range of 11-13%. This coincided with our observations.

The response of wheat seeds to heat treatments was also found to be dependent upon the physiological stage of the seeds. With age of seeds defined as the days from the end of dormancy to the beginning of heat treatment, the older the seeds were, the more serious was the damage from heat treatments. This is illustrated in samples 4 and 5 in Figure 3 where seeds which had been stored for more than 1 year showed a germination rate after HTAT of 26% less than the control immediately after heat treatment, and much greater decrease after 30 days. Newly harvested seeds appear relatively resistant to high temperature, and old seeds more susceptible to heat damage. Germination of old seeds decreased rapidly with the combination of high temperature and high CO_2 concentration in the HTAT treatment.

Average seedling length and fat acidity value were determined for selected samples with moisture content of about 12% (Table 4). Generally, seedling length decreased and fat acidity value increased with both heat treatments.

Table 4: Effects of High Temperature Treatment on Wheat Seedling Length and Fat Acidity Value

Variety	Treatment	Average Seedling Length (cm), Days after Treatment 35	Average Seedling Length (cm), Days after Treatment 63	Fat Acidity Value (mgKOH/100gDM), Before Treatment	Fat Acidity Value (mgKOH/100gDM), After Treatment
Nanda-2419	CK[a]	---	2.14	-----	-----
	HTAR[a]	---	1.47	10.72	11.68
	HTAT[a]	---	1.12	10.72	10.70
Shite-14	CK	1.41	2.33	-----	-----
	HTAR	1.37	1.50	9.78	10.82
	HTAT	.96	.76	9.78	11.09
Bima-1	CK	1.85	---	-----	-----
	HTAR	1.14	---	9.77	10.19
	HTAT	.69	---	9.77	11.07

[a]CK - Check; HTAR - Heat treatment aeration; HTAT - Heat treatment airtight.

Peroxidase activity of freshly harvested and old seed with moisture content of about 12% was measured before and after heat treatment (Table 5). Freshly harvested seed showed signficantly higher peroxidase activity than the old seed, reflecting the difference in the physiological stage of the seeds. The data also indicated that the enzyme activities of both freshly harvested and old seed were decreased following heat treatment, suggesting that high temperature interferes with the oxidation-reduction system of the seed.

Our research indicates that grain moisture content, composition of interseed atmosphere and physiological stage of the seed affect longevity of heat treated wheat seeds. If heat treatment is proposed as a pre-storage treatment for insect control of wheat designated for seed, the following suggestions are given: 1) apply heat treatment shortly after harvest while seeds are still dormant. 2) if newly harvested seeds have already passed dormancy, dry seeds to below 10% before heat treatment. 3) do not use heat treatment on seed-wheat harvested in a previous year.

Table 5: Peroxidase Activity of Wheat after Heat Treatment

Wheat (Jinmai-2148)	Treatment	Peroxidase Activity (Units/ml of Crude Extract)
New seeds	Untreated (CK)	25.2
New seeds	Heat treated (HTAR)	20.7
Old seeds	Untreated (CK)	17.3
Old seeds	Heat treated (HTAR)	14.4

Application of Heat Treatment to Rice, Soybeans and Peas

Studies on application of heat treatments to rice, soy beans and peas in China will be briefly reviewed here. Heat treatment experiments with newly harvested rice were conducted in Kiangxi (Grain Bureau, 1958) and Zhejiang (Zhou and Wan, 1959) provinces. The grain temperatures under sunlight reached 50-53°C, and the grain moisture content after sun-heating was 10%. The grains were put in the bin while still warm and grain temperature was retained above 40°C for several days. No insect infestation was observed in the heat-treated rice during the six-month experimental period although the untreated control rice was infested severely. Seed germination and cooking tests showed no significant differences between the treated and untreated samples. Since information on heat treatment of rice is limited, no general recommendation on heat treatment for rice has been given.

A large scale study on the use of sun-heating to dry soybeans provided some information in applying heat treatment to soybeans (Plant Physiology Institute, 1956). Results indicated that sun-heating effectively reduced the moisture content of soybeans, without detrimental effects to oil extraction or food quality, when the soybeans reached 43-46°C and immediately were put into storage while still warm. Soybeans for seed or for soybean sprouts should be cooled down before storage.

In some provinces heat treatment has been applied to peas on a small scale (two or three ton lots) to kill the pea weevil (Bruchus pisorum) which infests peas in the fields (Cereal Service Group, 1961). Not sunlight but energy derived from the seed respiration is used as the heat source. Peas with moisture contents of 13-20% were put into storage and respiratory heat was retained by surrounding the bin with insulation material. The pea temperature rose above 40° C in 1-12 days depending upon the pea moisture content. Most pea weevils were killed. The insulation material was taken off after several days and the pea temperature gradually decreased to ambient. Best retention of pea quality was obtained when pea moisture content was about 15% and insulation material was removed after 4 days of heat treatment.

CONCLUSIONS AND PERSPECTIVES

The advantage of using heat treatment over chemical treatment to control insects is its safety for human beings. The procedures now in use are rather laborious and are weather dependent, but it is feasible in China's countryside today, and has been adopted by Chinese farmers as an effective method for safe storage preventing both mould and insect damage. For large scale storage, of course, it is necessary to develop other effective ways for pre-storage heat treatment of grains. A bin designed for aeration equipped with heaters could be an option, and a grain dryer had been used to supply heat for the purpose of insect control (Evans and Dermott, 1979). We believe that with more effective technology, the method of controlling both mould and insects in stored cereals with heat treatment can be developed for large scale storage of cereal grains.

The experimental data reviewed in this paper provide information on heat treatment, seed dormancy and seed aging. Heat treatment shortens the post-harvest dormancy period of freshly harvested wheat seeds as shown by the increased germination of the heat-treated seeds, and accelerates the aging process of germinable wheat as shown by the decreased germination of heat-treated seeds. If we consider the dormancy period and the aging process as continuous stages of the seed life cycle, the over all effect of heat treatment would be to shorten the "life span" of the seed.

Aging may also refer to the gradual loss of optimal baking quality during long storage of wheat. The extensive commercial use of heat-treated wheat along with the laboratory tests indicate that heat treatment does not affect the baking quality. This suggests that the aging process altering baking quality goes at a much slower rate than does the loss of seed viability. In other words, in the aging process the loss of germination occurs long before the decrease of baking quality. Germination, in this case, does not appear to be a good criterion of baking quality of wheat.

ACKNOWLEDGEMENTS

We gratefully acknowledge the assistance of the Grain Market Research Center of USDA in providing mixograph and baking quality test results for our treated wheat grain samples.

REFERENCES

Cereal Science Group of Beijing Agriculture University (1961). Storage and Processing of Agricultural Products (in Chinese). Agriculture Press, Beijing, China, 67-68.

Chao, Tung-Fang (1956). Effect of hot-storage on the chemical composition and viability of the wheat grains (in Chinese with English summary). Experimental Biology Proceeding, 5(1), 23-33.

Evans, D.E. and T. Dermott (1979). The potential of a fluidized bed heating system for the insect disinfestation of grain. Contributions to the Symposium on the Protection of Grain Against Insect Damage During Storage at Moscow USSR, July 1978. CSIRO Div. of Entomology, Canberra, Australia, 32-38.

Grain Bureau of Jiangxi Province (1958). An experiment about heat treatment of rice grains (in Chinese). Bulletin of Cereal Science and Technology, 2, 16-19.

Grain Bureau of Shandong Province (1960). Planting experiment with wheat seeds subjected to heat treatment (in Chinese). Bulletin of Cereal Science and Technology, 7, 9-10.

Grain Ministry of China (1958). Storage of Cereal Grains (in Chinese). Finance and Economics Press of China, Beijing, China, 229-248.

Honold, G.R. and M.A. Stahmann (1968). The oxidation-reduction enzymes of wheat. IV. Qualitative and quantitative investigations of the oxidases. Cereal Chem. 45(2), 99-106.

Liu, Reizheng and Yiqiang Xiong (1962). Effect of heat treatment on germination of wheat seeds (in Chinese). Agriculture Science of China, 6, 47-50.

Liu, Reizheng and Cheng Wang (1981). Experimental data about heat treatment of wheat seeds (unpublished).

Plant Physiology Institute in Academia Sinica, and Grain Ministry of China (1956). The effect of sun-drying on oil extraction of soybeans (in Chinese). Grain Business, 13, 21-24.

Shik, Sheng-han (1962). Chi Min Yao Shu (An agricultural encyclopedia of the 6th century)(Chinese and English). Science Press of China, Beijing, China.

Zhou, Fengxian and Jinxin Wan (1959). Heat treatment as a method of insect control for rice (in Chinese). Bulletin of Cereal Science and Technology, 7, 21-22.

REMOVAL BY PROCESSING OF NATURALLY OCCURRING TOXICANTS AND ANTINUTRIENTS

Irvin E. Liener
Department of Biochemistry, College of Biological Sciences
University of Minnesota, St. Paul, MN 55108, USA

ABSTRACT

Foods of plant origin are known to contain a wide variety of substances which, if not destroyed by proper processing, may adversely affect the nutritive properties of the protein. Fortunately those which are protein, such as the protease inhibitors and lectins that are widely distributed in legumes and cereal grains, are destroyed by heat treatment and hence pose little risk to the consumer when such foods are properly processed. Many plants contain glycosides which, upon hydrolysis by endogenous enzymes, release compounds which may be toxic. Examples of such glycosides are the goitrogens of the cabbage family and the cyanogens of cassava, lima beans, and fruit kernels. Suitable processing techniques or traditional methods of preparation often take advantage of the enzymatic release of these toxic constituents in order to detoxify such foods. The destruction of a number of other toxic factors may be facilitated by the use of enzymes of microbial origin as exemplified by: the removal of flatulence-producing factors in beans, the hydrolysis of the phytate of soybeans and wheat, and the destruction of free gossypol in cottonseed.

KEYWORDS: Natural toxicants; protease inhibitors; lectins; goitrogens; cyanogens; flatulence; gossypol; phytate.

INTRODUCTION

Although proteins of plant origin offer considerable promise for alleviating the shortage of food protein now facing many segments of the world's population - a situation which will certainly become more acute if the expansion of the world's population continues unabated - many plants are known to contain substances which have an adverse effect on the nutritional qualities of the protein (Liener, 1980). Fortunately, those toxicants which are protein in nature can be readily inactivated by the heat treatment involved in household cooking or commercial processing. In other cases, potential toxicants remain relatively innocuous as long as they are not released by enzymes which frequently accompany these substances in the plant tissue. Paradoxically, advantage can sometimes be taken of the action of these endogenous enzymes to effect the removal of these toxicants by appropriate methods of food preparation or processing techniques. In this presentation, examples will be cited whereby natural toxicants have been effectively eliminated by some of the aforementioned procedures.

INACTIVATION OF PROTEIN TOXICANTS

Proteinase Inhibitors

Inhibitors which are capable of inactivating mammalian digestive enzymes such as trypsin and chymotrypsin are widely distributed in nature; legumes and cereal grains are particularly rich in such inhibitors. Since these inhibitors are proteins, they are readily denatured and hence inactivated by the application of heat. It is the relative ease with which these inhibitors can be inactivated by cooking that has no doubt contributed to the popularity of legumes as a staple component of the diet in many countries of the world.

Because of its economic importance, the soybean has received the most attention with respect to the effect of heat treatment on trypsin inhibitor activity, and, in general, the extent to which this activity is destroyed by heat is a function of the temperature, duration of heating, particle size, and moisture conditions - variables that are closely controlled in the industrial processing of soybean oil meal in order to botain a product having maximum nutritive value. An example of the relationship that exists between the destruction of the trypsin inhibitor and a concomitant improvement in nutritive value is shown in Figure 1. In general most commercially available soybean products, whether intended for animal or human consumption, such as soy protein isolates, concentrates, tofu, texturized meat analogues, etc., generally contain no more than 10% of the original activity present in raw soybeans. This is a level of activity which is believed to be well within the threshold level necessary to cause pancreatic enlargement in rats (Rackis et al., 1975; Churella et al., 1976), one of the characteristic physiological responses of most animals to the ingestion of trypsin inhibitors. What has been found to be true for trypsin inhibitors in soybeans applies equally as well for the trypsin inhibitors which are present in other beans such as kidney beans, navy beans, lima beans, broad beans, etc.

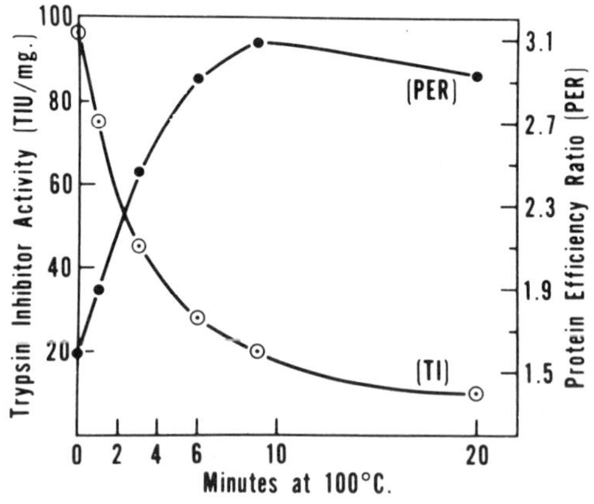

Figure 1: Effect of heat treatment on trypsin inhibitor activity and protein efficiency ratio (PER) of soybean protein. Taken from Anderson et al., 1979.

It has been recently reported that the addition of thiol-containing compounds such as N-acetylcysteine to soyflour and lima beans decreases the amount of heat treatment necessary to inactivate the trypsin inhibitors (Lei et al., 1981; Friedman et al., 1982, 1982b). The action of thiols has been postulated to involve the formation of mixed disulfide bonds between added thiols and the inhibitors whereby the structural integrity of the inhibitors essential for activity is disrupted. The economic feasibility of this approach as well as its toxicological implications remain to be studied.

Lectins

It has been recognized for many years that most legumes and cereals contain substances called phytohemagglutinins or lectins, which have the unique property of agglutinating the red blood cells of various species of animals and the more general property of combining with glycoprotein receptor sites of cell membranes. The toxicity of such lectins toward animals is quite variable, however, and ranges from the extremely toxic lectin of castor bean (ricin) to the one present in soybeans which is relatively innocuous (Turner and Liener, 1975).

Since these lectins are proteins, they are readily denatured by appropriate heat treatment which accounts for the marked improvement in the nutritive value of kidney beans and black beans fed to rats as test animals (Table 1). The extreme toxicity of purified lectins from these beans as compared with the heat-inactivated lectins is shown by the data presented in Figure 2. Several cases of human poisoning from raw or inadequately cooked kidney beans has in fact been reported in England (Noah et al., 1980). It is clear that beans classified as *Phaseolus vulgaris* should receive sufficient heat treatment (at least 15 minutes in boiling water) before they can be considered safe for human consumption.

Table 1: Effect of Heat on Nutritive Value and Hemagglutinating Activity of Some Legumes[1]

Legume	Gain in Wt. (g/day) Raw	Gain in Wt. (g/day) Heated	Hemagglutinating Activity (units/g) Raw	Hemagglutinating Activity (units/g) Heated
Phaseolus vulgaris				
black bean	-1.94(4-5)[2]	+1.61	2450	0
kidney bean	-1.04(11-13)[2]	+1.48	3560	0
Cicer aritinum				
chick pea	+1.25	+1.16	0	0
Cajanus cajan				
Red grain	+1.33	+1.74	0	0
Phaseolus aureus				
mung bean	+1.05	+1.07	0	0

[1] Source: Honavar et al., (1962).
[2] 100% mortality observed during period (in days) shown in parenthesis.

Removal of Goitrogenic Factors

Many cruciferous plants such as the rapeseed (*Brassica napus*), mustard seed (*Brassica juncea*), and Abyssian kale (*Crambe abyssinica*) provide a potentially valuable source of protein, particularly for feeding animals. The protein content of the meal remaining after extraction of the oil is relatively high (30-50%), and

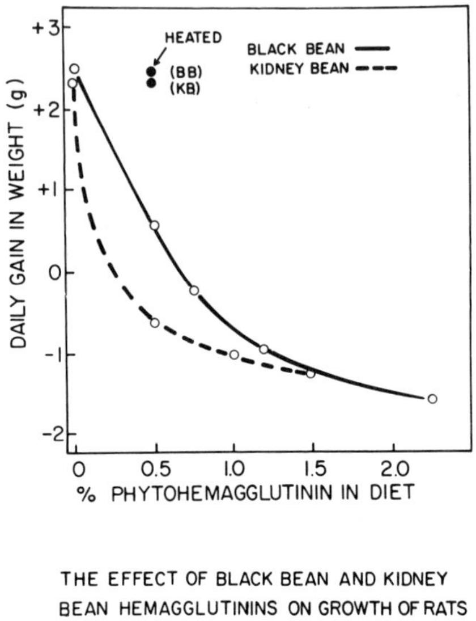

Figure 2: The effect of black bean and kidney bean lectin on growth of rats. Taken from Honavar et al., 1962.

the amino acid composition compares favorably with that of soybeans (Bell, 1955). The use of such oilseeds, however, is limited by the fact that they contain thioglycosides, also referred to as glucosinolates, which upon hydrolysis yield products that, unlike the parent glycoside, are goitrogenic and act as growth depressants. These goitrogenic products, primarily isothiocyanates and oxazolidine-thiones, are liberated from the parent glycosides by enzymes (β-thioglycosidases) normally present in the plant tissue, and these act on their substrates when the seed is crushed and moistened. Although a wide variety of thioglycosides are present in cruciferous seeds, the principal one in rapeseed and crambe is progoitrin which, when acted upon by myrosinase, yields an isothiocyanate that spontaneously cyclizes to a vinyloxazolidine-thione (Figure 3).

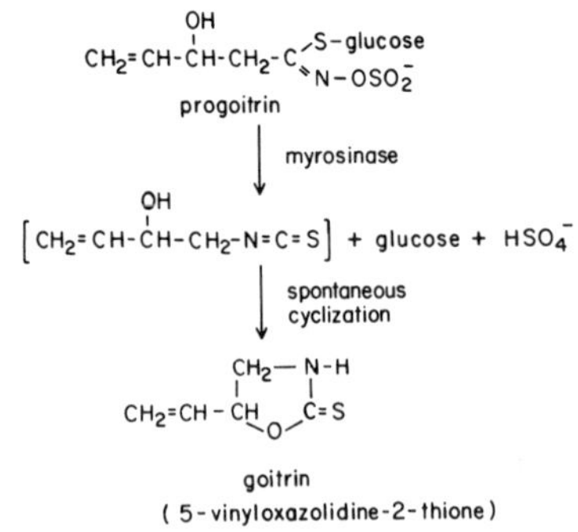

Figure 3: Structure of the goitrogenic glycoside present in rapeseed and the products produced by the action of the enzyme myrosinase.

The toxicity of rapeseed can be reduced in one or a combination of several different ways: (a) destruction of the enzyme by moist heat treatment (Eapen et al., 1968); (b) removal of the thioglycosides by extraction with hot water (Agren and Eklund, 1972), dilute alkali (Kozlowska et al., 1972), or acetone (Van Etten et al., 1965) or by decomposition with iron salts (Kirk et al., 1971); or (c) removal of goitrogenic end products.

Although procedure (a) prevents the further breakdown of the thioglycosides in the meal, the possibility remains that the glycosides which are allowed to remain in the meal could be subsequently hydrolyzed to toxic end products after ingetion by enzymes produced in the intestinal tract (Marangos and Hill, 1974; Oginsky et al., 1965). In one instance at least, enzyme-inactivated mustard seed cake was found to be toxic to cattle because kohlrabi, another cruciferous plant which contains myrosinase, was fed at the same time to the animals (Poulson, 1958). Although procedure (b), especially in combination with procedure (a), effects a marked reduction in toxicity, it does not eliminate any goitrogenic substances which may have been produced prior to processing.

In procedure (c), all of the thioglycosides are deliberately permitted to undergo conversion to goitrogenic end products which are subsequently removed. This conversion may be allowed to proceed by autolysis of the raw meal, or by adding the active enzyme, in the form of raw meal, to the processed meal. For example, Belzile and Bell (1966) noted a reduction in toxicity towards mice of raw rapeseed meal which had been allowed to autolyze and from which the goitrogenic products had been removed by extraction with buffer solutions (pH 6-9). Goering et al. (1960) observed that the addition of a small amount of raw rapeseed to solvent-extracted rapeseed meal followed by extraction with water produced a significant improvement in growth performance of rats. In the case of crambe seed, autolysis of the raw, moistened meal was followed by extraction of the goitrin with acetone containing 2-12% water or a ternary mixture of 53% acetone, 44% hexane, and 3% water (Tookey et al., 1965). Enzymatic hydrolysis followed by extraction with aqueous acetone provides a very effective means of detoxifying crambe seed meal.

Cassava

It has been known for over 150 years that a wide variety of plants are potentially toxic because they contain a glycoside which releases HCN upon hydrolysis (Montgomery, 1980). The glycoside itself is not toxic, but the HCN which is released upon hydrolysis by an endogenous enzyme may be toxic because its principal site of action is cytochrome oxidase, a key enzyme necessary for the survival of aerobic organisms. Those plants which contain high levels of such cyanogenetic glycosides and are commonly eaten by man or domestic animals include cassava (manioc), lima beans (*Phaseolus lunatus*), sorghum, and linseed (flax) (see Table 2).

Cassava (*Manihot esculenta* or *utilissima*) is a staple food item throughout the tropics where it is eaten as a boiled vegetable or prepared into a variety of dishes by traditional methods (Favier et al., 1971). The principal cyanogenetic glycoside present in cassava is linamarin, which is a β-glucoside of acetone cyanohydrin (Figure 4); the liberation of HCN is effected by an endogenous enzyme called linamarase. Since the peel is particularly rich in the glycoside, some reduction in potential toxicity is achieved by peeling and thorough washing of the crushed pulp with running water. A further reduction in toxicity is accomplished by the cooking process (boiling, roasting, or sun-drying) which serves to inactivate the enzyme and volatilize any HCN that may have been released. In the preparation of the native Nigerian dish called gari, the pulp is allowed to undergo fermentation, and because of the formation of acid, the spontaneous hydrolysis of the glycosides with the liberation of HCN is favored (Akinrele, 1964). Subsequent frying effects almost complete elimination of any remaining

HCN. Ikediobe and Oryike (1982) have recently shown that the addition during fermentation of exogenous sources of linamarase isolated from fungi not only causes a reduction in the fermentation time necessary to produce gari but also achieves an eight-fold reduction in its residual cyanide content.

Table 2: Cyanide Content of Plants Commonly Eaten by Man or Domestic Animals[1]

Plant	HCN Yield, (mg/100 g)
Lima bean (*Phaseolus lunatus*)	
Samples incriminated in fatal human poisoning	210-312
Normal levels	14.4-16.7
Sorghum	250
Cassava	113
Linseed meal	53
Black-eyed pea (*Vigna sinensis*)	2.1
Garden pea (*Pisum sativum*)	2.3
Kidney bean (*Phaseolus vulgaris*)	2.0
Bengal gram (*Cicer arietinum*)	0.8
Red gram (*Cajanus cajans*)	0.5

[1] Taken from Montgomery (1964).

Tapioca is the product obtained from cassava after thorough washing. Because of its high starch content (about 65%) and its high digestibility, tapioca has been recommended for human diets and for animal nutrition. Tapioca, however, may still contain sufficiently high levels of cyanide to preclude its use at levels higher than 10% for poultry feeding (Vogt, 1966). Because tapioca has not been subjected to any heat treatment, the hydrolytic enzymes still remain. It has been claimed, however, that these enzymes catalyze the condensation of HCN with aldehydic compounds to form cyanohydrins, so that the addition of glucose to unprocessed cassava causes the disappearance of HCN (Oke, 1966). It is therefore suggested that in the preparation of foods from unprocessed cassava liberal amounts of glucose should be added.

Despite the apparent effectiveness of traditional methods of preparing cassava dishes in reducing the cyanide content, there is evidence to indicate that a tropical disease known as ataxic neuropathy may be associated with the consumption of cassava (Osuntokun, 1970). One of the metabolic routes whereby cyanide is detoxified involves the formation of thiocyanate. Elevated levels of thiocyanate in blood plasma have been observed in individuals who suffered from ataxia neuropathy and who had a history of high consumption of cassava. Rats fed boiled or fermented cassava also showed high levels of thiocyanate is plasma and developed clinical signs of ataxia neuropathy. The mechanism whereby cyanide or its metabolic product, thiocyanate, is involved in the pathogenesis of this disease remains to be elucidated.

A high incidence of goiter in certain parts of Africa has likewise been associated with the consumption of cassava (Ekpechi, 1973). The goitrogenic effect of cassava has been attributed to the production of elevated levels of thiocyanate, a well-established goitrogen which interferes with the trapping of iodide by the thyroid.

Lima Beans

Serious outbreaks of poisoning in man and animals have frequently been attributed to the consumption of certain varieties of lima beans (Montgomery, 1980). The particular glycoside involved is identical to the one present in cassava (Figure 4) but is sometimes referred to as phaseolunatin. Rahman and others (1947) have described a simple procedure for reducing the cyanide content of lima beans to a negligibly low, nontoxic level. Water is added to the ground bean flour to a level of 25-50% which favors the autolytic release of cyanide. Subsequent drying and cooking of the flour with water for 20 minutes reduces the cyanide contents to levels where it is physically impossible to eat enough to reach an estimated toxic dose of 50 mg of HCN for man.

It is by no means clear why human intoxication has been occasionally reported in cases where the lima beans had been cooked and the hydrolytic enzymes presumably destroyed (Gabel and Kruger, 1920). Significant amounts of intact glycoside might still be present, for example, if conditions were such that very little autolysis had been allowed to occur prior to cooking. Since Gabel and Kruger (1920) observed that cyanide was present in the urine (36), it is possible that enzymes secreted by the intestinal flora may be responsible for the release of cyanide from any glycosides which remain after cooking (Winkler, 1958). Auld (1912-1913), however, contends that the enzymic hydrolysis of cyanogenetic glycosides is greatly inhibited by nearly all of the conditions prevailing in the digestive tract.

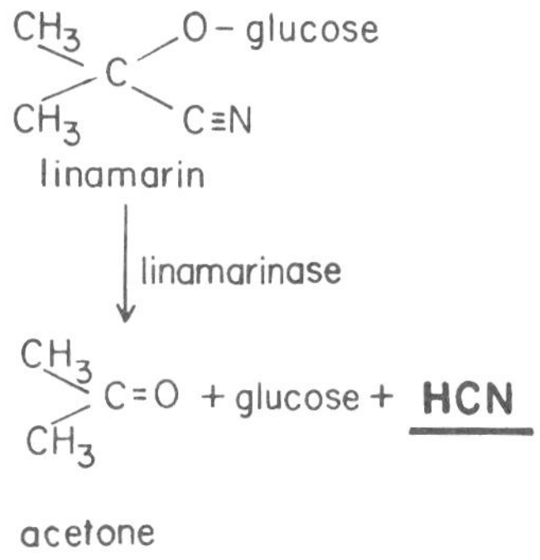

Figure 4: Enzymatic hydrolysis of linamarin, the cyanogenic glycoside of cassava and lima beans

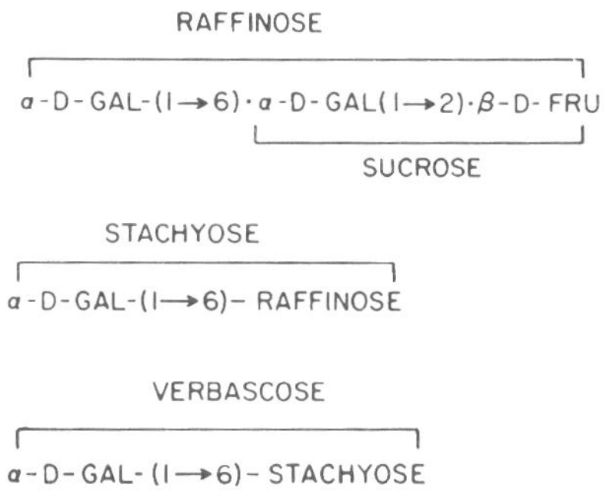

Figure 5: Structural relationship of the oligosaccharides involved in flatulence.

Flatulence-producing Factors

One of the important factors limiting the use of legumes in the human diet is the production of flatulence associated with their consumption (Rackis, 1975). Flatulence is generally attributed to the fact that man is not endowed with the enzymes (α-galactosidase and β-fructosidase) necessary to hydrolyze certain oligosaccharides which contain α-galacosidic and β-fructosidic linkages. As shown in Figure 5, these oiligosaccharides (raffinose, stachyose, and verbascose) are related by having one or more α-D-galactopyranosyl groups in their structure, where the α-galactose units are bound to the glucose moiety of sucrose. The intact oligosaccharides enter the lower intestine where they are metabolized by the microflora into carbon dioxide, hydrogen, and, to a lesser extent, methane.

It is the production of these bases which leads to the characteristic features of flatulence, namely nausea, cramps, diarrhea, abdominal rumbling, and the social discomfort associated with the ejection of rectal gas.

A number of studies have been made in an attempt to remove offending oligosaccharides by enzymatic degradation using, in most instances, crude enzyme preparations of fungal origin. Calloway et al. (1971) found that the treatment of soybeans with diastase A had only a negligible effect on the production of intestinal gas measured in human subjects despite the fact that stachyose and raffinose had been virtually eliminated. Although several studies have shown that the oligosaccharide content of soybeans can be reduced to negligible levels by treatment with mold enzymes, the flatulence-producing properties of the bean as measured in human subjects were not significantly altered by such treatment (Calloway et al., 1971). Becker et al. (1974) were able to lower the oligosaccharide content of California small white beans by allowing a slurry of the raw, ground bean to undergo autolysis at pH 5.2 and 55°C.

Such traditional soybean foods as tofu (soybean curd) and tempeh have little flatus activity (Calloway et al., 1971). In the case of tofu, oligosaccharides are are presumably eliminated during the course of its preparation, and, in the case of tempeh, enzymes produced by the mold (Rhizopus) during fermentation probably hydrolyze the oligosaccharides. As might be expected because of their low carbohydrate content (<1%), soy protein isolates are devoid of flatus activity (Rackis, 1975). It follows that textured meat analogs made from isolated soy protein are most likely free of flatus activity.

Phytic Acid

Phytata, a cyclic compound (inositol) containing six phosphate radicals, readily chelates such di- and trivalent metal ions as calcium, magnesium, zinc, and iron to form poorly soluble compounds that are not readily absorbed from the intestines (Oberleas, 1973). The ability of phytic acid to bind metal ions is lost when the phosphate groups are hydrolyzed through the action of the enzyme phytase. Although phytase activity has been shown to be present in the small intestine of various experimental animals (Spitzer and Phillips, 1945), its presence in the intestines of humans remains a controversial issue (McCance and Widdowson, 1942; Bitar and Reinhold, 1972). Even if phytase were present, its activity does not appear to be sufficiently great to prevent phytate from interfering with the utilization of calcium, zinc, and phosphorus from cereal-containing diets which are high in phytate (Bitar and Reinhold, 1972). This was found to be particularly true in Iran where the high incidence of a disturbance in mineral metabolism has been attributed to their life-long consumption of unleavened bread made from a high-extraction rate flour with a high phytate content (Rienhold et al., 1973). Reinhold (1975) found that the fermentation of whole wheat bread with baker's yeast caused a significant reduction in phytate content in three different whole wheat flours representing different extraction rates.

A low-phytate soybean protein isolate can be prepared from soybean flour, however, by allowing endogenous phytase to act on the phytate in a 6% suspension of the flour at pH 5 at a temperature of 65°C (Okubo et al., 1975). Hydrolysis of the phytate facilitates its separation from the bulk of the soybean protein which is then concentrated by ultrafiltration using a membrane which is permeable to phytate and its hydrolysis products but impermeable to protein. The product obtained by this method contains over 90% protein and only about 0.3% phosphorus. Brooks and Morr (1982) have recently reported that a combination of a cation- and anion-exchange process was capable of removing 96% to 97% of the phytate from soybean isolates.

The phytate content of California small white beans (Phaseolus vulgaris) can be greatly reduced by allowing a suspension of raw beans to autolyze (pH 5.2, 35°-55°C, 20-48 hr) or by adding a preparation of wheat germ phytase under the same conditions (Becker et al., 1974; Kon et al., 1973).

Gossypol

One of the most important factors limiting the utilization of cottonseed meal is gossypol, a phenolic compound which is peculiar to the pigment glands of the cottonseed where it constitutes about 20% to 40% of the weight of the glands. Although gossypol is toxic to non-ruminants, it is much less so to ruminants presumably because the gossypol becomes bound to protein while in the rumen (Reiser and Fu, 1962). The heat generated during the commercial production of cottonseed meal serves to cause 80 to 90% of the gossypol to become bound to the protein rendering it non-toxic. The amount of gossypol inactivated in this manner depends on the type of commercial processing involved (see review by Berardi and Goldblatt, 1980). Unfortunately the binding of gossypol to protein reduces the availability of the lysine because of the interaction of the aldehyde groups of gossypol with the epsilon amino group of lysine.

Gossypol forms stable chelates of low solubility with metal salts which reduces its toxicity and hastens its elimination from the body (Singleton, 1981). This property has prompted the use of iron salts to increase the tolerance of animals to cottonseed meals containing high levels of gossypol. Glandless, and hence gossypol-free, varieties of cottonseed have been bred and are beginning to be planted commercially, but it remains to be seen if pest and disease resistance and fibre yield and quality will enable the glandless seed to replace the glanded forms on a practical basis. An interesting alternative method for the detoxification of gossypol is suggested by the report that certain fungi are capable of effecting a 90% reduction of the free gossypol of cottonseed with concomitant elimination of its toxicity to rats and chicks (Baugher and Campbell, 1969).

REFERENCES

Agren, G. and A. Eklund (1972). Rapeseed protein fractions. III. The nutritive value of a detoxified protein concentrate from rapeseed (Brassica napus L.) by hydraulic pressing. J. Sci. Food Agr. 23, 1457-1462.

Akinele, I.A. (1964). Fermentation of cassave. J. Sci. Food Agr. 15, 589-594.

Anderson, R.L., J.J. Rackis and W.H. Tallent (1979). In: Soy Protein and Human Nutrition (H.L. Wilcke, D.T. Hopkins, and D.H. Waggle, eds.). Academic press, pp. 209-233.

Auld, S.J.M. (1912-1913). Cyanogenesis under digestive conditions. J. Food Sci. Agr. Sci. 5, 409-417.

Baugher, N.L. and T.C. Campbell (1969). Gossypol detoxification by fungi. Science 164, 1526-1527.

Becker, R., A.C. Olson, D.P. Frederick, S. Kon, M.R. Gumbman, and J.R. Wagner (1974). Conditions for the autolysis of alpha-galactosides and phytic acid in California small white beans. J. Food Sci. 39, 766-769.

Bell, J.M. (1955). The nutritional value of rapeseed oil meal. A review. Can. J. Agr. Sci. 35, 242-251.

Belzile, R.J. and J.M. Bell (1966). Growth-depressing factors in rapeseed oil meal. VII. Effects of myrosinase activity on toxicity following treatment with buffered solutions. Can. J. Anim. Sci. 46, 165-169.

Berardi, L.C. and L.A. Goldblatt (1980). Gossypol. In Toxic Constituents of Plant Foodstuffs (I.E. Liener, editor). Academic Press, New York, pp. 184-238.

Bitar, K. and J.G. Reinhold (1972). Phytase and alkaline phosphatase activities in intestinal mucosas of rat, chicken, calf, and man. Biochim. Biophys. Acta 268, 442-452.

Brooks, J.R. and C.V. Morr (1982). Phytate removal from soy protein isolates using ion-exchange processing treatments. J. Food Sci. 47, 1280-1282.

Calloway, D.H., C.A. Hickey and E.L. Murphy (1971). Reduction of intestinal gas-forming properties of legumes by traditional and experimental food processing methods. J. Food Sci. 36, 251-255.

Churella, H.R., B.C. Yao and W.A.B. Thompson (1976). Soybean trypsin inhibitor activity of soy infant formulas and its nutritional significance for the rat. J. Ag. Food Chem. 24, 393-397.

Eapen, K.E., N.W. Tape and R.P.A. Sims (1968). New process for the production of better-quality rapeseed oil and meal. I. Effect of heat treatments on enzyme destruction and color of rapeseed oil. J. Am. Oil Chem. Soc. 45, 194-196.

Ekpechi, O.L. (1973). Endomic goiter and high cassava diets in Eastern Nigeria. In Chronic Cyanide Toxicity (B. Hestel and R. McIntyre, editors). International Development Research Center, Ottawa, Canada, pp. 139-145.

Favier, J.-C., S. Chevassus-Agnes and G. Gallon (1971). Traditional technology of manioc. Influence on nutritive value. Ann. Nutr. Aliment. 25, 1-59.

Friedman, M., O-K.K. Grosjean and J.C. Zahnley (1982a). Inactivation of soya bean trypsin inhibitors of thiols. J. Sci. Food Agr. 33, 165-172.

Friedman, M., O-K.K. Grosjean and J.C. Zahnley (1982b). Comparative effects of heat and thiols in inactivating trypsin inhibitors from legumes in solution and in the solid state. Nutr. Rep. Int. 25, 743-752.

Gabel, W. and W. Kruger (1920). The toxic action of Rangoon beans. Muensch. Med. Wochschr. 67, 214-215.

Goering, K.J., O.O. Thomas, D.R. Beardsley and W.A. Carran, Jr. (1960). Nutritional value of mustard and rapeseed meals as protein source for rats. J. Nutr. 72, 210-216.

Hill, F.W. and K. Totsuka (1964). Studies on the metabolizable energy of cottonseed meals for chicks with particular reference to the effects of gossypol. Poultry Sci. 43, 362-370.

Honavar, P.M., C.-V. Shih and I.E. Liener (1962). Inhibition of the growth of rats by purified hemagglutinin fractions from Phaseolus vulgaris. J. Nutr. 77, 109-114.

Ikediobi, C.O. and E. Onyike (1982). The use of linamarase in grain production. Process Biochem. 17, 2-5.

Kirk, L.D., G.C. Mustakas, E.L. Griffin, Jr. and A.N. Booth (1971). Crambe seed processing. Decomposition of glucosinolates (thioglucosides) with chemical additives. J. Am. Oil Chem. Soc. 48, 845-890.

Kon, S., A.C. Olson, D.P. Frederick, S.B. Eggling and J.R. Wagner (1973). Effect of different treatments on phytate and soluble sugars in California white beans (Phaseolus vulgaris). J. Food Sci. 38, 215-220.

Kozlowska, H., F.W. Sosulski and C.G. Youngs (1972). Extraction of glucosinolates from rapeseed. Can. Inst. Food Sci. Technol. 5, 149-154.

Lei, M.-G., R. Bassette and G.R. Reeck (1981). Effect of cysteine on heat-inactivation of soybean trypsin inhibitors. J. Ag. Food Chem. 29, 1196-1199.

Liener, I.E. (1980) Toxic Constituents of Plant Foodstuffs. Academic Press, New York.

Marangos, A. and R. Hill (1974). Hydrolysis and absorption of thioglucosides of rapeseed meal. Proc. Nutr. Soc. 33, 90A.

Montgomery, R. (1964). Observations on the cyanide content and toxicity of tropical pulses. West Ind. Med. J. 13, 1-11.

Montgomery, R. (1980). Cyanogens. In Toxic Constituents of Plant Foodstuffs (I.E. Liener, editor). Academic Press, New York. pp. 143-160.

Noah, N.D., A.E. Bonder, G.B. Readi and R.J. Gilbert (1980). Food poisoning from raw red kidney beans. Brit. Med. J. 281, 236-237.

Oberleas, D. (1973). Phytates. In Toxicants Occurring Naturally in Foods. National Academy of Science, Washington, DC. pp. 363-371.
Oginsky, E.L., A.E. Stein and M.A. Greer (1965). Myrosinase activity in bacteria as demonstrated by the conversion of progoitric to goitrin. Proc. Soc. Exp. Biol. Med. 119, 360-364.
Oke, O.L. (1966). Chemical studies on some Nigerian foodstuffs. Nature 212, 1055-1056.
Okubo, K., A.B. Waldrop, G.A. Iacobucci and D.V. Myers (1975). Preparation of low-phytate soybean protein isolate and concentrate by ultrafiltration. Cereal Chem. 52, 263-271.
Osuntokun, B.O. (1970). Cassava diet and cyanide metabolism in Wistar rats. Brit. J. Nutr. 24, 797-800.
Poulsen, E. (1958). Poisoning by myrosinase-free mustard-seed cake in cattle. Nord. Vet. Med. 10, 487-489.
Rackis, J.J. (1974). Oligosaccharides of legumes: alpha-galactosidase and the flatus problem. Am. Chem. Soc. Symp on Physiological Effects of Food Carbohydrates, Atlantic City, New Jersey, pp. 207-222.
Rackis, J.J., J.E. McGhee and A.N. Booth (1975). Biological threshold levels of soybean trypsin inhibitors by rat bioassay. Cereal Chem. 52, 85-92.
Rahman, S.A., S.S. De and V. Subrahmanyan (1947). Effect of different treatments on the removal of hydrocyanic acid from the Burma bean. Current Sci. 11, 351-352.
Reinhold, J.G. (1975). Phytate destruction by yeast fermentation in whole wheat meals. J. Am. Diet. Assoc. 66, 38-41.
Reinhold, J.G., H. Hedayati, A. Lahimgarzadeh and K. Nasr (1973). Zinc, calcium, phosphorus, and nitrogen balances of Iranian villages following a change from phytate-rich to phytate-poor diet. Ecol. Food Nutr. 2, 157-162.
Reiser, R. and H.C. Fu (1962). The mechanism of gossypol detoxification by ruminant animals. J. Nutr. 76, 215-218.
Schab, R. and S. Yannai (1973). An improved method for debittering apricot kernels. J. Food Sci. Technol. 10, 57-59.
Singleton, V.L. (1981). Naturally occurring food toxicants: phenolic substances of plant origin common in foods. Adv. Food Res. 27, 149-242.
Spitzer, R.R. and P.H. Phillips (1945). The availability of soybean oil meal phosphorus for the rat. J. Nutr. 30, 117-126.
Turner, R.H. and I.E. Liener (1975). The effect of the selective removal of hemagglutinins on the nutritive value of soybeans. J. Ag. Food Chem. 23, 484-487.
Vogt, H. (1966). The use of tapioca meal in poultry rations. World Poultry Sci. 22, 113-125.
Winkler, W.O. (1951). Report on hydrocyanic glucosides. J. Assoc. Off. Agr. Chem. 34, 541-547.

WATER ACTIVITY AND INTERMEDIATE MOISURE FOODS

M. Karel
Department of Nutrition and Food Science
Massachusetts Institute of Technology
Cambridge, Massachusetts 02139, U.S.A.

ABSTRACT

Control of water activity is the principal factor in such important food preservation methods as dehydration, freezing, concentration, salting and production of intermediate moisture foods. Water activity appropriately describes water relations of foods, because it controls the equilibrium with respect to water between food components and between food and its environment. The dependence of water activity on food composition and on temperature is shown and methods given for predicting water activity of mixtures. Non-equilibrium situations due to phase changes in food (e.g. sugar recrystallization) and to other physical changes are described and their significance noted. A review is presented of the effects of water content and of water activity on food deterioration due to biological, chemical and physical mechanisms. An important application of the water activity concept is the production of intermediate moisture foods (IMF). A review is presented of principles and practice of marketing of "traditional" IMF in several areas of the world. Developments in the production of "novel" IMF utilizing less traditional water-activity-depressing solutes (e.g. glycerol, sorbitol) are described. Several important obstacles, requiring research and development, remain in the area of IMF development. These are reviewed and prospects for progress analyzed.

KEYWORDS: Water activity, intermediate moisture foods, food composition, water content, food deterioration, novel IMF, water activity depression.

WATER ACTIVITY AND FOOD PRESERVATION

Food preservation processes have as their goal the extension of shelf life of foods, to allow storage and convenient distribution without undue losses in quality. The first and most dangerous limitation of shelf life is due to the activity of microorganisms which are not only effective catalysts of chemical changes, but are also potentially health-threatening agents. Hence, food preservation processes are designed to eliminate the danger of microbial growth.

Several food preservation processes achieve this aim by lowering the availability of water to microorganisms. In freezing, water is immobilized in ice crystals, and thus made unavailable to microorganisms, which are exposed to the unfrozen concentrated solution. Salting, preservation by addition of sugars, and the more recent intermediate moisture technology depend on increasing the osmotic pressure by addition of solutes. Dehydration, concentration and sublimation remove water from the food, thus achieving the same aim.

The availability of water to microorganisms is best expressed in terms of water activity. Water activity, a, can be defined simply as the ratio of partial pressure of water in the food, p, to the vapor pressure of water at the same temperature, p^o (equation 1):

$$a = (p/p^o)_T \tag{1}$$

The importance of water activity arises from the fact that equilibrium between different phases with respect to water exists when the water activities of these phases are equal. For instance, when two food ingredients are within the same container they attain equilibrium when their water activities are equal, at which point their moisture contents would be vastly different. The interior of a microbial cell is subjected to potential dehydration when the solution surrounding the cell is at a lower water activity than the cell. The biochemical activity of the cell is then directed to increasing internal osmotic pressure to avoid a potentially lethal dehydration, and if the (a) of the surrounding is sufficiently low microbial growth ceases.

Different microorganisms have different values of limiting water activity (Scott, 1957; Christian, 1981; Troller and Christian, 1978).

Water relations of microorganisms may be summarized in a somewhat simplified manner as follows: (a) water activity, rather than water content, determines the lower limit of available water for microbial growth. Most bacteria do not grow below a = 0.91 and most molds cease to grow at water activities of 0.65, but the range of 0.70-0.75 is generally considered their lower limit; (b) environmental factors affect the level of water activity required for microbial growth; (c) some adaptation to low activities occurs; (d) the solutes themselves may have effects which complicate the effect of water activity per se; (e) water activity modifies sensitivity of microorganisms to heat, light, and chemicals. In general, organisms are most sensitive at high water activities (i.e., in dilute solution) and minimum sensitivity occurs in an intermediate moisture range; (f) minimum water activities for production of toxins are often higher than those for microbial growth (a phenomenon which may represent an important safety factor); (g) the methods used to test for microbial growth, including the conditions of rehydration, type of media used for dilution and plating, and the incubation temperature -- all can affect the observed minimum water activity (Mossel, 1976).

The effects of water on chemical reactions in foods are complicated. Water can act: (a) as a solvent for reactants and for products; (b) as a reactant (e.g. in hydrolysis reactions); (c) as a product of reactions (e.g. in condensation reactions in nonenzymic browning); and (d) as a modifier of the catalytic or inhibitory activities of other substances (e.g. water inactivates some metallic catalysts of lipid peroxidation).

The most recent findings concerning the role of water in chemical deterioration of foods include the following (Karel, 1973):

1. Non-enzymatic browning is initiated when the water content reaches a so-called "mobilization point" at which diffusion of reactants and of reaction products becomes possible. This mobilization point may coincide with exceeding the water content corresponding to "non-solvent" water, but is certainly above the so-called BET monolayer value (Duckworth, 1981; Simatos et al., 1981).
2. Enzymatic reactions respond to water content in a complex manner. Diffusion

is a factor, but most recent work indicates that effects of water on protein conformation in the vicinity of active sites of enzymes may be of importance (Karel, 1981; Silver and Karel, 1981).

3. Water affects strongly the course of reactions initiated by lipid free radicals. Lipid oxidation, as well as oxidative degradation of protein are dependent on water content (Karel and Yong, 1981; Karel, 1980; Karel, 1975a).

WATER ACTIVITY AND FOOD COMPOSITION

The relation between water activity and water content is given by water sorption isotherms whose determination is reviewed by Gal (1967) and Troller and Christian (1978). Typical isotherms in foods are S shaped, and can be approximated by a variety of mathematical relationships. For example, the BET equation (2), allows the calculation of the BET monolayer value.

$$\frac{a}{m(1-a)} = \frac{1}{m_1C} + \frac{C-1}{m_1C}a \tag{2}$$

where m = moisture content (g/g solids), C = constant, m_1 = monolayer value. The monolayer value should be understood as the amount of water strongly held on specific polar groups in food. Monolayer values for various food substances are generally in the range of 1 to 10% water (dry basis). The maximum heats of sorption in foods range typically from 1 to 8 Kcal/mole.

The role of water as solvent or mobilizer affects a variety of potential processes in foods. The determination of the amount of the total water that is available as a solvent is difficult. One method used involves mixing a water-soluble solute with the food, assuming perfect dilution, and the determination of concentration of the solute in a portion of the water withdrawn from the food for analysis. A variant of this method is to determine the concentration of the solute directly in the food provided that other food components do not interfere with determination of solute concentration in the aqueous phase. Duckworth and Kelly (1973) developed an NMR method for determination of solvent water. The pertinent property for commencing dissolution of a given solute was found to be water activity whose critical value depended only on solute and not on polymeric components present. For example, sucrose was found to begin to go into solution at an activity of approximately 0.88 in several water-polymer systems. At this activity the water contents of these systems were 0.34 g/g for agar, 0.27 for gelatin, 0.24 for starch, and 0.11 for cellulose. Glucose began to dissolve at an activity of approximately 0.85 and urea at an activity of approximately 0.45, independent of polymer studied.

Water capable of acting as a solvent is subject to depression of vapor pressure by solutes. Ideally, this vapor depression is given by Raoult's Law whereby the activity, a, would equal the mole fraction of water in solution. In foods there are usually substantial deviations from the ideal relationship, and a number of more complicated equations have been developed. A useful equation, assuming independent behavior of different solutes in a complex solution, is given as equation 3 (Ross, 1975)

$$a_f = (a_i)\,(a_1)\,(a_2)\,\ldots\,(a_n) \tag{3}$$

where a_f = final water activity of a food to which various solutes were added, a_i = initial water activity of a food, a_1 ... a_n = water activity of a solution containing a given solute at a concentration which could exist if all the water in the food were available to dissolve this solute only.

NONEQUILIBRIUM SITUATIONS IN WATER SORPTION

Many food components, and in particular low molecular weight carbohydrates and salts, may be present in one of several states: crystalline solids, amorphous solids, aqueous solution, and bound to other components. Sorption of water in such systems is complicated and may require considerations of kinetics as well as equilibrium or sorption. The recognition of the above phenomena is absolutely necessary for development of reliable data on behaviour of foods at different water contents. The history of samples is of paramount importance, especially for materials such as sugars and polymers which can readily be prepared in a glassy state and must be recognized as systems capable of undergoing changes in molecular state of aggregation as a result of absorption of water.

INTERMEDIATE MOISTURE FOODS

Historical Development

One of the oldest methods of food preservation is based on lowering of water activity to a level at which growth of microorganism is prevented or greatly reduced. In the 1960's a number of developments led to a renaissance of technology for production of foods with an intermediate moisture content which have been often referred to as intermediate-moisture foods (IMF). The modern IMF is a direct descendant of the traditional IMF. Some examples of traditional IMF are: products which have been dried with no addition of humectants (prunes, apricots, dates, figs, etc.), products to which sugar has been added (candied fruits, soft candies, marshmallows, jams, jelly, honey, pie fillings, and syrups), products which have been dried with added salt and sugars (country ham, pemmican), and bakery products such as fruit cakes and certain pie fillings. These traditional foods have moisture contents from 10-40% ("wet basis") and range in water activity from 0.65 to 0.9. Table 1 shows some traditional IMF of Japan and Table 2 some traditional IMF of Argentina. Traditional IMF of the United States were surveyed by Flink (1977). The modern IMF experienced the most dramatic growth in the pet food industry which has recently reached a total sales level of $3.4 billion. Of the US total in 1979 8.8% of the dollar value was in "moist" dog foods, 4.2% in "soft-dry" dog food and 4.0% in "moist" cat food.

Modern IMF have been developed according to the following technological principles: 1) lowering of water activity by addition of polyhydric alcohols, sugars, and/or salt; 2) retardation of microbial growth by addition of antimicrobial agents and 3) improvement of organoleptic properties through physical or chemical treatment.

Processing steps include means for 1) activity adjustment in the finished product, 2) microbiological stabilization by heating or use of chemical additives, 3) enzyme inactivation by blanching, 4) prevention of physical and chemical deterioration by addition of antioxidants, chelators, emulsifiers, or stabilizers, and 5) inclusion of appropriate nutrients.

The status of "modern" IMF in the 70's was surveyed thoroughly by Heidelbaugh and Karel, (1975) and Karel and Flink (1977). Recently, Erickson (1982) surveyed the field again, based mainly on the above reviews and other general reviews published in the important monograph by Davies et al. (1976). We shall concentrate therefore on most recent developments and on older studies, which have received less attention in previous reviews.

Table 1: Some Traditional IMF in Japan
(courtesy of Mr. M. Motoki of Ajinomoto)

Japanese Name	Type of Product	% Water Content	% Water Activity
Amanatto	Sugared Bean	23	
An	Bean Paste	53	
Chimaki	Rice in Bamboo Leaf	68	
Mushi Manju	Steamed Bun	37	
Neri Yokan	Bean Jelly	26	0.87
Yaki Chikuwa	Fish Cake	67	0.97
Konbu Tsukudani	Seaweed Confectionery	60	0.82
Shoyu	Soysauce	72	0.80 - 0.86
Ama Miso	Sweet Miso	49	0.75 - 0.80
Katsuo Kakuni	Tuna Preserve	30	0.88
Katsuo Shiokara	Tuna Tripe Preserve	67	0.72
Shio Sake	Cured Salmon	61	0.88
Sake Kunsei	Smoked Salmon	40	0.74
Kusaya	Dried Mackerel	30	0.70

Table 2: Some Traditional IMF in Argentina
(After Vigo et al., Rev. Agroquim. Tecnol. Alim. 21(1):91, 1979)

Product Types	Range of Water Activity			
Marmalades, fruit pastes	0.62 to 0.86			
Sweetened milk products	0.74 to 0.87			
Sausages	0.82 to 0.97			
Cheeses	0.66 to 0.97			
Pastries				
Torta galesa		0.76		
Budin ingles	Interior:	0.83	Surface:	0.76
Pan dulce	"	0.92	"	0.79
Pan lactal	"	0.97	"	0.95
Sauces				
Traditional	0.93 to 0.95			
Based on hydrolyzed vegetable proteins	0.69 to 0.70			

Recent Developments in IMF Technology

As noted by Flink (1977) the novel IMF in the US marketplace are limited to pet foods, selected bakery products and snacks. While not considered "novel" by Flink, various soy-based sauces and condiments are also in the IMF category. An important recent development has been the recognition by the US FDA of the important role of the water activity in food preservation. Under regulations promulgated by the Bureau of Foods of FDA, foods with water activities below 0.85 are exempted from the strict regulations applied to thermally processed low acid foods (Johnston, 1981). As a consequence, considerable interest developed in standardization of measurement of water activity (Stoloff, 1978) and in improvement of the safety factor in osmotically preserved foods.

Specific IMF products recently developed include the following:

a) Meat Products. The general state of preservation by control of water activity in semi-traditional meat products was reviewed by Leistner et al. (1981). Entirely novel products were developed on a laboratory scale by US Government agencies (Karel, 1976) and more recently by laboratories in Argentina, Brazil, and England (Chirife et al., 1980, Ledward et al., 1981).

b) Fish Products. Salt-cured fish have long been an acceptable component of human diet. The recently developed products are largely improvements on traditional technology, either by improvement in traditional salting and drying processes (Del Valle et al., 1973, Hsu et al., 1980), or by utilization of glycerol and other additives in addition to NaCl (Dymsza and Silverman, 1979).

c) Intermediate Moisture Cheeses and Soy-Based Cheese Analogs. Most cheeses have a substantial shelf life, but microbial growth can occur, and in particular there exists the possibility of growth of *Staphylococcus aureus*, and in at least some of the cheeses, which have a high pH, of *Clostridium botulinum* (Leung et al., 1976). To improve the microbial stability, intermediate moisture cheeses have been developed using various additives to reduce the water activity to less than 0.85. Leung et al. (1976) used nonfat dry milk and propylene glycol to reduce the water activity of cheddar cheese. Their additive also included potassium sorbate, NaCl, butter, and disodium phosphate. Kreisman and Labuza (1978) conducted storage stability studies on similar IM cheeses and reported that some formulations gave excellent stability.

Recently Motoki et al. (1982) conducted a development study on an intermediate moisture cheese analog made of soy protein isolate, casein, vegetable oil, several additives. A number of compositions were tried. The most acceptable product contained 12% sorbitol, had a water activity of 0.87, and had a shelf life of 3 months.

d) Cereal and Bakery Products

Traditional IMF exist in the bakery field. Many pies, cakes, and cookies are traditionally formulated with high sugar, and/or salt contents to provide long shelf life. Honey is another traditional water-activity-lowering ingredient. Recently, the feeding of astronauts in space provided a stimulus to formulation of IM baked products in both the U.S. and the Russian space programs (Karel, 1976). Arya et al. (1977) reported the development of chapaties with an activity of 0.88 to 0.92, and having 0.2% citric acid and 0.2% sorbic acid. This product is quite stable for 6 months. A General Foods patent (Shanbhag, 1981) reports the use of polyols to reduce the water activity of fruit to be used in cereal products. This allows the avoidance of moisture transfer from fruit to cereal, which is deleterious to the texture of both components. Pelaez and Karel (1980) reported the development of an IM tortilla containing various humectants in addition to corn flour, water and salt, which are the traditional ingredients. The composition with 5.56% glycerol, 0.2% sorbate and 3.75% corn solids had a water activity of 0.86 and was judged acceptable and stable. A product with similar charateristics is now being test marketed by McCormick Company.

e) Fruit Products, Candies and Confectioneries

Sweet IM products have long been used as part of shelf-stable food supplies. They include home-prepared as well as industrial products. Recent developments include improved utilization of osmotic drying using a variety of osmotic liquids (Contreras and Smyrl, 1981; Bongirwar and Sreenivasan, 1977; Moy et al., 1978; Lerici et al., 1977). Some of the techniques were perfected in our laboratory at MIT, including osmotic dehydration using mixed osmotic solvents (Hawkes and Flink, 1978) and development of an IM banana product (Luh and Palmer, 1973). Further studies on IM bananas were made by Levi et al. (1980).

Confectioneries have long been available in IM form in both Western and Oriental cuisines. Some Oriental IM pastries are listed in Table 3. Wang and Sun (1981) modified the traditional Ho-Hsui candies by using sorbitol and obtained a superior IM product with water activity lower than in the traditional candies. The utilization of sorbitol has become more promising because of recent pricing developments whereby the prices of the several potential humectants have become comparable (Table 4).

Table 3: Some Traditional Oriental IM Foods (after Wang and Sun, 1981)

Foods	Major Component	Moisture Content (%)	Water Activity
Gelatinized rice dough	Starch	26.6	0.88
Steamed starch powder	Starch	23.5	0.89
Baked starch cake	Starch	22.3	0.79-0.81
Candied sweet potato	Starch	16.8	0.72-0.73
Sweetened mung-bean starch	Starch	10.2	0.66-0.74
Candied pineapple	Fruit	16.1	0.72-0.73
Candied papaya cubes	Fruit	20.7	0.69-0.71
Candied mango slices	Fruit	20.5	0.60-0.62
Ho-Hsui candies	Agar-agar egg-white	37.0-19.4	0.70-0.85

f) Soft-Frozen, No Thaw Desserts

Another commercial development was the promotion of IM frozen desserts by Rich Products Corp. of Buffalo, New York. These desserts are simply frozen solute-rich formulations which do not contain enough water to result in crystal formation at the usual conditions of freezer storage (Anon, 1981).

THE POTENTIAL OF IMF IN FUTURE FOOD TECHNOLOGIES

Because of their convenience, energy savings and low capital investment needs, osmotic dehydration and IMF technology are likely to be widely applied.

Table 4: Prices of Osmotic Agents as of June 1982
(Based on Chemical Marketing Reporter, June 21, 1982)

Agent	Price U.S. $/Kg of solids	Price U.S. $/gram mole of ion or molecule in solution
Glycerol	1.80	0.16
Glucose	1.50	0.27
Corn syrup	0.46	-
Sorbitol	1.57	0.28
Sodium chloride	0.10	0.006
Sucrose	0.92	0.31

The commercial potential for IMF is large. These foods offer a combination of shelf stability, convenience, ease of nutrient content adjustment, and safety. The rapid penetration of the pet food market by IMF clearly proves the practical potential for these foods. The convenience, safety, acceptability, and nutrition provided by IM pet foods have been major factors in their success. Food intended for human consumption must be highly acceptable organoleptically as well as being in compliance with prevailing fads and being judged safe for a "long-term" consumption. To date, however, modern IMF has been slow to become a significant factor in human food systems primarily because of poor organoleptic acceptability. Specialized food systems such as space feeding and clinical nutrition do however demonstrate that the drawbacks in IMF for human consumption can be overcome by intensive application of technology. These drawbacks, nevertheless, still largely remain with regard to the mass marketing of IMF for human consumption.

RESEARCH AND DEVELOPMENT NEEDS

Areas in which research and development are required are:

1) Development of new formulations containing humectants or high organoleptic acceptability.
2) Development of new antimicrobial agents suitable for inclusion in IMF.
3) Control of storage-induced organoleptic changes, in particular those due to nonenzymatic browning.
4) Development of economic processes for large scale production of IMF.
5) Development of procedures to prevent microbial spoilage in situations in which local and transient changes in water activity produce potential for microbial growth.

Progress is being made in all of the above areas. As an example, we cite our recent work on prevention of microbial growth during transient changes in surface water activity. Microbial stability of intermediate moisture foods is affected by unsteady-state environmental conditions. This possibility has been mostly ignored by researchers working in the development of IMF's, who have tested them under conditions of constant temperature and relative humidity.

As shown in Table 5, many situations commonly encountered during production, storage, distribution and marketing have one consequence in common: the surface water activity deviates from the safe design value. This effect is attributable to condensed water at the surface due to fluctuations in headspace temperature. This presumably creates a high localized surface water activity unconnected with the food bulk composition. Thus, when looking at the overall problem of safety we have to consider the delicate balance between rate and/or extent of condensation, rate of reabsorption, and rate of microbial growth.

Table 5: Examples of Unsteady-State Environments

1. Temperature changes during the fabrication of IMF's:
   - Cooling of a packaged product.
   - Packaging of a warm product before its temperature reaches storage temperature.
2. Temperature changes during distribution:
   - Temperature differences between transporting vehicle and warehouse.
3. Temperature changes during storage:
   - Product temperature variations due to large day and night temperature differences.
   - Product temperature variations caused by a heating system turned on during warehouse operation and off during other times.
4. Temperature changes during marketing:
   - Temperature differences within a product caused by illumination.

The specific water activity and time length of these occurrences will be a function of environmental conditions as well as of food properties. It is necessary to emphasize that although each occurrence may be brief the total effect could be more than cumulative since: a) toxins may be produced; b) deterioration may occur; c) microorganisms may survive the period of lower safe a_w; and/or d) preservatives may be ineffective at higher cell numbers. There are other reasons why the stability of food surfaces should be analyzed separately. For example, the handling of a finished IM product (which may have received a heat treatment) may result in higher surface microbial loads than that of the bulk.

The solution to the surface microbial stability problem currently under investigation involves the use of a surface layer with an optimized capacity to resist microbial growth. Thus, if growth-allowing conditions appear on the surface microenvironment the surface properties will inhibit, or at least reduce, growth rate to such an extent that water diffusion will occur before a high cell concentration is achieved. There are currently two approaches being studied: 1) controlled diffusion of preservative(s) and, 2) controlled localized surface microenvironment pH.

REFERENCES

Anon (1981). Soft-frozen, no-thaw desserts. Food Eng. June, 45.

Arya, S.S., K. Vidyasagar and D.B. Parihar (1977). Preservation of chapaties. Lebensm.-Wiss.-u. Technol., 10, 208.

Bongirvar, D.R. and A. Sreeivasan (1977). Studies on osmotic dehydration of banana. J. Food Sci. Technol. (India) 14(3), 104.

Chirife, J., M.S. Vigo, M.H. Bertoni, P. Cattaneo, O.C. Scorza, S.A. Sasetru (1980). Retention of available lysine after long term storage of intermediate moisture beef formulated with different humectants. Lebensm.-Wiss.-u. Technol. 13, 44.

Chirife, J., O.C. Scorza, M.S. Vigo, M.H. Bertoni and P. Cattaneo (1979) Preliminary studies on the storage stability of intermediate moisture beef formulated with various water binding agents. J. Food Technol. 14, 421.

Contreras, J.E. and T.G. Smyrl (1981). An evaluation of osmotic concentration of apple rings using corn syrup solids solutions. Can. Inst. Food Sci. Technol. J., 14(4), 310.

Christian, J.H.B. (1981). Specific solute effects on microbial/water relations. In: Rockland and Stewart, 1981.

Davies, R., G.G. Birch and K.J. Parker (eds.) (1976). Intermediate Moisture Foods. Appl. Sci. Publ., London.

Del Valle, F.R., M. Padilla, A. Ruz and R. Rodriguez (1973). Pilot plant production of and large scale acceptance trials with quick-salted fish cakes. J. Food Sci., 38, 580.

Duckworth, R.B. (1981). Solute mobility in relation to water content and water activity. In: Rockland and Stewart, 1981.

Dymsza, H.A. and G. Silverman (1979). Improving the acceptability of intermediate moisture fish. Food Technol., 33(10), 52.

Erickson, L.E. (1982). Recent developments in intermediate moisture foods. J. Food Protec. 45(5), 484.

Flink, J. (1977). Intermediate moisture food products in the American marketplace. J. Food Process. Preserv. 1, 324.

Gal, S. (1967). Die Methodik der Wasserdampf-Sorptionsmessungen. Springer Verlag. Berlin.

Hawkes, J. and J. Flink (1978). Osmotic concentration of fruit slices prior to freeze dehydration. J. Food Process. Preserv. 2, 265.

Heidelbaugh, N.D. and M. Karel (1975). Intermediate moisture food technology. In: Freeze-Drying and Advanced Food Technology. S.A. Goldblith, L. Rey and W.W. Rothmayr (eds.) Academic Press, London, 619-641.

Hsu, W.H., J.C. Deng and J.A. Cornell (1980). Effect of salting time, dehydration temperature and dehydration time on quality of intermediate moisture millet roe. J. Food Sci. 45, 102.

Johnston, M.R. (1981). Private communication. (See also Code of Federal Regulations, Title 21, "Low-acid canned foods, acidified foods. Good manufacturing practice".)

Karel, M. (1973). Recent research and development in the field of low moisture and intermediate moisture foods. CRC Crit. Revs. Food Technol. 3, 329-373.

Karel, M. (1975a). Free radicals in low moisture systems. In: Water Relations of Foods, R. Duckworth (ed.) Academic Press New York, pp. 435-453.

Karel, M. (1975b). Physico-chemical modification of the state of water in foods. In: Water Relations of Foods, R. Duckworth, (ed.) Academic Press, London, pp. 639-657.

Karel, M. (1976). Technology and application of new intermediate moisture foods. In: Intermediate Moisture Foods. R. Davies, C.G. Birch and K.J. Parkers (eds.) Appl. Sci. Publ., London, pp. 2-28.

Karel, M. (1980). Lipid oxidation, secondary reactions, and water activity of foods. In: M. Simic and M. Karel (eds.). Autoxidation in Food and Biological Systems. Plenum Press, New York.

Karel, M. (1981). The significance of moisture to food quality. C.C. Tsen and C.Y. Li (eds.). Recent Advances in Food Science and Technology. Vol. II, 382-392. Hua Shiang Yuan Publ. Col., Taipei, Taiwan.

Karel, M. and S. Yong (1981). Autoxidation-initiated reaction in foods. In: L.B. Rockland and G.F. Stewart (eds.). Water Activity: Influences on Food Quality. Academic Press, New York.

Ledward, D.A., S.K. Lynn and J.R. Mitchell (1981). Textural changes in intermediate moisture meat during storage at 38°C. J. Texture Studies 12, 173.

Leistner, L., W. Rodel and K. Krispien (1981). Microbiology of meat and meat products in high and intermediate moisture range. In: Rockland and Stewart, 1981.

Lerici, C.R., M. Pepe and G. Pinnavaia (1977). La Disidratazione della frutta mediante osmosi diretta. Ind. Conserve 52, 125.

Levi, A., J.R. Ramirez-Martinez and H. Padua (1980). Influence of heat and sulfur dioxide treatments on some quality characteristics of IM bananas. J. Food Technol. 15, 557.

Luh, N.R. and J.K. Palmer (1975). Process for dehydrating bananas and banana product. U.S. Patent 3,879,568.

Mossel, D.A.A. (1976). Microbiological specifications for intermediate moisture foods with special reference to methodology used for assessment of compliance. In: Davies, Birch and Parker, 1976.

Motoki, M., J.A. Torres and M. Karel (1982). Development and stability of intermediate moisture cheese analogs from isolated soybean proteins. J. Food Process. Preserv. 6(1), 41.

Moy, J.H., N.B.H. Lau and A.M. Dollar (1978). Effects of sucrose and acids on osmovac dehydration of tropical fruits. J. Food Process. Preserv. 2, 131.

Pelaez, J. and M. Karel (1980). Development and stability of intermediate moisture tortillas. J. Food Process. Preserv. 4, 51-65.

Scott, W.J. (1957). Water relations of food spoilage microorganisms. Ch. in Vol. III of Advances in Food Research. Academic Press, New York.

Shanbhag, S.P. and A.S. Szczesniak (1981). Fruit and cereal products and process therefor. U.S. Patent 4, 256, 772.

Silver, M. and M. Karel (1981). The behaviour of invertase in model systems at low moisture contents. J. Food Biochem. 5, 283-311.

Simatos, D., M. LeMeste, D. Petroff and B. Halphen (1981). Use of electron spin resonance for the study of solute mobility in relation to moisture content in model food systems. In: Rockland and Stewart, 1981.

Stoloff, L. (1978). Calibration of water activity measuring instruments and devices: Collaborative study. J. Assoc. Off. Anal. Chem. 61(5), 1166.

Troller, J.A. and J.H.B. Christian (1978). Water Activity and Food. Academic Press, New York.

Vigo, M.S., J. Chirife, O.C. Scorza, P. Cattaneo, M.H. Bertoni and P. Sarrailh (1979). Estudios sobre alimentos tradicionales de humedad intermedia elaborados en al Argentina. Determinacion de la actividad acuosa, pH, humedad y solidos solubles. Rev. Agroquim. Tecnol. Aliment. 21(1), 91.

Villota, R., I. Saguy and M. Karel (1980a). An equation correlating shelf life of dehydrated vegetables with storage conditions. J. Food Sci. 45, 398-399, 401.

Villota, R., I. Saguy and M. Karel (1980b). Storage stability of dehydrated foods. Evaluation of literature data. J. Food Quality 3, 123-212.

Wang, H.H. and C.T. Sun (1981). A boundary condition for intermediate moisture foods. Recent advances in Food Science and Technology, Vol. II, Hua Shiang Yuan Publ. Co., Taipei, Taiwan.

NEW DEVELOPMENTS OF COOPERATION AMONG THE CHEMICAL INDUSTRIES TO IMPROVE CONTROL OF THE FOOD SUPPLY

Pierre Roessler

Member of the Board, Union Des Industries Chimiques, Paris

ABSTRACT

The increased future requirements from the chemical industry to meet increasing food production needs are emphasized. Such requirements include the need for dissemination of products, application methods, and specific knowledge suitable to the recipient area or country. This calls for innovative broad cooperation among disciplines and especially among various industrial groups. Some of the recent successful cooperation methods, both on a national and on an international scale are reviewed.

KEYWORDS: Cooperative activities, pesticides, plastics, technical institutes.

INTRODUCTION

Today, 2 billion people are employed in agriculture working 4.5 billion hectares, in extremely diverse soils, climates and socio-cultural environments. Between now and the year 2000 the number of people employed in agriculture will increase by a further 1/2 billion. Also today, chemistry, often associated with other related sciences, can offer an impressive range of chemical products and techniques, all of which contribute to improved management of quantities and qualities, from the seed stage through to the finished consumable product.

At the head of the list of such chemical products, are the two mainstays of productivity and crop protection: fertilizers, and plant and pest control products. In addition, there are less widely employed techniques, the advantages of which are growing: (1) soil conditioners, such as humidity retainers, (2) nitrification inhibitors to limit fertilizer breakdown, (3) synthetic chemicals and agents affecting reproduction of predatory insects, (4) plastics used in water management and soil protection. In addition chemicals are used for developments in the food industry, providing the even more essential functions of protection and preservation such as: (1) hydrolysis of milk lactose, (2) valorisation of the biomass by the use of enzymes, (3) protein enrichment, (4) high performance plastic sealants for protecting meat, (4) additives for preservation of foodstuffs.

The combining of the science of chemistry and the chemical industry with one of humanity's most essential and most complex activities, the production of foodstuffs, cannot be reduced to mere technical solutions or simple trade relations. Used in most of the links in the immense agricultural foodstuff chain, chemistry interacts with social-economic systems. It must also associate with other sciences: physics, genetics, mechanics, hydraulics. The application of chemistry in this vast ensemble raises a key question: how can the contribution provided by chemicals be effectively disseminated and assimilated?

The world food problem itself engenders the above question. According to the FAO, 80% of the growth required for food production will come from an increase in productivity, and therefore for a very large part from intensification of the use of chemicals. Between now and the year 2000, the production of fertilizers must be increased five-fold, and that of plant and pest control products by 2.6, and irrigated surface areas by 1.5. Crop losses due to predators and weeds are estimated to be one third of the world agricultural production, representing $75 billion (Dr. A. Robertson, Imperial Chemical Industries, 1980). Losses solely due to rodents and insects amount to 80 Mt for cereals, representing approximately the cereal deficit of the developing countries.

While chemical products and techniques are essential, there is also a considerable capacity for innovation and development in the chemical industry. However troubling barriers oppose the dissemination of techniques throughout the agricultural and foodstuff world: the low purchasing power of farmers in large areas of the world, the insufficiency of infrastructures of all types, and the lack of technical know-how.

We will endeavour to show that the procedures and methods of cooperation are as essential as the productions and techniques themselves, and that barriers may be overcome by a better organisation of technical applications of all types in the food industry and in agriculture.

CHEMICAL PRODUCTS: NEEDS AND INNOVATIONS

Pesticides and Cooperation in Adaptive Innovations

Taking pesticides as an example of a chemical product, there are numerous difficulties in the way of their general use: fragmentation of the market (how to reach 18 million farmers in Nigeria, each one of whom produces an average of 500 kilos of cereal per year?); insufficiency of purchasing power (for the same overland distance, the cost of transporting a pesticide in West Africa is eight times higher than that for a European country); frequent unsuitability of the storage equipment. On the whole, the cost of a pesticide in relation to the value of production is five times higher for an African producer than it is for a European producer. But pesticides could reduce post-harvest losses, currently estimated at 20% in developing countries, to 5%, and it is precisely the developing countries whose need is most vital in this type of technical assistance. First of all, the problem should be approached in a coordinated fashion, and from all aspects, going far beyond the simple application of a chemical technique.

Using pesticides in a diluted form faces the problem of scarcity of water in countries such as Senegal. Up to 1000 kg of water are required to dilute pesticides for the treatment of 1 hectare in our country. The search for a technological solution to reduce the water and work requirements was carried out by three companies working together: a pesticides manufacturer (Rhone-Poulenc Agrochimie), an agricultural equipment research centre (CEEMMAT, member of the GERDAT), and agricultural equipment manufacturers (Berthouc, Technova). This association was set up in 1978 to cooperate with the Senegal Agronomy Research Institute in an experimental programme and then in a general field programme.

This association enables the development of a technique for using pesticides in a more concentrated form (ultra light volume) for which new vaporizer equipment had to be designed. This new technique has since been adapted by the Senegal Institute and applied to ground-nut and cotton crops. This same technique was applied in the Sudan (aircraft insecticide spraying on cotton areas), and in Chad (soil insecticide treatment).

In these different cases, positive action was based on a two-fold cooperation: (1) between pesticide manufacturers, equipment manufacturers and technical centres for the application of a new treatment technique, and (2) between these operators and the rural development and research organisations in the recipient countries for the application of these techniques (experiments, training, servicing). This integration of vocations and cooperation between the possessors and the appliers of the technologies exemplifies the liaison which often appears to consitute the key to success in agricultural production projects.

Enhancing the Value of Foodstuff Products

Cassava is a basic foodstuff in many countries. Correction of its protein deficiency is therefore a major consideration in remedying a common scarcity in these countries. A protein enrichment process has been developed jointly between a technical university (Universite Technique de Compiegne, CNRS), a design company (Adour-Entreprise), and an engineering company (SPEICHIM), which has led to the isolation of a variety of amylolytic yeasts increasing the cassava amine acid content. The fermentation equipment required was then made available in different modules (1 ton/day to 1 ton/hour). For the installation of these units, cooperation is envisaged ranging from the cultivation of cassava producing areas, to ensure a regular supply, up to the manufacture on site of part of the fermentation equipment.

Integrated Approach and Cooperation in Agricultural Plastics

Large expanses of the developing world consist of semi-arid land subject to insufficient and irregular rainfall. Water storage and irrigation techniques which are accessible based on the revenues of these areas are particularly essential; the same applies to drainage techniques to combat soil salinity. In this area, concrete can only be used in exceptional cases.

Chemicals can contribute in several ways. For hillside dams, water storage, and also distribution channels, materials of the "Bidim" type (non-woven polyester) enable the installation of these systems by using a limited aamount of structural work and a larger degree of human resources. For distribution channels, plastic materials enable installations which do not require too much infrastructure, and which in addition encourage the emergence of a small plastics processing industry (extruders can also use the recovered waste from existing plastic materials). Clearly, average sized countries can develop a small plastic extrusion industry much more quickly and much more cheaply than they can develop an iron, steel or cement pipe industry.

A further application contributing to water management is also based on the highly diverse techniques used in agricultural plastics, enabling optimisation of the soil and sub-soil water, protection of the soil against erosion, speeding up of harvesting, and an increase in crop yields. Plastics can be used, drawing maximum benefit from the greenhouse principle, to affect both of the crucial factors - water and heat - and the crop sub-soil. Their use requires several techniques and several industries: from the chemist who develops plastic films with a well-defined yield strength, and behaviour under the sun's rays, to the contribution of hydraulics, the seed industry, and greenhouse and ventilation equipment manufacturers. Today, it is estimated that plastic films ʻor agriculture account for between 5 and 15% of the total production of plastic materials, depending on the country. Their application in semi-arid areas where there is a combined lack of water and wind erosion is obviously of interest. Establishing a crop-project under plastic needs a complete operational sequencing, from the draft study to the training of the farmers and, sometimes, the setting up of an extrusion industry to minimise costs. This requires using a system concept, so that the various associated techniques have the optimum effect: choice of seed, design of the water system, design of the greenhouses and other plant protection devices. These requirements clearly call for a coordinated effort so that the introduction of the technique into the rural environment concerned is achieved profitably.

The UNIDO, within the scope of its development aid objectives, has completed pioneer work promoting the applications of plastics in agriculture. In 1975-76, a group on assignment in Sahel spent several months presenting the use of agricultural plastics materials, giving demonstrations of the erection of tunnels and shade nets, and through films showing the interaction of agricultural plastics and the environment: fertilizers, pesticides, etc. Continuing its pilot role, UNIDO has since launched a programme establishing regional centres in five countries: Upper Volta, Cyprus, Jordan, Mexico and Egypt. The objective of these centres is to achieve the general use of agricultural plastics techniques, using local plastics processing factories, and, obviously, to carry out a qualitative assessment of plastic material requirements.

Plastic material chemical plants are contributing towards these projects by sending specialists and providing materials for experimental purposes. They also contribute to projects in various other countries which have begun the training and project design work which is basic to developing the use of plastics in their agriculture. It can thus be seen that cooperation can be achieved at various levels: agronomics, personnel training, agricultural production and industrial cooperation for the production of films, plastic tubes and greenhouses.

The Role of Technical Institutes in Agricultural Support Operations

The combination of several techniques and industries for the management of agricultural production should result in a more or less close cooperation between the various factors, based on the type of actual situations and the timing required for continuous action. This point can be illustrated from the efforts in France to improve adaptation of technical solutions to agricultural problems and needs in tropical countries.

In 1970, the eight specialist agricultural research and promotional institutes joined together to form GERDAT* which was thus able to henceforth offer a significant study and assistance capacity. Today this consists of 1500 engineers and technicians, of which 600 are actually in the field outside France. By necessity, the activities cover a whole range of agricultural requirements: research studies, assessments, experiments, training, promotion, documentation. The GERDAT works in close cooperation with two other institutes. One is the National Agronomy Research Institute, which employs 700 agents, and an important part of the activity of which is devoted to international cooperation. The other is the Overseas Scientific and Technical Research Office (ORSTOM) whose function is basically international cooperation, and which today operates 60 facilities, centres or missions, employing 2700 agents. These research organisations play a very important role in the development and dissemination of techniques, many of which are based on chemicals, in extremely diverse agricultural production, physical and social-economic contexts.

* GERDAT: (Groupement d'Etudes et de Recherches pour le Developpement de l'Agronomie Tropicale)
Centre Technique Forestier Tropical
Institut d'Elevage et de Medecine Veterinaire des pays Tropicaux
Institut Francais de Recherches Fruitieres Outre-Mer
Institute Francais du Cafe, du Cacao, et autres Plantes Stimulantes
Institut de Recherches Agronomiques Tropicales et des Cultures Vivieres
Institut de Recherches sur le Caoutchouc en Afrique
Institut de Recherches du Coton et des Textiles Exotiques
Institut de Recherches pour les Huiles et Oleagineux
Centre d'Etudes et d'Experimentation du Machinisme Agricole et Tropical

The need to marry disciplines and to group together mankind's skills can also be seen at an international level, where close cooperation is developed between organisations in different countries. In 1970, an information group (GASGA: Group for Assistance on Systems relating to grain after harvest, associating the Institutes of Australia, Canada, France, Holland, United States, Great Britain and Italy) was formed among eight organisations belonging to eight different countries to exchange information and, in many cases to develop together, post-harvest protection techniques. Technical seminars are organised on such themes as the proper use of pesticides, loss evaluation and measurement techniques, foodstuff processing techniques, storage systems and their implementation. Naturally, this inter-Institute cooperation results in mutual cooperation directed towards the tropical regions, in supplying expertise, technician training and even so far as conducting field projects.

CONCLUSIONS

National and international cooperative operations are required to meet the complexity of agricultural situations for the development of techniques requiring the marrying of complementary disciplines, and for their adaptation to the needs of rural societies. The desired objectives include, first of all, overcoming the technique dissemination barriers as much as possible, and, in the long term improving conditions. The challenge thrown down to science and modern industry is to take what is designed in a high-cost environment (technology, raw materials, wages) and to apply it in contexts where income is often under $1000 as compared to $5000-$10,000. Such high cost technologies must therefore be constantly improved to lower costs, to simplify methods, and to ease application limitations.

The case of agricultural plastics illustrates how the cost reduction objective can be achieved. Through its application centre programme, the objective of UNIDO is to get around the poverty trap problem and to make agricultural plastics available to the maximum number of farmers. Experiments are being conducted to find the most economic ways in which plastic technology can be used, by lightening the materials, recycling recovered plastics etc. All such actual improvements require research responsive to the recipient environment. The progress attained by UNIDO enables it to attract to the projects promoted by the agency, major financial aid from the World Bank, the FAO and the UNDP.

A more general objective is to contribute to the necessary autonomy of the development of the farmer in his agricultural activity and in his society. Neither international aid, obviously necessary, nor food trading, can supplant the objective of increased self-sufficiency.

Chemical research and the chemical industry have the ability to participate in a wide range of cooperative ventures whose very objectives are effectiveness in meeting the problems of agriculture and food production.

OVERVIEW OF ASSESSMENT AND CONTROL

R.A. Edwards

School of Food Technology
The University of New South Wales
P.O. Box 1, Kensington, New South Wales, Australia

ABSTRACT

The importance of chemistry in relationship to the assessment and control of the food supply is reviewed historically. The level of assessment and control of quality will vary significantly in different countries reflecting nutritional policy, food legislation and trade. Nonetheless the principles of quality control are common, whether in the area of new materials, of postharvest processing, or of distribution and marketing. Pressing needs in these areas are outlined, including the development of analytical procedures and instruments. Priorities are suggested with a strong emphasis on postharvest protection.

KEYWORDS: food legislation, food quality, quality control.

INTRODUCTION

The genesis of chemistry in man's history can no doubt be related to the ancient dream of the Alchemist to convert base metals to gold - a dream that despite its potent economic incentive remains unfulfilled. Nonetheless, the pervasive role of chemistry may be seen as an integral and component part of all of civilized man's achievements that today we take for granted. Indeed, in no aspect of our lives does chemistry, its understanding, development and application not occupy a central role. In no other area is the role of chemistry more relevant to human needs than in its application to the pressing food problems of the world.

The earliest application of chemistry to the food supply was in the Middle Ages, and, at least in Western society, is to be seen in the control of adulterants to foods of that time that represented a hazard to health or that constituted deception and fraud in the marketplace. Early references to these initial forms of food assessment and control were directed toward the chalk that was used to whiten bread made from inferior and discoloured flour, and to the inorganic pigments that were used to falsify the colour of wines and other beverages, often with disastrous consequences. It was the use of chemistry to detect such deceptive and unsafe practices that gave rise to food assessment and control by legislation, a development that was the forerunner of food acts and regulations at a national level, which is today evident in those international endeavours that make up the work of the Codex Alimentarius Commission and its related agencies. Such developments, which seek to facilitate food trade, continue to embrace those two basic tenets of food control, namely the safety of food and the avoidance of consumer deception in the marketplace. From these early beginnings, food chemistry was used in the long and laborious endeavours to define foods in terms of those chemical components that are a unique and definitive description of their nature and diversity.

It was not until the 17th century, as laboratory procedures improved and the development of the scientific method became a reality, that chemistry began to emerge as a true science. Its emergence reflected a general but growing interest in the composition of the entire world, and it was natural that food chemistry and analysis would have their beginnings at that time. Initial studies sought to separate foods into their component parts, with early attention being given to the pigments that provided foods with their definitive colours, but it was not until the 18th century that such simple components were isolated and subjected to further study. By the 19th century foods were recognized as being composed of essentially four component parts, namely water, carbohydrates, protein and fats, with the remainder being estimated by difference as "other substances" that had to wait for many more years for a more definitive classification. Nevertheless, it was this simple understanding of the basic constituents of food, together with a growing development in methods of chemical analysis, that was to result in a formulation of the foundations of the science of nutrition. It is somewhat surprising, in fact, to note that the earliest rudimentary account of the nutritive value of foods (Percy and Vacqueline, 1818) did not appear until about 164 years ago, and was restricted to analyses for moisture, dried matter, extractive matter and ash in a range of French foods. It was not until about 100 years ago that the first systematic sets of food composition tables were published in Germany (Konig, 1878) and in the United States (Atwater and Woods, 1896).

The 19th century also saw a continued development in organic chemistry, analytical chemistry and physical chemistry, all of which were essential to the growth of food chemistry and the development of methods of food analysis. The related sciences of biology and microbiology also saw a rapid development at this time, and these sciences, too, were required before further progress could be made in food chemistry and food preservation. This was made possible through an appreciation of the complex nature of cells - the common patterns that cells may have through their common heritage, and the different patterns that distinguish between species, families and even varieties.

As modest, in hindsight, as those early developments were, they gave rise in the early 1900's to the promulgation in many Western countries of pure food acts and regulations that greatly influenced the development and direction of food chemistry in subsequent years. As in earlier times, these early food regulations focussed upon consumer protection in the form of safety-in-use and avoidance of deception, but they were to develop substantially further in response to the advances made in the application of industrial processes to foods. It was during the early 1900's that rapid developments in industrial food processing occurred, being made possible by the earlier invention of mechanical refrigeration and the subsequent application of cold storage, and later freezing, as a preservation technique of major significance. In the same era was developed Mege - Mourier's concepts of margarine manufacture as a basis for the edible oil industry, and the traditional methods of the cereal miller gradually gave way to the modern power-driven, roller-milling systems made possible by the Industrial Revolution. Such developments, together with the development of the open-top, tinplate canister at the turn of the century, set the stage for the application of the scientific method to food processing and preservation to cater to the needs of a rapidly growing industrial population who provided the markets for mass produced foods in Western societies.

It remained only for the phenomenal growth in our knowledge of biochemistry within the last 30 years to complement those basic and developing sciences of chemistry, biology, microbiology and nutrition to complete our current understanding of metabolic pathways in plants, animals and man, in genetics and related sciences, to provide us with our current imperfect understanding of foods and the control of food supplies.

Just as the genesis of the earliest food laws stimulated the development of food chemistry in the 17th Century, so the gains made in the chemical understanding of foods and their relation to human health, began to stimulate changes in the scope of food law.

FOOD LEGISLATION

Until 80 years ago, "pure food" meant food that was free from visible and perhaps active wild life such as maggots and weevils, their body parts or their droppings, or other gross adulterants. Early in this century, the emphasis moved to the invisible components, to the microorganisms causing food spoilage or sickness, and to the production of food that contained its original content of low molecular weight components such as the vitamins and minerals. In more recent times, potential and actual food law has become concerned with contaminants present in low to very low and perhaps even to molecular concentrations in foods; the pesticides, herbicides, rodenticides and disinfectants that are currently essential to large scale agriculture and mass food processing techniques but which have no place even in trace amounts of foods. In a similar category are the plastic monomers that migrate from packages to foods, the heavy metal of polluted waterways that contaminate marine foods and a variety of other unintentional contaminants whose control poses a constant problem and a challenge. It can be argued that it is because of the force and vigour of our application of science to the food supply, that food law and the associated control measures of food processors have been forced to move from the gross and visible contaminants and components, to the molecular level at least for those ingredients that impact upon health in either a positive or negative manner. The control of the food supply has become considerably more complex in achievement because of both mass handling techniques and our modern ability to see and quantify materials that were once unknown to us.

This reference to food legislation, and its changing nature in association with our growing knowledge of food composition, is given because food legislation is the ultimate form of quality control exercised by governments. As a legal requirement, it should be an expression of the minimum compositional standard that is both sufficient to protect public health, and attainable by the food production skills of farmers and food processors of that region. Accordingly, it can be expected that the level of food quality control and assessment that is exercised by the food producers of one country may be different in detail to that of another country where the level of industrial development, the food problems, the food priorities and economic factors, may be quite different.

FOOD QUALITY

Before considering the control of food quality, it is necessary to appreciate that the term quality, as applied to foods, is a portmanteau term that can mean many different things to many different people even within any one country.

To the farmer, quality usually means the cultivar or variety of food plant or breed of animal that provides the highest yield of raw product, and that puts the most money into the farmer's pocket.

To the food processor, quality usually means the raw material that gives the highest yield of processed article, or that contains the highest amount of a key ingredient, or perhaps that will absorb the most water, or that generates a market interest and represents a profitable return.

To the nutritionist, quality means nutritive value and acceptability within the nutritional priority needs and social fabric of the society in which the nutritionist serves. Thus, there may be widely different priority needs as, for example, in an affluent and obese society where food is abundant and the choice wide, compared to a deprived society where foods, energy and essential nutrients are in undersupply.

To the public health authority, quality above all means safety, together with the minimum compositional standard identified by law, and food presented in a manner that avoids deception or misrepresentation in the marketplace.

To the consumer, quality usually means foods of the highest aesthetic appeal in terms of palatability, in appearance, colour, taste, odour, flavour and texture; and that is in accordance with established dietary habits, food prejudices, taboos and other social constraints - and all at the lowest possible cost.

Thus, in any country, there are different and often conflicting needs and quality attributes of foods that must be assigned a relative importance that is reflected in that country's nutritional policy, and in its food legislation and trade. The very great diversity in the condition and the needs among and within the world's population groups makes it necessary to relate food quality to the needs of such specific groups.

QUALITY CONTROL

Viewed as above, the use of quality control procedures can be directed in widely different ways in different societies. Despite such differences, the principles of quality control are the same and equally applicable in any situation. Their aims are to achieve as good and as consistent a standard in the finished product as is compatible with the market for which the product is designed, and for the price at which it will sell.

The scope of quality control embraces the control of raw material quality, involving their selection in terms of quality attributes that fit them for the intended purpose. This selection can be, and usually is, the most crucial aspect of any quality control program, for it is rarely the case that any gross quality deficiency can be corrected economically in subsequent stages of use. Too frequently in the past, the selection of plant crops in terms of variety, cultivar and genetic make-up has been focussed only upon the quality required by the farmer - namely a high yield of harvested material. However, if the high yield also results in poor postharvest characteristics, or in a poor appeal to consumers, then it will have little merit for it will neither transport nor sell. It is important for us to recognize and cater to the different meanings of quality to different groups in any society, and to balance yield characteristics with other quality attributes of importance to food processors, to nutritionists, to public health authorities, and above all, to consumers.

Other raw materials beside the food crop that require control include the water that may be required in further processing, the containers or packaging materials, and all other ingredients such as sugar, salt, spices and associated materials that may be component parts of the finished product. The principles of control of such materials are just as relevant to dendeng, the sugared and spiced meat delicacy of Indonesia, as they are to the canned caviar of the Northern Hemisphere or the chilled beef of Australia. The products and technology of such materials may all be quite different, but the principles of quality control of raw materials are identical.

Quality control procedures are also required for the processes that are applied postharvest. This area of quality control is, of necessity, dependent on the nature of the product and the process applied, but in most cases is concentrated on the control of time and temperature, either high or low, on sanitation and the avoidance of contaminants, and on the choice of conditions of unit processes that may include salting, smoking, drying, fermenting, mixing, milling, size reduction and many others that represent a process of modification or preservation of the raw material. There is the need to control all of these unit processes to their optimum level to retain or enhance the nutritional and aesthetic qualities inherent in the raw product during its projected market life. With such a wide diversity of possible postharvest processes that can be applied, it is the principles in the controlling of the process that must be considered. These include a listing of the sequence of steps in the process by means of a flow diagram, listing the deviations possible in each step, identifying those deviations that affect quality, setting limits to such deviations, and establishing controls to achieve such limits.

Finally, quality control needs to be exercised over the handling and storage operations involved in the distribution, marketing and retail sale of the processed product. This control is entirely dependent upon the level of storage, transport and handling facilities available in the region where the product is marketed. In particular, the distribution and associated transport and storage of perishable foods in many developing countries is not well advanced, and can be a major limitation to effective distribution. Because of this limitation, there are many processes of Western countries, e.g. chilling and freezing, that are often quite inappropriate processes for developing countries and, moreover, represent relatively expensive distribution systems that do not meet the requirement of low cost food supplies. In this way, the available distribution, marketing and storage facilities can be, and should be, the determinants of the type of postharvest process that is applied to the raw material.

Too frequently, the solutions to food problems of developing countries have been seen as requiring an application of the technological processes that have been found appropriate in Western or developed countries. While there are, no doubt, particular and specific situations where such approaches may be justified, many foods require a different form of technology - commonly termed an appropriate technology - that recognizes a different food usage pattern, different distribution and storage facilities based upon local and regional rather than national patterns, and a different use of preservation techniques that is within the economic means of consumers. There is, indeed, an enormous range of unique and traditional foods, many based upon drying, salting and fermentation processes, that represent a sophisticated but nonetheless appropriate technology that is the result of centuries of development, and that are in keeping with the aesthetic, social and economic means of the people of such regions. Very few of these traditional foods and processes have been the object of scrutiny by the scientific method, a scrutiny that should be designed to extend the storage life, reduce the level of postharvest losses, decrease the costs of the food and, if appropriate, to enhance its nutritional status.

CHEMICAL ANALYSIS

There is the need to exercise quality control in the three broad areas of food production as outlined above, and this may be frequently achieved most rapidly and economically by some form of analytical measurement made upon the food itself. It may be made in terms of water content, on fat, carbohydrate or protein content, or on salt, sugars and spice levels, or it may be made in terms of a composite measure of water activity or total plate count in microbiological and related terms. In products intended for specific use it may be focussed upon nutrient balance and specific nutrients or ingredients that indicate intolerance or essential quality factors.

In all of these chemical indices of quality, the need is for analytical procedures and equipment that are simple, economical, rapid, reliable, preferably of low initial and operating costs, and with a continuous availability of spares, replacement parts and accessories. Such qualities are not features found in many of the sophisticated analytical instruments developed in recent years. Indeed, many such instruments are utterly dependent on an extensive and complex back-up service. In the absence of such, their high performance life is very brief indeed. For this reason alone, such sophisticated instruments do not serve well the many control or research laboratories of a more humble but nonetheless important nature, that are to be found in many developing countries. There is, therefore, a pressing need for a new generation of analytical instruments, no doubt based upon our current advanced procedures coupled with the revolutionary silicon chip and computer technologies, to provide the low cost and the reliability in operation that is now required. The end of the decade will undoubtedly see the development of simplified, high-speed and automated analytical procedures and instruments, not only in food analysis, but also coupled to sampling and process control procedures. Such developments will also further revise our current understandings of the control of food supplies in developing countries.

PRIORITIES

The future priorities for the application of quality control and assessment to food supplies must be set both in the areas of food production and postharvest protection.

In terms of increased food production, we must look to chemistry, biochemistry, plant physiology and the broad area of agricultural science to increase crop productivity in many ways. In terms of plant crops, we can expect that advances in somatic and molecular genetics, currently in their infancy, will flow through to an increased tolerance to environmental stress; to improved protection from disease, insects and weeds; and to increased levels of desired nutrients and reduced levels of toxic ingredients, without a lowering of the yields that we have currently achieved.

In respect to animal crops, we can expect advances in genetics and our current systematic selection to improve animal breeds in a related manner. However, it clearly would be a mistake to make use of Western animal selection criteria without recognizing the multiplicity of traditional roles that animals occupy in many developing countries. These include, apart from meat and milk production, a diversity of other significant roles such as a source of power in land preparation, in crop cultivation and in road transport. In many regions, animals play an integral part in the life, the culture and the social customs of the people, being a source of wealth, of pride, of mystique and even of entertainment, which are important factors not to be ignored.

In regard to marine resources, our controls are likely to remain far less direct, and may be more closely related to international agreements on the harvest of the seas, as well as on the ability to mount an advanced technology in harvesting methods.

Even with all such possible gains, a focus upon the production of more food is not enough nor, indeed, is increased production a useful end in itself. The first priority is to focus a great deal more attention on the reduction of postharvest losses caused by insects, by chemical agencies and by microbial spoilage during food production, during food processing and especially during food distribution, storage and marketing. Without exaggeration, we can benefit more, in the next decade, by an appropriate conservation of what we now produce than by the use of energy-expensive methods to produce more, whether the crop should be of a plant, animal or marine source. With postharvest losses of many important crops in developing countries amounting to an estimated 30%, it is clear that increased crop production will compound these losses unless appropriate postharvest protective action is developed.

The role of chemistry and chemical analyses in these respects is to provide rapid, reproducible, low cost procedures for the evaluation of food constituents, for the evaluation of nutrients, differentiating between the chemical and biological availability of nutrients, for toxicants and contaminants, and methods that will allow measurement of small quantities of food components with greater reliability and precision. Increased understanding of the chemistry of the postharvest metabolism of plant materials will lead to simple but effective methods for the control of postharvest wastage in the absence of refrigerated storage. One such area, of current active interest, is the role of Ca^{++} in retarding or increasing senescence in harvested fruit, a phenomeon that has been shown (Wills, 1982) to delay postharvest senescence in a variety of tropical products. Such gains in our chemical understanding of plant metabolism will undoubtedly lead to simple, low-cost procedures that will extend storage life sufficiently to allow distribution and marketing to more distant areas with substantially decreased losses.

The chemical methods and analyses that will be developed to accompany such advances should well be non-destructive yet still meaningful, be part of an integrated and continuous quality control program, and be suitable for large scale use in food production, processing and marketing activities.

The collaboration of chemists with others equally concerned with the food supply - the technologist, the nutritionist, the food processor, the sociologist, the legislator and the politician - will be required if chemistry is to influence the food supply as dramatically in the future as it has undoubtedly done in the past. It is also as important in the future, as in the past, to recognize and cater to the aesthetic, the social, the nutritional and the emotive roles that foods play in all of our lives, for without such consideration our gains through chemistry will not achieve their full potential. Therefore, a high priority must be assigned to the need for the world's scientists to work together, to learn from each other, and to increasingly recognize the problems and the limitations that apply to their solutions, if they are to upgrade the economic and nutritional status of deprived populations in our society.

REFERENCES

Atwater, W.O. and C.W. Woods (1896). The chemical composition of American food materials. US. Off. Exp. Stn., Exp. St. Bull. 28.

Koning, J. (1878). Chemie der menschlichen Nahrungs-und Genussmittel. Berlin: Springer.

Percy, P.F. and N.L. Vacqueline (1818). Sur la qualite nutritive des aliments compares entre eux. Bull. Fac. Med. Paris: 75-91.

Wills, R.B.H. and S.I.H. Tirmazi (1982). Inhibition of ripening of avocados with calcium. Scientia Horticultural 16: 323-30.

EFFECTS ON NUTRITIONAL QUALITY OF FOOD LEGUMES FROM CHEMICAL CHANGES THROUGH PROCESSING AND STORAGE

R. Bressani, J.E. Braham, and L.G. Elias
Institute of Nutrition of Central America and Panama (INCAP)
P.O. Box 1188, Guatemala City, Guatemala, Central America

ABSTRACT

To demonstrate the effect on nutritive value arising from chemical changes induced by storage and processing of food, legumes were chosen because of their importance in human diets. Chemical components, most of them nutrients, found in foods as produced are affected by production, storage, processing and consumption conditions. The understanding of the chemical changes taking place is essential to maximize nutrient utilization. Improper storage will increase the hard-to-cook condition in common beans already initiated during post-harvest processing. Dry or wet processing techniques, if properly carried out, will inactivate anti-physiological substances and increase nutritive value. Improper processing will result in low digestibility of the protein. Germination and fermentation result in higher levels of vitamins and increased availability of nutrients, but germination effects have given contradictory results. Inadequate storage of processed food legumes reduce their nutritional value and thus the nutritional quality of diets. Recommendations for research include the understanding of the hard-to-cook problem, the chemical nature of low protein digestibility, and resistance to insect attack.

KEYWORDS: processing; storage; food legumes; nutritive value.

INTRODUCTION

Storage and processing of foods are technologies which man has always used as a pre-requisite for insuring availability of the food supply, and in many cases as a necessary step before consumption for such reasons as increasing stability, improving flavor, decreasing the possibilities of intoxication, and introducing functionality. All these end points of storage and processing of food influence to variable degrees the nutritional value of the food.

Chemistry plays an important role in the food chain from production to consumption, and the nutrients or chemical compounds synthesized by the plant and deposited in the seed interact with each other during storage and processing to give specific nutritional values. This suggests the need for geneticists and production specialists to work closely with food chemists and technologists as well as with biochemists and nutritionists to insure the maximum nutritive quality and acceptability to the consumer.

This paper selects common beans as the commodity whose nutritional value is affected by a variety of factors during storage and processing. Areas of research needed as a means to increase the availability, use and nutritive value of a food of great nutritional significance to people in developing countries are also pointed out.

POST-HARVEST PROCESSING

Processing For Storage

After harvest, common beans usually contain relatively high amounts of moisture which have to be removed before storage. It is a common practice by small farmers in developing countries to expose the grain to solar radiation. The exposure time is more or less well controlled, otherwise the beans will become hard. This has been confirmed by experimental studies (Table 1). As solar exposure time increased, there was an expected decrease in moisture content, and a decrease in water absorption and cooking time. Changes in other chemical constituents also took place, the significance of which is not well understood. However they probably have significant implications in the hard-to-cook condition often developing in common beans and other food legumes subjected to storage (Sefa-Dedeh et al., 1979; Burr and Kon, 1966, Molina et al., 1976)

Table 1: Physical and Chemical Characteristics of Recently Harvested Common Beans Exposed to Solar Radiation Prior to Storage (Garcia and Bressani, 1982)

Solar radiation cal	Moisture %	Water Absorption %	Cooking quality[1] %	Soils in cooking broth g	Soluble pectins %
0	15.2	33.4	20.0	3.98	5.77
91	13.7	30.4	18.0	3.60	5.46
390	12.0	20.4	8.0	2.00	4.69
1416	10.7	7.4	4.0	1.78	4.43

[1] Broken grains

The Hard-to-cook Phenomenon

Storage of common beans under improper conditions increases the hard-to-cook problem. In the example shown in Figure 1, common beans were stored for 0, 3 and 6 months at 35 °C and 85% relative humidity (Bressani, 1982). At the end of storage, samples were cooked under standard procedures at atmospheric pressure and their hardness measured by the Instrom texturometer. A standard cooking hardness as g-force of 90 was chosen on the basis of current home practices in Guatemala. As seen in Figure 1, it required 150 min. and 170 min. at atmospheric pressure to cook soft the samples stored for 0 months and 3 months respectively, but those stored for 6 months were still uncooked at 210 min. This hard-to-cook condition has significant economic implications both through lack of acceptability and energy cost increases. Furthermore, excessive cooking, which is often applied to soften the grain, will decrease protein quality as shown in Figure 1 for Phaseolus vulgaris. The loss in protein quality is the result of a lower protein availability as well as a loss in lysine, which becomes inactivated through the well known Maillard reaction effect shown by various investigators (Bressani et al., 1963; Almas and Bender, 1980). An interaction between storage and processing can also affect the nutritional value of legume foods. In some instances a high cooking time is needed to obtain a maximum possible nutritive value from the hard-to-cook beans. Table 2 shows that even after 240 min. cooking time digestibility of hard-to-cook beans was significantly lower compared to recently harvested beans. The factors involved are temperature, time, relative humidity, and the moisture content of the seed (Burr and Kon, 1966). The mechanism of the hard-to-cook condition is not known.* The evidence available suggests an increase in the bound protein probably taking place in the seed coat and aleurone layer. However, the cotyledons also lose the capacity to absorb water due to changes in pectins and calcium ions, and very often they develop a gray color, suggestive of carbohydrate protein reactions (Varriano-Marston and Omana, 1979; Sefa-Dedeh et al., 1978).

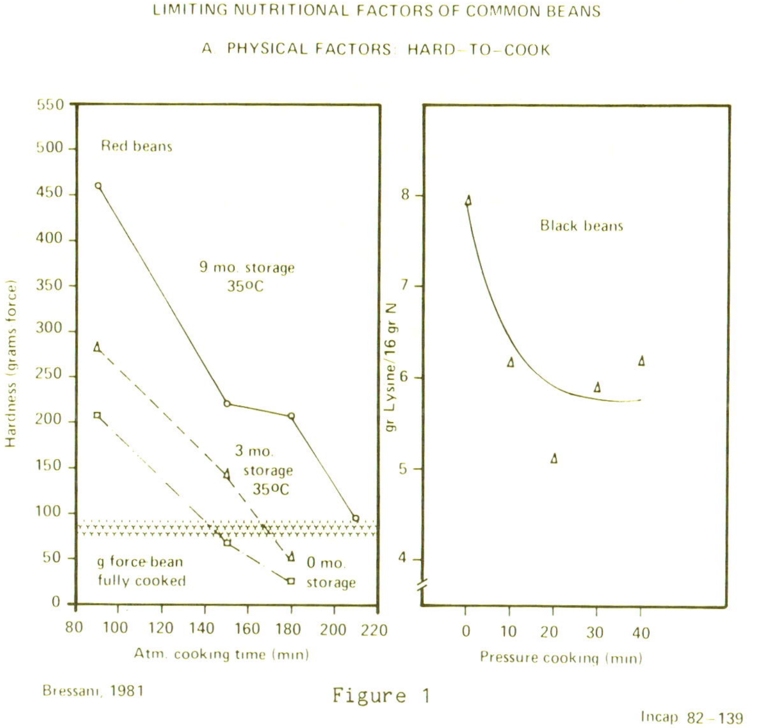

Figure 1

* Editor's Footnote

During discussion following presentation of this paper, it was noted by Prof. D. Boulter that recent work had clarified the mechanism leading to hard-to-cook beans (c.f. Jones, P.M.B. and D. Boulter, Hardbean Symptoms: Their Cause and Relevance to Textural Deteriorations, J. Food Sci. (in press); Jones, P.M.B. and D. Boulter, The Analysis of Development of Hardbean During Storage of Black Beans (Phaseolus vulgaris L), Qualitas Plantarium: Plant Foods for Human Nutrition (in press)).

Table 2: Effect of Storage and Processing on the Digestibility Values of Hard-to-cook Beans

Cooking time min.	Apparent Digestibility
60	48.2
120	55.7
180	58.3
240	57.3
60	69.3
Recently harvested beans	

Table 3: Protein Quality of Chickpea and Pigeon Pea Infested by Insects

Food Legume	PER	% decrease
Chickpea dhal (control)	2.21	-
Chickpea dhal (infested)	1.83	17.2
Pigeon pea (control)	2.04	-
Pigeon pea (infested)	1.66	18.6

Insect Infestation

Studies carried out in India (Parpia, 1972) shown in Table 3 indicate that chickpeas or pigeon peas lost about 18% of their protein quality due to insect infestation. The loss may be due to contamination from uric acid, as well as from increases in fat acidity and microbial contamination, and even losses of grain fractions. Resistance to insect attack has been reported for common and Faba beans (Pabon et al, 1976; Tahhan and Hariri, 1981). Thus, natural resistance to insect infestation may exist. Identification of the nature of this resistance would help alleviate storage problems reduce chemical treatment for insect control, maintain nutritive value and insure increased efficiency of processing.

PROCESSING

Milling

Although not universally used, in certain regions of the world food legumes are milled to remove the seed coat before the application of thermal processing. Storage conditions have significant quantitative and qualitative effects on legume milling. Both wet and dry milling may be carried out at the household or commercial level (Siegel and Fawcet, 1976). The wet method usually takes longer, and although yields are good, the cotyledons (dhal) become hard-to-cook, requiring longer cooking times. The dry method of dehulling has the disadvantage of producing high milling losses due to breakage of cotyledons and powdering. Furthermore, the loosening of the seed coat in this process is not adequate, much less when the drying operation of beans before storage has not been adequate. In addition, if the grain is insect infested, milling yields decrease (Parpia, 1972). Milling techniques have been developed to maximize the yield of the edible fractions and data have been obtained suggesting that the genetic make up of the cultivars also play an important role in the yield of edible fractions.

Roasting

Roasting is an interesting processing technique since it has a special attribute of developing attractive flavors in the food. It also induces important functional properties which should be compatible with nutritional value. An example of roasting for cooking Vicia faba is shown in Table 4. This product, converted into a fine flour, is utilized as a drink for young children in some developing countries (Bressani et al., 1981). About 15 minutes of roasting at 200°C was the optimum time for maximum protein quality. Chemical analysis for available lysine showed an expected decrease in this amino acid, which explains the lower quality as roasting time increased. The roasting process has been applied to processing of common beans (Table 5, Yadav and Liener, 1978) using a bed heat exchange dryer operated at 190-200°C for 20 to 30 second.

Table 4: Effect of Roasting Time on the Protein Quality of Vicia Faba

Roasting time min. @ 200°C	PER
0	0.86
7.5	1.00
15.0	1.12
22.5	1.06
25.0	0.75

Table 5: The Protein Quality of Autoclaved and Dry Roasted Navy Beans

Process	Trypsin inhibitor	Hemagglutination units/g x 10^{-3}	P.D. %	PER
Autoclaving, min				
0	16.5	35.5	44.3	-
15	2.5	0.2	66.0	1.69
30	0	0	66.4	1.46
60	0	0	62.8	1.15
Dry roasted 20-25 sec 196-200°C	4.1	0.2	69.2	1.92

Based on the parameters measured, roasting resulted in a product as good or higher in protein quality as that processed by the common wet cooking procedure under pressure. Thus well controlled processing conditions remove the danger of reducing the food's nutritional potential.

Heat treatment is also very useful in preserving the cooking characteristics of whole food legumes which become hard-to-cook upon storage, as indicated above. In one study, whole black beans were heat-treated for 2, 5 and 10 minutes at 121°C and for 10, 20 and 30 minutes under steam (98°C). The materials so treated were then stored at 25°C and 70% relative humidity and samples were withdrawn at 3, 6 and 9 months for cooking quality evaluation (Molina et al., 1975; Molina et al.,

1976). The results indicated that the process ensured that the beans retained acceptable cooking characteristics as compared to the untreated control, which was more hard-to-cook as time of storage increased.

Cooking

Wet cooking is probably the most common procedure used to prepare food for consumption, both at home and industrially. Common beans must be cooked for consumption for the purpose of inactivating the presence of antiphysiological factors (Liener, 1969; Tobin and Carpenter, 1978). This is shown in Figure 2 where the effect of processing on inactivating trypsin inhibitors and hemagglutinins is given on the left, while the improvement in protein quality is given on the right. The antiphysiological factors are destroyed in about 90 minutes, which results in an increase in protein quality. However, exceeding cooking time results in a progresive loss in nutritive value due to losses of lysine. The wet cooking process may be carried out under pressure which reduces the time of exposure as compared to cooking at atmospheric pressure.

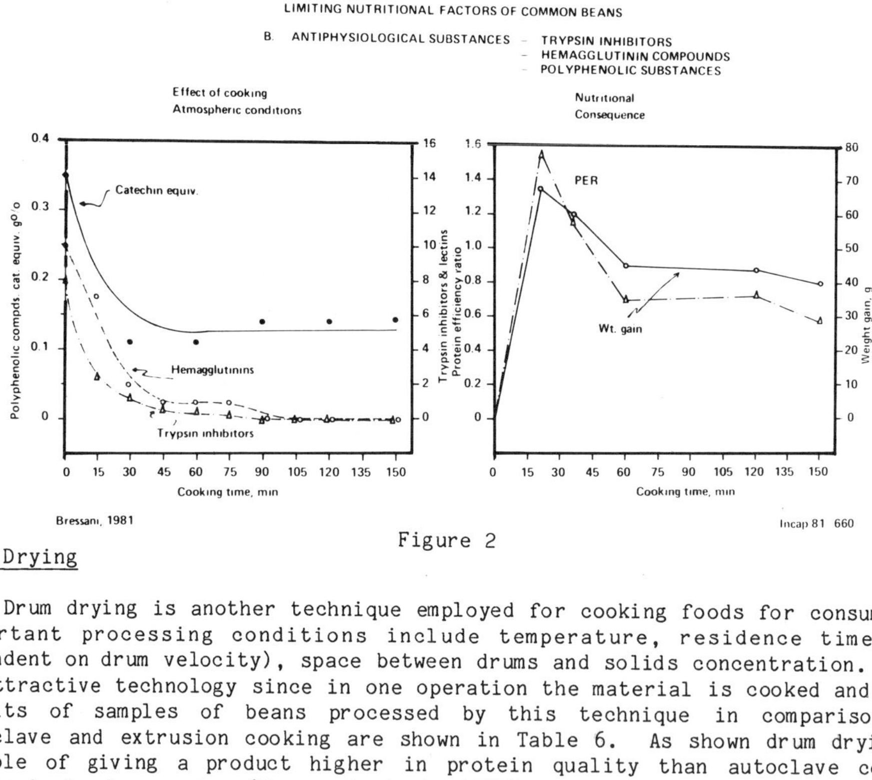

Figure 2

Drum Drying

Drum drying is another technique employed for cooking foods for consumption. Important processing conditions include temperature, residence time (both dependent on drum velocity), space between drums and solids concentration. It is an attractive technology since in one operation the material is cooked and dried. Results of samples of beans processed by this technique in comparison with autoclave and extrusion cooking are shown in Table 6. As shown drum drying was capable of giving a product higher in protein quality than autoclave cooking, particularly for cowpeas (Bressani et al, 1977).

Extrusion Cooking

One of the most recent processing techniques in food processing is extrusion cooking. This process may be carried out with some water addition. The data in Table 6 show an increase in protein quality taking place by extrusion cooking, over samples from autoclave cooking and drum drying. The possible reasons for the increase by extrusion cooking is that this procedure caused a greater inactivation of antiphysiological factors and increased the susceptibility of the protein to a

Table 6: Effect of Cooking of Food Legumes by Autoclaving, Drum-drying and Extrusion (Bressani et al., 1977)

Process		Cowpea/common beans 50/50	Cowpea
Autoclave	Wt. gain, g/28 days	41.00 ± 6.80	44.00 ± 4.20
	PER	2.33 ± 0.18	1.47 ± 0.09
Drum-drying	Wt. gain, g/28 days	28.00 ± 0.40	52.00 ± 2.60
	PER	1.38 ± 0.08	1.97 ± 0.52
Extrusion	Wt. gain, g/28 days	48.00 ± 4.90	75.00 ± 5.20
cooking	PER	1.54 _0.04	2.12 ± 0.06

more complete hydrolysis, or that there is a change in the carbohydrate fraction which favors a better protein utilization (Bressani et al, 1977).

Germination

It is generally known that germination has a marked effect on improving the nutritional quality of legumes. A dramatic increase in ascorbic acid and of other vitamins in legume seeds has been observed during germination. However, many contradictory results are reported in the literature. Most indicate that trypsin inhibitors activity is retained in germinated food legumes without affecting nutritive value, an observation deserving more research. Likewise, starch is broken down, and flatulence factors as well as polyphenolic content are reduced (Chen et al., 1975; Fordham and Wills, 1975; Everson et al., 1944; Kakade and Evans, 1966; Noor et al., 1980; Chandrasekhar and Jayalakshmi, 1978; Chattopadhyay and Banerjee, 1953; Singh and Jambunathan, 1981; Khaleque et al., 1983).

A significant decrease in the nutritive value of common beans has been found (Elias et al, 1973) as shown in Table 7. Protein efficiency ratio decreases significantly with respect to germination time which in part can be due to a decrease in the total sulfur amino acid content, since apparent protein digestibility showed no major changes until the end of the germination period. Although not shown in this Table, the protein fraction extracted with 5% KCl increased during germination, while soluble proteins in water and in 70% ethanol remained unchanged. The latter should be the object of further investigation, since available amino acid composition will depend on the protein fraction that prevails at different stages of the germination period. The differences in results reported from food legume germination may be associated with food legume species, an aspect which should be studied.

Fermentation

The processing of food by fermentation, an ancient practice, is used quite extensively in various parts of the world, particularly in the Orient. Some general observations made include increases in B_{12} as well as in other vitamins of the B-complex group. Likewise, there are increases in protein quality, increased availability of various nutrients and removal of antiphysiological substances. Of particular significance is the supplementary effect induced by the microbial growth on the substrate (Dworschak, 1982). Although fermentation is usually carried out on the food legume itself, from the nutritional point of view, fermentation of mixtures of legumes and cereal grains have attractive perspectives. An example of this is shown in Table 8 (Wang et al., 1968).

Table 7: Changes in the Nutritive Value of Common Beans (Phaseolus vulgaris) During Germination

Days of germination	Total sulfur amino acids g/16 g N	Ave. wt. gain in 4 weeks, g	Protein efficiency ratio (PER)	Apparent digestib. %
0	1.44	28	0.99	67
3	1.25	24	0.86	64
6	1.21	15	0.59	67
9	1.15	4	0.26	60

Table 8: Weight Gain, Food Consumption and Protein Efficiency Ratio of Rats Fed Fermented or Unfermented Grains as Protein Sources

Protein sources	Weight gain, g	Food consumption, g	PER
Casein	98.0 ± 6.6[1]	347 ± 13[1]	2.81 ± 0.10[1]
Wheat (control)	37.6 ± 2.7	295 ± 13	1.28 ± 0.05
Wheat (fermented)	55.0 ± 1.6[2]	322 ± 7[2]	1.71 ± 0.05[2]
Soybeans (control)	76.5 ± 2.3	353 ± 10	2.17 ± 0.03
Soybeans (fermented)	72.9 ± 3.3	321 ± 12[2]	2.27 ± 0.05
Soybean/Wheat (control)	97.1 ± 3.2	389 ± 8	2.49 ± 0.04
Soybean/Wheat (fermented)	94.2 ± 2.2	338 ± 12[2]	2.79 ± 0.04[2]

1 Standard error
2 Significantly different (P / 0.05) from corresponding unfermented grain.

Mixtures of soybean with wheat (in a 60/40 ratio on the basis of protein) give a food of a protein quality higher than that of the single components (Bressani, 1974). If these foods individually or combined are fermented through the use of a microorganism, such as Rhizopus oligosporus, there are important increases in protein quality (Table 8). The fermented product in every case showed a higher ptoein quality increase which depends on the protein quality of the food to be fermented. For example, wheat quality increased much more than soy quality, because the protein of the microorganisms contain lysine, deficient in wheat, but is deficient in methionine, which is also deficient in soybean protein.

STORAGE OF PROCESSED FOOD LEGUME PRODUCTS

Processed bean products include canned beans as well as powders. However, studies of the effect of storage on nutritive value are few. The powders if not properly stored may become damaged rather easily, and one example for soybean flour, is shown in Table 9. Samples, classified by color were taken for chemical and biological analysis. Available lysine was acceptable in the light colored samples, but it was significantly low in the dark colored samples. The loss in available lysine reflected itself in the loss of protein quality of the product (INCAP, unpublished data). Similar results have been reported to take place with other food products, particularly milk powder often shipped by developed to developing countries for purposes of distribution in supplementary feeding programs. Losses of foods through improper handling and storage conditions obviously have significant economic and nutritional implications which may be reduced or eliminated by improving storage conditions. Similar observations have been made with bean powders, made by moist cooking in the autoclave followed by deydration and stored in paper and polyethylene bags at ambient temperature. In biological trials, decreased protein quality of the product (Elias et al., 1973) resulted from prolonged storage and increases in free fatty acid content.

Table 9: Effect of Storage on the Nutritive Value of Soybean Meal (INCAP, unpublished data.)

Soybean meal color	Available lysine g/16 g N	Prot. Dig. %	NPR
Light yellow	5.82	86.6	3.89
Yellow/Brown	5.34	88.7	3.43
Light brown	4.45	83.5	2.58
Dark brown	1.78	26.1	0.80
Casein	-	-	4.53

OTHER PROCESSING PROBLEMS WITH COMMON BEANS

Recently, the results of various studies have indicated that common beans contain polyphenolic compounds which interfere with protein digestibility and protein quality (Bressani and Elias, 1979), as indicated in Figure 3 for human studies (Bressani et al., 1981).

The partition of polyphenolics during and after cooking has been studied (Bressani et al., 1981) as shown in Table 10. Of the total polyphenolics in raw beans, expressed as tannic acid, the amounts remaining after cooking are shown along with the amounts in the cooking liquor. If no destruction of tannins takes place, it can then be assumed that the balance became bound. Bound tannins would be those polyphenolic compounds not measured by present analytical methods, since they have reacted with amino groups of the protein and are thus not extracted by methanol, and would be responsible for the decrease in protein digestibility. To understand the effect of the polyphenolic compounds on digestibility, it is necessary to establish the fate and role of these compounds during cooking, an area which should receive increased attention, including improved analytical techniques for such compounds.

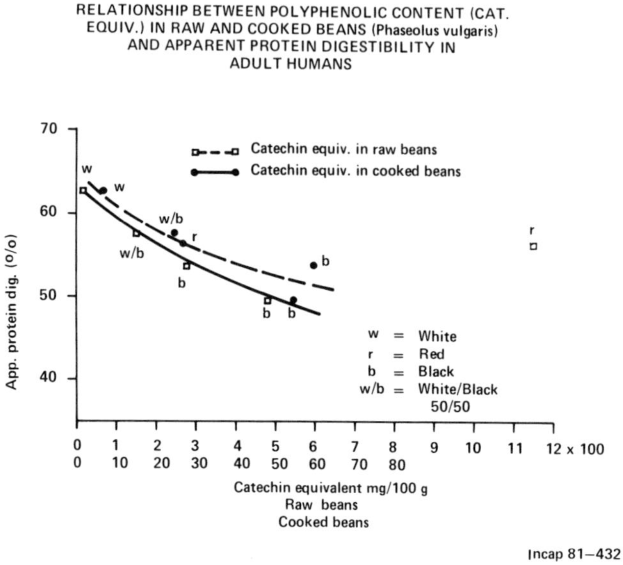

Figure 3

A second problem is related to the possible presence in beans of proteins resistant to enzymatic action even after "appropriate cooking". This fraction found in cooked beans decreases protein digestibility as shown by the regression equation in Figure 4. In this case the alkali soluble fraction in cooked beans, amounting to 9-20% of the total, negatively influenced protein digestibility in rats. Similar results have been obtained in dogs (Bressani et al, 1977) and, even more important, in human studies (Bressani et al, 1982) in which a negative correlation ($Y = 95.81 - 1.17 X$, $r = -0.94$) was also found. This is also another area which deserves additional research because the nutritional potential of bean protein is diminished by these two factors.

Table 10: Partition of Polyphenolic Compounds in Beans Upon Cooking

	Bean Color		
	Black	White g/500 g	Red
Raw bean	4.50	1.80	7.35
Cooked bean	2.72	1.20	2.75
Cooking liquor	0.86	0.28	0.86
		% Distribution	
Raw bean	100.0	100.0	100.0
Cooked bean	60.4	66.7	37.4
Cooking liquor	19.1	15.5	11.7
Bound (?)	20.5	17.8	50.9

Figure 4

RELATIONSHIP BETWEEN SALT SOLUBLE PROTEINS IN COOKED BEANS AND THEIR TRUE PROTEIN DIGESTIBILITY

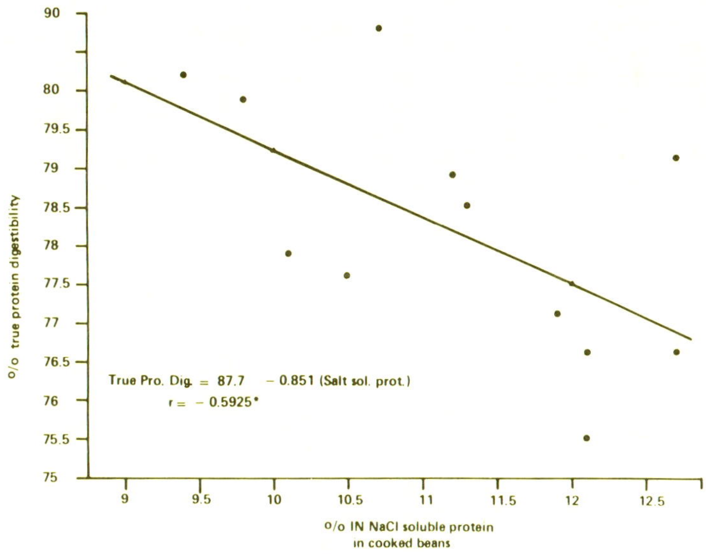

Incap 82–209

REFERENCES

Almas, K. and A.E. Bender (1980). Effect of heat treatment of legumes on available lysine. J. Sci. Food Agr. 31, 448-452.

Bressani, R. (1974). Complementary amino acid patterns. Chap. 16, p.149-166. In: Nutrients in Processed Foods: Proteins. Ed. P.L. White and D.C. Fletcher. American Medical Association. Publishing Sciences Group, Inc. Acton, MA.

Bressani, R. (1982). Research needs to up-grade the nutritional quality of common beans (Phaseolus vulgaris). In: Proceedings of the workshop "Current Aspects on Legumes as a Food Constituent with Special Emphasis on Lupines", San Diego, CA. August 1981.

Bressani, R. and L.G. Elias (1979). The nutritional role of polyphenols in beans. In: Polyphenols in Cereals and Legumes. Ed. J.H. Hulse. IDRC-145e. IDRC, Ottawa, Canada.

Bressani, R., E. Hernandez, D. Navarrete and J.E. Braham (to be published) Protein digestibility of common beans (Phaseolus vulgaris) in adult human subjects.

Bressani, R., L.G. Elias, and J.E. Braham (to be published). Reduction of digestibility of legume protein by tannins. Workshop "Physiological effects of legumes in the human diet". XII International Congress of Nutrition. San Diego, CA, August 1981.

Bressani, R., L.G. Elias and R. Gomez Brenes (1981). Valor proteinico del haba (Vicia faba) y su utilizacion en sistemas de consumo alimentario para seres humanos. INCAP Annual Report. p.72.

Bressani, R., L.G. Elias, M.T. Huezo y J.E. Braham (1977). Estudios sobre la produccion de harinas precocidas de frijol y caupi, solos y combinados mediante coccion-deshidratacion. Arch. Latinoamer. Nutr. 27, 247-260.

Bressani, R., L.G. Elias y M.R. Molina (1977). Estudios sobre la digestibilidad de la proteina de varias especies de leguminosas. Arch. Latinoamer. Nutr. 27, 215-231.

Bressani, R., L.G. Elias and and A.T. Valiente (1963). Effect of cooking and of amino acid supplementation on the nutritive value of black beans (Phaseolus vulgaris). Brit. J. Nutr. 17, 69-78.

Burr, H.K. and S. Kon (1966). Factors influencing the cooking rate of stored dry beans. p. 50-52. Report of 8th Dry Bean Research Conference, Bellaire, Mich. August 11-13, ARS-USDA 74-41, 1966.

Chandrasekhar, U. and K. Jayalakshmi (1978). Evaluation of protein quality of sprouted, roasted and autoclaved legumes on albino rats. Ind. J. Nutr. Diet. 15, 414-421.

Chattopadhyay, H. and S. Banerjee (1953). Effect of germination on the biological value of proteins and the trypsin inhibitor activity of some common Indian pulses. Ind. J. Med. Res. 41, 185-189.

Chen, L.H., C.E. Wells and J.R. Fordham (1975). Germinated seeds for human consumption. J. Food Sci. 40, 1290-1294.

Dworschak, E. (1982). Effect of processing on nutritive value of food: Fermentation. p. 63-76. In: Handbook of Nutritive Value of Processed Food. Vol.1. Food for Human Use. Ed. M. Rechcigl, Jr. CRC Press.

Elias, L.G. y R. Bressani (1981). Efecto del tiempo de almacenamiento sobre las caracteristicas tecnologicas y nutricionales del frijol comun (P. vulgaris). INCAP Annual Report.

Elias, L.G., R. Bressani y M. Flores (1973). Los problemas y los potenciales en el almacenamiento y procesamiento de las leguminosas de grano comestibles en America Latina. p.32-48, 1973. Seminario "El Potencial del Frijol y de Otras Leguminosas de Grano Comestibles en America Latina" Feb. 26 - Mar. 1, 1973, CIAT, Cali, Colombia.

Elias, L.G., A. Conde, A. Munoz and R. Bressani (1973). Effect of germination and maturation on the nutritive value of common beans (P. vulgaris) p. 139-152. In: Nutritional Aspects of Common Beans and Other Legume Seeds as Animal and Human Foods. Ed. W.G. Jaffe, Proc. Meeting held Nov. 6-9, 1973, Ribeirao Preto, S.P. Brasil.

Everson, G.J., H. Sttenbock, D.C. Cederquist, and H.T. Parsons (1944). The effect of germination, the stage of maturity and the variety upon the nutritive value of soybean protein. J. Nutr. 27, 225-229.

Fordham, J.R., C.E. Wells, and L.H. Chen (1975). Sprouting of seeds and nutrient composition of seeds and sprouts. J. Food Sci. 40, 552-556.

Garcia Soto, A. y R. Bressani (1982). Efecto de la radiacion solar sobre algunas caracteristicas fisico-quimicas del grano de frijol (P. vulgaris). Observaciones preliminares. Turrialba (in press).

Khaleque, A., L.G. Elias, J.E. Braham and R. Bressani (1982). Studies on the development of infant foods from plant protein sources. Part 1. Effect of germination of chickpea (Cicer arietinum) on the nutritive value and digestibility of protein. J. Sci. Food and Agr. (in press).

Kakade, M.L. and R.J. Evans (1966). Effect of soaking and germination on the nutritive value of navy beans. J. Food Sci. 31, 781-783.

Liener, I.E. (1969). Toxic Constituents of Plant Foodstuffs (1969). Academic Press, New York, USA.

Molina, M.R., M.A. Baten, R.A. Gomez Brenes, K.W. King and R. Bressani (1976). Heat treatment: a process to control the development of the hard-to-cook phenomenon in black beans (Phaseolus vulgaris). J. Food Sci. 41, 661-666.

Molina, M.R., G. de la Fuente and R. Bressani (1975). Interrelationships between soaking time, cooking time, nutritive value and other characteristics of the black bean (Phaseolus vulgaris). J. Food Sci. 40, 587-591.

Noor, I., R. Bressani, and L.G. Elias (1980). Changes in chemical and selected biochemical components, protein quality and digestibility of mung bean (Vigna radiata) during germination and cooking. Qual. Plant. Foods Hum. Nutr. 30, 235-244.

Pabon, R.L.A., C.J. Aguirre y J.A. Reyes Q. (1976). Resistencia de 17 variedades comeciales de frijol (P. vulgaris L.) en almacenamiento al ataque del gorgojo pintado de los granos (Zabrotes subfasciatus Boh). Acta Agron. 26, 39-47.

Parpia, H.A.B. (1972). Utilization problems in food legumes. p.281 In: Nutritonal Improvement of Food Legumes by Breeding. Proc. Symp. PAG/FAO July 3-5.

Sefa-Dedeh, S. D.W. Stanley and P.W. Voisey (1978). Effects of soaking time and cooking conditions on texture and micro structure of cowpeas (Vigna unguiculata). J. Food Sci. 43, 1832-1838.

Sefa-Dedeh, S. D.W. Stanley and P.W. Voisey (1979). Effect of storage time and cooking conditions on the hard-to-cook defect in cowpeas (Vigna unguiculata). J. Food Sci. 44, 790-796.

Siegel, A. and B. Fawcett (1976). Food legume processing and utilization with special emphasis on application in developing countries. IDRC-TS 1, 1976. IDRC, Ottawa, Canada.

Singh, U. and R. Jambunathan (1981). Studies on Desi and Kabuli chickpea (Cicer arietinum L.) cultivars: levels of protease inhibitors, levels of polyphenolic compounds and in vitro protein digestibility. J. Food Sci. 46, 1364-1367.

Tahhan, O. and G. Hariri (1981). Infestation of faba bean seeds by Bruchus Drutipes Bandi (Coleoptera Bruchidae). p.58. FABIS Newsletter No. 3, April 1981. ICARDA.

Tobin, G. and K.J. Carpenter (1978). The nutritional value of the dry bean (Phaseolus vulgaris). A literature review. Nutr. Abst. and Revs. 48, 919-936.

Varriano-Marston, E. and E. de Omana (1979). Effect of sodium salt solution on the chemical composition and morphology of black bean (P. vulgaris). J. Food Sci. 44, 531-536.

Yadav, N.R. and I.E. Liener (1978). Nutritional evaluation of dry-roasted navy bean flour and mixtures with cereal proteins. p.401-413. In: Nutritional Improvement of Food and Feed proteins. Ed. M. Friedman. Adv. Exptl. Medicine and Biol. Vol. 105. Plenum Press, New York.

Wang, Hwa, L., D.I. Ruttle and C.W. Hesseltine (1968). Protein quality of wheat and soybean after Rhizopus oligosporus fermentation. J. Nutr. 96, 109-114.

THE RAPID SCREENING OF FOOD FOR CONTAMINANTS AND TOXICANTS

H.B.S. Conacher

Food Research Division, Health Protection Branch
Health and Welfare Canada
Ottawa, Ontario, Canada K1A 0L2

ABSTRACT

A limited review of the various approaches taken for the rapid screening of toxicants and contaminants in foods in Canada over the past decade is presented. From an initial consideration of the need for, and desirable characteristics of, rapid screening methods in the regulatory setting, examples are given of the various techniques adopted in the area of pesticides and mycotoxins. The screening procedures are categorized into three classes: those requiring only inexpensive equipment, such as thin-layer chromatography (TLC), those requiring moderately expensive equipment, such as GLC, and those requiring highly expensive systems, such as tandem mass spectrometry. The advantages and disadvantages of each technique are discussed.

KEYWORDS: Rapid Screening Techniques; Toxicants and Contaminants, Pesticides; Mycotoxins; Thin-Layer Chromatography; Gas-Liquid Chromatography; Radioimmunoassay; Mass Spectrometry.

INTRODUCTION

The general public is more aware than ever before of the potential dangers to health associated with the presence of contaminants and toxicants in foods. They generally look to their governments not only to develop the appropriate legislation setting maximum permissable levels of such compounds in their foods, but also to institute the necessary control procedures to ensure compliance with these legislated levels.

In connection with both the development of the legislation and the subsequent enforcement programs, analytical chemical data are of primary importance. When such data are used for such regulatory purposes, which can have far-reaching impacts on society, the analytical methodology involved must be specific, precise, and accurate, and generally applicable in more than one laboratory. Unfortunately many of these 'regulatory' methods, although reliable and credible, can often be tedious, time consuming, labor-intensive, and expensive. For example, using the dioxin methodology shown in Figure 1, only approximately 5 samples can be analyzed in one week by two individuals. They are thus not readily applicable to, or desirable in, routine surveillance of the food supply. In this latter connection statistically designed surveys require large numbers of samples. Furthermore, in many instances where perishable items may be held in customs, or in situations where illness or death has resulted, data are always required as quickly as possible. It is in such instances that rapid screening methods play an important role in the regulatory program.

Figure1. METHODOLOGY FOR TETRACHLORODIBENZODIOXIN IN FISH

Extraction → Defatting → Florisil → HPLC → GLC/MS

In Canada, as in many other countries, we have been involved for a number of years in the development, evaluation and application of a wide variety of these screening procedures in the analysis of foods. In this report some of our past and present activities in this area are described and some of the various advantages and disadvantages of each technique noted. Although these activities have covered the whole spectrum of toxicants, contaminants, additives, etc., only examples from the pesticide and mycotoxin areas are presented here.

DISCUSSION

The typical trace organic/inorganic analysis consists of the four operations of collection and storage, extraction, the determinative steps and data acquisition and processing. In the development of rapid screening procedures it is in the second and third steps where considerable simplification and/or time saving can be introduced. However the first step, which is common to both 'regulatory' and screening procedures, cannot be ignored as a significant source of error (Hertz, 1978). Also, since many modern techniques involve computerized data acquisition and processing combined, the last two steps may be consolidated.

Desirable characteristics of rapid screening procedures include wide application, simplicity and speed, and a reasonable level of detection - certainly a level of detection below the maximum permissable level. Although a limited number of false positive findings can be tolerated (since they can be subsequently verified or negated by application of the 'regulatory' method) false negatives are not acceptable since food containing violative contaminant levels could enter the food supply. Such techniques can be compound-specific, for example, the radioimmunoassay (RIA) procedures for benomyl (Newsome and Shields, 1981a), chlorinated dibenzodioxins (Albro and co-workers, 1979) and chlorinated biphenyls (Newsome and Shields, 1981b); class specific, for example, the thin-layer chromatography-enzyme inhibition technique (TLC-EI) for organophosphorus and carbamate pesticides (Mendoza, 1981); or non-specific with respect to class of compound, for example, the screening of foods for mutagens using the Ames Salmonellae procedure (Stoltz and co-workers, 1982).

In some cases the screening procedures are devised to give only an indication of presence and at best a semi-quantitative response, while in others an accurate measurement can be obtained. Class-specific TLC-EI procedures would fall into the former category while those, for example, based on RIA might fall into the latter. The actual methods themselves can range from those utilizing only inexpensive equipment such as TLC apparatus, through those using moderately expensive equipment such as gas-liquid chromatographs, scintillation counters etc., to those utilizing highly expensive equipment such as fully computerized mass spectrometers.

SIMPLE INEXPENSIVE SCREENING PROCEDURES

The simplest screening procedures are probably those based on column or thin-layer chromatography (TLC) and find ready application in both the pesticide and mycotoxin areas.

Pesticides

Out of many TLC techniques described for the rapid screening of pesticide residues in foods (Mendoza, 1981) two have been studied extensively in our laboratories - a TLC enzyme inhibition technique (TLC-EI) (Mendoza, 1972; 1974) and a TLC technique for photosynthesis-inhibiting herbicides (TLC-PI) (Lawrence, 1980). The TLC-EI technique, as the name might imply is used with compounds that inhibit enzyme activity such as the organophosphorus compounds and the carbamates.

The extracted pesticide residues are resolved on a TLC plate (generally Silica Gel G, although others can be used), sprayed with an enzyme solution, and then with a substrate solution. The principle of the technique relates to the fact that the substrate will not be hydrolyzed where the enzyme and the pesticide interact leading to colourless spots where the pesticides are located against an intensely coloured background. In some cases the enzyme-substrate reaction product(s) must be coupled to produce a coloured compound (eg. 1-naphthol from 1-napthyl acetate is coupled with a diazonium salt to produce a highly coloured azo compound) whereas with others the reaction products themselves are coloured (eg. indophenol from indophenyl acetate) and need no coupling reaction. Sulfur analogs (P = S) of the organophosphorus compounds generally must be converted to their more active oxygen (P = O) analogues before the enzyme is applied to the plate. Bromine water and/or vapour is used to achieve this conversion.

Some of the sources or kinds of esterases and types of substrates that can be used with this technique are shown in Table 1. Differences in level of detection

Table 1: Sources of Esterases and Types of Enzyme Substrates Used With the TLC-EI Technique

Sources of Esterases	Substrates
Livers of	Indoxyl Acetate
- Cattle	5-Bromoindoxyl Acetate
- Pig	Indophenyl Acetate
- Chicken	Indophenyl Acetate
- Monkey	1-Napthyl Acetate
- Rat	
Human Blood Plasma or Serum	
Horse Blood Serum (available commercially)	
Bees' Heads	
Trypsin-Beef Pancreas (available commercially)	
Acid Phosphatase (available commercially)	

can be observed depending on the esterase, substrate and type of treatment adopted (Mendoza, 1972).

Some typical retention values (relative to carbaryl) and levels of detection for some commonly used carbamates are given in Table 2. In general an amount of extract equivalent to 100 mg sample is applied to the plate. It is therefore apparent that if the level of detection on the plate is 10 ng or less in the system adopted then one can readily screen for residues at 0.1 ppm. This is the minimum level for pesticide residues in foods that is generally aimed for in Canada. For the carbamates listed in Table 2, only aldicarb would not be detected under such a screening procedure.

Table 2: Detection Levels (NG) and Retention Values for Some Commonly Used Carbamates Determined by TLC-EI Technique Using Pig Liver Esterase and Indophenyl Acetate (Mendoza and Shields, 1973).

Chemical	Relative Retention Value	Detection Level (NG)
Carbaryl	100	0.05
Formetanate (HCL)	21	5.0
Methomyl	30	10.0
C-8353	55	5.0
Aldicarb	88	25.0
Meobal	119	0.5
Mesurol	124	0.5
Promecarb	141	0.5
Butacarb	158	0.5

The TLC-PI technique is applicable to herbicides such as triazines, phenyl ureas, phenyl carbamates, uracils and acyl anilides (Lawrence, 1980). It is based on the fact that these herbicides inhibit electron transport in the plant cell chloroplasts and thus inhibit photosynthesis. The extent of this inhibition can be measured in vitro using the artificial electron acceptors such as potassium ferricyanide or 2,6-dichloroindophenol (DCIP) which change their absorption maxima when reduced. In the present technique the reaction is carried out on a TLC plate as indicated in Figure 2.

The developed plate is sprayed with a mixture of spinach or bean leaf chloroplasts and the redox indicator, and the plate exposed to white fluorescent light. If no herbicide is present the DCIP is completely converted by the chloroplasts to its colourless reduction product. However in the presence of an inhibiting herbicide this reduction is prevented and the blue colour of the DCIP persists. Thus herbicides become visible as blue spots on a pale yellow-green background. Chloroplasts are readily isolated from fresh bean leaves or spinach, and may be refrigerated (2-4°C) for up to 3 days with retention of activity.

Minimum detection levels of most inhibiting herbicides are in the 0.1-0.5 ng range and the selectivity is such that the bipyridilium, phenoxy acid and wild oat herbicides are negative. In samples such as potatoes, carrots and corn with the minimum of cleanup (Figure 2), detection limits are of the order of 10 ppb. The procedure is extremely useful for the semi-quantitative estimation of many

Figure 2 TLC-PHOTOSYNTHESIS INHIBITION TECHNIQUE

Sample (25G) → TLC → Spinach Chloroplasts + Redox Indicator (2,6-Dichloroindophenol) → Soft White Fluorescent Light → Blue Spots On Yellow-Green Background

1. 100 ML Acetone
2. 5G Equiv. H_2O NaCl
3. 3x40 ML CH_2CL_2
4. 0.5 ML
5. 10μL

photosynthesis inhibiting herbicides which comprise as much as 40-45% of all commercial herbicides (Kovac and Henselova, 1976).

Mycotoxins

Rapid screening for the simultaneous analysis of several mycotoxins in the same sample (multimycotoxin analysis) of an agricultural commodity is an important and developing area of analytical research. At present attempts are being made to apply high-performance liquid chromatography (LC) to this problem but, although sensitivity is better than for TLC, much additional effort is needed. Whereas a blue and yellow fluorescent spot on a TLC plate can readily be detected and differentiated by the eye, simultaneous instrumental analysis by LC may require more than one detector or detector setting. Thus TLC generally remains the procedure of choice for multimycotoxin detection.

Several systems for one dimensional TLC of mycotoxins have been developed (Scott, 1980). The solvent system toluene-ethyl acetate-formic acid (6+3+1), for example, has proved of wide applicability. In combination with the Stoloff extraction procedure (Stoloff et al, 1971) it has found frequent use in our laboratories in the monitoring of contaminated grains and/or cultures. Typical RF values, levels of detection and means of visualization for some of the more common mycotoxins in such a system are illustrated in Table 3.

Some of the more recent methods of multimycotoxin analysis whether of standards alone (Gorst-Allman and Steyn, 1979) or of extracts of feeds and foodstuffs (Gimeno, 1979; Takada et al., 1979) appear to have become more complex with respect to the number of solvent systems recommended or number of spray reagents to be used. A comprehensive review of multimycotoxin analysis has recently been published (Steyn, 1981).

In the case of the developments of rapid screening techniques for individual mycotoxin analysis most attention has to date been devoted to aflatoxin. A comparison of three types of screening methods for detecting aflatoxin in corn is presented in Table 4.

Table 3: Typical RF Values, Levels of Detection and Visualization. Aids for Some Common Mycotoxins on Silica Gel G[a].

Mycotoxin	RF[b]	Limit of Detection (NG)	Visualization (Scott, 1980) L = longwave UV; S = shortwave UV
Citrinin	0-0.48	5	Yellow Fluorescence (L)
Aflatoxin G_2	0.08	0.04	Blue, Green Fluorescence (L)
Aflatoxin B_2	0.18	0.04	Blue Fluorescence (L)
Aflatoxin G_1	0.19	0.1	Blue, Green Fluorescence (L)
Aflatoxin B_1	0.32	0.1	Blue, Green Fluorescence (L)
Patulin	0.37	10	MBTH, Heat, Yellow Brown Fluorescence (L)
Penicillic Acid	0.43	10	Acidic Anisaldehyde Heat, Blue Fluorescence (L)
Zearalenone	0.59	20	Green-Blue Fluorescence (S)
Ochratoxin A	0.60	2	Blue-Green Fluorescence (L+S)
Sterigmatocystin	0.71	40	Brick-Red Fluorescence (L)

[a] Tol./EtAc/90% HCOOH (6:3:1)

[b] Durackova et al. (1976)

Table 4: Comparison of Screening Methods for Aflatoxin (Romer, 1976).

Characteristic	TLC	Minicolumn	Black Light
Analysis Time	30-60 Min	30-60 Min	5-10 Min
Ease of Analysis	Least Ease	Medium Ease	Most Ease
Sample Types	Corn Grain and Nuts	Corn Ingredients Mixed Feeds	Corn Only
False Negatives	Few	Few	Few
False Positives	?	Few	Many
Mycotoxins Measured	Aflatoxin Zearalenone T-2 Toxin	Aflatoxin	Aflatoxin

It is apparent that the black light test (which simply involves a crushing of the sample and examination under long wave ultra violet light for the bright yellowish-green fluorescence of aflatoxin contaminated corn) is the best of the three methods because of the short analysis time and the ease of performance. However it is limited by the fact that it gives too many false positives. Because of this limitation the black light test is generally used together with the minicolumn test to effectively screen for aflatoxin in corn in the feed mill. The Shannon and Shotwell (1979) extraction and minicolumn procedure for corn is most appropriate and permits the detection of 10 ppb aflatoxins. It is possible that the use of cellulose columns (Jemmali, 1973) would allow introduction of 50% H_2SO_4 and further indication of the presence of aflatoxin via the yellow fluorescence. A Handbook on the rapid detection of mycotoxins has recently been issued by OECD (Jan. 1982).

MODERATELY EXPENSIVE SCREENING PROCEDURES

Equipment used in procedures fitting into this category ranges in value from $20,000 to $50,000, and includes instruments such as gas-liquid chromatographs, spectrophotometers, scintillation counters, etc.

Pesticides

Modern agricultural practices have generated the possibility of having a great number of individual chemical pesticides in a large variety of foods. As a result attention has focussed on gas-liquid chromatographic (GLC) multiresidue methods to obtain maximum data on a single sample of food. For instance, the current general GLC screening method used in our laboratories can accommodate over 130 different pesticide residues in plant materials. Although still used extensively, this type of screening procedure is by no means rapid and most foods generally do not contain more than half a dozen different residues in any case. Therefore in this area of pesticide residue analysis we are presently emphasizing development of more rapid screening procedures for specific compounds or related (structurally) compounds. Recent examples of this would be the use of composite sampling in a recent study of pyrethrin residues in fruits (Ryan et al., 1982) and in the development and application of an RIA procedure for the fungicide benomyl (Newsome and Shields, 1981).

For composite sampling to be applicable, generally the right combination of a high tolerance, low detection level and an expected low incidence of positives, is required. This proved to be the situation with pyrethrins where the need arose to quickly assess residues in 130 samples of fruits. Application of the current Health Protection Branch general screening procedure (Analytical Methods for Pesticide Residues in Foods, 1973) would have involved 130 repetitions of acetone extraction, partitioning, florisil fractionation and GLC/EC. Because of the low detection limits (10 and 30 μg/g for pyrethrins I and II respectively) the 1 ppm tolerance, and the expected (and subsequently found) low-incidence, it was possible to combine fractions of the acetone extracts of various fruit batches to provide a single composite extract. Routinely, fractions from 5 fruit sample extracts were combined, and the composite submitted to the partition florisil and GLC steps. In this manner, it was possible to reduce the processing time of 130 samples to that normally required for 30 to 40. The obvious disadvantage of this technique is that if the composite result indicates a level above the tolerance, then each member of the composite may have to be sampled individually.

Radioimmunoassay (RIA) is a technique which has been widely used in clinical chemistry but which until comparatively recently has not been applied to food contaminants. The principle of the technique follows the equations:

$$L^* + A \rightarrow L^*A$$

$$L + L^* + A \rightarrow LA + L^*A$$

where L = liquid (pesticide), L^* = labelled liquid, and A = antibody.

In the absence of unlabelled ligand (pesticide) a fixed percentage of the labelled ligand is bound to the antibody. In the presence of unlabelled ligand there is competition which results in proportionally less of the labelled ligand being bound. The extent of this reduction in binding is an indication of the unlabelled ligand (pesticide) present. In general the technique lends itself more to the analysis of hydrophilic rather than lipophilic compounds.

With regard to the application of this technique to benomyl it was found possible to conduct a survey of 150 samples in approximately 1/5 the time that would have been required for the more common physico-chemical procedures. Some of the advantages and disadvantages associated with application of the technique to benomyl are indicated in Table 5.

Table 5: Assessment of RIA Procedure for Benomyl on Fruit and Vegetables.

Advantages	Disadvantages
Relatively Specific	Need for Scintillation Counter
Rapid	Preparation of Radioligand
Wide Applicability	Generation of Antibodies
Minimum Clean-up	Use of Radioactive Materials
Good Level of Detection	Reliance on Adsorption Phenomena

Some of these disadvantages can be overcome by use of the enzyme-immunoassay technique (Jhangiani et al., 1982) which eliminates the need for the use of radioactive material and counting facilities, and may only use a simple colorimeter. This technique is presently under investigation in our laboratories.

EXPENSIVE SCREENING PROCEDURES

Instrumentation of this category falls in the >$50,000 range and in the case of the more sophisticated mass spectrometers can reach the $500,000 level. The most exciting developments in this area are in the application of tandem mass spectrometry to trace analysis (Bonner, 1982), particularly in the potential of such systems for the rapid screening of foods for trace contaminants. Such an instrument, the Sciex™ TAGA™ 6000, is presently undergoing intensive studies in our laboratories to access its potential for the rapid monitoring of foodstuffs for a wide variety of chemicals.

The system is presently being applied to the screening of vomitoxin, a trichothecene mycotoxin, which has been recently detected in some varieties of Eastern Canada wheat. Some preliminary data comparing results (on the same sample extracts) from this tandem mass spec. technique with those from the 'regulatory' procedure (Scott et al., 1981) are presented in Table 6.

With the tandem screening procedure all 10 wheat extracts, plus appropriate standards, could be analyzed in approximately 1 - 1-1/2 hours as compared to 2 days for the 'regulatory' method. In general, with one or two exceptions, there is good agreement between the 2 techniques indicating the strong potential of this technique for the rapid screening of vomitoxin. The technique is also being applied to other mycotoxins and to selected pesticides.

Table 6: Comparison of Vomitoxin Levels in Wheat[a] by Taga™6000 and GLC/EC Vomitoxin (PPB).

Wheat	GLC/EC	TAGA™ 6000 MS/MS[b]
1	838	484
2	173	178
3	373	329
4	426	286
5	549	386
6	271	299
7	465	320
8	493	498
9	843	779
10	597	535

[a] The same cleaned-up sample extract was used for mass spec. and for GLC analysis.

[b] Calculated using m/z 175.

SUMMARY AND CONCLUSIONS

A variety of techniques developed and/or evaluated in the laboratories of the Canadian Health Protection Branch over the past decade for the rapid screening of toxicants and contaminants in foods have been described and attempts made to point out the various advantages and disadvantages of each.

Although the procedures range in requirement from those utilizing only inexpensive TLC approaches to those utilizing highly sophisticated mass spectrometry systems it is considered that there is still a role for all in the regulatory environment.

What may initially appear to be a prohibitive capital investment may soon pay for itself in terms of increased specificity, accuracy, precision and overall rapid monitoring ability and capability. On the other hand, the increasing availability and lure of sophisticated instrumentation should not preclude further study and/or application of the lower priced chromatographic methods, particularly in situations where they are effective.

REFERENCES

Albro, P.W., M.I. Luster, K. Chae, S.K. Chaudhary, G. Clark, L.D. Dawson, J.T. Corbett and J.D. McKinney (1980). A Radioimmunoassay for Chlorinated Dibenzo-p-Dioxins. Toxicol. Appl. Pharmacol. 50, 137-146.

Analytical Methods for Pesticide Residues in Foods (1973). Health Protection Branch, Health and Welfare Canada, Ottawa, Ontario, Canada.

Bonner, R. (1982). Chemical Analysis Problems Yield to Quadrupole MS/MS. Canadian Research, Sept., 31-36.

Durackova, Z., V. Betina and P. Nemec (1976). Systematic Analysis of Mycotoxins by TLC. J. Chromatogr. 116, 141-154.

Gimeno, A. (1979). Thin Layer Chromatographic Determination of Aflatoxins,

Ochratoxins, Sterigmatocystin, Zearalenone, Citrinin, T-2 toxin, Diacetoxyscirpenol, Penicillic Acid, Patulin, and Penitrem A. J. Assoc. Offic. Anal. Chem. 62, 579-585.

Gorst-Allman, C.P. and P.S. Steyn (1979). Screening Methods for the Detection of Thirteen Common Mycotoxins. M. Chromatogr. 175, 325-331.

Hertz, H.S., W.E. May, S.A. Wise and S.N. Chesler (1978). Trace Organic Analysis. Anal. Chem. 50, 4, 429A-436A.

Jemmali, M. (1973). Sensitive Cellulose Column for the Rapid Detection of Aflatoxin in Agricultural Products. J. Assoc. Offic. Anal. Chem. 56, 1512-1513.

Jhangiani, R.K., G.D. Chase and T. Barry (1982). Review of Radioimmunoassay and Gas Chromatography-Mass Spectrometry as Comparative Analytical Techniques. F.D.A. By-Lines 12, 4, 245-255.

Kovac, J. and M. Henselova (1977). Detection of Triazine Herbicides in Soil by a Hill-Reaction Inhibition Technique After TLC. J. Chromatogr. 133, 420-422.

Lawrence, J.F. (1980). Simple, Sensitive, and Selective Thin Layer Chromatographic Technique for Detecting Some Photosynthesis Inhibiting Herbicides. J. Assoc. Off. Anal. Chem. 63, 758-761.

Mendoza, C.E. (1972). Analysis of Pesticides by the Thin-Layer Chromatographic-Enzyme Inhibition Technique. Residue Reviews, 43, 105-142.

Mendoza, C.E. and J.B. Shields (1973). Determination of Some Carbamates by Enzyme Inhibition Techniques Using Thin-Layer Chromatography and Colorimetry. J. Agr. Food Chem. 21, 178-184.

Mendoza, C.E. (1974). Analysis of Pesticides by the Thin-Layer Chromatographic-Enzyme Inhibition Technique. Residue Reviews 50, 43-72.

Mendoza, C.E. (1981). Thin Layer Chromatography. In 'Pesticide Analysis', ed. K.G. Das Marcel Dekker, Inc., New York, 1-43.

Newsome, W.H. and J.B. Shields (1981a). A Radioimmunoassay for Benomyl and Methyl 2-Benzimidazolecarbamate on Food Crops. J. Agric. Food Chem. 29, 220-222.

Newsome, W.H. and J.B. Shields (1981b). Radioimmunoassay of PCBs in Milk and Blood. Intern. J. Environ. Anal. Chem. 10, 295-304.

Romer, T. (1976). Methods of Detecting Mycotoxins in Mixed Feeds and Feed Ingredients. Feedstuffs 48 (16), 18-46.

Ryan, J.J., J.-C. Pilon and R. Leduc (1982). Composite Sampling in the Determination of Pyrethrins in Fruit Samples. J. Assoc. Offic. Anal. Chem. 65 (4), 904-908.

Scott, P.M. (1980). Analysis of Mycotoxins Other Than Aflatoxin. In 'Thin-Layer Chromatography: Clinical and Environmental Applications'. Eds. J.C. Touchstone and D. Rogers. Wiley, New York, 251-274.

Scott, P.M., P.Y. Lau and S.R. Kanhere (1981). Gas Chromatography with Electron Capture and Mass Spectrometric Detection of Deoxynivalenol in Wheat and Other Grains. J. Assoc. Off. Anal. Chem. 64, 1364-1371.

Shannon, G.M. and O.L. Shotwell (1979). Minicolumn Detection Methods for Aflatoxin in Yellow Corn: Collaborative Study. J. Assoc. Offic. Anal. Chem. 62, 1070-1075.

Steyn, P.S. (1981). Multimycotoxin Analysis. Pure and Appl. Chem. 53, 891-902.

Stoloff, L., S. Nesheim, L. Yin, J.V. Rodricks, M. Stack and A.D. Campbell (1971). A Multimycotoxin Detection Method for Aflatoxins, Ochratoxins, Zearalenone, Sterigmatocystin and Patulin. J. Assoc. Off. Anal. Chem. 54, 91-97.

Stoltz, D.R., B. Stavric, D. Krewski, R. Klassen, R. Bendall and B. Junkins (1982). Mutagenicity Screening of Foods 1. Results with Beverages. Environmental Mutagenesis 4, 477-492.

Takeda, Y., E. Isohata, R. Amano and M. Uchiyama (1979). Simultaneous Extraction and Fractionation and Thin Layer Chromatographic Determination of 14 Mycotoxins in Grains. J. Assoc. Offic. Anal. Chem. 62, 573-578.

RAPID, SIMPLE TESTING FOR NUTRIENTS AS AN AID TO PLANT BREEDING

Bienvenido O. Juliano

Chemistry Department, The International Rice Research Institute
P.O. Box 933, Manila, Philippines

ABSTRACT

Chemists involved in plant breeding programs have developed reliable methods for screening for protein and the first limiting amino acids lysine, tryptophan, and sulfoamino acids. These screening methods are described and their effective use, including micro tests for processing quality is discussed. Appreciation for the interdisciplinary nature of nutritional and acceptable characterization of crops is increasing. Chemists are in a unique position to catalyze research activities to better understand the interrelation between nutrient properties and the postharvest quality of the various crops. Such research will form the basis of more rapid, simple microtests for nutritional, processing, and cooking and eating qualities of consumable products in plant breeding programs to ensure that consumers can avail themselves of such products with minimum additional cost.

KEYWORDS: Nutrients, screening, breeding, available nutrients, protein content and quality, acceptability.

INTRODUCTION

This paper briefly describes test for nutrients, related important factors and future trends in plant breeding programs. Nutrients in this paper refer to compounds or substances involved in the nutritional quality of the crop. Toxicants, antinutrients, and contaminants are discussed by Conacher (1983) in these proceedings.

Although the major criterion of plant breeding programs is the increase and stabilization of yield, consumer acceptability and nutritional value of a new variety relative to the ones it replaces have recently received greater attention. Consumer quality determines the price of the crop, and farmers are fully aware that a high-quality, low-yielding traditional variety fetches a higher market price than a poorer quality, high-yielding variety. By contrast, farmers and processors seldom receive a premium in return for growing a more nutritious variety. Hence, such a variety should yield at least as much as the variety it is to replace.

Improvement of the nutrient yield of a crop has been approached by increasing nutrient content while maintaining yield, or by increasing yield while maintaining nutrient content, or both, or by improving nutrient quality. The first limiting essential amino acid to man in cereals is lysine, and sulfoamino acids, cysteine and methionine in food legumes, but also tryptophan in corn and in pigeon pea (Cajanus sp.). Earlier attempts to increase the protein content of corn, the Illinois high-protein selections, did not result in better nutritive value relative to 8-10% protein corn, because the increase in protein occurred mainly in the zein (prolamin or alcohol-soluble) fraction, which has only trace lysine (Mitchell et al., 1952). The discovery of opaque-2 (Mertz et al., 1964), and floury-2 (Nelson et al., 1965) high-lysine mutants in corn triggered interest in screening world collections of seeds and selection in other crops for varieties with improved amino acid balance. Higher lysine mutants have been discovered in barley (Munk et al., 1970), sorghum (Singh and Axtell, 1973), and wheat (Mattern et al., 1970; Vogel et al., 1973; Johnson and Mattern, 1980), but not in rice (Juliano et al., 1973; IRRI, 1980). The MA13 rice mutant from USDA (Schaeffer, 1981) was found to have only 0.3% higher lysine than its parent (IRRI, 1982). All these cereals except rice are rich in prolamin (alcohol-soluble protein), which occurs in endosperm protein bodies. Mutants are selected for a reduction in prolamin synthesis and an increase in matrix or amorphous protein. By contrast, rice protein is mainly body protein with little matrix protein (Bechtel and Juliano, 1980), 80% glutelin and less than 5% prolamin. Prolamin occurs mainly in the lamellar-type protein bodies (Tanaka et al., 1980). Among legumes, the storage proteins are mainly globulins, legumins, and vicilin (Derbyshire et al., 1976).

TESTING METHODS

Several compilations of recommended procedures for screening for nutritional quality have been published for specific crops such as corn (maize) (Villegas and Mertz, 1971; Mertz et al., 1975), sorghum (Guiragossian et al., 1977), and wheat (Johnson and Mattern, 1980), cereals (PAG, 1975; IAEA, 1977), legumes (PAG, 1973; Evans and Boulter, 1974; Hulse et al., 1977), and protein foods in general (Pellett and Young, 1980).

Most methods require a clean sample of 5 grams of flour of uniform fineness. Some crops such as rice need removal of hull or seed coat. Low-oil seeds are ground in a UD or Cyclotec cyclone mill or a Christy-Norris or Falling Number hammer mill. Larger grains require preliminary grinding in a burr type mill such as Falling Number or Buhler Laboratory mill before being passed through the cyclone mill (Hulse et al., 1977). Higher-oil seeds, such as maize, sorghum, and oats, tend to clog the cyclone grinder and an impact mill such as the Krups Model 75, the Moulinex mill or even a hand grinder is suitable. The ground sample may be defatted with hexane or petroleum ether, a procedure that also helps remove pigments that may interfere with some colorimetric tests (Villegas and Mertz, 1971). The defatted sample can then be milled in the UD or Cyclotec cyclone mill.

Because of the limitations of sample size during early generations, the large number to be screened between growing seasons (two crops a year in the tropics), and the limited supply of highly trained technicians, screenings for nutrients must be rapid, economical of material, and simple, but sensitive enough to detect differences in nutrient levels. These methods are not the most accurate for the nutrient being measured. The screening tests applicable in cereal field stations include total N (Hulse et al., 1977), dye binding capacity (DBC) for lysine (Mossberg, 1969), ninhydrin screening (Mertz et al., 1974), and lysine and tryptophan in corn (Villegas and Mertz, 1971) (Table 1). Crude-protein content has the most obvious nutritional implication. Although the classic macro, semi-macro or micro Kjeldahl digestion method followed by colorimetric or titrimetric assay of the ammonia produced remains the reference procedure; DBC (Mossberg,

1969) or near infrared reflectance (NIR) spectroscopy (Williams, 1975) can also be used. The latter two methods need to be calibrated against Kjeldahl protein. The factor used to convert Kjeldahl N to crude protein is usually 6.25 for most proteins (based on 16% N in protein), but is 5.95 for rice and 5.7 for wheat.

Table 1: Checklist of Nutrients Tested in Breeding Programs

Crop	Nutrient	Method
Cereals	Crude protein or total N	Kjeldahl
	Lysine	Dye binding capacity or colorimetry (corn only)
	Tryptophan (corn only)	Colorimetry
Legumes	Protein	Kjeldahl
	Sulfoamino acids (methionine)	Microbiological or gas chromatography
	Tryptophan (pigeon pea only)	Colorimetry
	Cystine plus methionine	Liquid or gas chromatography
	Lysine	Dye binding capacity
Oil seeds	Oil content	Nuclear magnetic resonance
Root crops	Dry matter	Sp. gravity or moisture content
	Crude protein	Kjeldahl

The DBC is based on the quantitative binding of the diazo dye, 1-phenylazo-2-napthol-6-sulfonic acid monosodium salt (acid orange 12 or orange G) by the basic amino acids (lysine, arginine, and histidine) and the free amino terminal of flour protein. The unbound dye remaining in the solution is measured colorimetrically after filtration or centrifugation. Since only the level of lysine and arginine is increased in high-lysine cereals relative to histidine, the differing ratios of DBC to protein can be utilized to select these genotypes, preferably run at similar protein or DBC values. The presence in samples of excessive chlorophyll, fiber and certain tannins (polyphenols) can give rise to erroneously high DBC values (PAG, 1975). DBC was used to identify the hiproly barley mutant (Munck et al., 1970) and to screen rice for higher lysine cultivars (Juliano et al., 1973).

The ninhydrin screening method is based on the unusually high levels of free amino acids in high-lysine mutants of corn, sorghum, and mutant 1508 barley (Mertz et al., 1974). The test may be applied either to a sectioned surface of single kernels or to an extract of the flour, and the intensity of the purple color of the reaction product between free amino acids and ninhydrin at 570 nm is used as indicator of amino acid level, and indirectly of lysine content.

The colorimetric procedures for lysine and tryptophan are applicable primarily for corn (Villegas and Mertz, 1971), although they have also been applied successfully to rice. Lysine and tryptophan are present in corn in a roughly constant ratio of 4:1. Consequently, when tryptophan is determined an estimate of lysine is obtained. The defatted flour is digested with papain and tryptophan color is developed by means of reagents that when combined produced glyoxylic acid (Opienska-Blauth et al., 1963; Hernandez and Bates, 1969). In the case of lysine, the papain digest is treated with a solution of 2-chloro-3,5-dinitropyridine (Villegas and Mertz, 1971; Tsai et al., 1972). The lysine method works only on nonpigmented cereals, and high tanning sorghums give low values. In contrast, the trytophan method is applicable to pigmented grains. Tryptophan colorimetric assay recommended for pigeon pea (Hulse et al., 1977) is alkaline hydrolysis and colorimetric reaction in the dark with dimethylaminobenzaldehyde (Spies, 1967).

Another screening method used on corn at CIMMYT is the turbidimetric determination of zein extract of 70% ethanol-0.5% sodium acetate extract of corn flour (Paulis, et al., 1974).

Palter and Kohler (1969) reported on a survey hydrolysis procedure for lysine screening by lysine decarboxylase.

APPLICATION TO CEREAL BREEDING PROGRAM

The Protein Advisory Group of the United Nations (PAG, 1975) recommended that the screening for dye binding capacity (DBC) be started on F_3 seeds from F_2 plants in a high-nutrient cereal breeding program other than corn. Only the 50% with high DBC are analyzed for Kjeldahl protein. The 10-20% with high DBC for protein content are verified for lysine by DBC at a constant protein basis, e.g., 60 mg. In corn, colorimetric analysis is made of tryptophan content of F_3 seeds, and only the high 50% are run for Kjeldahl N. Lysine content of corn is verified on the high 10-20% by the 2-chloro-3,5-dinitropyridine method. Based on these analyses of F_3 seeds, only the high 10-20% (0.5-2% of materials) are replanted in F_3 and F_4 generations.

In the F_5 and later generations, complete amino acid analyses are made by ion-exchange chromatography and the high 10-20% with the best amino acid balances undergo biological evaluation in rats by relative protein value (PAG, 1975). Actual protein fractionation of corn samples using the Landry-Moureaux (1970) procedure is used to verify the low zein (prolamin) content of high-lysine lines at CIMMYT.

Some breeders, however, feel that early generation F_2 and F_3 materials are preferably selected primarily for grain size and yield (disease and pest resistance) as well as other desirable agronomic characteristics (PAG, 1975). Nutrient screening may well commence with F_4 and F_5 materials, their numbers now reduced drastically to a more manageable size.

FOOD LEGUME BREEDING PROGRAM

For legumes, screening assays for sulfoamino acids analyzed only methionine (Hulse et al., 1977). Microbial methods using Streptococcus zymogenes have been used for total and available methionine (Ford and Hewitt, 1979), but require specialized techniques (Hulse et al., 1977). The other method involves treatment with cyanogen bromide, which converts methionine to methyl thiocyanate, which is then analyzed by gas chromatography (Finlayson and MacKenzie, 1976). However, 8-glutamyl-S-methylcysteine in some bean varieties also yields methyl thiocyanate on cyanogen bromide treatment (Hulse et al., 1977). Analysis of total sulfur content, particularly by energy dispersive X-ray fluorescence spectrometry (Knudsen et al., 1981), is much simpler than sulfoamino acids analysis. In contrast to brown rice, where 71-80% of total sulfur was cysteine plus methionine

(IRRI and R.B. Clark, University of Nevada, unpublished data, 1982), sulfoamino acids fraction of total sulfur was more variable at 61-95% (mean of 78%) in cowpeas (Evans and Boulter, 1974), at 41-68% (mean of 55%) in chickpea, and at 59-85% (mean of 76%) in pigeonpea (Jambunathan and Singh, 1981).

Analysis of cystine and methionine ideally requires performic oxidation before acid hydrolysis and ion-exchange chromatographic analysis of cysteic acid and methionine sulfone (Spackman et al., 1958; Evans and Boulter, 1974).

Hulse et al. (1977) recommended screening of F_2 and succeeding generations of legumes for total N. The high 20% are analyzed for sulfoamino acids in all species except pigeonpea, which is analyzed for tryptophan. The high 20% (4% of materials) are then verified for methionine and cystine by ion-exchange or gas-liquid chromatography. At F_5 and subsequent generations, the lines are analyzed for lysine by DBC and the high 20% verified for lysine by ion-exchange chromatography (Mattern et al., 1970; Villegas et al., 1968). The subsequent high 20% undergoes further evaluation for antimetabolites and cooking quality, and only the high 20% undergoes biological evaluation for relative protein value in rats.

Oilseeds, such as peanuts, are also routinely screened for oil content by nuclear magnetic resonance.

RELATED CONSIDERATIONS

Several factors have to be considered in relation to testing for nutrients in plant breeding programs in order for such testing to succeed.

1. Environmental. Screening nurseries must be planted under uniform conditions, and if possible, in an environment that allows the greatest expression of genetic differences, and where effects on nutrient level are reduced to a minimum. Unfortunately, these environmental factors are still not well understood. Manipulation of management and cultural practices, particularly nitrogen fertilizer application, increases both yield and protein content of rice up to a threshold value characteristic of a variety and season, beyond which any further increase in protein content results in a decrease in grain yield (Gomez and De Datta, 1975). Many high-protein entries with decent yields tend to be of short growth duration.

2. Genetic and Physiological. The effective exploitation of high-nutrient mutant lines depends upon an understanding of the number of genes involved, their linkage relationship with other genes, and whether or not they have adverse pleiotropic effects on other characters of agronomic interest (IAEA, 1982). The physiological basis of their high nutrient content would also be valuable information. Unfortunately, most of these mutants are not linked with morphological properties of the crop. In a few cases where they are, the opaqueness of endosperm has to be removed to increase yield (grain weight) and improve resistance to postharvest fungal infestation.

Heritability estimates provide a measure of the proportion of the total variance in a character that is of genetic origin. Very few heritability estimates have been reported for protein content in crops, but they tend to be complex relative to nutrient quality mutants. In rice, only about 25-50% of the protein variability is estimated to be genetically controlled (Jennings et al., 1979). Another consideration is that cereal endosperm is triploid tissue but legume cotyledon is diploid.

In many cereals, the physiological basis of high-protein grain seems to involve the more efficient translocation of vegetative N to the developing grain

(Perez et al., 1973; Scholtz, 1982). However, the current semidwarf varieties already have low straw production or a high harvest index close to 0.50 and as much as 70-80% of total foliar N is in the seed. Thus, the prospects of further increasing seed protein by this mechanism may be limited. Further increased protein has to come from greater N uptake by the plant itself (Scholtz, 1982). Protein, and particularly the limiting amino acid lysine, required more glucose substrate to synthesize as compared to polysaccharides like starch (Bhatia and Rabson, 1976; Mitra et al., 1979).

3. Nutritional. The higher nutrient level sought must be nutritionally advantageous under typical methods of processing (such as rice milling) and based on typical consumer diets of the region. A staple food is seldom consumed alone, except as weaning food. In cassava-based diets, legumes provide protein in which sulfoamino acids contents are important. In cereal-legume diets, high sulfoamino acids content of cereals and high lysine content of legumes are important. The relationship between protein content and protein quality has been studied: in cereals, lysine content of protein usually decreases as protein content increases from genetic and environmental factors to about 10% protein in rice (Juliano et al., 1973) and to 14% protein in wheat (Mattern et al., 1970; Vogel et al., 1973) after which lysine value remains constant. In legumes, sulfoamino acids are also reported to decrease with an increase in protein content in field bean (Adams, 1973).

In cereals such as milled rice with little antinutrition factors, protein content is a good index of protein quality based on amino acid score and rat trials (Bressani et al., 1971; Eggum and Juliano, 1973; Hegsted and Juliano, 1974) (Table 2). The slightly lower nutritional quality of high (11%) protein IR480-5-9 milled rice as compared to 8% protein IR32 was confirmed by preschool children balance studies in Peru (MacLean et al., 1978) and in the Philippines (Roxas et al., 1979) using similar protocols. Huang (National Taiwan University, 1978) also found the IR480-5-9 rice to have the same protein quality in young men as a local 8% protein rice. Thus, the higher protein content of milled rice was of nutritional advantage. Results of long-term feeding trials in preschool children on rice-based diets fed either average (8%) protein or high (11%) protein rice did not consistently demonstrate the nutritional advantage of the high protein rice, as other factors such as zinc may become limiting (Roxas et al., 1980; Pereira et al., 1981).

4. Consumer Acceptance. Higher nutrient content should not have adverse effct on consumer acceptance. The price, appearance, cooking qualities, and flavour all influence consumer acceptance. Because starch and protein constitute the major nutrients of cereal endosperm, legumes, and tubers, they probably greatly influence the properties of grain such as grain hardness (Simmonds et al., 1973; Juliano, 1972). High protein content of rice due to N fertilizer application was reported to give flaky cooked rice using electric rice cookers (Ishima et al., 1974). The harder texture of high-protein rice may be explained by the slower rate of cooking due to the water-insoluble nature of rice protein; by adding more water to achieve similar water content of cooked rice, the rices have similar softness values (Juliano et al., 1965). Other examples of acceptability problems is the poor tortilla quality produced from some opaque-2 corn and the poor chapatti quality produced from Triticale. Carotene content of durum wheat is also important for macaroni color quality.

The importance of tests for consumer acceptability is well recognized in cereals (PAG, 1975) and legumes (Hulse et al., 1977). They included milling and baking properties for wheat (PAG, 1975) and amylose content, alkali spreading value, and gel consistency as indicators of cooked-rice texture for rice (Juliano, 1979), and cooking time for legumes (Hulse et al., 1977). At IPC, potatoes are

Table 2: Summary of Protein and Nutritional Data on Various Milled Rices. IRRI, 1970-1977

Property	Intan	IR8	IR22	IR22	IR8	IR11-3-15-8	IR-480-5-9	BPI-76-1
Protein (% N x 6.25)	6.0	7.7	7.9	10.2	10.2	11.6	11.8	15.2
Lysine (g/16 g N)	4.1	3.6	3.8	3.9	3.5	3.6	3.3	3.2
Methionine (g/16 g N)	3.3	2.9	2.2	2.3	2.3	2.4	2.5	1.7
Amino acid score[a/] (%)	74	65	68	70	64	66	61	58
Rat assay(%)								
N growth index[b/]	76	71	-	-	67	-	-	55
Relative nutritive value[c/]	72	74	75	74	65	64	55	50
Relative protein value[d/]	109	84	-	-	-	-	76	56
Net protein utilization[e/]	75	70	-	69	65	71	64	66

[a/] Based on 5.5 g lys/16 g N as 100%. [b/] Bressani et al. (1971). Based on a value of 75% for casein and corrected for differences in Kjeldahl factors. [c/] Hegsted and Juliano (1974). [d/] By B.E. McDonald, Dept. of Foods and Nutrition, University of Manitoba, 1974 (PAG, 1975). [e/] Eggum and Juliano (1973).

screened for specific gravity readings only as an index of total solids of tubers (O.T. Page, personal communication, 1982) in view of the good quality of potato protein (IPC, 1974). At IITA (1981), gari quality (panel evaluation) and garification rate are related to percentage dry matter of the cassava tuber. Gari processing involves peeling off the cassava skin, grating the peeled roots, dehydrating and fermenting, sieving off the chaff, and frying the pulp.

5. Protein and Energy Requirements. The field of protein and energy requirements has been in a state of flux and influences the relative emphasis on nutrient breeding. Protein requirements were obviously underestimated by the 1973 FAO/WHO recommendations. As a result of data of the protein-energy requirements research network of the UNU World Hunger Program, the FAO/WHO/UNU Expert Consultation on Protein-Energy Requirements, held in Rome, October 1981, increased the recommended safe allowance of protein for adults from 0.57 to 0.75 g/kg body weight, expressed in terms of high-quality protein (Food Nutr. Bull. 4(1): 42-43 (1982)). This recommendation should increase the protein deficiency level of diets.

THE FUTURE

Advancement in rapid, simple testing for nutrients in selective breeding will continue as our understanding of the relationship between seed structure and function, nutrient content and distribution, and acceptability and nutritive value improves. Such studies are interdisciplinary among nutritionists, botanists, food scientists and technologists, and chemists. Improvement in methodology will need inputs from analytical chemists. Nondestructive methods for analysis such as NIR spectrophotmetry (Norris, 1982) will continue to be developed, including those for specific amino acids (Panford et al., 1981).

Better understanding of the related topics discussed above would also be helpful in the breeding program for improved nutrients: environmental effects on nutrient content, understanding the genetics of nutrient content and its effect on grain chemistry, availability of nutrient being assayed, and possible effect on acceptability of the crop and physiological basis of high nutrient content.

Newer breeding methods using mutation of fertilized egg cell (Satoh and Omura, 1981) or protoplast or DNA fusion may result in mutation with higher protein content or quality with altered protein morphology and composition.

A much closer cooperation among chemists involved in nutrient screening in various institutions would be desirable to insure that the most practicable procedures are used in most breeding programs and that data are comparable.

ACKNOWLEDGEMENT

The author thanks E.I. Ortega of CIMMYT, El Batan, Mexico; D.O.T. Page, Centre Internacional de la Papa, Lima, Peru; R. Jambunathan, ICRISAT, Hyderabad, A.P., India; and S. Tsou and M.H. Yang, AVRDC, Tainan, Taiwan, China for sending information on methods used in the respective centers.

REFERENCES

Adams, M.W. (1973). On the quest for quality in the field bean. Nutritional improvement of food legumes by breeding, PAG, N.Y., 143-149.

Bechtel, D.B. and B.O. Juliano (1980). Formation of protein bodies in the starchy endosperm of rice (Oryza sativa L.): a reinvestigation. Ann. Bot. 45:503-509.

Bhatia, C.R. and R. Rabson (1976). Bioenergetic considerations in cereal breeding for protein improvement. Science 194:1418-1421.

Bressani, R., L.G. Elias and B.O. Juliano (1971). Evaluation of the protein quality of milled rices differing in protein content. J. Agric. Food Chem. 19:1028-1034.

Conacher, H.B.S. (1983). Rapid, simple testing for contaminants and toxicants. CHEMRAWN II Technical Proceedings. Pergamon Press Canada Ltd.

Derbyshire, E., D.G. Wright and D. Boulter (1976). Legumin and vicilin, storage proteins of legume seeds. Phytochem. 15:3-24.

Eggum, B.O. and B.O. Juliano (1973). Nitrogen balance in rats fed rices differing in protein content. J. Sci. Food Agric. 24:921-297.

Evans, I.M. and D. Boulter (1974). Chemical methods suitable for screening for protein content and quality in cowpea (Vigna unguiculata) meals. J. Sci. Food Agric. 25:311-322.

Finlayson, A.J. and S.L. MacKenzie (1976). A rapid method for methionine determination in plant materials. Anal. Biochem. 70:397-402.

Ford, J.E. and D. Hewitt (1979). Protein quality in cereals and pulses. 1. Application of microbiological and other in vitro methods in the evaluation of rice (Oryza sativa L.), sorghum, (Sorghum vulgare Pers.), barley and field beans (Vicia faba L.). Brit. J. Nutr. 41:341-352.

Gomez, K.A. and S.K. De Datta (1975). Influence of environment on protein content of rice. Agron. J. 67:565-568.

Guiragossian, V.Y., S.W. Van Scoyoc and J.D. Axtell, ed. (1977). Chemical and biological methods for grain and forage sorghum. Department of Agronomy, International Programs in Agriculture, Purdue University, West Lafayette, Indiana, 232.

Hegsted, D.M. and B.O. Juliano (1974). Difficulties in assessing the nutrutional quality of rice protein. J. Nutr. 104:772-781.
Hernandez, H.H. and L.S. Bates (1969). A modified method for rapid tryptophan analysis in maize. CIMMYT Res. Bull. 13, 7.
Hulse, J.H., K.O. Rachie and L.W. Billingsley, ed. (1977). Nutritional standards and methods of evaluation for food legume breeders. International Development Research Centre, Ottawa, Canada, IDRC-TS7e, 100
IAEA (International Atomic Energy Agency) (1977). Nutritional evaluation of cereal mutants. Proceedings of the Advisory Group Meeting, Vienna, 1976, Vienna, 187.
IAEA (International Atomic Energy Agency (1982). Conclusions and recommendations of the research co-ordination meeting of FAO/IAEA/GSF/SIDA Coordinated Research Programme on the Use of Nuclear Techniques for Cereal Grain Protein Improvement. Induced mutants for cereal grain protein improvement, IAEA, Vienna, pp. 207-213.
IITA (International Institute of Tropical Agriculture) (1981). 1980 annual report. IITA, Ibadan, Nigeria, 51-64.
IPC (International Potato Center) (1974). Potato protein quality. Report of the potato quality planning conference, CIP, Lima, Peru, 1973. CIP, Lima, Peru, 57.
IRRI (International Rice Research Institute) (1979). Proceedings of the workshop on chemical aspects of rice grain quality. IRRI, Los Banos, Laguna, Philippines, 390.
IRRI (International Rice Research Institute) (1980). Improvement of the protein content of rice. Summary report 1966-1979, Los Banos, Laguna, Philippines, 20.
IRRI (International Rice Research Institute) (1982). Annual report for 1981. Los Banos, Laguna, Philippines (in press).
Ishima, T., H. Taira, H. Taira and K. Mikoshiba (1974). Effect of nitrogeneous fertilizer and protein content in milled rice on organoleptic quality of cooked rice (in Japanese). Rep. Nat. Food Res. Inst. (Tokyo) 29:9-15.
Jambunathan, R. (1980). Improvement of the nutritional quality of sorghum and pearl millet. Food Nutr. Bull. 2(1): 11-16.
Jambunathan, R. and U. Singh (1981). Relationship between total sulphur and sulphur amino acids in chickpea (Cicer orietinum L.) and pigeonpea (Cajanus Cajan [L.] Millsp.). Qual. Plant. Plant Food Hum. Nutr. 31:109-117.
Jennings, P.R., W.R. Coffman and H.E. Kaufmann (1979). Rice improvement. IRRI, Los Banos, Laguna, Philippines, 186.
Johnson, V.A. and P.J. Mattern (1980). Genetic improvement of productivity and nutritional quality of wheat, 1966-1979. Final report of research findings. University of Nebraska-Lincoln Institute of Agriculture and Natural Resources, Dept. of Agronomy, 110.
Juliano, B.O. (1972). The rice caryopsis and its composition. Rice chemistry and technology, D.F. Houston, ed. American Association of Cereal Chemists, Inc., St. Paul, Minn., 16-74.
Juliano, B.O. (1979). The chemical basis of rice grain quality. Proceedings of the workshop on chemical aspects of rice grain quality. IRRI, Los Banos, Laguna, Philippines, 69-90.
Juliano, B.O., L.U. Onate and A.M. del Mundo (1965). Relation of starch composition, protein content, and gelatinization temperature to cooking and eating qualities of milled rice. Food Technol. 19:1006-1011.
Juliano, B.O., A.A. Antonio and B.V. Esmama (1973). Effects of protein content on the distribution and properties of rice protein. J. Sci. Food Agric. 24:295-306.

Juliano, B.O., C.M. Perez, S. Barber, A.B. Blakeney, T. Iwasaki, N. Shibuya, K.K. Keneaster, S. Chung, B. Laignelet, B. Launay, A.M. del Mundo, H. Suzuki, J. Shiki, S. Tsuji, J. Tokoyama, K. Tatsumi and B.D. Webb (1981). International cooperative comparison of instrument methods for cooked rice texture. J. Texture Studies 12: 17-38.

Knudsen, D., R.B. Clark, J.L. Denning and P.A. Pier (1981). Plant analysis of trace elements by X-ray. J. Plant Nutr. 3:61-75.

Landry, J. and T. Moureaux (1970). Heterogeneity of the glutelins of the grain of corn: selective extraction and composition in amino acids of the three isolated fractions (in French). Bull. Soc. Chim. Biol. 52:1021-1037.

Mattern, P.J., J.W. Schmidt and V.A. Johnson (1970). Screening for high lysine content in wheat. Cereal Sci. Today 15: 409-411.

MacLean, Jr., W.C., G.L. Klein, G. Lopez de Romana and G.G. Graham (1978). Protein quality of conventional and high-protein rice and digestibility of glutinous and non-glutinous rice by preschool children. J. Nutr. 108:1740-1747.

Mertz, E.T., L.S. Bates and O.E. Nelson (1964). Mutant gene that changes protein composition and increases lysine content of maize endosperm. Science 145:279-280.

Mertz, E.T., P.S. Misra and R. Jambunathan (1974). Rapid ninhydrin-color test for screening high-lysine mutants of maize, sorghum, barely and other cereal grains. Cereal Chem. 51:305-307.

Mertz, E.T., R. Jambunathan and P.S. Misra (1975). Simple chemical and biological methods used at Purdue University to evaluate cereals for protein quality. International Programs in Agriculture, Agric. Exp. Sta., Purdue Univ. Bull. No. 70, 25.

Mitchell, H.H., T.S. Hamilton and J.R. Beadles (1952). Relation between the protein content of corn and the nutritional value of the protein. J. Nutr. 48:461-476.

Mitra, R.K., C.R. Bhatia and R. Rabson (1979). Bioenergetic cost of altering the amino acid composition of cereal grains. Cereal Chem. 56:249-252.

Mossberg, R. (1969). Evaluation of protein quality and quantity by dye-binding capacity: a tool in plant breeding. New approaches to breeding for improved plant protein. International Atomic Energy Agency, Vienna, 151-160.

Munck, L., K.E. Karlssen, A. Hagberg and B.O. Eggum (1970). Gene for improved nutritional value of barley seed protein. Science 168:985-987.

Nelson, O.E., E.T. Mertz and L.S. Bates (1965). Second mutant gene affecting the amino acid pattern of maize endosperm proteins. Science 50:1469-1470.

Norris, K.H. (1982). Multivariate analysis of raw materials. CHEMRAWN II.

Oparka, K.J., P.J. Gates and D. Boulter (1981). Regularly aligned mitochondria in aleurone and sub-aleurone layers of developing rice caryopses. Plant Cell Environ. 4: 355-357.

Opienska-Blauth, J., M. Charezinksi and H. Berbec (1963). A new rapid method of determining tryptophan. Anal Biochem. 6:69-76.

PAG (Protein Advisory Group of the United Nations System) (1973). Nutritional improvement of food legumes by breeding. Proceedings of a symposium held at FAO, Rome, 1972, ed. by M. Milner. United Nations, N.Y., 389.

PAG (Protein Advisory Group of the United Nations System) (1975). PAG guideline (No. 16) on protein methods for cereal breeders as related to human nutritional requirements. PAG Bulletin 5(2):22-48.

Palter, R. and G.O. Kohler (1969). Survey hydrolysis procedure for lysine analysis. Cereal Chem. 46:22-26.

Panford, J.A., P.C. Williams and K.R. Preston (1981). Determination of lysine and other amino acids by near-infrared reflectance spectroscopy (abstr.). Cereal Foods World 26:495.

Paulis, J.W., J.S. Wall and W.F. Kwolek (1974). A rapid turbidimetric analysis for zein in corn and its correlation with lysine content. J. Agric. Food Chem. 22:313-317.

Pellett, P.L. and V.R. Young, ed. (1980). Nutritional evaluation of protein foods. United Nations University World Hunger Programme Food and Nutrition Bulletin Suppl. 4, Tokyo, 154.

Pereira, S.M., A. Begum and B.O. Juliano (1981). Effect of high protein rice on nitrogen retention and growth of preschool children on a rice-based diet. Qual. Plant. Plant Foods Hum. Nutr. 31:97-108.

Perez, C.M., G.B. Cagampang, B.V. Esmama, R.U. Monserrate and B.O. Juliano (1973). Protein metabolism in leaves and developing grains of rices differing in grain protein content. Plant Physiol. 51:537-542.

Resurreccion, A.P. and B.O. Juliano (1981). Properties of poorly digestible fraction of protein bodies of cooked milled rice. Qual. Plant. Plant Foods Hum. Nutr. 31: 119-128.

Roxas, B.V., C. Ll. Intengan and B.O. Juliano (1979). Protein quality of high-protein and low-protein milled rices in preschool children. J. Nutr. 109:832-839.

Roxas, B.V., C. Ll. Intengan and B.O. Juliano (1980). Effect of zinc supplementation and high-protein rice on the growth of preschool of children on a rice-based diet. Qual. Plant. Plant Foods Hum. Nutr. 30:187-193.

Satoh, H. and T. Omura (1981). New endosperm mutations induced by chemical mutagens in rice, Oryza sativa L. Japan J. Breed. 31:316-326.

Schaeffer, G.W. (1981). Mutations and cell selections: increased protein from regenerated rice tissue cultures. Environ. Expt. Bot. 21:333-345.

Scholtz, F. (1982). Problems and possibilities in genetic improvement of protein yields in cereals. Induced mutants for cereal grain protein improvement. IAEA, Vienna, 71-79.

Simmonds, D.H., K.K. Barlow and C.W. Wrigley (1973). The biochemical basis of grain hardness in wheat. Cereal Chem. 50:553-562.

Singh, R. and J.D. Axtell (1973). High lysine mutant gene that improves protein quality and biological value of grain sorghum. Crop. Sci. 13:535-539.

Singh, U. and R. Jambunathan (1981). A survey of the methods of milling and consumer acceptance of pigeonpeas in India. Proceedings of the International Workshop on Pigeonpeas, Vol. 2, ICRISAT, Patancheru, A.P., India, 419-425.

Spackman, D.H., W.H. Stein and S. Moore (1958). Automatic recording apparatus for use in the chromatography of amino acids. Anal. Chem. 31:1190-1206.

Spies, J.R. (1967). Determination of tryptophan in protein. Anal. Chem. 39:1412-1416.

Tanaka, K., T. Sugimoto, M. Ogawa and Z. Kasai (1980). Isolation and characterization of two types of protein bodies in the rice endosperm. Agric. Biol. Chem. 44:1633-1639.

Tanaka, Y., A.P. Resurreccion, B.O. Juliano and D.B. Bechtel (1978). Properties of whole and undigested fraction of protein bodies of milled rice. Agric. Biol. Chem. 42: 2015-2023.

Tsai, C.Y., L.W. Hansel and O.E Nelson (1972). A colorimetric method for screening maize seeds for lysine content. Cereal Chem. 49:572-579.

Villegas, E., C.E. MacDonald and K.A. Gilles (1968). Variability in the lysine content of wheat, rye and triticale. CIMMYT Res. Bull. No. 10, 32.

Villegas, E. and E.T. Mertz (1971). Chemical screening methods for maize protein quality at CIMMYT. CIMMYT Res. Bull. 20, 14.

Vogel, K.P., V.A. Johnson and P.J. Mattern (1973). Results of systematic analysis for protein and lysine composition of common wheats (Triticum aestivum L.) in the USDA World Collection. Univ. Nebraska-Lincoln Coll. Agric. Expt. Sta. Res. Bull. No. 285, 27.

Williams, P.C. (1975). Application of near infrared reflectance spectroscopy to the analysis of cereals and oilseeds. Cereal Chem. 55:561-576.

MULTIVARIATE ANALYSIS OF RAW MATERIALS

Karl H. Norris
Instrumentation Research Laboratory
Beltsville Agricultural Research Centre, USDA,
Beltsville, Maryland, 20705, U.S.A.

ABSTRACT

Techniques have recently been developed for providing a rapid prediction of the composition of agricultural products. The basic technology involves computer analysis of spectral transmittance or spectral reflectance curves of the product being measured. The major constituents of starch, protein, water, and oil can all be measured using near-infrared diffuse reflectance spectra. Measurements are made very rapidly with little or no sample preparation. Spectral transmittance or reflectance data in the visible region are used to predict chlorophyll, anthocyanins, carotenoids, and other components having absorption bands in this region. These measurements are made on sample slices, homogenates, or intact samples.

Examples of this technology that are discussed include: protein and moisture content of ground wheat; oil content of intact sunflower and soybean seeds; fat, moisture, and protein content of meat; chlorophyll and lycopene content of intact tomatoes; and moisture and protein content of bulk samples of wheat without grinding. The composition of such samples is predicted with a precision equal to or better than that which is obtained by normal chemical procedures.

KEYWORDS: Raw materials, rapid analysis, composition, diffuse reflectance spectroscopy, diffuse transmittance spectroscopy, oil, moisture, protein.

INTRODUCTION

The concept of direct spectrophotometric measurement of moisture content of grains was introduced by Norris and Hart (1965). The initial development used diffuse transmittance techniques, but diffuse reflectance techniques have advanced at a much faster pace (McClure et al., 1977; Norris et al., 1976; Hymowitz et al., 1974; and Watson, 1977) and are now accepted techniques for rapid prediction of the composition of grains and oilseeds.

NEAR INFRARED REFLECTANCE (NIR)

Diffuse reflectance spectrophotometry in the 1000 to 2600 nm region is used to predict the major constituents of starch, protein, oil, and water in agricultural products. This is possible because each of these constituents has definitive absorption properties in this region, and these absorptions can be sensed with diffuse reflectance. The spectra for these constituents are shown in Figure 1. The data are plotted as log (1/R) where R is the diffuse reflectance. Reflectance spectra plotted as log (1/R) are very similar to absorption spectra. The reflectance varies in a non-linear mode with changes in concentration of an absorber while the relation between log (1/R) and concentration is sufficiently linear to permit linear data treatment. Other data transformations, such as the Kubelka-Munk function, provide linear data, but these have not had much application for grain and seed analyses. The spectra show that the different constituents have absorptions which seriously overlap each other, complicating the problem of measuring any one of the constituents. However, data treatment procedures have been developed for such problems. The data treatments include: multiterm log (1/R), multiterm Δlog (1/R), curve fitting, and single-term derivative ratios of log (1/R). All of these procedures start with spectral data on a set of calibration samples of known composition. The relationship between the spectral data and the constituent data are determined by linear regression analysis to develop a calibration equation for each constituent. For multiterm log (1/R), the equation is in this form:

$$C = K_0 + K_1 \log (1/R_1) + K_2 \log (1/R_2) + K_3 \log (1/R_3) + K_4 \log (1/R_4) + K_5 \log (1/R_5) + K_6 \log (1/R_6) \quad (1)$$

where C is the concentration of the constituent being measured, R_1, R_2, R_3, R_4, R_5 and R_6 are the reflectance values at the six measured wavelengths, and K_0 through K_6 are the respective coefficients for the regression equation. The six wavelengths normally include one wavelength at the peak absorption wavelength for each of the major constituents in addition to one or two reference wavelengths. These wavelengths are 1940 nm for water, 2100 nm for starch, 2180 nm for protein, 2308 nm for oil, and reference wavelengths at 1680, and 2230 nm.

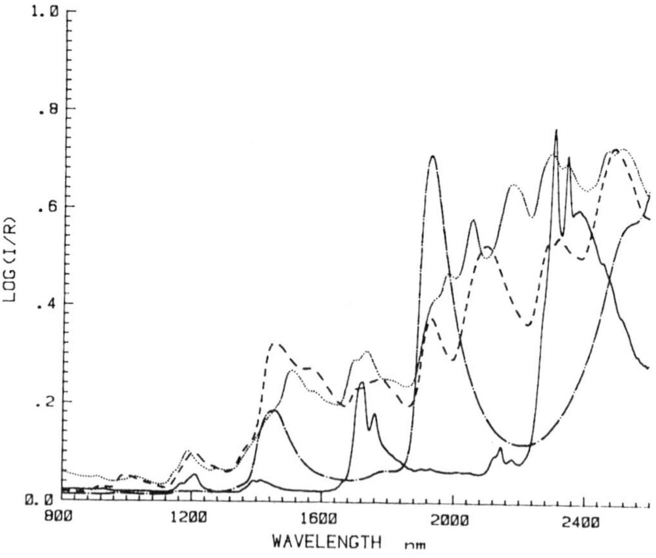

Figure 1: Spectral reflectance of major components of agricultural products. Oil ————, Protein ·········, Starch - - - - -, Water —— . —— .

For multiterm $\Delta\log(1/R)$, the same wavelengths may be used in an equation of this form:

$$C = K_0 + K_1 [\log(1/R_1) - \log(1/R_2)] + K_2 [\log(1/R_3) - \log(1/R_4)] + K_3 [\log(1/R_5) - \log(1/R_6)] \, . \quad (2)$$

The derivative ratio treatment can be with either the first or second derivative. The equations are in this form:

$$C = K_0 + K_1 \frac{d \log(1/R_1)}{d \log(1/R_2)} \quad (3) \qquad C = K_0 + K_1 \frac{d^2 \log(1/R_1)}{d^2 \log(1/R_2)} \quad (4)$$

The wavelengths for R_1 and R_2 are chosen by correlation searches of all wavelengths, but as shown by Norris (1982), the numerator wavelength is chosen to provide a derivative term best sensing the constituent being measured. The denominator wavelength is chosen to provide a derivative term which senses a summation of all the constituents.

Curve-fitting procedures use spectral data over a range of the spectrum where all the major constituents make a contribution. The procedure developed by Hruschka and Norris (1982) uses spectral curves of samples as components in a multiple regression analysis.

Practical NIR instruments for routine analyses are now available from at least five firms. These use interference filters to isolate the wavelengths to be measured. These instruments use one of the above four equations for their calibration. Some of the instruments provide additional wavelengths which can be used to measure other constituents, but the most successful applications have been in measuring protein, oil, and moisture in grains and oilseeds. The measurement requires about 2 minutes per sample with most of the time required for grinding the sample and packing it into a sample cell. The actual instrument readout occurs approximately 10 to 30 seconds after the sample is inserted into the instrument.

The type of performance attainable with NIR is illustrated in Figures 2 and 3. The protein content of 50 samples of Hard Red Spring wheat were predicted with a standard error of 0.14% protein, and the moisture content of the ground samples were predicted with a standard error of 0.09% moisture. These results were obtained using the data treatment of equation 3 with a numerator wavelength of 2160 nm and a denominator wavelength of 2270 nm for protein prediction. For moisture prediction, the numerator was at 2027 nm and the denominator at 2270 nm. The instrument was calibrated with a separate set of 50 samples of wheat of known moisture and protein content. The errors in protein prediction in this test are lower than normal because the chemical data are the average of 16 different protein determinations on each sample. If the NIR predicted proteins are compared to single protein determinations by the Kjeldahl procedures, the standard error is about 0.18% protein. The error in the NIR prediction is limited by the error in the chemical procedure used to test the NIR, unless multiple determinations are averaged to reduce the error in the chemical procedure.

The moisture prediction results of Figure 3 are the results on ground wheat samples for both the NIR data and the data for the standard procedure of oven drying. The NIR technique predicts the moisture of the ground sample very accurately, but it should be noted that considerable error can be introduced in using these data to predict the moisture content of the grain before grinding. Law and Tkachuk (1977) have reported on the use of diffuse reflectance measurements on whole grain wheat to predict moisture content.

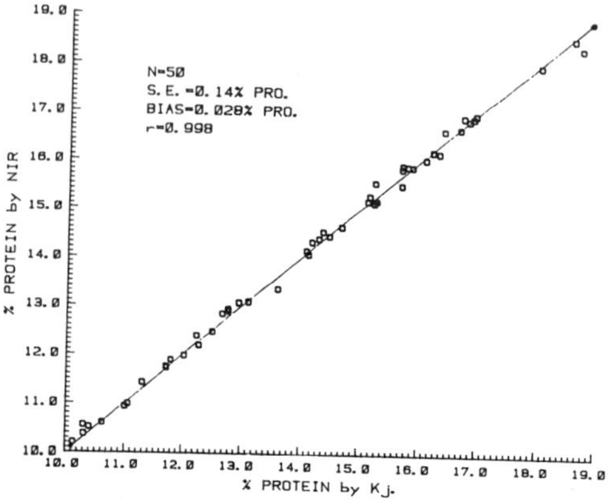

Fig. 2: Scatter plot of predicted protein vs. protein by chemical procedures for ground wheat.

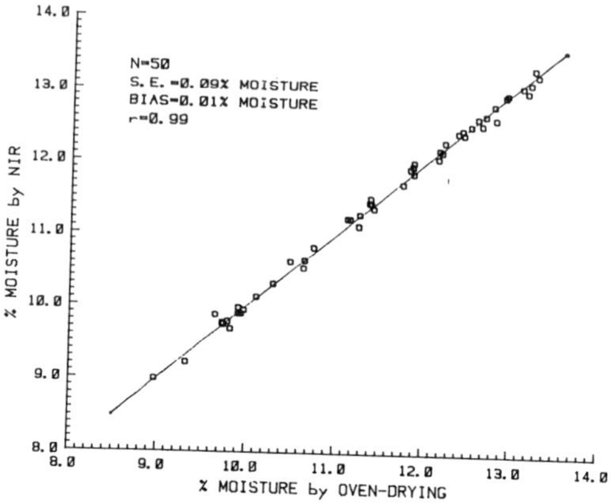

Fig. 3: Scatter plot of predicted moisture vs. moisture by oven-drying procedure for ground wheat.

A test of NIR in predicting the protein content of ground samples of brown rice did not give results equivalent to those obtained on wheat. The results (Figure 4) show a standard error of 0.31% protein using the average of two determinations by the Kjeldahl procedure. The standard deviation between the two Kjeldahl procedures was 0.15% protein, so this should not be the source of the error. The samples used in this test were from an international collection of rice, and it is possible that the protein-to-nitrogen ratio variability causes an error. The NIR does not sense nitrogen as does the Kjeldahl procedure, but senses protein, so if the protein-to-nitrogen ratio varies from sample to sample, an error will be introduced. A variation of 15% in the protein-to-nitrogen ratio between extreme samples would create the error observed in this test.

The test with rice samples served another useful purpose, because in the first test, a different set of Kjeldahl data were used and the errors were much larger as shown in Figure 5. The same samples were reanalyzed in duplicate without any change in the NIR data to obtain the results of Figure 4. These results demonstrate the value of NIR as a means of checking laboratory data. The instrument was calibrated with a set of 124 samples of rice for both predictions. The calibration equation used the first derivative ratio with the numerator at 2150 nm and the denominator at 2270 nm.

NEAR-INFRARED TRANSMITTANCE (NIT)

Although transmittance techniques were used for the initial work with near-infrared analyses, the requirement of working with a very thin sample has limited the applications. This requirement for thin-samples of grains and oilseeds occurs because the light-scattering from the particles, combined with the absorption coefficients of the constituents, causes the samples to have transmittance values too low for measurement. However, as we move to shorter and shorter wavelengths, the absorption coefficients become much lower, and in the 800 to 1100 nm region, the transmittance is high enough to make useful measurements through thick samples. The major constituents of protein, starch, water, and oil still have defined absorption bands in this region, although the overlapping is more serious (Figure 6).

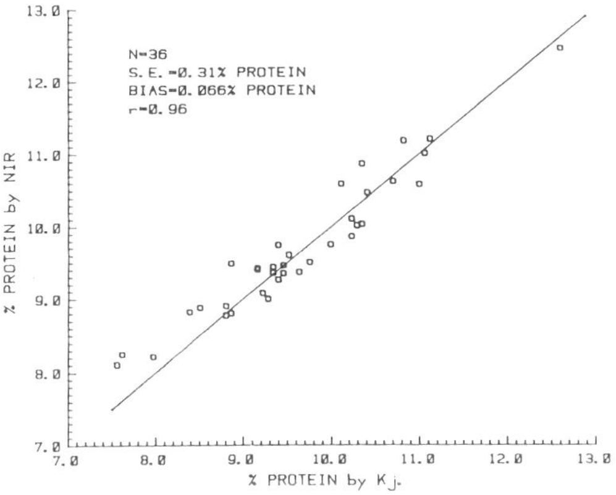

Fig. 4: Scatter Plot of predicted protein vs. protein by chemical procedures for ground rice.

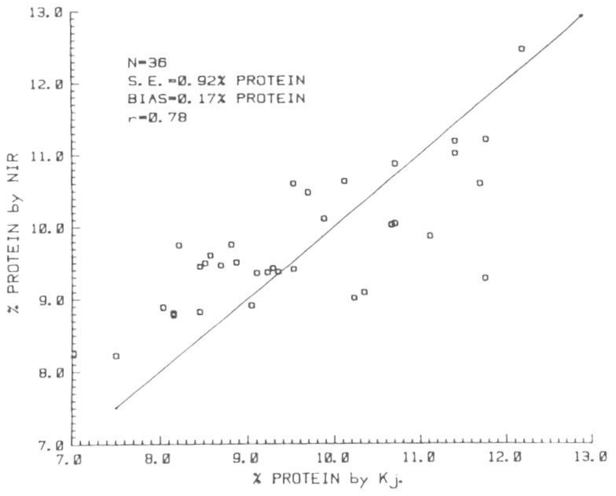

Fig. 5: Scatter plot of predicted protein vs. protein from initial chemical procedures for ground rice.

Large-area, low-noise silicon photo detectors, combined with high-efficiency monochromators, make it possible to measure the transmittance of 10 cm of clear water or oil, 2 cm of high moisture tissue such as a potato, 2 cm of particles of whole grain, 1 cm of homogenized meat, and 0.3 cm of ground grain or other dry product in the 800 to 1100 nm region. This spectral region has been explored for measuring the fat, moisture, and protein content of raw pork, the oil content of individual oil seeds, and the protein and moisture content of bulk samples of whole wheat.

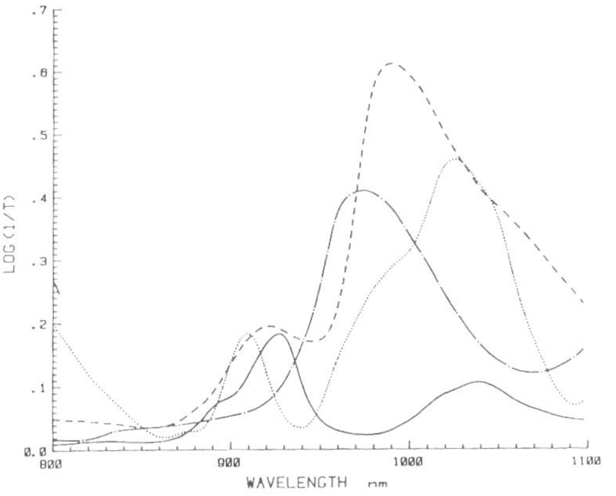

Fig. 6: Spectral transmittance properties of major components of agricultural products.
Oil ———, Protein ·····,
Starch - - -, Water — · — .

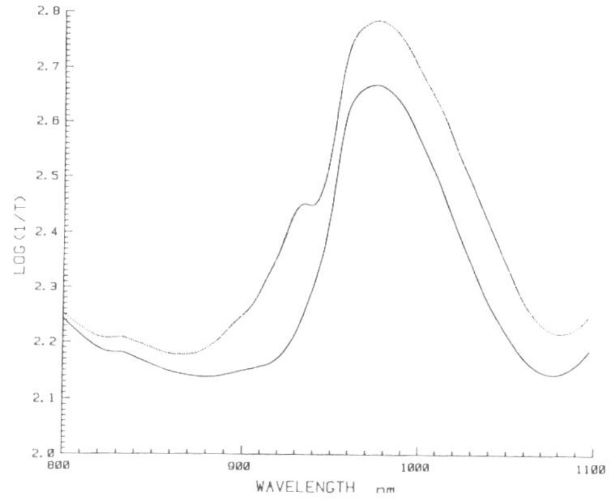

Fig. 7: Typical log (1/T) spectra for homogenized raw pork. Low-fat sample ———, High-fat sample ····· .

Raw pork samples, representing different cuts from an animal and from different animals, were ground to a paste in a food processor to provide a very uniform sample for analysis. Spectral transmittance measurements were made on 1-cm thick samples for 38 such samples for calibration. The same measurements were made on 19 other pork samples to test the performance of the calibration equation. Typical log (1/T) spectra for the pork samples are shown in Figure 7. A calibration equation was developed using the second-derivative data treatment similar to equation (4). The numerator wavelength for fat was 931 nm and the denominator wavelength was 1070 nm. For moisture, the numerator wavelength was 1043 nm with the denominator at 1070 nm, and for protein the numerator was 929 nm with the same denominator. The calibrations were tested by predicting the fat, moisture, and protein content of the set of 19 samples not included in the calibration samples. Excellent results (Table 1) were shown for all three components.

Table 1: Prediction Errors for 19 Pork Samples.

Constituent	Mean %	S.D.[a] %	S.E.[b] %	Bias[c] %	r^2[d]
Fat	8.25	3.83	0.44	-0.027	0.990
Moisture	7.2	3.03	0.66	-0.040	0.956
Protein	15.9	3.05	0.50	0.011	0.975

[a] Standard deviation of chemical data.
[b] Standard error of prediction, standard deviation of differences between prediction and chemical data.
[c] Bias, difference between the means of prediction and chemical results.
[d] Coefficient of determination for predicted results.

Spectral tansmittance measurements on intact individual soybean seeds were made to develop a non-destructive procedure for measuring oil content. Oil contents on an individual seed basis were not available, but the oil content of the samples from which the seeds were drawn were available. Therefore, although individual seeds were measured, the performance was evaluated on the basis of an average of 20 seeds from each lot. The spectra from the 20 seeds from each lot were averaged to obtain one spectrum for each lot for analyses. Ten lots of seeds were measured to develop a calibration equation, and a separate set of 7 lots were used for prediction. The typical log (1/T) spectra for soybean seeds are shown in Figure 8. A calibration equation using a second derivative ratio in the form of equation (4) was developed with the numerator at 1145 nm and the denominator at 965 nm. This calibration produced the results of Table 2 in predicting the unknown samples. These results indicate that it is possible to predict the oil content of individual soybean seeds without harming the seed in any way. It should be possible to also predict the protein and moisture content with the spectral data, but this was not tested.

Table 2: Prediction Errors for Oil Content of 7 Soybean Seed Lots and 7 Sunflower Seed Lots.

Seed	Mean %	S.D. %	S.E. %	Bias %	r^2
Soybean	18.8	3.55	0.52	-0.02	0.987
Sunflower	41.9	2.67	0.41	0.085	0.983

A similar test was performed on individual sunflower seeds, again averaging 20 seeds per sample lot. Ten sample lots were used to develop a calibration which was tested with a prediction set of 7 sample lots. The prediction results in Table 2 show that it is possible to predict the oil content of individual sunflower seeds. These individual seed measurements required 60 sec. to scan the spectrum for each seed, but the measurement could be made in less than 10 sec. per seed.

Spectral transmittance measurements on 2.1 cm thick samples of whole grain, Hard Red Spring wheat, were measured in the 800 to 1100 nm region. A calibration equation was developed for protein prediction using the second-derivative-ratio data treatment similar to equation (4). The numerator wavelength was 1016 nm, and the denominator was 952 nm. This equation was tested on a prediction set of 49 samples, with the results shown in Table 3. The standard error is considerably higher on this whole grain test compared to the results on the ground grain. However, this measurement can be made much faster because the sample preparation time is eliminated.

Table 3. Prediction errors for protein content of whole wheat.

	Mean %	S.D. %	S.E. %	Bias %	r^2
Protein	15.55	2.39	0.28	0.014	0.988

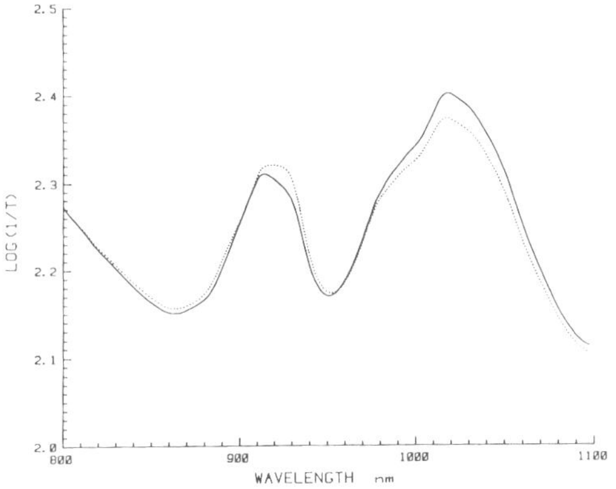

Fig. 8: Typical log (1/T) spectra for individual intact soybean seeds. Low-oil, high-protein seed ——— , High-oil, low-protein seed ······ .

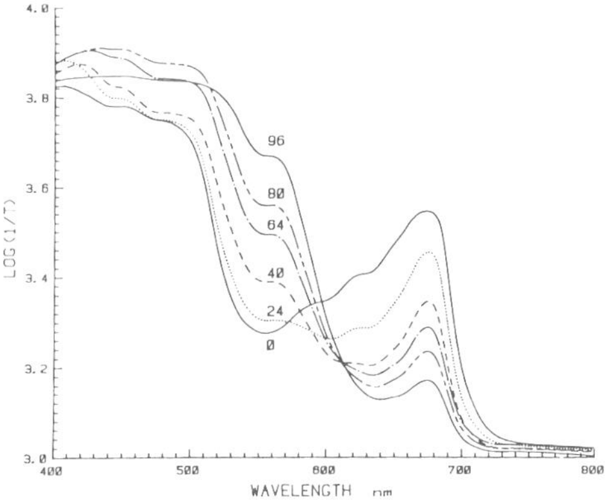

Fig. 9: Changes in log (1/T) spectra as a tomato ripens. Numbers on the curves are the hours stored at 22°C starting with zero at the mature green stage.

VISIBLE TRANSMITTANCE

The data treatment techniques developed for the near-infrared region can also be applied in the visible region. Thus, it is possible to measure the pigment concentration of fruit and vegetable tissues with very little sample preparation. Transmittance techniques are recommended over reflectance for these measurements because most often the pigment is distributed through a large part of the tissue. Reflectance measures the constituents at or near the surface of a sample while transmittance measures the constituents distributed through the total volume of tissue within the path of the measuring radiation. Reflectance measurements are widely used to evaluate the color of agricultural products, and many products are sorted automatically with reflectance techniques. Color-measuring instruments are often used to predict pigment concentrations, but such instruments were designed to measure color rather than composition, and they do not represent the optimum design for composition analyses.

The availability of high-sensitivity, low-noise, end-window, multiplier-type phototubes makes it possible to measure the spectral transmittance properties of intact fruits and vegetables. The spectral changes of a ripening tomato are shown in Figure 9. As the tomato ripens, the chlorophyll absorption (675 nm) decreases and the lycopene absorption (570 nm) increases. Pigment changes such as these can be measured on a wide range of products. The curve for the mature-green tomato illustrates one of the greatest problems of spectral analyses on fruits and vegetables. The chlorophyll absorption peak in the 675 nm region is flattened by fluorescence interference. It is difficult to overcome the fluorescence limitation although Massie and Norris (1975) described a dual-wheel, filter monochromator to minimize the fluorescence.

Light-transmittance techniques have been developed to predict the chlorophyll, lycopene, and beta-carotene content of intact tomatoes (Watada et al., 1976); chlorophyll content of peaches (Sidwell et al., 1961); and chlorophyll content of apples (Yeatman and Norris, 1965). Spectral measurements in the visible are widely used to study pigment changes in plant tissues, but the application to routine analysis of food products has not developed. The technology is available, and it is anticipated that it will be applied.

CONCLUSIONS

Technology has been developed to provide a rapid prediction of the major constituents of agricultural products. The constituents which can be predicted include: starch, protein, oil, sugar, moisture, chlorophyll, and carotenoids. Additional work is needed to develop procedures for other constituents. Instrumentation is now available for most of the measurements listed above, and new instruments are being developed. These instruments make it possible to give a direct readout to six constituents in a ground product in less than 30 seconds.

REFERENCES

Hruschka, W.R. and K.H. Norris (1982). Least-squares curve fitting of near infrared spectra predicts protein and moisture content of ground wheat. Applied Spectroscopy 36(1), 261-265.

Hymowitz, T., J.W. Dudley, F.I. Collins and G.M. Brown (1974). Estimation of protein and oil concentration in corn, soybean, and oat seed by near-infrared light reflectance. Crop Sci. 14, 210-214.

Law, D.P. and R. Tkachuk (1977). Determination of moisture content in wheat by near-infrared diffuse reflectance spectrophotometry. Cereal Chem. 54(4), 874-881.

Massie, D.R. and K.H. Norris (1975). A high intensity spectrophotometer interfaced with a computer for food quality measurements. Trans. ASAE 18, 173-176.

McClure, W.F., K.H. Norris and W.W. Weeks (1977). Rapid spectrophotometric analysis analysis of the chemical composition of tobacco. Part 1: Total reducing sugars. Beitrage zur Tabakforschung 9(1), 13-18.

Norris, K.H. and J.R. Hart (1965). Direct spectrophotometric determination of moisture content of grain and seeds. In: "Humidity and Moisture", 4, "Principles and Methods of Measuring Moisture in Liquids and Solids", Reinhold Publishing Corp., New York.

Norris, K.H., R.F. Barnes, J.E. Moore and J.S. Shenk (1976). Predicting forage quality by infrared reflectance spectroscopy. J. Animal Sci. 43(4), 889-897.

Sidwell, A.P., G.S. Birth, J.V. Ernest and C. Golumbic (1961). The use of light transmittance techniques to estimate the chlorophyll content and stage of maturation of Elberta peaches. Food Tech. 15(2), 75-78.

Watada, A.E., K.H. Norris, J.T. Worthington and D.R. Massie (1976). Estimation of chlorophyll and carotenoid content of whole tomato by absorbance techniques. J. Food Sci. 41, 329-332.

Watson, G.A. (1977). Near infrared reflectance spectrophotometric analysis of agricultural products. Anal. Chem. 49(9), 17-21.

Yeatman, J.N. and K.H. Norris (1965). Evaluating internal quality of apples with new automatic fruit sorter. Food Tech. 19, 123-125.

THE IMPORTANT DIFFERENCE BETWEEN CHEMICAL ANALYSIS AND BIOLOGICAL AVAILABILITY

A.E. Bender

Department of Nutrition, Queen Elizabeth College
University of London, Campden Hill, London W8 7AH, England

ABSTRACT

The nutrients present in food are usually incompletely absorbed, and the amount available may be influenced, either enhanced or depressed, by other ingredients of that food or of the remainder of the diet. Chemical/physical measurements of nutrients are reproducible and relatively precise but rarely reveal availability. Animal assays provide information of the amounts of nutrients available to the test animal, but methods are lengthy and the results invariably cover a wide range. Extrapolation to man presents additional problems. However, a limited amount of information has been obtained from human experiments. While chemical/physical methods must continue to provide the greater part of the required information, all such determinations must be verified by bioassay when novel foods, new varieties and novel methods of processing are involved. If the food is expected to make a significant contribution to the diet, final verification on human subjects will be needed.

KEYWORDS: nutrients; availability; bioavailability; nutrient analysis; nutritional value.

GENERAL CONSIDERATIONS

The objective of measuring the nutrient content of a food is usually to ascertain how much is available to the consumer - apart from comparisons before and after processing, and similar considerations. This is probably the most difficult problem facing the nutritionist, and viturally impossible to achieve. Two indexes are used - chemical/physical measurement and animal bioassay but, in both cases, these have to be extrapolated to man.

The 'true' assay is the biological one in which the defined response of the test animal is measured. If vitamin E prevents resorption of the fetus, then the amount of food required to prevent resorption of the fetus compared with the standard amount of the pure substance is, by definition, a true measure of the vitamin E potency of the food. Such bioassays are long and laborious and consequently extremely expensive. For example, in the bioassay of vitamin E discussed below (Weiser and Vecchi, 1981) the authors used six groups of test animals each on a different dose of standard, and another six groups, each on a different dose of the material under test. It required 3 months on a vitamin-E-free diet to deplete the animals, and depletion was verified by mating at 4-weekly intervals, and when no living embryos could be detected after day-20 of pregnancy then depletion was complete and the assay could start. Five days after mating, the depleted animals were allotted to their respective dosage groups and continued for a total of 20 days of pregnancy, after which they were sacrificed and examined for fetuses. The test depends on the protection provided by the test substance against resorption of the fetus, i.e. the vitamin E potency, and the result is calculated from the dose-response relation of test substance compared with standard. In all biological work there is considerable variation between animals, and extremely large variations invariably come to light when inter-laboratory assays are carried out; so as well as being lengthy and expensive, the assay is very imprecise.

Chemical and physical measurements of nutrients are precise, usually rapid and reproducible but suffer two drawbacks. The colour developed with the particular reagent, or the physical measurement such as absorption of light of a given wavelength, depends on the presence of a specific chemical grouping in the molecule and so is not specifically measuring the biologically active material. Secondly, such measurements may be enhanced or repressed by the presence of other substances present in the food.

When dealing with a pure substance in relatively high concentration such as a pharmaceutical preparation, the chemical or physical properties provide a precise measure of the material present. In a food, however, there are three different considerations. Firstly, the active substance may be present in one or more of a variety of chemical forms. For example, there are three forms of vitamin B6, more than ten forms of vitamin A, and eight major forms of vitamin E. In modern nomenclature, the term vitamin is used as a generic descriptor for substances with defined biological properties while the substance itself, e.g. retinol or cryptoxanthin or beta-carotene, is referred to by its chemical name. Secondly, a micronutrient will be present in a food or tissue in extemely small amounts, of the order of parts per million or less, and this means that there are vast amounts of numerous other substances present. Thirdly, there may be substances present which enhance or inhibit the physical/chemical reaction. This is particularly a problem in microbiological assay which may respond precisely and reproducibly in pure solutions, but the growth of the organism may be stimulated or repressed by the other substances present in the test material.

This comparison of the relative advantages of physical/chemical methods with biological assay concerns measuring the amount of the nutrient present - it does not provide any information about the amount available to the consumer. It is rarely possible to measure this, the only exceptions being research carried out on human subjects. There are a very limited number of experiments designed to determine the amount of the nutrient required by man such as the Sheffield experiments to determine the requirements for vitamin A and vitamin C (Hume and Krebs, 1949; Bartley et al., 1953) and work of the type carried out by Layrisse et al. (1969) to determine the availability of iron in food. These human experiments reveal very large variations between subjects, as indeed do laboratory animal assays: yet these are the 'true' estimations. The human trials provide the information required, i.e. the proportion of the nutrient actually available, and the animal assays measure the potency of the substance relative to a pure substance (although such measurements then have to be extrapolated to man). So, the results show very large variability of 'true' values while chemical/physical methods provide precise and reproducible results which may not be the true values. Hence the dilemma.

These problems can be exemplified by considering vitamins A and E, selected minerals, protein and energy.

Vitamin A

Vitamin A is the generic descriptor for a variety of chemical substances which possess a specified biological activity, including retinol and its derivatives, α-, β-, γ- carotene, cryptoxanthin and some other minor carotenoids. Enormous effort has been devoted in a very few experiments in attempting to determine human requirements for vitamin A. Fortunately, the average results of the few investigations show good agreement with extrapolations made from animal experiments. But the work indicates that it would not be possible to use human subjects to assay the vitamin content of a food. The best we can do is to use an animal bioassay in which the biological potency of the food is compared with a standard amount of the vitamin and to assume, in the instance of vitamin A, with some confidence, that the same comparison is valid in man.

In the Sheffield experiment (Hume and Krebs, 1949) determination of requirements involved measuring the amount of vitamin A that was available in the foodstuffs provided but later, with more sophisticated methods, the turnover of labelled retinol was used and this does not provide information on availability. The advantages of physical/chemical methods listed above are usually considered to outweigh the drawbacks when compared with the problems involved in bioassay, but there are unrecognised problems exemplified by vitamin A. Until 1971, it was considered that canning of fruits and vegetables had no effect on vitamin A potency because colorimetric estimation of the extracted carotenoids showed no change on processing. Sweeney and Marsh (1971), however, showed that heat isomerises the carotenoids and changes their biological potencies (Table 1) and concluded that green vegetables, containing largely beta-carotene, lose 15-20% of their vitamin potency on processing, while red and yellow vegetables, containing largely alpha-carotene, lose 30-35% of their potency. Bioassay would have revealed this (but probably with less precision). Chemical work involves separation of the different chemical forms before measurement - also a long and laborious procedure but presumably far more precise and reproducible.

Table 1: Relative potencies of isomers of carotene (Sweeney and March, 1971)

All -trans β- carotene	100
neo -β- carotene-B	53
all - trans -α- carotene	53
neo -β- carotene - U	38
neo -α- carotene - B	16
neo -α- carotene - U	13

One of the more intractable problems in assaying the content of a nutrient, whether chemically or biologically, is that of extracting it (Bender and MacFarlane, 1965). There are reports in the literature of apparent increases in the carotene content of foods because it was less completely extracted from the raw food than after the cell walls had been broken down by heat. Similarly, there are considerable potential errors in the assay of folic acid (microbiological) because the food has first to be autoclaved to permit extraction of the vitamin. A measurement of bioavailability carried out by feeding the food itself would apparently yield the correct value, yet the findings of the Sheffield experiment belie this. By measuring blood levels in depleted human subjects it was found that the proportion of carotene absorbed from carrots varied from 25% to 50% depending on the physical form of the food, namely on how finely it was minced.

Vitamin E

Vitamin E presents an excellent example of the complex nature of the problems involved. The term vitamin E is the generic descriptor for all substances possessing specific biological activities including the prevention of resorption of fetuses in the rat and the protection of red blood cells from haemolysis. There are eight substances with the biological property, namely α, β, γ and δ - tocopherols and α, β, γ and δ-tocotrienols. For chemical estimation it is necessary to separate the isomers, usually by thin-layer chromatography, and then measure the amount of each isomer present and finally to calculate these in terms of α-tocopherol equivalents.

As recently as 1981 it became possible to separate the various diastereomeric forms of α-tocopherol itself and they include one form with R-configuration at all three assymetric centres in the molecule (namely C2', 4' and 8'); this is termed RRR-α-tocopherol acetate and is one of the main substances of commercial importance. A second is RS-R-R (or 2-ambo-alpha) tocopherol. A third is all-racemic-α-tocopherol, RS-RS-RS. These three forms were compared for their biological potency by Weiser and Vecchi (1981) using the procedure outlined above. The potency of all-rac-αTA: RRR-α-TA was 1:1.5; the potentcy of 2-ambo-α-TA: RRR was 1:1.34, and that of all-rac-α-TA: 2-ambo was 1:1.1. The US Pharmacopoeia XX, 1980, quotes dL-α-TA as 1 USP unit, compared with dl-α-tocopherol at 1.1 and d-α-tocopherol at 1.49 USP units. This presents new problems.

Iron

The availability of iron in foods is a well-established problem, although its extent is not always realised (Lock and Bender, 1980). The WHO Report (1972) and standard textbooks (Davidson et al., 1975), corrected later (Davidson et al., 1979), provide a figure showing the proportion of total iron in various foods found to be available from human experiments using doubly-labelled iron (Figure 1). This shows, by a divided rectangle, the proportion of iron absorbed and implies that the rectangle indicates the mean and standard error or deviation. The original publications (Layrisse and Martinez-Torres, 1971) show that they are, in fact, the antilogs of the log standard error on each side of the geometric mean - a statistical device to deal with the enormous variation in the results. The mean absorption of iron ranges from 1% from rice, 3% from beans, to 8% from soya and 20% from veal muscle. However, the individual values cover the enormous ranges shown in Figure 2.

The absorption of iron depends on three groups of factors. 1) The chemical form of the iron in the food. Inorganic iron is absorbed most readily, haeme iron is well absorbed compared with plant foods, ferrous salts more readily than ferric. The proportion absorbed also varies with the amount of iron in the diet. 2) Other foods consumed at the same time. Absorption is enhanced by ascorbic acid, an acid medium, and by fructose and amino acids which chelate with the iron. Absorption is decreased by phytates, oxalates and phosphates. 3) The physiological state of the subject, including iron reserves, concentration of acid in the stomach and intestinal motility.

None of the proposed *in vitro* methods of measuring bioavailability, has found general acceptance nor, in view of the numerous factors other than the food itself which influence the amount absorbed, can any method be verified by a human trial. The amount releasable from a food may be useless as an index of the amount absorbed if the other groups of factors listed above affect absorption. Moreover, measurements carried out directly on human subjects, even were they possible under routine conditions, show such enormous variations as to render them virtually valueless. Layrisse and Martinez-Torres (1971) used labelled iron and found that

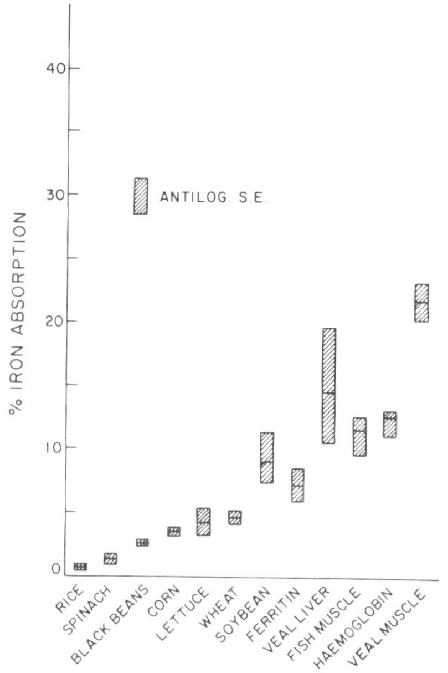

Fig. 1: % Iron Absorption from Foods.

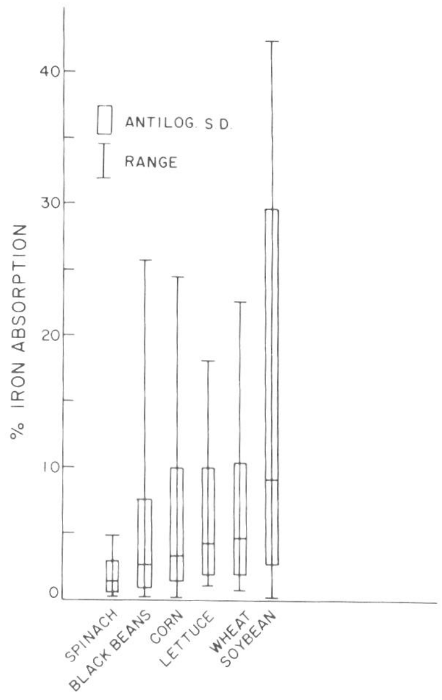

Fig. 2: % Iron Absorption from Foods.

in a group of 28 individuals under controlled conditions, the amount of iron absorbed from soybeans ranged from 2% to 42.2%!

The problems of iron availability, which are widely appreciated, are not very different from those for vitamin A which are not generally appreciated. This also exists in foods in a variety of chemical forms, its absorption is influenced by the presence of other nutrients in the diet and the physiological state of the individual, but the figure for retinol equivalents, calculated from chemical estimation of retinol and carotenoids stated in food composition tables, is generally accepted as the amount that is available. For vitamin A, we tend to regard dietary and physiological factors quite separately from the chemical forms of the vitamin, whereas for iron, we combine all the factors.

Proteins

It is usual procedure to measure total nitrogen in foods, and rarer to measure true proteins. It is also usual to measure the total amount of each amino acid present although it is realised that some part of the amino acids may not be biologically available. This is certainly well established for lysine when the Maillard complex results in discrepancies between chemical and biological assays. There is an accepted chemical method for available lysine but not for other amino acids and little, indeed, is known of the availability of other amino acids (Hurrell and Carpenter, 1976; Almas and Bender, 1980). Consequently it is accepted practice to depend on one of the methods of biological evaluation of protein quality rather than chemical tests. However, it is not often realised that the lengthy biological assay, usually regarded as the ultimate test for the 'true' value of a protein, provides only a single figure which depends solely on the limiting amino acid and supplies no information about the other amino acids.

Energy

Most of the problems associated with availability relate the micronutrients and proteins, but there is even a problem of estimating the energy available from foods. The procedure falls into two parts: measurement of the fat, carbohydrate and protein; and conversion into energy. The problems of measuring lipid content of foods are well known and, although not always employed, methods involving hydrolysis are required to yield total lipid.

The greatest problem is that of carbohydrate which some workers determine 'by difference' while others measure total, and yet others measure the amount available (i.e. dietary fibre separately from available carbohydrate measured as monosaccharide). This is still presenting a problem despite the fact that it has been known for so long. For example, Ashley and Heck (1981) stated that the energy values for fruits and vegetables were 'remarkably' similar in Swiss and German food composition tables but that these values were 'strikingly' higher than those from British tables, where the carbohydrate was always lower. The authors then proceed to state that this 'seems' to reflect the way in which the data are expressed - in the British tables the carbohydrates are reported in terms of equivalent monosaccharides and indigestible dietary fibre, whereas the total carbohydrates (available and unavailable) are reported in the Swiss and German tables. That such confusion can arise in the literature in 1981 when the procedure was clearly explained in the original British tables in 1940 serves to illustrate the difficulties.

With the change to SI units, the former energy conversion factors for carbohydrates, proteins and fats (respectively, 4, 4 and 9 according to Atwater and 4.1, 4.1 and 9.3 according to Rubner, the former allowing for incomplete absorption from the diet) were changed. Tables 2A and B show those that have come into common usage, with existing minor differences. Little direct information of the conversion factors for human beings is available since the classical work of Atwater in 1903. These were derived from specified diets and there is no evidence that the same figures (and their subsequent approximations as joules) pertain to different diets or to single foods under analysis. Southgate and Durnin (1970) re-examined the figures using two diets with and without fruits and vegetables which provide much unavailable carbohydrate. Their attempt to determine the figures for digestible or metabolisable energy of the foods did note how this differs from gross energy, which is the heat of combustion in vitro without allowing for the losses in urine and faeces. The authors pointed out the variation between subjects in their ability to digest pentosans and cellulose, presumably due to variations in the intestinal microflora. Obviously variations of this kind will be greater on diets rich in dietary fibre, thus producing larger errors in energy estimates.

Table 2A: Atwater Energy Conversion Factors Currently is Use (kJ)

	Germany	United Kingdom
Fat	38	37
Middle-chain length triglycerides	34	-
Protein	17	17
Carbohydrate	17	16
Ethanol	30	29
Organic acids	13	-

Table 2B: Comparison of Energy Content of Foods Calculated by Three Methods (Kcal)

Food	4.1;9.3;3.75[a]	4;9;4.2[b]	FAO 1947 Merrill and Watts (1955)[c]
Bread, brown	242	245	251
Wholemeal flour	339	336	327
Butter	793	768	748
Milk	66	65	68
Beef steak	177	172	177
Apple	45	45	55
Bananas	77	76	103
Currants, black	29	28	79
Oranges	35	35	49
Beans, raw butter	266	264	350
Potatoes	87	86	92
Groundnuts	603	586	576

a - protein x 4.1; fat x 9.3; available carbohydrate as monosaccharides x 3.75
b - protein x 4; fat x 9; available carbohydrate as starch x 4.2
c - specific factors for different foods (Food and Agriculture Organisation).

Energy-yielding components of foods and computation of calorific values. 1947. UN, FAO, Washington DC; Merrill, A.L. and Watts, B.K. (1955) Energy Value of Foods - Basis and Derivation. US Dept. of Agriculture, Agr. Handbook No. 74. Washington D.C.

Quoted from Paul, A.A. and Southgate D.A.T. 1978. McCance and Widdowson's The Composition of Foods. 4th Ed. MRC Spec. Rpt. Ser. No. 297. HMSO. London.

SOME BASIC PROBLEMS

In addition to the fundamental difficulties discussed above, there are problems still remaining in chemical/physical methodology. Micronutrients are present in extremely small quantities so that methods of estimation must be either highly specific so as to quantify the nutrient in the presence of vast amounts of numerous other substances, or some degree of purification is required with the subsequent application of a less specific test. Amino acids, for example, are separated by chromatography and then a general reaction is applied for the amino-group. The unsatisfactory nature of much of our physical/chemical methodology is underlined by the continuing development of new and improved methods.

Assays for thiamin, riboflavin and niacin are well-established and in common use. Both chemical and microbiological methods are used and the choice is often based on convenience. However, the microbiological assay is reputed to be reproducible only to about ±10% in the hands of the same operator and ±25-30% (or worse) between laboratories. The chemical estimation of thiamin also presents problems. It is reported that polyphenols present in the food interfere with the determination, and methods of reducing this effect have been suggested (Ratanaubolchai and Panijpan 1980), yet others have failed to verify that such interference exists (Horman and Brambilla, 1981).

SUBSTANCES THAT INTERFERE WITH AVAILABILITY

There are two related problems in attempting to measure the availability of nutrients - the presence in the food of substances that reduce absorption, and the presence of interfering substances in other ingredients of the diet. The latter is the concern of the nutritionist. The former, however, may be considered to lie within the realm of the analyst.

Taking iron as an example, it is possible to determine total iron present and also the other substances known to influence its absorption and so possibly achieve an index or forecast of the bioavailability. This has not been done, but it has been suggested for the effect of phytate in reducing the absorption of minerals where there appears to be an inverse relation between the amount of phytate and the proportion of minerals such as iron, calcium, and zinc that are absorbed. Indeed, it was partly as the result of such calculations that the British Department of Health (DHSS 1980) suggested the level of enrichment of soya products with zinc and iron. Support for the effect of phytate comes from the work of Welch and House (1982) who showed that mature soybeans (120 days harvest) rich in phytate, allowed 60% absorption of zinc while the 84 day bean with one third the amount of phytate allowed 90% absorption.

Selenium

A parallel problem is presented by selenium, recently shown to be essential to man since it is part of the enzyme system glutathione peroxidase. First, relative to food supplies, it must be realised that there is a wide range of selenium contents of foods which reflects the concentration in the soil. Thus cereals make little contribution to the intake of selenium in the diet in New Zealand, where the levels are low, but make a major contribution in the Canadian and US diets where the selenium content is high. The chemical form of selenium in food influences its availability and it is possible to measure the availability by restoration of glutathione peroxidase activity in selenium-depleted chicks. By this method it was shown (Gabrielson and Opstvedt, 1980) that availability relative to selenite was 48% (range 39-60) from capelo fish meal, 34% (32-36) from mackerel fish meal, 17% from soya, 26% from corn gluten, and that 78% of selenomethionine was absorbed. Douglas et al. (1981) showed that selenium from tuna fish is only 50% available compared with that from beef kidney and wheat.

CONCLUSIONS AND FUTURE DEVELOPMENTS

We are faced with the development of new varieties of traditional foods which will provide higher yields and other advantages; with the introduction of novel foods, and with the development of novel methods of food processing. Novel foods are those not previously eaten in significant amounts in the community in question, and so include molds and yeasts which have been part of the diet for many centuries but in only small amounts, and foods commonly consumed in other areas of the world but not previously known to the community in question - such as durian fruit common in the SE Pacific, spirulina consumed in N. Africa, and even deer, horse and snails regularly consumed in some communities. The problem here is illustrated by favism - the faba bean is toxic to those with a genetic disorder but not to other population groups. Thus, the introduction of a food already in common use elsewhere does not make it safe - nor nutritionally adequate - in other areas.

Changes in processing methods also can seriously affect the nutrient content of food as has been detected by laboratory analysis. One example is the damage effected on vitamin C in instant potato (now rectified by adding vitamin C to levels in amounts greater than found in the raw material). Laboratory analysis before and after processing may be misleading when the nutrient cannot be completely extracted for analysis before processing and more is extracted after heat breakdown of cellular structure, and by destruction of substances that inhibit or even enhance the analytical results. Carotene, for example, can be isomerised to substances of reduced biological activity. For accepted processes such as air-drying, vacuum drying, blanching, freezing, canning, etc., there is no evidence that any significant change in nutrient intake has resulted, although such changes may have been masked by the concomitant provision of a much greater variety of foods through these very processes. Before any new processes are adopted, it is necessary to ascertain whether any such nutritional changes do take place. Here the prime example is extrusion cooking which appears to have drastic effects on some of the ingredients of the foods.

New knowledge may deepen the problems. We were once content with our knowledge of vitamin D2 and D3, and believed that foods contained the natural form D3, cholecalciferol (CC). Now we know that milk, for example, contains a range of metabolites, 25 HOCC, 24,25 diHOCC, 25,26 diHOCC, 1,25 diHOCC; that fish, a traditional source of vitamin D, contains esters of C20:5 and C22:6 fatty acids, and that chicken liver contains 1,24,25 triHOCC. Although we do not know whether vitamin D esters need to be hydrolysed before absorption, it has been shown that the esterified form is less potent that the free vitamin D (Rambeck et al., 1981) and that the palmitate is less active than the butyrate. The discovery of a water-soluble sulphate of vitamin D led to problems regarding the analysis, which was always carried out on the fat fraction, and consequently on the requirements of infants whose intake had been based on such analyses.

Some analytical methods have been so thoroughly examined over the years that they are accepted as providing a reasonable index of available nutrients and are certainly adequate for comparisons before and after processing, and to determine the stability of nutrients on storage. For example, past efforts to establish the factor required to convert the ultra-violet absorption of vitamin A into biological potency, along with correction factors to allow for interference, led to this method being accepted as a measure of the vitamin.

The acceptance of such methods applied only to foods that have been thoroughly investigated and results found to be reliable. 'New' foods present a new problem. New varieties and novel foods will be the principal means of increasing the food available for the growing populations. These may contain substances which interfere with the standard assay, and similarly novel methods of processing may introduce errors. Furthermore, such new foods may introduce materials into the diet which interfere with the bioavailability of nutrients present in the traditional foods or they may replace traditional foods of known nutritional value.

Comparisons of nutrient contents of new foods or those processed by novel means can be made by standard physical/chemical/microbiological techniques, but this can only be regarded as a preliminary step. They must be verified by trials on animals with metabolic needs and pathways similar to those in man. Furthermore, if the food is to become a major part of the diet then human trials will be necessary, despite the drawbacks described. At least comparisons can be made with the traditional foods that they may replace.

REFERENCES

Almas, K. and A.E. Bender (1980). Effect of heat treatment of legumes on available lysine. J. Sci. Food Agric. 31, 448-452.

Ashley, D. and E. Heck (1981). Differences among UK, Swiss and German food composition tables. Internat. J. Vit. Nutr. Res. 51, 190.

Bartley, W., H.A. Krebs and J.R.P. O'Brien (1953). Vitamin C requirements. Spec. Ser. Med. Res. Count. London No. 280.

Bender, A.E. (1982). Evaluation of protein quality: methodological considerations. Proc. Nutr. Soc. In press.

Bender, A.E. and A.J. MacFarlane (1965). Determination of β-carotene in a roller dried food. J. Soc. Anal. Chem. 90, 536-540.

Davidson, S., R. Passmore, J.F. Brock and A.S. Truswell (1975). Human Nutrition and Dietetics, 6th Ed., Churchill-Livingstone.

Davidson, S., R. Passmore, J.F. Brock and A.S. Truswell (1979). Human Nutrition and Dietetics, 7th Ed., Churchill-Livingstone.

DHSS (1980). Novel foods intended to simulate meat. Dept. of Health and Social Security Report No. 17. HMSO London.

Douglas, J.S., V.C. Morris, J.H. Soares and O.A. Lavender (1981). Nutritional availability to rats of selenium in tuna, beef kidney and wheat. J. Nutr. 111, 2180-2187.

Gabrielson, B.O. and J. Opstvedt (1980). Availability of selenium in fish meal in comparison with soybean meal, corn gluten meal and selenomethionine relative to selenium in sodium selenite for restoring glutathione peroxidase activity in selenium-depleted chicks. J. Nutr. 110, 1096-1100.

Herman, I. and E. Brambilla (1981). Evidence refuting the antithiamine effect of chlorogenic and caffeic acids. Internat. J. Vit. Nutr. Res. 51, 385-390.

Hume, E.M. and H.A. Krebs (1949). Vitamin A requirements. Spec. Rep. Ser. Med. Res. Coun. London No. 264.

Hurrell, R. and K.J. Carpenter (1976). An approach to the rapid measurement of 'reactive lysine' in foods by dye binding. Proc. Nutr. Soc. 35, 23A-24A.

Layrisse, M., J.D. Cook, C. Martinez-Torres et al. (1969). Food iron absorption: a comparison of vegetable and animal foods. J. Hematol. 33, 430-443.

Layrisse, M. and C. Martinez-Torres (1971). Food iron absorption: iron supplementation of food. Prog. Haemat. 7, 137-160.

Rambeck, W.A., H. Weiser, R. Haselbrauer, R. and H. Zucker (1981). Vitamin D activity of different vitamin D3 esters in chicken, Japanese quail and in rats. Internat. J. Vit. Nutr. Res. 51, 353-358.

Ratanaubolchai, K. and B. Panijpan (1980). Determination of thiamine modification by polyphenols using $K_3Fe(CN)_6$, $HgCl_2$ and CNBr. Internat. J. Vit. Nutr. Res. 50, 3-9.

Southgate, D.A.T. and J.V.G.A. Durnin (1970). Calorie Conversion Factors. An experimental reassessment of the factors used in the calculation of the energy value of human diets. Brit. J. Nutr. 24, 517-535.

Sweeney, J.P. and A.C. Marsh (1971). Effect of processing on provitamin A in vegetables. J. Amer. Dietet. Assoc. 59, 238-243.

Weiser, H. and M. Vecchi (1981). Stereoisomers of a-tocopheryl acetate: characterisation of the samples by physico-chemical methods and determination of the biological activities in the rat resorption-gestation test. Internat. J. Vit. Nutr. Res. 51, 100-113.

Welch, R.M. and W.A. House (1982). Availability to rats of zinc from soybean seeds as affected by maturity of seed, source of dietary protein and soluble phytate. J. Nutr. 112, 879-885.

World Health Organisation (1972). Tech. Rep. Ser. No. 503, Geneva.

World Health Organisation (1973). Trace elements in human nutrition. Tech. Rep. Ser. No. 452, Geneva.

PROBLEMS OF CORRELATION AND DEFINITION OF ANALYTICAL TECHNIQUES

Werner Baltes
Institute of Food Chemistry, Technical University Berlin
Strasse des 17. Juni 135, D-1000 Berlin 12
Federal Republic of Germany

ABSTRACT

The growing, international interdependence of food production and commerce requires approved and sound methods of food analysis for consumer protection. Both differences in food laws and traditionally developed procedures hinder the introduction of such methods. Modern, instrumental procedures alone cannot offer a remedy, because their application often hinders error recognition. Therefore, it seems necessary to equalize internationally the analysts' level of knowledge by encouraging better, personal rapport among them. Inter-laboratory studies can also be useful for correlating various analytical techniques and, thereby, achieve standardized methods that will be generally recognized.

KEYWORDS: Analysis; standardized analytical methods; inter-laboratory studies; correlation of analyses.

INTRODUCTION

Hardly any other product within man's sphere needs so much control of purity, authenticity, and quality as food. Therefore, it is no surprise that from historical times, laws were made regarding production and marketing of food. Thus, it is known that the Babylonian emperor, Hammurabi, more than 3600 years ago, created the legal foundations for the punishment of beer adulteration and of profiteering. Moreover, there existed food control organizations in ancient Rome and Hellenistic Greece.

Concerning the problems of definition and correlation of analytical techniques, the following points are important: 1) The selection of analytical methods depends on the laws of the country concerned. 2) Food analysis began in a time when only weak support from analytical chemistry was available. As a result, certain traditions in the examination of food became established. Replacing these methods will not be easy. 3) Modern chromatographic and spectrographic methods enable us to determine almost all components of food. However, the possibilities of error become in no way smaller, but larger. 4) Various countries have been trying recently to limit the possibilities of error and to guarantee correlation of results by stipulating official methods on a national basis. 5) Incomparably much more difficult will be the problem of correlating analytical methods on an international basis, because ultimately, the political bodies and not the experts have the final say.

According to the Code of Ethics for International Trade in Food (Codex Alimentarius Commission, 1979): "Food legislation and food control infrastructures are not sufficiently developed in many countries to enable adequate protection of their food imports and prevent the dumping of substandard and unsafe foods." These countries will have a lot of difficulty in establishing an up-to-date system for food control and analysis. Moreover, many problems are assessed differently. For example, DDT was banned in the U.S.A. and in most European countries because of its persistence; the basis for this action being a recommendation from the FDA. The official laboratories in these countries can detect this compound very quickly. However, the situation is different in many developing countries in which DDT has helped greatly in the fight against malaria. It has often been observed that food from these countries was heavily contaminated with this pesticide. Nevertheless, one can be sure that DDT had not been recently used. However, plants absorb DDT from the ground where, because of its high persistence, it remains unchanged for years. In the Federal Republic of Germany, the pesticide problem became legally regulated through an ordinance stipulating the maximal amounts of plant pesticides which may be present in food. These amounts lie on the average between 50 ppb and 1 ppm. Importers from malaria-infested areas demand certificates from approved laboratories in these countries which state that the allowed residue levels are not exceeded. However, such laboratories are rare. A further problem which especially affects East Asia is the occurrence of histamine and mercury residues in fish preserves, about which American and European officials are very sensitive. Therefore, importers of such goods demand certificates guaranteeing that the allowed amounts of these contaminants have not been exceeded. There are seldom laboratories which possess the equipment as well as the know-how to do such analyses. Their task becomes even more difficult since each importing country specifies its own analytical methods.

CHEMICAL FOOD ANALYSIS

Chemical food analysis goes back to around the year 1900. At that time, there were neither chromatographic nor spectographic methods, rather the chemists in those days had to deal with methods which we today consider to be very inexact. In spite of this, their results were remarkable. A good example is the analysis of fats, which was carried out with the help of exactly defined, so-called reference numbers. Many of these methods are still used today because they are accurate and require little investment in equipment. One example is the Butyric-Acid Number derived from the Reichert-Meissl Number (DGF). The Butyric-Acid-Number describes the sum of the volatile acids dissolved in an aqueous solution saturated with potassium sulfate and caprylic acid. Although one does not exactly determine the amount of butyric acid, the Butyric-Acid-Number is nevertheless capable of determining, with sufficient accuracy, the purity of butter fat compared to, for example, butter-margarine mixtures. This procedure employs a special apparatus in which all measurements are exactly fixed. This analysis is done in many laboratories in middle-European countries where there is a legal ban on the mixing of butter with margarine. Naturally, one can also use a chromatographic method or do a sterine analysis. One can even separate triglycerides and perform mass-spectrometric analyses. However, assuming proper performance, the Butyric-Acid Number method is less prone to error than, for example, the gas-chromatographic methods. Another example of a traditionally used method is the procedure for the determination of reducing sugars in food. In the U.S.A., the method of Munson and Walker (AOAC, 1980) is followed, which uses a sodium hydroxide-tartrate solution in the presence of copper sulfate. In the method of Luff-Schoorl, sodium carbonate is added instead of sodium hydroxide. Both methods are empirical and the conditions must be carefully specified.

However, there is no correlation between these methods.

Today, physical measurement methods have prevailed everywhere in chemical food analysis, and one can assume that newly developed methods will be tested very quickly. Unfortunately, in such methods, especially since the amounts of substance to be determined are becoming smaller, errors are even more difficult to recognize. Such errors can cause a catastrophe when the instruments are linked to an automatic computer. Test manuals should be made available to the user to help him to attain optimal functioning of the particular instrument. The trained analyst can generally recognize malfunctions because he works mostly with an internal standard. However, embarrassing misinterpretations can happen. For instance, in Germany there was quite an uproar, together with heavy financial losses for the company involved, when a scientist using a polarographic method, mistook tin for lead in a dry soup mix and even published the results. Similar alarms arose repeatedly through misinterpretations of gas chromatograms in the analysis of pesticides. By employing automatic analyses, we will place greater demands on the operator's critical capability. Increased employment of electronics in no way compensates for human knowledge.

There is a flood of analytical methods for food. Even in the national sphere, each laboratory has its own special analytical methods. This was adequate as long as trading was restricted to a small area, and each analyst could maintain direct contact with the producer. However, the concentration of food production and the increased international interdependence of the food trade require the establishment of official methods. There are now few such collections, e.g., the Official Methods of Analysis of the AOAC.

In their latest version, the food laws of the Federal Republic of Germany have called for the organization of an official collection of analytic procedures for the testing of food, tobacco products, cosmetics and other commodities. This collection is currently being organized. These methods will obtain legal status, and the official analytical laboratories will be required to use these methods for legal purposes or provide an explanation for selecting another method.

Furthermore, in order to possibly guarantee a high degree of reliability of the collection, methods involving quantitative determinations will include specifications for the statistical assessment of the results. The smallest random error occurs whenever the same person analyzes identical material in the same laboratory, with the same methods and instruments. From examinations according to ISO 3534 (1977), one obtains values for the repeatability of a method. However, the determination of inter-laboratory reproducibility of a method is necessary. One obtains these values by testing identical material with the same method but in different laboratories, i.e. by different people, with different instruments and, consequently, under different conditions. This means conducting inter-laboratory studies.

There is presently a recommendation from the ISO (ISO 5725-1981) concerning inter-laboratory studies. Firstly, one needs a laboratory which is equipped to produce test materials and which is capable of assuming supervision of such a study. Together with an experienced statistician, it is important that there be a committee of experts which includes, at best, one representative from each laboratory involved in the study. The composition of the test materials, whose analysis the methods concern, shall include about 6 different orders of magnitude. No fewer than 8 laboratories should be involved to provide an adequate number of parallel tests.

Outlying values should be reported because their number provides information about statistical significance. Therefore, according to the ISO recommendations, one should consider a method with more than 5% outliers as being statistically insignificant. Naturally, that depends on the measurement range. The results should be so reported that the reproducibility as well as the repeatability are given depending on the level of material present. The smallest concentration of a substance which can be detected is another relevant factor.

In the Federal Republic of Germany, many laboratories are currently working under the direction of the Federal Department of Health on inter-laboratory studies to aid in the preparation of the official collection of analytical methods. There are, moreover, many international attempts. In August 1981 in Helsinki, an IUPAC symposium on the topic "Collaborative Inter-Laboratory Studies in Chemical Analysis" (Egan and West, 1982), represented a first attempt to coordinate the current inter-laboratory studies being conducted by many international organizations. The following organizations, among others, are concerned: the International Office of Cocoa and Chocolate (IOCC); the International Sugar Confectionery Manufacturers' Association (ISCMA); the International Commission for Uniform Methods of Analysis (ICUMSA); the International Office of Wine (OIV); the International Dairy Federation; the International Federation for Cereal Chemistry (ICC); the Nordic Committee on Food Analysis; the European Monitoring and Evaluation Programme (EMEP); the Clincial Chemistry Division; the Collaborative International Pesticides Analytical Council (CIPAC); and the International Organization for Standardization (ISO). At this meeting, it became obvious that still greater efforts for international collaboration were needed, and it was recommended that IUPAC should adopt the standardized definitions compiled by the ISO (ISO 3534-1977). Furthermore, the necessity to establish standards for the international acceptance of analytical methods was emphasized. Thought was also given to the formation of an international organization somewhat like the AOAC. One international problem is the formation of a sample bank which can issue reference materials on demand. The lack of standardized products has hindered again and again the execution of inter-laboratory studies. Of importance is the listing of suitable laboratories which can participate in such studies. It is clear that poorly equipped laboratories with little know-how produce more outliers than institutes having high standards. This especially affects residue analysis in which the results are sometimes reported with errors up to 100%. In the closing statements of the IUPAC symposium, standards of accreditation of involved laboratories were demanded. Last but not least, much could be achieved if more opportunities could be created for international meetings of food chemists.

Another very important problem is that of sample taking. For example, how can one sample 100 tons of one product originating from 130 different producers and packaged in bundles weighing up to 5 kg? Statistical guidelines are needed to adequately assure a representative sampling. This problem is probably more difficult to solve than the evaluation of analytical techniques.

CONCLUSIONS

In conclusion, some possible solutions for the above problems are offered.

1. The IUPAC symposium in Helsinki revealed the desire for the formation of a larger organization to handle this task. Three alternatives could be: (a) the IUPAC; (b) a consolidation of chemical societies, i.e., the American Chemical Society with the Federation of European Chemical Societies, and the Federation of Asian Chemical Societies, and (c) formation of a new, overlapping society. Probably the first two alternatives are more likely to succeed because these societies already have the necessary experience. The new organization should bring the national groups more closely together,

organize joint symposia, and set up analytical committees. It should work with the appropriate U.N. bodies, e.g., FAO/WHO.

2. If it is possible, one should offer for each analysis not one, but two or three alternative methods which, of course, must correlate with each other. The fact that laboratories are equipped differently would thus be taken into account.

3. Just as one standardizes analytical methods, one must set standards for the personnel involved in this work to ensure an adequate level of knowledge and skill.

4. The recognition of specially accredited laboratories must be promoted. In particular cases such already exist, e.g., for the examination of wine. Such laboratories should constantly prove themselves by participating in inter-laboratory studies whereby outliers would eventually be eliminated. In the less industrialized countries, laboratories should be created which specialize in a few analyses of importance locally. Personnel training for these laboratories could take place at recognized institutions.

5. Such actions naturally cost money which many national treasuries do not have. The possibility of obtaining U.N. recognition of these efforts, and thereby financial aid, should be explored.

Finally, it is important to recognize that while synthetic chemistry can fight world hunger by supplying agricultural chemicals, analytical chemistry is equally essential since it helps to secure a food supply of good quality.

REFERENCES

AOAC, 1980, Official Methods of Analysis of the Association of Official Analytical Chemists (W. Horwitz, Ed.), 13 Edition, Washington, DC.

Codex Alimentarius Commission, 1979: Code of Ethics for International Trade in Food. FAO/WHO of United Nations, CAC IRCP 20-1979.

Egan, H. and West, T.S. (1982): Collaborative Inter-Laboratory Studies in Chemical Analysis, Pergamon Press.

ISO 5725 - 1981: Precision of Test Methods - Determination of Repeatability and Reproducibility by Inter-Laboratory Tests.

ISO 3534 - 1977: Statistics - Vocabulary and Symbols. ISO Standards Handbook 3 (1981). International Organization for Standardization. ISO Central Sekretariat, Case postale 56, CH 1211, Genève 20, Schweiz.

OVERVIEW OF THE POTENTIAL AND PROSPECTS IN GENETIC ENGINEERING OF PLANTS

Lawrence Bogorad

Department of Cellular and Development Biology
The Biological Laboratories, Harvard University
16 Divinity Avenue
Cambridge, Massachusetts 02138, USA

ABSTRACT

Features of bacterial genes and of genes in plastids, mitochondria and nuclei of eukaryotic cells are reviewed. Procedures for manipulating genes for genetic engineering are summarized.

Genetic engineering tasks are considered at three different levels of complexity: introduction of a foreign gene which is expressed at all times in all tissues and organs; introduction of a gene bearing control elements that limit its expression to specific stages or times in development; introduction of sets of genes for multi-gene traits. Some of the difficulties at each of these levels are discussed.

The need for continuing basic research is emphasized as a prelude to the realization of the many potential applications of genetic engineering to food production that can now be seen.

KEYWORDS: Gene manipulation, bacterial gene, nuclear gene, genetic engineering, gene expression, genetic diversity.

INTRODUCTION

The objective of genetic engineering and related biotechnologies is to change the hereditary makeup of an organism. This is accomplished by permanently installing in the cell new genetic material -- new genes. The techniques of genetic engineering permit us to consider realistically introducing genes of one organism into any other organism -- species to species and kingdom to kingdom. In order to proceed with a discussion of the potentials and limitations of the broad range of newly available and imagined biotechnology for improving agriculture, particularly plants, it is necessary to first review the features of genes.

Operationally, genes are units of heredity that specify biochemical activities that may be manifested as morphological features. The experiments of Mendel provide a classical example -- genes that determine flower petal color in peas. In the simplest and most direct form, a gene is a transmissable hereditary element containing information that specifies an enzyme protein. For example, an enzyme that catalyzes a step in the production of a violet flower pigment. In contrast to such single gene traits, many characters are controlled by complex sets of genes.

Genes are made of DNA. DNA is a sequence of deoxyribonucleotides connected to one another by phosphate ester bonds linking their deoxyribose moeities. Each of the two strands of the DNA double helix is comprised of a backbone chain of deoxyribose residues linked to one another by phosphate ester bonds with a purine or pyrimidine residue protruding from each deoxyribose. The sequence is generally written as a series of A's, T's, G's, and C's. That is, a series of adenine, thymine, guanine and cytosine residues. One chain of A's, G's, T's and C's is complemented by another set of bases attached to a chain of deoxyribose sugars. The purine A forms hydrogen bonds with the pyrimidine T and the purine G forms hydrogen bonds with the pyrimidine C. The famous DNA double helix is thus comprised of two chains of deoxyribose sugars, each sugar of a single chain is linked to the next through a phosphate ester bond. Protruding from each sugar is the base A, G, T or C. The other chain of the double helix has the same sort of sugar-phosphate-sugar backbone but in each position a complementary base is present, i.e. a base that can form hydrogen bonds to the base at that position on the other strand of the helix.

THE CENTRAL DOGMA: GENE TO ENZYME PROTEIN

By a process called transcription the enzyme RNA polymerase catalyzes the formation of a chain of ribonucleotides complementary to one of the strands of the DNA double helix. The information contained in the gene, i.e. in DNA, is thus rewritten in a sequence of ribonucleotides, i.e. RNA. The portion of the RNA that carries information for a sequence of amino acids, using a triplet ribonucleotide code, is translated into a polypeptide by cellular machinery including special micro-organelles called ribosomes. These two steps, transcription and translation, convert the information present in the gene into a protein; some proteins serve structural roles in the cell while others are enzymes. The enzyme catalyzes a reaction such as a step in the synthesis of a flower pigment. Gene expression is the sum of activities starting with transcription, including translation, and in some cases, the consequences of the activity of the enzyme specified by the gene.

THE STRUCTURE OF A BACTERIAL GENE

The simplest sort of gene for a protein consists of three major regions. The first, a promoter, is the sequence of nucleotides recognized by the RNA polymerase (the enzyme for transcription). The polymerase binds to the DNA chain here and is directed to a place at which transcription will start. The second region of the gene is the portion that is transcribed into RNA. This latter region ends at the third region of the gene, a sequence of deoxyribonucleotides that signals the RNA polymerase to drop off of the DNA and thus to stop transcription.

Most of the stretch of DNA that is transcribed is a series of nucleotides that, when read out in the proper triplet code, specifies the sequence of amino acids in the final protein product of the gene. This amino acid coding region, or reading frame, is preceded almost immediately by a short sequence of nucleotides that serves as a site for binding to the ribosome on which translation of the RNA transcript into an amino acid chain will occur. A transcript of this sort is called a messenger RNA (mRNA). This is the form in which information contained in a gene is made available for the synthesis of a protein.

THE STRUCTURE OF A NUCLEAR GENE

Genes in the chromosomes in the nucleus of cells of higher organisms (eukaryotes) differ from bacterial genes in having different types of promoter sequences, in lacking distinguishable ribosomal binding sequences and in frequently having introns. Introns are sometimes very long stretches of DNA which contain no coding information (although in a few cases some amount of amino acid coding information has been found in introns), but interrupt reading frames coding for proteins. The RNA transcript of an intron-containing gene is a series of alternating coding and intron sequences. Before the transcript becomes a useful mRNA, the introns are excised.

Eukaryotic cells contain genes not only in their nuclei but also in mitochondria, the sites of respiration and oxidative phosphorylation, and, in higher plants, genes are present in plastids as well. Plastids may take the form of photosynthetic chloroplasts, starch-storing amyloplasts, carotenoid crystal-containing chromoplasts, etc.

Genes in mitochondria, at least in fungi and animals, are strikingly different from those in plastids and nuclei in that some codons for amino acids are not the same -- a single mRNA would specify different proteins when translated by the mitochondrial system than by the plastid or cytoplasm. DNA sequences flanking coding regions of plastid genes resemble bacterial genes more than nuclear genes. Each of these three gene-containing compartments of a eukaryotic cell has a unique set of enzymes for replicating DNA, for transcription, and for translation. Considering the difference in codon usage in mitochondria versus plastids and the nuclear-cytoplasmic system, it seem unlikely that genes from any other genome would be faithfully transcribed and translated there. Whether genes of plastids can replicate, be transcribed and translated in the nuclear-cytoplasmic system remains to be determined.

To summarize: Amino acid coding regions of prokaryotic genes are continuous; many eukaryotic genes are interrupted by non-coding DNA sequences called introns. There can be several introns in a single gene and the total length of intron sequence often exceeds the length of the coding sequences in a gene. Some amino acid codons used in mitochondrial genes (at least from fungal and animal mitochondria) are different from those in plastid and nuclear genes. Sequences flanking the amino acid coding regions of nuclear genes are much different from those in plastid genes. These differences will probably limit the expression of genes from prokaryotic cells in eukaryotic ones unless the regions that flank the coding sequences can be modified properly. Similar modification would probably be needed to have plastid genes function in nuclei or nuclear genes in plastids. Mitochondria of yeast use some codons differently from Neurospora mitochondria; human mitochondria exhibit other differences. If such variability is general, there are likely to be severe limitations to using mitochondria or mitochondrial genes in engineering.

MANIPULATING GENES

There are different levels of complexity of genetic engineering tasks, but they have in common the need to have and know about the DNA to be introduced and a mechanism for introducing it into an organism. That is, the gene of interest must be physically identified and isolated from the rest of the genome of the donor, it must be multiplied so that adequate amounts of purified DNA are available to permit determining its sequence for further manipulation, and it must be introduced into the host cell and there into a specific genome. The process of introducing a foreign gene into a cell and thus altering its genotype and phenotype is called transformation -- a cell into which foreign DNA has been successfully introduced and expressed is designated a "transformed" cell.

Getting a gene separated from all the other genes of its own genome requires that it be located. This has been one of the most difficult steps, but newly emerging technologies and improvements on established techniques promise to make that task easier. Let me describe an approach that can be used. Enormous refinements in the techniques for determining the amino acid sequence of proteins have been made. It is now possible to determine the sequence of the amino acids in a protein with minute amounts of purified protein. But how to isolate even a small amount of pure protein? Among the important procedures are gel electrophoresis, high performance liquid chromotography and selection by monoclonal antibodies. Once an even relatively small segment of the amino acid sequence has been determined, knowledge of the nucleotide code for the amino acid sequence enables one to predict the probable nucleotide sequence in the gene that codes for this segment of the polypeptide and a small DNA molecule of the predicted sequence can be synthesized. A DNA sequence as short as 14 or 16 nucleotides has proven to be adequate for the next step in the operation. (A sequence of three nucleotides codes for a single amino acid; a 15 nucleotide-long sequence of DNA could code for 5 amino acids.) This synthesized short sequence of nucleotides complementary to one of the strands in the coding region of the gene can be made radioactive and used as a molecular hybridization problem to find the gene. Because this DNA sequence is complementary to one of the strands of the gene, it will form hydrogen bonds with the gene and reveal the gene.

How the gene is located and isolated requires some explanation. First, the total DNA of the genome in which the gene of interest lies is isolated from the cell and digested into smaller fragments using restriction endonucleases, enzymes that recognize particular DNA sequences and sever the double stranded DNA at a location specified by the DNA sequence. Secondly, all of the fragments of the DNA of that genome are ligated enzymatically -- the enzyme is a "ligase" -- into modified bacterial viruses or, for example, into a plasmid. (A bacterial plasmid is an autonomously replicating DNA molecule much smaller than the main chromosome of the bacterium. Each plasmid has an origin of replication of its own and plasmids designed for cloning foreign DNA contain, as distinguishing marks, genes for enzymes that destroy antibiotics. Plasmids can be reintroduced into bacterial cells and those cells transformed with the plasmid carrying a gene for antibiotic destruction can be identified by their resistance to that antibiotic; transformed cells can be distinguished from non-transformed cells when grown in the presence of the antibiotic that is lethal to cells not carrying the gene for the destructive enzyme.) Thirdly, after fragments of the DNA to be cloned are ligated into plasmids, the plasmid DNA is used to transform a population of bacterial cells, and cells that have taken up DNA are identified by their antibiotic resistance. Each bacterial cell is mostly likely to take up and maintain a single plasmid and thus to contain clones of a single fragment of the foreign genome. Fourthly, which of the transformed bacterial cells contains the cloned gene of interest is determined by probing the cloned DNA fragments by using the synthetic radioactive DNA complementary to a segment of the gene that is being sought; since the synthetic DNA is complementary to the gene, it will form hydrogen bonds with the gene and the gene can be identified by the radioactivity of the synthetic probe adhering to it. This is a great oversimplification of the steps and all of the technical details are omitted, but this sort of approach works. Finally, if the gene is to be cloned into a plant or animal, it must be removed from the bacterial plasmid by a restriction endonuclease and be ligated into another vector. The use of the Ti plasmid, carried by Agrobacterium tumifaciens, as a means of introducing genes into dicotyledonous plants is described by Dr. Schell in the next paper.

To summarize: Steps in genetic engineering include (a) isolation from the donor of DNA containing the gene which is to be manipulated, (b) digestion of the DNA using a restriction endonuclease, (c) ligation of the DNA fragments so generated into a vector that can replicate in a bacterium (in the example given, a plasmid was used), (d) transformation of bacterial host cells with chimeric plasmids and cloning the DNA in the bacteria, (e) identification of the gene of interest (in the example, by hybridization with a radioactive synthetic DNA sequence complementary to a portion of the gene), and (f) if the gene is to be introduced into a plant or aminal, the cloned gene must be transferred into an appropriate vector, (g) the gene in the appropriate vector is used to transform the target organism.

LEVELS OF GENETIC ENGINEERING

The procedures outlined in the previous section describe one of the simplest sorts of genetic engineering: the cloning and replication of a gene from a foreign source, bacterial or enkaryotic, in a bacterial host. In some cases, the gene may be not only replicated, but also transcribed and translated and protein with some enzymatic activity may be produced from the cloned DNA.

If the gene is from a higher organism, it is likely to be necessary to modify the regions flanking the amino acid coding segment of the gene so that it will be transcribed by the bacterial RNA polymerase. It may also need additional modification so that it will be translated by the ribosomes of the bacteria, which are different from those in a eukaryotic cell's cytoplasm. These modifications are generally made by using restriction endonucleases to separate the noncoding regions of the gene from the nucleotides specifying the amino acid sequences and relegating the latter into a position behind the promoter or promoter and ribosome recognition sequences borrowed from a bacterial gene. The foreign gene is put into a new set of clothes so it will be taken for a bacterial gene and be expressed. Manipulations of these sorts have been used successfully to have human growth factor, interferons and insulins be produced in bacteria from animal genes. Bacteria cannot process out introns so animal genes from which introns have been deleted are sometimes used -- DNA copies of mRNA's can serve this function.

Genetic engineering at the same level, the introduction of a single gene which is either always on, i.e. expressed as a protein, or always off (not expressed), can be carried out -- or will be very soon -- with dicotyledonous plants and certain animals. In the case of dicotyledonous plants, although the limit is not known, DNA for at least 10 or so genes can be incorporated into a plant chromosome using the Ti plasmid system.

A much more complicated level of genetic engineering — a level of which we are not yet capable -- is to introduce a gene that will be expressed in only one tissue or at only one time during development -- in seeds but not roots, during germination but not during flowering. A leaf cell is different from a petal cell not because they contain different complements of genes, but because different sets of genes are being expressed in the two cell types -- different sets of genes are on in petal cells than in leaf cells. The DNA sequences that flank the amino acid coding region and perhaps some sequences within the coding region are tough to determine when an RNA polymerase recognizes and transcribes a particular gene. We know the genetic code for amino acids but not the code for regulation of gene expression. Scientists studying these problems believe they know what answers are needed; they are being sought in energetic research programs.

There is a still more complicated level of genetic engineering that remains a challenge. Most agronomic traits of plants are not coded for by single genes but by genes that are at several genetic loci. These loci could be sites of genes for a series of enzymes in a biosynthetic path. In a more complicated relationship, some of these could be sites of genes for enzymes and others that somehow regulate the expression of the genes for the enzymes. In the latter case, we do not know enough about the genes themselves or how they act on one another to understand how to use them for engineering. The genetic engineering technology, in these cases, is ahead of the basic knowledge required. We know so little that it is not possible even to guess how long the time lag will be in getting this information.

In summary: We can visualize three levels of genetic engineering. The first is to introduce into an organism a gene which is always expressed, as in the case of transformation of bacteria to produce interferon or insulin. In principle, there is no limit to the number of genes that can be introduced, but the organism might balk at producing large amounts of proteins for which it has no use. At the second level of genetic engineering is the introduction of a gene bearing control elements that limit its expression to specific stages or times in development or restrict its expression in different organs. This is a problem for which considerable basic research needs yet to be carried out. At a third level is the introduction of multi-gene traits into an organism. Here, only mapping by transmission genetics tells us that sets of genes are functionally interrelated; a great deal of basic information remains to be obtained before such genes can be used for engineering.

OTHER BIOTECHNOLOGIES

A marvel of genetic engineering is that it has created a single pool of genes from genomes that have become reproductively isolated from one another as a result of the forces of speciation. Two examples that are agriculturally important are: First, the cloning of the gene for a protein of the foot and mouth disease virus in *E. coli* for the production of the protein by the bacteria -- the protein can be used as a cheap, safe vaccine -- and, second, cloning of a DNA copy of the potato spindle viroid in *E. coli* to provide a molecular probe to test plants to see if they carry the viroid.

The techniques outlined above describe how to go about moving one or a few genes at a time from one genome to another, and thus to increase, in a limited way, the genetic diversity in the recipient organism. There are other biotechnologies that reveal existing diversity or, in ways other than those described above, increase genetic diversity for further breeding.

Plant cells from a leaf, for example, can be dissociated from one another by enzymatically digesting away cell wall pectin and cellulose to produce protoplasts. Naked cells of some plants regenerate walls and, by a succession of changes of hormones in the medium, the cells can be stimulated to multiply and differentiate and to finally start to regenerate whole new plants. James Shepard showed that when such experiments are carried out with protoplasts from leaves of potato plants, the regenerants are not all alike. They differ, for example, in the shapes and sizes of tubers produced by the plant, in resistance to drought, in resistance to some pathogens, and in other features. There is some diversity among potato plants grown in the field and it is thought that the procedures used to produce the protoplast and to regenerate plants may contribute to exposing more of the normal genetic diversity. For whatever reason, soma-clonal variation seems to be important as a means of obtaining plant material with new traits that can be bred into a crop.

Another approach to increasing genetic diversity is by fusing protoplasts of two different species. Under these circumstances, some chromosomes are lost and others or fragments of others may be retained. Viable cells form walls and can be regenerated into whole plants. This is a way to increase the gene pool of the recipient plant in a random way. Further selection and breeding would be required to produce plants with the desired properties. Another paper in these proceedings deals with wide crosses -- another way of increasing the pool of genes available for development of crop plants.

Thus, there are available now, in addition to methods for the directed introduction of identified bits of genetic material using the tools of recombinant DNA technology and genetic engineering, other biotechnologies that expose inherent variation or have the potential of adding large number of undefined genes to the gene pools or regeneratable species. In the latter cases, selection and screening methods are required to identify varieties with the desired traits. In the former, the trait is decided upon and introduced. But, as already pointed out, the limitation is in the regulation of the introduced gene.

Finally, certain cell and tissue culture technologies are already contributing to commercial agricultural practice. The potential contribution of cell and tissue culture to crop improvement are discussed elsewhere in these proceedings (Crocomo, 1983).

PROSPECTS FOR GENETIC ENGINEERING TO PROMOTE PLANT PRODUCTIVITY IN AGRICULTURE

Two examples of agriculturally important advances made by genetic engineering have already been mentioned -- production of vaccines and production of a DNA probe for infection by a viroid. But what's ahead, both in learning about molecular biology of plants and associated microorganisms and in using the knowledge for increasing productivity?

As already outlined, the simplest kinds of genetic engineering involve introducing into relatively simple organisms, like bacteria, genes that are always on. An example of an agriculturally useful step at this level would be the introduction of a gene for hydrogenase into otherwise desirable nodulating strains of Rhizobium that lack this enzyme. Nodules formed on roots of legumes by rhizobia that lack hydrogenase liberate hydrogen gas into the atmosphere. The energy from photosynthetic products of the plant is used to produce hydrogen gas rather than to fix nitrogen. The waste is very great. By contrast, bacteria with the enzyme hydrogenase recapture the hydrogen gas and use it to reduce ferredoxin, which can contribute electrons to dinitrogen reductase. Other sorts of relatively simple genetic engineering of microorganisms that associate with plants can and have been easily imagined. The tools of molecular biology and genetic engineering have already been used to increase enormously our understanding of the organization of nitrogen fixation genes in Rhizobium and should help us comprehend mechanisms for the control of expression of these genes in legume nodules. The same tools should, in the near future, reveal the organization of nitrogen fixing genes in actinomycetes that fix nitrogen in association with the roots of certain woody species. Here, too, genetic engineering possibilities for improvement of the strain have a good prospect of succeeding.

In a more general way, interest in bacteria of the rhizosphere as biological control agents and in promoting growth of plants in other ways, is rising rapidly -- mostly because detailed analyses of their genomes can be accomplished by use of currently available techniques and this opens the way to genetic engineering to improve their agricultural usefulness. The analysis of plant diseases caused by bacteria and fungi seems on the verge of a revolution, again, because our relatively recently gained knowledge of molecular genetics promises to lead us to the genes involved in pathogenicity.

The study of molecular biology of nodule-forming bacteria shows another direction for the future. Up to now, the study of the legume-Rhizobium association has been concentrated on the bacterial genes, an exception is the isolation and identification of the plant genes for the leghemoglobins that are produced as part of the process of nodule formation. Sure to follow will be the identification of additional plant gene products that appear in the course of the establishment of the nodule. This pattern of the analysis of the genes of a symbiont or a pathogen followed by identification of plant genes that are activated in response to an infection is likely to be repeated over and over again in research of the coming decade. This may well be a route by which the products of genes at several loci that all affect resistance to a given disease can be identified physically. This would open the way to genetic engineering with genes that have complex interactions with one another. Surely, the insights to be gained from these lines of research will have a profound effect on how we think about plant disease and resistance to disease in the last decade of the century.

The other side of making genetic improvements to promote plant productivity by genetic engineering of symbionts and pathogens is by altering the plants themselves. A 1981 review of "Impacts of Applied Genetics" by the United States Office of Technology Assessment states that the three major goals of crop breeding are:

1. To maintain or increase yields by selecting varieties for: pest (disease) resistance, drought resistance, increased response to fertilizers, and tolerance to adverse soil conditions.

2. To increase the value of the yield by selecting varieties with such traits as: increased oil content, improved storage qualities, improved milling and baking qualities, and increased nutritional value, such as higher levels of proteins.

3. To reduce production costs by selecting varieties that: can be mechanically harvested, reducing labor requirements, require fewer chemical protectants or fertilizers; and can be used with minimum tillage systems, conserving fuel or labor by reducing the number of cultivation operations.

Can we meet any of these goals by the use of genetic engineering techniques and other biotechnologies currently available? We can (or soon should be able to) use genetic engineering techniques to transfer single genes or groups of genes that are constitutively expressed. Production costs could be reduced by producing a plant variety resistant to a specific herbicide. A specific example would be by introducing a single gene that codes for a herbicide-destroying enzyme (e.g., the basis of resitance to atrazene in maize) or a herbicide-resistant receptor protein to replace a sensitive one (e.g., as may be the case in pigweed varieties that mutate to resistance to atrazene). One can imagine isolating from a maize plant or another source a gene for an enzyme that degrades the herbicide, introducing it into a plant that is sensitive to the herbicide and thus creating a crop plant for which weeding is less necessary by using a chemical herbicide safely. We don't know yet whether the introduction of such a single gene, constitutively expressed in the plant, will affect the agronomic values of the crop. If such properties are affected, a further breeding program would be necessary before the modified crop is released for use, but it is possible that such an introduction would have no deleterious effect on yield. Another possible kind of single constitutive gene introduction that can be imagined is in forage crops. If it were economically worthwhile, one could imagine introducing a gene for a protein whose presence would alter the nutritional quality of all the cells of the plant, including above-ground parts eaten by foraging domesticated animals.

More attractive, of course, is the prospect of changing the nutritional quality of grains by altering the protein composition or the amino acid makup of their storage proteins; this would require more than we can do now -- not only the introduction of a new gene or a modified form of one already present, but also, knowing about the control elements on the gene, so that it is expressed in the seed and not in all parts of the plant. This is likely to be done first by attaching a modified gene to some of the sequences flanking transcripts of genes that already occur only in seeds.

One view of the genetic engineer's future parts kit is: (a) sets of DNA sequences which limit gene expression only to roots, or only to leaves, or only to fruits, or only to seeds, etc.; plus (b) sets of reading frames for proteins obtained from anywhere (but mitochondria) to be connected with the control elements; plus (c) the proper kinds of transcription termination signals; plus (d) vectors for introducing the engineered gene into the plastid or nuclear genome. Then, perhaps in 5-10 years' time, one will be able to design such genes precisely enough to introduce them for expression at a particular time in the life history of the crop. Considering that the tools of genetic engineering began to be forged only about six years ago, it seems very likely that, with the passage of 15 additional years, today's fantasies as well as unimagined possibilities will be at or near reality.

To summarize: Genetic engineering of bacteria that promote or impede plant growth can be carried out now. Genetic engineering tools available now also permit the introduction of single genes or of many genes into plants. Little is known about portions of genes that are required to control their expression in specific tissues or at specified stages of development. Further along is the prospect of transferring sets of physically distant genes that may all be required to confer a useful trait such as resistance to a specified disease. We do not know how introduction of foreign genes will affect the agronomically significant properties of a plant nor, therefore, how extensive breeding programs may have to be to develop useful crops this way. We also lack a vector system for introducing genes into monocotyledonous plants including cereals. But is is most important to recognize that deficiencies in our basic knowledge are among the major barriers to genetic engineering of crop plants.

CONCLUDING COMMENTS

Knowledge of molecular genetics plus the development and exploitation of genetic engineering tools have brought us to define the information needed to move to higher levels of manipulation of crop plant genes in order to increase productivity. Accomplishment of the technically simplest genetic engineering tasks will help us understand how to go forward but a great deal of basic research on plants, their symbionts and their pathogens must be acquired if we are to climb quickly to the higher and most productive level of genetic engineering capability. We need to understand mechanisms of ion uptake to devise plants that need lower levels of nutrients; we need to understand how plant hormones work to modify growth and development of crops; we need to understand the structure of plant genomes and genes; we need to know about gene expression regulatory mechanisms; etc. Yet, basic research in molecular genetics and related biochemical problems underlying agriculture is on a very small scale in the world. In the United States, where I am most familiar with the magnitude and limitations of the research support, financial support lags far behind the capacity of people to carry the research forward and far behind the level needed to sustain the modest momentum we see now.

One of the interesting aspects of the genetic engineering revolution that is surely before us is the capacity to spread benefits of this high technology not only through the most developed countries but also to those that are developing. One of the promises of genetic engineering is that we may be able to build into the organism more properties that increase productivity to expand on the benefits bestowed by plant breeders. This conserves the world's resources, but especially aids locales where the administration of fertilizers and agricultural chemicals may be difficult for institutional, cultural or economic reasons. But to what extent regional or national efforts will be needed to adapt the technology to local situations deserves careful attention. How the world will organize itself to learn quickly what is needed, to implement the technology and to spread the benefits is not simple. New on this scene are many commercial efforts. How will governments, the existing regional research establishments, and commercial interests join forces to foster the exploitation of the new technologies?

ACKNOWLEDGEMENTS

The preparation of the paper and of related research in the author's laboratory has been supported by research grants from the U.S. National Science Foundation, the Competitive Research Grants Office of the U.S. Dept. of Agricuture and the National Institute of General Medical Science. It has also been supported in part by the Maria Moors Cabot Foundation of Harvard University.

REFERENCES

Bogorad, L. (1979). The Choroplast, Its Genome and Possibilities for Genetically Manipulating Plants. In: Genetic Engineering, J.K. Setlow and A. Hollaender, 1, 181-203. Plenum Publishing Corporation, New York.

Impact of Applied Genetics: Micro-organisms, Plants, and Animals. Office of Technology Assessment Report HR-132. U.S. Government Printing Office, Washington, DC (1982).

Maniatis, T., E.F. Fritsch and J. Sambrook (1982). Molecular Cloning. A Laboratory Manual. Cold Spring Harbor Laboratory, Cold Spring Harbor, New York.

Shepard, J.F., D. Bidney and E. Shahin (1980). Potato Protoplasts in Crop Improvement. Science 208: 17-24.

Crocomo (1983). The Potential Contribution of Cell and Tissue Culture to Crop Improvement. Chemrawn II Technical Proceedings, Pergamon Press.

APPLICATIONS OF GENETIC ENGINEERING TO PLANT AND ANIMAL PRODUCTION

Joachim Schroder and Jeff Schell
Max-Planck-Institut fur Pflanzenzuchtung
D-5000 Koln 30 FRG

ABSTRACT

The main limitations of conventional plant and animal breeding are the length of time required to produce new, useful varieties and the relatively small gene pool available for sexual crossings. In plants, the length of time can be reduced by employing haploid tissue cultures and plants. Elimination of undesirable phenotypes and selection of genetic variants is thereby simplified, and the diploid state can be restored, providing the breeder with suitable material for further crossings. Somatic hybridization of plant cells is a first step to enlarge the available gene pool. Protoplasts from sexually incompatible plants can thus fused to form a new genetic unit which contains the genomes of both parent plants. The direct introduction of genes as DNA molcecules with help of gene vectors represents a system of minimal interference with existing, stable genomes. With dicotyledonous plants, the Ti-plasmid system is available for direct transfer of genes, and improvement of crop plants can be expected in the near future. Gene vectors for monocotyledonous plants and preferably also for chloroplasts and mitochondria need to be developed. With animals, recent developments indicate that genetic engineering has become a feasible project, although many fundamental problems remained to be solved. Both with plants and animals, the chief limitation for progress in genetic engineering is insufficient knowledge on structure, function, and regulation of genes.

KEYWORDS: haploid tissue cultures, somatic cell hybridization, crown gall, Ti-plasmids, T-DNA, gene vectors, genetic engineering.

INTRODUCTION

Conventional plant and animal breeding has been of great importance in the last decades for the production of better and more useful plants and animals, and there is little doubt that it will continue this role in the future. However, it also has limitations. It takes a long time to produce new and better varieties (with plants for example between eight and 30 years), and the gene pool available for new combinations is limited by the fact that under normal conditions only closely related species can be crossed sexually. The vast pool of potentially valuable genes in other organisms cannot be tapped by conventional breeding techniques.

Therefore much of the current interest is focussed on the development of techniques which either shorten the time required for breeding or which enlarge the gene pool available for production of new and useful gene combinations. The last few years have seen important developments in both of these areas, and although much is still in the experimental stage and many fundamental questions still have to be resolved, it is nevertheless worthwhile and timely to assess the possible applications.

PLANT TISSUE CULTURES

Haploid Cells and Plants

In contrast to higher animals, many plants can be regenerated from tissue cultures or even from single cells. With the advancement of these techniques it also became possible to establish tissue cultures from haploid male or female gametes and to regenerate haploid plants. Since these cells or plants contain only one set of chromosomes instead of two, they offer significant advantages over diploid plants in several steps necessary in plant breeding. Recessive traits which are otherwise not detected because of complementation by the second chromosome are immediately visible in the phenotype of haploid plants, and thus can be detected much faster than by crossing of diploid plants. If undesirable they can be recognized and eliminated directly in the haploid stage, and this leads to a considerable reduction in the time required to select for optimal gene combinations. Another advantage of haploid cells and plants is that the screening for genetic variants, for example resistence to pathogens, is simplified. Since haploid cells often return spontaneously to diploidy in culture or can be induced to do so by suitable treatments, the original diploid chromosome number can be restored. In these cells the two chromosome sets are identical, and this material provides the plant breeder with good material for further improvement by conventional plant breeding techniques.

Successful application of haploid cells requires that it is possible to produce protoplasts which are capable of growth and division and which can be regenerated to plants. The breeding of plant species amenable to these procedures can be expected to be accelerated considerably by use of haploid cell techniques.

Hybridization of Somatic Cells

After removal of plant cell walls, the naked cells (protoplasts) often fuse spontaneously or they can be induced to do so, even with protoplasts from other plants. Since it has been shown that with a certain frequency not only the cytoplasms but also the nuclei merge to form a new unit which combines both genomes and which is capable of growth and division, this technique can be used to produce genetic combinations from distantly related plants which cannot be crossed sexually. However, in many cases it has been found that the new combination was not stable. Progressive loss of chromosomes has been observed in culture, sometimes to the extent that only one or a few chromosomes of one parent plant were left in the hybrid, and the properties of the finally stabilized hybrids cannot be predicted. Accordingly it is also not possible to predict the genotype and phenotype if plants can be regenerated from such hybrids. To obtain plants with specific, desirable properties it is necessary to apply large and time-consuming screening or selection programs, and it has to be shown that the phenotype is stable.

Hybridization of somatic cells can be performed only with plants from which viable protoplasts capable of plant regeneration can be obtained. It seems unclear whether this approach to fuse complete genomes will be useful in the near future for crop plants. It is possible, however, that the full potential of these techniques can be realized when they are used in combination with genetic engineering methods.

GENE TRANSFER IN PLANTS

The development and practical use of so-called gene transfer methods or "gene vectors" represents a major breakthrough. These systems aim for a minimal disturbance of existing stable genomes by introduction of a single gene or a set of few genes for desirable properties. The most developed of a number of potential gene-vector systems has been derived from a natural system for genetic manipulation of plants.

Agrobacterium tumefaciens is a gram-negative soil bacterium which infects wounds in dicotyledonous plants and in gymnosperms, and it induces tumorous outgrowths (crown galls). Cells from such tumors grow and divide - in contrast to most normal plant cells - in simple media without added growth hormones like auxins and cytokinins; they are hormone-autotrophic. Since growth and differentiation in plants are largely controlled by these hormones, this property of the cells appears to be one of the major factors in tumorous growth. The cells also produce a number of low molecular weight compounds (opines), often unusual aminoacid derivatives, which are not detected in normal cells, and these substances specifically support the growth of the inducing Agrobacteria. Thus the biological system essentially represents a parasitic relationship in which the Agrobacteria force the plants to divert some of their anabolic capacity towards the production of opines. The capacity of the bacteria to induce tumorous growth and opine synthesis resides in large plasmids (Ti-plasmids, Fig. 1) which are

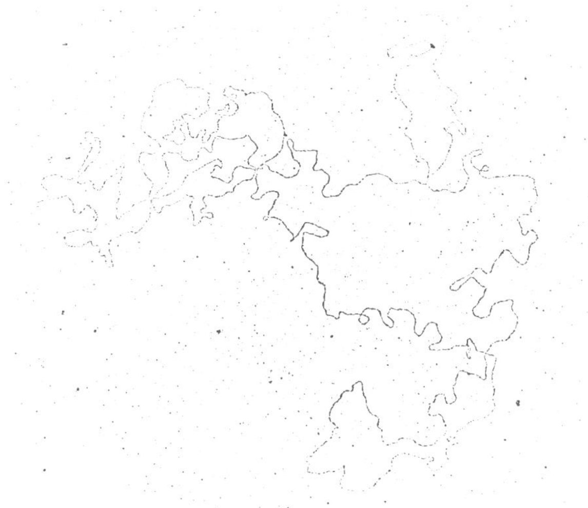

Fig. 1: Electronmicrographic picture of a tumor-inducing plasmid (Ti-plasmid) isolated from Agrobacteria (15 000 x magnification).

found in all pathogenic Agrobacteria. A large body of evidence shows that the bacteria use the Ti-plasmids as a tool for genetic engineering of plants. All tumor cells contain a specific part of the plasmid (the T-DNA) which had been transferred from the bacteria into the plant cells and which is stably incorporated, maintained and transmitted in the chromosomal DNA of the nuclei. The T-DNA contains defined genes which are expressed in plant cells, and the actions of the gene products are responsible for tumorous growth and opine synthesis. Thus, the interaction between Agrobacteria and plants can be described as a "genetic parasitism" or "genetic colonization" of plants which benefits the Agrobacteria (Fig. 2).

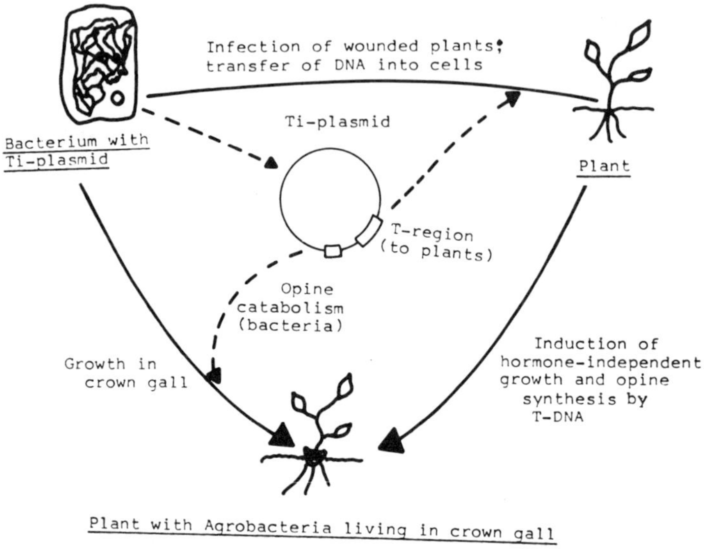

Fig. 2: Ti-plasmids as natural gene vectors: Genetic colonization of plants by Agrobacterium tumefaciens. After infection of a wound by the bacteria, a specific part of the Ti-plasmid is transferred into the nuclei of plant cells where it is stably incorporated and maintained in the genome. The transferred DNA (T-DNA) contains genes which are responsible for tumorous growth of the infected plant cells and synthesis of various compounds (opines) which specifically support the growth of the Agrobacteria. The capacity of the bacteria to catabolize the opines produced in the tumor cells is also encoded on the Ti-plasmids.

The key to this interaction is the Ti-plasmid; it is a natural gene vector. Intensive research has been undertaken in the last years to investigate whether Ti-plasmids can also be used to introduce other, well-defined genes into plants. The results clearly show that Ti-plasmids containing additional genes in the T-DNA will transfer these genes unchanged into plant cells where they become integrated into the chromosomes and where they are stably maintained and transmitted to progeny of the cells. If the genes responsible for tumorous growth are still present and active, however, such plant cells usually cannot regnerate to normal plants since the tumor genes affect hormonal control of differentiation processes. These genes were identified and mapped in the T-DNA, and it was shown that they have no function in the actual process of gene transfer from bacteria to plants nor in the integration of the transferred DNA into the plant genome. It was therefore possible to eliminate these genes without affecting the genetic engineering capacity of the Ti-plasmid. Plant cells containing such modified T-DNA were shown to regnerate into normal plants which were fertile and transmitted the transferred gene to progeny in a Mendelian fashion.

In the last few years, more than 15 model genes have been transferred via Ti-plasmids into plant cells, and therefore a generalization seems to be justified that Ti-plasmids will transfer any gene into plants. It can be expected that its application will yield in the short term (5-10 years?) plants with improved biological pest resistances, for example against fungi and viruses. The chief limitation to fast results appears to be that disease resistance genes - as defined by the plant breeder - are very often not defined by their biochemical mechanism, and without this basic knowledge it is very difficult to isolate the genes as DNA molecules. Since understanding of the biochemistry of disease resistances is crucial, active support of basic research in this field appears to be important. One can also envision significantly new strategies for a more efficient use of chemical pest control methods. For example, soil organisms contain genes for degradation of herbicides, and these genes can be isolated, suitably engineered for eucaryotic cells, and then introduced into important crop plants. This would supply the plants with resistance against specific herbicides often used in agriculture.

A wide open field for genetic engineering techniques is the use of plant cell cultures for the production of valuable proteins, enzymes and high value drugs. Plant cells can now be grown like microorganisms in large suspension cultures using fairly standard fermentation procedures, and many attempts have been made to produce with them rare chemicals or proteins. However, these substances are usually formed by complex biosynthetic pathways which are not necessary for rapid growth of the cells and expression actually represents a metabolic load which slows growth down. Such cells are fairly quickly lost since the standard culture conditions select for rapidly growing cell types. The problem is therefore to change the regulation of the genes for valuable proteins or drugs in such a way that they can be activated under controlled conditions when high cell density has been reached by rapid growth. Genetic engineering techniques are undoubtedly critical and powerful tools to achieve these goals.

The natural host range of Agrobacteria is very broad and encompasses probably all dicotyledonous plants and most gymnosperms. Accordingly, Ti-plasmids are useful gene vectors for all of these plants. However, monocotyledonous plants are not attacked by Agrobacteria, and so far it seems not possible to transfer genes via Ti-plasmids into these plants. Since this group contains the most important crop plants it remains of high priority either to adapt the Ti-plasmid system to these plants or to develop gene vectors for chloroplasts and mitochondria, since Ti-plasmids transfer genes exclusively into the nucleus. One of the problems with direct transfer of genes appears to be that the transferred genes are often not active in the host cells. Further basic research on the structure, function, and regulation of plant genes is essential for the practical application of gene transfer techniques.

GENE TRANSFER IN ANIMALS

The prospects for genetic engineering in animal husbandry are not yet as advanced as in plants. One of the reasons is that - unlike plants - animals cannot be regenerated from single somatic cells. It is therefore necessary either to introduce genes directly into isolated eggs which are then reimplanted into the animal womb, or to introduce generically altered complete cells into embryos at the blastula stage. It has been demonstrated that they can be integrated into the differentiating embryo and, with a certain probability, can enter the germ line. Very important progress in this field has been recently reported in studies with mice. It was possible to show that genes introduced via these procedures were retained in adult animals, that they were expressed (e.g. skin color) and that the genes were transmitted to progeny. One of the problems appears to be that foreign genes are inactivated when introduced into embryogenic cells and stay inactive since there is no developmental program for activation later in life. However, although many fundamental questions remain to be solved, the progress in this field indicates that genetic engineering in animal husbandry has become a conceivable project.

REFERENCES

Plant Tissue Cultures

Perspectives in Plant Cell and Tissue Culture (International Review of Cytology, Supplement 11A). Ed. I.K. Vasil, Academic Press 1980.

Plant Improvement and Somatic Cell Genetics (XIII. International Botanical Congress). Eds. I.K. Vasil and W.R. Scowcroft, Academic Press 1982.

Variability in Plants Regenerated from Tissue Culture. Eds. E.D. Earle and Y. Demarly, Praeger Publishers (New York) 1982.

Gene Transfer in Plants

Gene Transfer in the Plant Sciences. Ed. N.J. Panopoulos, Praeger Publishers (New York) 1981.

J. Schell (1982). The Ti-plasmids of Agrobacterium tumefaciens. In, Encyclopedia of Plant Physiology, New Series, 14B, Nucleic Acids and Proteins in Plants II, 455-474. Eds. B. Parthier and D. Boulter, Springer Verlag Berlin, Heidelberg.

J. Schell, M. Van Montagu, J.P. Hernalsteens, M. Holsters, H. De Greve, J. Leemans, P. Dhaese, and L. Willmitzer, (1982). The use of Ti-plasmids as gene vectors for plants. In, From Genetic Engineering to Biotechnology - The Critical Transition, 41-52. Eds. W.J. Whelan and S. Black, J. Wiley and Sons Ltd.

J. Schell, M. Van Montagu, M. Holsters, J.P. Hernalsteens, J. Leemans, H. De Greve, L. Willmitzer, L. Otten, J. Schroder, Ch. Shaw (1981). The development of host vectors for directed gene transfers in plants. In, Developmental Biology Using Purified Genes (ICN-UCLA Symposia on Molecular and Cellular Biology, Vol. XXIII), 557-575. Ed. D.D. Brown, Academic Press.

Gene Transfer in Animals

K. Illmensee, K. Burki, P.C. Hoppe, and A. Ullrich: Nuclear and gene transplantation in the mouse. In, Developmental Biology Using Purified Genes (ICN-UCLA Symposia on Molecular and Cellular Biology, Vol. XXIII), 607-619. Ed. D.D. Brown, Academic Press.

R. Jaenisch (1980). Germ line integration and Mendelian transmission of exogenous type C viruses. In, Molecular Biology of RNA Tumor Viruses, 131-162. Academic Press.

R. Jaenisch (1982). Insertion of leukemia virus genomes into the germ line of mice: a model system to study gene expression during development and differentiation. In, Expression of Differentiated Functions in Cancer Cells, 21-28. Ed. R.F. Revoltella, Raven Press, New York.

B. Mintz, and C. Cronmiller (1981). METT-1: a karyotypically normal in vitro line of developmentally totipotent mouse teratorcarcinoma cells. Somatic Cell Genetics 7, 489-505.

THE APPLICATION OF WIDE CROSSES TO PLANTS

L.H. Shebeski
International Institute of Tropical Agriculture
Ibadan, Nigeria

ABSTRACT

Wide crosses are defined as those in which the chromosomes of the two contributing species do not normally pair in the hybrid. Wide crosses are made either to increase genetic variability in existing cultivated species or to develop new crops species.

Four major constraints to successful completion and/or utilization of wide crosses (incompatability, embryo absortion, hybrid sterility, gene transfer), and methods to minimize constraints are discussed. Through improved technology, hybrids have been made between species previously considered non-crossable. Genes, primarily those governing diseases or insect resistance, have been transferred from alien to cultivated species. Relatively few have been put to practical use. Impact of transferred genes on increasing food crop production has been negligible, but the potential value for stabilizing and improving crop production is considerable.

Only one of many new allopolyploids that have been synthesized, Triticale, has become established as a new cultivated food crop. Its potential is discussed.

Research on wide crosses is not well supported and potential values not sufficiently appreciated. Lack of knowledge of the mechanisms of speciation, of the factors involved in the stabilization of new species, and why some species hybrids cannot be made, are the most limiting impediments. Overcoming these impediments requires added personnel, adequate financing, and an official dedication to advancing wide cross research.

KEYWORDS: wide crosses, hybrids, gene transfer, triticale, speciation.

WIDE CROSSES APPLIED TO PLANTS

Conventional plant breeding, which has been so effective in increasing the production of cultivated food crops, particularly during the last three decades, basically is the process of creating a genetically variable population of plants with subsequent selection amongst the variants. Key to the process is the crossing of plants which differ genetically.

A cultivated crop plant has between one and ten million genes, each of which holds information that specifies the structure of a product RNA and, often, a protein. In addition, each has nucleotide sequences involved in recognizing on-off signals (Freeling, 1982). The importance of individual genes or combinations of genes to plant development and hence to plant breeding, should not be underestimated.

It takes relatively few genic differences between two cultivars that are crossed to provide an enormous amount of genetic variability. If two plants that are to be crossed differ by as few as 30 of the million or more genes present in each plant (assuming independent segregation), the number of possible genetically different plants in the first segregating generation (3^{30}) is so large that in order to provide for the chance presence of each of these genotypes (4^{30}), the entire cultivated area of the earth would have to be planted at a million plants per acre each year for more than 300 years.

Despite the unbelievably large variability that can so easily be induced by hybridization, if neither of the two parental cultivars possess a gene or genes conferring resistance to a specific disease, none of their progeny can be expected to be resistant to that disease. It is in this context that I wish to review some of the important aspects of the application of wide crosses to agricultural food crops.

Wide crosses are usually considered to be those that are made between plants from different species or genera, as compared with crosses between plants from the same species. However, because a number of closely related species can be intercrossed as readily as the cultivars within a species (Knott, 1981), and hence genetic transfers can readily be made by conventional plant breeding methods, "wide crosses" are here defined as those in which the chromosomes from the two contributing species or genera normally do not pair in the hybrid. By this definition, wide crosses are usually intergeneric and often difficult to make, and the transfer of useful genes from the donor to the recipient cultivated species requires special techniques. There are numerous published reports on the successful transfer of useful genes from wild to cultivated species but the majority of these are on crosses between closely related species that are relatively easy to make and in which the hybrids are fertile and chromosome pairing and segregation tend to be normal. These, by definition, should not be considered wide crosses.

Wide crosses are made for one of two reasons: either to increase genetic variability in existing crop species (Knott, 1981) or to develop new and useful crop species that will be more productive or have greater adaptability than existing crop species. Either application <u>initially</u> requires the successful completion of a hybrid between widely diverse genotypes and, as implied earlier, certain difficulties may be encountered in any attempt to produce such hybrids.

The first major constraint which has plagued plant breeders interested in making "wide crosses" is that of incompatibility. Pollination is completely ineffective and fertilization does not occur. According to Bates (1974) little is known about this fertility barrier. A number of methods have been and are being tried to induce fertilization in very wide crosses. These include hormonal treatments, the transplantation of embryos from the seed of one parent species to the seed of the other species, bud pollination, the use of bridging species, and more recently <u>in vitro</u> fertilization and ovule/egg culture. As techniques improve, ever wider crosses have become possible.

A second and equally critical constraint to the successful completion of wide crosses is a post-fertilization phenomenon termed by Bates "<u>hybrid necrosis</u>". The embryo aborts before a viable seed is produced, or if a hybrid seed is produced and germinates, the resulting seedling does not survive. Obviously, this constraint must also be overcome before a transfer of genetic material from one species or genus to another or the development of a new species will be possible. The wider the cross the more likely will this constraint be encountered.

The failure of embryos to develop normally because of endosperm defects has long been recognized as a problem in a number of diverse species and species hybrids. Experiments on the artificial culturing of embryos were described as early as 1904, but it was not until the work of Tukey (1933) on the culturing of sweet cherry embryos that the usefulness of embryo culture as a rescue technique began to be more widely recognized and utilized. According to Jensen (1981), embryo culture "is perhaps the oldest practice in in vitro culture, and paradoxically, the most neglected one". He suggested "that utilizing the new culture techniques currently employed with cells and protoplasts would give a new boost to embryo culture". Nevertheless, embryo culture has made it possible to produce hybrid plants from crosses between distantly related species or genera that previously had not been considered crossable. Two examples will be cited to indicate both the difficulties that are encountered and some of the advances that are being made in the science and techniques of embryo culture.

The cultivated tomato, Lycopersicon esculentum, does not produce viable seed when crossed with the wild species Lycopersicon peruvianum, as Thomas and Pratt (1982) report that only about 1 in 10,000 embryos develop enough to be able to continue their development into plants on appropriate culture media. They are now experimenting with a new embryo callus technique. Rather than excizing and culturing embryos, they state that when whole undeveloped seeds are placed on a culture medium suitable for the growth of a callus, up to 1 in 10 such seeds forms a callus. All plants subsequently regenerated from the callus were derived from hybrid embryos rather than from endosperm or maternal cells in the seed.

The first report on the successful production of a hybrid between barley and wheat was viewed with skepticism. Many previous attempts by a number of plant breeders (myself included) to complete this wide cross were unsuccessful. In the early 1970s Anthony Kruse, working in Denmark, developed a new simple embryo rescue technique (Jan et al., 1982). The excised embryos of the hybrid, instead of being transferred to a standard culture medium, were placed on immature endosperm excized from the developing seed of one of the parent species. Australian workers subsequently found that the barley endosperm provided the best nurse tissue. Despite using these improved techniques, Jan et al. (1982) found it difficult to produce barley-wheat hybrid plants. In two years, more than 50,000 wheat ovaries were pollinated by barley but only about 440 showed embryo development out of which just 20 were true barley-wheat hybrids. The production of the barley-wheat hybrid was the first step in a project which has as its objective the transfer of a gene which confers resistance to an aphid transmitted virus that causes serious yield losses to both barley and wheat. Genetic resistance to the virus has been found in barley but not in wheat.

The barley-wheat hybrid plants produced by Jan and co-workers were completely male sterile. Sterility is a third constraint to the successful completion of wide crosses. Invariably, hybrid plants from wide crosses will be sterile because of the lack of chromosome pairing during the sexual stage of the life cycle of the hybrid plant. It is essential as a first step in the synthesis of a new seed-bearing crop species that fertility be restored. Blakeslee and Avery (1937) reported on methods of inducing the doubling of chromosomes in plants by the application of a carefully timed and appropriate concentration of the drug colchicine. This discovery opened a new era, not only for conventional plant breeding, but for the synthesis of new fertile crop species. However, newly synthesized allopolyploid species are not necessarily highly fertile from the start, and they may require considerable genetic adjustment before they become sufficiently diploidised to be fully fertile.

When the purpose of making a wide cross is to provide for the transfer of a useful gene, as in the transfer of the virus resistance gene from a chromosome of barley to a chromosome of wheat, the initial sterility of the hybrid, which usually can partially be overcome by backcrossing to the recipient species, is not as much a constraint as is the actual subsequent transfer of the useful gene. The transfer of the useful gene can be completed only if an appropriate piece of chromatin from the donor species can be translocated to a chromosome of the recipient cultivated species. Ideally, the alien chromosome segment should be as small as possible because the introduction of foreign germplasm into the finely balanced genetic system of a cultivated species usually causes a major disturbance (Knott and Dvorek, 1976). A procedure for inducing such translocations by using radiation to break chromosomes into segments for subsequent reattachment to other chromosomes was described and successfully used by Sears (1956), when he transferred a gene for leaf rust resistance from Aegilops umbellulatum to a Triticum aestivum chromosome. This was a major advance. Despite the difficulty in making such transfers the procedure subsequently was used successfully by a number of workers. In most of the reported transfers, plant breeders were attempting to broaden the available genetic base of major food crops by transferring genes for resistance to diseases. Genes for resistance to disease are readily identifiable, and recognition is a requisite to successful transfer.

Recently, Knott (1981) reviewed the literature on the transfer of genes conferring resistance to a number of diseases and an insect pest from different genera to wheat. Briefly, wheat had been successfully crossed with seven different species belonging to three genera, and from the resulting hybrids, genes governing resistance to six different diseases and to the greenbug were subsequently transferred to wheat. But despite the successes of plant breeders in making intergeneric gene transfers, the evidence so far, according to Knott, is that although resistances from an alien source can be useful they are not necessarily long lasting, and transfers can be deleterious. Flavell (1981) supports this view, stating that crosses between distantly related species have not commonly led to agriculturally successful cultivars, primarily because too many deleterious genes are introduced that are difficult to remove. Gustafson (1982), while admitting to a few instances (very few) where wide crosses have led to improved disease resistance, indicated that in most instances their need was not critical and "that plant breeders have been making tremendous progress generally without the need for wide crosses". Robbelen (1978) also questions the usefulness of wide crosses, particularly with respect to the transfer of quantitative traits, suggesting that "the addition of foreign variation does nothing but spoil the achieved high level of performance" of the cultivated crops.

Therefore, when the objective for making wide crosses is to increase the genetic variability of existing crop species, the question may well be asked ... "has anything come out of wide crosses that should result in an increased world food supply?" The consensus of opinion is definitely "no". Should work continue on the transfer of specific genes or blocks of genes from alien species or genera to our major food crops? In my opinion, most definitely yes. In the long history of man's continuing struggle to provide an adequate food supply for all, time and again his cultivated crops have been decimated by the ravages of disease or insect pests. I personally recall the great rust epidemics of 1936 and 1951 that swept from Mexico to Canada, wiping out or greatly reducing the yields of wheat on the better part of a continent. Each epidemic was caused by the appearance of a new virulant race of the disease organism. Each of the cultivated crops in all parts of the world has been similarly and periodically plagued - late blight of potatoes, blast in rice, and streak virus in maize. Currently, devastating losses to cassava are taking place in Africa because approximately 10 years ago the mealybug was accidentally brought there from South America. Losses in the production of cassava caused by the mealybug this year are estimated to be over a billion dollars, and these losses are expected to continue or even escalate until suitable resistance to the mealybug or its destruction by biological control can achieved.

Earlier I referred to the program at the University of California designed to transfer resistance to a yellow dwarf virus from barley to wheat, because resistance to the virus had not been found in the existing wheat germplasm banks. I fully expect that there will be many occasions in the future where resistance to new diseases or new virulant strains of old diseases or other pests will not be found in the existing variability of a cultivated crop. The time to search for and transfer additional sources of resistance is before rather than after need arises. There can be no finer insurance for the continuing protection of cultivated crops than by widening the genetic base for resistance to the diseases and insects that attack them.

At least one of the International Research Institutes, CIMMYT, supports by its actions the view I have just expressed. In the 1982 CIMMYT Review, wide crosses are referred to in two sections. According to the Review, "Crosses between maize and two alien genera, sorghum and Tripsacum, are being pursued to determine the feasibility of using potentially useful genes from these genera for maize improvement. The aim is to make maize a more environmentally stable crop with better insect and disease resistance and drought and water logging tolerance". In wheat, CIMMYT is continuing its research efforts to incorporate disease and environmental stress tolerance characteristics from several genera. Wheat has been hybridized with Elymus giganteus, with at least three species of Agropyron, and with Secale cereale in its Triticale program. The wide cross research programs at CIMMYT are well conceived and most worthy of continued strong support.

Earlier, I indicated that the second of the two major reasons for making wide crosses is to develop new and useful crop species that will be more productive or have greater adaptability than existing crop species. In a recent article on "polyploidy in plant breeding", Simmonds (1980) was very positive about the value of developing new species of crop plants. He indicated that one new crop species, Triticale, has recently become established in agriculture and more are in prospect. He described the work taking place at the Scottish Plant Breeding Station at Edinburgh, and in Sweden, on the development of Radicole (also called Raphanobrassica) with the objective of combining the hardiness of Kale with the disease resitance of radish. The new species is intended to compete with forage rape but it is still plagued in its development by "infertility".

Infertility also plagued the development of Triticale particularly in the early stages of its development, and as stated by Simmonds, "one lesson learned from Triticale was that, whatever the textbooks used to say, new allopolyploids are not necessarily highly fertile from the start; they may need a great deal of genetic adjustment to become sufficiently diploidized". Simmonds (1980) concluded his article by stating that the two examples, Triticale and Radicole, are "the most striking ones available, but they can represent no more than a fraction of the possibilities inherent in our crops at large. Systematic attempts to exploit a new allopolypolidy have been few - not surprisingly, perhaps, because the work tends to be laborious and success is far from guaranteed: long-term commitment is always necessary. Legumes, cucurbits, jutes and various fruits and ornamentals are among the most obvious possibilities". Although sharing Simmonds optimism about the potential of wide crosses for the development of a few useful new crop species, I wish to caution, for two reasons, that it is not realistic to expect easy or quick success from a "wide" hybridization program: first, the vast majority of newly synthesized species do not appear to have the potential productive capacity worthy of exploitation; and second, for those that appear to have the potential, the difficulties and time that is required to develop them from initial synthesis to an accepted cultivated crop.

During the past three decades, numerous wide crosses have resulted in the synthesis of new species. These include hybrids between Elymus and Triticum; Hordeum and Triticum; Brassica and Raphanus, Festuca and Lolium; Zea and Tripsacum. In our work at the University of Manitoba, we successfully synthesized hexploid, octoploid, and decaploid Triticale; the E genome of Agropyron elongatum was combined with the AB genomes of different species of tetraploid Triticums; similarly the Haynaldia genome and several Aegilops genomes were combined with the genomes of several species of Triticum. But in only one of all these newly synthesized species were any hybrids produced that would excite the imagination, or evoke the desire, to develop the hybrid into a new and useful crop species. However, the hybrids have to be produced, before value judgements can be made about potential usefulness. A case in point is the hexaploid Triticale, which should not be confused with the octoploid Triticale first reported in 1883 and on which so much pioneering developmental research was done by Dr. Muntzing in Sweden during the 1940's and the early 1950's. The hexaploid Triticale, unlike the octoploid Triticale, required the use of embryo culture and the use of colchicine for its successful synthesis. One of the first hexaploid Triticales - combining tetraploid durum wheat with rye - was reported by O'Mara in 1948.

We grew O'Mara's Triticale in Winnipeg in 1954 and were tremendously impressed with its yield potential. I wish to stress the word "potential" and return to it. A 5-foot row of O'Mara's Triticale was grown between corresponding rows of its two component parental species, durum and rye. The Triticale produced far more robust and vigorous plants -- much larger spikes, and well over 50% more spikelets per spike than either of the component species, or any of the neighbouring well adapted bread wheats. Thus it was easy to speculate that the new species had a yield potential at least 50% above that of wheat. The fact that fertility was incomplete, that seeds tended to shrivel upon maturity, was not really important. Surely, these faults and other weaknesses could be corrected with time by breeding. It was the apparent potential that provided the incentive to embark on the improvement of the hexaploid Triticale. The establishment of a Research Chair in our Faculty by the Bronfman Family Foundation at that time, provided a scientist and a technician to undertake the Triticale improvement program. That was more than 25 years ago. We soon recognized that in order to make progress with this newly synthesized species, one scientist and one technician was too small a team. Before a breeding program could be launched, variability had to be introduced by synthesizing amphiploids containing many different combinations of tetraploid wheats and rye. We recognized that to make progress would require the combined efforts of breeders, cytologists, physiologists, pathologists and chemists. But it was most difficult indeed to attract research support. The traditional granting agencies had advisors who were well aware of Dr. Muntzing's lack of success in launching octoploid Triticales. In fact if it were not for the Research Chair, it is doubtful if we could have continued our program. Two other groups, Kiss in Hungary and Sanchez-Monge in Spain, had also embarked on the development of hexaploid triticales in the early 1950s but they were handicapped in their work as we were with lack of finances and scientific manpower.

Fortunately, Dr. Bourlaug of CIMMYT became aware of our work in the early 1960s and initiated a breeding program for the development of Triticale on the scale required to make reasonable progress. Currently, according to Zillinsky et al (1980) "CIMMYT's advanced Triticale lines are as productive as the best wheats when both are grown under optimum production conditions. However, international testing data have shown that Triticale can yield twice as much as wheat on acid soils, in semi-tropical highland areas, in certain rainfed, sandy soils, and in some areas affected by serious wheat disease problems. Areas where some or all of these production conditions exist include southern Brazil, the Himalayan foothills, highland areas throughout East Africa, parts of Europe, the Central Plateau of Mexico, and the Andean Region of South America."

To indicate the potential impact of this "wide cross" species on world food production, Gustafson (1982) wrote that in Poland and Brazil there are at least 18 to 20 million hectares of acid soils not suitable for wheat where triticale can be grown, raising yields from nothing to at least one to two tons per hectare. That would result in a yearly increase of 20-40 million tonnes of cereals just from these two countries alone.

It has taken more than 30 years to bring the hexaploid triticale from its initial synthesis, its partial sterility, shrivelled seed, and low yield to its present stage of development of full fertility, fairly good seed characteristics, and yields that not only equal the best wheat yields, but far surpass them on many of the acid and light soils found in different parts of the world. A similar time frame for the development of any other newly synthesized species showing productive potential will surely be required. During the next two decades, based on the current deployment of scientific manpower and funding, far more effort will be made on a world base to improve wheat, rice, or maize than there will to improve triticale, even though the number of countries that are developing triticale breeding programs is on the increase. Despite this differential in activity favouring the established principal cereal food crops of man, I forecast that triticale will increasingly begin to outyield wheat in virtually every region of the world where wheat is grown.

The pandora's box of still unexplored "wide crosses" indicated by Simmonds (1980) raises the question of what actions are needed to realize potential opportunities. What problems need solutions? What are the knowledge gaps? The major impediments are the old standards of hybrid incompatability, chromosome doubling, and plant regeneration through tissue culture. The basic knowledge gaps stem mainly from the fact that not enough people are working on wide crosses. The gaps include lack of knowledge of the mechanisms of speciation, the factors involved in the stabilization of new species, and why some species hybrids cannot be made. Overcoming these impediments will take added personnel, money, and an official dedication to advancing wide cross research at the plant level.

REFERENCES

Bates, L.S. (1974). Wide crosses. Proc. Symp. World-wide maize improvement in the 70's and the role of CIMMYT. El Batan, Mexico.

Blakeslee, A.F. and A.G. Avery (1937). Methods of inducing doubling of chromosomes in plants by treatment with colchicine. Journal of Heredity, 28: 393-411.

de Wet, J.M.J. (1978). Principles of evolution and cereal domestication. Proc. Conf. Broadening Genet. Base Crops, Wageningen, 1978.

Flavell, R.B. (1981). An overview of the role of plant genetic engineering in crop improvement. Genetic Engineering for Crop Improvement, Working Papers/Rockefeller Foundation.

Freeling, M. (1982). Maize anaerobic genes. California Agriculture, 36(8).

Gustafson, J.P. (1982). Research Geneticist, USDA, Univ. of Missouri, Colombia MO. Personal communication.

Jan C.C., C.A. Qualset and J. Dvorak (1982). Wheat-barley hybrids. California Agriculture, 36(8).

Jensen, C.J. (1981). Use of cell and tissue culture techniques in plant breeding and genetics. Genetic Engineering for Crop Improvement, Working papers/The Rockefeller Foundation, 1981.

Knott, D.R. and J. Dvorak (1976). Alien germplasm as a source of resistance to disease. Annual Review of Phytopathology, 14, 1976.

Knott, D.R. (1981). Novel approaches to wheat breeding. Paper presented at the annual ASA meeting, Atlanta.

Robbelen, G. (1978). Transfer of quantitative characters from wild and primitive forms. Proc. Conf. Broadening Genetic Base Crops, Wagenengin.

Sears, E.R. (1956). The transfer of leaf-rust resistance from Aegilops umbellulata to wheat. Brookhaven Symp. Biol. 9, 1-22.

Simmonds, N.W. (1980). Polyploidy in plant breeding. Span, 23(2).

Thomas R., and Pratt, D. (1982). Embryo callus hybrids. California Agriculture, 36(8).

Tukey, H.B. (1933). Artificial culture of sweet cherry embryos. Jour. of Heredity, XXIV.

Zillinsky, F., Skoomand, B. and van Bentum, G.P.W. (1980). Triticale: adoption, production and uses. Span, 23(2).

IMPROVED CONVENTIONAL STRATEGIES AND METHODS FOR SELECTION AND UTILIZATION OF GERMPLASM

Donald N. Duvick
Director, Plant Breeding Division
Pioneer Hi-Bred International, Inc., Johnston, IA, 50131, U.S.A.

ABSTRACT

Plant breeding is an art as well as a science. Its success depends on the skill and good judgment of individual plant breeders, trained to utilize a judicious blend of scientific knowledge and plant breeding experience for production of improved cultivars. Plant breeders use continually updated combinations of effective hybridization/selection procedures and modern technological aids. Chief goals are yield and stability of yield, durable pest resistance, and useful genetic diversity. Genetic yield gains of major field crops in the United States have averaged about 1% per year during the past 40 years and show no sign of leveling off. The genetic component has comprised about 50-70% of total yield gains. It is hypothesized that research inputs, rather than biological ceilings, control rates of genetic improvement in cultivar performance. Plant breeding's most needed research input today is additional fundamental knowledge about plant biology. It is recommended that funding be increased for support of the basic biological sciences that serve plant breeding.

KEYWORDS: Plant breeding; yield gains; pest resistance; genetic diversity; hybrids; molecular biology; genetic engineering; recurrent selection; pedigree breeding.

HISTORY AND PROGRESS TO DATE

Plant breeding, as an art, preceded and made possible the development of agriculture. Plant breeding, as a science, is less than 100 years old. Its foundation on modern scientific principles followed the rediscovery of Mendel's laws in the first year of the 20th century. Initial progress in plant breeding was rapid and the technology of plant breeding was simple. As years went by, progress continued but breeders required increasingly large inputs of energy-intensive technology and increasingly complex breeding schemes. In recent years computers and electronic data collecting devices have added yet another dimension to the plant breeders' program of cultivar selection and improvement.

Wide-scale success of the new cultivars, combined with improved farming technology and improved transportation facilities, led to large-scale and intensive cultivation of a few cultivars of a few species. This practice intensified problems with recurring disease and insect epidemics, thereby forcing plant breeders to add disease, insect and nematode resistance to their list of highly important traits for selection and improvement.

The trend toward specialization on a few crop species increased the need for reliability of yields. Farmers no longer had the insurance offered by diversified crop and livestock farming, and needed to be sure that the two or three crop species on their farm would be reasonably successful every year. Thus reliability of yield moved near the top of the plant breeder's list of important traits for breeding and selection.

Agriculture experts have feared that progress in improving crop yield potential could not continue. However, genetic improvements in yielding ability, in disease and insect resistance, and in stability of yield have continued to appear. Recent summaries of genetic yield gains for U.S. wheat, maize, soybeans, sorghum and cotton have indicated continuous gains over the past 30 to 50 years, at rates of about 1% per year, using the mean over years as the base. Total yield gains have averaged nearly 2 percent per year; therefore, genetic improvements account for 50 percent or more of the total gain in yield. (Duvick, 1981; Meredith, 1981; Miller and Kebede, 1981; Schmidt, 1981; Specht and Williams, 1981).

In no crop is the curve for genetic yield increase starting to level out. In all crops, expression of gains in genetic yield potential does not necessarily require changed agronomic practices, such as increased amounts of nitrogen fertilizers. However, the yield potential of the newer cultivars is greatly enhanced by modern agronomic practices, whereas yield potential of the older cultivars is enhanced only slightly or sometimes even decreased by such modern cultural practices as high plant density and high nitrogen fertilizer applications. Where measurements have been made, most of the newer cultivars also have greater stability of yield in regard to environmental stresses such as heat, drought, excessive moisture or unusually cool weather. The new cultivars are tougher (Russell, 1974; Duvick, 1977; Evans, 1980).

Gains in pest resistance are not easily quantified since changes in pest biotype and the continual appearance of entirely new pest species give a continually changing base. Breeders, up to now, usually have been able to meet the needs for disease, insect and nematode resistance sufficiently to allow the new cultivars to express their enhanced yielding ability (Duvick, 1981). However, in some crops, especially cotton, insecticide applications are an indispensable aid to reliable production. In other crops, insecticide or fungicide applications are required to maintain present-day rotation practices, although different rotation patterns could eliminate the need for most of these chemical aids.

Herbicides seem to have become indispensable in crop production in the economically advanced countries, allowing earlier planting and novel plant spacing patterns. They also are the indispensable basis for the present-day movement to reduced tillage and zero-tillage. But herbicides as such can be nearly as much trouble to plant breeders as new insect or disease problems, because of most herbicides' ability to injure some of the non-target cultivars. Screening of cultivars for genetic resistance to new, highly potent, herbicides is becoming as important as screening the same cultivars for genetic resistance to prevalent disease and insect pests.

PLANT BREEDING GOALS

At present, plant breeders of major farm crops such as wheat, maize, soybeans, sorghum, alfalfa and cotton have three primary goals: 1) increased yield potential, 2) increased pest resistance, and 3) increased genetic diversity. Cultivars with increased yield potential must at the same time be reliable; their yields should not drop drastically in unfavorable seasons. Therefore, stability of yield is always implied in any statement about high yields. Cultivars with increased pest resistance must have durable long-lasting types of resistance that will not be overcome by changes in race or biotype of disease, insect or nematode. Cultivars that are genetically diverse - that originate at least in part from unrelated parents - and that have differing genetic responses to disease, insect and weather stresses are needed. Such genetic diversity will help prevent nationwide vulnerability to epidemics or weather patterns that disastrously affect a specific genotype.

NEW STRATEGIES

Improved strategies in plant breeding may be classified under three major headings: 1) new breeding schemes, 2) new technological aids to plant breeding, and 3) new biological concepts.

New Breeding Schemes

Although most plant breeders still rely on pedigree selection for the major part of their breeding effort, an increasingly large amount of work is devoted to recurrent selection of one kind or another. Recurrent selection, as used here, indicates any cyclic program of population improvement, of relatively few generations per cycle, and in which individual plant lineages usually are not available. Pedigree selection refers to longer cycle breeding schemes in which plant lineages are recorded and in which, usually, well-tested cultivars are used to start a new cycle of improvement. Recurrent selection has been advocated, and used, for about 30 years. To date its results, even in maize in the US where it has been used longest, have not been impressive in terms of numbers of useful inbred lines or cultivars. But practical, innovative variations on the basic design increasingly are coming into use; some of the longer term recurrent selection programs are finally maturing; and it seems likely that this technique will be more productive in the future. Recurrent selection should be especially useful for producing large - even though infrequent - step-wise increases in yielding ability, for developing new kinds of useful multigene pest resistance, and for introgressing exotic germplasm in order to increase the genetic diversity in useful breeding materials.

Hybrid seed production using cytoplasmic male sterility (CMS) and (usually) fertility restoration (FR) from nuclear genes has been advocated for wheat, cotton, alfalfa and soybeans for 20 years or more. To date the system does not work in any of these crops, at least to the extent that successful hybrids can be produced and sold for a profit. However work continues and breeders expect success eventually for wheat and cotton at least.

Hybrid seed production using gametocides (pollen control agents) depends on products of the chemical companies. No pollen control agent has yet had wide scale commercial success, although several wheat gametocides are being tested extensively.

In the self-pollinated crops such as wheat, large scale production of commercial hybrid seed using a pollen control agent will require that it be virtually 100% effective in all environments, and that it should have a broad time-span for effective application. Even a very small percentage of pollen shedding plants in the female could contaminate the remaining male sterile female plants to the point that the seed could not legally be sold as hybrid. A narrow time-span for useful application of the pollen control agent would mean that several days of wet or otherwise unfavorable weather could totally prevent application of the control agent and thus cause abandonment of thousands of hectares of seed fields, with concomitant loss of much needed hybrid seed, and seed company profits.

It bears repeating, that achievable heterosis in wheat, cotton and soybeans is not very large, compared to that in maize and sorghum. Therefore the cost of producing hybrid seed in these self-pollinating crops must be kept to a minimum, since farmers will not be willing to pay a high price for a small yield gain.

New Technological Aids to Plant Breeding

This area has seen much innovation during the past 10-15 years and holds much promise, in the short term, for further advances. New machinery, pesticides, improved transportation and digital computers have been important foundations to most of the technological advances. Specialized drilling and planting machines, combine harvesters, transplanting machines, and electronic weighing devices have multiplied many-fold the number of breeding selections that can be grown and observed by each breeder. Additionally these new tools greatly increase the precision of planting, measuring, and recording data. Computers have revolutionized data summarization and analysis. They will be even more useful in the future as interactive capabilities are improved to allow individual breeders to analyze subsets of pooled data. Further advances will come from use of portable computers for data entry and data transfer.

New Biological Concepts

Three classes of new biological concepts promise to help plant breeding in the future.

Durable (long-lasting) genetic resistance to diseases, insects and nematodes is a clearly defined goal for many plant breeders today. A body of theory about this concept is developing, along with data to test the theories. The concept rests on the belief that incomplete genetic resistance lasts longer than complete resistance and that often, but not always, durable resistance is best provided by polygenic rather than by monogenic systems. Recurrent selection and introgression of foreign germplasm are regarded as two useful breeding tools for development of durable resistance in those germplasm pools that lack it.

Genetic diversity is perhaps oversold as a preventative of disease and insect epidemics and as providing a strong buffer against environmental perturbations. Nevertheless, judicious application of this principle can benefit the farmer. Cultivar blends of barley or of wheat have been shown to provide more resistance to disease epidemics than did any of the components planted individually (Wolfe and Barrett, 1980). Genetic diversity in time, meaning sequential replacement of leading cultivars by new ones of different genotype, is another useful although usually unrecognized form of genetic diversity (Simmonds, 1962; Duvick, 1981). Introgression of exotic germplasm into adapted, high performance gene pools can introduce generalized but non-specific genetic diversity. The usefulness of this latter technique is limited, however, by the lack of knowledge about kinds of diversity available in germplasm collections and by lack of any workable standard for measurement of useful genetic diversity.

Molecular biology, and cell and tissue culture, are the basis of the new movement known as genetic engineering. Genetic engineering is based on the concept that DNA is the prime mover of all biological organisms, and that mastery of DNA manipulation means mastery of biology. Practitioners of this new kind of science already have pushed the boundaries of basic biological knowledge well past the limits once thought to be fixed forever. As plant breeders develop an understanding of these new facts and concepts they will revise their ideas of what can and should be done in traditional plant breeding. New breeding schemes and new sources of germplasm will be used. Individual pieces of the technology that supports this new kind of science will be found useful for speeding up or giving greater precision to traditional plant breeding practices. There is little likelihood of near term success in developing most of the novel products promised to us by some proponents of genetic engineering for plants. Nevertheless, the new sciences that comprise genetic engineering will be immensely useful to plant breeders and probably will be indispensable in helping plant breeders to maintain - or increase - present rates of progress in turning out new, improved cultivars by conventional means.

CHIEF IMPEDIMENTS TO PROGRESS

Narrow-minded Theoreticians and Plant Breeders

Plant breeding, as practiced by the most successful plant breeders, still is largely an art, even though faith in scientific principles supports the breeder's decisions. Individual breeders use highly personalized and continually modified combinations of hybridization/selection procedures. They have found that the science of probabilities is their most efficient tool. It allows them to manipulate breeding populations, selecting for traits governed by large but unknown numbers of genes with numerous but unknown interactions and with intricate but unknown biochemical pathways of expression. Recurrent selection, as a breeding tool, is based explicitly and rigorously on the science of probabilities.

Proponents of recurrent selection have ignored pedigree selection's potential and have left it out of their calculations of ways to achieve maximum progress in breeding and selection. Proponents of pedigree breeding have ignored recurrent selection because of its paucity of practical successes and have therefore unnecessarily limited the scope of their breeding populations and the kinds of useful genetic recombinations available to them. The weakness of recurrent selection is that, used improperly, it continually destroys newly created optimal genetic linkages. The weakness of pedigree selection is that, used improperly, it fails to construct useful new genetic linkages. Recurrent selection makes new gene combinations; pedigree selection identifies them with precision and saves them. The two approaches complement each other. Unfortunately, theoreticians have not recognized this complementarity, or having done so, they have not been courageous enough to try to build a body of theory encompassing both approaches. Lacking the theory, careful experimentation to correct and improve it cannot be done. Lacking the experimentation, development of practical breeding schemes that utilize both approaches to maximum advantage will be slow and inefficient.

Need for Better Understanding of Durable Genetic Resistance

Durable genetic resistance to disease, insects and nematodes is a definable goal so long as the definition means that the resistance has protected the crop for many years. Unfortunately this proof takes a long time to establish. We need to know, biochemically, molecularly, cytologically, and physiologically, why some genetic resistance is durable and why some is not. Why do pests frequently produce new virulent genotypes against some forms of genetic resistance, but not against others? Are there any general rules, or will each case be unique? If we find answers to some of these questions will they also help in designing durable pesticides?

The goal of the plant breeder is to eliminate all need for pesticides; since that goal can never be reached, practically, some pesticide use always will be necessary. If we understood durable genetic resistance better than we do now, we might be able to decide where to concentrate our plant breeding efforts in developing genetic resistance, and where to let pesticides provide protection, at least temporarily.

Lack of Knowledge About Fundamentals of Plant Growth and Metabolism

Molecular biology, and cell and tissue culture, hold great promise as aids to plant breeding but unfortunately a large gap in understanding exists between molecular genetics and cultivar performance. Plant breeders and molecular geneticists alike are just now realizing how badly we need knowledge of plant gene action, plant biochemistry, plant physiology and even plant ecology and coevolution, in order to use rationally the precise information now coming to use from the new molecular studies. Lack of knowledge about these intermediate processes is a severe impediment to further progress. Lacking this knowledge, genetic engineering cannot replace traditional plant breeding. Even with this knowledge, genetic engineering probably cannot replace traditional plant breeding, but it certainly could be much more useful to plant breeders if the intermediate processes were better understood.

Poor understanding of genetic diversity and how to use it.

Genetic diversity is widely advocated but poorly understood. We don't know how to measure it in practical, precise units. We don't know how to increase it in our breeding populations, efficiently and usefully. We don't know where exactly to get it, because our germplasm banks have as yet done very little to describe or catalogue their collections in regard to useful traits. We don't even know how effective it really is in providing protection against epidemics and environmental stresses, because with few exceptions only anecdotal or theoretical statements, rather than experimental or statistical data, have been used to show its usefulness in protecting crop plants. This gap in our knolwedge is a series impediment to progress in plant breeding.

ESTIMATES OF FUTURE PROGRESS

Curves of progress in plant breeding, crop by crop, are in reality "envelope curves", composites of a series of S-curves, each reaching its asymptote at the line of the "envelope curve", also called the "curve of need". Stated in other terms, as each technological input has reached the limit of its usefulness, another new technological aid to plant breeding has come forward to take its place, thus sustaining the overall forward momentum at about the same rate. The rate is established - and maintained - by the strength of demand for progress in that crop.

In maize breeding, for example, early improvement depended on use of replicated yield trials plus simple mass selection or unsophisticated variations of it; these techniques were supplemented by the invention of the double cross hybrid; then came ingenious schemes to increase efficiency of finding the best four-inbred combinations among the thousands that could be made; then came inputs of technical expertise from plant pathologists and entomologists; then selection of hybrids better able to use large amounts of nitrogen fertilizer; then mechanical aids such as research plot planters and combines; then the move to single cross hybrids; then use of computers; and now - probably - use of sophisticated laboratory techniques for analysis, understanding and prediction. During this entire 50-year period, summarized here very sketchily, inputs in numbers of skilled researchers also have been advanced continually. In the case of my own company, I have found that during the 50-year period in which our maize hybrids have made genetic yield gains at a linear rate of 1.5% per year, the numbers of our maize breeding personnel likewise have been increased at a rate of about 4% per year. Thus each successive increment of gain was more expensive, as measured in numbers of researchers. We seem nevertheless to have increased the researcher inputs at a rate sufficient to keep up a constant rate of breeding progress. The strength of the demand for improved maize hybrids has controlled the speed - the rate - of hybrid improvement. This was possible because enlightened company management understood the need for and utility of increasing the company's research efforts through the years.

As noted earlier, most major US crops have made genetic yield gains of about 1% per year. Projections of the regression curves indicate progress will continue at this rate for at least the next 10 to 20 years. Is this rather slow but constant rate of gain the consequence of a "demand" that asked for no more than 1% per year, or are there biological limits to the size of annual rates of gain? I believe these rates of gain are not limited by biological constraints - today. I believe that if society demanded faster rates of improvement, inputs would be increased proportionately and rates of gain would be increased. Every plant breeder knows that the best cultivars are not perfect and every breeder knows of genetic material that could improve the best cultivars, no matter whether yield, pest resistance, quality, or any other trait, is concerned. However the public must be cautioned that the simplest advances take, on average, 10 years from inception of breeding effort to placement on the farm in quantity. Longer lead times are required for complicated or untested advances to be implemented. Thus, simply increasing research budgets today will not give results before 1990, or more realistically before 2000. And, ominously, decreasing budgets today will bring on slow-downs that may not be seen before 1990 or 2000.

RECOMMENDATIONS

To improve conventional plant breeding's rate of progress in developing improved cultivars, we must increase investment in traditional plant breeding and in the basic biological sciences which support plant breeding. The US Variety Protection Act, and similar breeders' rights laws in most other countries with advanced economies, have stimulated private investment to the point that it generally will provide the funds for plant breeding of most - but not all - food and fiber crops. This investment will continue as long as breeding gains, and seed company profits, can be realized. Continued breeding gains will depend in part on mechanical innovations, which private firms probably can produce, and in part on utilization of new basic biological knowledge, which private firms probably cannot produce. Public institutions must provide this knowledge; they must decide to do it; they must be funded sufficiently to allow them to do it. At this time great indecision is to be seen in many public institutions - especially the universities - in the U.S. They are losing traditional sources of funds and are looking to short-term solutions, most of which involve dropping some long-term research in favor of short-term, saleable research. Major funding institutions - usually federal and state bodies - must recognize this problem. They must fund the university research departments to the extent needed to support the kinds of basic biological research which will be useful to plant breeding.

Fundable research needs lie not only in the new fields of molecular biology and cell and tissue culture, but also - and emphatically - in the fields of plant physiology, genetics, plant pathology, entomology, and related plant biology disciplines. Major effort must be made to integrate all of these research programs with plant breeding. Researchers with expertise in both plant breeding and one or more of the above fields must be trained. These broadly trained researchers will become group leaders, cross-discipline interpreters and innovators, helping to bring the results of basic studies to fruition in practical plant breeding programs and finally in the farmers' fields.

REFERENCES

Duvick, D.N. (1977). Genetic rates of gain in hybrid maize during the past 40 years. Maydica XXII, 187-196.

Duvick, D.N. (1981). Genetic contribution to yield gains of hybrid corn during the past 50 years. Agron. Abstracts. American Society of Agronomy, Madison, WI.

Duvick, D.N. (1981). Genetic diversity in major farm crops on the fram and in reserve. Presented to the 13th International Botanical Congress, Sydney, Australia, August 18, 1981.

Evans, L.T. (1980). Response to challenge: William Farrer and the making of wheats. Jour. Austral. Inst. Agric. Sci. 1980, 3-13.

Meredith, W.R. Jr. (1981). Genetic contributions to yield gains: Cotton. Agron. Abstracts. American Society of Agronomy, Madison, WI.

Miller, F.R. and Y. Kebede (1981). Genetic contributions to yield gains in sorghum over the past 30 years. Agron. Abstracts. American Society of Agronomy, Madison, WI.

Russell, W.A. (1974). Comparative performance for maize hybrids representing different eras of maize breeding. Proc. 19th Annual Corn and Sorghum Research Conference. 29, 81-101.

Schmidt, J.W. (1981). Genetic contribution to yield gains in wheat (Triticum species) in the United States. Agron. Abstracts. American Society of Agronomy, Madison, WI.

Simmonds, N.W. (1962). Biological Reviews 37, 422-465.

Specht, J.E. and J.H. Williams (1981). Genetic technology contributions to soybean productivity - restrospect and prospect. Agron. Abstracts. American Society of Agronomy, Madison, WI.

Wolf, M.S. and J.A. Barrett (1980). Plant Disease 64; 148-155.

CHEMICAL, BIOLOGICAL, GENETIC, BIOLOGICAL AND AGRONOMIC APPROACHES TO IMPROVED OR ALTERNATE TECHNOLOGIES TO PROVIDE FIXED NITROGEN

Ralph W.F. Hardy

Central Research and Development Department, Experimental Station
E.I. du Pont de Nemours and Company, Wilmington, DE, USA 19898

ABSTRACT

Initiated by a chemist's seminal advance, the knowledge base of the chemistry, biochemistry, genetics, biology and agronomy of N_2 fixation has expanded greatly during the past two decades and continues to expand. A large increase in fixed N input is required to support the estimated growth in crop production to meet global food and feed needs. The major current technologies and their products or processes are synthetic nitrogen fertilizer based on chemical and engineering skills, fertilizer N responsive crops, N_2-fixing legumes and recycling wastes. Improved or alternate technologies are desirable on an economic, fossil energy and self-sufficiency basis but are not essential. Potential improved or alternate technologies for the near, mid- and long-term are described. These perceived technologies will have modest impact in the near term but could produce major impact in the long term.

KEYWORDS: Fertilizer Nitrogen, Biological Nitrogen Fixation, Grain Yield and Nitrogen, Advances in Nitrogen Research, Rhizosphere Microbial Associations, Cost of Symbiotic N_2 Fixation, Molecular Genetics of N_2 Fixation, Technologies for Nitrogen Input, Nitrogenase, *Nif* Genes.

INTRODUCTION

The importance of adequate fixed nitrogen for crop production is established. For example, the continuously increasing input of fertilizer nitrogen into cereal grains is strongly correlated with their increasing yield (Hardy and Havelka, 1975). This almost linear correlation is true for less developed and more developed countries for the 3rd quarter of the 20th century. The data form a continuum except that the average cereal grain yield and average nitrogen fertilizer use on these crops in more developed countries is about 20 years ahead of that in less developed countries. The significant apparent opportunity is to extend this relationship of increasing fixed nitrogen input and cereal grain yield in both economic environments. It is, of course, recognized that inputs of improved cultivars, better cultural practices, and increased use of agrochemicals and irrigation as well as increased fertilizer nitrogen use have contributed to these yield increases.

What are the sources of fixed nitrogen and their past, present and future contribution? Biological and abiological N_2 fixation (Table 1) (Burns and Hardy, 1975) were estimated at 255 M tonnes in 1975. A small estimated increase of 15 Mt in biological N_2 fixation by expanded legume cropping accompanied by the development of industrial N_2 fixation produced an estimated 80 Mt increase from 1900 to 1975. By 2000 A.D. an increase of up to 175 Mt, a major quantitative change in the global N cycle, may occur mainly in industrial fixation to meet the estimated 160 Mt of fertilizer nitrogen required to meet crop production needs. New technologies to provide fixed nitrogen are not anticipated by that time. It is impossible to suggest the contributions from biological and industrial fixation processes by 2025 since the useful technologies will probably be much different but it is too early to predict the specific technologies. My comments will provide an informational base with which to evaluate possible future technologies. Economic, environmental, health and possibly political factors will control the selection of the technologies to be used. Chemistry of animate and inanimate systems will continue as in the past to be a key skill in generation of the possible technologies.

Table 1: Estimated Annual Amounts of Total N_2 Fixed (Mt) (1975 Estimate from Burns and Hardy, 1975)

	1900	1975	2000	2025
Biological	160	175	190	?
Abiological				
Lighting	10	10	10	10
Combustion	5	20	30	?
Industrial	0	50	200	?
	175	255	430	?

The impressive expanding base of knowledge especially in the area of biological N_2 fixation is surveyed in this paper and the expanding need for fixed N estimated. Characteristics of established technologies to provided fixed N make it desirable but not essential that improved or alternate technologies be found. Possible future technologies are described with projected time of impact.

THE EXPANDING KNOWLEDGE BASE

The past 20 years have been most productive in improved knowledge of nitrogen fixation processes. These advances have resulted from engineering, chemical, biochemical, genetic, biological and agronomic investigations. The list of advances (Table 2) in almost all cases is impressive. In 1960 there was no information on the molecular aspects of biological N_2 fixation; in 1982 molecular understanding of the chemical, biochemical and genetic area is well advanced (Hardy, 1977).

Recent books (Bothe and Trebst, 1981; Burns and Hardy, 1975; Chatt et al., 1980; Gibson and Newton, 1981; Hardy et al., 1979; Hardy and Gibson, 1977; Hardy and Silver, 1977; Hollaender et al., 1977; Newton and Lyman, 1976; Newton and Orme-Johnson, 1980; Newton et al., 1977; Sprent, 1979; Stewart and Gallon, 1980; and Subba Rao, 1980) and reviews (Ausubel, 1982; Bauer, 1981; Brill, 1980; Emerich and Evans, 1980; Gutschick, 1981, 1982; Hardy, 1977; Hardy, 1980ab; Hardy et al.,

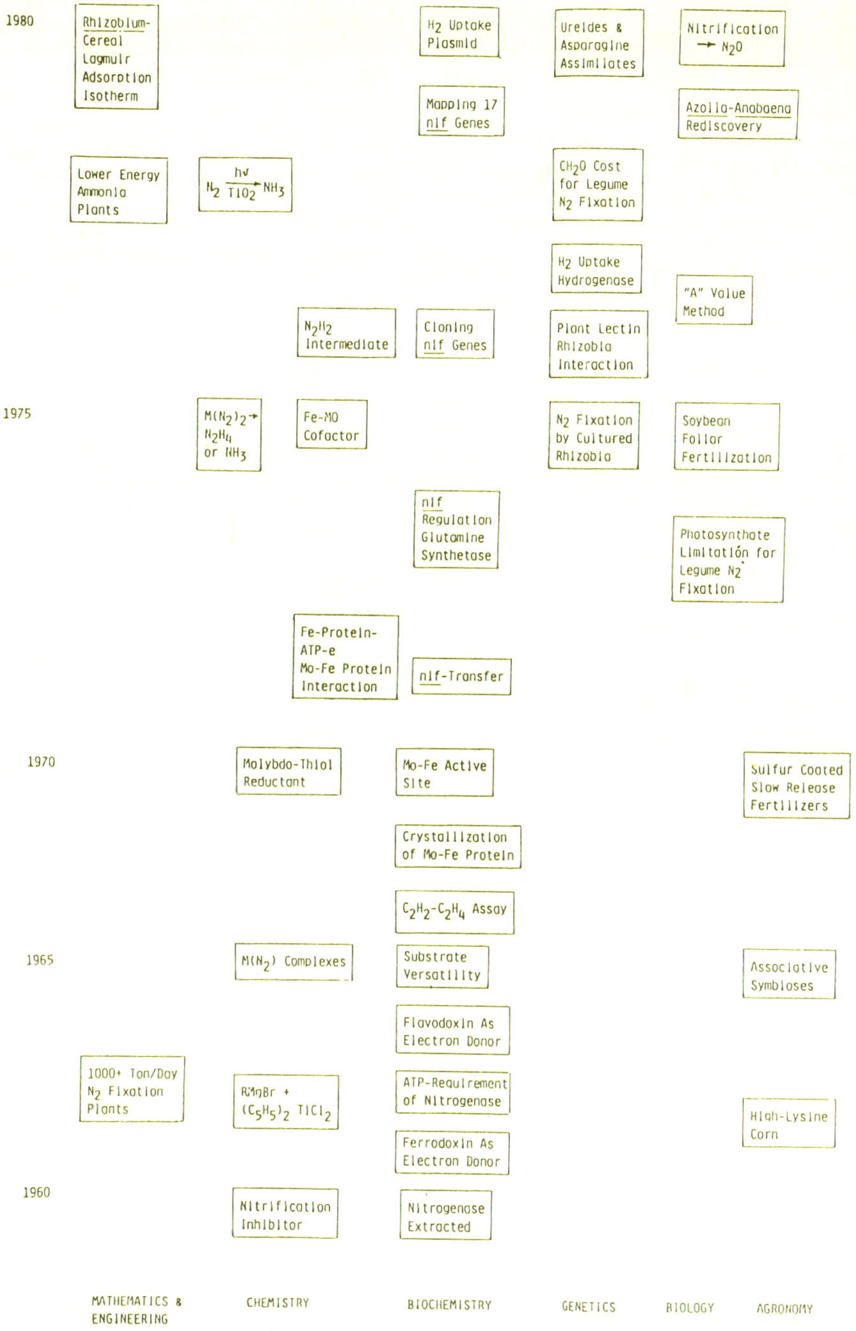

Table 2 Significant Advances in Nitrogen Fixation Processes Since 1960

1983; Hardy and Havelka, 1975, 1977; Havelka et al., 1982; Lamborg and Hardy, 1982; La Rue and Patterson, 1981; Mortenson and Thornley, 1979; Phillips, 1980; Postgate, 1982; Roberts and Brill, 1981; Robson and Postgate, 1980; and Shanmugam et al., 1978) on N_2 fixation provide much additional information to supplement the following overview.

Explosions in knowledge usually can be traced to one or more seminal advances. Biological N_2 fixation is no exception and indicates the importance of different approaches and backgrounds to such advances. In the late 1950's, a chemist, Dr. James E. Carnahan, who had worked in the catalyst area in the Du Pont company was urged to tackle a challenging problem outside of traditional catalyst research. He selected the area of biological N_2 fixation. His objective was to learn the molecular secrets of the biological N_2 fixation system that functioned under mild conditions of temperature and pressure with the hope that such information might lead to, for example, an improved catalyst for industrial reactions of molecular nitrogen. At this time no one had successfully extracted the N_2-fixing enzyme, already named nitrogenase, in an active form from any living cell. Within a few years, Carnahan successfully extracted nitrogenase that was so active that the product NH_3 could be measured by titration (Carnahan et al., 1960). Why did a chemist succeed where many biochemists had failed? He selected an anaerobic organism, Clostridium pasteurianum which had not been as extensively explored in attempts to extract nitrogenase. He systematically tested many methods of cell culture to find one that worked. Of more importance was his addition of molar quantities of a substrate, pyruvate, which subsequent work revealed was needed in such quantities to fuel the highly inefficient nitrogenase.

This example supports the concept of balanced funding of research for both scouting and concentration efforts, and scientist diversification. It is critical to support scouting research by individual highly creative scientists since the seminal advance from these efforts will open refractory but important fields to major advances in knowledge from which may come important new technologies. Furthermore, scientists with backgrounds and experience outside a field may be more effective than those within.

In addition to a seminal advance, the facilitation of dialog among scientists stimulated advances in N_2 fixation. An informal meeting at Butternut Lake in Wisconsin in 1964 brought together about 20 scientists working on the biochemistry of N_2 fixation and this was followed by similar meetings at Segehen, California, and Sanibal Island, Florida. In 1968, chemists joined the biochemists. The next major step was the initiation of a biannual international meeting on N_2 fixation in 1974 at Pullman, Washington, involving chemists, biochemists, geneticists, biologists and agronomists (Newton and Lyman, 1976). These meetings have grown to seveal hundred scientists with the information expansion paralleling the scientist involvement (Newton et al., 1977; Newton and Orme-Johnson, 1980; Gibson and Newton, 1981). The 5th international meeting will be held in the Netherlands in 1983. Nitrogen fixation has been selected consistently from the early 1970's as a high priority area for agricultural research. For example, a recent scientist survey gave an expected contribution of 142 kg/ha of corn from biological N_2 fixation by 2100 A.D. (Brady, 1982).

Engineering

Engineering input led to large nitrogen fertilizer synthesis plants in the 1960's. These plants make 1000-1500 tons of ammonia daily and have displaced smaller less economic plants. Natural gas provides H_2 for reduction of N_2 to NH_3 as well as energy to operate the plant at the required high pressures and temperatures. About 33 M BTU of natural gas are needed per ton of ammonia but recent engineering advances may save about 4-6 M BTU/ton by retrofitting existing plants to recover heat from flue gases to preheat the combustion air (Buividas and LeBlanc, 1982). A saving of up to 20% in capital costs and 23% in natural gas was achieved with a new catalyst that operates at a lower pressure (Anon, 1982b).

Chemistry

Molecular nitrogen although believed to be almost inert was fixed under mild conditions by scientists in the USSR, US, and Canada in the 1960's. A chemically defined complex with two nitrogen molecules attached to a transition metal can be reduced to hydrazine and ammonia; however, a strong reductant is required (Chatt et al., 1980). These systems are not expected to provide the base for a useful industrial process but may provide information on the activation of molecular nitrogen that might be coupled to an oxidative fixation process. Industrial as well as the biological N_2 fixation are reductive requiring expensive H_2 or its equivalent. The major cost of fertilizer nitrogen is the increasing cost of the natural gas to provide H_2 and energy. The product of industrial N_2 fixation is ammonia.

Fertilizer is most frequently applied as ammonia or in many developing countries as an ammonia derivative, urea. Soil microorganisms rapidly oxidize the ammonia or ammonia from urea to nitrate. The International Fertilizer Development Center (IFDC) and others are seeking new or modified fertilizer products that would be used more efficiently (Anon, 1982a). Fixation of N_2 by an oxidative vs. a reductive process would eliminate the need for H_2 and provide the same ultimate soil nitrogen form, nitrate. An oxidative process using N_2 and O_2 (air) to produce dilute aqueous solutions of nitrate is thermodynamically feasible. Chemical reserch on N_2 fixation should seek catalysts for such an oxidative reaction that could provide fertilizer nitrogen with zero direct energy input. Research in this area needs to be increased. Improved catalysts to activate molecular nitrogen for oxidation are the key need.

Biochemistry

Biological N_2 fixation is catalyzed by a binary enzyme, nitrogenase, consisting of two brown metalloproteins. The larger which is believed to be the dinitrogenase is an $\alpha_2\beta_2$ tetramer containing 2 Mo, 32 Fe and 30 sulfide atoms part of which has been purified as a Mo-Fe cofactor; the smaller, which is believed to reduce dinitrogenase, is an α_2 dimer containing 4 Fe and 4 sulfide atoms. Sophisticated physical methods, e.g., EXASF, are proving useful in the characterization of these proteins. There is only minimal structural variation in the nitrogenase proteins from different organisms. In a few cases, the small protein is phosphorylated to an inactive form in the presence of NH_3. The nitrogenase proteins are extremely sensitive to O_2 and this liability limits both their location within the living cell and their use as a catalyst in a chemical process. Nitrogenase is a most sluggish enzyme with each molecule converting only about 50-100 molecules of N_2 to ammonia each minute.

The biological process is a reductive one, e.g.,

$$N_2 + 3H_2 \text{ ---> } 2NH_3 + \text{Energy}$$

analogous to the industrial Haber-Bosch process. Unfortunately, nature provided us with only a reductive process and no oxidative one, or if it did, scientists have yet to find it.

Nitrogenase generates an even more potent reductant than H_2 at great metabolic cost by hydrolysis of ATP. Four ATP's are consumed for every two electrons transferred by nitrogenase. These electrons are provided to nitrogenase by a ferredoxin or flavodoxin at the reduction level of H_2. Furthermore, nitorgenase is not absolutely specific for the substrate N_2 and protons are reduced to H_2 simultaneously with the reduction of N_2, e.g.,

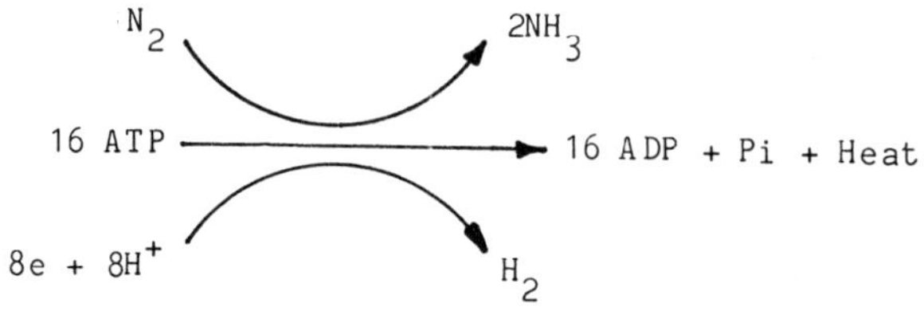

Much energy is wasted as heat since the reaction of N_2 and H_2 to produce ammonia is itself energy producing. The coproduct, H_2, may be scavenged and oxidized to recapture some of the energy used in its formation if an enzyme called an uptake hydrogenase is present. Unfortunately nitrogenase in vitro is excessive in its energy requirements and it is doubtful that such an inefficient system will provide information useful for the design of more efficient chemical catalysts.

A variety of other triple bonded or potentially triple bonded small molecules containing N, C and O are also reduced by nitrogenase making it one of the most versatile reductases known. These substrates include nitrous oxide, azide, nitriles, isonitriles, and actylenes. Reduction of acetylene to ethylene with gas chromatographic analysis has proved to be a most useful quantitative assay for N_2 fixation in vitro and in vivi (Hardy et al., 1968, 1973). Further elucidation of the biochemistry of nitrogenase will probably come from molecular genetic studies.

Genetics

Molecular genetics is providing powerful tools to expand our understanding of biological N_2 fixation. Genetic advances (Table 2) during the past decade dominates the research advances on N_2 fixation. Some hope, and some predict, that these tools will allow, in time, the incorporation of functional N_2-fixing genes into crop plants to enable them to become self-sufficient in fixed nitrogen. However, current activities in genetics are at the understanding stage - the practical, hopefully, will come later.

The collective genes for N_2 fixation are called *nif*. They have been most intensively studied in *Klebsiella pneumonia* where 17 have been mapped between *his* and *shi* P (Brill, 1980; Gibson and Newton, 1981; Postgate, 1982; Roberts and Brill, 1981). The indicated function of the *nif* genes are shown in Table 3. The *nif* K and D genes code for the α- and β-subunits respectively of the Mo-Fe protein into which the Fe-Mo cofactor is inserted while *nif* H codes for the Fe protein. A few of the structural genes have been sequenced thereby providing the amino acid sequence of the proteins.

These *nif* structural genes in the N_2-fixing blue-green alga *Anabaena* (Mazure et al., 1980; Mazur and Chui, 1983; Rice et al., 1982) have provided fundamental understanding of the nitrogenase structural proteins and identified a molecular relationship between *Anaebaena* and higher plant chloroplasts. Based on the sequence of the *nif* K gene, its protein product which is the precursor of the β-subunit of the Mo-Fe protein is 512 amino acids long with a molecular weight of 57, 583 daltons and contains 6 cysteine residues, three of which may be ligands to which FeS clusters bind.

Although the *nif* K and D genes from *Anabaena* are separated in the chromosome by 11,000 nucleotides, they have homologous promoter-like sequences suggesting coordinate control. Even more striking is the almost identical homology of the

Table 3: The nif Genes and Their Function Based on Studies on Klebsiella pneumoniae and Extension to Anabaena (Hardy et al., 1983; Roberts and Brill, 1981). Genes in brackets are less well defined.

Regulation	Pronitrogenase Structure	Modification	Active Nitrogenase Components	Electron Donors
nif A,L	nif K	nif B,N,E,(Q,C,J) →	β-Subunit of Mo-Fe Protein	nif F,J
	nif D	Fe-Mo-CO → →	α-Subunit of Mo-Fe Protein	
	nif H	nif M,S(V,U) →		

nif K promoter to a promoter-like sequence preceding the start of the large subunit of ribulose bisphosphate carboxylase (Rubisco) from maize (McIntosh et al., 1980). This molecular similarity supports the blue-green algal evolutionary origin of chloroplasts and suggests that incorporation of algal nif genes into chloroplasts may be appropriate.

The nif genes are regulated by NH_3 and O_2. Products of gln genes including gln A which specifies glutamine synthetase are believed to be responsible for NH_3 regulation of nif. Another system is responsible for O_2 regulation of nif. In addition, products of nif A and nif L regulate nif as well as possibly others. Regulation is complex and this is not surprising for such an energy-consuming process.

The genes for nitrogen fixation are limited exclusively to lower life forms -- the reason, if any, for this limitation is unknown. These nif genes have been transferred to and function in Escherichia coli, Salmonella typhumurium, Serratia marcescens, Erwinia herbicola, Pseudomonas fluorescens but did not function in Proteus mirabilis and Agrobacterium tumefaciens. Recently molecular geneticists have moved nif up the evolutionary scale to yeast, but the genes failed to fucntion. Undoubtedly, in time, scientists will learn how to move these genes in a functional form to higher forms such as plants and maybe even animals.

Of greatest near-term interest to agriculture is the symbiotic nitrogen fixation system of legumes. The genes for N_2 fixation, host-range specificity and nodulation appear to be located on a large plasmid in Rhizobium. It is not known if some of these genes are inserted into the plant's chromosomes analogous to the process of tumor production by Agrobacterium (Ausubel, 1982). The excess enthusiasm of some molecular geneticists has led them to make premature claims of the development of super mutants of Rhizobium; if such were the case in the laboratory, they were not substantiated by field trials.

Biology and Agronomy

Within the last few years, Rhizobium in many cases have been manipulated so as to fix nitrogen in culture independent of legumes. However, the rates are low and the spectrum of organisms does not include all Rhizobium. This limits most studies to the more complex plant-microorganismal system than the more simple Rhizobium culture. Fixed nitrogen, nitrate, and ammonia produce multiple inhibitions of symbiotic nitrogen fixation. They inhibit the infection process, the nodulation process, and the quantity of nitrogen fixed by the system as well as promoting senescence of the symbiotic system.

The large in vitro cost of N_2 fixation described above may be a major barrier to expanded biological N_2 fixation in high-yield crops and has stimulated an examination of the in vivo cost of N_2 fixation in the legume symbiotic system (Schubert and Wolk, 1982).

We have measured simultaneously the respiration of nodulated and non-nodulated root systems and their C_2H_2 reduction in intact soybean plants over the period from initial nodulation to pod fill (Hardy et al., 1983; Heytler and Hardy, 1979). The nodulating isoline of Clark was compared with its non-nodulating isoline to provide an appropriate control for respiration not associated with nodule function. The cost of nodule function was further subdivided into maintenance and N_2-fixing activities by comparison of nif^- and nif^+ Rhizobium nodules using the nonfixing mutant SM-5 provided by Winston Brill.

The overall nodulated root respiratory cost of N_2 fixation was 100+ g (CH_2O)/g N_2 fixed at low rates of N_2 fixation declining to 12 at high rates. Subtraction of the respiratory cost for the nif^- Rhizobium nodulated root from the nif^+ Rhizobium nodulated root gives a net cost for N_2-fixing related costs -- operation of nitrogenase and export of NH_3. This net cost of N_2 fixation decreases from 100^+ g CH_2O/g N_2 fixed at low rates of N_2 fixation to about 6 at high rates. At high rates of N_2 fixation, approximately 25% of the nodulated root respiration supports activities of the root, 25% supports non-N_2-fixing-associated activities (maintenance) of the nodule and 50% supports N_2-fixing-associated activities.

The incremental cost at high rates -- the most favourable case -- approaches that of in vitro nitrogenase while the cost at lower rates greatly exceeds the in vitro cost due to the cost of nodule maintenance. Clearly in vivo symbiotic N_2 fixation is a major consumer of carbohydrate. These results are consistent with the parallel response of symbiotic N_2 fixation, and photosynthate production (Hardy and Havelka, 1977). Decreased nodule maintenance cost or improvement of nitrogenase energetics, a most challenging proposition, probably is needed to make extension of biological N_2 fixation attractive for high yield agriculture.

The biological system in many cases evolves hydrogen during its fixation of nitrogen as discussed under biochemistry. The energy of the hydrogen is lost to the plant if it does not possess a hydrogen uptake system. In greenhouse tests, soybeans inoculated with Rhizobium possessing a hydrogen uptake capability outyield those inoculated by Rhizobium without this capability. No difference was found in field tests presumably because endogenous soil Rhizobium dominated (Emerich and Evans, 1980). Legumes and most cereals tend to be photosynthetically inefficient. They have the wasteful process of photorespiration that decreases their new dry matter accumulation by 25-50%. This photosynthetic inefficiency of legumes makes the cost of biological nitrogen fixation of even greater concern.

The symbiotic system undergoes premature senescence. Biological nitrogen fixation stops increasing and may even decline while significant soybean seed growth is still occurring. The amount of nitrogen fixed will vary with the fixed nitrogen content of the soil. In high fertility soils, approximately 25% to a maximum of 50% of the nitrogen in legume plants arises from biological nitrogen fixation. There is an inadequate amount of nitrogen fixation for high yield soybeans. The symbiotic biological nitrogen fixation system is difficult to optimize because of the multiplicity of the strains of Rhizobium that exist, the specificity of bacterial strain-plant cultivar interactions, and the impact of soil and climatic variables. Nitrogen fixation can be increased by legume breeding program. The standard alfalfa cultivar fixed 109 kg N/ha.yr while the cultivar selected for increased N_2 fixation fixed 148 kg N/ha.yr (Viands et al., 1982).

Conditions for use of biological systems such as bacterial inocula are more demanding than for inanimate ones. Manufacture, storage, handling, and application of inocula may all result in partial or complete less of biological activity while an inanimate chemical additive is not similarly affected. In addition, competition for nodule formation between applied Rhizobium with improved N_2-fixing activities and endogenous ones as suggested above for the added organisms containing hydrogen uptake capabilities. Interest in the rhizosphere association of N_2-fixing microorganisms with crop plants exploded in 1974 when Dobereiner reported the association of the N_2-fixing bacteria Azospirillum with roots of tropical plants (Dobereiner and Day, 1976). The potential for high rates of N_2 fixation measured under special conditions of preincubation of roots was reported. Most controlled greenhouse or field experiments were unable to show a N_2 fixation or yield benefit from inoculation with Azospirillum.

Recent multi-year field experiments on several grain and forage crops by Okon and his colleagues in Israel have shown significant yield increases from inoculation with Azospirillum (Kapulnik et al., 1981; Cohen et al., 1980). Recently, we have explored, jointly with Dr. Okon, the Azospirillum rhizosphere association under controlled growth room conditions. Setaria italica was used as the model plant and was inoculated with Azospirillum brasilense and grown under limited N. N_2 fixation measured by C_2H_2 reduction peaked at the booting stage and the integrated activity during the plant life cycle was estimated as about 20% of the plant's N content at maturity. $^{15}N_2$-Experiments confirmed a small amount of N_2 was fixed in the model system. The transfer of this fixed nitrogen to the plant is very slow presumably requiring bacterial decomposition and mineralization. Factors other than the contribution of newly fixed nitrogen to the plant must account for the improved field yield (Okon, Y., Heyter, P.G., Hardy, R.W.F., unpublished results). One of these factors may be an enchancement of ion uptake. The uptake of NO_3^-, K^+ and Pi by corn was increased about 50% by inoculation with A. brasilense while other bacteria produced no effect. The growth of corn and sorghum plants on limited nutrients was increased 16 to 34% by Azospirillum root inoculation (Lin, W., Okon, Y., Hardy, R.W.F., unpublished results). Based on these model experiments, nutrient uptake enhancement probably contributes more to beneficial yield responses of Azospirillum inoculation than contribution of newly fixed nitrogen.

THE NEED AND ESTABLISHED TECHNOLOGIES

A variety of factors will increase the world need for food, feed, and biomass and consequently fixed nitrogen. Population is growing at about 2% annually. Growing affluence is increasing the fraction of animal vs. plant food in the diet which accelerates the need for plant product growth at a greater rate than population growth because of the inefficiency of the plant to animal conversion. These animal feeds will require a greater increase in high protein crops such as grain legumes than cereal grains. New foods such as single cell proteins probably will make a negligible contribution to world food. The crop production increase will come mostly from yield enhancement rather than expansion of crop acreage. Development of economic processes to convert biomass to commodity chemicals (Ng et al., 1983) and selected energy substitutes, such as ethanol as an octane enhancer in gasoline, will place additional demands on crop production.

The above assumptions suggest an estimated need to increase cereal grain production at 3% annually and grain legume production at 6% annually. Recycling of the nitrogen in the harvested part of these grain crops to the cropping fields is minimal. The above projected increase in cereal grain production from 1300 M tonnes in 1979 and 2600 M tonnes by 200 A.D. will increase the nitrogen content in the harvested crop fom 39 M tonnes in 1975 to 78 M tonnes in 2000. The above projected increase in grain legume production from 130 M tonnes in 1975 to 520 M tonnes by 2000 A.D. will increase the harvested nitrogen from 10 M to 39 M tonnes. This increase in harvested nitrogen of about 70 M tonnes in these grain crops coupled with the need for fixed nitogen by root, forage and biomass crops and an average efficiency of use of fixed nitrogen of about 50% emphasizes the major need for additional fixed nitrogen by the end of this century.

This need may be met by an expansion of existing technologies, improvement of existing technologies, and/or development of new technologies. Twelve technologies that span biblical times to the present are listed in Table 4. The key skills from which these technologies were developed range from empirical observation to chemistry, microbiology, engineering, plant breeding, agronomy and plant physiology. The most significant current inputs are synthetic nitrogen fertilizers, legume symbiotic N_2-fixing systems, nitrogen responsive crops and nitrogen recycling. They all have disadvantages. Industrial nitrogen fertilizer manufacture requires capital expensive plants and use large amounts of fossil energy -- about 25% of fossil energy consumed by agriculture up to the farm gate in developed countries is for nitrogen fertilizer manufacture. A more economic and energy-saving technology would be advantageous. Furthermore, specific countries may need to import fertilizer nitrogen. Legume symbiotic N_2 fixation consumes large quantities of plant carbohydrate to support the highly inefficient process. A process that consumed less plant material may enable the plant to convert more of its carbohydrate to harvestable yield. Nitrogen-responsive crops such as the high-yielding wheats and rices require large fixed nitrogen additions. Crops that are more effective in scavenging fixed nitrogen may have lower requirements for fixed nitrogen to produce the same yield. Recycling is impractical when the nitrogen product, e.g., manure, must be transported long distances.

Newer technologies are being evaluated. They include intercropping (Wiley, 1983), nitrogen transformation effectors, e.g., N-Serve , slow release nitrogen fertilizers (Anon, 1982a), foliar fertilization and new legume crops (Haq, 1983). some may provide improvements but with modest impact.

FUTURE IMPROVED OR ALTERNATE TECHNOLOGIES

Future technologies are also considered in Table 4. They are organized into highly probable short-term, less probable mid-term, and more speculative long-term to provide an indication of the possibility of realization and timing of impact. The short-term examples include improved energy efficient N fertilizer plants, legume cultivars and *Rhizobium* species matched for enhanced N_2 fixation and yield, inceased harvest index for nitrogen in certain food legumes where the overall harvest index may be as low as 20%, and improved methods for effective *Rhizobium* inoculation. The overall impact is expected to be modest.

Less probable, mid-term impact technology include improved *Azolla-Anbabaena* symbiosis for paddy rice, mycorrhizal inocullants, and improved *Rhizobium* to simplify inoculation which could be important in developing countries where errors in inoculation are probably even more prevalent than in the developed countries.

Table 4. Technologies for Nitrogen Input

Skill	Technology - Product/Process
Established	
Empirical Observation	Recycling of N-Containing Wastes
" "	Legumes as Green Manures
" "	*Azolla-Anabaena* (Goddess of Azolla)
Chemistry	Mined Nitrates as Fertilizer N
Soil Microbiology	*Rhizobium* Inoculation of Legumes
Chemistry/Engineering	Synthetic N Fertilizer
Plant Breeding	N-Responsive Crops
Agronomy	Multiple Cropping and Intercropping of Legumes and Nonlegume Crops
Chemistry/Soil Microbiology	N Transformation Effectors
Chemistry	Slow Release N Fertilizers
Plant Physiology/Chemistry	Foliar Fertilization
Agronomy	New Legume Crops, e.g., Winged Bean
Highly Probable Short-Term Impact	
Engineering	Improved Energy Efficient N Fertilizer Plants
Agronomy/Microbiology	*Rhizobium* Species/Legume Cultivar Matches
Microbiology	Improved *Rhizobium* Inoculation Methods
Plant Breeding	Increased Harvest Index for Nitrogen
Microbiology/Genetics	*Rhizobium* with H_2 Uptake Gene
Plant Physiology/Microbiology	*Azospirillum* Inoculation
Chemistry/Agronomy	Nitrogen Fertilizer Responsive Legumes
Less Probable Mid-Term Impact	
Plant Physiology/Microbiology	Improved *Azolla-Anabaena* Symbiosis for Paddy Rice
Microbiology	Mycorrhizal Inoculants
Microbiology	Improved *Rhizobium*
Genetics	Promiscuous *Rhizobium*
Genetics/Chemistry	Symbioses Insensitive to Fixed N
More Speculative Long-Term Impact	
Genetics/Microbiology/Plant Breeding	Non-*Rhizobium* N_2-Fixing Associations for Nonlegume Crops
Chemistry/Genetics/Physiology	Increased Photosynthesis
Chemistry/Genetics/Physiology	Improved Nodule Efficiency
Microbiology/Agronomy	Naturally Occurring but Undiscovered N_2-Fixing Systems, e.g., *Rhizobium-Trema*
Genetics	Extension of Rhizobial N_2-Fixing Symbiosis to Nonlegume Crops
Chemistry	Synthetic N Fertilizers by Zero-Direct Energy Input
Genetics	Synthetic N_2-Fixing Microsymbionts, e.g., Mycorrhiza, *Agrobacterium*, *Erwinia*, Viruses
Genetics	Transfer of Genetic Information for N_2 Fixation and Associated Genes to Crop Plants
Chemistry/Genetics	Synthetic Gene That Codes for Small Stable (Oxygen and Temperature), High Turnover, Absolute Substrate Specificity (No H_2 Production), Zero-Direct Energy Requiring (No ATP), N_2-Fixing Enzyme with Appropriate Regulators

Rhizobium symbioses insensitive to fixed nitrogen also may provide mid-term impact. A modest impact is expected.

The more speculative, long-term impact technologies will provide by far the greatest contribution to the nitrogen input problem, if they are realized. The skills are chemistry, genetics, microbiology and plant breeding. A major increase in photosynthate available to the nodule would have a strong benefit analogous to the multifold increase in N_2 fixation experimentally demonstrated by foliar CO_2 enrichment of field-grown soybeans. Improved efficiency of the nodule could be significant. Effective non-*Rhizobium* nitrogen-fixing associations or ion uptake enhancers for nonlegume crops could be important. Mycorrhizal associations are important in scavenging ions for plants and one might suggest that mycorrhizal associations containing functional *nif*. Some have suggested that pathogens with the capability of nitrogen fixation but made nonpathogenic would provide effective systems for incorporating nitrogen fixation capabilities into crops. Examples might include *Agrobacterium* currently being studied as a general system to deliver foreign genes to plants and *Erwinia*. Some hope that yet-to-be discovered N_2-fixing systems exist in nature that could be readily commercialized for nitrogen input. The discovery in the last decade of the *Rhizobium-Trema* association and the rediscovery of the *Azospirillum*-grass association document the possibility of the existence of such systems.

Some suggest that the *Rhizobium* nitrogen-fixing symbiosis may be extended to nonlegume crops. No doubt this will occur in time but specialized genes both in the plant and in the bacteria are critical to the formation of a symbiotic nitrogen-fixing system. Already chemists have produced a novel enzyme, β-thiolactamase, by chemical modification of the β-lactamase gene so that a cysteine replaces a serine at the active site. This synthetic enzyme is active but with different characteristics than the natural one (Sigal et al., 1982). The ideal syntehtic nitrogenase gene will code for a small, stable, high turnover, absolute substrate specificity, zero direct energy input nitrogen-fixing enzyme with appropriate regulation of fixed nitrogen. Such an accomplishment will clearly demonstrate man's superior capability over that of nature at the present stage of evolution. One would hope that such a system would code for a nitrogen-fixing enzyme with an oxidative rather than a reductive reaction, in other words, zero direct energy input.

SUMMARY

The need for fixed nitrogen for crop production is massive but solvable even with current technology. As well as the direct impact of the technology(ies) on food and feed, there are indirect impacts of the specific technologies on health, energy, capital and environment. The dynamic science of N_2 fixation is highly productive but there could be more aggressive efforts to develop the best improved or alternate technologies. The current technologies should be described as mostly commodity in character while the movement to a future high technology is possible but major technology changes are not imminent. The size of the nitrogen need makes multiple technologies possible and even desirable.

REFERENCES

Anon (1982a). Nitrogen research. In: IFDC Annual Report 1981. International Fertilizer Development Center, Muscle Shoals, Alabama, 23-29.

Anon (1982b). Improved ammonia process unveiled. Chem. Eng. News, 60, 47, 28.

Ausubel, F.M. (1982). Molecular genetics of symbiotic N_2 Fixation, Cell, 29:1-2.

Bauer, W.D. (1981). Infection of legumes by rhizobia. Ann. Rev. Plant Physiol., 32:407-449.

Bothe, H. and A. Trebst (eds.) (1981). Biology of Inorganic Nitrogen and Sulfur. Springer-Verlag, New York.

Brady, N.D. (1982). Chemistry and world food supplies. Science, 218:847-853.

Brill, W.J. (1980). Biochemical genetics of nitrogen fixation. Microbiol. Rev., 44: 449-467.

Broughton, W.J. (ed) (1981). Nitrogen Fixation. Vol. 1: Ecology. Claredon Press, Oxford.

Buividas, L.J. and J.R. LeBlanc (1982). Lower energy ammonia with retrofits. The Fertilizer Industry Roundtable, Washington, D.C.

Burns, R.C. and R.W.F. Hardy (1975). Nitrogen Fixation in Bacteria and Higher Plants. Springer-Verlag, Heidelberg.

Carnahan, J.E., L.E. Mortenson, H.F. Mower and J.E. Castle (1960). Nitrogen fixation in cell-free extracts of Clostridium pasteurianum. Biochem. Biophys. Acta, 44: 520-535.

Chatt, J., L.M. da Camara Pina and R.L. Richards (eds.) (1980). New Trends in the Chemistry of Nitrogen Fixation. Academic Press, London.

Cohen, E., Y. Okon, J. Kigel, I. Nur and Y. Henis (1980). Increases in dry weight and total nitrogen content in Zea Mays and Setaria italica associated with nitrogen fixing Azospirillum spp. Plant Physiol., 66:246-249.

Dobereiner, J. and J.M. Day (1976). Associative symbioses and free-living systems. In: Proceedings of the 1st International Symposium on Nitrogen Fixation. W.E. Newton and C.J. Wyman (eds.). Washington State University Press, Pullman, 518-538.

Emerich, D.W. and H.J. Evans (1980). Biological nitrogen fixation with an emphasis on legumes. In: Biochemical and Photosynthetic Aspects of Energy Production. Academic Press, London, 117-145.

Gibson, A.H. and W.E. Newton (eds.) (1981). Current Perspectives in Nitrogen Fixation (Proc. 4th Int. Symp. Nitrogen Fixation, Canberra). Australian Academy of Science, Canberra.

Gutschick, V.P. (1981). Evolved strategies in nitrogen acquisition by plants. The American Naturalist, 118: 607-637.

Gutschick, V.P. (1982). Energetics of microbial fixation of dinitrogen. Adv. Biochem. Eng., 21: 109-167.

Hardy, R.W.F., R.C. Burns, and R.D. Holsten (1973). Applications of the acetylene-ethylene assay for measurement of nitrogen fixation. Soil Biochem. Biol., 5:47-81.

Hardy, R.W.F. and A.H. Gibson (eds.) (1977). A Treatise on Dinitrogen Fixation. Section IV. Agronomy and Ecology. John Wiley and Sons, New York.

Hardy, R.W.F. and U.D. Havelka (1975). Nitrogen fixation research: a key to world food? Science, 188:633-643.

Hardy, R.W.F. and U.D. Havelka (1977). Possible routes to increase the conversion of solar energy to food and feed by grain legumes and cereal grains (crop production): CO_2 and N_2 fixation, foliar fertilization and assimilate partitioning. In: Biological Solar Energy Conversion. A. Mitsui et al. (eds). Academic Press, New York, 299-322.

Hardy, R.W.F., P.G. Heytler, and R.M. Rainbird (1983). Studies of novel nitrogen inputs. In: Better Crops for Food. Ciba Foundation Symposium No. 97, London.

Hardy, R.W.F., R.D. Holsten, E.K. Jackson and R.C. Burns (1968). The acetylene-ethylene assay for N_2 fixation: laboratory and field evaluations. Plant Physiol., 43: 1185-1207.

Hardy, R.W.F. and W.S. Silver (eds.) (1977). A Treatise on Dinitrogen fixation. Section III. Biology. John Wiley and Sons, New York.

Hardy, R.W.F. (1976). Potential impact of current abiological and biological research on the problems of providing fixed nitrogen. In: Proceedings of the 1st International Symposium on Nitrogen Fixation. W.E. Newton and C.J. Nyman (eds). Washington State University Press, Pullman, 693-717.

Hardy, R.W.F. (1977). Increasing crop productivity: agronomic and economic considerations on the role of biological nitrogen fixation. In: Report of the Public Meeting on Genetic Engineering for Nitrogen Fixation. U.S. Government Printing Office, Washington, 77-106.

Hardy, R.W.F. (1980a). Futuristic system for N input for crop production. In: Proceedings Bio-Energy '80 World Congress and Exposition. The Bio-Energy Council, Washington, 280-290.

Hardy, R.W.F. (1980b). Translating basic research on biological nitrogen fixation to improved crop production in less-developed countries - a user's view. In: Linking Research to Crop Production. R.C. Staples and R.J. Kuhr (eds.). Plenum Press, New York, 137-151.

Hardy, R.W.F., F. Bottomley and R.C. Burns (eds.) (1977). A Treatise on Dinitrogen Fixation. Sections I and II. Inorganic and Physical Chemistry and Biochemistry. John Wiley and Sons, New York.

Haq, N. (1983). New food legume crops for the tropics. In: Better Crops for Food. Ciba Foundation Symposium No. 97, London.

Havelka, U.D., M.G. Boyle and R.W.F. Hardy (1982). Biological nitrogen fixation. In: Nitrogen in Agricultural Soils. F.J. Stevenson (ed.). Agronomy, 22:365.

Heytler, P.G. and R.W.F. Hardy (1979). Energy requirement for N_2-fixation by rhizobial nodules in soybeans. Plant Physiol., 63S:84.

Hollaender, A., R.H. Burris, P.R. Day, R.W.F. Hardy, D.R. Helinski, M.R. Lamborg, L. Owens, and R.C. Valentine (1977). Genetic Engineering for Nitrogen Fixation. Plenum, New York.

Kapulnik, Y., S. Sarig, Y. Okon, J. Kigal and Y. Henis (1981). Yield increases in summer cereal crops in Israel fields inoculated with Azospirillum. Exp. Agric., 17: 171-178.

Lamborg, M.R. and R.W.F. Hardy (1982). Microbial effects. In: Increasing Atmospheric pCO_2 and Crop Production. E. Lemon (ed.). Am. Assoc. Adv. Sci., Washington.

La Rue, T.A. and T.G. Patterson (1981). How much nitrogen do legumes fix? Adv. in Agron., 34: 19-38.

Mazur, B.J. and C.-F. Chui (1982). Sequence of the gene coding for the -subunit of dinitrogenase from the blue-green alga anabaena. Proc. Natl. Adad. Sci. USA, 79: 6782-6786.

Mazur, B.J., D. Rice and R. Hasselkorn (1980). Identification of blue-green algal nitrogen fixation genes by using heterologous DNA hybridization probes. Proc. Natl. Adad. Sci. USA, 77: 186-190.

McIntosh, L., C. Powlson, and L. Bogorad (1980). Chloroplast gene sequence for the large subunit of ribulose diphosphate carboxylase of maize. Nature, 288: 556-560.

Mortenson, L.E. and R.N.F. Thorneley (1979). Structure and function of nitrogenase. Ann. Rev. Biochem., 48: 387-418.

Newton, W.E. and C.J. Lyman (eds.) (1976). Proceedings of the 1st Interntional Symposium on Nitrogen Fixation Vol. 1 and 2. Washington State University Press, Pullman.

Newton, W.E. and W.H. Orme-Johnson (eds.) (1980). Nitrogen Fixation. Vol. 1. Free-Living Systems and Chemical Models; Vol. 2 Symbiotic Associations and Cyanobacteria. University Park Press, Baltimore.

Newton, W.E., J.R. Postgate, and C. Rodriguez-Barrucco (1977). Recent Developments in Nitrogen Fixation. Academic Press, London.

Ng, T., R.M. Busche, C.C. McDonald, and R.W.F. Hardy (1983). The production of chemicals. Science, 219: 000-000.

Phillips, D.A. (1980). Efficiency of syumbiotic nitrogen fixation. Ann. Rev. Plant Physiol., 21: 29-49.

Postgate, J.R. (1982). Biological nitrogen fixation: fumdamentals. Phil. Trans. R. Sci. Lond. B., 296: 375-385.

Rice, D., B.J. Mazur, and R. Hasselkorn (1982). Isolation and physical mapping of nitrogen fixation genes from the cyanobacterium Anabaena 7120. J. Biol. Chem., 257: 131257-13163.

Roberts, G.P. and W.J. Brill (1981). Genetics and regulation of nitrogen fixation. Ann. Rev. Microbiol., 35: 207-235.

Robson, R.L. and J.R. Postgate (1980). Oxygen and hydrogen in biological nitrogen fixation. Ann. Rev. Microbiol., 34: 183-207.

Schubert, K. (1982). The Energetics of Biological Nitrogen Fixation. Workshop Summaries - I. Am. Soc. Plant Physiologists, Rockville, M.D.

Shanmugam, K.T., F. O'Gara, K. Anderson, and R.C. Valentine (1978). Biological nitrogen fixation. Ann. Rev. Plant Physiol., 29: 263-276.

Sigal, I., B.G. Harwood and R. Arentzen (1982). Thiol- -lactamase: replacement of the active-site serine of RTEM -lactamase by a cysteine residue. Proc. Natl. Acad. Sci. US, 70:0000-0000.

Sprent, J.I. (1979). The Biology of Nitrogen-Fixing Organisms. McGraw-Hill, London.

Stewart, W.D.P. and J.R. Gallon (1980). Nitrogen fixation. Academic Press, London.

Subba Rao, N.S. (1980). Recent Advances in Biological Nitrogen fixation. Arnold, London.

Viands, D.R., D.K. Barnes and G.H. Heichel (1982). Nitrogen fixation in alfalfa-response to bidirectional selection for associated characteristics. USDA Tech. Bull. 1643.

Willey, R.W. (1983). Intercropping. In: Better Crops for Food. Ciba Foundation Symposium No. 97, London.

THE ROLE OF GROWTH REGULATORS AND HORMONES IN ENHANCING FOOD PRODUCTION

Louis G. Nickell
Velsicol Chemical Corporation
341 East Ohio Street, Chicago, Illinois 60611, U.S.A.

ABSTRACT

Plant growth regulators are expected to play an important role in the efforts to double the world food supply by the end of this century. Although the search for plant growth regulators differs in numerous ways from that for pesticides, it lends itself to more (crop) specific, concentrated, and directed efforts that can be utilized on a smaller scale by smaller companies and countries. The diversity of effects already established for plant growth regulators promises a continuing optimistic future for them with considerable savings in energy and costs. Increasing the yield of a crop for a given area of land through their use means incremental gain, the only outlay for which is usually the cost of the plant growth regulator and its application and sometimes, extra processing costs.

KEYWORDS: plant growth regulators, hormones, weed control, phytohormones.

INTRODUCTION

The need to increase the world food supply substantially by the end of the century, maybe even to double it, poses one of the biggest challenges that this world has ever seen (Hudson, 1976). Agricultural scientists believe that this challenge can be met, and it is expected that plant growth regulators will play an increasingly important role in meeting this challenge.

The use of plant growth regulators may be the cause of the most important quantitative yields yet achieved in agriculture (Agricultural Production Efficiency, 1975). The principal aim of the agrochemical industry has been to provide chemicals that control competition to the crop, i.e., the weeds, insects, fungi, and nematodes that reduce yield or quality or that interfere with harvesting. Product performance has been judged in simple terms, e.g., in the case of herbicides it is the death of the weed and adequate margins of safety for the crop and the farmer. Plant growth regulators are used to modify the crop by changing the rate or pattern or both of its response to the internal and external factors that govern all stages of crop development from germination to vegetative growth, reproductive development, maturity, and senescence or aging, as well as post harvest preservation.

Controlling the internal metabolism of a given crop should make it produce more sugar, more protein, more oil, more latex, or more or better fruit than it would under the best conditions without treatment. Enchancement of productivity of crops grown for processing could be achieved by selecting for high levels of processing components, e.g., insoluble solids, flavor, acidity, color, etc.

CHEMICALS FOR PLANT GROWTH REGULATION

The concept of the use of chemicals to regulate plant growth is not new. In fact, the use of chemicals to control various physiological processes in plants is older by several decades than their use as weed killers. The use of indolebutyric acid to induce rooting in an extremely wide variety of plants dates back to the 1920's (Avery and Johnson, 1947). Although its use has been expanded in the intervening decades to an enormous number of plant species, it is surprising that a more effective compound has not yet been found (Nickell, 1982b). The use of maleic hydrazide to inhibit sprouting in potatoes and onions and for the retardation of grasses is decades old, as is the use of a number of chemicals to induce fruit set in apples as well as to thin apples (Wittwer, 1971). Weed control is considered by many to be only one facet of plant growth regulation. The discovery in the early 1940's by Zimmermann and Hitchcock that 2,4-D has a differential effect on grasses as opposed to broadleaves would not have had its immediate significance had there not been a World War and labor for traditional weeding of crops not available in the technologically advanced countries.

Plant growth regulators, other than nutrients, usually are organic compounds. They are either natural or synthetic compounds and are applied directly to a plant to alter its life processes or structure in some beneficial way so as to enhance yield, improve quality, or facilitate harvesting. Plant hormones, i.e., phytohormones, are plant produced growth regulators and, therefore, are naturally-occurring plant substances. The term "plant growth regulators", however, applies to phytohormones as well as to synthetic compounds (Nickell, 1978, 1982c).

Growth regulating compounds which are produced by plants generally are classified as one of four types of hormones: auxins, gibberellins, cytokinins, and inhibitors. Many authorities would include ethylene as a fifth class (Nickell, 1978). These materials are produced by plants but can be extracted and applied to a different plant to affect that second plant's growth and/or development. Many synthetic compounds have physiological activity, and a number of these are important in maximizing crop production.

SEARCHING FOR PLANT GROWTH REGULATORS

Because of the success with, the immediate economic use of, and the relative simple and straightforward methods of searching for herbicides, the weed-killing aspect of plant growth regulation has overshadowed its other uses for the last several decades. Because the niches for uses of herbicides are rapidly becoming saturated, and with the successful introduction of plant growth regulators in some important crops and their commercial uses (such as in sugar cane for ripening and in rubber for increasing the length of time of latex flow between tappings), industry has increasingly turned its interest to the search for plant growth regulators within the last few years. However, it should be pointed out that the task of discovering a plant growth regulator is substantially more demanding that that involved in the search for a new herbicide. Estimates of the complexity and the costs connected with this complexity are estimated to be 10 to 1000 times greater than that for herbicide research (Sacher, 1982). Nevertheless, the absence of effective plant growth regulators for most of the major crops makes this field extremely attractive -- wtih large potential payoffs. For success, long term commitments are necessary, both to the research and development programs and to the time periods now necessary for approval by regulatory bodies.

Because of the long term commitment necessary as well as the high expense, much strategic thinking as well as gathering of information needs to enter into a plant growth regulator program when comparing it with the search for a herbicide. For example, in a screening program for herbicidal activity, one can screen a chemical and let the results of the screen suggest on what crop or crops the chemical might be used for the control of which weeds. There is no screen such as this in connection with plant growth regulators. First, one must decide the crop for which to develop a plant growth regulator. In reaching this decision, a considerable amount of market research information is necessary, covering such factors as: a) value of the crop, b) history of chemical use for the crop, c) what specifically does one want to do to make a plant growth regulator of value to that crop. Next, when the crop is chosen, one must know or should develop an extensive knowledge of the physiology and biochemistry of that crop. Next, one must answer how the plant growth regulator is to enhance the value of the crop after its treatment. That is, is it to change some metabolic process, to change the physiognomy of the crop, or to increase yield per se. If the latter, is the economic value of the crop in its vegetative or reproductive state? Is this value in a developing organ such as a fruit or in a root? Do you expect to apply the plant growth regulator as a foliar spray, directly to the soil, or through an irrigation system? To have its maximum economic effect, when should the chemical be applied? The complexities in searching for plant growth regulators for a given crop have been amply presented by Archer and his colleagues (1982) in their study of the opportunities for the use of plant growth regulators in maize. In this study, they clearly demonstrate the need for knowledge of the life cycle of the crop involved and the need to determine the rate-limiting factors and their interactions in order to determine when a plant growth regulator must be applied to have its maximum effect or, in some cases, to have any effect.

APPLICATIONS OF PLANT GROWTH REGULATORS

The value of being able to control the flowering of plants cannot be over-emphasized. The ability to induce flowering at the proper time in order to obtain maximum return for marketing of the flowers is apparent to everyone. The fact that flowering in certain crops causes a decrease in yield is not so apparent outside the field of agriculture. The ability to prevent flowering, therefore, is extremely important with crops such as sugar cane (Nickell, 1978, 1982a). Interestingly enough, more recent studies have shown a relationship in certain crops such as sugar cane between the lack of flowering and resistance to disease. For example, in Mauritius the incidence of yellow spot in sugar cane has been found to be related to non-flowering. Because of this importance in the control of flowering, it is expected that plant growth regulators will be increasingly used by plant breeders as tools to enable them to bring into flowering earlier those plants which are either difficult to flower or those which take a long time to flower such as forest trees.

Environmental conditions are the greatest limiting factors in crop production. Our ability to both increase yields per unit area in existing agricultural lands as well as our ability to convert now non-productive areas is dependent upon our ability to control the crop processes to fit into the existing natural limitations or to increase the resistance of the crop to those limitations. Both of these facets can be attacked by the use of plant growth regulators. For example, it is well established that chlormequat can increase the resistance of some plants to cold, drought, and salt damage. Another example is the ability of gibberellic acid to enable sugar cane to increase its yield during the cold winter weather when normally there is little production (Nickell, 1978a).

Growth regulators are already used to change the shape of fruits as well as the appearance of fruits by increasing or decreasing certain colors. It is quite likely that changes in taste characteristics can also be affected in a similar manner, affecting in a positive way that nebulous marketing factor called quality. More recent findings show that treatment with plant growth regulators has a positive effect in either helping control diseases and pests or in making the crop less attractive or more resistant to them (Smith et al., 1974, 1975).

The use of ethephon to boost the yield of latex from rubber trees has become standard plantation practice in recent years. This compound increases the length of time for the flow of latex between tappings and thereby increases by as much as 100% the commercial yield of dry rubber from commercially important tree varieties. Use of ethephone obviates frequency cutting of the tree bark thus helping preserve the life of the tree (Bridge, 1980). Rubber made from the guayule plant, which grows in southwest United States and Mexico, is equal in every respect to that made from the traditional Hevea rubber trees. Inexpensive triethylamines sprayed about 3 weeks before harvest increase yields 200 to 600%. Use of such chemicals could increase overall yields 30 to 35% and cut the growing time by 1 to 2 years (Yokoyama, et al., 1977). The oleoresin content of pine trees can be increased by treating the trees with the herbicide paraquat prior to harvesting (Parham, 1976; Schwarz and Ryan, 1980). As much as 20-fold increases in the extractive content of treated pines have been reported. One of the most important commercial developments in recent years has been the use of chemicals as ripeners for sugar cane. Many chemicals have been found which will increase the sucrose content of cane at harvest. The most commonly used compounds commercially are glyphosine, its mono-substituted derivative glyphosate, and ethephon (Nickell, 1982a). The increased sugar yield produced by treatment with ripeners is 5 to 25% depending on the variety of sugar cane treated and on prevailing weather and soil conditions. Changes in cultural conditions to adjust to the use of ripeners probably should increase sugar yields still further. Newer groups of compounds are considerably more active than those originally considered as commercial candidates.

The discovery of naturally-occurring molecules which have biological activity will continue to add to the arsenal of agriculture (Maugh, 1981). Those which have sufficient activity in their own right, such as triacontanol and brassinolide, will be easier to register for use because they are naturally occurring. Those with insufficient activity will serve as leads to the chemists in the preparation of more active compounds. The developing field of allelopathy should be another source both of chemical configurations and of ideas.

Other areas of potential value are (1) the reevaluation of biologically active compounds used for other purposes (herbicides, fungicides, insecticides, etc.) for plant growth regulator activity, and (2) the combination of current plant growth regulators to determine (a) possible synergism, (b) and/or a shifting of the biological spectrum. Examples of the former include the cytokinin activity of the systematic fungicide benomyl (Benlate) (Skene, 1972; Thomas, 1974) and the induction of abscission by the antifungal antibiotic cycloheximide (Actidone) and the insecticide carbaryl (Sevin).

THE FUTURE

In summary, it is important to emphasize the already established diversity of effects of plant growth regulators and to speculate upon the potential for increasing this diversity. So many uses have already been shown that one has to be optimistic about their future. Thus, flowering can be hastened, flowering can be slowed down, flowering can be prevented; plants can be made to grow taller, plants can be made to grow shorter; side shoots can be made to develop more rapidly, side shoots can be prevented from developing at all; flowers can be made to fall off prematurely, flowers can be made to stay on; fruits can be made to fall off prematurely, fruits can be made to stay on; maleness can be increased, femaleness can be increased; sprouting may be prevented, sprouting may be induced; rooting can be initiated, rooting can be prevented; senescence can be initiated early, senescence can be delayed; metabolic products such as sugar can be increased, or metabolic products can be decreased. These examples amply demonstrate the versatility of plant growth regulators and one can easily see why they will undoubtedly have a substantial part in future increases in agricultural production. Once they are available, the use of plant growth regulators will have an impact on the economic production of crops both because of their saving in energy and their saving in cash outlay. Increasing the yield of a crop for a given area of land means an incremental gain, the only output for which is the cost of the plant growth regulator and its cost of application. No extra land must be purchased, no extra land must be tilled, generally no extra fertilizer must be added, no extra weed control or other pest control measures must be taken, no additional harvesting need be made. There will be extra processing costs because there will either be additional plant material or additional economic product such as sugar, oil, or protein added to the crop by the proper use of the appropriate plant growth regulator.

Because of the specialization and inherent differences in searching for plant growth regulators as compared to herbicides mentioned earlier, the search for plant growth regulators is not the exclusive domain of the developed country or the large industrial concern. In local areas there are principal crops. A smaller program concentrating on fewer crops and with a direction of effort can be as successful as a large program where a large number of crops are studied.

Experts seem to agree that the future of plant growth regulator research must take a multidisciplinary approach, including an understanding of the basic biological processes and how they regulate the productivity of crops. Such an approach provides both a cross-fertilization of ideas and a synergistic effect whereby the results achieved will be more than those from a number of individual unidisciplinary approaches taken together.

REFERENCES

Agricultural Production Efficiency (1975). National Academy of Sciences, Washington, D.C., pp. 199.

Archer, M.C., D.J. Hansen, and D.F. Loussaert (1982). Corn yields -- rate-limiting factors and opportunities in plant growth regulation. In: Chemical Manipulation of Crop Growth and Development, (J.S. McLaren, ed.), pp. 253-265. Butterworths, London.

Avery, G.S., and E.F. Johnson (1947). Hormones and Horticulture. McGraw Hill, New York.

Bridge, K. (1980). Ethrel-Its use in natural (Hevea) rubber. Proc. Plant Growth Regulator Working Group 7: 207-211.

Hudson, J.P. (1976). Future roles for growth regulators. Outlook on Agriculture 9: 95-98.

Maugh, T.H. (1981). New chemicals promise larger crops. Science 212: 33-34.

Nickell, L.G. (1978). Plant growth regulators. Controlling biological behavior with chemicals. Chem. Eng. News 56(41): 18-34.

Nickell, L.G. (1928a). Plant growth regulators in the sugar cane industry. In: Chemical Manipulation of Crop Growth and Development, (J.S. McLaren, ed.), pp. 167-189. Butterworths, London.

Nickell, L.G. (1982b). Plant Growth Regulators -- Agricultural Uses. Springer-Verlag, Berlin, Heidelberg, pp. 173.

Nickell, L.G. (1982c). Plant-growth substances. Encyclopedia Chem. Technol. (3rd Ed.) 18: 1-23.

Parham, M.R. (1976). Stimulation of oleoresin yield in conifers. Outlook on Agriculture 9: 76-81.

Sacher, R.M. (1982). Strategies to discover plant growth regulators for agronomic crops. In: Chemical Manipulation of Crop Growth and Development, loc. cit., pp. 13-15.

Schwarz, O.J., and F.J. Ryan (1980). Paraquat induced oleoresin loading in pine. Proc. Plant Growth Regulator Working Group 7: 199-206.

Skene, K.G.M. (1972). Cytokinin-like properties of the systemic fungicide benomyl. J. Hort. Sci. 47: 179-182.

Smith, M.C., B.D. Smith, S. Halfacre, J. Collins, and M. Churchill (1974). Resistance of cultivated plants to pests. Long Ashton Res. Stn. Rpt. for 1974, pp. 116.

Smith, M.C., B.D. Smith, D.A. Kendall, S. Halfacre, and M. Churchill (1975). Resistance of cultivated plants to pests. Long Ashton Res. Stn. Rpt. for 1975, pp. 112-113.

Thomas, T.H. (1974). Investigations into the cytokinin-like properties of benzimidazole-derived fungicides. Ann. Appl. Biol. 76: 237-241.

Wittwer, S.H. (1971). Growth regulants in agriculture. Outlook on Agriculture 6: 205-217.

Yokoyama, H., E.P. Hayman, W.J. Hsu, S.M. Poling, and A.J. Bauman (1977). Chemical bioinduction of rubber in guayule plant. Science 197: 1076-1077.

THE POTENTIAL CONTRIBUTION OF CELL AND PLANT TISSUE CULTURE TO CROP IMPROVEMENT

Otto J. Crocomo and Neftali Ochoa-Alejo*

Center for Biotechnology in Agriculture (CEBTEC/FEALQ) and
Center for Nuclear Energy in Agriculture (CENA/USP)
13400-Priacicaba, SP, Brasil

**Permanent address: Center for Biological Research (CIB/CONACYT)*
P.O. Box 128, 23060, La Paz, BCS, Mexico

ABSTRACT

Plant tissue culture involves the culture of cells, tissues and organs of plants in aseptic conditions in culture media that allow for their growth and proliferation. Tissue culture is now a basic technique for plant propagation and an excellent auxiliary tool in the breeding of economically important plant species. The methodology allows for a single plant cell to proliferate, producing amorphous cell conglomerates called callus which in turn, under appropriate conditions, differentiate into tissues, organs or intact plants. Five main types of plant tissue culture technology are considered: callus culture, cell suspensions, organ culture, meristem culture and protoplasts culture. Through these methodologies the plant material can be manipulated in such way to obtain clones or to obtain genetic variants. Plant cell and tissue culture technique, with all its implications, will play an important future role in the improvement of economic crops. A complete application of this technology to plant molecular genetics and to basic investigations will require the establishment of research needs and priorities, which are presented and discussed.

KEYWORDS: Plant Tissue Culture, Protoplasts, Callus Culture, Cell Suspension Culture, Organ Culture, Meristem Culture, Clonal Propagation, Disease Elimination, Cryopreservation, Hybridization.

INTRODUCTION

Plant tissue culture involves the culture of cells, tissues and organs of plants in aseptic conditions in culture media that allow for their growth and proliferation. In the "new biology" that has been florishing for some time now, tissue culture is a basic technique for plant propagation and an excellent auxiliary tool in the breeding of economical important plant species; it offers a good system for the understanding of the biochemical, physiological and genetic processes occurring in plant cells; it opens and enlarges the routes for the potential applications of the biochemical and genetic technologies of molecular biology in plants. The methodology allows for a single plant cell to proliferate producing amorphous cell conglomerates called callus which in turn, under appropriate conditions, differentiate into tissues, organs or intact plants.

When grown as free suspensions in liquid medium, plant cells can be manipulated as microbial cells are, allowing for their utilization in the production of specific natural compounds. The preservation, conservation and distribution of germplasm is made possible using cells and tissues cultured in vitro if these are maintained at very low temperatures in the process of cryopreservation. When plant cells are submitted to enzymatic hydrolysis to dissolve their cell wall, protoplasts will result. Protoplasts offer a very good system for transfer of genetic information through somatic fusion, transplant of organelles and microorganisms and the incorporation of exogenous DNA.

THE STATE OF THE ART

Five main types of plant tissue culture technology are considered here: callus culture, cell suspensions, organ culture, meristem culture and protoplasts culture. Through these methodologies the plant material can be manipulated to obtain clones or to obtain genetic variants.

Callus Cultures

In order to obtain callus a small segment (the explant) of a highly organized and specialized tissue is transferred to semisolid synthetic medium to induce the formation of a non-differentiated cell mass with a non-organized structure. The explant source can be any part of the plant such as stem, root, leaf mesophyll, cotyledon, shoot tip, floral tissue, vascular cambium, storage parenchyma, reproductive tissues, bulb scales, endosperm, embryo axis, medullary ray, vascular parenchyma, etc. (Narayanaswamy, 1977). Juvenile tissues are considered to be the most appropriate. In general, induction of callus from each type of explant depends upon specific conditions either of the culture medium composition or of environmental factors. The size, form and source of explant are important factors in callus induction (Evans et al., 1981). Cells at the callus stage should be transferred (subcultured) to fresh medium periodically in order to promote good cell proliferation.

Through the manipulation of the ratio and balance of hormones and nutrients callus can resume differentiated structures and functional organization. In the callus, organ primordia can thus be produced de novo which later give rise to shoots, roots, flowers, buds or embryoids. Plant regeneration from callus can be achieved either through organogenesis or through embroyogenesis. In organogenesis plant regeneration takes place in culture media the composition of which determines induction of shoots and regeneration of roots (Skoog and Miller, 1957). Plants can be regenerated much more easily from dicotyledoneous than from monocotyledoneous species (Sharp and Evans, 1981).

Cell Suspension Cultures

Cell suspension cultures are initiated by transferring plant explants or callus to a liquid medium, usually maintained in a rotary shaker to help dispersion of the cells and exchange of gases. The suspension consists of cell aggregates, cell clusters and single cells. In general, cells in liquid medium have a higher growth rate than in semisolid medium, as the cells are in intimate contact with the culture medium constituents. Cells can be cultured either in finite volumes of medium (batch propagation) or in semi-continuous cultures in which periodically the culture medium is replaced by a fresh medium. Continuous cultures can be established in which the medium is withdrawn and replaced continuously by the same volume of fresh medium (King, 1980; Crocomo et al., 1981). Cells in suspension can be a good source for plant regeneration either through organogenesis or through embryogenesis (Evans et al., 1981). Usually the capacity of callus or cell suspension cultures to regenerate plants declines as the number of subcultures increases (Torrey, 1967).

Organ Culture

This technique involves the culture of any plant organ such as roots, shoots, embryo axis, anthers, microspores, ovaries, ovules, etc., which can be inoculated in liquid or semisolid medium. Organ cultures provide organized, differentiated tissues grown under defined conditions.

Embryo Axis Culture - Embryo axes at different stages of development can be removed from the maternal tissue and cultured *in vitro*. Embryo axes from at least 40 plant families have been successfully cultured (Murashige, 1978). Several factors influence the development of mature or immature embryo axes *in vitro* (Raghaven, 1980), carbohydrates and nitrogen compounds being amongst those considered most necessary. A balanced ratio between cytokinins and auxins promotes growth and differentiation of the embryos (Raghavan and Torrey, 1963; Crocomo et al., 1979). In general, the more immature the embryo the more complex will be the requirements for culture media.

Anthers/Microspores Culture - the use of anthers/microspores in culture has been recently reviewed (Vasil, 1980). Anthers at the uninucleate microspore stage are excised from the stamens of sterilized buds and inoculated in liquid or semisolid media. The entire anther with enclosed microspore or the expressed microspore can be cultured. The physiological state of the anthers is important for successful culture (Nitsch, 1981); young plants maintained in optimal conditions of temperature, light and nutrition, are the best source. Embryogenesis is enhanced when anthers are pretreated at low temperature (ca. 2-10°C, during 2-30 days). The nutritional requirements for anthers are different from those required by the expressed microspores (Vasil, 1980). The majority of plants regenerated via anther culture are haploids. However, in some plant species they can be diploid or polyploid (Keller and Stringam, 1978). The plants regenerated from another callus exhibit genetic variation (Vasil, 1980). The cells or haploid plants can be diploidized by treatment with colchine resulting in homozygous cells or plants (Burk et al., 1972).

Meristem Culture

This technique pertains to the aseptic culture of meristems of both apical and axillary buds to obtain complete plants. The meristem is a dome of cells undergoing intense division being ca. 0.1mm in diameter and 0.25mm in length. Plants obtained are genetically uniform as they originate from preexisting or newly formed meristems, skipping the callus stage. Apical or axillary meristems are protected by developing leaves or scales which are excised after the shoot or the defoliated segments are disinfected, the explant being transferred to a semisolid or liquid medium. The composition of the medium is important to enhance the number of shoot buds and to allow for the formation of roots. In general, high concentrations of cytokinins (10-30 mg/1) stimulate the process of bud proliferation and induce the growth of dormant meristem (Murashige, 1977).

Protoplast Culture

Isolation and Culture - Protoplasts are naked cells which possess the potential to regenerate cell walls, to grow and to divide. Protoplasts have to be isolated in aseptic conditions. The source can be either cell cultures or plant tissues; the protoplast is subsequently cultured to produce small colonies and regenerate whole plants (Vasil and Vasil, 1980).

Mechanical procedures and most frequently enzymatic digestion are employed to obtain protoplasts (Cocking, 1960). High yielding protoplasts can be obtained through the use of cell wall degrading enzymes (cellulase, hemicellulase and pectinase). Success in the isolation of protoplasts depends upon the physiological stage of the plant material, the source of the required enzymes and osmotic stabilizers (Vasil and Vasil, 1980; Gamborg et al., 1981; Crocomo and Ochoa-Alejo, 1981). Protoplasts can be cultured in semisolid or liquid medium similar to that for cell cultures (Eriksson et al., 1978). Solid media can be used when the protoplasts are plated onto or embedded in agar; alternatively liquid media can be used as suspensions, hanging droplets or in microculture chambers (Bajaj, 1977; Eriksson et al., 1978; Vasil and Vasil, 1980). The concentration of protoplasts in the initial inoculum is an important factor for cell division to occur (Bajaj, 1977). Rapid cell wall regeneration can be observed in protoplast cultures in appropriate conditions followed by cell division after 2-7 days in culture. Small colonies are formed after 1-3 weeks which undergo callus formation, or they can be transferred to a liquid medium to form cell suspensions. So far successful regeneration of plants from protoplasts has been reported only for a restricted group of species, among them the Solanaceae (Vasil and Vasil, 1980) and recently the Leguminosae and Gramineae (Conger, 1981b).

Protoplast Fusion and Culture - Protoplasts can be fused to form multinucleated cells or produce hybrid cells when derived from two different sources. Fusion can be spontaneous or induced by different substances,, including polyethyleneglycol (PEG) (Kao et al., 1974). PEG causes protoplast agglutination in small clusters. Fusion takes place during the dilution of PEG with alkaline solutions containing high calcium concentrations. PEG is utilized in the implantation of organelles, microorganisms and in the incorporation of DNA bearing liposomes (Uchimiya, 1979; Crocomo and Ochoa-Alejo, 1981). Hybrid cells resulting from fusion can regenerate the cell wall and divide (Cocking, 1981). Hybrid cells are selected through mechanical methods or genetic complementation (Cocking, 1978). The fusion products usually are cultured in complex media. Hybrid plants have been regenerated from hybrid protoplasts from several species such as *Nicotiana*, *Daucus*, *Datura* and less commonly from intergeneric somatic hybrids such as *Datura-Atropa*, *Lycopersicon-Solanum* (Kao, 1980), *Brassica-Arabidopsis* (Gleba and Hoffmann, 1978).

APPLICATIONS

Current and potential applications of tissues culture methods in crop improvement are in the following areas: 1) clonal propagation; 2) disease elimination; 3) long term storage of germplasm exchange; 4) production of haploids; 5) generation and selection of genetic variants; 6) secondary products; 7) gene transfer by wide hybridization; 8) plant molecular genetic engineering.

Clonal Propagation

A common use of plant tissue culture is cloning or asexual propagation of plants. The aim of clonal propagation is to produce genetically identical plants (Lawrence, 1981). This technique can be used for plants including forest, horticultural and agronomic species (Vasil and Vasil, 1980; Bottino, 1981; Conger, 1981b; Hughes, 1981; Mott and Zimmerman, 1981; Skirvin, 1981). Clonal propagation can be accomplished mainly by: a) formation of adventitious meristems; b) enhancing axillary branching; c) adventive embryo or embryoids formation (Murashige, 1977). Clonal propagation advantages lie mainly in the way with which plant multiplication can be achieved and the number of plants that can be produced in a relatively short time with space saving. Furthermore, this technique can be applied for maintaining heterozygosity, to bypass sexual sterility or incompatibility. In some plant breeding programs, clonal propagation can be used to increase the supply of limited plant material and reduce the time required to introduce the selected genetic traits.

Disease Elimination

One of the major obstacles to crop productivity has been the occurrence of plant diseases caused by fungi, bacteria, viruses, mycoplasms and nematodes. Effective control methods have been developed for most of the diseases except those caused by viruses and mycoplasm-like organisms. One of the most important applications of tissue culture technique has been the use of meristem tip culture to eliminate virus from infected plants (Smith and Oglevee-O'Donovan, 1979; Walkey, 1978). Viruses exist at low frequencies in meristematic cells and often are not transmitted to progeny plants derived from the meristem in culture. The meristem tip culture has its major application in vegetatively propagated species where parent plants have been virus infected (e.g. cassava, potato, sugar cane, strawberries, sweet potato, cauliflower, etc.)

The size of the meristem tip is an important factor governing the regenerative capacity of meristems and also in increasing the probability of recovering virus-free plants. The success of eliminating viruses also depends upon the characteristics of the host-virus combinations.

Long Term Storage of Germplasm and Germplasm Exchange

Successful plant breeding is based on the amount of genetic diversity available for insertion of new genes into plants. Germplasm is needed as a resource for improving characteristics of plants and as a means to guarantee supplies of known plant derivatives and potential new ones. The need to preserve all potentially valuable germplasm is balanced by the time, labor and space requirements for its maintenance for a large number of economic plant species. For vegetatively propagated crops, preservation of germplasm is a process which is labor intensive and expensive. Maintenance of seed propagated crops is not always feasible due to the facts that: a) some plants do not produce seeds; b) some seeds remain viable for a limited time; c) some seeds are very heterozygous; d) seeds of certain species deteriorate rapidly owing to seed-born pathogens (Kartha, 1981a).

Plant tissue culture in combination with cryopreservation techniques offers great promise for both the storage of plant materials in a disease-free condition, and the exchange of germplasm between countries. Despite recent advances, the cryopreservation techniques are still at an experimental stage. Cultured plant cells, protoplasts, somatic embryos, pollen, whole seeds, shoot tips, vegetative buds and meristems from a few species have been frozen and recovered with different degrees of success (Bajaj and Reinert, 1977; Withers, 1980; Kartha, 1981a). It seems that cultured plant cells are not good systems for germplasm preservation because of the possible genetic changes which may occur during the growth phase before and after freezing. Thus, meristems can offer a better system for the long-term storage of germplasm in a genetically stable condition. Somatic embryos may eventually represent a useful form of cultured plant tissue for longterm storage with acceptable genetic stability. At present, liquid nitrogen freeze-preservation is probably the only method allowing for the long-term storage of cells and tissues in a completely non-dividing state.

Production of Haploids

Haploids are sporophytes with the chromosome number of gametophytes. Haploids can be induced artificially either by using the male or the female gametophyte. In vitro culture of anthers has created much interest and been successful for haploid production in an increasing number of crop species (Han and Qiquan, 1981). The most common use of haploids involves the production of instant homozygous lines by chromosome doubling. The breeding programs which lead to pure line varieties of self-fertilized plants are extremely time consuming. By using haploid cells or plants a large number of inbreds can be obtained in one step by chromosome doubling. The homozygous line may be used as inbred lines for hybrid production, for fixing gene combinations at the end of a backcrossing program, or to stabilize other unique germplasm in homozygous form (Reinert and Bajaj, 1977).

Germplasm and Selection of Genetic Variants

Phenotypic variation expressed at the cell, tissue and regenerated plant levels has been reported for most plants cultured in vitro (Larkin and Scowcroft, 1981). The phenotypic variability observed in regenerated plants reflects either pre-existing cellular genetic differences or tissue culture-induced variability. The new variants have morphological, physiological as well as qualitative and quantitative characteristics bearing chromosomic, nuclear and cytoplasmic differences from the donor plant. Spontaneous and induced variants from tissue and cell culture systems are excellent sources of plant genetic variations which may be useful for the selection of novel genotypes, saving field space and time. Of course, in order to make effective and short time impact on breeding programs, the selected traits must be stable and expressed throughout successive regenerations. Somaclonal variants are having some impact on the improvement of sugar cane and potato cultivars and in the breeding of new ornamental varieties (Heinz et al., 1977; Liu and Chen, 1978; Shepard et al., 1980; Larkin and Scowcroft, 1981). A large number of mutants have been selected from cells and tissue culture (Widholm, 1977; Maliga, 1980). Disease resistant, stress tolerant, herbicide resistant and biochemical mutants have been obtained (Henke, 1981; Dix, 1981; Crocomo and Ochoa-Alejo, 1982).

Secondary Products

About 30 years ago it was suggested that plant tissue cultures could be used as an alternative to whole plants as a source of potentially useful compounds. This idea came from the demonstration that plant cells can be grown in suspension in liquid medium in the same way as microorganisms (Nickell, 1956). Generally, the plant products of commercial interest are the secondary metabolites. There are three main groups of products: essential oils, glycosides, and alkaloids. Many of these substances are used in the pharmaceutical, food, flavor and perfume industries. The majority of commercial secondary metabolities are obtained by extraction from plants grown in tropical and subtropical regions of the world. Selected secondary products and their properties or uses from wild or semi-cultivated Brazilian species were recently reported by Crocomo et al. (1981).

Advantages expected for cell culture systems over the conventional cultivation of whole plants are that (a) useful substances could be produced under controlled conditions; (b) cultured cells would be free of contamination; (c) cells of any plant could be multiplied to yield specific metabolites; and (d) cell growth could be controlled automatically and metabolic process could be regulated rationally, all contributing to the improvement of productivity and the reduction of labor and costs (Tabata, 1977).

Gene Transfer by Wide Hybridization

Sexual crosses are usually limited to closely related species because of a variety of biological barriers associated with the pollination process and embryo development. The possibility of bypassing sexual pathways in hybridization could be of basic interest in plant improvement programs and for the creation of entirely new plant types. Hybridization can be used to transfer genetic capacities and this can be accomplished through embryo axis culture or somatic hydridization. Embryo axis culture has a particular value when wide hydridization are involved (interspecific, intergeneric, etc.). The technique overcomes the problem of cross sterility due to collapse of endosperm and subsequent embryo abortion. Progeny from wide crosses can often be recovered by excision and culturing of the immature embryo axis. Raghavan (1977) has reviewed the obtainment of interspecific and intergeneric hybrids via embryo axis culture. Promising results, mostly between cultivated and wide related specifies have been obtained with this technique, especially among cereal crops and legumes. When normal fertilization does not occur, then _in vitro_ fertilization and ovule/egg culture may be useful (Stewart, 1981).

The second method for the production of wide hybrids between species is by fusion of protoplasts from different sources with subsequent regeneration of an intact hybrid plant. Somatic hybrid plants have been produced in sexually compatible species in the following genera: _Nicotiana_, _Datura_, _Petunia_, _Daucus_ and _Solanum_ (Kao, 1980). Intergeneric plants, such as _Lycopersicon esculentum_ + _Solanum tuberosum_, _Datura innoxia_ + _Atropa belladona_, _Daucus carota_, _Aegopodium podagria_, _Daucus carota_ + _Petroselinum hortense_, and _Arabidopsis thaliana_ + _Brassica campestris_, which may be difficult to achieve by sexual crosses, have been obtained through protoplast fusion. Unfortunately none of these plants has resulted in crop improvement as yet (Dudits et al., 1980; Kao, 1980; Melchers, 1980). Hybrid cells involving many diverse species have already been created, for example, between soybean and pea, tobacco and soybean, and carrot and barley, although plants have not been regenerated from such fusions. Thus, the production of plants of one parental type by including genetic traits of the other parent seems to be a goal for the future.

Plant Molecular Genetic Engineering

Dramatic advances in the area of cell and molecular biology have engendered new enthusiasm in the field of plant genetics. Genetic transformation through DNA uptake implies that DNA from one source is taken up, incorporated into the recipient cell in a stable form, and expressed as new stable characteristics. Plant protoplasts are considered to be the most logical and efficient recipients for introducing foreign genetic information. Numerous attempts have been made to induce transformation in higher plant through uptake of DNA and organelle implantation into protoplasts (Crocomo and Ochoa-Alejo, 1981). Data on nuclei, chloroplasts, mitochondria and chromosome transplantation into the protoplast by induced fusion with PEG have been reported by several authors but, unfortunately, the ability of organelles to survive and to express their genetic information in recipient protoplasts has not be unequivocally demonstrated (McDaniel, 1981). At present, genetic engineering technology makes possible routine transfer of genes between different prokaryotes and between eucaryotes and prokaryotes.

ACTUAL LIMITATIONS

A major fundamental problem which severely limits the application of tissue culture technology to the major feed and food crops remains one of inability or inefficiency of plant regeneration from cells and tissues, which has been particularly difficult with grasses, cereals and woody species. Evans et al. (1981), have listed species for which plant regeneration has been reported, including most vegetables, fruit and nut trees, flowers, foliage crops, wheat, oats, rice, corn, barley, sorghum and sugar cane. This is due mainly to the lack of knowledge concerning the biochemical and molecular mechanisms which trigger plant regeneration from tissue culture. Empirical manipulations of chemicals in the nutrient media is the only way through which plant regeneration has been achieved.

Meristem culture techniques have found their place in clonal propagation, disease elimination, and in germplasm preservation. However, few (16) important vegetatively propagated food crops have been freed of systemic viral pathogens (Kartha, 1981b). At present the application of cryopreservation techniques for germplasm storage is very limited. Kartha (1981a) has reported a list of only 9 species where plant organ storage has been achieved. Research in this area is still needed.

Plant cells propagated in culture can undergo considerable variation in phenotype, either spontaneously or as a result of specific environmental situations. Cultured plant cells can exhibit a wide variety of phenotypes many of which are maintained in the progeny of individual cells (Binns, 1981). Genetic and cytogenetic instabilities from cell and callus culture may represent serious obstacles when the objective is to reproduce asexually, at a reliable and defined level of fidelity, a given individual plant genotype. The mechanisms by which such clonal variation arises and is maintained are unknown (Larkin and Scowcroft, 1981).

Variants cells in tissue culture are selected by their unusual phenotype. However, phenotypic changes may be the result either of genetic or epigenetic alterations. Mutations are permanent, sexually transmissible, hereditary alterations in the primary structure of DNA. Epigenetic changes, on the other hand, are distinguished only by operational criteria, since their molecular basis is unknown. Since the major aim of plant breeders is to regenerate recoverable genetic variation, epigenetic variants are often an obstacle. Therefore, extensive investigation is needed in order to confirm that variants selected at the cellular level still express those same traits at the whole plant level. The potential application of haploids include production of stocks for genetic and cytogenetic analysis, selection of mutants at the haploid level, facilitation of interspecific hybridization, and rapid production of homozygous breeding lines. Unfortunately efficient methods for producing large number of haploids are unavailable for many agronomic species. On the other hand, reduction in vigor and genetic instability in haploid and double haploid populations represent basic problems (Conger and Collins, 1981).

The use of protoplasts in somatic hybridization and genetic manipulation is dependent on the instability to regenerate plants from them. One of the most significant problems is the failure to regenerate plants from protoplasts of the two most important families: Leguminoseae and Gramineae (Vasil and Vasil, 1980). Although protoplasts of widely separated species can be induced to fuse, the recovery of hybrid plants from the fusion products is a rare event (Schieder and Vasil, 1980). Other important limitations include lack of selection systems to isolate the hybrids, incompatibility of different genomes, and the sterility and instability of somatic hybrids.

Molecular genetic engineering in plants is an emerging field whose potential is primarily long range. The problems involve the isolation of biologically adequate organelles, identification and isolation of specific genes or gene groups, and their incorporation and expression in appropriate tissues of the multicellular plants (Kado and Kleinhofs, 1980).

PROSPECTS FOR THE FUTURE

The most important clue is the development of procedures to determine the efficient regeneration of intact and fertile plants from either single cells, callus or protoplasts of economic plant species. The genetic and developmental control of totipotency and its consequence, which is the regeneration of a whole plant, should be understood taking into account the transference of the conditions for successful manipulation from one species to another. Primary requirements must be satisfied, such as the capability a particular plant species has to favor in vitro culture initiation (which can be related to the type of explant), identification of nutrients, control of morphogenetic responses (which can be achieved through balance of growth factors) and selective conditions.

A recent Workshop on priorities in biotechnology (National Academy of Sciences, 1982) pointed out eight research areas identified as prospects for development specially in developing countries. Due to regional constraints, the research areas were divided into three groups representing immediate, mid-range and long-range development opportunities. Suggested for immediate application were clonal propagation, disease elimination, germplasm exchange, gene transfer by wide hybridization. The mid-range application list included: variant selection and haploid production. For the long-range, secondary product production from cell suspension cultures, and molecular genetic engineering in plants were envisaged.

Once a strategy for the total program of improvement for a given plant species is established, it is possible to envisage the role of plant cell and tissue culture as an important auxiliary tool to be used in breeding and/or other research programs such as crop production/ management. It should be emphasized that a program involving tissue culture must be pursued in collaboration with plant breeders and other plant scientists such as biochemists, physiologists, microbiologists and molecular biologists the latter mainly when use of recombinant DNA techniques is needed. To complete the program evaluation and testing of the developed plant materials would require the collaboration of production agronomists as well as extension workers.

High frequency of plant regeneration is crucial for any breeding program using plant tissue technique. Evans et al., (1981) have classified the successful reports in the literature pertaining to plant regeneration either by organogenesis or somatic embryogenesis. Vegetable and tree species were predominant in a list of 26 crop and 14 non-crop species capable of plant regeneration via somatic embryogenesis. Organogenic pathway was identified as the route through which 20 grass and cereal crops and hybrids can regenerate plants. The crop species include sugar cane, barley, millet, wheat, oat, rice, sorghum, corn, etc. Forty-three non Gramineae crop species have been found capable of regenerating plants either directly or indirectly through an intermediate callus stage. Thus, 103 species (63 via organogenesis and 40 via somatic embryogenesis) are recorded, of which 46 species are classified as of agronomic interest.

A few examples among several found in the literature show that future use of this technique is promising. Conger (1981a) stressed that in vitro techniques for cloning or mass propagation are definitely more advanced than other culture techniques. Some species such as orchids are more amenable to this technique. It is estimated that as many as 4 million orchid Cymbidium could be produced from a single explant in a year. Asparagus can be successfully propagated by in vitro technique; it is estimated that 300,000 transplantable plants may be produced from a single shoot apex in a year and indeed 70,000 plants can be produced by one person working 200 days/years. This level of production depends upon adequate supply of aseptic stock plants and the culture of 500 bud segments by one person in one day. Very recently, Phillips (1982) reported that through the modern techniques of the "new biology", corn growers in the USA expect to double corn production in the next decade.

REFERENCES

Bajaj, Y.P.S. (1977). Protoplasts isolation, culture and somatic hybridization. In: Applied and Fundamental Aspects of Plant Cell, Tissue and Organ Culture. (J. Reinert and Y.P.S. Bajaj, Eds.). Springer-Verlag, Berlin. pp. 467-496.

Bajaj, Y.P.S. and J. Reinert (1977). Cryobiology of plant cell cultures and establishment of gene-banks. In: Applied and Fundamental Aspects of Plant Cell, Tissue and Organ Culture. (J. Reinert and Y.P.S. Bajaj, Eds.). Springer-Verlag, Berlin. pp. 757-777.

Binns, A.N. (1981). Developmental variation in plant tissue culture. Env. Exp. Botany 21: 325-332.

Bottino, P.J. (1981). Vegetable crops. In: Cloning Agricultural Plants via in vitro Techniques (B.V. Conger, Ed.). CRC Press, Inc., Boca Raton, Florida. pp. 141-164.

Burk, L.G., G.R. Swynn and J.F. Chaplin (1972). Diploidized haploids from aseptically cultured anthers of Nicotiana tabacum. A colchicine method applicable to plant breeding. J. Heredity 63: 355-360.

Cocking, E.C. (1960). A method for the isolation of plant protoplasts and vacuoles. Nature 187: 927-929.

Cocking, E.C. (1978). Selection and somatic hybridization. In: Frontiers of Plant Tissue Culture. (T.A. Thorpe, Ed.). International Association for Plant Tissue Culture, Calgary. pp. 151-158.

Cocking, E.C. (1981). Opportunities from the use of protoplasts. Phil. Trans. R. Soc. Lond. B. 292: 557-568.

Conger, B.V. (1981a). Cloning Agricultural Plants Via in vitro techniques. (B.V. Conger, Ed.). CRC Press, Inc., Boca Raton, Florida. pp. 1-4.

Conger, B.V. (1981b). Agronomic Crops. In: Cloning Agricultural Plants via in vitro techniques. (B.V. Conger, Ed.). CRC Press, Inc., Boca Raton, Florida. pp. 165-215.

Conger, B.V., and G.B. Collins (1981). A Discussion on Potential in vitro Applications for Agronomic Crops. Env. Exp. Botany 21: 407-413.

Crocomo, O.J., E. Aquarone and O.R. Gottlieb (1981). Biosynthesis of Secondary Products in vitro. In: Plant Tissue Culture. Methods and Applications in Agriculture. (T.A. Thorpe, Ed.). Academic Press, New York. pp. 359-372.

Crocomo, O.J., L.A. Gallo, G.S. Tonin and N. Sacchi (1979). Developmental Control of Phaseolus vulgaris using embryo axis culture. Energ. Nucl. Agric., Piracicaba. 1: 55-58.

Crocomo, O.J., and N. Ochoa-Alejo (1981). The Technology of Genetic Characters Transfer in vivo. In: Proceedings of the Symposium on Tissue Culture of Economically Important Plants. Singapore 28-31 April, 1981. (A.N. Rao, Ed.).

Crocomo, O.J., and N. Ochoa-Alejo (1982). Selection for Herbicide Tolerance in Tissue Culture Derived Plantlets. In: Treatise on Plant Tissue Culture. (D.A. Evans, W.R. Sharp, P.V. Ammirato, and Y. Yamada, Eds.). MacMillan, Moorestown, NJ, in press.

Dix, P.J. (1980). Environmental stress resistance. Selection in Plant Cell Cultures. In: Plant Cell Cultures: Results and Perspectives. (F. Sala, B. Parisi, R. Cella, and O. Ciferri, Eds.). Elsevier, Amsterdam. pp. 183-186.

Dudits, D., G. Hadlaczky, G. Lazar and Z. Haydu (1980). Increase in Genetic variability through somatic cell hybridization. In: Plant Cell Cultures: Results and Perspectives. (F. Sala, B. Parisi, R. Cella, and O. Ciferri, Eds.). Elsevier, Amsterdam. pp. 207-214.

Eriksson, T., K. Glimelius, and A. Wallin (1978). Protoplast Isolation, Cultivation and Development. In: Frontiers of Plant Tissue Culture. (T.A. Thorpe, Ed.). International Association for Plant Tissue Culture, Calgary. pp. 131-139.

Evans, D.A., W.R. Sharp, and C.E. Flick (1981). Growth and behavior of cell cultures: Embryogenesis and organogenesis. In Plant Tissue Culture. Methods and Applications in Agriculture. (T.A. Thorpe, Ed.). Academic Press, New York. pp. 45-113.

Gamborg, O.L. and J.P. Shyluk (1981). Nutrition, media and characteristics of plant cell and tissue cultures. In Plant Tissue Culture. Methods and Applications in Agriculture. (T.A. Thorpe, Ed.). Academic Press, New York. pp. 21-44.

Gamborg, O.L., J.P. Shyluk and E.A. Shahin (1981). Isolation, Fusion, and Culture of Plant Protoplasts. In: Plant Tissue Culture. Methods and Applications in Agriculture. (T.A. Thorpe, Ed.). Academic Press, New York. pp. 115-153.

Gleba, Y.Y. and F. Hoffman (1980). "Arabidobrassica": A novel plant obtained by protoplast fusion. Planta 149: 112-117.

Han, H. and S. QiQuan (1981). Advances in Plant Cell and Tissue Culture in China. Adv. Agron. 34: 1-13.

Heinz, D.J., M. Krishnamurthi, L.G. Nickell and A. Maretzki (1977). Cell, Tissue and Organ Culture in Sugar cane Improvement. In: Applied and Fundamental Aspects of Plant Cell, Tissue and Organ Culture. (J. Reinert and Y.P.S. Bajaj, Eds.). Springer-Verlag, Berlin. pp. 3-17.

Henke, R.R. (1981). Selection of Biochemical Mutants in Plant Cell Cultures: Some Considerations. Env. Exp. Botany 21: 347-357.

Hughes, K.W. (1981). Ornamental species. In: Cloning Agricultural Plants Via _in vitro_ techniques. (B.V. Conger, Ed.). CRC Press, Inc., Boca Raton, Florida. pp. 5-50.

Kado, C.I. and A. Kleinhofs (1980). Genetic Modification of Plant Cells through uptake of foreign DNA. Int. Rev. Cytol. Supplement 11B; 47-80.

Kao, K.N. (1980). Expression of foreign genetics material through protoplast fusion, through uptake of prokaryotic cells and cell organelles. In: Plant Cell Cultures: Results and Perspectives. (F. Sala, B. Parisi, R. Cella and O. Ciferri, Eds.) Elsevier, Amsterdam. pp. 195-205.

Kao, K.N., F. Constabel, M.R. Michauluk and O.L. Gamborg (1974). Plant protoplasts Fusion and Growth of Intergeneric Hybrid Cells. Planta. 120: 215-227.

Kartha, K.K. (1981a). Meristem Culture and Cryopreservation-Methods and Applications. In: Plant Tissue Culture. Mehtods and Applications in Agriculture. (T.A. Thorpe, Ed.). Academic Press, New York. pp. 181-211.

Kartha, K.K. (1981b). Tissue Culture Techniques for virus elimination and germplasm preservation. In: Genetic Engineering for Crop Improvement. (K.O. Rachie and J.M. Lyman, Eds.). Rockefeller Foundation. USA pp. 123-138.

Keller, W.A. and G.R. Stringam (1978). Production and utilization of microspore-derived haploid plants. In: Frontiers of Plant Tissue Culture (T.A. Thrope, Ed.). International Association for Plant Tissue Culture, Calgary. pp. 113-122.

King, P.J. (1980). Cell proliferation and growth in suspension cultures. Int. Rev. Cytol. Supplement 11A: 25-53.

Larkin, P.J. and W.R. Scowcroft (1981). Somaclonal variation - a novel source of variability from cell cultures for plant improvement. Theoret. Appl. Genet. 60: 197-214.

Laurence, R.H. (1981). _In Vitro_ plant cloning systems. Env. Exp. Botany 21: 289-300.

Liu, M-C. and W-H Chen (1978). Tissue and Cell Culture as Aids to Sugar cane Breeding.2. Performance and yield potential of callus derived lines. Euphytica 27: 273-282.

Maliga, P. (1980). Isolation, characterization and utilization of mutant cell lines in higher plants. Int. Rev. Cytol. Supplement 11A: 225-250.

McDaniel, R.G. (1981). Plant genetic engineering: Possibilities for organelle transfer. In: Genetic Engineering for Crop Improvement. (K.O. Rachie and J.M. Lyman, Eds.). Rockefeller Foundation, USA. pp. 185-207.

Melchers, G. (1980). The somatic hybrids between tomatoes and potatoes (topatoes and pomatoes). In: Plant Cell Cultures: Results and Perspectives. (F. Sala, B. Parisi, R. Cella and O. Ciferri, Eds.). Elsevier Amsterdam. pp. 57-58.

Mott, R.L. and R.H. Zimmerman (1981). Trees: Round-Table summary. Env. Exp. Botany 21: 415-420.

Murashige, T. (1977). Clonal crops through tissue culture. In: Plant Tissue Culture and its Bio-Technological Application. (W. Barz, E. Reinhard and M.Y. Zenk, Eds.). Springer-Verlag, Berlin. pp. 392-403.

Murashige, T. (1978). The impact of plant tissue culture on agriculture. In: Frontiers of Plant Tissue Culture. (T.A. Thorpe, Ed.). International Association for Plant Tissue Culture, Calgary. pp. 15-26.

Narayanaswamy, S. (1977). Regneration of plants from tissue cultures. In: Applied and Fundamental Aspects of Plant Cell, Tissue and Organ Culture. (J. Reinert and Y.P.S. Bajaj, Eds.). Springer-Verlag, Berlin. pp. 179-206.

NAS (1982). Workshop on Priorities in Biotechnology Research for International Development, NAS, Washington, DC, U.S.A. July 26-30, 1982 (in press).

Nickell, L.G. (1956). The continuous submerged cultivation of plant tissue as single cells. Proc. Natl. Acad. Sci. USA. 42: 848-850.

Nitsch, C. (1981). Production of isogenic lines: Basic technical aspects of androgenesis. In: Plant Tissue Culture. Methods and Applications in Agriculture. (T.A. Thorpe, Ed.). Academic Press, New York, pp. 241-252.

Phillips, R.L. (1982). Integration and synthesis. Conference on Genetic Engineering of Plants. August 15-19, 1982, Davis, Calif., USA. (In Press).

Raghavan, V. (1977). Applied Aspects of Embryo culture. In: Applied and Fundamental Aspects of Plant Cell, Tissue and Organ Culture. (J. Reinert and Y.P.S. Bajaj, Eds.). Springer-Verlag, Berlin. pp.375-397.

Raghavan, V. (1980). Embryo Culture. Int. Rev. Cytol. Supplement 11B: 209-240.

Raghavan, V. and J.G. Torrey (1963). Growth and Morphogenesis of Globular and older embryos of Capsella in Culture. Am. J. Botany 50: 540-551.

Reinert, J. and Y.P.S. Bajaj (1977). Anther Culture: Haploid production and its significance. In: Applied and Fundamental Aspects of Plant Cell, Tissue and Organ Culture. (J. Reinert and Y.P.S. Bajaj, Eds.). Springer-Verlag, Berlin. pp. 251-267.

Schieder, O. and I.K. Vasil (1980). Protoplast fusion and somatic hybridization. Int. Rev. Cytol. Supplement 11B: 21-46.

Sharp, W.R. and D.A. Evans (1981). Plant Tissue Culture: The Foundation for Genetic Engineering in Higher Plants. In: Genetic Engineering for Biotechnology. (O.J. Crocomo, F.C.A. Tavares and D. Sodrezeieski, Eds.). PROMOCET, SP, Brasil. pp. 23-28.

Shepard, J.F., D. Bidney and E. Shahin (1980). Potato protoplasts in crop improvement. Science 28: 17-24.

Skirvin, R.M. (1981). Fruit crops. In: Cloning Agricultural Plants Via in vitro Techniques. (B.V. Conger, Ed.). CRC Press, Inc., Boca Raton, Florida. pp. 51-139.

Skoog, F. and C.O. Miller (1957). Chemical regulation of growth and organ formation in plant tissues cultivated in vitro. In: Biological Action of Growth Substances. 11th. Symp. Soc. Exp. Biol. 11: 118-131.

Smith, S.H. and W.A. Oglevee-O'Donovan (1979). Meristem-tip culture from virus-infected plant material and commercial implications. In: Plant Cell and Tissue Culture. Principles and Applications. (W.R. Sharp, P.O. Larsen, E.F. Paddock and V. Raghavan, Eds.). Ohio State University Press, Columbus. pp. 453-460.

Stewart, J.M. (1981). In vitro fertilization and embryo rescue. Env. Exp. Botany 21: 301-315.

Tabata, M. (1977). Recent advances in the production of medicinal substances by plant cell cultures. In: Plant Tissue Culture and its Biotechnological Application. (W. Barz, E. Reinhard and M.H. Zenk, Eds.). Springer-Verlag, Berlin. pp. 3-16.

Torrey, J.G. (1967). Morphogenesis in relation to chromosomal constitution in long-term plant tissue cultures. Physiol. Plantarum 20: 265-275.

Uchimiya, H. (1979). Progress in plant protoplast technology. Isolation, culture and genetic manipulation. Reports Inst. Agric. Res. Tohoku Univ. 30: 29-53.

Vasil, I.K. (1980). Androgenetic haploids. Int. Rev. Cytol. Supplement 11A: 195-223.

Vasil, I.K. and V. Vasil (1980). Isolation and culture of protoplasts. Int. Rev. Cytol. Supplement 11B: 1-19.

Walkey, D.G.A. (1978). In vitro methods for virus elimination. In: Frontiers of Plant Tissue Culture. (T.A. Thorpe, Ed.). International Association for Plant Tissue Culture, Calgary. pp. 245-254.

Widholm, J.M. (1977). Selection and characterization of biochemical mutants. In: Plant Tissue Culture and its Bio-technological Application. (W. Barz, E. Reinhard and M.H. Zenk, Eds.). Springer-Verlag, Berlin. pp. 112-122.

Withers, L.A. (1980). Preservation of germplasm. Int. Rev. Cytol. Supplement 11B: 101-136.

PHOTOSYNTHETIC ACTIVITY AND PARTITIONING

L.T. Evans
CSIRO Division of Plant Industry, P.O. Box 1600
Canberra City, A.C.T. 2601, Australia

ABSTRACT

Photosynthesis, translocation, partitioning, growth and storage constitute an integrated set of processes linked by many interactions, and each should be viewed within that context. Although the duration of photosynthetic activity has increased during crop improvement, maximum photosynthetic rate has not done so, probably because of the need for compromise between greater leaf area and faster photosynthesis. Also, rubisco, the major photosynthetic enzyme, has been highly conserved in evolution, and the selection of forms with lower $K_m(CO_2)$ or oxygenase activities seems unlikely. Past increases in yield potential have come from changes in the partitioning of assimilates and nutrients due to reduced growth of non-harvested organs, more prolonged and faster storage, and enhanced competitiveness of the storage organs; i.e. from changes in the regulatory processes rather than in the photosynthetic processes. The prospects for further change, and the relevant research priorities, are considered.

KEYWORDS: photosynthesis, partitioning, translocation, rubisco, storage, senescence, photorespiration, competition.

INTRODUCTION

The process of photosynthesis and of the subsequent partitioning of its products are among the most crucial in relation to the yield potential of crops. However, they provide a stark contrast in that humankind has not improved on nature with respect to photosynthesis, but has profoundly changed the partitioning of assimilates in crop plants in the course of their domestication and improvement. Two quite different strategies might therefore be envisaged for the future. One would be to continue along the path that has contributed most of the rise in crop yield potential so far, namely, to increase still further the partitioning of assimilates into the harvested organs. The other assumes that further progress in that direction will be limited, and that we should now concentrate on increasing the efficiency of photosynthesis.

This paper focuses, therefore, on what factors may have prevented even indirect selection for higher rates of photosynthesis in the past, on the processes that have already contributed to greater partitioning into the harvested organs, and on the likelihood of extending these still further. There is need, however, to consider the processes of photosynthesis, translocation, partitioning, growth and storage as an integrated whole, since they are linked by numerous interactions (Gifford and Evans, 1981). In some conditions, the rate of photosynthesis may drive the rate of storage, and faster photosynthesis at one stage may also enhance the storage capacity at a later stage, e.g. by allowing more ears or grains to develop, thereby creating a greater demand for assimilate later in the life cycle of the crop.

Greater demand during the grain growth phase may either enhance or reduce leaf photosynthesis, depending on the nature of the storage products. If they have a high nitrogen or mineral content, as in legume seeds, greater demand will lead to a faster remobilization of nitrogen out of the leaves and, therefore, to earlier senescence and a faster fall in photosynthetic rate. On the other hand, where grain growth is mainly based on carbohydrate storage, as in the cereals, and where nitrogen fertilizers eliminate the need to mobilize nitrogen out of the upper leaves, leaf photosynthetic rate can be higher the greater is the storage demand. Where storage is slow, carbohydrates accumulating in the leaves may cause a feedback inhibition of photosynthesis, leading to early senescence, as occurs in wheat (King et al., 1967; Thomas and Stoddart, 1980).

Thus, the rates of photosynthesis and storage are interlinked in several ways, involving both feed-forward and feedback reactions which can complicate comparisons of photosynthetic rates. Similar interactions occur between translocation and storage, underlining the need for us to consider the processes of photosynthesis, translocation, partitioning and storage within an integrated framework, and not each in isolation if we are to understand how to raise yield potential. Current understanding is greatest for photosynthesis, least for storage, and a more balanced approach, with relatively more emphasis on what happens in the growing and storing organs, will be needed through the rest of this century.

PHOTOSYNTHETIC RATE

Selection for greater photosynthetic rates in crop plants would seem, at first sight, to be a logical and direct route to greater yield potential. Crop yields often bear a close relation to irradiance at certain stages, and exposure to greater CO_2 concentrations at those stages may increase yields substantially. Photosynthesis, at some stage in many crops, almost certainly limits potential yield and we might expect indirect selection for greater photosynthetic rate to have occurred during the domestication and improvement of crops. However, there is no evidence that any selection for increase in the maximum light-saturated CO_2 exchange rate (CER) per unit leaf area has occurred to date in wheat, maize, sorghum, pearl millet, sugar cane, cotton or cowpea (Gifford and Evans, 1981). In fact, the highest rates for four of these crops are found among their wild relatives. Similarly, in rice there appears to have been no marked increase in CER in the course of evolution and improvement (Cook and Evans, 1983). The only crops in which higher CER may have been associated with higher yields are beans (Peet et al., 1977) and soybeans (Buttery et al., 1981), but the evidence is by no means conclusive.

To support this assessment, we need to make an important distinction between rankings among cultivars when CER is at its maximum, and rankings for photosynthetic rate at some later stage. Among the wheats, for example, CER is highest among the wild diploids, but these fall rapidly with the result that several weeks after anthesis the highest rates of photosynthesis are found among the modern wheats (Evans and Dunstone, 1970; Austin et al., 1982). Maximum CER in cowpeas was not related to yield potential, but later measurements of photosynthetic rate were (Lush and Rawson, 1979), just as in soybeans CER during flowering was not related to yield whereas those during pod filling were (Buttery et al., 1981). Thus, correlations between CER and yield probably reflect differences in the duration rather than in the maximum rate of photosynthesis.

Plants can be selected for higher CER, so it is important to understand why this has not occurred during selection for greater yield potential. Two of several possible reasons are considered here.

Area Rate Compensation

The first concerns the partitioning of the resources available for leaf growth between greater area and greater investment per unit leaf area, expressed as more nitrogen or dry weight per unit leaf area (the latter being called specific leaf weight or SLW). Insofar as there is often only minor variation between cultivars in the proportion of biomass invested in leaves and in the rate of leaf appearance, (Cook and Evans, 1983), there tends to be an inverse relation between leaf area and SLW. In turn, both SLW and nitrogen dm^{-2} of leaf - especially the latter - tend to be closely and positively related to CER, as in rice and wheat, whether the variation is due to cultivar, species, nutrient level or environmental conditions (Khan and Tsunoda, 1970b; Cook and Evans, 1983). Consequently, there tends to be a negative relation between the area of a leaf and its CER, as in wheat (e.g. Evans and Dunstone, 1970; Khan and Tsunoda, 1970a; Gale et al., 1974; Planchon and Fesquet, 1982; Austin et al., 1982).

Selection for higher CER may, therefore, simply result in a compensatory reduction in leaf size, as in alfalfa (Hart et al., 1978) and peas (Mahon, 1982). The best compromise between leaf size and CER depends on stage of growth and on agronomic and environmental conditions. Rapid leaf area expansion is needed in the young crop to cover the ground, increase light interception, and exclude weeds. Later, when the crop canopy has closed, smaller, more upright leaves with greater CER would be advantageous. There is little evidence of such a change in strategy during development, but selection for it might be worthwhile. In the meantime, high yielding cultivars of both wheat and rice tend to be found near the middle of the leaf area/CER or SLW curve (e.g. Cook and Evans, 1983), presumably reflecting an optimum compromise. In fact, when selections for higher SLW or CER have been made, as in alfalfa and peas, it has been found that yield and crop growth rates tend to be more closely related to leaf area, being higher in the lines selected for low CER (Hart et al., 1978; Mahon, 1982).

Agronomic trends, however, seem likely to favour high CER plants increasingly in the future. Denser plantings should reduce the premium on leaf area expansion, while higher levels of nitrogen fertilizer application to non-legume crops will make it desirable to shift the flattening of the CER response to increasing leaf nitrogen content, evident for wheat in Figure 1, to higher levels.

Thus, selection for higher CER remains a valid objective, but it will be ineffective unless the counterproductive associations with smaller leaf area are taken into account.

Photosynthetic Efficiency

Photosynthesis involves the coordinated action of a large number of enzymes and processes, but the primary target for the improvement of CER has been ribulose-1, 5-bisphosphate carboxylase/oxygenase, referred to below as rubisco. This is probably the most abundant enzyme in the world, commonly representing 30-50% of soluble leaf protein, and reportedly up to 73% in an octoploid tall fescue (Joseph et al., 1981). Rubisco already constitutes such a major fraction of leaf protein that further increase in this proportion is likely to be limited. Seasonal irradiance and nitrogen nutrition levels have no effect on this proportion in wheat (Figure 2), but in tall fescue it supposedly increased with increasing ploidy (Joseph et al., 1981), so there may be some scope for further selection.

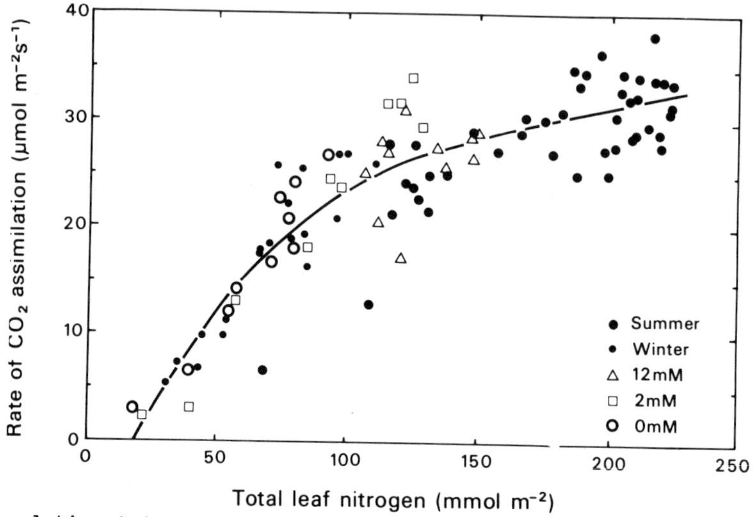

Fig. 1 The relation between the net photosynthetic rate of flag leaves of wheat (measured at a quantum flux density of 1800 µE m^{-2} s^{-1}, 340 µ bar CO_2 and 23°C) and total leaf nitrogen content for plants grown at various seasons and levels of nitrogen. (J.R. Evans, 1983).

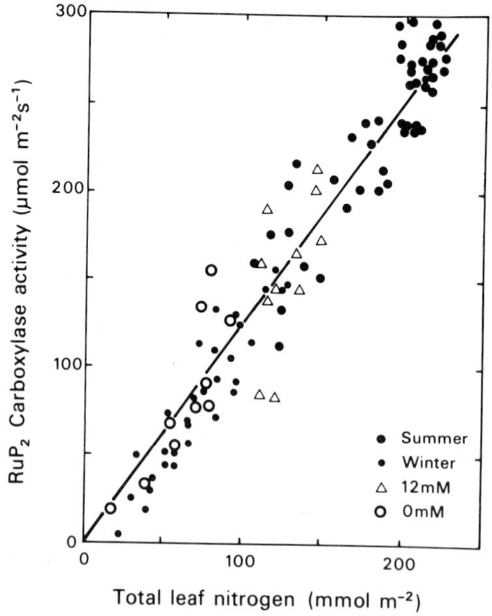

Fig. 2 The relation between rubisco activity and nitrogen content in leaves of wheat grown at various seasons and levels of nitrogen. (J.R. Evans, 1983).

Several investigations have suggested that CER is limited by the activity of rubisco (Lorimer, 1981), but comparison of Figures 1 and 2 suggests that, while this may be true in the lower part of the range, it is less so for wheat leaves with more than 100 m mol nitrogen m^{-2} and rubisco activities of more than 130 mol $m^{-2}s^{-1}$ (Evans, J.R., 1983).

Earlier work with a decaploid line of tall fescue suggested that its higher CER was associated with a rubisco of considerably higher specific activity, but with better techniques this has now been shown not to be the case (Joseph et al., 1981). However, Seeman and Berry (1982) report a substantially higher specific activity for spinach rubisco compared with that from soybean. Two earlier reports of rubisco with a lower $K_m(CO_2)$ in tetraploid ryegrass have not been substantiated (McNeil et al., 1981), but some taxonomic variation in $K_m(CO_2)$ has been reported (Yeoh et al., 1981).

Rubisco acts not only as a carboxylase but also as an oxygenase, to an extent depending on the relative concentrations of CO_2 and O_2 at the active site, and with the result that the CER is substantially reduced by the process of photorespiration. Considerable efforts have therefore been made, in large screening programs with many different C_3 plants, to find variants with reduced photorespiration. The outcome of all this work is that, across the many lines, cultivars and species examined so far, there is a close positive relation between CER and photorespiration, and no evidence of genotypes with high CER due to slow photorespiration (Lorimer, 1981).

Given the selective advantage that would accrue to plants with higher CER due to reduced oxygenase activity *if* photorespiration did not play an adaptive role, and the great span of time that has been available for this evolution, these results suggest either that photorespiration has an important adaptive role(s) or that the topological requirements for the site of carboxylation inevitably confer oxygenase activity. Rubisco has been highly conserved throughout plant evolution, and displays substantial homology, particularly in the composition of the active site, across many species, including spinach, peas, barley and maize (Lorimer, 1981). The fact that the dimeric form of rubisco from the bacterium *Rhodospirillum rubrum* also has oxygenase activity in spite of its quite different composition suggests that there may be little chance of increasing photosynthetic rate in crop plants by eliminating photorespiration. Nevertheless, if forms of rubisco with substantially higher specific activity can be found, a high CER could be maintained with a reduced investment in rubisco, as Seeman and Berry (1982) have shown. The resources thus freed from photosynthesis could have a major impact on yield even in the absence of any rise in CER.

PARTITIONING

Partitioning of the photosynthetic and nutrient resources available to a crop can be considered at several levels: between the individual plants within the crop; between the organs of each plant; within those organs, as has already been discussed for partitioning between leaf area and SLW; and between compartments or compounds within cells. Thus, many quite different processes are involved in partitioning and the following discussion concentrates on those which influence the partitioning of assimilates among the various organs of the crop plant, and particularly into the harvested organs.

Since there has been no increase in CER or relative growth rate of crop plants during their domestication and improvement, past increases in yield potential have mostly come from increase in the proportion of biomass which is harvested (the harvest index). This was first shown to be the case by van Dobben

(1962) for Dutch wheat varieties, and pronounced rises in the harvest index of wheat, barley, oats, peanuts, soybeans and other crops have been found to match increases in their yield potential. For wheat and its wild relatives grown in the field, the harvest index ranged from 5-17% in the wild diploids to 51% in Hobbit, a modern cultivar (Austin et al., 1982). How much further it can be raised is a matter for conjecture, but it could possibly reach 63% in wheat (Austin et al., 1980), permitting 25% further increase in yield potential.

The harvest index varies greatly between crops and with environmental conditions, being the outcome of many interacting processes, some of which are now considered.

1. Reduced Investment in Other Organs, and Its Consequences

In the absence of any change in crop biomass, increase in yield potential must be at the expense of the non-harvested organs, and can be achieved either by direct selection for greater yield of the harvested organs or by selection for reduced investment in the other organs. A clear example of the latter is provided by wheat, where the fall in straw weight of British cultivars with progressive selection for shorter stems has been accompanied by an equivalent rise in grain yield (Austin et al., 1980). This matching rise and fall is probably fortuitous, because the savings on stem growth precede the period of grain growth, and only part of them would go to temporary reserves which are subsequently mobilized during grain growth. Some would have been used to support the development of more ears or more florets, thereby increasing the capacity of the shorter cultivars to mobilize their resources during grain growth.

Such an effect is shown more clearly by tropical maize populations selected for reduced height: these showed no reduction in biomass at flowering, but the reduction in stem weight of 1-2 t/ha was associated with an increase in ear weight of only about 0.1 t/ha at flowering, but of more than 3 t/ha at maturity (Fischer and Palmer, 1983). Clearly, although only a small part of the assimilate released by reducing stem growth was immediately invested in greater inflorescence growth, its ultimate impact on grain yield was considerably amplified.

Stem height and weight can probably be reduced still further in many crop plants without adverse effects on crop photosynthesis. Some reduction in leaf growth in the later stages of the crop life cycle may also be possible although, as we have seen, reductions in leaf area should be compensated by increases in SLW and CER. Root weight might also be reduced, especially where irrigation and fertilizer application is adequate. Likewise, the accumulation of mobilizable reserves for osmotic adaptation, recovery from pest attack, or prevention of stalk rots and lodging (e.g. in sorghum and maize), could be reduced in environments where stress, pests, diseases and lodging can be controlled by other means.

Important policy issues follow from these considerations. While further increase in the genetic yield potential via increased harvest index should be possible in most crops for the remainder of this century, such selection will be effective only to the extent that the crops are provided with improved agronomic support. For example, shorter stems and smaller leaves are of advantage only when weeds are controlled. Smaller root systems are possible only where water and nutrient supply is adequate. Reserves can be reduced only when stress is eliminated. If increased yield potential was due to faster photosynthesis and growth the situation would be different, but since it is due to reduced investment in the non-harvested organs, we should not expect "wonder wheat" and "miracle rice" to be wonderful and miraculous without high levels of agronomic inputs.

A related issue, with major policy implications, is whether higher yielding cultivars are likely to be more or less stable in their yields. With the requisite agronomic support their yields could be more stable, but to the extent that selection for greater harvest index means they must sail closer to the physiological wind, they could also be inherently more vulnerable to environmental variation. Here, too, is an area that deserves a more open-minded experimental approach.

2. Changed Balance in the Crop Life Cycle

In many environments, climatic factors largely determine when crop life cycles can begin and must end or pass certain critical stages such as the beginning of ear development, meiosis or grain set. Nevertheless, some adjustment of the relative lengths of the various phases of the life cycle has been possible as the level of agronomy has improved. Early irrigation, higher levels of fertilizer application and better weed control, for example, may accelerate the early stages of crop growth, thereby allowing the final storage phase to be extended, as may the use of grain drying equipment, or breeding for greater resistance to frost or drought. Greater or later applications of N and P fertilizers in these circumstances also allow the upper leaves to remain photosynthetically active for longer, by reducing the need to mobilize N and P out of them in order to support grain growth.

Thus, greater agronomic support has made it feasible to select plants for longer durations of photosynthetic activity by their leaves and of storage by their grains. Such changes have been observed in many crops, e.g. in wheat (Evans and Dunstone, 1970; Sofield et al., 1977a; and Austin et al., 1982), soybeans (Gay et al., 1980) and maize (Crosbie and Mock, 1981). To be fully effective, the durations of grain growth and of leaf photosynthesis should increase in parallel. In wheat, for example, grain growth ceases, even though flag leaf photosynthesis and export may still be active, when the chalazal zone is blocked by the deposition of lipids (Sofield et al., 1977b), a plant equivalent of arteriosclerosis. The timing of this cessation is presumably under hormonal control and readily subject to selection or chemical regulation, as is that for leaf senescence, provided environmental and agronomic conditions permit.

Increased duration of photosynthesis and storage could, within the limits set by environmental constraints, provide one of the most effective and immediate routes to greater yield potential, and research on the factors determining leaf senescence and grain growth duration deserve a higher priority than they have received to date.

3. Faster Storage

Yield depends on both the rate and the duration of storage, and increases in both components have contributed to past rises in yield potential, as in wheat (e.g. Sofield et al., 1977a) and maize (Crosbie and Mock, 1981). Very high rates of growth per grain are achieved in some legume crops, but these are coupled with short durations of growth, as in cowpeas (Lush and Evans, 1981). In such cases, grain growth rates per unit ground area may substantially exceed the crop growth rates, being sustained by the mobilization of N, P and carbohydrates out of the leaves, leading to their early senescence.

4. Enhanced Competitiveness of the Harvested Organs

We have already seen that yield potential can be enhanced when the growth of the non-harvested organs is reduced by genetic means, and also that there is a need to maintain a balance in the partitioning of resources between competing organs so that faster storage is not at the price of earlier senescence. The geometry of a crop may have a profound influence not only on the amount and distribution of its photosynthetic activity, but also on the partitioning of its resources. However, the mechanisms and rules which control the partitioning of assimilates among competing organs have been little studied, and more widely based research in this area is needed if we are to understand how to optimize growth habit and the characteristics of the storage organs.

Earlier experiments with wheat indicated that, for competing sinks of a similar kind, greater relative size and proximity to the source resulted in a disproportionate increase in the share of current assimilates received by a sink (Cook and Evans, 1978). In recent experiments (e.g. Figure 3), the extent of vascular connection to the source has also been shown to be important. Thus, as in human affairs, it pays to be large, close and with direct connections to the source. One expression of these effects may be the tendency towards closer aggregation of the storage organs into a larger, more competitive sink, evident, for example, in the evolution of sunflower, maize, sorghum and pearl millet.

However, other mechanisms probably come into play in the control of partitioning between organs of different kinds and at different stages of growth. Effects of the various plant hormones are probably crucial in this context, yet we have remarkably little understanding of how geometric and hormonal factors interact to determine the pattern of partitioning, which I regard as the physiological black box in most urgent need of attention if we are to understand the limitations to yield.

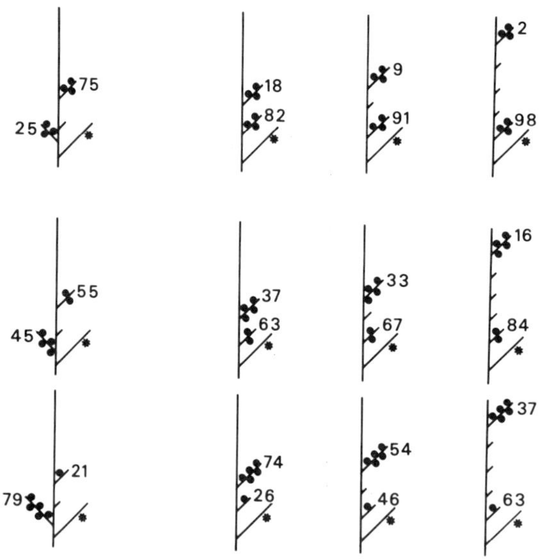

Fig. 3 Percentage distribution of ^{14}C, fixed by the awn of a basal floret, between grains in two competing spikelets of Highbury wheat. The awn (starred) was the only source of current photosynthate, and only 6 evenly sized grains were left in each ear at various positions. (M.G. Cook and L.T. Evans, unpublished data).

CONCLUSIONS

1. The steady increase in the genetic yield potential which has been achieved - and is still continuing - in many crops has been due to changes in partitioning rather than to any improvement in maximum photosynthetic rate so far. Regulatory processes have presumably been easier to modify than assimilatory processes.

2. Even for highly productive crops, further shifts in partitioning and further increase in the harvest index appear feasible, and seem likely to provide scope for substantial increases in yield potential until the end of this century. There is still scope for further reduction in the non-harvested organs provided there is strong agronomic support for the crops. With such support, it should also be possible to increase the duration of the storage phase of the life cycle still further and possibly, though less certainly, the rate of storage. Further modification of crop geometry to enhance partitioning to the sink organs should also be possible, at least in some crops.

3. However, by the end of this century, such changes as a source of greater yield potential, could well be approaching exhaustion. Thus, all possible ways of increasing the rate or efficiency of photosynthesis should be actively explored over the next 20 years, to provide the basis for further increase in yield potential in the 21st century. Although increased growth rates (whether of seedlings (RGR) or crops (CGR)) have not been realized so far, there may be some scope for direct selection for greater RGR even in the absence of any increase in CER, e.g. through reduced rates of maintenance respiration.

4. In the shorter term (this century), relatively limited increases in yield potential might come from selections for higher photosynthetic rate when these are (a) are not associated with smaller leaves but with greater investment in leaf growth; or (b) associated with smaller leaves in cultivars selected for closer planting with greater applications of nitrogen fertilizers; or (c) associated with a more pronounced developmental switch from larger early leaves to smaller upper ones.

5. In the longer term, there may be some hope of improving the efficiency of rubisco by either selection or genetic engineering for greater specific activity. We can hardly be optimistic about achieving what half a billion years of natural selection has not, but given the large proportion of leaf protein which is invested in rubisco, and the limitation it nevertheless often imposes on CER, the impact of effective selection for more efficient forms of rubisco could be very great indeed, even if CER is not raised.

6. However, rubisco is already the target of much physiological and genetic engineering research, and stimulation of research in that area is probably less needed and urgent than in the areas of (a) the control of leaf senescence and of the factors determining the duration and rate of storage processes in harvested organs; and (b) the control of partitioning of current assimilates, reserves and remobilized compounds among competing organs of the major crop plants.

REFERENCES

Austin, R.B., J. Bingham, R.D. Blackwell, L.T. Evans, M.A. Ford, C.L. Morgan, M. Taylor (1980). Genetic improvements in winter wheat yields since 1900 and associated physiological changes. J. Agric. Sci. Camb. 94: 675-689.

Austin, R.B., C.L. Morgan, M.A. Ford, S.G. Bhagwat (1982). Flat leaf photosynthesis of Triticum aestivum and related diploid and tetraploid species. Ann. Bot. 49: 177-189.

Buttery, B.R., R.I. Buzzell, W.I. Findlay (1981). Relationships between photosynthetic rate, bean yield and other characters in field grown cultivars of soybean. Can. J. Plant Sci. 61: 191-8.

Cook, M.G. and L.T. Evans (1978). Effect of relative size and distance of competing sinks on the distribution of photosynthetic assimilates in wheat. Aust. J. Plant Physiol. 5: 495-509.

Cook, M.G. and L.T. Evans (1983). Some physiological aspects of the domestication and improvement of rice (Oryza spp.). Field Crops Res. 6:(in press)

Crosbie, T.M. and J.J. Mock (1981). Changes in physiological traits associated with grain yield improvement in three maize breeding programs. Crop Sci. 21: 255-9.

van Dobben, W.H. (1962). Influence of temperature and light conditions on dry matter distribution, development rate, and yield in arable crops. Neth. J. Agric. Sci. 10: 377-389.

Evans, J.R. (1983). Nitrogen and photosynthesis in the flag leaf of wheat (Triticum aestivum L.), Plant Physiol. (submitted).

Evans, L.T. and R.L. Dunstone (1970). Some physiological aspects of evolution in wheat. Aust. J. Biol. Sci. 23: 725-41.

Fischer, K.S. and A.F.E. Palmer (1983). Maize in Potential Productivity of Field Crops under Different Environments. IRRI, Los Banos, in press.

Gale, M.D., J. Edrich, F.G.H. Lupton (1974). Photosynthetic rates and the effects of applied gibberellin in some dwarf, semi-dwarf and tall wheat varieties (Tritium aestivum). J. Agric. Sci. Camb. 83: 43-46.

Gay, S., D.B. Egli, D.A. Reicosky (1980). Physiological aspects of yield improvement in soybeans. Agron. J. 72: 387-91.

Gifford, R.M. and L.T. Evans (1981). Photosynthesis, carbon partitioning, and yield. Ann. Rev. Pl. Physiol. 32: 485-509.

Hart, R.H., R.B. Pearce, N.J. Chatterton, G.E. Carlson, D.K. Barnes, C.H. Hanson (1978). Alfalfa yield, specific leaf weight, CO_2 exchange rate and morphology. Crop Sci. 18: 649-53.

Joseph, M.C., D.D. Randall, C.J. Nelson (1981). Photosynthesis in polyploid tall fescue II Photosynthesis and ribulose 1,5-bisphosphate carboxylase of polyploid tall fescue. Plant Physiol. 68: 894-8.

Khan, M.A. and S. Tsunoda (1970a). Evolutionary trends in leaf photosynthesis and related leaf characters among cultivated wheat species and its wild relatives. Jap. J. Breed. 20: 133-140.

Khan, M.A. and S. Tsunoda (1970b). Growth analysis of cultivated wheat species and their wild relatives with special reference to dry matter distribution among different plant organs and to leaf area expansion. Tohoku J. Agric. Res. 21: 47-59.

King, R.W., I.F. Wardlaw, L.T. Evans (1967). Effect of assimilate utilization on photosynthetic rate in wheat. Planta (Berl) 77: 261-76.

Lorimer, G.H. (1981). The carboxylation and oxygenation of ribulose 1,5-bisphosphate: the primary events in photosynthesis and photorespiration. Ann. Rev. Pl. Physiol. 32: 349-83.

Lush, W.M. and L.T. Evans (1981). The domestication and improvement of cowpeas (Vigna unguiculate (L.) Walp.). Euphytica 30: 579-87.

Lush, W.M. and H.M. Rawson (1979). Effects of domestication and region of origin on leaf gas exchange in cowpea (Vigna unguiculata L.), Photosynthetica 13: 419-27.

McNeil, P.H., C.H. Fayer, D.A. Walker, I.F. Bird, M.J. Cornelius, A.J. Keyes (1981). Similarity of ribulose 1,5-bisphosphate carboxylases of isogenic diploid and tetraploid ryegrass (Lolium perenne L.) cultivars. Plant Physiol. 67: 530-4.

Mahon, J.D. (1982). Field evaluation of growth and nitrogen fixation in peas selected for high and low photosynthetic CO_2 exchange. Can. J. Plant Sci. 62: 5-17.

Peet, M.M., A Bravo, D.H. Wallace, J.L. Ozbun (1977). Photosynthesis, stomatal resistance, and enzyme activities in relation to yield of field grown dry bean varieties. Crop. Sci. 17: 287-93.

Planchon, C. and J. Fesquet (1982). Effect of the D genome and of selection on photosynthesis in wheat. Theoret. and Appl. Genet. 61: 359-365.

Seeman, J.R. and J.A. Berry (1982). Interspecific differences in the catalytic properties of RuBP carboxylase protein in relation to photosynthetic carbon fixation. Carnegie Inst. Washington Year Book 1982 (in press).

Sofield, I., L.T. Evans, M.G. Cook, I.F. Wardlaw (1977a). Factors influencing the rate and duration of grain filling in wheat. Aust. J. Plant Physiol. 3: 785-97.

Sofield, I., I.F. Wardlaw, L.T. Evans, S.Y. Zee (1977b). Nitrogen, phosphorus and water contentxs during grain development and maturation in wheat. Aust. J. Plant Physiol. 4: 799-810.

Thomas, H. and J.L. Stoddart (1980). Leaf senescence, Ann. Rev. Pl. Physiol., 31: 83-111.

Yeoh, H-H., M.R. Badger, L. Watson (1981). Variations in kinetic properties of ribulose-1,5-bisphosphate carboxylases among plants. Plant Physiol. 67: 1151-5.

NEW APPROACHES TO MEAT AND MILK PRODUCTION

Virgil W. Hays
Department of Animal Sciences
University of Kentucky
Lexington, Kentucky 40546-0215, USA

ABSTRACT

There remains tremendous potential for increasing animal products for human use with current resources by applying research developments. Microbial synthesis through genetic engineering or improved chemical methods should make practical the synthesis of essential amino acids to allow widespread supplementation of cereal grains. This would reduce the protein needs for nonruminants including swine, poultry and humans and greatly extend existing protein supplies. Both precise control of reproduction in food producing animals through use of hormones and other chemicals, and the use of growth hormones, estrogens and androgens, and antimicrobial drugs can result in significant improvements in overall production efficiency. The combined effects of treating crop residues to increase nutrient availability, the application of plant growth regulators to maintain plants in a vegetative and highly digestible state and the use of regulators of rumen metabolism such as the ionophore drugs to increase efficiency of energy utilization especially from fibrous feedstuff are all means of greatly increasing the productivity from existing resources.

These developments, combined with genetic improvement, improved nutrient balance and more effective disease control, provide for optimism about the future contributions of animal products to the world's food supply.

Keywords: drugs, hormones, amino acids, reproduction, growth, feed efficiency, estrus synchronization.

INTRODUCTION

The contributions chemistry has made to our present day efficiency of meat production are well recognized. The identification of essential nutrients for use as diet supplements represents a series of new discoveries and opportunities for increased efficiencies in our food chain. A series of papers were presented at the Agriculture and Food Chemistry sessions of the American Chemical Society's Centennial Meeting, April 4-9, 1976 (Teranishi, 1978) to highlight developments in production, processing and utilization with particular emphasis on the role of chemistry in that progress.

The opportunity to review all the great progress of the past may lead one to question what remains to be discovered in the future and also may lead some to a very pessimistic view relative to our ability to meet human needs now and in the future. However, we do appear to be on the verge of great breakthroughs. This paper concentrates on a few of the developments which could have a major impact on total production and the efficiency of animal production.

AMINO ACIDS

Adequate amounts of high quality protein is one if not the major limitation in our world food supply. Animals play a major role in providing high quality protein for human diets; but, animals can also compete with humans for the limited supply of protein. Availability of certain of the essential amino acids at a reasonable cost could greatly reduce the total protein needs for poultry, pigs and humans. The developments in genetic engineering would suggest that microbial synthesis of amino acids at a much reduced cost is not only possible but should be realized soon.

Two of the 10 amino acids essential for growth in nonruminants, lysine and methionine, are now available at prices that will permit their use in animal diets. Table 1 illustrates how the availability of three others could markedly reduce the need for supplemental protein in swine diets. In the U.S., soybean meal serves as a major protein source for animals, particularly swine and poultry. This simplified example shows that 150 kg (15%) of soybean meal/t of diet is required to supplement corn protein in order that the most limiting amino acid, lysine, is present at a level to meet the pig's need. That level of soybean meal provides an excess of several amino acids. In fact, the protein from corn alone meets the essential amino acid needs except for lysine, methionine, threonine, tryptophan and valine. In this example 150 kg of soybean meal or approximately 68 kg of soya protein is required to replace 7.2 kg of essential amino acids. For poultry or with the use of other combinations of feedstuffs, the order and magnitude of limitations would differ. From this illustration, however, it is apparent that the availability of amino acids for supplemental purposes would greatly extend our protein supplies. Direct supplements to human foods would further reduce the protein needs. In the near future microbial synthesis or improvements in chemical synthesis should lead to the availability of these limiting amino acids at costs to permit more extensive use as feed supplements.

The use of amino acid supplements to replace or reduce natural ingredients will necessitate other adjustments and may lead to vitamin or mineral dificiencies which will need to be corrected. In fact, such diet manipulations may lead to the discovery of the dietary essentiality of vitamins or minerals not now recognized as being essential.

REPRODUCTION CONTROL

Improvements in reproductive efficiency of food producing animals is probably the area of animal production that offers the greatest opportunities for gains in efficiency of production. Failures to show or return to estrus, failures to conceive, prenatal mortality and early postnatal mortality are all factors that greatly reduce the efficiency of production. More complete control over the reproductive processes would allow implementation of management and disease control practices to improve reproduction and reduce mortality. It is anticipated that within the near future, we will be able to exercise complete control of reproduction through the use of chemicals, primarily exogenous hormones or substances that regulate the secretion of endogenous hormones. In order to reap the maximum benefits we will need to synchronize the onset of estrus, synchronize

Table 1. Amino Acid Limitations of Corn-Soybean Meal Diet (14.0% Protein in Combination) or Corn Alone

Amino Acid	Corn + Soybean Meal[1]	Corn Alone[2]	Pig's Need[3]	Supplement Needed[4]
	%	%	%	kg
Arginine	.87	.45	.22	
Histidine	.32	.18	.18	
Isoleucine	.76	.45	.42	
Leucine	1.34	.99	.55	
Lysine	.60	.18	.60	4.2
Methionine + Cystine	.41	.18	.40	2.2
Phenylalanine	.71	.45	.30	
Threonine	.56	.36	.42	0.6
Tryptophan	.16	.09	.10	0.1
Valine	.66	.36	.37	0.1

[1]Approximately 15% soybean meal in diet.

[2]Amino acids provided by corn alone.

[3]Amino acid requirement of pigs from 55 to 100 kg body weight.

[4]Kg of amino acid needed per 1000 kg of corn to meet the requirement level.

time of ovulation, synchronize parturition and preditermine the sex of the offspring.

Some of the improved efficiencies that can result from reproduction control can be illustrated by the efficiencies in the poultry industry. We are not exercising great control of reproduction in chicks; but, simply because of the high rate of lay and the fact that we can hold the ova (egg) for several days, we can have a large number of chicks at a specific time. By filling a chick starter unit at one time with chicks of the same age we are able to reduce disease by not exposing the young chick to older birds and applying appropriate preventive measures, to provide the correct nutritional program at all times, and to provide necessary labor and management at the critical periods of the life cycle. We are not able to control sex; but, again poultry provides an excellent example of the benefits that could be derived from sex control. The male chicks from the layer strains are a total loss to the egg industry and very inefficient producers of meat, whereas the females of the broiler strains are less desirable than males in rate and efficiency of meat production.

Progress is being made in controlling reproduction in cattle and swine. Synchronizing of estrus and to some extent ovulation is being accomplished in cattle at present through the use of prostaglandin $F_{2\alpha}$. Relatively normal conception rates can be accomplished by inseminating at a specified time (80 hr) after administration of the substance (Table 2). Certainly, the system is not perfected as yet and research is in progress on estrogen-progestin combinations that also show promise. Researchers (Pursel et al., 1982) have developed partially effective synchronization programs in swine. More precision is needed before we can consider that estrus and ovulation synchronization is successfully accomplished; but, once accomplished, the timing of parturition should be immediately forthcoming with the application of prostaglandins or other substances.

Table 2. Conception Rates of Heifers and Cows Inseminated at 80 Hours or at Estrus Following Injection with Prostaglandin $F_{2\alpha}$[1]

Animal Type	Insemination Time[2]	Conception Rate[3]
Heifers	80 hr	47
Heifers	Estrus	70
Cows	80 hr	46
Cows	Estrus	52

[1]Adapted from Heersche (1982).

[2]Insemination at approximately 80 hr after injection with prostaglandin $F_{2\alpha}$ or 12 to 18 hr after onset of estrus.

[3]Percent conceiving following artificial insemination.

Precise timing of estrus and ovulation would allow maximum application of artificial insemination and more extensive application of embryo transfer, two techniques which provide our greatest opportunities for rapid genetic progress in both milk and meat production. In our U.S. dairy industry, the national average level of milk production is approximately 5500 kg of milk/cow/year. A very substantial number of the herds on production records programs and making extensive use of performance tested sires are producing above 9100 kg/cow/year. The difference in production between the average and the high producing herds is not entirely a result of genetic differences but these figures do illustrate that we have a far greater genetic potential for production than we are now accomplishing.

Effective methods for preselecting the desired sex of the offspring could have a great impact on total production efficiency. Progress is being made in two areas of research that could lead to the solution of this problem. One area is the separation of the X- and Y-chromosome bearing sperm. A second means of control could develop through chemical or immunochemical techniques to selectively permit the development of the embryo of the desired sex. Accomplishment by either method would lead to widespread application of sex control in our food producing animals; however, sperm separation would probably be more effective. Seidel (1981) predicts that we will accomplish or at least partially accomplish the separation of X- and Y-chromosome bearing sperm within the next 10 to 20 years.

In controlling sex in dairy production it would be desirable to choose a very high percentage of females and to be very selective as to which females would be the dams of potential sires. Sex control could essentially double the number of females from which to select replacement cows and also assure a high degree of precision in selecting only superior males for performance evaluation.

For beef production under extensive pasture or range conditions as is done in the U.S.A., Canada, Australia, Argentina, etc., significant gains in efficiency could be accomplished with greater application of crossbreeding. With sex control, we could opt for females in the first cross offsprings and these could serve as the dams for three-way cross calves. For the three-way or terminal cross, we could produce all male offspring because of their superior efficiency in producing beef. The reproduction strategies would differ among species or between purposes within a species (e.g. milk vs beef); but, for each species, there can be very significant improvements in efficiency of production from exercising complete control, including sex control of reproduction.

GROWTH PROMOTANTS

A number of chemical substances are being used to increase rate and efficiency of production in food producing animals. These effective agents can be broadly classified either as hormone type substances or antimicrobial agents. The hormones increase rate and efficiency of production by altering the anabolic or catabolic processes allowing the animal to synthesize tissue at a rate greater than their genetic potential. Whereas, antimicrobial agents (antibiotics produced by microbial synthesis or chemotherapeutics produced by chemical synthesis) increase the rate and efficiency of production by protecting against bacterial diseases or by altering the inhibitiory effects of microorganisms and allowing the animals to perform up to their genetic potential.

Hormone or Hormone-Like Substances

A number of hormonal substances are being effectively used as implants or as feed additives for the purpose of stimulating rate and efficiency of growth, particularly in beef animals. A summary of the response to some of these agents is presentedd in Table 3 where each of the hormones re: lt in substantial improvements in performance of grazing cattle or cattle finished in the feed lot. Some of these products have been used for extended periods of time, although Estrodiol-17β is a recently approved drug. In the U.S., diethylstilbestrol is no longer used; however, that drug is included because the extensive information available clearly demonstrates the magnitude and consistency of improvement that can be accomplished from the use of such substances.

Table 3. Effects of Anabolic Agents on Rate and Efficiency of Gain in Beef Cattle[1]

Anabolic Agent	Number of Experiments	Average Daily Gain, kg Control	Average Daily Gain, kg Treated	Feed/Gain Control	Feed/Gain Treated
Steers					
Diethylstibestrol	35	1.00	1.14	9.43	8.64
Diethylstilbestrol[2]	18	1.02	1.18	8.87	7.89
Zeranol	59	0.98	1.10	8.72	8.08
Estradiol benzoate	14	1.04	1.19	10.92	9.95
Estradiol-17β	5	0.99	1.15	9.19	8.52
Average improvement, %			14.5		8.6
Heifers					
Diethylstibestrol	12	0.86	0.93	8.82	8.36
Melengestrol acetate[2]	14	1.04	1.13	7.88	6.03
Zeranol	13	0.93	1.01	8.18	7.57
Estradiol benzoate + testesterone propionate	6	1.03	1.15	8.23	7.73
Average improvement, %			9.3		10.3

[1]Adapted from Overfield and Hatfield (1975) except for the Estradiol-17B data.

[2]Administered as a diet supplement, others implanted.

Estrogenic substances alone have been ineffective as a growth stimulant in swine and the androgenic substances have been unsatisfactory because of undesirable effects on meat quality. However, a combination of androgen and estrogen (methyl-testosterone and diethylstilbestrol) has been effective in improving rate and efficiency of gain and reducing carcass fat in pigs and is approved for use in some countries.

Recent developments in producing protein hormones by microbial synthesis should result in adequate quantities of growth hormone at a price sufficiently low to permit its use in animal production. There appears to be great potential for use of bovine growth hormone to increase growth rate and milk production. Though not adequately tested as yet, growth hormone should also be effective in improving efficiency of pig production. A great deal of research is needed to establish optimum dosage, length of response time and adjustments needed in feeding and management programs before widespread application can be expected. Also, a satisfactory delivery system is needed to provide the optimal level on a continuous or controlled intermittent basis; and, developing the delivery system will likely be far more difficult than establishing dosage and response time.

Machlin (1973) and Peel et al. (1981) provide evidence from short term trials that milk production may be increased by as much as 10 to 25%. The magnitude of the response to growth hormone is illustrated in Table 4. Machlin (1973) and Brumby and Hancock (1955) continued cows on treatment for 10 to 12 weeks and noted a sustained improvement from growth hormone throughout the test period. The potential benefits from administering exogenous growth hormone has been recognized for several years, but the advent of microbial synthesis has brought application much closer to realization. The accomplishment of microbial synthesis of insulin and growth hormones opens the door for a new era in regulating growth and lactation in farm animals.

Antimicrobial Agents

For more than 30 years, antimicrobial substances have played a major role in controlling diseases and as growth promotants in poultry, swine and cattle. A search for more effective drugs or more effective combinations of existing drugs is continuing. An example of the benefits, particularly in improving performance of young animals, is presented in Table 5. Such antimicrobial substances improve the efficiency of production by reducing mortality and morbidity in disease situations or by improving rate and efficiency of growth in the apparent absence of disease. Comprehensive summaries describing their effects and the various factors that influence their effects abound in the literature including reviews by Hays (1980) and Hays et al. (1981).

Table 4. Effect of Bovine Growth Hormone Injections on Milk Production by High Producing Dairy Cows[1]

Mammary Secretion	Absolute Change	Relative Change
	kg/day	%
Total milk	3.30	9.5
Lactose	0.24	14.6
Fat	0.28	22.7
Protein	0.05	4.7

[1]Adapted from Peel *et al*. (1981).

Table 5. Response of Starter Pigs in Field Tests or Experiment Station Tests To Antibacterials[1,2]

Location	% Improvement	
	Daily Gain	Feed/Gain
32 field tests	28.4	14.5
128 experiment station tests	16.9	7.0

[1]Hays, 1980.

[2]Drugs: Tetracycline-sulfamethazine (or sulfathiazole)-penicillin, tylosin-sulfamethazine, tetracycline and mecadox.

One relatively new application of the ionophore drugs (polyether antibiotics that have been used primarily as coccidiostats in poultry) is as a rumen metabolic regulator in beef cattle. For many years we have searched for ways (drugs, diet manipulations, etc.) to alter the rate and pathways of rumen fermentation. The ionophores do alter rumen fermentation and improve performance of beef cattle. At this time one cannot credit the total response from ionophores to their effects in the rumen as control of coccidiosis in cattle may also be of benefit. However, the improvements in feed conversion on high energy diets and the improvements in both rate and efficiency of gain on high fiber diets are of sufficient magnitude to have a great impact on beef production. Table 6 illustrates the benefits of the ionophores when added to high concentrate diets in which the average improvement in rate of gain was 3.6% and that for feed conversion was 6.5%. Even greater benefits can be expected for cattle on high roughage diets as illustrated in Table 7.

Table 6. Effect of Ionophores on Performance of Finishing Cattle on High-Concentrate Diets[1] (Summary of 59 Experiments)

Response Criteria	Treatment Control	Ionophore[2]	Percent Improvement
Daily gain, kg	1.10	1.14	3.6
Feed/gain	8.87	8.29	6.5

[1]Adapted from Wagner (1982).

[2]Ionophore and (no. experiments): Monensin, (32); Lasalocid, (16); Avoparcin, (5); Narasin, (4); and Salinomycin, (2).

Table 7. Summary of Response by Cattle to Monensin

Diet Type	Improvement in Performance, % Daily Gain	Feed Efficiency
Finishing Cattle		
High grain diets	0	11
High roughage diets	10	12
Growing cattle		
Green chop feed	33	22
Hay	14	15
Corn silage	6	15
Pasture, grazing	16	-

Possibly of even greater significance than the responses observed to date is the model the ionophores provide for altering rumen fermentation. In general they depress microbial protein synthesis and methane production and increase the amount of propionic acid produced without appreciably altering total volatile fatty acid production. The typical effects on propionic acid as related to acetic acid production are illustrated in Table 8 for two ionophores (monensin and lasalocid) which are presently used to improve efficiency of beef production.

The enhanced performance on diets high in fiber can greatly reduce the concentrate feeds required for finishing beef cattle. Though we do not have support data at the present time, it is anticipated that these or related compounds may have a similar beneficial impact on milk production by dairy cows and on meat and wool production by sheep and goats.

As mentioned above, the potential for improving milk production to levels of 9,000 to 10,000 kg/cow/year is realistic. However, to produce that level of milk, diets contain 55 to 60% high energy concentrates. Reid (1977) and Broughman (1982) estimate that milk production on all forage diets may be limited to 4,500 to 6,000 kg. Reid predicted that forages of 60 to 65% digestibility will support 5,000 kg/cow/year and that an increase to 70% digestibility would allow for production levels as high as 8,000 kg/year. The ionophores increase feed efficiency for the growing animal by about 10 to 20%, a magnitude that should have a tremendous impact on milk production by cows on all forage diets.

Table 8. Effect of Ionophores (Monensin and Lasalocid) on Acetic:Propionic Acid Ratio in Rumen Fermentation Media[1]

Ionophore Level, ppm	Ionophore Monensin	Lasalocid
0	2.78[2]	2.35
22	2.27	2.24
44	2.11	1.65
88	2.02	1.65
176	1.80	2.37

[1]Adapted from Bartley *et al*. (1979).

[2]Acetic to propionic ratio.

Table 9. Performance of Cattle Fed Crop Residue[1]

Diet Type	Control	NaOH Treated
72% rice straw		
Gain, kg/day	0.23	0.71
Feed/gain	38.4	16.1
80% corn cobs		
Gain, kg/day	0.30	0.73
Feed/gain	14.3	7.5

[1]Adapted from Klopfenstein (1978).

MODIFYING FEEDSTUFFS

It is well recognized that a very significant portion of the concentrate feeds used in livestock production could be used as human food. Significant gains in total food production can be accomplished by increasing the productivity from those materials that cannot be used directly by humans. A discussion of the potential impact of plant breeding and cropping systems on animal production is beyond the scope of this paper. However, two areas of research illustrate the tremendous potential for more effective utilization of the feedstuffs now available.

Treating Crop Residues

In the U.S., most of the 200 Mt of crop residues from corn, wheat and grain sorghum production is poorly utilized. Byerly (1966) estimated the world supply of such materials at more than nine times that in the U.S.. Certainly some is used effectively as feed, fuel and fiber, and it is essential that some be returned to the soil. However, tremendous potential remains for increased animal production from these crop wastes, especially if we can develop practical methods for improving digestibility. Treating crop residues with sodium hydroxide, for example, will increase the digestibility by as much as 20 to 30% (Klopfenstein, 1978), and such treatment will markedly improve utilization of fiberous feeds by ruminants. An illustration of the potential for greater productivity from such materials is presented in Table 9. Treating rice straw or corn cobs resulted in more than a doubling of the rate of gain by cattle fed diets high in fibrous feeds (72% rice straw or 80% corn cobs) and reduced the feed required per unit of gain by 50% or more. Sodium hydroxide treatment is not the final answer to effective utilization of low quality roughages, but such data do illustrate that a great deal of food could be produced from materials that are not now used effectively.

Regulation of Plant Growth

Many of our forage crops progress through a growth and maturation cycle which markedly affects the nutrient availability and acceptability to ruminants. This maturation process is necessary if seed production is a primary goal; but, for much of our land areas, it is desirable to maintain the plants in a vegetative state throughout the grazing or forage harvesting cycle.

For some time researchers have been evaluating the effects of a plant growth regulator, mefluidide, on the productivity of tall fescue (Festuca arundenacea) and on the performance of animals grazing the greated pastures. Ely et al. (1982) have treated fescue pastures in early spring with mefluidide (280 g/ha). The improved productivity of cattle grazing the treated pastures and the improvement in digestibility of the forage is illustrated in Table 10. Note that the mefluidide improved digestibility of the regrowth by an average of 6% and improved performance of cattle or productivity per ha by 17 to 18%. The differences between treated and nontreated pastures are even more pronounced during the hot summer weather when the quality of fescue is lowest. The total productivity of fescue (dry matter/ha) did not appear to be significantly affected.

Table 10. Effect of Mefluidide Treatment on Forage (*Festuca arundnacea*) Quality[1]

Item	Control	Mefluidide[2]	Improvement %
Performance by grazing steers (140 days)			
Total gain/steer, kg	67.0	79.4	18.5
Total gain/ha, kg	167.2	196.2	17.3
Dry matter digestibility by lambs, %			
Accumulated growth	46.1	46.5	0.8
Monthly regrowth	53.4	56.7	6.2

[1]Ely *et al.* (1982).

The pretreatment of the world's supply of low quality crop residues with enzymes, acids or alkali; the regulation of growth and maturation of plants grown for forage purposes with plant growth regulators; and the regulation of rumen functions to enhance utilization of fibrous feedstuffs with additives such as the ionophores are means of greatly increasing human food production.

Our success in meeting the world's food needs is highly dependent on a full knowledge of the chemistry of plant and animal growth. Furthermore, success is dependent on using that knowledge to manipulate and control those chemicals and chemical reactions needed to maximize productivity of animals in converting unusable or less desirable products into human foods.

LITERATURE CITED

Bartley, E.E., E.L. Herod, R.M. Bechtle, D.A. Sapienza and B.E. Brent (1979). Effect of monensin or lasalocid, with and without niacin or amicloral, on rumen fermentation and feed efficiency. J. Animal Sci. 49:1066-1075.

Broughman, R.W. (1982). Practical livestock-forage systems: Model to managers. In: (J.A. Smith and V.W. Hays, eds.), Proc. XIV International Grassland Congress, 48-53, Westview Press, Boulder, CO.

Brumby, P.J. and J. Hancock (1955). The galactopoietic role of growth hormone in dairy cattle. New Zealand J. Sci. Technol. A36:417.

Byerly, T.C. (1966). The role of livestock in food production. J. Animal Sci. 25:552-566.

Ely, D.G., T.W. Robb and K.P. Coffey (1982). Potential for increasing productivity of ruminants by treating forages with a plant growth regulator, mefluidide. Private Communication.

Hays, V.W., R.L. Bard, R.F. Brokken, S.A. Edgar, W.T. Elliott, M. Finland, W.D. Heffernan, C.A. Lassiter, H.W. Moon, S.A. Naqi, A.A. Paulsen, J.V. Shutze, H.S. Teague, R. Vinson, W.J. Visek, G. Vocke, H.D. Wallace, R.H. White-Stevens and D. Wolf (1981). Antibiotics in Animal Feeds. Council for Agric. Sci. and Techn. Rept. 88, Ames, IA.

Hays, V.W. (1980). Effectiveness of feed additive usage of antibacterial agents in swine and poultry production. The Hays Report. Rachelle Lab. Inc., Long Beach, CA.

Heersche, G.E. (1982). Artificial insemination of dairy heifers. Proc. National Dy. Cattle Workshop. USDA-ES, Washington, DC.

Klopfenstein, T. (1978). Chemical treatment of crop residues. J. Animal Sci. 46:841-848.

Machlin, L.J. (1973). Effect of growth hormone on milk production and feed utilization in dairy cows. J. Dy. Sci. 56:575-580.

Overfield, J.R. Jr. and E.E. Hatfield (1975). Anabolic agents for beef cattle. University of Illinois Beef Cattle Report AS-669, 48-51, Urbana, IL.

Peel, C.J., D.E. Bauman, R.C. Gorewit and C.J. Sniffen (1981). Effect of exogenous growth hormone on lactational performance in high yielding dairy cows. J. Nutr. 111:1662-1671.

Pursel, V.G., D.O. Elliot, R.B. Staigmiller and C.W. Newman (1982). Fertility of gilts fed Altrenogest with and without hormone injections to control ovulation. J. Animal Sci. 55: Suppl. 1:379.

Reid, J.T. (1977). Potential for increased use of forage in dairy and beef rations. Proc. 10th Research-Industry Conf. 165. Amer. Forage Grassland Council. Lexington, KY.

Seidel, G.E. Jr. (1981). Management of reproduction of cattle in the 1990's. Carl B. and Florence E. King Visiting Scholar Lectures. Special Report 77. Univ. AK.

Wagner, D. (1982). Ionophore comparisons for feedlot cattle in the future of the beef industry. National Beef Symposium and Oklahoma Cattle Feeders Seminar, N1-N9. Oklahoma State University, Stillwater, OK.

BIORATIONAL DESIGN OF CHEMICALS

Hans Geissbuhler, Urs Muller, Johannes Paul Pachlatko
and Hans-Rudolf Waespe
Agricultural Division, CIBA-GEIGY Ltd.,
CH-4002 Basel, Switzerland

ABSTRACT

Biorational approaches in the design of new bioactive chemicals must be pursued, perfected and accelerated to ameliorate the selectivity vis-a-vis non-target organisms and to assist in improving a diminishing success rate observed with conventional approaches. The analysis of the present state of the art demonstrates that biorational reflections have been of minor importance in the discovery and development of presently used chemicals. Four major categories of impediments have been identified: 1) the relatively high success rate of conventional approaches in the past, 2) the multitude or organisms and control factors involved in determining the efficacy and fate of chemicals, 3) the complexity and irrelevance of testing systems, 4) a continuing and significant lack of basic scientific knowledge. Measures to overcome these impediments are projected and the required priorities are defined. An area which needs particular attention is the physiology of crops, insect pests, diseases and weeds, under the subtropical and tropical conditions of developing countries.

KEYWORDS: Biorational Approach, Biorational Design, Optimisation, Pesticide, Bioactive Chemicals.

INTRODUCTION

Francis Bacon once said: "If a man will begin with certainties, he shall end in doubts; but if he will be content to begin with doubts and if he will be patient he shall end in certainties". We believe that this citation accurately reflects the pitfalls experienced in the past, and the basic attitude to be taken towards the biorational design of new chemicals which are to be successfully used for the improvement of food production in the future.

Let us first answer the question: why biorational design of chemicals? We believe that there are four reasons or incentives for pursuing, perfecting and accelerating this approach. In order of priority they are (1) to improve the selectivity of bioactive chemicals vis-a-vis socalled non-target organisms i.e., to reduce the incidence of unwanted or undesirable side-effects; (2) to assist in the improvement of a diminishing success rate observed with conventional approaches; (3) to facilitate the synthetic chemist's task of selecting, by more rational means, promising lead structures or molecules; and (4) to help the biologist by constantly improving the relevance of his battery of testing systems and processes.

For reasons of expediency and personal experience we shall limit the following discussion to chemicals which are to be used in crop protection and in the improvement of crop growth. However, we believe that most of the arguments are also applicable to the discovery and development of chemicals for the production and health of animals, the protection and amelioration of stored and processed food and for public hygiene, etc.

DEFINITIONS

The term "biorational" is utilized very ambiguously in current literature on the design and optimisation of bioactive chemicals. To be clear and comprehensible, therefore, we propose first to introduce a minimum set of indispensable definitions (Table I). By the term "biorational" we mean the intellectual and experimental exploitation of working hypotheses generated from biological information on specific target systems.

Table 1: Definition and description of biorational approaches (optimisation and design) in synthetic chemistry and their fit to appropriate biological testing levels.

Biology	Chemistry	Biorational
morphological	random (novel) analogy	Optimisation (incl. QSAR)
physiological	vague models	
biochemical	defined models: substrate - analogs	Design (incl. molecular modeling)
molecular	- antagonists - metabolites "de novo"synthesis	

In the search for new biologically active molecules the conceptual approaches of synthetic chemistry must be accompanied by appropriate biological testing procedures. Among these procedures or systems, four different levels are to be distinguished: (1) morphological (i.e., the macroscopic or microscopic observation of effects on whole organisms, such as insects, weeds, etc.); (2) physiological (the use of separate or isolated organs, organelles, tissue cultures, etc.); (3) biochemical (i.e., testing systems which reflect the mode of action of specific enzymes, hormones, semiochemicals, immunosystems, neurotransmitters, electron-carriers, etc.); and (4) molecular (i.e., models reflecting, in both chemical and physical terms, defined receptor sites of enzymes, membranes, etc.).

For the screening of randomly-selected (novel) chemicals and the synthesis of structural analogs of recognized bioactive entities, the morphological and physiological levels of biological testing are the most appropriate. Biorational reflections and models used at these levels are, by nature, relatively vague and best suited to the optimisation of identified lead structures. Regression and classification methods of quantitative structure-activity analyses (QSAR) are tools for rationalizing and accelerating this approach (Hansch, 1981; Magee, 1981). A typical example of biorational reflections during the optimisation of a lead structure is the discovery of the herbicide glyphosate, which, according to the literature, was worked out from the notion that tertiary amines are subjected to metabolic N-dealkylation in plants (Franz, 1982).

For the actual biorational design of new bioactive compounds, more specific and more stringent criteria must be met both by chemical synthesis and by biological testing. Options open to chemical synthesis include the exploitation of defined biochemical or molecular models in terms of substrate - analogs, - antagonists, - metabolites, etc. or of specific compounds of novel structure, designed on the basis of receptor site information. Modern methods which can profitably assist in rationalising this approach include various computerized molecular modeling procedures, which we shall briefly refer to later. Chemical structures conceived and synthesized along these lines must be examined with appropriate biochemical and/or molecular testing systems. An example which meets some of the criteria of biorational design mentioned above is the discovery and optimisation of synthetic insect juvenile hormone analogs (Bowers, 1971; Slama, 1981; Menn and Henrick, 1981).

Another example, which meets at least one of the criteria to the fulfilled in the biorational design of new chemicals, is the synthesis of sulfenylated insecticidal carbamates, which reflect the substrate metabolite (or "procide") approach and which have been designed to reduce the actual toxicity of their parent molecule (Fukuto, 1976).

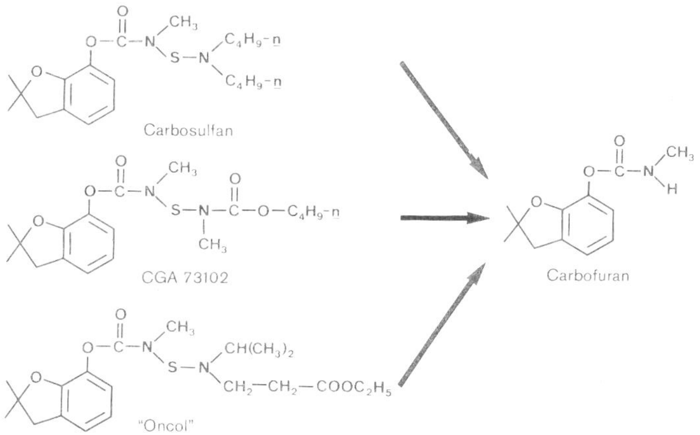

This and the preceding examples demonstrate that, in practice, not all approaches can be classified exactly according to our idealized set of definitions. However, as long as it is accepted that the boundaries between the different chemical and biological levels are fluid, the definitions fulfil their purpose of making evident the different methodological concepts.

Another approach, which does not fit into the scheme of our definitions as given earlier, but which is often referred to as being "biorational", is the exploitation of naturally-occurring products obtained from microorganisms, plants, etc. having biocidal or bioregulatory activities of an "a priori" undefined physiological and/or biochemical nature (Menn and Henrick, 1981). The most prominent example in this category is the design and optimisation of the synthetic pyrethroid insecticides which have been derived from lead structures provided by the natural pyrethrins (Elliott, 1977).

In the following discussions this "natural products" approach which is and will continue to be an essential part of any comprehensive synthesis concept, will not be further pursued. We also omit from our discussion the so-called biological control agents, which include, among others, bacteria, fungi and viruses with insecticidal or fungicidal activities.

The scheme of definitions given in Table I has not only been described to clarify such terms as biorational approach, biorational optimisation and biorational design, but also to demonstrate that these activities cannot profitably be pursued in isolation but only as part of an integrated and comprehensive chemistry concept, which will continue to include such major elements as random-, analog- and natural products chemistry. The overall concept must also be sufficiently flexible and dynamic to exploit or to inject biorational thinking in any of its component areas at any time.

STATE OF THE ART

In order to evaluate and demonstrate the present state of the art, we have attempted to subject the entire range of available structural classes of pesticides and growth regulators to a systematic analysis with regard to biorational reflections and approaches (optimisation and/or design) which were applied in their discovery and exploration. In judging the validity of such an analysis, we accept that published information rarely reveals the whole chain of chemical rationales and biological considerations which were followed by the respective scientists. On the other hand, we also suspect that, in some instances, newly introduced coompounds have subsequently been disguised with a biorational jacket, although their discovery took place under more fortuitous circumstances.

A short-hand version of our analysis is presented in Fig. 1. It lists the major structural classes of insecticides, fungicides, herbicides and plant growth regulators according to the time scale of their introduction. The scheme demonstrates that the large majority of structural classes of pesticides were discovered by conventional means, i.e. by random screening and subsequent chemical optimisation. Biorational reflections and approaches of some kind were exploited in no more than 14 of the more than 60 different classes of chemical structures which eventually reached the stage of practical application. The frequency of biorational approaches was found to be somewhat higher for insect control agents and plant growth regulators than for herbicides and fungicides. When looking at the frequency in terms of the 40-years' time scale, the scheme suggests a slight increase in the use of the biorational approach in the seventies as compared wth previous decades. However, this increase is hardly significant.

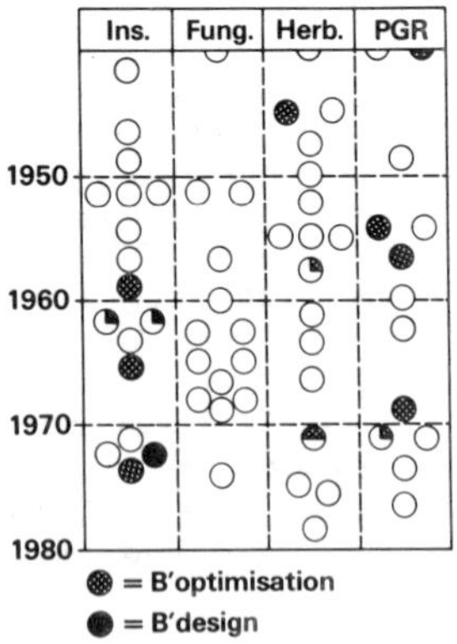

Fig. 1: Time scale of disclosure of different structural classes of pesticides and evaluation of biorational approaches (optimisation and design) used in their discovery.

In summarising the present state of the art, we conclude that biorational approaches and especially the biorational design have been and (at least for the moment) continue to be a minor element in the design of new bioactive chemicals.

IMPEDIMENTS/MEASURES TO OVERCOME THEM

What are the impediments which for the moment prevent a more extensive application of biorational approaches to the design of new and more selective bioactive molecules? We believe that these impediments can be grouped into four major categories: (1) the past success rate of conventional approaches; (2) the multitude and variety of target organisms and the multitude of control-(loss-) factors which affect the behaviour and fate of bioactive molecules under practical use conditions; (3) the complexity and/or irrelevance of biological testing systems; and (4) (last but not least) a considerable and continuing lack of basic scientific knowledge of the physiology and biochemistry of major crop plants, pest organisms and soil substrates. In what follows we shall briefly discuss these four categories of impediments and postulate a number of measures to overcome them.

Success Rate of Conventional Approaches. The yearly rate of world-wide introduction of new active ingredients as crop protection agents and plant growth regulators, as compiled from the latest edition of the "Pesticide Manual" (Worthing, 1979), is shown in Fig. 2. This compilation demonstrates that the rate of introduction reached its maximum level of 20 or more compounds per year in the late sixties. Since then, it has progressively declined to less than 10 compounds per year in the late seventies. Though efforts were and are being made to maintain an adequate success rate by increasing the number of chemicals subjected to conventional screening procedures, the responsible parties face increasing pressure to explore alternative strategies, including biorational approaches. Thus we believe that this impediment, which is intellectual rather than technical or scientific, will continue to be reduced and that a readiness to exploit biorational thinking will influence developments to an increasing extent.

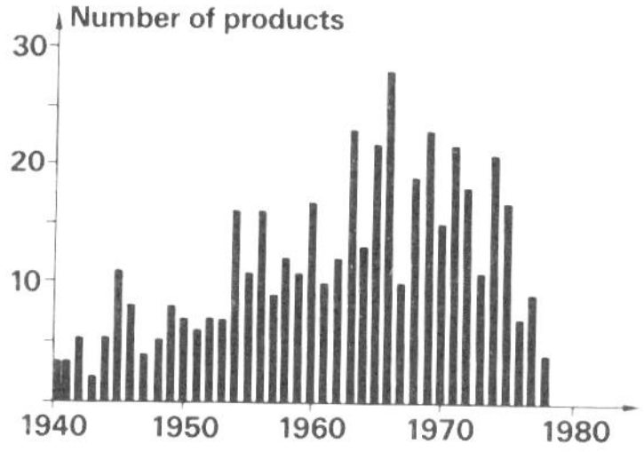

Fig. 2: Development of yearly rate of world-wide introduction of new pesticides and growth regulators.

Multitude of Organisms and Control Factors. The total of so-called "economically important" insect pests, crop diseases and weeds is estimated to comprise roughly 200 different species belonging to almost as many systematic orders, families and genera. In contrast to the pharmaceutical researcher, who can concentrate on the diseases and features of one single mammalian species, the human being, the agricultural scientist is faced with an immense variety of morphological features physiological mechanisms, biochemical systems etc., not only of the organisms to be controlled, but also of the crops to be protected or regulated.

Having to work on this multitude of organisms has two consequences: (1) the selection of appropriate biochemical or molecular models which cover a sufficient technical range of target-organisms in one particular area of indication is made difficult; and (2) potential candidate compounds, which are designed to be active on a specific system, may turn out to have such a limited range of practical applications that the economics of development and introduction are not justifiable: synthetic juvenile hormone analogs as well as certain synthetic pheromones are typical examples of this happening.

The measures to overcome this impediment are a careful selection of those testing models which have sufficient latitude in terms of organisms covered and a concentration of effort on those targets whose control is economically justified. The multitude of the control-(or loss-)factors which affect the behaviour and fate of a chemical in the target organism, in the crop plant to be protected and in the soil substrate is depicted in Fig. 3. Additional criteria which have to be taken into account, and which are, to a varying degree, independent of the control factors of a particular pest-crop-soil complex, are selectivity, and mammalian and eco-toxicity.

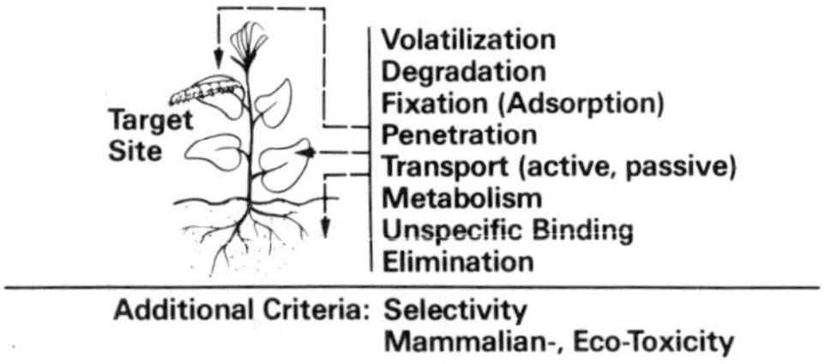

Fig. 3: Multitude of control-(loss-) factors affecting the fate of bioactive chemicals in the pest-crop-(soil-)complex.

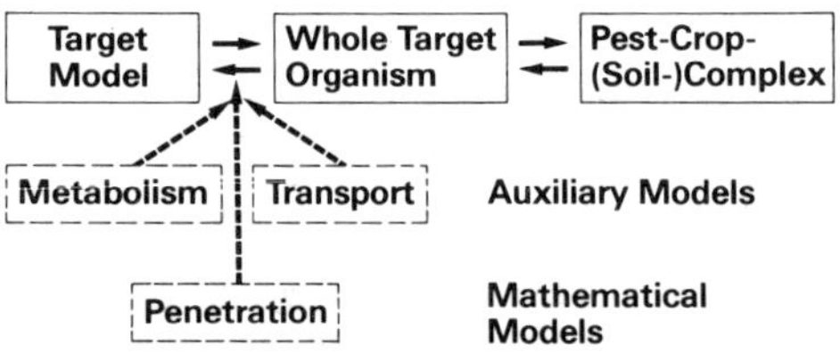

Fig. 4: Scheme illustrating the basic approach to cope with control-(loss-)factors in the biorational design of bioactive molecules.

A simple but typical example to demonstrate the composite effect of control factors is given in Table 2. It shows that certain chemical structures, such as diuron and atrazin, which are strong inhibitors of the photosynthetic Hill-reaction, are very active herbicides, whereas others, such as 2,5-dibromo-3-methyl-6-isopropyl-1,4-benzoquinone (DBMIB) or CGA 71'884, in spite of being excellent Hill-reaction inhibitors, failed to perform as herbicides under greenhouse or field conditions. On the other hand, compounds like bentazone, which have less inhibitory activity, turned out to be very useful herbicides.

Measures to overcome the impediments presented by the multitude of control factors are presented in Fig. 4. This scheme essentially suggests a three-step approach during the biorational design of new molecules:

(1) when using a biochemical or molecular target model or system, chemicals which have promising activity in this system must immediately be examined for effect on the whole target organism. Provided that the activity continues to be observed, examination must be extended to a suitable pest-crop-(soil-)complex. Information on activity losses from this process serves as feedback to the synthetic chemist.
(2) as more systematic knowledge is gained or becomes available, auxiliary models which reflect crucial control factors (such as metabolism or transport phenomena) may be inserted on the track between the primary target model and the whole organism.
(3) provided that the main physical-chemical parameters of a particular control

Table 2: Composite effect of control-(loss-)factors on the activity of herbicides. The effect is demonstrated by comparison of the inhibitory activity of the listed compounds in the photosynthetic Hill-reaction test with their herbicidal activity under greenhouse and field conditions.

CGA 71'884 = 1-(3-chloro-4-trifluoromethyl-phenyl)-3,4-dimethyl-5-hydroxy-2-oxo-pyrrole.
DBMIB = 2,5-dibromo-3-methyl-6-isopropyl-1,4-benzoquinone.

Structure	Hill-Reaction $-\log I_{50}$	Herbicidal Activity Greenhouse	Field
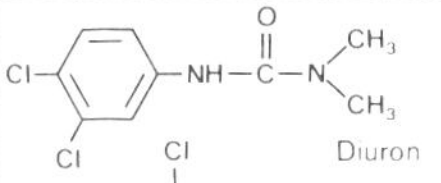 Diuron	6,7-7,5	Good	Good
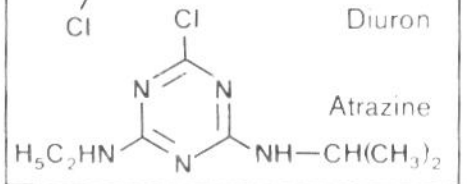 Atrazine	6,1-6,6	Good	Good
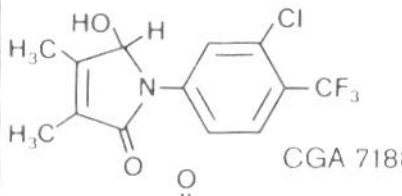 CGA 71884	6,2	Good	Poor
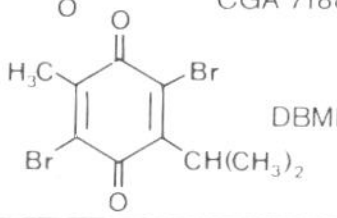 DBMIB	6,1	Poor	–
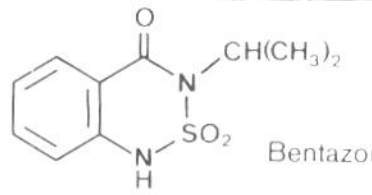 Bentazone	3,7-4,4	Good	Good

phenomenon (such as penetration, for example) can be defined, the auxiliary biological model can be replaced by a mathematical model, the use of which is facilitated by modern electronic data evaluation and storage systems (Schaper, 1982; Schonherr, 1979).

The use of mathematical models is most advanced in pharmaceutical chemistry (Hansch, 1981) but is also becoming increasingly visible in agricultural (Iwamura and Fujita, 1982) and toxicological (Veith, 1981) chemistry. Present efforts to exploit molecular modeling are based on different conceptual approaches and include, among others, "de novo"-modeling of biocides based on X-ray analysis of receptor sites (Gund et al., 1980), receptor mapping by combining the features of a series of active molecules (Humblet and Marshall, 1981), and mathematical and statistical analysis based on physico-chemical and steric parameters of active moieties without receptor recognition (Martin, 1978). To illustrate the first one of these approaches, Fig. 5 shows a view of the drug-receptor site interaction between methotrexate and the enzyme dehydro-folate-reductase of Escherichia coli. In agricultural chemistry more extensive application of molecular modeling is not limited by the availability of appropriate computerized mathematical models but by the lack of sufficiently purified and relevant key enzyme preparations.

Complexity/Non-Relevance of Testing Systems. For the biorational design of chemicals, efficient and accurate physiological and biochemical testing systems are essential. Until recently, such systems were technically rather complex, labour-intensive and often of questionable reliability and reproducibility. This technical complexity has deterred researchers from using them more frequently and on a more extended scale. Now this impediment of technical complexity is being rapidly overcome by continuous and impressive advances in analytical instrumentation and data handling. For example, modern automated continuous-flow analyzers are capable of reliably performing a multitude of enzyme assays and other biochemical reactions with minimum quantities of reagents, fast separation procedures and reaction times, ultra-sensitive measuring devices and unlimited flexibility in programming, data evaluation, printout and storage, etc. A typical example of these advances in biochemical assay procedures is demonstrated in Fig. 6, which shows a modern version of a flow-sheet for performing the automated photosynthetic Hill-reaction with plant chloroplasts.

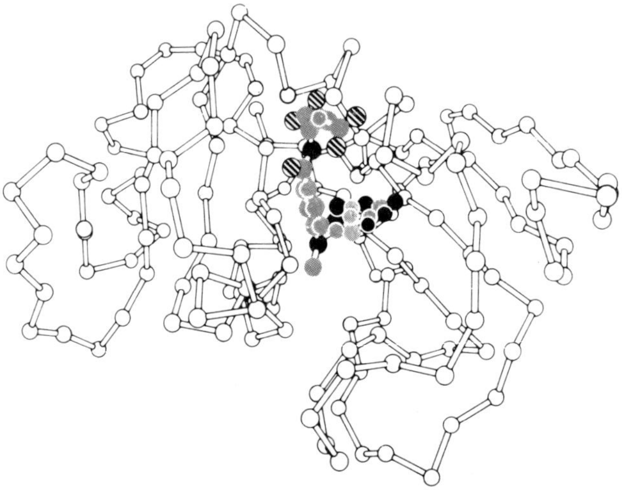

Fig. 5: Illustration of drug-receptor interaction by a stereo drawing of E. coli dihydrofolate reductase with a bound methotrexate molecule. (Matthews et al., 1978).
Methotrexate components: = nitrogen; oxygen, = carbon.

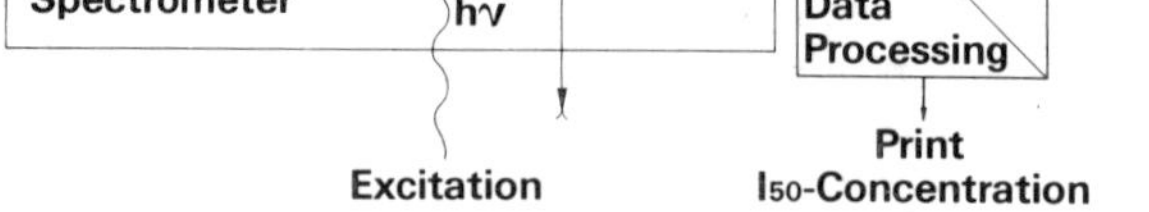

Fig. 6: Flowchart of a modern version of the Hill-reaction test designed to allow high sample throughput capacity.

Even more important than the technical feasibility and reliability of testing systems is their relevance. They must be relevant to the biological effects to be achieved, and these biological effects, in turn, must be relevant to the control procedures envisaged in practice. A typical example where, for lack of basic scientific knowledge, the biological and physiological testing procedures did not effectively meet the criteria of relevance, is provided by the various assays which were commonly used in the past for the examination and testing of synthetic juvenile hormone analogs (Fig. 7).

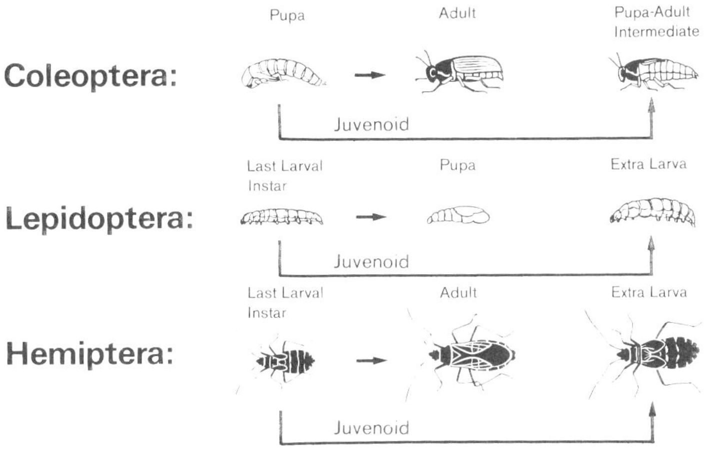

Fig. 7: Scheme summarizing routine procedures used for the biological testing of synthetic juvenile hormone analogs (juvenoids).

The juvenoids were routinely tested on the last larval stages or on pupae of various insect species. The objective was to prevent pupation or metamorphosis of these larvae and to thus cause their premature death. It was not foreseen that these larvae would turn into voracious unnaturally large "superlarvae", which, although they eventually died, continued to cause damage for an unacceptable period of time (Bowers, 1982). The point to be made by this example is to underline that, whatever testing system is used in the biorational design of new molecules, its relevance to the envisaged biological effects and to the potential use in practice must be carefully evaluated and clearly defined at a relatively early stage.

Lack of Scientific Knowledge. It is almost trivial to mention that efforts to develop and exploit biorational design procedures are seriously hampered by a significant and continuing lack of basic scientific knowledge concerning the physiology and biochemistry of insects, diseases, weeds and crop plants. In addition, basic information at the biochemical and molecular levels on the mode of action of presently used pesticides is, in general, not sufficient to be systematically exploited for biorational improvements and modifications. This is illustrated by Fig. 8, which summarizes the present state of knowledge of the mode of action of some major structural classes of herbicides (Fedtke, 1982). Although it is 10 to 30 years since their discovery, and we have a wealth of information on the physiology of herbicide action, biochemical information is definitely incomplete and information at the molecular level is practically non-existent.

When analyzing the reasons for this continuing lack of basic information, we might ask the somewhat provocative question of whether the efforts invested have been insufficient or whether they have partially been directed in the wrong way. We believe that there is a tendency to overdo physiological experimentation, and to use these organisms and systems which can easily be manipulated in laboratories but which are not necessarily relevant as models for the conditions prevailing in agricultural practice. A case to illustrate this point is the plant hormone indole acetic acid (IAA), of which we are still not able to describe the mechanisms of action accurately, at the molecular level, more than 40 years after its discovery. A computer survey of the literature on IAA revealed that out of the 6600 papers registered between 1969 and early 1982, no more than 163 dealt with the biochemistry of its mechanisms of action and only 85 with its receptor sites (Table 3).

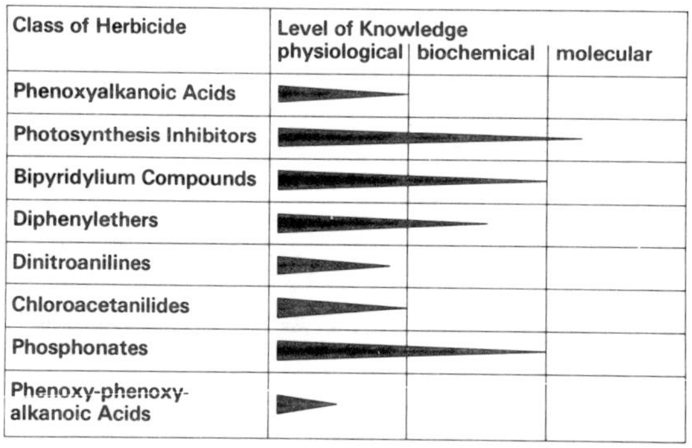

Fig. 8: State of knowledge on the mechanisms of action of some major classes of herbicides.

Table 3: Computer survey of literature regarding the number of papers published on the plant hormone indole acetic acid (IAA) with special reference to those publications dealing with its mechanisms of action and/or its interactions with receptor sites.

Keywords	Number of Hits		
IAA	6601		
Mode of Action	163	33 130	1969–1976 1977–1982
Receptor Site	85	13 72	1969–1976 1977–1982

In most areas of chemical crop protection and regulation we need a more determined and concerted effort to advance to the biochemical and molecular level of events and to more quickly translate this information into practical, biorational solutions. That this is feasible is illustrated by a classical and a more recent example:

The discovery and development of organic phosphorus esters and carbamates provoked rapid advances in defining their mode of action at the biochemical level and in localizing their receptor site on the cholinesterase enzyme (Fig. 9). This knowledge was repeatedly exploited in the design of structural modifications of these compounds (Linwood et al., 1966; Eto, 1974; Kuhr and Dorough, 1976). More recently, the discovery of the key role of the amino acid D-alanine in the formation of bacterial cell walls (see Fig. 10) was exploited for designing structural analogs of this substrate to be active as specific inhibitors of bacteria, and thus to control diseases in medicine and agriculture (Dingwall et al., 1981; Neuhaus and Hammes, 1981).

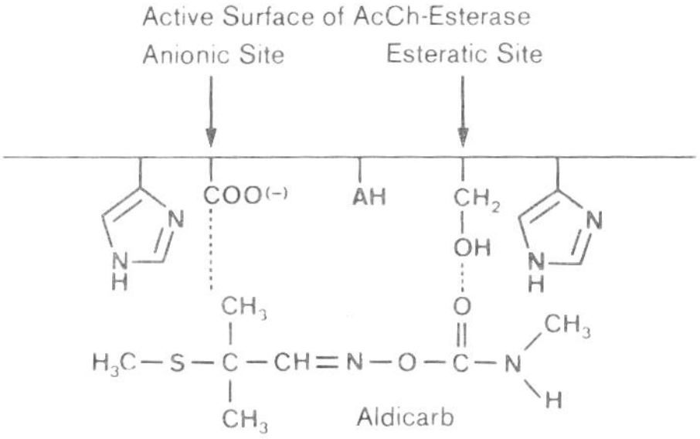

Fig. 9: Biorational design of oxim-carbamates (representative: aldicarb) by exploiting the physical-chemical features of the acetylcholine-esterase surface.

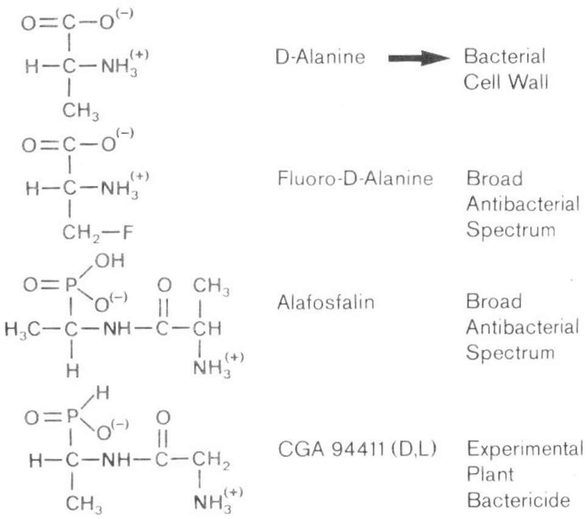

Fig. 10: Biorational design of bactericides by exploiting the structural analog approach, i.e. the key role of d-alanine in the formation of bacterial cell walls.

alafosfalin = 1-(2-amino- propionamido)-ethyl-phosphonic acid.

CGA 94'411 = 1-(2-amino-acetamido)-ethyl-phosphinic-acid.

TRENDS AND PRIORITIES

In looking back on the four major categories of impediments with which we are still confronted, the priorities to be observed in applying the measures to overcome them are almost self-evident. The first priority is to intensify the acquisition of basic scientific knowledge on the growth, development and reproduction of crop plants, insect pests, diseases and weeds and to define in more detail their respective interactions and natural defense mechanisms. This task is and must continue to be mainly academic. We suggest that more determined efforts are made to push forward to the biochemical and molecular levels of the phenomena and to concentrate on organisms and model systems which are relevant to the agricultural environment, even if these biological objects and tools prove to be more difficult to handle.

We believe that scientists working in subtropical and tropical regions have a particularly important mission to fulfil in the acquisition of basic information. It is common knowledge that the agronomic environment in these areas in terms of soil conditions, climate, crop and pest species and varieties, etc. is, in many ways different from that encountered in temperate zones. We urgently need more extended and detailed information on the peculiar features of subtropical and tropical crop ecosystems. This information will then allow an approach to pertinent problems by more rational biochemical and chemical means. The acquisition of physiological knowledge requires neither advanced instrumentation nor sophisticated biochemcial modeling procedures. A typical example, which illustrates the necessity of taking into account such physiological particularities, is provided by the rice ecosystem, which, in certain areas, includes the fern Azolla as an efficient "in situ" auxiliary source of nitrogen for crop growth (see Fig. 11). Unless we are aware of, and know in sufficient detail, the physiology of the rice/Azolla-Anabaena interactions, the design and development of sufficiently selective herbicides, which do not upset the balance of the system, will be jeopardized (Mathar et al., 1981).

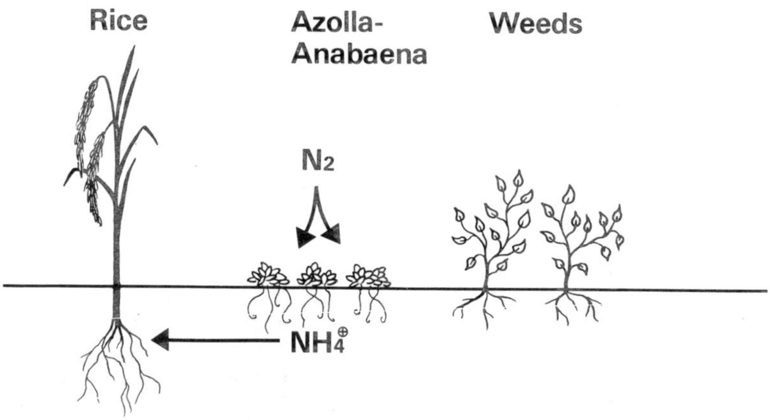

Fig. 11: Scheme of rice ecosystem comprising the crop species, the Azolla-Anabaena nitrogen-fixing component and weeds.

The extension of the scientific information base is essential for reducing further the complexity of biochemical and molecular testing systems and for improving their relevance to practical agronomic conditions. It is also indispensable for the development of more realistic biological, and eventually mathematical, models to cope with the multitude of control factors which affect the efficiency and fate of bioactive chemicals.

These improvements can only be achieved by closer cooperation and interfacing between chemists and biologists. It means that the synthetic chemist must be prepared to open his horizon towards biochemistry and mathematics, whereas the biologist, in addition to improving his biochemical background, must be able to evaluate the agronomic relevance of his testing systems and models.

In looking at the various areas of crop protection and crop growth, we believe that two of them deserve the highest priority in our efforts to intensify biorational design procedures. The first is insect control which is the one most hampered by problems of poor selectivity, undesirable side effects, resistance, etc., and which appears to be most affected by the diminishing success rate of conventional approaches. The second area is plant growth regulation, which, in spite of past efforts, has had, so far, only limited impact in terms of economically important practical applications. We accept that the two mentioned areas are more complex and more difficult in terms of biorational handles than either weed or disease control. However, we believe that priorities ought to be dictated by urgency and not by simplicity.

CONCLUSIONS

We are convinced that biorational approaches, and particularly the biorational design of new molecules, will come to be an essential and increasingly important part of the total and comprehensive system required for the discovery and optimisation of bioactive agricultural chemicals. To overcome present impediments, the basis of scientific knowledge must be extended and improved. The intelligent and efficient exploitation of this knowledge will depend on a determined and unbiased cooperative effort shared among chemists and biologists from academia, governments and industry. In focusing biorational approaches on agricultural problems in developing areas and regions, the particular features of subtropical and tropical ecosystems must be defined in precise physiological terms.

REFERENCES

Bowers, W.S. (1971). Insect Hormones and Their Derivatives as Insecticides. Bull. Wld. Hlth. Org., 44, 381-389.

Bowers, W.S. (1982). Endocrine Strategies for Insect Control. Ent. Exp. & Appl., 31, 3-14.

Dingwall, J.G., E.K. Baylis, C.D. Campbell, W. Pickles (1981). α -Aminophosphonous Acids: A New Class of Biologically Active Amino Acid Analogs. Phosphorous Chemistry. Proceedings of the 1981 International Conference. ACS Symp. Series 171, Quin, L.D., J.G. Verkade, Eds., Amer. Chem. Soc., Washington, D.C., 183.

Elliott, M. (1977). Synthetic Pyrethroids. ACS Symp. Series 42, Gould, R.F., Ed., Amer. Chem. Soc., Washington, D.C.

Eto, M. (1974). Organophosphorus Pesticides: Organic and Biological Chemistry. CRC Press. Cleveland, Ohio, 123-158.

Fedtke, C. (1982). Biochemistry and Physiology of Herbicide Action. Springer Verlag, Berlin.

Franz, J.E. (1982). Using a Biorational Approach, We Discovered Glyphosate. Farm Chemicals, 48-52.

Fukuto, T.R. (1976). Carbamate Insecticides. In "The Future for Insecticides", Metcalf, R.L., J.J. McKelvy, Jr., Eds., John Wiley & Sons, New York, N.Y., 313-342.

Gund, P., J.D. Andose, J.B. Rhodes, G.M. Smith (1980). Three-Dimensional Molecular Modeling and Drug Design. Science, 208, 1425-1431.

Hansch, C. (1981). The Physiocochemical Approach to Drug Design and Discovery (QSAR). Drug Dev. Res., 1, 267-309.

Humblet, Ch., G.R. Marshall (1981). Three-Dimensional Computer Modeling as an Aid to Drug Design. Drug Dev. Res., 1, 409-434.

Iwamura, H., T. Fujita (1982). QSAR-Studies in Pesticide Research in Japan. J. Pesticide Sci., 7, 289-299.

Kuhr, R.J., H.W. Dorough (1976). Carbamate Insecticides: Chemistry, Biochemistry and Toxicology. CRC Press, Cleveland, Ohio, 41-101.

Linwood, K.P., Jr., H.A. Stansbury, Jr., M.H.J. Weiden, (1966). The Synthesis and Insecticidal Properties of Some Cholinergic Trisubstituted Acetaldehyde O-(Methylcarbamoyl)oximes. J. Agr. Food Chem., 14, 356-365.

Magee, Ph.S. (1981). Discovering Bioactive Materials. Chem. Tech., 378-384.

Martin, Y.C. (1978). Quantitative Drug Design. Marcel Dekker, New York.

Mathar, A.S., S. Krishnamoorthy, L. Anovaradham (1981). Azolla Influence on Rice Yield. Int. Rice Res. Newsletter, 6, 23.

Matthews, D.A., R.A. Alden, J.T. Bolin, B.J. Filman, S.T. Freer, R. Hamlin, W.G.J. Hol, R.L. Kisliuk, J. Kraut, E.J. Pastore, L.T. Plante, N. Xuong (1978). Dehydrofolate Reductase from Lactobacillus casei. X-ray Structure of the Enzyme · Methotrexate · NADPH Complex. J. Biol. Chem., 253, 6946-6954.

Menn, J.J., C.A. Henrick (1981). Rational and Biorational Design of Pesticides. Phil. Trans. R. Soc. London, Ser. B295, 57-71.
Neuhaus, F.C., W.P. Hammes (1981). Inhibition of Cell Wall Biosynthesis by Analogues of Alanine. Pharm. Ther., 14, 265-319.
Schaper, J-J. (1982). Absorption of Ionisable Drugs: Nonlinear Dependence on log P, pKa and pH - Quantitative Relationships. Quant. Struct.-Act. Relat., 1, 13-27.
Schonherr, J. (1979). Transcuticular Movement of Xenobiotics. Advances in Pesticide Science Vol. III, 392-400. H. Geissbuhler, Ed., Pergamon Press.
Slama, K (1981). Juvenoids in Retrospect and Juvenogens in Propsect. In: Regulation of Insect Development and Behaviour. Sehnal, F., A. Zabza, J.J. Menn, B. Comborowski, Eds., Proc. Int. Conf., Karpacz, Poland, 853-868.
Veith, G.D. (1981). State of the Art Report on Structure-Activity Methods Development. Govt. Reports Announcements and Index, Issue 17.
Worthing, Ch.R. (1979). The Pesticide Manual, publ. The British Crop Protection Council, 6th Ed.

Author Index

Subject Index